MW00760665

Springer Series in

MATERIALS SCIENCE 81

Springer Series in
MATERIALS SCIENCE

Editors: R. Hull R. M. Osgood, Jr. J. Parisi H. Warlimont

The Springer Series in Materials Science covers the complete spectrum of materials physics, including fundamental principles, physical properties, materials theory and design. Recognizing the increasing importance of materials science in future device technologies, the book titles in this series reflect the state-of-the-art in understanding and controlling the structure and properties of all important classes of materials.

61 **Fatigue in Ferroelectric Ceramics and Related Issues**
By D.C. Lupascu

62 **Epitaxy**
Physical Principles and Technical Implementation
By M.A. Herman, W. Richter, and H. Sitter

63 **Fundamentals of Ion-Irradiated Polymers**
By D. Fink

64 **Morphology Control of Materials and Nanoparticles**
Advanced Materials Processing and Characterization
Editors: Y. Waseda and A. Muramatsu

65 **Transport Processes in Ion-Irradiated Polymers**
By D. Fink

66 **Multiphased Ceramic Materials**
Processing and Potential
Editors: W.-H. Tuan and J.-K. Guo

67 **Nondestructive Materials Characterization**
With Applications to Aerospace Materials
Editors: N.G.H. Meyendorf, P.B. Nagy, and S.I. Rokhlin

68 **Diffraction Analysis of the Microstructure of Materials**
Editors: E.J. Mittemeijer and P. Scardi

69 **Chemical–Mechanical Planarization of Semiconductor Materials**
Editor: M.R. Oliver

70 **Applications of the Isotopic Effect in Solids**
By V.G. Plekhanov

71 **Dissipative Phenomena in Condensed Matter**
Some Applications
By S. Dattagupta and S. Puri

72 **Predictive Simulation of Semiconductor Processing**
Status and Challenges
Editors: J. Dabrowski and E.R. Weber

73 **SiC Power Materials**
Devices and Applications
Editor: Z.C. Feng

74 **Plastic Deformation in Nanocrystalline Materials**
By M.Yu. Gutkin and I.A. Ovid'ko

75 **Wafer Bonding**
Applications and Technology
Editors: M. Alexe and U. Gösele

76 **Spirally Anisotropic Composites**
By G.E. Freger, V.N. Kestelman, and D.G. Freger

77 **Impurities Confined in Quantum Structures**
By P.O. Holtz and Q.X. Zhao

78 **Macromolecular Nanostructured Materials**
Editors: N. Ueyama and A. Harada

79 **Magnetism and Structure in Functional Materials**
Editors: A. Planes, L. Manõsa, and A. Saxena

80 **Ion Implantation and Synthesis of Materials**
By M. Nastasi and J.W. Mayer

81 **Metallopolymer Nanocomposites**
By A.D. Pomogailo and V.N. Kestelman

A.D. Pomogailo V.N. Kestelman

Metallopolymer Nanocomposites

With 239 Figures

Prof. Anatolii D. Pomogailo
Russian Academy of Sciences
Institute of Chemical Physics
Chernogolovka
Moscow Region 142432, Russia
E-mail: adpomog@icp.ac.ru

Prof. Vladimir N. Kestelman
KVN International
632 Jamie Circle
King of Prussia
PA 19406, USA
E-mail: kvnint@earthlink.net

Series Editors:

Professor Robert Hull
University of Virginia
Dept. of Materials Science and Engineering
Thornton Hall
Charlottesville, VA 22903-2442, USA

Professor Jürgen Parisi
Universität Oldenburg, Fachbereich Physik
Abt. Energie- und Halbleiterforschung
Carl-von-Ossietzky-Strasse 9–11
26129 Oldenburg, Germany

Professor R. M. Osgood, Jr.
Microelectronics Science Laboratory
Department of Electrical Engineering
Columbia University
Seeley W. Mudd Building
New York, NY 10027, USA

Professor Hans Warlimont
Institut für Festkörper-
und Werkstofforschung,
Helmholtzstrasse 20
01069 Dresden, Germany

ISSN 0933-033x

ISBN-10 3-540-20949-2 Springer Berlin Heidelberg New York

ISBN-13 978-3-540-20949-2 Springer Berlin Heidelberg New York

Library of Congress Control Number: 2004114850

Springer is a part of Springer Science+Business Media.
springeronline.com

Printed in Germany

Typesetting by the authors and TechBooks using a Springer LaTeX macro package
Cover concept: eStudio Calamar Steinen
Cover production: *design & production* GmbH, Heidelberg

Printed on acid-free paper SPIN: 10864309 57/3141/jl 5 4 3 2 1 0

To the memory of our fathers Dmitriy Ivanovich and Nikolay Yakovlevich,
participants in World War II and great toilers, with hope that our children
and grandchildren will be deserving of their memory

Foreword

The way from the micrometer to the nanometer scale is a further step in the miniaturization of active particles and functional components for technical applications. Physical and chemical properties change due to the occurrence of size-dependent quantum and confinement effects. Ceramics may become transparent like glass; glass can become tough like adhesive; metals may become coloured like dyes; magnetism can be switched on and off, etc. In order to understand this behaviour, fundamental research is necessary. In addition, it can be foreseen that nanomaterials and their technology will be one of the most important technologies of the 21st century.

The kinds of materials addressed are, in principle, the same as those encountered in macro- and microtechnologies. Examples are polymers, ceramics, metals, semiconductors, and insulators. But in place of the top-down approach, the bottom-up approach now becomes the key to devising new, for example, composite or hybrid materials. Nanotechnology exploits the same building blocks and structural elements as other technologies: atoms, molecules, solids. But the difference to the classical materials is that the structural elements consist of small and sometimes very small ensembles of atoms and molecules. In the classical case the number of surface atoms or molecules is small compared to the number of bulk particles. In the case of nanomaterials this situation is reversed.

One of the most important materials for future nanotechnology will be metal nanoparticles, due to the very unusual properties that they display. After some decades of fundamental research in this field, a summary of the state of the art is of great value. The monograph of A.D. Pomogailo and V.N. Kestelman is the first book that compiles detailed information about metallopolymer nanocomposites. These are entities that involve a combination of metal nanoparticles and macromolecules. Two excellent and well-known scientists have taken up the challenge and provided this detailed overview.

Before discussing the fundamental role of the polymer, the text first addresses stabilization of metal nanoparticles, fundamentals in classification/ organization and nucleation mechanisms. The chapters about synthesis methods include all details about physical, chemical and physico-chemical procedures for the synthesis of metal-polymer nanocomposites. Polymer-inorganic nanocomposites (sol-gel processes, intercalation hybrids, nanobiocomposites) are also described. Special attention is paid to properties and many applications are described. A large number of references will help readers to gain a more-detailed acquaintance with almost all the details and problems involved in research into metallopolymer nanocomposites. I am sure that this excellent book will be a great success and an important contribution to furthering the field of nanoscience and nanotechnology.

University of Bremen, Germany
Dieter Wöhrle
January, 2005

Preface

By the end of the twentieth century, considerable success in nanochemistry and nanotechnology had been achieved. As a result, two novel scientific areas of nanosize chemistry and functional materials science have appeared. The nanoworld is a world of quantum coherences and nonequilibrium processes of self-organization. Nanoparticles (with sizes in the range 1–100 nm) are at the boundary of the classic and quantum worlds. At present, great progress is being achieved in these investigations.

New properties of nanoparticles, nanolayers, superlattices and metal–polymeric composites have been discovered, and these are of great interest as very promising materials for microelectronics, holography and magnetic recording of information. This monograph deals mostly with the logic of investigations and the inner methodological and scientific interconnections in the field. It is hardly possible to present a survey of the whole of polymer nanocomposite science taking place throughout the world. This science is continually developing and many principal results and their interpretations are being revised. There are many reviews and books about specific properties, technologies and applications of nanomaterials; we have decided to devote this book to the basics of the science of metallopolymer nanocomposites. This is the first book of this kind, summarizing the results of more than 30 years of research in this area, from the technology of nanoparticles, problems of stabilization, and immobilization in the polymer matrix, to design of nanohybrid polymer–inorganic nanocomposites and different applications of these materials.

This book reflects the present state of the problems considered, and many ideas are still being elaborated. However, certain materials and structural aspects, and also biological problems, are not considered in detail here.

We hope that the book will be of interest to experienced and young specialists in nanochemistry and nanophysics.

Our thanks go to colleagues who supported this book, mostly to Profs. Dieter Wohrle, Franz Faupel, David Avnir, Kennett Suslik, Robert E. Cohen, Mercouri Kanatzidis, Ulrich Schubert, Dominique Bazin, Ralf Zee, Sergey Gubin, Rostislav Andrievskii, etc. Our colleagues in the laboratory have carried out a large number of experiments, most of which are represented in this book. Prof. Aleksander Rozenberg did his share for Chaps. 2 and 11 and offered his valuable remarks. Dr. Gulzhian Dzhardimalieva has made an important contribution to the synthesis of metallopolymer nanocomposites by polymerization of metal-containing monomers. We would like to express our special appreciation of her enormous help and contribution to this book.

Undoubtedly, as usual for a first multidisciplinary monograph, there may be room for improvement. We would greatly appreciate any comments and suggestions.

Chernogolovka, Russia, KVN International, USA *Anatolii Pomogailo*
January, 2005 *Vladimir Kestelman*

Contents

List of Abbreviations

A	Hamaker's constant
AcAc	acetylacetone
AFM	atomic-force microscopy
AIBN	azobis(isobutironitrile)
AOT	bis(2-ethylhexyl) sulphosuccinate
ATP	adenosine triphosphate
ATRP	atom transfer radical polymerization
C_m	volume concentration
Cerasomes	organoalkoxysilane proamphilites
COD	cyclooctadiene
CSDVB	copolymer of styrene with divinylbenzene
CVD	chemical vapor deposition
D	diffusion coefficient
D	fractal dimensionality
D_s	fractal dimensionality
Dipy	dipyridyle
DMAAm	N,N-dimethacrylamide
DMFA	dimethyl formamide
DMSO	dimethyl sulfoxide
DNA	deoxyribonucleic acid
EDTA	ethylenediaminetetraacetate
ESCA	electron spectroscopy for chemical application
EXAFS	extended X-ray absorption fine structure
FC	fractal cluster
Fe_3O_4	magnetite
Fe_3S_4	greigite
FMR	ferromagnetic resonance
FTIR	Fourier transform infrared spectroscopy
HDPE	high density polyethylene
HEMA	2-hydroxyethyl methacrylate
HOPG	highly oriented pyrolytitic graphite
HSA	human serum albumin
HTSC	high-temperature superconducting ceramics
IPN	interpenetrating polymer networks
ITO	indium tin oxide electrode

L	ligand
LB	Langmuir-Blodgett films
LCG	layered interstitial compounds in graphite
LCST	low critical solution temperature
LDLPE	low density linear polyethylene
LDPE	low density polyethylene
M_j	nanoparticle consisting of j atoms
MAO	methylalumoxane
MC	Monte Carlo method
MCC	micelle-forming critical concentration
MCM-41	hexagonal materials (Mobil codes)
MCM-48	cubic materials (Mobil codes)
MMA	methylmethacrylate
MMD	molecular-mass distribution
MMO	methane monooxygenase
MMT	montmorillonite type clays
$MO_n(OR)_m$	oxoalkoholates
MRI	magnetic resonance imaging
MX_n	metallocomplex
(MV^{2+})	methylviologene (dimethyl-4,4-bipyridine)
NADH	nicotine-amide reduced
NC	coordination number
NLO	nonlinear optical
NP	nanoparticle
NVI	N-vinylimidazole
OGC	oxygen-generated center
ORMOCER	Organically Modified Ceramics
ORMOSIL	Organically Modified Silicates
P	cross-linked polymer
P*i*PA	poly(N-*iso*-propylacrylamide)
P2VPy	poly(2-vinylpyridine)
P4VPy	poly(4-vinylpyridine)
PA	polyamide
PAA	poly(acrylic acid)
PAAm	poly(acrylamide)
PAMAM	polyamidoamine (dendrimer)
PAn	PANI polyaniline
PAN	polyacrylonitrile
PB	polybutene
PBD	polybutadiene
PBIA	polybenzimidazole
PBIA	polyheteroarylene
PC	percolation cluster
PCL	poly(ε-caprolactone)
PDMS	poly(dimethylsiloxane)

PE	polyethylene
PEO	polyethylene glycol
PE-gr-P4Vpy	polyethylene grafted poly(4-vinylpyridine)
PEI	polyethyleneimine
PEO	polyethyleneoxide
PES	polyestersulfone
PETP	polyethylene terephthalate
Phen	phenantroline
PhTMOS	phenyltrimethoxysilane
PI	polyisoprene
PM	polymer matrix
PMAAc	poly(methylacylic acid)
PMMA	poly(methylmethacrylate)
PMPhSi	poly(methylphenylsilane)
PMVK	poly(methylvinyl ketone)
POSS	polyhedral oligosilsesquioxane-based macromers
PP	polypropylene
PPG	polypropylene glycol
PphO	poly(2,6-dimethyl-1,4-phenylene oxide)
PPO	polypropyleneoxide
PPy	polypyrrole
PS	polystyrene
PSE	polymeric solid electrolytes
PTFE	poly(tetrafluoroethylene)
PU	polyurethane
PVA	poly(vinylalcohol)
PVAc	poly(vinyl acetate)
PVB	poly(vinylbutyral)
PVC	polyvinylchloride
PVDC	poly(diallyldimethyl ammonium chloride)
PVIA	poly(1-vinylimidazole)
PVK	poly(vinylcarbazole)
PVP	poly-N-(vinyl-2-pyrrolidone)
Py	pyridine
QD	quantum dots
S_{sp}	specific surface
S	solvent
SAS	surface active substance
SAXS	small-angle X-ray scattering
SCA	specific catalytic activity
SEM	scanning electron microscopy
S-layers	surface layers
SSA	specific surface area
STM	scanning tunnel microscopy
TEM	transmission electron microscopy

TEOS	tetraethoxysilane
T_g	glass transition temperature
TGA	thermal gravimetry analysis
THF	tetrahydrofuran
Ti-MCM-41	Ti-materials (Mobil codes)
T_m	melt temperature
TMOS	tetramethoxysilane
TN	turnover number of catalyst
t-RNA	t-ribonucleic acid
XPS	X-ray photoelectron spectroscopy
ZSM-5	zeolites
(α, β and γ)-PP	crystal form of isotactic PP
γ-Fe_2O_3	maghemite
φ_{cr}	volume portion of filler
4Vp	4-vinylpyridine

Part I

Structure and Organization of Metal Nanoparticles in Polymer Matrices

It is supposed that properties of solids do not depend on their dimensions, but this assumption is true only for objects which contain macroscopic amounts of atoms. Recent research has shown that many physico-chemical properties of particles on the nanometer-scale differ greatly from those on a macroscopic scale. This is because of the following three factors. First, the size of nanoparticles is comparable with the Bohr radius of an exciton; this determines their optical, luminescent and redox properties. Secondly, being small in size, surface tension effects are expected to increase. This makes surface atoms very active and determines their considerable contribution to the thermodynamic characteristics of solids. Thirdly, the sizes of nanoparticles themselves are also comparable to those of molecules. These factors make the kinetics of chemical processes with nanoparticles specific.

Most of the existing routes for fabrication of nanoparticles are based on condensation and dispersion approaches. In the first case, nanoparticles are spontaneously assembled from single atoms during a phase transformation. According to the second method, nanoparticles are formed in the course of dispersion (disintegration) of a macroscopic phase. Such approaches allow metal-containing particles of different dispersion to be obtained. The latter determines a variety of physico-chemical properties of low-dimensional systems. Dimensional phenomena in the chemistry and physics of nanoparticles and their structural organization are essential for the creation of new technologies such as composition information media. The conception of nucleation, phase formation and self-organization of such systems are being extensively studied. Size control in the synthesis of nanoparticles is a kinetically driven process, where the ratio between the rates for nucleation and for growth is responsible for the final nanoparticle size. The particles formed are characterized by a high energy due to a large portion of surface atoms having non-compensated bonds. Therefore, for practical applications nanoparticles need to be stabilized. Polymers adsorbed on nanoparticles reveal a stabilizing ability. The mechanisms of such stabilization are not clear because a number of processes occur and a variety of factors are responsible for the stability of nanoparticles. However, using high-molecular-weight compounds as stabilizers enables researchers to produce metallopolymer nanocomposites already at the stage of nanoparticle synthesis. In such materials thermodynamically stable three-dimensional structures are formed due to cohesion of particles within a dispersion phase. However, a series of problems emerge here: How do the properties of the individual particles evolve into those of the phase (the so-called size effect)? What relations are there between an individual macromolecule and macroscopic matter? The role of molecular and supramolecular self-organization in such processes also becomes important.

1 Nanoparticles in Materials Chemistry and in the Natural Sciences (Introduction)

The major discoveries of the twentieth century can be mainly related to nuclear power engineering, space exploration, development of nanomaterials, semiconducting and microprocessor-based technology. All sciences have made fundamental contributions to the breakthrough in science. Speaking in support of nanomaterials it can be said that the nanoworld is extremely broad and it is practically impossible to find any field in the natural sciences that is not in some way or another connected with nanostructures. Terms such as nanophase, nanohybrid, nanocrystalline and nanoporous materials, nanochemistry, nanophysics, nanostructures, nanocrystals, nanophase geometry, nanosize hierarchy and architecture, nanostructured organic networks, molecular and nano-level[1] design and finally nanotechnology[2] are most frequently cited currently in the scientific literature as applied to small-scale and small-dimensional phenomena. A peculiar place is occupied in nanochemistry by the particles participating in various biological processes, including supramolecular functional systems to which belong enzymes, liposomes and cells. Fullerenes and nanotubes relate to this group too. These materials are used in chemistry in a number of new reactions, catalytic and sensor systems, fabrication of new compounds and nanocomposites with previously unknown properties. In physics they constitute some of the materials for microelectronics, structures with nanogeometry for information recording systems and are used to transform various natural radiations. In biology and medicine they are being used as novel medicinal preparations and carriers of these preparations to human organs. The relationship between materials science

[1] Nanos means a dwarf in Greek. A nanometer (nm) equals 10^{-9} m. Materials whose individual crystallites or phases making up their structural base (small forms of a substance) are below 100 nm in size at least in one dimension are called nanostructures. Although the limit is subjective and is accepted for convenience, beginning from this size, the exact portion of the regions bordering on disordered structures is becoming perceptible. It correlates with the characteristic size of one or another phenomenon (see later). As far as limiting values of the characteristic values for various physical properties are concerned, different metals and their oxides differ; this ensures a certain conventionality of 100 nm values.

[2] Unfortunately we have to admit that the above notions are not always used appropriately, e.g. in the marketing strategy for various brands.

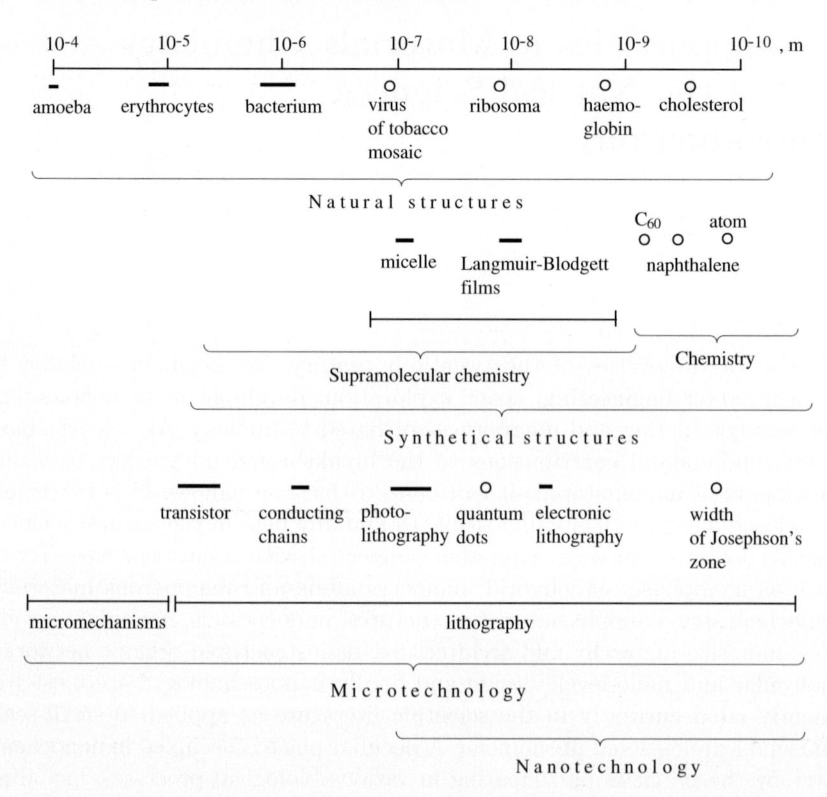

Fig. 1.1. Size correlation between synthetic and natural structures

and the natural sciences becomes more and more evident (Fig. 1.1 illustrates size correlation).

The assertions that science and engineering in the twenty-first century will acquire a nano and angstrom character are becoming a reality. In traditional technologies the limits of miniaturization of separate elements (e.g. density of arranging crystals in microelectronics) have already been attained. This motivates the search for some alternative procedures. Fabrication of modern chips is based on so-called planar techniques, being a combination of nanolithography (formation of surface drawings as lines and dots at the nanolevel) and etching. To reduce their size still more new recently developed lithographic procedures are used, particularly electronic, ion-beam and X-ray methods and dry etching, including plasma-chemical, reactive, ionic, etc. These techniques reach below the 100 nm size of elements in optoelectronic chips.

The part of materials science that studies nanophases differs from the traditional area by not only the creation of new materials but also by the necessity of designing corresponding instruments for manipulating such materials. Among the most promising nanotechnologies of metal materials and their products one can mention, in the first place, micro and nanometallurgy, laser treatment of material surfaces whose thickness is limited to a few hundreds and tens of nanometers, and various kinds of nanoceramics.

The transfer to high technologies necessitates the development of fundamentally new structural materials whose functional parameters are conditioned by the properties of microdomains and processes at the atomic and molecular level, in monolayers and nanovolumes.

1.1 Classification of Nanoparticles by Size

Ultradispersed particles can be subdivided by size into three types [1]: nanosize or ultradispersed to about 1–(30–50) nm particles,[3] highly dispersed of size (30–50)–(100–50) nm and micron size particles – flocculi of 100–1000 nm. The later consist, as a rule, of individual particles or their agglomerates representing either mono or polycrystals of fractal type. The first two types of particles are colloidal and the latter are coarse-dispersed ones.

Aerosol particles of metals (~50 nm) prepared in normal conditions are spherical or nearly spherical due to the high surface energy of fine particles [2].

The most frequently used terms are *ultrasmall particles* and *nanocrystals* to denote nanoparticles of metals with diameter from 2–5 to 50 nm, as well as colloidal crystallites [3] and subcolloidal particles. The upper threshold of the size of semiconducting nanocrystals in polymer matrices [4] is the condition which gives the optical homogeneity of compositions (no scattering by the environment at particle size below a quarter of the wavelength of light). The lower boundary is conditioned by the existence of crystalline particles at the interface between the crystalline phase and the quasi-molecular phase. The terms *molecular aggregation* and *crystallite clusters* are used less often [5].

There are two types of nanoparticles:

1. Clusters of 1–10 nm, the particles of an ordered structure possessing commonly 38–40, and sometimes more, metallic atoms (e.g. Au_{55}, Pt_{309}, a family of palladium clusters consisting of 500–2000 atoms).
2. Nanoparticles with diameter 10–50 nm and consisting of 10^3–10^6 atoms.

Nanoparticles are also classified by the number, N, of atoms they consist of [6–10]. Clusters are subdivided into very small, small and large. Table 1.1 shows the diameter, $2R$, for corresponding N_a atoms and the relation of N_s to N_v atoms.

The surface and the bulk appear to be inseparable for very small nanoparticles. For nanoparticles containing 3000 metallic atoms the relation $N_s/N_v \approx$

---

[3]
$$D = \Sigma n_i - D_i/\Sigma n_i \ ,$$
where n_i is the number of crystallites with diameter in $D_i+\Delta D$, $\overline{D} = D_i+(\Delta D/2)$ ($\overline{D} = D_{av}$). As a rule, it is necessary to estimate not less than 200 particles.

Table 1.1. Classification of nanoparticles and clusters by size

Very small	Small	Large
$2 < N \leq 20$	$20 < N \leq 500$	$500 < N \leq 10^7$
$2R_{\mathrm{Na}} \leq 1.1\,\mathrm{nm}$	$1.1\,\mathrm{nm} \leq 2R_{\mathrm{Na}} \leq 3.3\,\mathrm{nm}$	$3.3\,\mathrm{nm} \leq 2R_{\mathrm{Na}} \leq 100\,\mathrm{nm}$
Surface and inner volume are inseparable	$0.9 \geq N_s/N_v \geq 0.5$	$0.5 \geq N_s/N_v$

20% is true, whereas compact metallic particles ($N_s/N_v \ll 1$) are reached only at $N \gg 10^5$. Hence, particles can be subdivided by their size into the following four groups (domains):

I – molecular clusters ($N \leq 10$);
II – clusters of a solid body ($N \leq 10^2 \leq 10^3$);
III – microcrystals ($10^3 \leq N \leq 10^4$);
IV – particles of dense substances ($N > 10^5$).

Figure 1.2 illustrates the main stages of how an individual atom transforms into a bulk metal through a cluster, nanosize and colloidal particles (active metals) according to [11].

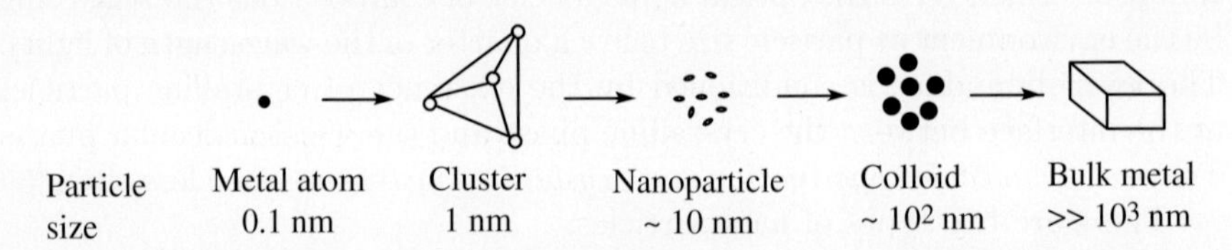

Fig. 1.2. The main stages of the transformation metal atoms into a bulk metal

Moving along the axis of sizes from an individual atom in a zero valence state (M^0) until we reach a metallic particle showing all the properties of a dense metal, the system passes through a number of intermediate stages, among which the main stages are considered to be clusterization and formation of nanoparticles.

In the idealized form the relationship between the number of atoms in clusters and their share of surface atoms, in providing a dense hexagonal packing, looks as shown in Table 1.2 [12, 13].

The structure of small and medium size homometallic clusters (3, 4, 5, 6, 10, 20 and 38 atoms) has been examined by X-ray analysis. The metal skeleton was found to consist of the following structural constituents [14]:

Table 1.2. Full-shell magic number clusters

Number of Shells	Number of Atoms in a Cluster	Share of Surface Atoms, %
1	13	92
2	55	76
3	147	63
4	309	52
5	561	45
6	923	
7	1415	With 3.2 nm diameter

cycle, carcass and polyhedron. After the transition from clusters to nanoparticles the non-monotonic behavior becomes less prominent. Yet the symmetry in the distribution of forces in them is violated because of uncompensated bonds between the atoms in the subsurface layers (having neighbors predominantly on one side). These atoms constitute a substantial portion of the total number of atoms in a particle. Thus, there are $\sim 10^{23}$ atoms in $1\,\text{cm}^3$ of Ni monocrystals, in contrast to $1\,\text{cm}^2$ of the surface where there are only $2 \cdot 10^{15}$ atoms. This means that the share of the surface atoms of a cubic crystal is $\sim 10^{-7}$. The number of surface atoms can be increased substantially if we enlarge the total surface (or dispersion degree) of the particles. For example, the share of atoms found on nanoparticle surfaces whose mean diameter is 1 nm (~12–15 atoms) is 100%, whereas the share of 5 nm size nanoparticles reaches only 15%.

Nanoparticles 2 nm in size contain comparable amounts of surface and inner atoms. The atomic packing in the center of the core corresponds to a face-centered cubic lattice but on the surface it has an icosahedron structure. The violation of equilibrium and symmetry in the force and mass distribution for clusters of this type results in the formation of hollows and shear deformations. In this case the coordination numbers of the surface atoms are smaller than those for inner atoms. The electronic shells of the former atoms are highly polarized and the bonds are oriented in only one direction, into the particle. This alters the physical and electrochemical properties of the nanoparticles as compared with large particles and bulk metals.

The properties of these ligand-deficient atoms turn out to be similar to the properties of the surface atoms of crystals. They are arranged at the interface and the atoms near crystal faces enter into interaction and are displaced. Furthermore, the translational symmetry is violated and the number of neighboring atoms is much lower, and anisotropy and inharmonic features of oscillations are stronger [2,15–18]. This results in a raised capability of adsorption, ionic and atomic exchange, and contact interaction between structural elements. On the other hand, the interpretation of the behavior of such particles is complicated because it is difficult to distinguish between their bulk and surface properties. It is crucial that surface atoms contribute most to the thermodynamic characteristics of solids and nanoparticles and largely ensure structural transitions and determine the melting point.

The term *metallic* reflects only the composition, but not the nature, of constituent particles that are intermediate between the metal and its individual atoms [15]. As they say, clusters are metal embryos. This simple scheme demonstrates the mechanical increment of the number of metal atoms participating in the construction of j-mers. In contrast to a known paradox of the ancient Greek philosopher Eubulide from Mileth[4] about a heap that demonstrates an objective dialectic of how a quantity transforms into quality from the standpoint of improbability of cognition, the formation of a metal embryo (a heap) from individual atoms can be described quantitatively. The association of electrons in a forming nucleus is spontaneous and in essence is similar to the formation of molecules from separate atoms. Special attention is paid to all these processes including the formation of a new phase that follows various models of homogeneous and heterogeneous nucleation, dynamics of particle growth and their structural formations. The real picture of particle nucleation and growth in a new phase at the micro- and macrolevels is extremely complex and should represent a single physico-chemical process where some stages might be of chain origin. This includes many interrelated stages, among which are the chemical transformation reactions (the source of the building material), mass transfer (diffusion mobility and transport of condensing particles into the assembly site), sorption phenomena observed at adsorption and desorption, as well as reaction of particles on nuclei surfaces, crystallization, etc. Some of these stages are heterogeneous which means that they are inhomogeneous in space, especially on the solid-phase surface or in its bulk.

The thermodynamic approach allows the examination of conditions of the generation of nuclei in the new phase, estimating their critical size and finding factors for their control. The kinetic equations are commonly used to interpret experimental results and define the distribution function of nanoparticles by size. The equations based on macroscopic approximations of the known kinetic models are able to specify the velocity and mechanisms of formation (coagulation) and decay of j-nuclei structures. Often, a statistical approach is used or numerical modeling of Lennard–Jones atoms and clusters by the

[4] Neither a grain can make up a heap, nor two grains can make a heap, and the addition of one more grain does not form a heap either; so it is doubtful whether a heap can be formed if one grain is added at a time, if each of them is not a heap.

methods of molecular dynamics and Monte Carlo methods, or a computer experiment is run.

1.2 Structural Organization of Nanoparticles

Nanoparticles are the objects of supramolecular chemistry which studies the synthesis of molecules and molecular ensembles capable of self-organization following their high reactivity and ability to self-assemble. The final formations have a loose-branching and most often fractal structure. They are called fractal clusters, fractal aggregates or fractal filaments [19]. These formations are characterized by diminishing mean density of the bulk of the substance as they grow while preserving self-similarity (scale invariance). These fractal structures are frequently formed in thin film metal–polymer nanocomposites. They are also critical objects in representing percolation structures appearing when isolated clusters join into a large conducting structure. The fractal phenomenon is a peculiar criterion for visualizing the level of the system disorder. The basic forces for the formation of these artificial systems are related to nonvalent interactions like in living nature.

Ligand-protected particles form, as a rule, solid powders which cannot be used immediately since cold or hot pressing and sintering are inapplicable due to intensive recrystallization. The great advantage of these 30 nm stabilized nanoparticles is that they become soluble in specific solvents beginning from nonpolar solvents (pentane) and ends with water. Nevertheless, the organization of such particles is a serious problem Nanoparticles with almost ideal properties can become useless unless the problem is solved. Their structure can be three- (commonly cross-linked by spatial molecules of various lengths), two- (e.g. self-organized ligand-stabilized particles on the solution surface, often with participation of linking blocks) or one-dimensional (quantum dots, quantum wires and even quantum cables). Three-dimensional (3D), two-dimensional (2D), and one-dimensional (1D) materials are now in extensive use today (Table 1.3).

1.3 Dimensional Phenomena in the Chemistry and Physics of Nanoparticles

The degree of particle dispersion significantly affects their activity which varies with particle growth. Properties of nanoparticles as a function of their size are a fundamental problem in the chemistry and physics of the ultradispersed state.

Nanoparticles exhibit so-called dimensional effects when their structural parameters are commensurable in at least one direction with the correlative radius of one or another chemical or physical phenomenon. This happens,

Table 1.3. Typical nanomaterials

	Size (approximate)	Materials
1D nanocrystals and clusters (quantum dots)	1–10 nm in diameter	Metals, semiconductors, magnetic materials, Langmuir–Blodgett films
Other nanoparticles	1–100 nm diameter	Ceramic oxides
Nanowires	1–100 nm diameter	Metals, semiconductors, oxides, sulfides, nitrides
Nanotubes	1–100 nm diameter	Carbon, layered metal, chalcogenides
2D arrays (of nanoparticles)	A few nm^2 to 1 μm^2	Metals, semiconductors, magnetic materials, polymer films
Surfaces and thin films	1–100 nm thick	Various materials
3D structures (superlattices)	A few nm in all three dimensions	Metals, semiconductors, magnetic materials, consolidated material, nanostructured materials
Nanoparticles in polymers	1–100 nm	Metal–polymer nanocomposite

e.g. with the free path length of electrons for electrokinetic properties or phonons, with coherent length in a superconductor, the size of a domain for magnetic characteristics, the size of a Frank–Reed loop for dislocation sliding, or the length of mechanical defects of dislocation or disclination type. They are characterized by quantum-dimension effects. The classical physical laws are replaced by the rules of quantum mechanics. When the size of a solid or liquid particle diminishes down to 100 nm and less, quantum-mechanical effects start to become more and more noticeable. The effects are displayed in the variation of the quantum-crystalline structure of the particles and their properties. At least three reasons responsible for these effects can be mentioned [20]. First, is commensurability of the size with one of several fundamental values of the substance or characteristic length of some process in it that evokes various dimensional effects. Second, the expanded specific surface and raised surface energy of nanoparticles at a limited number of atoms and uncompensated electronic links affect their lattice and electronic subsystems. And, third, the severe conditions of their formation (high or low temperatures and process rates, exposure to powerful radiation sources, etc.) transform nanoparticles into a non-equilibrium (metastable) state. These factors determine the specifics of the atomic structure of separate nanoparticles and of the atomic and crystalline structure of a nanomaterial as a whole. A strictly periodic lattice typical of a crystal does not answer the minimum of energy of a nanoparticle. Its stability corresponds to an inhomogeneous strain and, in the case of multicomponent materials, to an inhomogeneous distribution of components and phases by the particle radius. A great variety of imperfections have been detected in the atomic structure of nanoparticles, including variations in the interatomic distance, increasing mean square shifts of atoms, defectiveness, microdistortion, amorphization, inhomogeneous strain; structural, concentration and phase inhomogeneity of nanoparticles; stabilization of high-temperature phases, etc. The specific heat, susceptibility, conductivity and other critical characteristics of metals change significantly as nanoparticles reach the nanoscale. Moreover, their specifics are imposed by a non-monotonic dependence of the basic characteristics, including melting point, pressure of recombination of the crystalline structure, ionization potentials, binding energy per metal atom, change in interatomic distances, optical and magnetic parameters, electronic conductance, electron–phonon interactions, etc. upon cluster size and the number of atoms M in it.

Formally [6,10], nanoparticles of various sizes are treated as a link between the classical objects of the chemistry and physics of solids (Fig. 1.3 and Table 1.4).

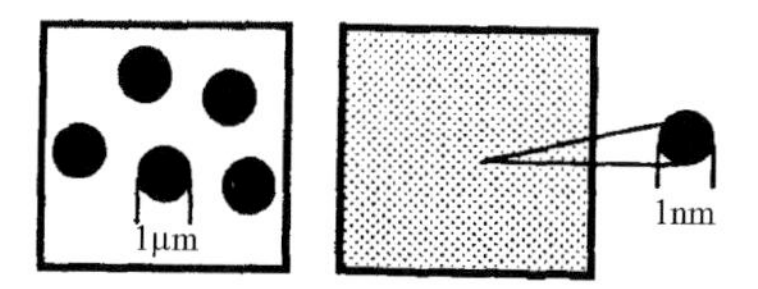

Fig. 1.3. Size correlation between micro- and nanoparticles

Table 1.4. Classification of nanoparticles by fields of investigation

Chemistry			Nanoparticles			Physics of Solids
Atom	$N = 10$	$N = 10^2$	$N = 10^3$	$N = 10^4$	$N = 10^6$	Compact substance
	0.1	1 nm	2 nm 3 nm	5 nm 7 nm	10 nm	100 nm

Proceeding from the above, dimensional effects can be conventionally subdivided into two groups. Inner size effects are connected with variation on the surface and in the bulk. They are exhibited by the electronic and structural parameters of nanoparticles, including binding energy, chemical activity, ionization potential, melting point and crystalline structure. The effects of

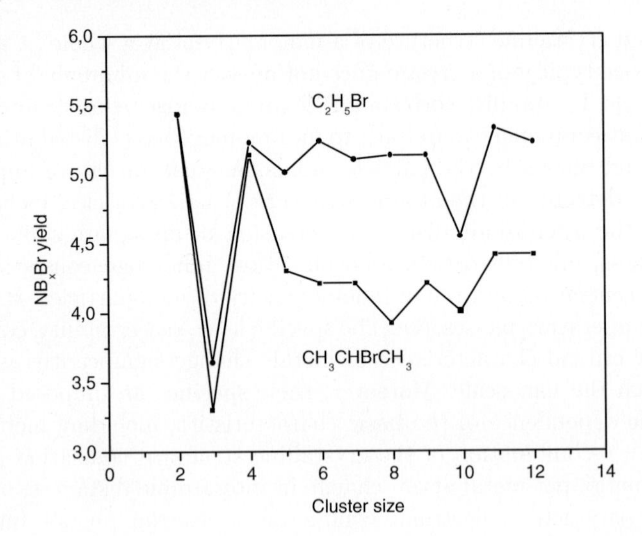

Fig. 1.4. Size dependence of the reactivity of niobium clusters (Nb_x) to alkyl bromides as the number of atoms (x) in the niobium clusters was varied

external size attributed to collective properties or the lattice excitation are displayed in the variation of optical spectra of nanoparticles [21,22]. This can be proved for the example of niobium clusters Nb_x (x = 1–12) with either isopropyl bromide or ethyl bromide (Fig. 1.4). The yield of Nb_xBr for x = 4 is a maximum due to the high activity of particles with uncompensated energy of surface atoms which then reduces in the case of isopropyl bromide because of the intensified steric factors in the niobium cluster.

In Fig. 1.5a, [23, 24] Fig. 1.5b, Fig. 1.6, [25, 26], Fig. 1.7, Fig. 1.8 [28] and Fig. 1.9 [27] typical size-dependent properties of some semiconductors and nanocrystals are shown. The drop in melting temperature T_m with size reduction has been known for a long time. Recently, this has been proved experimentally for only comparatively low-temperature two-phase systems, including In-Al (Fig. 1.6a). The melting enthalpy of indium particles (ΔH_M^E) was shown to depend on particle chemical activity which lowers in the following order (in the example of low-temperature co-condensation on cold surfaces [22]): metal atom > cluster > nanoparticle > nanoparticle aggregate > bulk metal (Fig. 1.9). Actually, the dependence bears an extreme rather than monotonic behavior.

1.4 Nanoparticles and Materials on their Base

It is important that in contrast to homogeneous materials, nanoparticles having an extremely developed interfacial surface show excess energy due to which they are often called energy-saturated systems. Peculiar to them

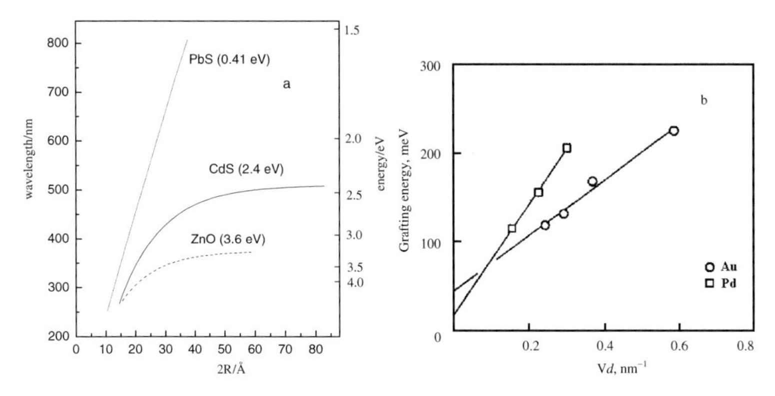

Fig. 1.5. Size dependence of the wavelength of the absorption threshold in semiconductors (**a**) and of the nonmetallic gap (local density of states) with the volume of metal nanocrystals (**b**)

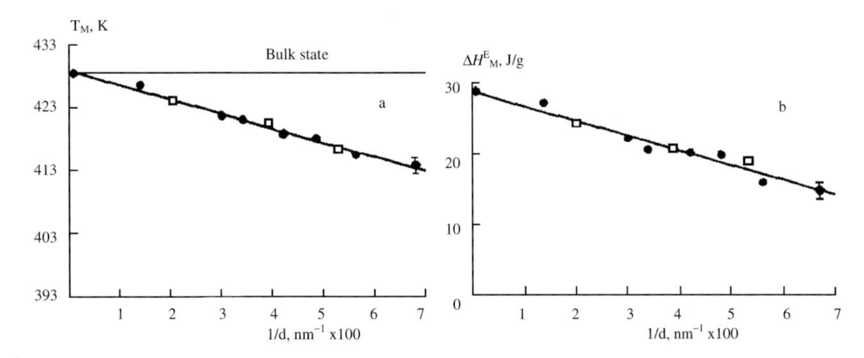

Fig. 1.6. The size dependence of melting temperature (**a**) and enthalpy (**b**) of indium particles in an aluminum matrix

properties serve the base for relating them to the fifth aggregate state of the substance [29]. These materials are the subject of a new developing scientific discipline – physico-chemistry of nanoparticles (sometimes called nanochemistry or physics of clusters). Successes in this field are closely related to the development of synthesis methods (Table 1.5).

For example, scanning probe microscopy (SPM), and its variants such as scanning tunneling (STM), atomic force (AFM), and magnetic (SMM) microscopy, allow us not only to visualize the structures formed on the atomic and subatomic levels but to carry out the transfer and localization of single molecules into the solid surface [30–34]. The range of production methods of nanoparticles is extremely broad. They can be grouped into physical (first-order phase transformations in the absence of chemical reactions) and chemical. Among commonly used physical methods employing aggregation of free molecules and ions into clusters, the most widespread are condensation procedures. The condensation procedure includes assembly of nanoparticles

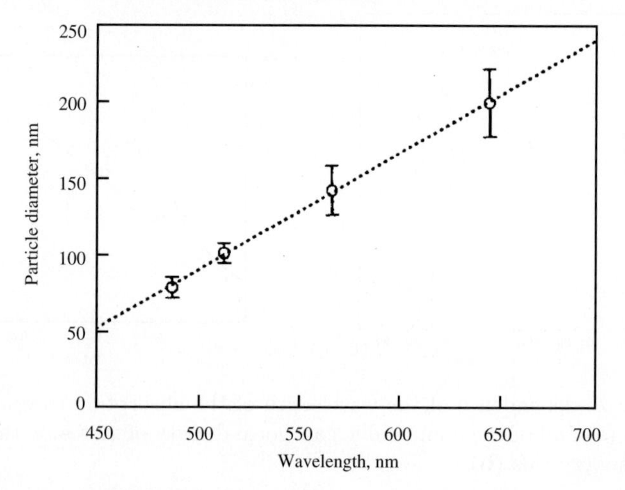

Fig. 1.7. Plot of particle size as a function of excitation wavelength. Particle sizes are the average values of five or more nanoparticles measured at each wavelength

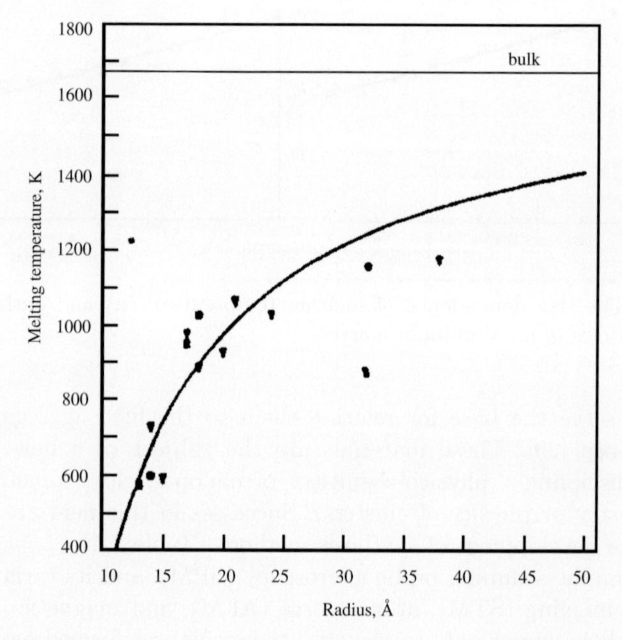

Fig. 1.8. The variation of the melting temperature in CdS nanocrystals with size

from individual metal atoms, or molecular and cluster groups. Upon reaching certain dimensions and a definite physical interface the atoms associate and after a transient state become nuclei of a new phase. This results in the formation of an equilibrium and stable (like in micelle formation) structure.

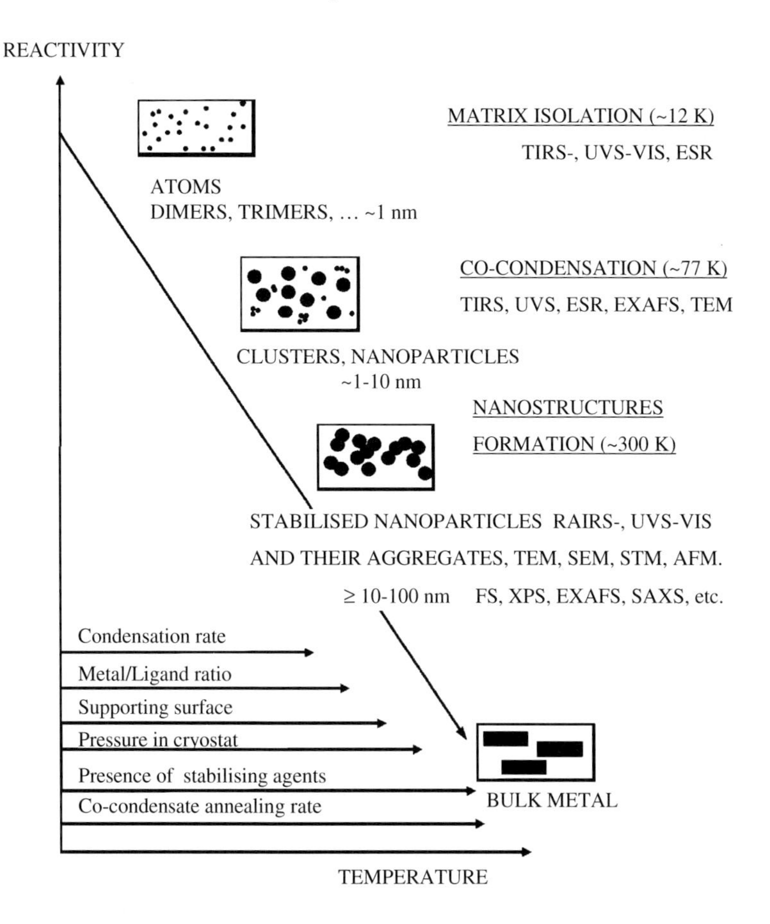

Fig. 1.9. The relative chemical activity of metal species versus their size

Modifications of the procedure will be analyzed in detail in the chapters to follow. Note that the majority of them are complementing rather than competing methods.

Less frequently, dispersion methods are employed that consist of disintegration of the macroscopic phase of the substance (coarse-dispersed particles in colloidal mills, agitators, ultrasonic, high-frequency and electric-arc crushing, etc.). They all belong to forcible methods since perceptible work is spent on breaking the links, although sometimes a spontaneous dispersion of substances is feasible under specific conditions.

Among chemical methods the reduction of metal compounds in solutions in the presence of various stabilizers has been in extensive use for more than a hundred years. The reducers can be hydrogen and hydrogen-containing compounds, e.g. tetrahydroborates, citrates of alkali metals, hypophosphates, alcohols, metal–organic compounds, high-energy radiation, etc. The kinetic

Table 1.5. Nanoparticles and materials on their base

Scale (approx.)	Synthetic Methods	Structural Tools	Theory and Simulation
0.1–10 nm	Covalent synthesis; reduction procedures; sol-gel synthesis; intercalation methods	Transmission infrared spectroscopy (TIRS); UV and visible electronic spectroscopy (UVS-VIS); nuclear magnetic resonance (NMR); electron-spin resonance spectroscopy (ESR); diffraction methods; scanning probe microscopy (SPM)	Electronic structure
< 1–10 nm	Self-assembly techniques; Langmuir–Blodgett technique; biomineral-ization	Scanning electron microscopy (SEM); transmission electron spectroscopy (TEM); atomic force microscopy (AFM); X-ray photoelectron spectroscopy (XPS); extended-ray absorption fine structure (EXAFS)	Molecular dynamics and mechan-ics
100 nm–1 µm	Dispersion; microen-capsulation; biosorption; processing, modifica-tions		Coarse-grained models, hopping, etc.

parameters of the reduction reactions depend upon the nature of the agent to be reduced and the conditions thereof. Chemical methods include various modifications of electrochemical synthesis, thermal decomposition of unstable covalent metal-containing compounds (including metal–organic compounds) and molecular complexes. Common features of the processes of nanoparticle formation are the high rate of their nucleation and low growth velocity. This is the condition of their promising use and search for the means of exercising various types of synthesis. Their general drawback is the formation of polydispersed colloids within a wide range of particle distribution by size. At the same time, many of the methods aimed at production of nanoparticles furnish the chance for researchers to choose a synthesis procedure in response to concrete purposes and requirements.

It is worthwhile mentioning a new direction in physico-chemistry of nanoparticles – the chemistry of gigantic clusters. A number of synthesis methods for compounds with metal–metal links, whose nuclearity reaches several hundreds, have been elaborated lately.[5]

It has been noted earlier that the severe conditions of synthesis, and the large specific area S_{sp} of nanoparticles, characterized also by small-size morphological elements, can induce variations in their physico-chemical properties and even violation of the atomic structure [36]. Extremely high (or low) temperatures and velocities of the processes, and various external effects (e.g. fast condensation or quenching) assist in the formation of nonequilibrium so-called frozen states in growing j-nuclei particles.

The developed interface, excess energy of the surface atoms (e.g. change in Helmholtz energy ΔF under dispersion of a homogeneous metal of volume V till the particle radius r is $\Delta F(r = 3\sigma V/r)$, where σ is the energy increase of the system at the formation of the new interface of unit area) and lowered activation energy of various reactions promote high activity of the particles. This is why energy-saturated nanoparticles obtained in strongly nonequilibrium conditions enter into highly intensive interactions with components of the medium they are formed in. Even when they are formed in an inert atmosphere, they spontaneously enlarge and result in a powder of usual dispersivity.

Nevertheless, the very high activity of nanoparticles can sometimes be helpful, although it creates the tedious problem of their stabilization during transporting and storage.

However, strong passivation of their activity, especially under catalysis, may be undesirable. Passivation should preserve the high chemical activity of nanoparticles. Nanoparticles in the absence of stabilizers are typical lyophobic colloids with low aggregate stability. So the choice of an efficient means of controlling chemical passivation of nanoparticles is considered as one of the most important trends. The solution of this problem is facilitated by the high sorption capacity of such particles towards surfactants and a number of high-molecular compounds. As a result of steric stabilization the formations transform into a structure with a nucleus encapsulated into a sort of lyophilic steric barrier consisting of a continuous layer of solvate polymer chains. Sometimes only a minimum amount of polymer stabilizer is needed to

---

[5] Strictly speaking, these huge clusters are not treated as individual compounds but include particles of different size (e.g., Pd_{561} contains 570 ± 30 atoms) and have a narrow range of distribution (like the molecular-mass distribution of polymers). Their formulas are averaged by the statistical ensemble, e.g. $Pd_{561}Phen_{60}(OAc)_{180}$ produced by $Pd(OAc)_2$ reduction by hydrogen in the presence of 1,10-phenantropine [35] or Pd_{1415} [15]. By their essence the clusters are nanoparticles (Pd_{561} is 2.6 nm in diameter) covered by a ligand protecting them from the environment.

overcome the coagulation threshold. Often, especially when nanocomposites are produced, the share of nanoparticles is low in contrast to the protecting cover. The role of polymers cannot be reduced to only screening of particles as it will be a strongly simplified approach to the problem of stabilization. In reality its mechanism is extremely intricate and many problems of polymer and nanoparticle interactions remain unsolved.

In this monograph we have tried to present the current state of three aspects in the physico-chemistry of nanoparticles critical for understanding the structure and properties of nanocomposites. This also relates to adsorption and chemisorption of macromolecules on nanoparticle surfaces from solutions, generation of interfaces, and phenomena of surface conductivity, specific interactions dependent upon the chain origin and length, its conformation, and composition of copolymers. Hydrophobic polymers in polyelectrolytes similarly charged with nanoparticles are inclined to associate ionic groups and form domains as microphases of ion regions.

Physico-chemical properties of polymer surfactants which present macromolecules with both hydrophobic and hydrophilic fragments are to be considered as stabilizers of the ultradispersed state in all the processes cited above. Furthermore, the topochemistry of interactions and the morphology of future nanocomposites are to be taken into account as well. The stabilization mechanism is based on structural and mechanical constituents of stability in dispersed systems, spatial networks of the coagulation structure type and formation of adsorption-solvate structured films on nanoparticle surfaces.

Block copolymers based on polar and nonpolar comonomers are found in some solvents in the form of inverted micelles. Their structure is separated at the microphase level. Nanoparticles are formed in the nuclei from a polar monomer like in a micelle (a peculiar nanoreactor – a domain) and they are covered by blocks of a nonpolar monomer (crown) serving as a stabilizer (a hairy ball structure). In this case, we deal with the realization of the ligand-controlled synthesis of nanoparticles. Modern trends in this field are the stabilization of concentrated and highly concentrated systems, surface stabilization in Langmuir–Blodgett (LB) films, dendrimer-template nanocomposites and matrix insulation processes based on the formation of a polymer stabilizer in the presence of dispersed nanoparticles.

Nanocomposite materials are a polymer matrix with nanoparticles and clusters randomly distributed (a totality of adjoining discrete particles in the form of aggregates of indefinite shape and size). The production method does not differ formally from the traditional approaches to nanoparticle synthesis without polymers. These procedures will be covered in this book. Nanoparticles in such systems serve as a dispersed phase whereas the polymer matrix serves as a dispersed medium. Nanoparticles are microencapsulated in the polymer shell. Recently, composites whose nanoparticles are localized only on the surface of powders, fibers, films and nanocomposites on the base of nanoparticles having 3D, 2D and one-dimensional structures have become widely used.

Nanocomposites can be produced by various methods: by generating nanoparticles in a specially prepared polymer matrix, less often by polymerization or polycondensation of corresponding precursors and very rarely by creation of the materials in a single stage with simultaneous shaping of both nanoparticles and the polymer shell. The last method is complicated by strict requirements on the velocities of the processes. In fact, these two completely different processes have the same kinetics including nucleation, growth and termination of material chains.

In spite of the abundance of such methods and their modifications, not many of them are complete or used in industry. So the methods of metal coating of polymer surface by evaporation are widely used. Nanocomposites with high content of metal particles can be obtained by thermal decomposition of the corresponding metal carbonyls (mostly in thermoplastics). Still, the broadest sphere of activity in the field lies in chemical reduction of metal compounds in the presence of a polymer matrix. This is, however, a complex multistage process which includes the formation of unstable low-covalent forms of metals undergoing spontaneous decay and metal isolation. Along with stabilization of these highly reactive particles in situ nascendi the polymer matrix comparatively often controls the size and sometimes shape of the forming nanoparticles, the rate of their crystallization, the formation of percolation structures, and self-organized growth of cluster formations. Sometimes the macroligand is the factor that helps to register the formation and even to isolate intermediate particles or clusters of specific sizes.

In fact all methods and their modifications for producing nanocomposites of nonmetallic type fit the synthesis of multimetallic materials. During multimetallic nanocomposites production various complex physico-chemical transformations occur that result in inhomogeneities and various phases at the microlevel. The processes have been comprehensively studied on the example of metal–polymer systems with bimetallic inclusions of Pd/Pt, Pd/Cu, Pd/Au, etc., triplet systems like Y–Ba–Cu, and the like. Preliminary synthesis of polymers is especially convenient in synthesis of structurally homogeneous materials. It employs the polymer matrix with metal salts dispersed in it at the molecular level. This procedure is also used to prepare multicomponent steels of M50 type. However, a number of problems concerning the matrix structure and composition effect on the physico-chemistry of nanoparticle nucleation and growth, and the formation of individual phases of metals and their alloys in multicomponent systems, are still awaiting solution.

The polymer matrix remains indifferent in the reactions of nanocomposite formation. They are frequently accompanied (especially in subsurface layers) by polymer destruction and abrupt reduction of molecular mass, cross-linking and formation of spatial structures (sometimes by isomerization of polymer units). This results in the loss of matrix solubility, fracture of the polymer crystalline phase followed by its transfer into the amorphous state and rupture of separate links (more often halogenide, ether and epoxide links). Nevertheless, a critical factor that ensures the degree of interaction between the polymer and nanoparticles is the formation of complexes and even chemical metal compounds and salts of arrhenic, bis-arenic and some other π-complexes of metals and chelate cycles of various composition with participation of polymer functional groups. When it has either been proved or presumed that the matrix and nanoparticles interact with each other, the systems are called polymer-immobilized. This is especially evident in the example of polymer-immobilized clusters and polycyclic formations with the identified structure including the heterometallic types (individual clusters in contrast are nanoparticles). They are produced in a specific manner (chiefly by phosphor-, nitrogen- and oxygen-containing groups) from functionalized polymers through polymer-analogous transformations, leaving the polymer chain and cluster skeleton intact. It looks prospective to produce materials by poly- and copolymerization of cluster-containing monomers (by the simultaneous formation of the matrix and nanoparticles similarly to nanocomposites).

An original variant of producing nanoparticles stabilized by a polymer matrix presents a combined pyrolysis of polymers and metal-containing precursors. The origin of precursors introduced into the polymer matrix may exert a strong effect on its thermal decomposition. High-temperature thermolysis (depolymerization) makes monomers and some other products, and is accompanied by graphitization of the matrix and nanoparticles. Conditions under pyrolysis (gaseous medium, ligand environment, metal nature) might influence thermolysis and lead to formation of different nanoparticles including metal, oxide, carbide, and sometimes nitride nanoparticles which are encapsulated into either a cross-linked polymer matrix or carbon-containing shell. Nanoparticles of transient metals and their lowest-order oxides with a deficient structure turn out to be the most efficient inhibitors of high-temperature thermal and thermooxidative destruction of polymers. Approaches that are based not only on the pyrolysis of polymer-immobilized polynuclear metal complexes but on the controlled thermolysis of metal-containing monomers with a series of stages seem to be most promising too. These stages also include the formation of metal–polymers presenting self-regulating systems in which synthesis of the polymer matrix, nucleation and growth of nanoparticles are running simultaneously.

For a long time synthesis of new monomers under optimum kinetic regimes has been the dominant method of obtaining novel materials. Significant achievements have been attained in this field. The potential of this method is however bounded. It is getting harder to reach the desired results. Recently the investigation has moved to the search for modification of known in the art procedures and materials. In particular, these approaches have been put into the base of developing hybrid polymer–inorganic nanocomposites. What is common in these materials and the topic of this book is the nanometer parameter, distances between networks and layers formed by the polymer and inorganic ingredients and the nanosize of the forming particles including metal-containing particles.

A rapidly developing area is sol-gel production of hybrid materials as well as intercalation of polymers and nanoparticles into layered systems in terms of the chemistry of intracrystalline structures like the guest–host. It is important that these procedures employ substances and reagents that are ecologically safe and convenient for deriving highly dispersed materials. Hydrolysis that presupposes condensation of metal and metalloid alcoholates in polymer-forming compounds or in polymer solutions is widely used. Interactions between phases of inorganic and polymer components here play the leading part. Special attention is paid to network polymer–inorganic composites whose components have strong covalent and ionic chemical bonds. Initiated by practical demands intensive elaboration of nanosize (soft template systems) and polymetallic ceramics based on sol-gel synthesis is underway today.

It is of special interest to investigate polymer intercalation into porous and layered nanostructures. Intracrystalline cavities of the host represented by inorganic oxides (silicate layers), metal chalcogenides with a regulated system of nanosize pores and channels are filled with atomic or molecular guest structures which are clusters, nanoparticles, coordinate structures of CdS type, or polymer units synthesized in situ. The most challenging type of intracrystalline reactions is inclusion of monomer molecules into the guest pores followed by their post-intercalation conversion into the polymer, oligomer or hybrid sandwich products. This method is used currently to fabricate polyconjugate conducting nanocomposites based on polyaniline, polypyrrole and others. Thus induced encirclement of the nascent polymer inclusion in self-assembling polylayered nanocomposites is characterized by high degree of properties.

Clusters and nanoparticles in LB films turn out to be promising materials for molecular designs, especially with respect to developing self-organized hybrid nanocomposites. Metal–polymer LB films and their properties are actively investigated now. One more aspect in understanding nanostructures is recognition that they are bridges to modeling surface and biological processes as objects of supramolecular chemistry and structural chemistry of intermolecular links [36]. Although only the first steps have been taken in this direction, one can perceive the far-reaching vistas in applying the described techniques for generation of self-organized layers at the molecular level.

Hybrid nanomaterials are common in the living nature. Interactions between metal-containing particles, biopolymers (proteins, nucleic acids, polysaccharides) and cells are critical in enzymatic catalysis, geobiotechniques, biohydrometallurgy and biomineralization. Perfection of the processes, modeling, and creation of synthetic analogs allow us to approach living organisms. This is especially so for polynuclear metal enzymes, as well as biosorption and biomineralization processes. Thus, microbes work hard like gold-diggers extracting nanoparticles from gold and ore by biological leaching and then enlarge them till they are visible gold grains. Clusters and nanoparticles serve as models for a series of biological concepts that are the objects of biomimetics and bioinorganic chemistry. Practically all approaches to the production of nanocomposites analyzed in this book (polymer-immobilized cluster and nanoparticles, sol-gel syntheses, layered intercalation compounds, LB films, etc.) are intensely used for biopolymers and biocomposites. Profound successes have been attained in modeling polynuclear negemic complexes of iron (methane monooxigenase components), in designing photosystems for oxygen isolation from water similarly to enzymatic processes, in biochemical nitrogen fixation, use of nanoparticles in diagnostics of pathogenic and genetic diseases through their aggregation in complementary oligonucleotides, in the creation of metal–enzyme preparations, and many others.

When microorganisms are combined with colloidal metals, they concentrate bacteria. These are preceded by their adsorption on the cell surface and is followed by assimilation and precipitation of the metal–bacterial mass. This procedure can become a reliable selective method for removing bacteria (owing to selective affinity to metals) and collective recrystallization coarsening of metal particles (microbiological geotechnology). These phenomena are based on the same colloidal and chemical interactions as in synthetic polymers. Biogenous formations frequently encountered in nature can be formed in the course of biomineralization and bioconcentration with the strict control of the matrix (template) over nucleation and growth of materials with a perfected structure. This leads to generation of materials with a complex hierarchic structure of biological nanocomposites. Although it is extremely difficult to reproduce intricate and accurate bioinorganic reactions under laboratory conditions, a set of complex processes, including microencapsulation of active enzymes into a sol-gel matrix, production of biosensors, enzymatic electrodes, and components of optical meters have already been successfully performed.

The fields where clusters and nanoparticles encapsulated into the polymer matrix can be adopted are, in fact, unbounded. We will enumerate only most important among already realized processes.

Nanoparticles are commensurable in size with Bohr's radius of excitons in semiconductors. This governs their optical, luminescence and redox properties. Since the intrinsic size of nanoparticles is commensurable with that of a molecule, this ensures the kinetics of chemical processes on their surface [37, 38]. Current investigations are concentrated also on the study of boundary regions between nanoparticles and the polymer, because the interfaces are responsible for the behavior of adsorption and catalysis.

The majority of investigations in nanoparticles bear an interdisciplinary character as far as methodologies of a number of scientific domains, including physico-chemistry, materials science, biotechniques, nanotechnology, etc. are required to get further insight in the subject. The science of nanocomposites has sprung up in the recent decades (see [39–42,44]) at the junction of various fields of knowledge and almost at once started to yield practical results. Its intensive advance, enrichment with new notions and wide spectrum of knowledge presented, until recently, a barrier to making an at least provisional review of the results attained. This monograph is aimed at generalizing the state of the art in this progressing boundary sphere of science.

References

1. Y.I. Krasnokutsky, B.G. Vereschak: *Production of Refractory Compounds in Plasma* (Kiev 1987)
2. Y.I. Petrov: *Clusters and Small Particles* (Nauka, Moscow 1986)
3. R. Rossetti, L. Brus: J. Phys. Chem. **90**, 558 (1986)
4. I.A. Akimov, I.Y. Denisyuk, A.M. Meshkov: Optics and Spectroscopy **72**, 1026 (1992)
5. D. Braga, F. Grepioni: Acc. Chem. Res. **27**, 51 (1994)
6. K.J. Klabunde: *Chemistry of Free Atoms and Particles* (Acad. Press, New York 1980)
7. H.S. Halwa: In: *Handbook of nanostructured Materials and Nanotechnology. Vol. 1. Synthesis and Processing, Vol. 2. Spectroscopy and Theory, Vol. 3. Electrical properties, Vol. 4. Optical properties, Vol. 5. Polymers and Biologycal Materials* (Academic Press, New York 1999) N4
8. U. Kreibig, M. Volmer: *Optical Properties of Metal Clusters* (Springer, Berlin 1995)
9. Nakoto Takeo: *Disperse Systems* (Wiley-VCH, 1999)
10. K.J. Klabunde: *Free Atoms, Clusters and Nanoscale Particles* (Acad. Press, San Diego 1994)
11. *Active Metals* ed. by A. Furstner (VCH, Weinheim 1996)
12. G. Schmid: Endeavour, New Series **14**, 172 (1990)
13. C.N.R. Rao, G.U. Kulkarni, P.J. Thomas, P.P. Edwards: Chem. Soc. Rev. **29**, 27 (2000).
14. S.P. Gubin: *Chemistry of Clusters* (Nauka, Moscow 1987)
15. G. Schmid: Chem. Rev. **92**, 1709 (1992)
16. I.D. Morokhov, V.I. Petinov, L.I. Trusov: Russ. Phys. Rev. **133**, 653 (1981)
17. *Clusters, Atoms, and Molecules. Springer Series in Chem. Phys.* ed. by E. Heberland **52** (1994)
18. S.L. Lewis: Chem. Rev. **93**, 2693 (1993)
19. M. Folner: *The Kinetics of New Phase Formation* (Mir, Moscow 1986)
20. V.F. Petrunin: Nanostruct. Mat. **12**, 1153 (1992)
21. U. Kreibig, Y.I. Petrov, M.A. El-Sayed: J. Phys. Chem. **95**, 3898 (1991)
22. G.B. Sergeev, I.I. Shabatina: Surface Science **500**, 628 (2002)
23. A. Henglein: Ber. Bunsen-Ges. Phys. Chem. **99**, 903 (1995)
24. P.J. Thomas, G.U. Kulkarni, C.N.R. Rao: Chem. Phys. Lett. **321**, 163 (2000)
25. H.W. Sheng, Z.Q. Hu, K. Lu: J. Mater. Res. **11**, 2841 (1966).

26. R.A. Andrievsky, A.M. Glezer: Physics of Metals and Metal Science **88**,50 (1999)
27. A.P. Alivistatos: Endeavour **21**, 56 (1997)
28. S.R. Emory, W.E. Yaskins, S. Nie: J. Am. Chem. Soc. **120**, 8090 (1998)
29. H. Muler, C. Opitz, L. Skala: J. Mol. Catal. **54**, 389 (1989)
30. Y. Wada: Jpn. J. Appl. Phys. **39**, 3825 (2000)
31. C.P. Collier: Science **289**, 391 (2000)
32. G.P. Lopinski, D.D. Wayner, R.A. Wolkov: Nature **408**, 48 (2000)
33. H. Siegenthaler: In: *Scanning Tunneling Microscopy, Vol. 28* (Springer, New York 1992) p. 231
34. G.J. Fleer, M.A.C. Stuart, J.M.H. Scheutjens, T. Cosgrove, B. Vicent: *Polymers at Interfaces* (Chapman and Hall,London 1993)
35. M.N. Vargaftik, V.P. Zagorodnikov, I.P. Stolarov, I.I. Moiseev, D.I. Kochuney, V.A. Likholobov. A.L. Chuvilin, K.I. Zamaraev: J. Mol. Catal. **53**, 315 (1989)
36. M.N. Vargaftik, I.I. Moiseev, D.I. Kochubey, K.I. Zamaraev: Faraday Disc. **13**, 92 (1991)
37. J.-M. Lehn: Angew. Chem. Int. Ed. Engl. **27**, 89 (1989), **29**, 1304 (1990); Macromol. Chem. Macromol. Symp. **69**, 1 (1993)
38. A.D. Pomogailo: *Catalysis by Polymer-Immobilized Metal Complexes* (Gordon & Breach Sci. Publ., Amsterdam 1998)
39. A.D. Pomogailo: Plat. Met. Rev. **38**, 60 (1994)
40. A.M. Natanson, Z.R. Ulberg: *Colloidal Metals and Metal-polymers* (Naukova Dumka, Kiev 1971)
41. A.M. Natanson, M.T. Buk: Russ. Chem. Rev. **41**, 1465 (1972)
42. B.K.G. Theng: Clay Miner. **18**, 35 (1970)
43. A.D. Pomogailo, A.S. Rozenberg, I.E. Uflyand: *Nanoparticles of Metals in Polymers* (Khimiya, Moscow 2000)
44. M. Kohler: *Nanotechnologie* (Wiley-VCH, Weinheim, New York 2001)

2 Nucleation Mechanism of Nanoparticles and Metal Clusters, their Structure and Chief Properties

The physico-chemistry of metal nanoparticles and their nucleation processes have been comprehensively studied and quantitatively described most fully for the cases of homogeneous and heterogeneous formation in the gaseous and liquid phases. Despite the diversity of synthetic methods of producing nanoparticles they can fit into two characteristic classes. The methods belonging to the first class are sometimes called dispersion or top-down methods and these are physical processes. They are based on first-order phase transformations in the absence of chemical reactions during which block metal is atomized to nanoparticles.

The second class presents a rather numerous group of bottom-up methods that employ chemical approaches to the assembly of nanoparticles from either mononuclear metal ions or kernels of a lower inclination to nucleation. To eliminate agglomeration of such particles at storage and transportation, a great variety of stabilizers, including donator ligands, surfactants, and polymers, etc., are used.

The formation of nanoparticles proceeds in stages among which the main are nucleation and growth of nuclei, rupture of material chains and formation of a new phase. It is difficult to identify the stages distinctly although the formation of a new phase is a well-explored physical domain.

Chain nucleation (stage of formation of active M_1 particles), growth of the new phase particles and generation of transformation products (P, P*) can be treated from the viewpoint of kinetics as a chain process (or *bimolecular*)

$$\begin{aligned} &\mathrm{M} \xrightarrow{k_0} \mathrm{M}_1 + \mathrm{M}_j && (monomolecular) \\ &\mathrm{M} + \mathrm{M}_j = \mathrm{M}_1 + \mathrm{P}^* && (bimolecular) \end{aligned} \tag{2.1}$$

and chain extension (*cluster growth*)

$$\mathrm{M}_j + \mathrm{M} \xrightarrow{k_j} \mathrm{M}_{j+1} \quad (1 < j < \infty)\,. \tag{2.2}$$

The reverse reactions, e.g. dissociation of clusters, take place along with chain nucleation and growth

$$\mathrm{M}_j \xrightarrow{k_j^-} \mathrm{M}_{j-n} + \mathrm{M}_n \qquad (n \geq 1) \tag{2.3}$$

interaction between chain carriers (including coagulation)

$$\mathrm{M}_j + \mathrm{M}_n \xrightarrow{k_{j,n}} \mathrm{M}_{j+n} \qquad (j, n > 1) \tag{2.4}$$

and destruction of the chains as a result of interaction of the growing cluster and stabilizer molecule (or matrix S^*)

$$\mathrm{M}_j + S^* \xrightarrow{k_s} \mathrm{M}_j S^* \,. \tag{2.5}$$

To describe phase formation, it is necessary to consider large series of successive elementary reactions and to solve a large number of differential equations. Quantitative characteristics of elementary reactions during nucleation and growth of metal-containing nanoparticles are almost unknown so far.

The major distinctive feature of nanoparticle formation in the kinetic respect is the high rate of metal-containing phase nucleation as opposed to its low growth rate. It serves as a criterion for determining topological means of synthesizing nanoparticles.

In this chapter we shall dwell upon common representations, general trends and methods of depicting nucleation processes, structure and growth of new gas and condensed phases as the start of initiating high-dispersed metal-containing particles and their further transformations.

2.1 Characteristic Features of Nanoparticle Nucleation

The formation of nanoparticles from a single metal atom in a zero valence state M_1^0 and their conversion into a compact metal M_∞ proceeds through generation of intermediate ensembles (clusters, complexes, aggregates) consisting of j atoms M_j $(1 < j < \infty)$ and can be presented by the following scheme (without consideration of back reactions)

$$\mathrm{M} \xrightarrow[k_1]{\mathrm{M}} \mathrm{M}_2 \xrightarrow[k_2]{\mathrm{M}} \mathrm{M}_3 \xrightarrow[k_3]{\mathrm{M}} \ldots \xrightarrow[k_{j-1}]{\mathrm{M}} \mathrm{M_j}. \tag{2.6}$$

The balance equation of the process looks like

$$n\mathrm{M}_1^0 \to \lim \sum f(j, L)\mathrm{M}_j \,, \tag{2.7}$$

where $f(j, L)$ is distribution function of atom ensembles by their size and number of atoms in the ensemble (in case a new phase is formed in the condensed matrix).

Let us proceed from the atomic point of view. As the number of atoms in an ensemble grows, some stable state is reached. The mean frequency of an atom joining the ensemble becomes equal to the mean frequency of separation, making their further joining with the ensemble unproductive. This ensemble is called a critical nucleus of the new phase [1–3].

In classical thermodynamics the emergence of a new phase presents a phase transition of the first order when the density and thermodynamic functions (energy E, Gibbs' energy G, enthalpy H, entropy S), except the thermodynamic potential Φ, change jumpwise at the point of transition. Nucleation of a new phase is critical to the formation of an interface within the maternal phase bulk. The interface bounds the amount of another phase, named a critical nucleus, capable of progressive spontaneous growth.

Nucleation in real supercooled liquids and gases in a metastable state may be invisible for a long time. Various kinds of glass such as overcooled melts are an example of such matter. They are able to exist in an amorphous state not crystallizing for a long time. This means that the reasons of stability of metastable systems towards appearance of a new phase are hidden behind difficulties in nucleation.

The origin and new phase of nucleation (homogeneous, heterogeneous, electrochemical) and growth processes are described theoretically by different methods (thermodynamic, statistical, macro- and microkinetic). Problems in the classical theory of nucleation formulated by J. Gibbs [4] have been extended by Folmer, Becker and Doring, Zeldovich and Frenkel [2, 4–13].

Heterogeneous nucleation is considered to be a process during which interactions resulting in formation of new phase nuclei are running in contact with either heterogeneities found in the maternal phase or with the surface. Otherwise, nucleation within the maternal phase bulk are called homogeneous and can be perfectly modeled by the phenomenon of drop formation in the volume of supersaturated vapor.

The description of nucleation on the surface and in the bulk are equivalent since in the latter case it is considered in two-dimensional and in the former in three-dimensional regions [10].

Homogeneous phase formation is characterized by the appearance of new phase nuclei within the metastable homogeneous system. This can be referred to as transition of the substance into the thermodynamic stable state through a sequential reversible association.

The metastable phase is found in a quasi-equilibrium state where the rate of formation of spontaneously growing associates of a supercritical size is low and the equilibrium statistical distribution of associates by their size down to the critical one is not violated and is invariable in time. The associates can be presented as spherical microdrops of radius R_j. For common liquids in capillaries, the drops show intrinsic properties provided the capillary is under vapor pressure P_j expressed by the Gibbs–Thompson formula

$$P_j = P_\infty \exp\left(2\gamma\, \Omega / k_\mathrm{B} T R_j\right) \tag{2.8}$$

where P_∞, γ are the pressure of the supersaturated vapor and surface tension (specific free surface energy, J/cm^2), $\Omega = M/N_A\,\rho$, M is the atomic mass of the substance, ρ is its density, N_A is Avogadro's number, and $k_\mathrm{B} = 1.38 \cdot 10^{-23}$ J/K is Boltzmann's constant.

This model of nucleation is sometimes known as the capillary model [13].

In the general case, the change of Gibbs energy, ΔG, in the system (rated by one atom) under formation of a new phase nucleus can be presented by the sum of two components. The first of them denotes system properties in the macrovolume and makes allowance for the lessening of the Gibbs energy, $\Delta G_j)_{\text{macr}}$, during formation of the new phase nucleus. This formation incorporates j atoms and is characterized by the difference, $\Delta \mu$, in the chemical potentials between the metastable maternal phase and the environment (e), and between the new stable phase and the nucleus (n)

$$(\Delta G_j)_{\text{macr}} = -j\,(\mu_e - \mu_n) = -j\,\Delta\mu \tag{2.9}$$

where μ_e and μ_n are the chemical potentials of the environment and nucleus, correspondingly for one atom and $\Delta\mu$ is the variation of chemical potential (J/atom).

The value $\Delta\mu$ can be expressed through the parameters being measured. The simplest equations for $\Delta\mu$ for a number of cases of isotropic homogeneous phase formation characterize vapor condensation as well as crystallization of single-component particles from a solution, melt or during boiling.

Let the nucleus of the new phase consisting of j atoms be spherical with radius $R_j = (3j\Omega/4\pi)^{1/3}$, so it can be written that $(\Delta G_j)_{\text{micr}} = 4\pi R_j^2$. Then the total change of Gibbs' energy during formation of a single spherical nucleus will be

$$\Delta G_j = (\Delta G_j)_{\text{macr}} + (\Delta G_j)_{\text{micr}} = 4\pi R_j^3 \Delta\mu/3\Omega + 4\pi R_j^2 \gamma\,. \tag{2.10}$$

This condition is valid for the following radius of the critical nucleus

$$R_{j,\,\text{cr}} = 2\Omega\gamma/\Delta\mu\,. \tag{2.11}$$

The number of atoms in this critical nucleus will be, correspondingly,

$$j_{\text{cr}} = 4\pi R_j^3 \Delta\mu/3\Omega = 32\pi\,\Omega^2(\gamma/\Delta\mu)^3/3\,. \tag{2.12}$$

The energy barrier to be overcome at formation of a new phase will be

$$\Delta G_{j,\,\text{cr}} = 16\pi\,(\Omega/\Delta\mu)^2\gamma^3/3\,. \tag{2.13}$$

The energy barrier constitutes one third of the surface energy of the critical nucleus according to Gibbs.

The system potential will diminish for an ensemble of atoms having $R_j < R_{j,\,\text{cr}}$, so aggregates less than $R_{j,\,\text{cr}}$, will disintegrate, whereas aggregates with $R_j > R_{j,\,\text{cr}}$ show growth that lessens the energy of the system. The nuclei will be just these particles.

Let us estimate the characteristic size of a critical nucleus. To condense from vapor, high temperatures and supersaturation are needed for metal-containing particles. These conditions can be modeled by the method of impact tubes. So, for iron [14] at $T = 1600\,\text{K}$, $s = P/P_0 = 3040$, $\gamma = 1.8 \cdot 10^{-4}$

J/cm^2, we have $\Delta\mu = k_A T \ln(P/P_0) \approx 9 \cdot 10^{-20}$ J. At $\Omega = 10^{-23}$ cm^3 we get $R_{j,\mathrm{cr}} = 2\Omega\gamma/\Delta\mu \approx 0.47$ nm. The critical nucleus of this size contains about 33 atoms of Fe.

Data on the size of critical nuclei of some metals and the number of their constituent atoms are presented in Table 2.1 ($\Delta H/T_0$, surface energies and experimental values of overcooling ΔT_{cr} are taken from [4, 14]). It is evident that the sizes of the critical nuclei are within 1–2 nm (the number of atoms is 10^2–10^3) which is higher than during crystallization from vapor.

Table 2.1. Radius of critical nucleus $R_{j,\mathrm{cr}}$ and number of constituent atoms j_{cr} during metal crystallization from the melt

Metal	$\gamma \cdot 10^7$, J/cm^2 [8]	$\Delta H_m/T_0$, J/(mol·K) [8]	ΔT_{cr}, K [8]	$\Omega \cdot 10^{23}$, cm^3	$R_{j,\mathrm{cr}}$, nm	j_{cr}
Fe	204	8.6	295	1.21	1.17	553
Co	234	8.9	330	1.12	1.07	457
Ni	255	10.1	319	1.13	1.07	453
Pd	209	8.4	332	1.51	1.36	696
Pt	240	10.45	370	1.54	1.15	418
Cu	177	9.6	236	1.21	1.14	512
Ag	126	9.2	227	1.75	1.27	489
Au	132	9.5	230	1.73	1.26	483
Pb	3.33	7.95	80	3.08	1.94	991
Sb	101	22.1	135	3.39	1.38	324
Al	93	11.65	130	1.67	1.23	466
Sn	54.4	14.1	105	2.76	1.22	275
Hg	31.2	9.7	79	2.52	1.24	316

In case the results of nucleation are solid crystalline products, particularly, metal-containing ensembles, the formation of a thermodynamically equilibrium crystalline nucleus will follow anisotropic behavior. This is due to crystallographic inequality of the surface energy in the boundary as the surface area, density and disposition of atoms in the surface layer are changing. During nucleation elements of the interface and interfaces themselves first emerge and then disappear. It is to be borne in mind that it is strongly problematic and almost impossible to identify volume and surface properties of nuclei with the size approaching atomic size. A large portion of surface atoms j_s, in total j ($j_s/j \sim j^{-1/3}$) in a nucleus, make the properties of a particle with a solid metal nonadditive. The geometrical interpretation of minute particles is also problematic. The boundary of small spherical particles of j atoms can be presented as a spherical surface formed by the centers of outer atoms whose volume corresponds to that of an infinite crystal. The boundary of an ensemble consisting of j atoms can also be treated as an extremely fine transition layer between the medium and ensemble nucleus differ in thermodynamic properties [13]. The width of the interface is estimated as 1.0–1.5 nm.

The major concepts of the isotropic model of nucleation in vapor phase condensation do not change while considering isotropic formation of a new phase in a solution or melt.

A different behavior is observed when the new phase is formed in a solid matrix for which the spatial anisotropy of properties is commonly characteristic, e.g. when one phase transforms into another. The formation of a new phase nucleus, having its own specific volume and structure different from the maternal, brings stresses [15] observed as the lattice strain U_{ik}^0.

Total variation of the Gibbs energy of the system during solid-phase nucleation is

$$\begin{aligned}\Delta G_j &= (\Delta G_j)_{\mathrm{macr}} + (\Delta G_j)_{\mathrm{contr}} + (\Delta G_j)_{\mathrm{surf}} \\ &= -j[\Delta\mu - \lambda(U_{ik}^0)^2 \Omega] + \gamma\,(j\,\Omega)^{2/3}. \end{aligned} \tag{2.14}$$

Nucleation of the new phase takes place only in case $\Delta\mu \geq \lambda(U_{ik}^0)^2\Omega$. This becomes possible (using the above cited parameters λ, Ω, $U_{ik}^0 \approx 3{\cdot}10^{-2}$) when $\Delta\mu \geq 2{\cdot}10^{-21}$ J. If $\Delta\mu \approx \Delta H \Delta T/T_0$, where ΔH is the transient heat $\leq 7{\cdot}10^{-21}$ J; T_0 is the transition temperature; $\Delta H \Delta T/T_0 = \Delta T = T_0 - T$ is the overcooling, then the new phase is able to form within the bulk of the solid maternal phase, providing the relative overcooling is $\Delta T/T_0 \geq 0.3$.

Solids with an ideal crystalline lattice were discussed above. Real crystalline bodies always have one or another defect, including packing, dislocation or structural complexes resulting from impurities, interstitial atoms, domain or grain boundaries. This might elevate the thermodynamic potential of the crystal. The formation of new phase regions might lead to gains in energy as compared to formation of such regions in the ideal lattice. The energy needed to form a critical nucleus in an ideal lattice is high enough. That is why considerable deviations from conditions of phase equilibrium are to be reached for nucleation under such conditions. The presence of various flaws lessens this energy and catalyzes the nucleation process so strongly in some cases that nucleation of the new phase could run without overcoming any noticeable energy barriers at activation energy of the order of the diffusion activation energy. Nucleation of a new phase on linear defects (dislocations) is often a very involved phenomenon in solid body transformations.

These models of homogeneous nucleation make it possible to estimate the level of critical saturation, size and quantity of the critical nucleus.

Heterogeneous nucleation occurs in various multicomponent systems with spatial macroinhomogeneities at the interface (foreign inclusions and surfaces, substrates, including polymer substrates, additives or already formed crystalline particles). These inhomogeneities raise the probability of heterogeneous nucleation of a new phase on them.

The formation of metallic silver particles as nanosize agglomerates joined into chains within the volume of photolysis silver bromide crystals can be attributed to silver crystallization prevailing along the lines of round grain boundaries. The condensation effect analogous to heterogeneous one could bring ionized particles in Wilson's chamber.

Nucleation on an ideally wettable surface ($\theta \to 0$) will run exclusively on the solid phase surface. This fact is critical for surface metallization of functionalized polymers.

Thus, the formation of a new phase on spatial inhomogeneities, particularly on the surface, is much more advantageous than homogeneous nucleation in the bulk. The potential barrier height $(\Delta G_j)_{\mathrm{cr,het}} \approx 0.2(\Delta G_j)_{\mathrm{cr,hom}}$ found in formation of the heterogeneous critical nucleus is nearly five times lower than that in nucleation in the bulk.

The peculiarities of heterogeneous nucleation with allowance for the surface energy anisotropy at the crystalline nucleus–surface–medium interface are to be taken into consideration during nanoparticle formation on the polymer surface. The dimensions of the critical nuclei of the new metal-containing phase during both homogeneous and heterogeneous condensation are below 10 nm.

Electric crystallization of metals and alloys is an important means of new phase nucleation observed on the cathode during electrolysis of solutions or melts of certain salts. Its mechanism helps us to understand electrochemical reactions for producing polymer-immobilized metal nanoparticles. The variation of the chemical potential of the system $\Delta\mu_{\mathrm{el}}{}^*$ in electrocrystallization is expressed through variation of the electrochemical potential

$$\Delta\mu^*_{\mathrm{el}} = \Delta\mu + zF\Delta E = \Delta\mu + zF\eta_k \,, \tag{2.15}$$

where z is the number of electrons participating in electrochemical equilibrium; F is Faraday's number; ΔE is the deviation of the electrode potential to the positive side of its equilibrium value equal to half the difference of the escape work of the base metal electrons and of the precipitating metal. It is characterized by an overvoltage $\eta_k = k_{\mathrm{B}}T\ln(C_{\mathrm{ad},0}/C_{\mathrm{ad},\eta})$ where $C_{\mathrm{ad},0}$ and $C_{\mathrm{ad},\eta}$ are concentrations of adsorbed atoms in conditions of equilibrium and oversaturation.

The frequency with which atoms separate and join the nucleus cluster is conditioned by the rate of electrochemical reaction and curvature of the nuclear surface, by the ion discharge velocity on the cluster surface that affects the magnitude $\Delta\mu$. So, electrochemical synthesis may become a promising direction in designing nanostructural materials.

The thermodynamic approach to the present problem has turned out to be helpful in determining the conditions of nucleation and evaluating its critical size for systems in various aggregate states.

2.2 Kinetic Features of New Phase Formation

In thermodynamic models of homogeneous nucleation an ensemble of j metal atoms was treated as a functional diffusion process of either a contracting or augmenting nucleus along its size axes. A nucleus with atoms equals to $j = j_{\text{cr}}$ is found in metastable equilibrium with the environment. To form a new phase nucleus, the system has to overcome an energy barrier $\Delta G_{j,\text{cr}}$ (2.13). According to the theory of fluctuations [24] the probability of the event at which the system potential will differ from the mean one by $\Delta G_{j,\text{cr}}$ is proportional to $\exp(-\Delta G_{j,\text{cr}}/k_{\text{B}}T)$. The number of nuclei $N_{j,\text{cr}}$ or their density $n_{j,\text{cr}} = N_{j,\text{cr}}/V$ (V is the volume of the system) at thermodynamic equilibrium with single atoms N_1 at density n_1, can be defined by Boltzmann's principle [2] by the equations

$$N_{j,\text{cr}} = N_1 \exp(-\Delta G_{j,\text{cr}}/k_B T) \,. \tag{2.16}$$

If $\Delta\mu$ is small, then $\Delta G_{j,\text{cr}} \sim (\Delta\mu)^{-2}$ (see (2.13)) is large which makes the probability of critical nuclei formation negligible and the formation of thermodynamically favorable large nuclei practically improbable. The function $f(\Delta G) = N_1 \exp(-\Delta G_{j,\text{cr}}/k_{\text{B}}T)$ reflects the distribution of precritical nuclei. The formation of critical nuclei appears to be probable when $\Delta\mu$ becomes high enough ($\Delta G_{j,\text{cr}}$ is low) and the addition of one or a few metal atoms results in their irreversible growth.

The velocity of homogeneous formation of critical nuclei $W_{j,\text{cr}}$ is proportional to the density and frequency $\nu_{j,\text{cr}}$ of metal atoms joining the critical nucleus:

$$W_{j,cr} = \nu_{j,cr} n_{j,cr} = \nu_{j,cr} n_1 \exp(-\Delta G_{j,cr}/k_B T) \,. \tag{2.17}$$

So $\nu_{j,\text{cr}}$ is a function of the frequency v of the passage of metal atoms through the interface and density of atoms n_s in the vicinity of the critical nucleus.

Nucleation is a stepwise macrokinetic process that might follow one or another mechanism limiting the nucleation stage. A kinetic or diffusion-free mechanism can govern the limiting regimes of nucleation [2]. This occurs when the velocity of atomic ensemble growth is conditioned by the interface processes, particularly, by adsorption, surface mobility, chemical reactions, and other phenomena. In the general case, when the growth of a nucleus is controlled kinetically, its linear size will be proportional to its growth time.

If the characteristic time of surface processes is small (in contrast to their velocity), the regime of nucleation will be governed by the flow velocity of atoms to the interface, in particular, by diffusion. So, their formation mechanism will be diffusive and their growth will be diffusion controlled. The process kinetics can be described by a corresponding diffusion equation.

As a result of the extending surface of the nucleus with its growth, and consequently, increasing diffusion field, the linear dimensions of the formed atomic ensemble increases, too, as $t^{1/2}$. Intermediate kinetic regimes of the new phase nucleation and their alteration during condensation are also probable.

Tests were carried out to study the condensation of Pb particles [17] from an oversaturated vapor in impact tubes at temperatures from 990 to 1180 K and various oversaturation values ($s = P/P_0 = 30-683$). The growth kinetics of particles with mean size $R_j(t)$ during $t \approx 10 - 10^{-2}$ s (including nucleation stages) have been examined.

The particles grew mainly due to addition of atoms to the nucleus but not as a result of nuclei coagulation. This fact is supported by the Gaussian pattern of the curve for particle distribution by size. At small condensation times (to $\sim 10^{-4}$ s) the dependence of $R_j(t)$ is linear, i.e. nuclear growth is controlled kinetically. With increasing condensation time the concentration of atoms around nuclei drops due to the depletion and growth of Pb particles follows the diffusion regime. Hence, the velocity falls while $\overline{R}$ rises with increasing time as $t^{1/2}$ (Fig. 2.1). So, $R_{j,\text{cr}}$ for Pb estimated at $T = 1110$ K, $s = 50$–240, $\Delta H_{\text{vol}} = 2.9 \cdot 10^{-19}$ J ($\Delta H_{\text{vol}} = 177.2$ kJ/mol), $\Omega = 3.1 \cdot 10^{-23}$ cm^3, $a = 4.95 \cdot 10^{-8}$ cm [26], $\gamma_{\text{Pb}} = 2 \cdot 10^{-3}$ J/cm^2 for $\gamma = \Delta H_{\text{vol}}/6a^2$, gives $R_{j,\text{cr}} \approx$ 18 nm.

The nucleation velocity in condensed liquid media (solution, melt), including polymer-containing substances, is similar to nuclear condensation from vapor, which is proportional to the square density of dissolved atoms and their flow to the surface of the nucleus being formed. The directed flow of atoms in a condensed phase is an activated process since breaking through

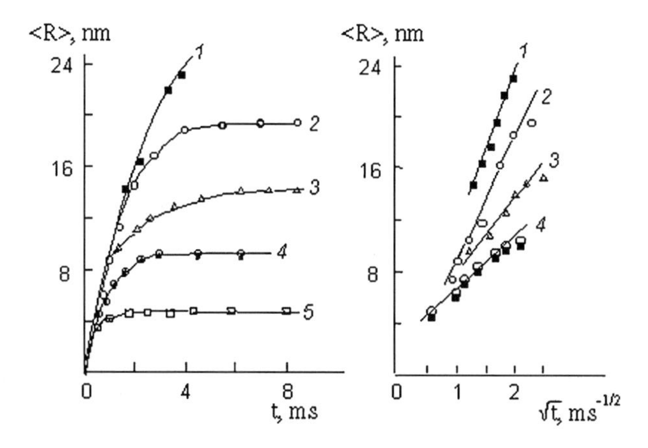

Fig. 2.1. Growth kinetics of mean radius $\langle R \rangle = \mathrm{R}_j(t)$ of Pb particles upon their appearance in the shock wave under various temperatures and oversaturation values s. 1 – $T = 1181$ K, $s = 30$; 2 – $T = 1140$ K, $s = 54$; 3 – $T = 1140$ K, $s = 60$; 4 – $T = 1123$ K, $s = 94$; 5 – $T = 990$ K, $s = 683$ (according to data from [17])

potential barriers is accompanied by the rupture or violation of the neighboring ties with the medium (e.g. with a solvent) under transformation into a nucleus. With solutions the mechanism will be that of atomic diffusion to the nuclear surface and embedding inside. The process in melts includes rearrangement of the neighboring order as atoms transform into the nucleus. Its activation energy can be characterized as the viscous flow activation that is also accompanied by a relative atomic shift.

For crystallization of metallic iron from solution (e.g. under reduction of aqua solutions of iron-containing salts) at $\Omega = 1.2 \cdot 10^{-23}$ cm^3, $\gamma \approx 2 \cdot 10^{-4}$ J/cm^2, $s = 4$, $n_l \approx 3 \cdot 10^{20}$ cm^{-3} (approximately one atom of Fe per 100 molecules of solvent), $v \approx 10^{13}$ s^{-1}, $a = 3 \cdot 10^{-8}$ cm, $T \approx 300$ K, we have $B \approx 10^{21}$ cm^{-3} s^{-1} [18]. As is seen, crystallization velocities from solution are comparable with those of condensation from a saturated vapor in the overcooling regime.

Melts show much lower ΔG_{tr} values ($\Delta G_{\mathrm{tr}} \approx (0.33\text{–}0.5)\Delta H_{\mathrm{m}}$) than solutions according to [18]. For iron crystallization from the melt at $T_{\mathrm{m}} = 1810$ K, $\Delta T = 295$ K [4], $\Omega = 1.2 \cdot 10^{-23}$ cm^3, $\gamma \approx 2 \cdot 10^{-4}$ J/cm^2, $v \approx 10^{13}$ s^{-1}, $a = 3 \cdot 10^{-8}$ cm, $\Delta H_{\mathrm{m}} = 1.66 \cdot 10^4$ J·mol^{-1}, $\Delta G_{\mathrm{tr}} \approx 0.5$, $\Delta H_{\mathrm{m}} = 1.4 \cdot 10^{-21}$ J·mol, we get $B \approx 1.7 \cdot 10^{39}$ cm^{-3} s^{-1}, i.e. nucleation velocities from the melt in overcooling are considerably higher than from solution. This can be attributed to both higher concentrations of atoms in the melt and to lower ΔG_{tr} values.

The velocity of heterogeneous nucleation. In real situations one cannot find regular smooth surfaces at the atomic level unless they are intentionally machined. So far, the formation of a new phase occurs on surface defects such as cracks, chips, growth stages, dislocation exits, violations of translation symmetry of the crystalline lattice, or vacancies. All of these could become nucleation centers whose nucleation energies $\Delta G^*_{j,\mathrm{cr}}$ can be lower than on regular surface areas.

The reduced energy of nucleation over strongly adsorption-active surfaces (as polymers with functional groups are) might elevate the probability of nucleation. That is why the nucleation velocity will be higher on the surface in the case of the homogeneous mechanism.

The analysis of nucleation velocity is based on the approximation of when a stationary distribution of atomic ensembles by size $N_0(j)$ is attained for a given oversaturation in a medium (vapor, melt or solution). By reaching oversaturation (overcooling) there is some stationary distribution by size, so a certain time is needed to attain a new stationary distribution $N(j)$ corresponding to the newly sprung one.

The evolution of a stationary size distribution at a system transition from one state to another was considered by Zeldovich [8] and Frenkel [9] within the drop model. These works are the foundation of the nonstationary theory of condensation. Statistical examination of the process yields a time-dependent function of distribution by size $N(j, t)$. The time to reach a stationary value in the condensed phase at $\tau \approx 10^{-4}$ s is $t \approx 1.5 \cdot 10^{-3}$ s which is less than the typical condensation time.

The stage of new phase nucleation was described by neglecting coagulation of the forming particles. However, situations are feasible when coagulation arises at the nucleation stage.

Allowance for coagulation will vary the character and evolution of nucleation. Some known kinetic models of homogeneous isothermal clusterization are as follows:

$$A_j + A_l \longrightarrow A_{j+l} \quad \text{– see Smolukhovsky's model,}$$

$$A_j + A_l \longrightarrow A_{j+l-m} + A_m$$
$$\qquad\quad | \longrightarrow A_{j-l} + A_l \quad \text{– Mezlak's model,}$$

$$A_j + A_l \longleftrightarrow A_{j+l} \quad \text{– quasi-chemical model by Scillard–Farkash,}$$

$$A_j + A_l \longleftrightarrow A_{j+l} \quad \text{– Zeldovich's model,}$$

$$A_j + A_l \longleftrightarrow A_{j+l-m} + A_m \quad \text{– Dubrovsky's model which is a refinment of Scillard–Farkash.}$$

The models make allowance, in the kinetic equation of clusterization [19], for the processes of association and disruption of clusters along various channels. Particularly, Smolukhovsky's and Mezlak's models were used in coagulation and decomposition problems for dispersed systems, whereas the Scillard–Farkash and Zeldovich models have often been used in problems of condensation in gases [8, 20].

Effect of nuclear temperature on the new phase formation process. Only one feature of the nucleus has so far been taken into consideration while treating nucleation, namely the number of constituent atoms. Nevertheless, the temperature of the nucleus becomes of no less importance at high velocities of new phase nucleation in formation and in the relationship with the environment, especially in condensed media (e.g. in polymers). When a cluster of nuclei is formed, a condensation energy of perceptible magnitude is isolated from atoms. The energy is so strong that it can surpass the kinetic energy of vapor atoms and become comparable with the energy spent on evaporation of atoms (atomization energy of a massive body per atom). The generated condensation energy is expended on excitation of the internal degrees of freedom of the cluster and elevation of its temperature relative to the environment.

A still more complex situation might occur at nucleation in the condensed phase. This could happen during nucleation of the new phase either on the surface or within the polymer matrix bulk (metal application on a polymer substrate, formation of metal-containing particles in the polymer matrix). In this case the temperature of the cluster nucleus will be determined by relaxation of the condensation energy along the inner degrees of freedom of a particle and by relaxation interrelations between the nucleus and the matrix as well. The last depends on the local structure of the matrix in the region of nucleus formation. Possible overheating of the local sites of the matrix could result in its structural changes in the near-nucleus region, including irreversible changes.

Hence, from the phenomenological viewpoint the application of nucleation models within the thermodynamic approach to condensation phenomenon makes possible the qualitatively forecast of the appearance of critical nuclei and corresponding oversaturation leading to macroscopic condensation. A significant quantitative discrepancy exists between the theoretical forecasts and experiments that are specifically expressed in condensation velocities reaching there a few orders of magnitude [21]. This happens because transfer of properties of a massive object onto small atomic ensembles (nuclei) consisting of tens of atoms is most often unacceptable in principle since the properties of a fine particle cannot be divided into volume and surface properties. In this respect the statistical approach shows doubtless advantages.

The statistical concept of nucleation describes new phase nucleation and growth by solving the equation of the macroscopic system state. It describes the dependence of thermodynamic potentials of the system on temperature, pressure, volume, and number of particles using the parameters of interatomic interactions. As applied to condensation from saturated vapors, the statistical approach makes provision for an accurate enough consideration of nucleation in a non-ideal gaseous medium containing clusters. One of the main advantages of the approach is the rejection of the capillary approximation and transfer of macroscopic characteristics on ensembles with few atoms.

The statistical approach to the problem of new phase nucleation considers a medium with volume V containing a vapor with interacting atoms. Interacting atoms form multiatomic clusters in dynamic equilibrium, so the vapor can be represented as a gas consisting of various particles where N atoms are distributed between different clusters having j–j^* atoms supposing the clusters are somehow individually distributed by size and balanced

$$N = \sum_{j=1}^{j*} jN_j \,. \tag{2.18}$$

The total statistical sum of the vapor Q_N with allowance for all individual distributions of clusters by size is

$$Q_N = \sum_N \prod_j \frac{(Z_j)^{N_j}}{N_j!} \,, \tag{2.19}$$

where Z_j is the statistical sum of individual ensembles having j atoms.

It is presumed that clusters like molecules of an ideal gas are found in material and energy equilibrium with the metal vapor $jm_1 \leftrightarrow M_j$ and $\mu_j = j\mu_l$. The volume V in which the interaction of vapor atoms is considered exchanges energy and particles with a very large system that imposes its temperature and chemical potential values on the small system.

If we treat a cluster as a molecule of an ideal gas with j atoms and $3n$ degrees of freedom (three translation, three rotation and $3n-6$ oscillation), its energy can be written as the sum of the translation (forward) motion E_{tr} and internal motion E_{in} of atoms in the cluster relative to the mass center. The cluster will then be the sum of rotation E_{rot} and oscillation energies E_{osc}

$$Z(j,T) = Z_{tr}(j,T) \cdot Z_{rot}(j,T) \cdot Z_{osc}(j,T) . \quad (2.20)$$

Clusters differ in number, origin of constituent atoms and their spatial orientation (in case it forms on the substrate, in particular on the polymer matrix surface). They are characterized by only one parameter – the potential energy ε_j which is less than the sum of the energies ε_1 of individual constituent atoms.

The statistical sum for the system of clusters formed on the substrate surface will be

$$Z = \sum_{N_j} \prod_{j} \frac{1}{N_j!} \left[\beta_j N_1 \exp\left(\frac{\varepsilon_j}{k_B T} \right) \right]^{N_j} \quad (2.21)$$

where β_j is the number of probable orientations of the cluster.

In this case, the velocity of origination of critical nuclei will be

$$I_{cr} = N_{j,cr}\omega_j = \omega_j N_0 \left(\frac{N_1}{N} \right)^{j_{cr}} \exp\left(-E_{j,cr}/k_B T\right) , \quad (2.22)$$

where ω_j is the velocity of individual atoms joining clusters. The velocity is evaluated by proceeding from molecular-kinetic considerations.

Numerical solutions of the equations perfectly agree with experiment.

Numerical simulation of phase formation. Thermodynamic and kinetic features of the new phase clusters are simulated on the basis of numerical experiments using the molecular dynamics (MD) and Monte Carlo (MC) methods. The main aim of the MD method is to calculate the location and velocity of each atom subjected to forces imposed by the gradient of potential energy during times of the order of $\sim 10^{-14}$. In such investigations the potential energy $U(j)$ is expressed as the sum of paired potentials $u(r_{nm})$ described in the general case by Mie's equation (Fig. 2.2).

$$\begin{aligned} U(j) &= \sum_{n=1}^{j} \sum_{m>n}^{j} u(r_{nm}) \\ &= \sum_{n=1}^{j} \sum_{m>n}^{j} \frac{\varepsilon_e}{l-k} \left(\frac{l^l}{k^k} \right)^{\frac{1}{l-k}} \left[\left(\frac{\sigma}{r_{nm}} \right)^l - \left(\frac{\sigma}{r_{nm}} \right)^k \right] \end{aligned} \quad (2.23)$$

where r_{nm} is the distance between the nth and mth atoms; ε_e is the energy value at potential minimum; $\sigma = r_e(l/k)^{l/(l-k)}$ is the interatomic distance at

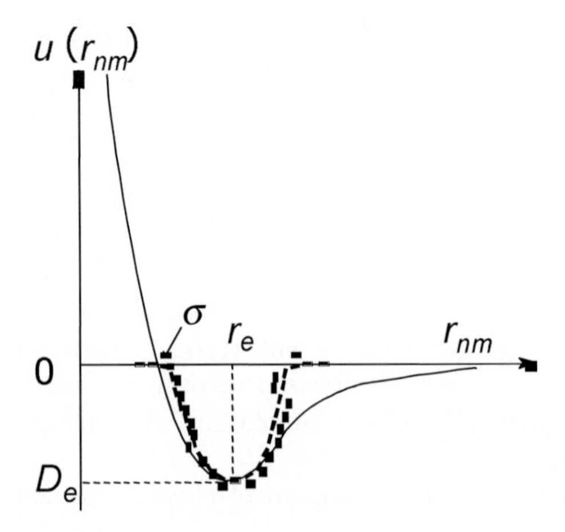

Fig. 2.2. The dependence of u_{nm}: the relation of σ to r_e (*solid* line – Lennard–Jones potential, *dash* line – short-lived interaction potential corresponding to the model of solid spheres)

zero potential; and r_e is the interatomic distance at minimum potential. For $l = 12$ and $k = 6$ this is the Lennard–Jones potential (12:6 potential).

An approximate averaged value of the required physical parameter can be found by the MC method and is based on a large set of random atomic assemblies by a sequential stepwise displacement of a single atom to a limited distance (step-by-step motion strategy). The potential energy of the system and average potential energy of clusters are estimated at each step by sorting out the assemblies.

The MD and MC methods supplement each other when studying Lennard–Jones clusters and could fit well the description of clusters of inert gases. Metal clusters have been considered in a few model experiments [22, 23].

The numerical methods employ the cluster model so as to comply with the physical notion of a cluster as a combination of linked particles (atoms, molecules) whose interaction potential breaks abruptly at distance $2r_e$, and a sphere with radius r_e drawn from the center of each particle intersects at least one of the spheres of the other particles. Two critical parameters such as the distance r_{cr} (geometrical criterion) and critical energy of linkage E_{cr} (energy criterion) are to be introduced here to adequately correlate with the mechanism of cluster reactions.

Elementary processes with a given interaction potential and kinetics of cluster formation and growth were studied by computer experiments. The MD method has proved [24] that it is impossible to observe the state of critical nuclei origination when condensation processes are slow (at small oversaturation values) or with a small number of particles (about 1000 j) as in computer experiments. Investigations of condensation of 512 Lennard–Jones atoms in an adiabatic system (number of particles N, system volume V and energy

E are constant) with density $\rho = 0.085\sigma^{-3}$, $T = (0.78–0.93)\varepsilon/k_B$, where ε and σ are the Lennard–Jones potential parameters, and periodic boundary conditions, have shown that the share of evaporated j-mers (dimers, trimers) is ~20% of the total amount of monomers and clusters.

An infinite system can be modeled by a finite number of particles. The approach is justified also when studying elementary events of evaporation and condensation of individual clusters and microrops surrounded by their vapor.

The dynamics of microparticles has lately become the subject of investigations with numerical methods. The motion and coagulation of particles are computer-modeled within the lattice diffusion-bounded models of particle aggregation [25–30]. The procedure is simpler as compared to MD and MC methods. The investigations have yielded the following results. First, in contrast to an equidistant distribution, aggregation was already observed at the beginning of particle motion due to nonequivalent location of the original atoms in their cells. Clusters start to grow in more densely filled areas although variations in particle number do not change the distribution structure of clusters because of random alteration of more and less dense sites. Second, the rate of cluster accumulation in the early stages is higher for an equidistant distribution but with time it asymptotically approaches that of a random distribution. Third, although the kinetic characteristics of systems are different, it is possible to reduce distributions by size and particle number in various times to a universal kind by a common transformation of scale factors.

Recently complex investigations of thermodynamic and kinetic properties of nuclei arising in spontaneous condensation with and without chemical reactions have become more frequent.

2.3 Phase Formation in Chemical Reactions

The real picture of nucleation and growth of particles of a new phase during chemical reactions is rather intricate. It includes a set of numerous interrelated processes, principally the following.

First, chemical reactions of transformations. These can be thought of as the source of building material for the new phase. They may consist of one or many reactions with participation of either single or several reagents. They might proceed on the surface or in the bulk. During chemical transformations highly reactive particles capable of condensation reach a certain local concentration at which they associate and form the new phase.

Second, mass transfer processes. These may be transfer of reagents into the reaction zone (if few compounds take part in the reaction) or carriage of products of chemical interaction capable of aggregation to the condensation zone, or removal of some products of reaction not participating in crystallization from the zone in case the new particles grow on the surface, particularly, of the polymer.

Third, sorption processes. These could make a contribution to nucleation and growth of the new phase particles. The processes are observed in adsorption of synthesized particles, reagents or products of interaction on the surface of growing nascent clusters and desorption from the surface. Chemisorption interactions on the growing particle surfaces are interrelated with the processes of stabilization (conservation, see Chap. 3).

The formation of the new phase during chemical transformations can be presented as a combination of two sequential macrokinetic stages: 1. the chemical reaction itself as a result of which particles capable of condensation are synthesized; 2. physical processes of particle assembly (condensation). Their relationship can be described through characteristic times that are the inverse of their efficient velocities: the characteristic time of chemical reaction τ_{chem} and the characteristic time of condensation τ_{cond}. The following limiting cases are probable. If $\tau_{\text{chem}} \ll \tau_{\text{cond}}$, then the nucleation velocity will mainly be conditioned by the condensation process, while the chemical factor will be bound to the creation of the original spatial distribution of crystallizing particles in the condensation zone and areas of oversaturation. This situation can arise in high rates of chemical reactions, particularly in severe operation regimes (impact waves, powerful radiation flows) when interactions between particles are weak, e.g. in gaseous phase chemical transformations. Another case is when $\tau_{\text{chem}} \gg \tau_{\text{cond}}$. Here nucleation and growth of particles of the new phase is fully under the control of chemical transformations.

However in most cases τ_{chem} and τ_{cond} are proportional. Particle synthesis and condensation nucleation and growth of the new phase include transfer and sorption processes. It is difficult to consider the integrity of models of new phase formation. Temperature rise assists in accelerating the synthesis of crystallizing particles on the one hand, but lessens the probability of their association into clusters due to raised tendency to disordering. Association into clusters could at the same time be alleviated due to diffusion mobility of condensing particles and redistribution of links at crystallization.

One of the features of kinetic approaches to treating the origination of a new phase is the presentation of clusters as a multiatomic molecular formation with intrinsic intramolecular and intermolecular interactions. Thanks to this fact the apparatus of statistical mechanics can become useful in describing properties of clusters.

There are two extreme tendencies in constructing kinetic models. One of them admits an assumption on the quasi-equilibrium behavior of chemical processes resulting in nucleation of the new phase. It is bounded by the region of minor deviations from the equilibrium state like in the case with the classical theory nucleation. This may happen at slowly running chemical reactions due to low concentration of reagents. The other tendency is, vice versa, connected with rejection of quasi-equilibrium chemical transformations, and the formation of the new phase is treated as an irreversible nonstationary process, especially in the initial stages.

Investigation results of the initial stages of isothermal condensation ensure accurate estimates of process characteristics such as induction period t_{in}, condensation time t_{c}, mean number of atoms in a cluster in the course of and upon termination of condensation, variance of distribution function of clusters by size, mean size $\langle j \rangle$, dispersion and the description of variation kinetics of the degree of oversaturation. The approach takes into account dimensional effects but neglects heat generation induced by condensation energy during cluster growth [14].

A model of nonisothermal homogeneous condensation was built in the context of the quasi-chemical approach with allowance for heat generation during condensation [31]. The model was useful in the analysis of homogeneous processes of nucleation and condensation in gaseous and condensed phases, including polymer matrices. The most efficient appeared to be when the monomer to diluent ratio is $\geq 10^{-2}$:10^{-3}, when particle heating and energy relaxation conditions start to be critical for the whole condensation process.

It is accepted in the above considered quasi-chemical models of condensation that chemical synthesis reactions do not affect nucleation or any further stages of the process. A chemical reaction is an instantaneous source of initial concentration of the condensing substance. Experimental study of iron clusters behind the front of an incident shock wave in the disintegration of $Fe(CO)_5$ gave evidence that this supposition was not always justified [14, 32, 33].

Condensation of iron vapors behind the shock-wave front. The accumulation kinetics of Fe_j clusters during disintegration of 0.1–2.0% (mol) of $Fe(CO)_5$ in argon behind the shock-wave front has been studied for the first time on the basis of laser beam diffraction over density inhomogeneities created by iron clusters [14]. It appeared that for the $j\text{Fe} \rightarrow \text{Fe}_j$ reaction displaying energy increment (at normal conditions) $\Delta G_j^\circ = \Delta H_j^\circ - T\Delta S_j^\circ$ and $\Delta G_1^\circ \equiv 0$, the empirically found dependence $\Delta H_j^\circ(j)$ [18] is $\Delta H_j^\circ / j \approx 6.6 \cdot 10^{-19}(1 - j^{1/4})$ J/(Fe atom). The difference from analogous dependencies predicted by the classical drop model and that of clusters with dense packing $\Delta H_j^\circ / j \sim j^{2/3}$ [34] was significant. Comparison of the calculated $\Delta G_j^\circ / k_{\text{B}}T$ for $T = 1600\,\text{K}$, $s = 3040$, $\gamma = 1.8 \cdot 10^{-3}$ J/cm^2 [17] with the expected behavior of the dependence of $\Delta G_j^\circ(j)$ forecast by classical theory has shown the following. The appearance of a maximum in the region $j \approx 6$–7 on the dependence of $\Delta G_j^\circ / k_{\text{B}}T$ on j cannot be attributed to critical conditions induced by the dimensional dependence of cluster energy according to classical theory, but it reflects the real kinetics of cluster formation.

The Scillard–Farkash scheme was used [17] to describe the formation kinetics of Fe_j ($j \leq 80$) clusters in condensation of iron vapors diluted by argon at two different oversaturations ($s = 3040$; 30400).

The main kinetic processes occurring during cluster formation are monomer addition and separation (Fe atom). The time to reach a stationary regime of cluster formation increases with reducing oversaturation and rising j,

$$\mathrm{Fe}_j + \mathrm{Fe}_m + \mathrm{Ar} \underset{-\ k_{j,m}}{\overset{k_{j,m}}{\rightleftharpoons}} \mathrm{Fe}_{j+m} + \mathrm{Ar}$$

where $k_{j,m}$, $k^{-}_{j,m}$ are constants of direct and back reaction velocities $j = 1, \ldots, \infty$, and $m = 1, 2, \ldots, 5$.

Flows of clusters to the sides of increasing and decreasing along size axes to equalize j_{cr} is taken as the criterion of an instant macroscopic condensation in the stationary state (at $j \approx 70$ and $t > 5 \cdot 10^{-7}$ s) with certain j_{cr}. At $j > j_{\mathrm{cr}}$ the value C_j does not depend on j and t. The calculated velocities of stationary flow of clusters J_{st} at various oversaturation values of Fe vapor depend strongly on the degree of oversaturation like in the classical theory. But it exceeds by almost 12 orders of magnitude the classical values J_{st}. The dependencies of $C_j(t)$ for various j_{cr} values and $s = 3040$ (Fig. 2.3) pass a minimum as j grows. The minimum location rises along j with increasing j_{cr} from $j \approx 14$ ($j_{\mathrm{f}} \leq 30$) to $j \approx 24$ ($j_{\mathrm{f}} = 70$); consequently the minimum concentration drops from 10^{-8} to 10^{-25} cm^{-3}.

Evaluation of kinetic dependencies of Fe atom expenditure during $\mathrm{Fe(CO)_5}$ disintegration behind the shock wave front using atomic-absorption spectroscopy and heat generation kinetics in condensation [33] has proved that disintegration of iron pentacarbonyl ran very quickly during times τ_{f} (less

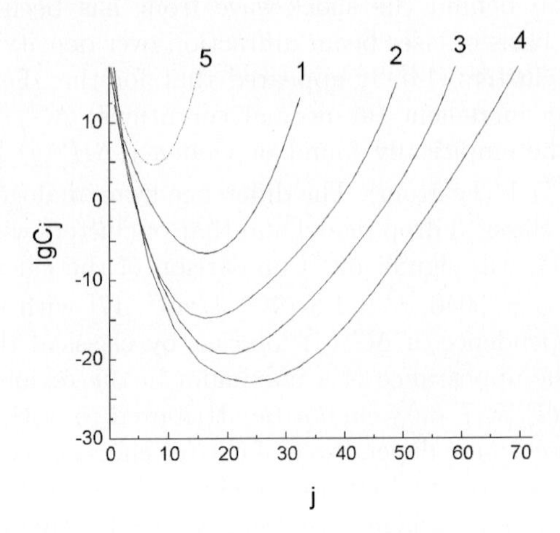

Fig. 2.3. Dependence of the concentration C_j of clusters Fe_j on j at forced equilibrium and $s = 3040$. 1 – $j_{\mathrm{f}} \leq 30$; 2 – $j_{\mathrm{f}} = 50$; 3 – $j_{\mathrm{f}} = 60$; 4 – $j_{\mathrm{f}} = 70$; 5 – j_{f} according to the classical theory for $\gamma = 1.8 \cdot 10^{-7}$ J/cm^2 and $T = 1600$ K [32]

than 1 μs) and merged with the front gradient region. Oversaturation with Fe atoms reaches $s = 10^2 – 10^7$. The observed heat generation in the condensation process is recorded during time $t > \tau_f$. The concentration C_{Fe} behind the shock-wave front increases and then drops due to consumption of Fe atoms on cluster formation followed by leveling cluster–Fe atom content. The rate of consumption of Fe atoms in the initial stages of transformation can be described by the law of the first-order velocity $W_{Fe} = dtC_{Fe} = -k_o C_{Fe}$, where k_o is the velocity constant. During further stages of condensation the velocity of Fe atoms transformation is decelerated due to enlargement of the clusters. This happens as a result of coagulation and competing reactions of disproportioning and decay of clusters accompanied by isolation of Fe atoms under elevated temperatures.

Consideration of the disintegration of $Fe(CO)_5$ followed by nucleation from the standpoint of a common physico-chemical process allowed the generalization of the multilevel kinetic scheme that includes models of thermal decomposition of iron penacarbonyl, nucleation and coagulation of Fe_j clusters. The kinetic parameters of transformation are listed in Table 2.2.

The kinetic approach based on a real pattern of chemical synthesis of nanoparticles and constants of elementary reaction velocities is very effective.

2.4 Self-Organization of Metal-Containing Nanoparticles (Fractal Structures)

Nanoparticles formed in the course of synthesis in real chemical systems, including metal-containing systems, display strong reactivity that impels them to self-organize. The inclination to aggregation was obvious in the absence of stabilizing factors. The resulting products are characterized mostly by the fractal structure. Objects of this kind are often called fractal clusters (FC) or fractal aggregates. Fractals are similar objects invariant with regard to local dilations, i.e. they have the same form at different magnifications. The notion of fractals as similar multeities has been introduced by Mandelbrot. The fractal dimension, D, of an object embedded into d-dimensional Euclid space changes from 1 to d. Most often the fractal objects are classed into deterministic (those which are constructed on several basic laws) and statistical (they are generated by random (incidental) processes) objects. Their structure is usually friable, branching and carries information about the integrity of physical phenomena accompanying their association close in size to nanoparticles in disordered systems. FC can be observed in experiments producing metal-containing nanoparticles, especially in the formation of metal–polymer composites. The estimation of the fractal dimensions is carried out on microphotographs using the method of grids [44]. The image is covered by a grid with square cells. The number N of the cells which are covered the image is estimated at the different sizes of the cell. Then the fractal dimension is calculated from the logarithmic dependence of N and d. Recently

Table 2.2. Parameters of the Arrhenius equation $k = A\exp(-E_a/RT)$ and heat of disintegration reactions and iron vapor condensation within the disintegration pattern of $Fe(CO)_5$ [33]

No	Reaction	A	E_a, kJ/mol	$-\Delta H$, kJ/mol
I. Disintegration of $Fe(CO)_5$				
1	$Fe(CO)_5 \rightarrow FeCO + 4CO$	$2{\cdot}10^{15}$	167.2	–514.1
2	$FeCO + M \rightarrow Fe + CO + M$	$6{\cdot}10^{14}$	85.7	–75.2
3	$Fe + CO + M \rightarrow FeCO + M$	$5{\cdot}10^{14}$	0	75.2
II. Formation and evaporation of small clusters				
4	$FeCO + FeCO \rightarrow Fe_2 + 2CO$	$6{\cdot}10^{14}$	8.4	–20.9
5	$Fe + FeCO \rightarrow Fe_2 + CO$	$6{\cdot}10^{14}$	0	62.7
6	$Fe_2 + FeCO \rightarrow Fe_3 + CO$	$6{\cdot}10^{14}$	0	104.5
	$Fe_3 + FeCO \rightarrow Fe_4 + CO$	$6{\cdot}10^{14}$	0	146.3
8	$Fe_4 + FeCO \rightarrow Fe_5 + CO$	$6{\cdot}10^{14}$	0	188.1
9	$Fe + Fe + M \rightarrow Fe_2 + M$	$4{\cdot}10^{15}$	0	125.4
10	$Fe_2 + Fe + M \rightarrow Fe_3 + M$	$5{\cdot}10^{14}$	0	175.6
11	$Fe_3 + Fe \rightarrow Fe_4$	$5{\cdot}10^{11}$	0	230.0
12	$Fe_4 + Fe \rightarrow Fe_5$	$3{\cdot}10^{13}$	0	255.0
13	$Fe_2 + M \rightarrow Fe + Fe + M$	$1{\cdot}10^{15}$	133.8	–133.8
14	$Fe_3 + M \rightarrow Fe_2 + Fe + M$	$1{\cdot}10^{16}$	183.9	–183.9
15	$Fe_4 \rightarrow Fe_3 + Fe$	$1{\cdot}10^{13}$	230.0	–230.0
16	$Fe_5 \rightarrow Fe_4 + Fe$	$2{\cdot}10^{14}$	255.0	–255.0
17	$Fe_6 \rightarrow Fe_5 + Fe$	$2{\cdot}10^{15}$	271.7	–271.7
18	$Fe_7 \rightarrow Fe_6 + Fe$	$5{\cdot}10^{15}$	280.1	–280.1
19	$Fe_8 \rightarrow Fe_7 + Fe$	$1{\cdot}10^{16}$	284.2	–284.2
20	$Fe_9 \rightarrow Fe_8 + Fe$	$4{\cdot}10^{16}$	292.6	–292.6
III. Coagulation of small clusters				
21	$Fe_2 + Fe_2 \rightarrow Fe_4$	$5{\cdot}10^{13}$	0	271.7
22	$Fe_2 + Fe_3 \rightarrow Fe_5$	$1{\cdot}10^{14}$	0	355.3
23	$Fe_2 + Fe_2 \rightarrow Fe_3 + Fe$	$3{\cdot}10^{14}$	0	41.8
24	$Fe_2 + Fe_3 \rightarrow Fe_4 + Fe$	$3{\cdot}10^{14}$	0	104.5
25	$Fe_2 + Fe_4 \rightarrow Fe_6$	$3{\cdot}10^{14}$	0	397.1
26	$Fe_3 + Fe_3 \rightarrow Fe_6$	$3{\cdot}10^{14}$	0	459.8

Table 2.2. (Cont.)

No	Reaction	A	E_a, kJ/mol	$-\Delta H$, kJ/mol
27	$Fe_3 + Fe_4 \rightarrow Fe_7$	$3{\cdot}10^{14}$	0	501.6
28	$Fe_4 + Fe_4 \rightarrow Fe_8$	$3{\cdot}10^{14}$	0	564.3
IV. Heterogeneous reactions in k-phase ($j \geq 5$)				
29	$Fe + Fe_j \rightarrow Fe_{j+1}$	$2.3{\cdot}10^{13}$	0	401.3
30	$FeCO + Fe_j \rightarrow Fe_{j+1} + CO$	$2.3{\cdot}10^{13}$	0	334.4
31	$Fe_2 + Fe_j \rightarrow Fe_{j+2}$	$2.3{\cdot}10^{13}$	0	702.2
32	$Fe_3 + Fe_j \rightarrow Fe_{j+3}$	$2.3{\cdot}10^{13}$	0	940.5
33	$Fe_4 + Fe_j \rightarrow Fe_{j+4}$	$2.3{\cdot}10^{13}$	0	1128.6

Note: Dimensionality A for bimolecular transformations – $cm^3/mol \cdot s$; for trimolecular – $cm^6/mol^2 \cdot s$. For coagulation of k-phase particles $Fe_n + Fe_m \rightarrow Fe_{n+m} (n, m \geq 5)$; $k^+_{n,m} = Z_{1.1}(n^{1/3} + m^{1/3})^2 (n+m)^{1/2}/4(2nm)^{1/2}$; $Z_{1.1}$ – collision factor.

many monographs and reviews have been published on this rapidly developing scientific subject [6, 35–37]. Aspects of generation and properties of fractal systems and their analogues have been extensively studied and discussed from the viewpoints of physics, chemistry and biology starting from burning and bursting processes and up to human physiology [38].

We shall dwell upon just the main postulates and notions on the nature and origin of FC to the extent needed for understanding the genesis of formation of new phase particles, which also concerns polymer-containing systems.

Computer modeling has been carried out by many authors [6, 35, 36]. It has enabled the development of models with variable parameters, approximation of the generation mechanism of FC, elaboration of a ready object incorporating all details, and has made it possible to analyze its properties and growth behavior.

Modeling of fractal structure formation. Among the major properties of FC one can name the decreasing mean density of the substance within the volume occupied by a cluster with its growth. Fume is a characteristic example. The aggregate can be presented as a collection of original particles (formed during j-atomic cluster synthesis) of approximately similar size that are rigidly bonded in contact sites, and whose main volume portion is empty space. This means that the original particles are not densely packed. Photographs of experimentally observed FC and a cluster obtained by computer assembly are shown at Fig. 2.4.

A fractal cluster as a physical object appears to be a system whose minimum cell size r_0 (in our case it is a particle as a structural element) and

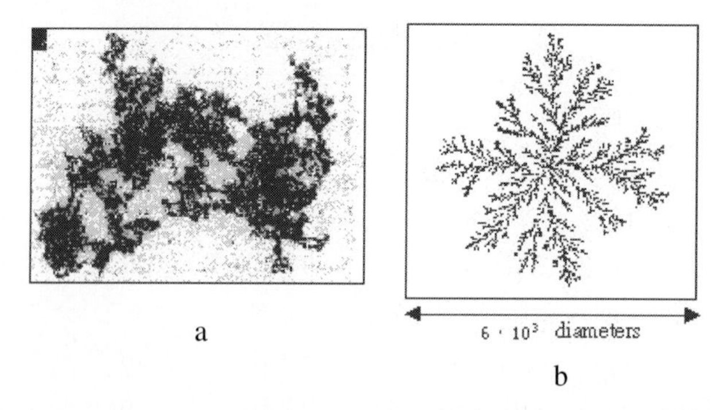

Fig. 2.4. (a) Microphotograph of a silver fractal cluster ($D = 1.9$) obtained during cryochemical synthesis during co-condensation of metal vapors and isopropanol (227 nm in 1 cm) [48]. **(b)** Computer model of a 2D fractal cluster consisting of 106 original particles. The cluster is formed as a result of diffusion-limited aggregation [41]

maximum size r_φ are associated with the full size of the FC. It also possesses the property of self-similarity (scale invariance) and satisfies the following relation

$$N(r_0) \sim (r/r_0)^D \tag{2.24}$$

where $N(r)$ is the number of particles in a volume with linear size $r(r_0 \leq r \leq r_\mathrm{f})$ and D is a parameter called the fractal dimensionality.

Fractal dimensionality shows that the system is imperfect. It always satisfies the inequality $D < d$, where d is the dimensionality of Euclidean space where the FC has been formed (for a surface $d = 2$; for a volume $d = 3$). The property of self-similarity of randomly placed particles of a physical FC is understood statistically as follows. If most of the equal-volume pieces in different parts of a cluster are removed, then they will contain on average a similar number of original particles. This fact does not contradict the main property of FCs (decreasing mean density of particles in a cluster during its growth) since a simultaneous increase in the volume of voids occurs. Large voids do not affect the mean density of particles within a small volume but influence the mean density of particles in a cluster as a whole.

The gyration radius R_r of cluster size r_f may be found from the relation $R_r = (\langle R^2 \rangle)^{1/2}$, where $\langle R \rangle$ is averaged by particle distribution in a cluster distance from a considered point of the cluster to its mass center. The experimental value of fractal dimensionality is determined by (2.24). It is found from the number of particles contained in clusters of various sizes or in different-size parts of a single cluster.

Models of FC take account of critical properties of motion and agglutination of clusters and particles during their growth [6]. These parameters are the type of association (cluster–particle or cluster–cluster), motion features (linear or Brownian), and the character of joining, depending on the probability of their sticking together at contact.

There are two main models of FC formation. The first model of diffusion-limited aggregation (DLA) [40] presumes that FCs grow as a result of association at a certain probability of cluster sticking to particles exercising Brownian movement in space. The second model of cluster–cluster aggregation (CCA) [41] is based on the supposition of a stage-by-stage formation of FCs. At the early stage of formation particles collide during motion along various trajectories and stick to each other forming many clusters of small size. At the second stay they aggregate into large clusters. It is assumed that in this case more friable FC are formed as compared to those following the DLA model since it is more difficult to fill voids in this kind of cluster assembly.

Various computer algorithms for FC formation are possible within the framework of the approaches considered. They are based on different kinds of modeling of the motion path of each test particle from its generation until it merges with the cluster during FC growth. However, the final results of both approaches practically coincide. The generalization of computer studies for named FC assembly models is presented in Table 2.3.

Computer experiments make it possible to define the effect of various factors and to achieve a rather full idea of the diversity of FCs in real situations.

Table 2.3. Fractal dimensionality of clusters in the different models of particle aggregations

Aggregation Model The Type of Association	The Kind of Particle Motion	Fractal Dimensionality $d = 2$	$d = 3$
Cluster–particle	straight trajectories	2	3
	Brownian motion (DLA-model)	1.68 ± 0.02	2.46 ± 0.05
Cluster–cluster	straight trajectories	1.54 ± 0.03	1.94 ± 0.08
	Brownian motion (CCA-model)	1.44 ± 0.04	1.77 ± 0.03
	(RLCA-model, small probability of sticking)	1.60 ± 0.01*	2.11 ± 0.02*

* For polydisperse systems

Structure and kinetics of fractal cluster growth. The structure and properties of FC growth are affected by many factors, including type of association (particle–cluster or cluster–cluster), movement of associating particles, degree of penetration inside the growing aggregate, length and thickness of branches, anisotropy level of the cluster being built, and structural features of its substrate.

The structure of FCs growing on a substrate is affected by its properties which are more strongly displayed, the larger the cluster and the smaller the fluctuations of its parameters during growth [32, 42]. When modeling a two-dimensional DLA cluster, it turned out that the character of the growth of its branches depended upon the symmetry of its substrate. A slight asymmetry of the substrate in the model results in increasing growth rate of a branch in a set direction.

Structural properties of a cluster are also characterized by the mean coordination number [43] that is calculated as an average number of the nearest neighbors of the cluster comprising particles. It slightly depends on the peculiarities of the FC assembly model. It is 2.19–2.22 for a 2D cluster and 2.2–2.25 for a 3D one. The next branching in such a FC is observed after every 4–5 joinings.

Fractal structures were mostly observed experimentally during condensation processes of metal vapors. The first FCs with fractal dimensionality 1.60 ± 0.07 [44] were investigated in condensation of Fe and Zn vapors. Electron-microscopic studies of cobalt FCs obtained in condensation in an Ar atmosphere (~1.33 kPa) showed that the main volume in the condensate was occupied by the pores (relative volume of cobalt is just 10^{-2}–10^{-4}) [45]. Self-organized processes were observed in the co-condensation of palladium and paraxylenene [46]. At a concentration of Pd less than 1%(vol.) and over 8%(vol.) a regular structure is formed with dimension about 2. It was found that aggregation of metal nanoparticles in the cured polymer matrix could proceed on several mechanisms [47]. It is observed that the fractal dimensions of the percolation clusters and aggregates presented in the matrix are close to each other.

FC formation includes the stage of generation of multiatomic clusters which during the next stage aggregate into FCs. The associate is affected by the solution concentration, the pH of the medium, and the presence of surfactants and stabilizers, including macromolecular ones. It is possible to regulate the time of FC formation [6, 35, 36] by changing the acidity of the solution during generation of metal nanoparticles in chemical transformations. This effects the mechanism of FC formation. For short times of formation (about a few minutes and less) the cluster acquires fractal dimensionality 1.75 ± 0.05 which corresponds to the CCA model. Increased time of cluster growth (from hours to days) due to changed solution acidity raises D to 2.05 ± 0.05. This proves the change of CCA mechanism for the RLCA regime of growth. Clusters formed in the RLCA mechanism are usually more resistant to variations in fractal dimensionality as compared to those formed in the CCA regime. Alteration of formation regimes (see Table 2.3) can also be observed when the origin of the solvent is changed. This is evident in the formation of fractal aggregates during cryochemical synthesis of metal-containing nanoparticles. The production of silver nanoparticles by cryochemical co-condensation of metal vapors with isopropyl alcohol, acetonitrile or toluene [48] may be used as an example. Silver fractal clusters with different fractal dimensionality depended on the organic solvent used (isopropyl alcohol – 1.9, acetonitrile – 1.7 and toluene – 1.5). The mean particle size in a cluster was, correspondingly, 16.0, 21.0 and 9.4 nm.

In the general case the dependence of the association rate constant versus the characteristic size of colliding particles can be written as

$$k \sim R^{2\omega} \tag{2.25}$$

where the parameter ω takes into account the structure and motion of clusters in space.

As the FC grows, the mean distance between clusters becomes commensurable with their sizes and further aggregation results in generation of a single large cluster. As with aggregation in colloidal solutions, this stage simulates sol to gel transfer and displays its own specifics [49].

The physical process of sol transformation into gel is reflected in the supposition of cluster–cluster association. It is presumed that if clusters are small, their agglomeration is connected with diffusion movement in the medium. Large clusters grow as a result of capturing particles when moving in the force of gravity.

FCs present a porous substance with maximum size $\overline{R}$. FCs can be obtained from small particles at a rather high material concentration in the medium. When the time of cluster formation is not long enough, removal of the building material from the zone of aggregation could hamper and even prevent FC formation. The presence of charged particles in solution or gaseous phase (in laser or electric-arc evaporation) hinders the formation of fractal aggregates. This means a reduced probability of particles sticking together on contact according to the model considered. During reduction by trisodium salt of citric acid in $NaAuCl_4$ solution spherical aurum nanoparticles are formed with a narrow distribution by size and about 14.5 nm mean diameter [50]. The small size of Au particles is attributed to their probable stabilization through sorption on the surface of citric acid ions. In this way nanoparticles acquire a charge that hampers their further growth. The introduction of a small amount of pyridine decreases the surface charge and assists agglutination. The aurum FCs formed are a few micrometers in size with fractal dimensionality $D = 1.77 \pm 0.10$.

An interesting peculiarity of the process of the formation of fractal aggregates is the generation of electric and magnetic charges on aggregating particles under the action of electric and magnetic fields. Upon reaching a high density of original particles, e.g. in evaporation of metals by electric explosion, cylindrical associates are formed in the gas-plasma system at the initial stage of aggregation. Such agglomerates are called chain aggregates and are found in systems of aggregating particles displaying high magnetic activity. For example, aggregates of ferrite nanoparticles, being thermolytic products of Fe and Co oxalate co-crystallizers as well as Ba and Fe formates, also form chain structures.

Surface fractal clusters. Fractal cluster formation on the surface and two-dimensional FCs are very important due to the role in catalysis [51] and the production of metal-containing fine films.

Two extreme variants of surface FC assembly are possible. The first one is connected with agglomeration of separate particles isolated on the substrate surface. The second variant is the formation of fractal surface aggregates as a result of deposition on the surface. It is used mostly for electrodeposition of metals in thin amorphous films including metal–polymer films. FCs (Fig. 2.5) can be formed during electrolysis on the substrate serving as an electrode [52]. The metal under the electrode voltage is isolated as a FC until the potential difference surpasses some critical value. In this case the fractal dimensionality is 1.66 ± 0.03 which corresponds to the DLA model. A higher potential difference leads to variation in the deposition character and increasing fractal dimensionality of the cluster. Zinc and copper deposition on the anode during electrochemical reduction of their salts in a cell with central cathode and annular anode, when disoriented fractal growth transformed into the generation of FCs with radial geometry, was observed in [53] (Fig. 2.5).

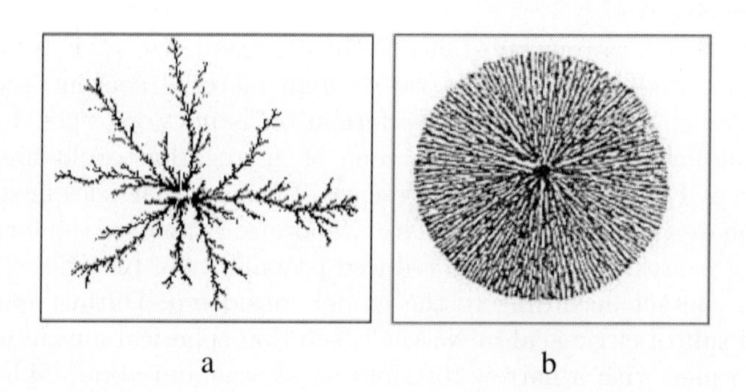

Fig. 2.5. (a) Fractal zinc cluster formed during electric deposition on electrode surface [52]. **(b)** Dendrite aggregate of zinc with radial geometry obtained during electrolysis of 0.03 M solution of $ZnSO_4$ at a potential difference of 10 V [53]

Fractal patterns arise in thin film structures during transformation of the amorphous phase into the crystalline phase. These FC display fractal dimensionality close to 1.7. An example of a similar fractal formation in nontraditional objects is crystallization in thin Al/Ge films in the course of self-propagating high-temperature synthesis (SHS) [54]. It is a combustion process during which chemical transformations are supported by heat generated during reactions. Such clusters forming in the bulk have $D = 1.60 \pm 0.1$ in the 2D case and $D = 1.80$–2.56 in the 3D case.

Surface FC could be formed on rough surfaces of solids mostly of porous objects, including those of polymers.

Fractal filaments are similar in structure to aerogels whose porous volume is strongly rarefied, the specific surface is high (300–1000 m^2/g) and the density is rather low. Almost all of the volume of formation is occupied by voids or pores, bonded particles forming the aerogel skeleton that constitutes just a small portion of the total volume. A small fragment of the aerogel with a representative amount of separate particles present a FC. Aerogels consisting of various natural particles, including oxides of Al, V, Fe, Sn, SiO_2 were mostly obtained in solutions. Fractal filaments were produced from a number of metals by exposure [55] to laser radiation of 10^6–10^7 Wt/cm^2. The mechanism of filament generation is the same in all the conditions described. Small filiform structures display fractal properties. The number of individual particles $n(r)$ in a sphere of radius r with one of the particles of radius r_0 in its center is equal to

$$n(r) = (r/r_0)^D \,. \tag{2.26}$$

The relation is true when $R \gg r \gg r_0$, where R is the radius of the FC or the maximum radius of pores in the structure. For the structures considered $D = 1.8$–1.9. This value is lower than the fractal dimensionality in aerogels $D \approx 2.1$. It follows from (2.26) that the relation of the material mean density in a fractal filament $\overline{\rho}$ to that of a separate particle ρ_0 is $\overline{\rho}/\rho_0 = (r_0/R)^{3D}$. This relation in reality is about 10^{-3}, which is commensurable with the density of atmospheric air and is characteristic of fractal formations. Chemical composition plays a significant role in the formation of such structures.

Percolation clusters as objects with fractal properties. The percolation cluster is another physical object showing fractal properties. The notion of a percolation cluster is connected with the phenomenon of conductivity in metal-filled composite materials, first of all in metal–polymer materials (Chap. 11). According to theory, the volume concentration of particles increases C_n $(0 \leq C_n \leq 1)$ and isolated conducting particles of one of the phases of a strongly inhomogeneous system are disordered inside a non-conducting phase. These particles gradually tend to form small aggregations and then to form conducting isolated clusters. The mean size of these clusters is conditioned by the correlation length ζ. Conducting chains arise between separate clusters at some critical value of concentration $C_{n,\mathrm{cr}}$, called the percolation threshold. They may include both individual particles and their aggregates. Isolated clusters join through connecting chains into a new large conducting cluster named the percolation cluster (PC). Starting from some density of the substance in a PC density becomes infinite. When the concentration reaches $C_{n,\mathrm{cr}}$, the system transforms stepwise into a conducting state. Further PC growth as C_n increases is connected with absorption of individual particles and isolated clusters, the appearance of additional chains and a monotonic rise of conductivity of the composite.

The presence of a percolation threshold is a general feature intrinsic to disordered systems and lattice structures. It has been shown [56] that $C_{n,\mathrm{cr}}$ is independent of the lattice type but is related to the space dimensionality. For the 2D case $C_{n,\mathrm{cr}} = 0.45 \pm 0.02$; for the 3D case $C_{n,\mathrm{cr}} = 0.15 \pm 0.01$. If the system is disordered and particles of the conducting phase are randomly distributed, the percolation threshold can vary in a wide range and is determined either by the distribution function of these particles or space dimensionality.

Percolation and fractal clusters differ in their physical nature. To obtain an infinite PC, it is necessary that the density exceeds some critical value. In contrast, the FC density can be as small as desired. As for their geometry, there is a certain correlation between structures of conducting systems, in particular, between the fractal dimensionality and percolation parameters. The analysis of experimental concentration dependencies of metal-filled polymer films (based on epoxy resin ED-20 with addition of nickel carbonyl particles) and those of geometrical characteristics of the forming particle aggregates in the vicinity of percolation threshold [57] showed the coincidence of fractal geometry of PC and FC. This agrees with the familiar relation between PC parameters and fractal dimensionality of the initial D aggregates

$$D/2 = \theta/\iota - (d - 2) , \tag{2.27}$$

where ι is the critical factor of the correlation length, and d is the dimensionality of the system.

In conclusion it should be emphasized that fractal phenomena are a peculiar indicator of the degree of disorder in a system. Fractal properties of objects and processes of different physical nature are exhibited in a single mathematical description. FC should be treated from the viewpoint of the new phase formation as an intermediate state of matter at phase transitions and a certain stage of phase transformations. At the same time, FC as a solid body state presents special scientific and practical interest.

2.5 Brief Account of Major Production Methods of Metal-Containing Nanoparticles

Methods for producing nanoparticles are numerous and diverse. The researcher is furnished with unlimited potentialities in the choice of both already known and constantly perfecting optimum approaches that take into account efficiency, ecological safety, energy consumption, physical and chemical properties of the nanoparticles, and the aim and problems of the final product. Let us consider the chief production principles of nanoparticles in the absence of polymers, recognizing that the polymer matrix might add its own specifics to metal composite formation processes.

Methods of molecular beams consist in emission of atoms (molecules or clusters) from a compact source that is a metal target heated to a high temperature in vacuum and condensation of these atoms on a substrate mostly cooled to low temperatures [58–60]. Modifications of the method differ in beam intensity (low-intensity methods include beams with $J = 10^{12}$–10^{14}, high-intensity methods have $J = 10^{16} - 10^{18}$ particles$\cdot$cm$^{-2}\cdot$s^{-1}) and in evaporation methods of the material (most perfect is electronic beam heating). Advantages of the method of molecular beams are the possibility of accurate regulation of beam intensity (material feeding velocity to condensation zone), simultaneous use of evaporated beams of several materials or their alloys and readiness of adoption in industry for growing epitaxial layers and lattices at the nanolevel (10–100 nm). A modification of the method is production of metallic clusters in beams under supersonic emission of the particles from the nozzle [61–65]. During motion of the thus-formed beam, clusters containing up to 10^6 atoms are generated. The described method employs laser evaporation of metals and acceleration of charged cluster ions by an outer electric field aimed at changing the trajectory of their motion. Another method of metal model evaporation is ion bombardment by high-energy fluxes obtained under bombardment by inert ions of gases or some metals [66]. The method of shock waves forms nanoparticles as a result of shock wave transit and strong oversaturation with metal vapors induced by high-temperature decomposition of metal-containing compounds with the vapor elasticity (carbonyl, metal-organic) [67]. An original modification of the method is evaporation of metals in vacuum followed by condensation on a solid surface [68], including that in a cryogenic reactor (production of matrix-stabilized clusters and nanoparticles during condensation of metal atoms on low-temperature matrices) and cathode spraying. In the latter case the negative electrode (cathode) is destroyed in a gas discharge under bombardment with positive ions. Substrates are placed on the anode for depositing metal atoms evaporated from the cathode. A stationary glowing discharge is ignited between the cathode and anode in the inert gas medium. The magnetotron regime of cathode spraying which employs crossed electric and magnetic fields enabling dispersion of high-melting metals and alloys is widely used [69]. Good results have also been obtained by the aerosol method (gas evaporation). It consists of metal evaporation [58, 60] in a rarified atmosphere of inert gas under low temperatures followed by vapor condensation in a reactor. Its modification is the levitation crucible-free method [70] that includes evaporation of a large suspended metal or alloy drop heated by a high-frequency magnetic field in a circulation setup with a laminar flow of inert gas. Another version is production of metal nanoparticles using an electric burst of conductors in the inert gas atmosphere [71].

Low-temperature plasma is an important means of synthesizing nanoparticle powders. It employs modern plasma installations incorporating plasma generators, a reactor and a device for hardening reaction products [72, 73]. The latest design of a low-temperature plasma apparatus is the plasmatron electric discharge generator. Its modifications are an electric arc, glowing discharge, barrier discharge and high-frequency discharge. Plasma-chemical processes are easily controlled and optimized. By using plasma technology high-dispersed powders of practically any metals and alloys can be obtained. Still most frequently applied modifications are plasma-activated chemical deposition in the gaseous phase (PECVD) based on volatile metal-containing organic compounds and metal–polymer nanofilms.

Dispersion is a widely used method of producing high-dispersed powders, suspensions, emulsions and aerosols by dispersion of coarsely dispersed (>100 μm) metal particles. Dispersion metals containing even colloidal systems are produced by mechanical or acoustic (supersonic) dispersion. Mechanical dispersion is used for loading (energy) to solid bodies and relaxation (distribution) of absorbed energy within the material bulk [74–76]. Investigation of physico-chemical transformations occurring in solids at deformation is the subject of mechanical chemistry. There are three stages of deformation in this case. The first one is structural disordering leading to amorphization of the substance without destruction of its crystalline structure. The second one is the structural mobility impelling multiplication and motion of dislocations as well as onset and propagation of cracks. The third one is structural relaxation related to shift of defects and system inclination to return to the equilibrium state. There are two limiting cases here: 1. deformation of the substance in a permanent field of mechanical stresses (e.g. under uniaxial tension); 2. effect of variable mechanical loads (when grinding by mills, friction, ultrasound in solutions).

Sometimes various mechanical effects are simultaneously acting in grinding devices, such as crushing (compression), splitting, free and constraint impact, attrition, rupture. Most promising grinding devices are planetary, differential, centrifugal and colloidal mills and disintegrators. They meet modern technological requirements, include thermostating, sealing, inert atmosphere, and ensure atomizing in the presence of polymers.

Although dispersion of metal containing particles under the effect of ultrasonic waves has not found wide application in industry, it is often used in laboratory practice. This process is called sonochemical synthesis or sonolysis. It proceeds in a dispersed medium consisting of water, organic solvents and surfactants. It is stipulated by sonic oscillations of medium and high intensity. Exposure to ultrasonic waves in a dispersed medium results in breakage of continuity due to a drop of local pressure P below its critical value P_{cr}, i.e. to acoustic cavitation. Contraction of the cavitation bubble of either gas or vapor inside the liquid is accompanied by a sound pulse which leads to an abrupt pressure rise (like at burst) and emergence of a compression pulse exerting a dispersion effect on the particles [77–80]. Ultrasonic wave frequency and intensity, time of treatment, initial dispersity of the metal, and the presence of surfactants, affect the efficiency of sonochemical synthesis.

Chemical methods of producing nanoparticles. These methods are most applicable at present procedures for obtaining nanoparticles. Already Faraday reduced metal salts in the presence of stabilizers resulting in formation of metal colloids [81]. However, major investigations took place only in the last 40–50 years (see Chap. 1). Hydrogen and some hydrogen containing compounds (ammonium, hydrazine and its derivatives, hydrogen sulfide, hydrogen peroxide and aliphatic alcohols), hydrides of light metals, boron and aluminohydrades, some amine and hydrazine boranes, hypsophosphates, formaldehyde, tartaric and oxalic acid salts, hydroquinone, dextrins and many inorganic compounds, such as CS_2, CO, NO, $SnCl_2$, were used as reducers. In this case metals are reduced in both gaseous and condensed phases. The process of nanoparticle formation is influenced by the choice of reducer–oxidizer combination, environment, its pH, temperature, concentration of reagents, diffusive and sorption characteristics of reacting particles and stabilizing substances. Methods of heterogeneous reduction in gas–solid systems under high temperatures and the systems of metals (Li, Na, K) and metal–organic compounds, including boron, zinc and organoaluminium) are employed less frequently.

The reduction methods and methods producing bimetallic colloids where nanoparticles look like a nut whose core is formed of one metal, while the shell by another, are widely used. Such a structure allows one to control properties of the core-forming metal. This approach is most often used when manufacturing colloids having spherical particles of the core–shell type, namely (Pd)Cu, (Pd)Pt, (Au)Pt with narrow dispersity and mean diameters (the core diameter is in brackets): 2.0 (1.5), 3.5 (2.5), 22.5 (4.5) [82,83]. It is possible to obtain colloids presenting either a mixture of metal phase nanoparticles or a core–shell structure.

The general picture of formation kinetics of the new phase in the bulk might be applied to the description of nanoparticle nucleation and growth in redox reactions. Chemical reduction of metals in solutions occurs with acceleration [84]. There are three characteristic transformation regions on the diagram of the being reduced metal consumption dependence in a solution (η) versus time (Fig. 2.5): the induction region from 0–τ_{in}, and the regions of transformation acceleration and attenuation. The induction period is connected with the initial stages of the chemical reduction reaction during which stable particles of a solid product are formed. At the end of the period 1 nm size stable particles appear as a result of sequential incrementing of small particles by metal atoms:

$$
\begin{aligned}
&M^+ + H \leftrightarrow M + H^+ \\
&M + M^+ \leftrightarrow M_2^+ \\
&M_2^+ + H \leftrightarrow M_2 + H^+
\end{aligned}
$$

....

$M_l + M_m \leftrightarrow M_j$

Small particles $M_j (j \leq 100)$ are unstable and their thermodynamic behavior in solutions is determined by the difference ΔE of equilibrium potential values E_{redox} of the solution: $\Delta E = E - E_{redox}$. At $\Delta E > 0$ the particle grows, while at $\Delta E > 0$ it dissolves and at $\Delta E \approx 0$ the state of thermodynamic equilibrium is reached. Upon termination of the induction period metal cluster nuclei cease their formation and further reduction occurs without increase in their number but with increment of particle mass. As it has been already mentioned in Sect. 2.3, growth of metal-containing particles presents a heterogeneous reaction during which kinetic regularities are conditioned not only by correlation of the rates of sequential stages but by mass transfer processes and sorption as well.

High-energy radiation-induced reactions are characterized by generation of new highly reactive particles (electrons, atoms, radicals, unstable excited particles, etc.) which are strong reducers. The reactions can be classified as photochemical (photolysis, <60 eV), radiolytic (radiolysis, 60–10^4 eV) and irradiation by a flow of high-velocity electrons (>10^4 eV). Plasma-chemical methods stand close to them. Chemical processes stimulated by high-energy radiation show the following characteristics [85]: high volume and surface energy density; nonequilibrium properties displayed in non-Maxwell and non-Boltzmann functions of particle distribution by the forward motion energies and occupancy of quantum levels; approach and overlapping of characteristic times τ of physical, physico-chemical and chemical processes; the leading role of highly reactive particles in chemical transformations; multichannel and nonstationary processes of the reacting system (Table 2.4).

Production of nanoparticles using photo and radiation-chemical methods proceeds in dispersed media (aqua, aqua-alcohol, organic solutions of reduced compounds). In this case organic compounds serve as secondary reducers which are short-lived radicals. For example, photochemical reduction turns out to be a phototransfer of electrons directly at photon absorption as a result of excitation of the charge transfer band (CT) by light quanta along the ligand–metal chain.

This method is used frequently and its significance has expanded considerably due to the use of laser radiation for stimulating chemical reactions. Along with silver and aurum colloids, photoreduction of light-sensitive metal complexes of Cu, Rh, Pd, Fe and some other nanoparticles in the presence of surfactants in aqua and aqua-alcohol solutions became popular. Under radiation-chemical reduction γ radiation (^{60}Co source) is used more often [85–87]. Radiation-chemical gain, i.e. number of active, emerged, disintegrated and recombined particles (molecules, ions, radicals, etc.) per 100 eV of absorbed energy, depends upon a diversity of physical and chemical factors, and is a function of absorbed radiation dose and varies within a wide range (from 10^{-6} to 10^8). Radiolysis in a condensed phase is conventionally

Table 2.4. Some physico-kinetic characteristics of radiation effects

Characteristics	Methods Photolysis	Radiolysis	High-velocity Electrons
Reactive particles	Oscillation excited molecules, ions and electrons	Electrons, ions, electron-excited particles	Electrons, ions, electron-excited particles
Initial processes	$h\nu$+ M $\rightarrow$ M*$\rightarrow$ products	e^-,$h\nu$+ M$\rightarrow$M$^+$, e^-, (M$^+$)*$\rightarrow$ products	e^-+ M $\rightarrow$ M$^+$, M*, (M$^+$)*$\rightarrow$ products
Hierarchy of characteristic process times	$\tau_o < \tau_c \leq \tau_e$	$\tau_c \approx \tau_{dif}$	$\tau_c \approx \tau_t$
Generation velocity of reactive particles ($\upsilon\cdot 10^{-15}$, cm^3/s)	10^{-2} -10^{-3}	$\sim 10^{-3}$	$\sim 2 \cdot 10^4$

Notation: c – chemical reaction, o – oscillation relaxation, e – electronic relaxation, t – transfer, dif – diffusion.

divided into three stages [85], namely a physical stage that is connected with formation of the initial intermediate products (high-velocity $<3\cdot 10^{-16}$ s), a physico-chemical stage attributed to spatial distribution of the initial intermediate products as a result of transfer processes (10^{-8}–10^{-7} s) and a chemical stage which includes chemical transformations inside a homogenized system lasting as long as the mean transformation time. Following the quick single-electron reduction of metal ions they become disproportional and form either simple atomic and ion clusters or intermediates. Their interaction leads to the appearance of subcolloidal particles containing 10–20 atoms. These processes terminate in an hour upon a γ pulse. Besides nanoparticles of Ag, Au, Cu, Pd, Hg, In, it is possible to obtain by such methods colloids of bimetallic particles with the core–shell structure, including (Ag)M, M = Cu, Pb, Au, etc. [88]. Stimulation by a flow of high-velocity electrons differs just in high generation velocity of reactive particles which is almost 10^6 as high as under γ irradiation.

Thermolysis of metal-containing precursors (salts of inorganic and organic acids, metal colloids, covalent metal-containing compounds, including metal-organic and molecular complexes) is widely employed in production of nanoparticles of metals and/or their oxides. Thermolysis of the precursors can be carried out in gaseous, liquid and solid states. Thermal decomposition in the gaseous phase is performed at various regimes of chemical vapor deposition – CVD synthesis in conditions of low-temperature plasma. The main requirement for precursors is rather high vapor concentration, and high volatility when the compound transforms into the vapor phase without noticeable decomposition. Thermolysis in a liquid medium employs high-temperature solvents (e.g. paraffin) or their suspensions [89]. Platinum metal (Pt, Pd, Ru, Os, Rh, Ir) compounds can potentially fit with a wide range of ligands for this purpose. Decomposition in a solid phase can be exercised both in vacuum or in gaseous atmosphere, including gaseous products of transformation (self-generated atmosphere). A general feature for the decomposition mechanism of metal-containing precursors is breakage of the metal–ligand bond. Mean Gibbs' dissociation energy $\Delta\overline{G}$(M–L) values are from 60 to 300 kJ·mol^{-1} for various metals and differ much for various types of precursors. The reactions of metal disproportionation, β-hydride elimination and others are important along with homolysis of M–L bonds.

Direct thermal decomposition of ion metal-containing precursors usually runs under high enough temperatures. The thermal transformation mechanism in the solid phase resulting in high-dispersed metal-containing particles was comprehensively studied in the example of decomposition of mono- and polycarboxylates of transition metals incorporating nonequivalent in energy links M–O, C–O, C–C. The velocity of their thermolysis is affected by the chemical and physical prehistory of the salts, e.g. dispersity of the original substance and its variation during the process, dehydration regime, and nature of the thermolytic medium (inert gases, dynamic vacuum, air, self-generated atmosphere). The dimensions of the forming nanoparticles can be varied by changing the regimes of thermolysis of transition metal carboxylates in conditions of self-regulating atmosphere and within 10–100 nm limits reaching below 10 nm size using special treatment methods [90].

2.6 Metal Clusters as Nanoparticles with Fixed Dimensions

There are metal nanoparticles of disordered structure with 1–100 nm dimensions having 10^5–10^6 metal atoms, and clusters of ordered structure 1–10 nm in size incorporating 20–40 and sometimes more atoms.

The chemistry of cluster compounds is a novel scientific direction at the interface of inorganic, organoelemental and colloidal chemistry, surface chemistry, physics of ultra-dispersed particles and electronic materials science. Clusters in terms of chemistry are metal compounds whose molecules contain a skeleton of metal atoms surrounded by ligands. These metal atoms are at such a distance from each other (less than 0.35 nm) that direct metal–metal interactions occur. As applied to nanocomposites, the terms cluster materials, nanoclusters or cluster metal compounds are more often used. Cluster formations exhibit a nonmonotonic dependence of their properties to a greater degree as compared to nanoparticles. Examples are ionizing potentials, energies of the links per metal atom, interatomic distances, optical and magnetic properties, electron conductivity, electron–phonon interactions, and so on) upon cluster size, i.e. on the amount of metal atoms in it [91, 92]. This is a crucial fact conditioning the existence of so-called magic numbers – the discrete set of the number of atoms (N) corresponding to the formation of clusters with the lowest energy value. The numbers equal to 13, 55, 147, 309, 561, 923, 1415, 2057, 2869 and so on are calculated by the formula [93, 94]

$$N = 1/3(10n^3 + 15n^2 + 11n + 3)$$

where n is the number of layers around a central atom. Thus, a minimal dense-packing core consists of 13 atoms: one central atom and 12 atoms of the first layer. The formation of a molecular cluster core proceeds in accordance with the conception of dense atomic packing of bulk metal. The dense packing core is the 12-apical polygon (cube-octahedron, icosahedron, anticube-octahedron). The structure of small metallic clusters (3, 4, 5, 6, 10, 20, 38 atoms) has been determined by X-ray diffraction analysis. Metal skeletons of clusters consist mainly of metal chains, cycles and polyhedrons encircled by an ordered structure of ligands.

One of the major conditions of its success is the stability of the skeleton which, in turn, is determined by electronic, thermodynamic and kinetic aspects. A significant role is played by ligand shells (including polymer shells). They avert interaction between individual metal particles that may lead to their sticking and formation of a compact metal and, on the other hand, they hamper interaction between the core and active reagents able to break the core due to easily polarizable metal–metal links, their high activity and comparatively low strength (100–150 kJ per electron participating in M–M interactions). It allows us to raise the activation energy of the disintegration reaction of clusters. Ligands are subdivided by their influence on the cluster core into weak field ligands (halogenide-ions, carboxylates, amines and phosphines) and strong field ligands (carbonyls, π acceptors). Ligands might dictate the behavior of cluster interactions too. Chloride clusters (e.g. with octahedral core $[Mo_6Cl_8]^{4+}$) are most often isolated as islet complexes, carbonyl clusters of the type $Os_3(CO)_{12}$, $Rh_6(CO)_{16}$ enter into strong interactions pulling together cluster fragments into some supercluster. Many cluster compounds of high nuclearity (Pd_{38}, Au_{55}, Rh_{55}, gigantic clusters whose nuclearity reaches hundreds of units, e.g. $Pd_{561}Phen_{60}O_{60}$ (Phen - phenanthroline) [95] and even Pd-1415 [96]) crystallize as monocrystals. Such large clusters are not individual compounds but include particles of various sizes and composition within a narrow distribution range (like molecular-mass distribution of polymers).

These large clusters represent nanoparticles covered by a ligand shell that protects them against the environment. High nuclearity clusters can be not only homometallic but heterometallic as well, e.g. of the composition $[Ni_{38}Pt_6(CO)_{48}H_2]^{4-}$ (diameter of the metallic skeleton is 1.16 nm and approximately spherical) [97]. The general principle of the production of heterometallic clusters is substitution of the atoms of one metal in highly symmetrical metal polyhedrals by some other metal while preserving the total amount of cluster valent electrons and ligand type. The most used heterobi-, heterotri and heterotetra metal clusters are produced by the methods of addition, substitution, elimination and recombination. Centering by μ_3-bridges consisting of the main subgroup elements (C, S, P) improves their stability. Problems of synthesis of such clusters, their structure and classification based on the topological and geometrical principles were analyzed elsewhere [98]. In [99] carbon-containing bimetallic nanoparticles can be produced by hydrogen-induced reduction of molecular carbonyl clusters at 673 K. The nanoparticles were about 1.5 nm in size. This demonstrates the evolution of metal clusters in nanoparticles.

The analysis has shown that in spite of a wide spectrum of available methods for synthesizing nanoparticles, the choice and elaboration of new procedures for obtaining nanoparticles with a narrow range of distribution by size, control over their synthesis, i.e. a purposeful competition between processes of nanoparticle nucleation and growth still remain actual problems. These are fundamental problems of the physico-chemistry of the dispersed state. Of no less importance is the question of stabilizing highly reactive forming particles which is connected with the methods of production.

References

1. I.D. Morokhov, L.I. Trusov, S.P.Chizhik: *Ultra-dispersed Metallic Media* (Atomizdat, Moscow 1977)
2. P. Barret: *Kinetics of Heterogeneous Processes*(Gauthier-Villars, Paris 1973)
3. I.Y. Frenkel: JETP **6**, 87 (1939)
4. J. Gibbs: *Thermodynamic Works* (Goskhimizdat, Moscow 1950)
5. F. Folmer: *Kinetics of the New Phase Formation*(Nauka, Moscow 1986)
6. B.M. Smirnov: *Physics of Fractal Clusters* (Nauka, Moscow 1991)
7. B.M. Smirnov: Russ. Phys. Rev. **161**, 6, 141 (1991); **163**, 7, 51 (1993)
8. Z.Y. Zeldovich: JETP **12**, 525 (1942)
9. I.Y. Frenkel: *The Kinetic Theory of Liquids* (Nauka, Moscow 1975)
10. B. Delmon: *Kinetics of Heterogeneous Reactions* ed. by V.V. Boldyrev, (Mir, Moscow 1972)
11. *Chemistry of Solid State* ed. by V. Garner (Izdatinlit, Moscow 1961), p. 213
12. D. Hirs, G. Pound: *Evaporation and Condensation* (Metallurgy, Moscow 1966)
13. Y.I. Petrov: *Clusters and Fine Particles* (Nauka, Moscow 1986)
14. S.H. Bauer, D.J. Frurip: J. Phys. Chem. **81**, 1015 (1977)
15. C. Kittel: *Introduction into the Physics of Solids* ed. by A.A. Gusev (Nauka, Moscow 1978)
16. L.D. Landay, E.M. Livshits: *Theoretical Physics. Vol. 5. Statistical Physics* (Nauka, Moscow 1978)
17. D.J. Frurip, S.H. Bauer: J. Phys. Chem. **81**, 1007 (1977)
18. A.A. Chernov: *Crystallization Processes. Modern Crystallography, Vol. 3. Formation of Crystals* ed. by B.K. Veinstein, A.A. Chernov, L.A. Shuvalov (Nauka, Moscow 1980)
19. V.G. Dubrovsky: Colloid. J. **52**, 243 (1990)
20. S. Kotake, I.I. Glass: Progr. Aerospace Sci. **19**, 129 (1981)

21. G.M. Pound: Metallurgical Trans. A **16**, 487 (1985)
22. G. D'Agostino: Philos. Mag. B **68**, 903 (1993)
23. G. D'Agostino, A. Pinto, S. Mobilio: Phys. Res. B **48**, 14447 (1993)
24. V.M. Bedanov: Mol. Phys. **69**, 1011 (1990)
25. *Kinetics of Aggregation and Gelation* ed. by F. Family, D.P. Landau (Elsevier, New York 1984)
26. P. Meakin: Faraday Disc. Chem. Soc., **83**, 113 (1987)
27. F. Family: Faraday Disc. Chem. Soc. **83**, 139 (1987)
28. S.L. Narashimhan: Phys. Rev., A **41**, 5561 (1990)
29. A.F. Shestakov, V.N. Soloviev , V.V. Zagorsky, G.B. Sergeev: J. Phys. Chem. **68**, 497 (1994)
30. V.M. Samsonov, S.D. Muraviev, A.N. Bazulev: J. Phys. Chem. **74**, 1971, 1977 (2000)
31. A.V. Krestinin: Russ. J. Chem. Phys. **5**, 240 (1986)
32. C. Tang: Phys. Rev. A **31**, 1977 (1985)
33. A.V. Krestinin, V.N. Smirnov: Chem. Phys. **9**, 418 (1990)
34. G. Seifert, R. Schmidt, H.O. Lutz: Phys. Lett. A **158**, 231, 237 (1991)
35. E. Feder: *Fractals* (Mir, Moscow 1991)
36. T. Vichek: *Fractal Growth Phenomena* (World Sci., Singapore 1989)
37. *Fractal Phenomena in Disordered System* ed. by R. Orbach, In: *Annu. Rev. Mater. Sci.* **19**, 497 (1989)
38. A.L. Goldberg, D.R. Regny, B.D. West: In the World of Science, Sci. Am. **4**, 25 (1990)
39. H. Spindler, G. Vojta: Z.fur Chemie **12**, 421 (1988)
40. T.A. Witten, L.M. Sauder: Phys. Rev. Lett. **47**, 1400 (1981)
41. P. Meakin: Phys. Rev. Lett. **51**, 1123 (1983)
42. J.P. Eckmann, P. Meakin, I. Procaccia, R. Zeitak: Phys. Rev. A **39**, 3188 (1989)
43. R. Jullien, M. Kolb: J. Phys. A **17**, L639 (1984)
44. S.R. Forrest, T.A. Witter: J. Phys. A **12**, L109 (1979)
45. G.A. Niklasson, A. Torebring, C. Larsson, C.G. Granqvist, T. Farestam: Phys. Rev. Lett. **60**, 1735 (1988)
46. M.Yu. Yablokov, S.A.Zavyalov, E.S. Obolonkova: Russ. J. Phys. Chem. **73**, 219 (1999)
47. V.V Vysotskii, T.D. Pryamova, M.V. Shamurina, T.M. Shuman: Colloid J. **61**, 473 (1999)
48. F.Z. Badaev, V.A. Batyuk, A.M. Golubev, G.B. Sergeev, M.B. Stepanov, V.V. Fedorov: Russ. J. Phys. Chem. **69**, 1119 (1995)
49. H.J. Herrmann, P. Staufer, D.P. Landau: J. Phys. A **16**, 1221 (1983)
50. D. Weitz, M. Oliveria: Phys. Rev. Lett. **52**, 1669 (1984)
51. J. Mai, A. Casties, W. von Niessen: Chem. Phys. Lett. **196**, 358 (1992)
52. M. Matsushita, M. Sano, Y. Hayakawa, H. Honjo, Y. Sawada: Phys. Rev. Lett. **53**, 286 (1984)
53. L.M. Sander, D. Grier: Annu. Rev. Mater. Sci. **19**, 502 (1989)
54. V.M. Myagkov, L.E. Bykova: JETP Lett. **67**, 317 (1998)
55. A.A. Lushnikov, A.E. Negin, A.V. Pakhomov: Chem. Phys. Lett. **175**, 138 (1990)
56. B.I. Shklovsky, A.L. Afros: *Electronic Properties of Alloyed Semiconductors* (Nauka, Moscow 1979)
57. I.A. Chmutin, S.V. Letyagin, A.V. Shevchenko, A.T. Ponomarenko: Polym. Sci. Ser. A **36**, 699 (1994)

58. R. Uyeda: Progr. Mater. Sci. **35**, 1 (1991)
59. N. Ramzei: *Molecular Clusters* ed. by B.P. Adiasevich (Izdatinlit, Moscow 1960)
60. Y.I. Petrov: *Clusters and Fine Particles* (Nauka, Moscow 1986)
61. O.F. Hagena: Surface Sci. **106**, 101 (1981); Rev. Sci. Instrum. **3**, 2374 (1992); Zs. Phys. D **4**, 291 (1987); **17**, 157 (1990); **20**, 425 (1991)
62. K. Kimoto, I. Nashida, H. Takahashi, H. Kato: Jap. J. Appl. Phys. **19**, 1821 (1980)
63. R.S. Bowles, J.J. Kolstad, J.M. Calo, R.P. Andres: Surface Sci. **106**, 117 (1981)
64. M.L. Alexandrov, Y.S. Kusner: *Gasodynamic Molecular Ionic and Clusterized Beams* (Nauka, Leningrad 1989)
65. J. Muhlbach, K. Sattler, P. Pfau, E. Reckhagel: Phys. Lett. A **87**, 415 (1982)
66. H.H. Andersen, B. Steumn, T. Sorensen, H.J. Whitlow: Nucl. Instrum. Meth. Phys. Res. B **6**, 459 (1985)
67. D.J. Frurip, S.H. Bauer: J. Phys. Chem. **81**, 1001, 1007 (1977)
68. S.H. Norrman, T.G. Andersson: Thin Solid Films, **69**, 327 (1980)
69. G. Schmid, N. Klein, B. Morum, A. Lehnert: Pure Appl. Chem. **62**, 1175 (1990)
70. M.Y. Gen, A.V. Miller: Surface **2**, 150 (1983); Invention Certificate no. 814432
71. V.P. Ilyin, G.V. Yablunovsky, N.A. Yavorovsky: *Clusters in Gaseous Phase* (Novosibirsk, 1987) p. 132
72. L.S. Polak, G.B. Sinyarev, D.I. Solovetsky: *Chemistry of Plasma* (Nauka, Novosibirsk 1991)
73. F.A. Saliyanov: *Foundations of the Physics of High-Temperature Plasma of Plasma Apparatuses and Technologies* (Nauka, Moscow 1997)
74. P.Y. Butyagin: Russ. Chem. Rev. **40**, 1935 (1971); **53**, 1769 (1984)
75. P.A. Rebinder: *Selected Works. Surface Phenomena in Dispersed Media* ed. by G.I. Fux (Nauka, Moscow 1978)
76. G. Heinike: *Tribochemie* (Akad. Verlag, Berlin 1984)
77. K.S. Suslick: *Sonochemistry. In: Kirk-Othmer Encyclodpedia of Chemical Technology*, 4th edn. (John Wiley & Sons, Inc., New York 1998) **26**, 516
78. K.S. Suslick: Science, **247**, 1439 (1990)
79. E.B. Flint, K.S. Suslick: Science, **253**, 1397 (1991)
80. K.S. Suslick, T. Hyeon, M. Fang, J.T. Ries, A.A. Cichowlas: Mater. Sci. Forum **225-227**, 903 (1996)
81. M. Faraday: Philos. Trans. R. Soc. London, **147**, 145-153 (1857)
82. M. Harada, K. Asakura, N. Toshima: J. Phys. Chem. **97**, 5103 (1993)
83. G. Schmid, H. West, H. Mehles, A. Lehnert: Inorg. Chem. **36**, 891 (1997)
84. V.V. Sviridov, T.N. Vorobieva, T.V. Gaevskaya, L.I. Stepanova: *Chemical Deposition of Metals in Water Solutions* (University Publ., Minsk 1987)
85. L.T. Bugaenko, M.G. Kuzmin, L.S. Polak: *Chemistry of High Energies* (Khimiya, Moscow 1988)
86. A. Henglein: J. Phys. Chem. **97**, 5457 (1993)
87. B.G. Ershov, N.L. Sukhov: Radiat. Phys. Chem. **36**, 93 (1990)
88. P. Mulvaney, M. Giersig, A. Henglein: J. Phys. Chem. **96**, 10419 (1992); **97**, 7061 (1993)
89. M.J. Hampden-Smith, T.T. Kodas: Chem. Vap. Deposition, **1**, 8 (1995)
90. A.S. Rozenberg, G.I. Dzhardimalieva, A.D.Pomogailo: Dokl. Phys. Chem. **356**, 66 (1997)
91. F.A. Cotton, R.A. Walton: *Multiple Bonds between Metal Atoms* (Wiley, New York 1982)

92. S.P. Gubin: *Chemistry of Clusters* (Nauka, Moscow 1987)
93. P. Chinni: Gazz. Chim. Ital. **109**, 225 (1979); J. Organomet. Chem. **200**, 37 (1980)
94. B.K. Teo, N.J.A. Sloan: Inorg. Chem. **24**, 4545 (1985)
95. M.N. Vargaftik, I.I. Moiseev, D.I. Kochubey, K.I. Zamaraev: Faraday Discuss. R. Soc. Chem. **92**, 13 (1991)
96. G. Schmidt: Chem. Rev. **92**, 1709 (1992)
97. A. Geriotti, F. Demartin, G. Longoni: Angew. Chem. Intern. Ed. **24**, 697 (1985)
98. S.P. Gubin: Russ. J. Coord. Chem. **20**, 403 (1994)
99. M.S. Nasher, A.I. Frenkel, D.L. Adler, J.R. Shapley, R.G. Nuzzo, J. Am. Chem. Soc. **119**, 7760 (1997)

3 Principles and Mechanisms of Nanoparticle Stabilization by Polymers

The high chemical activity of nanoparticles with a developed surface is the reason for various, often strongly undesirable spontaneous processes. Nanoparticles are sensitive to impurities, binding in high solution concentrations during boiling or agitation under the action of radioactive energy. The occurring processes are commonly irreversible. So far, one of the most important goals is to raise the stability of nanoparticles during storage and transportation. Various stabilizers are used with this aim. Previously they were low-molecular organic compounds (carbonic acids, alcohols, amides) and natural polymers (gelatin, gum arabic, agar-agar, starch, cellulose and so on). At present more frequently synthetic polymers are employed. They solve the problems of nanoparticle stabilization and introduction of a polymer ingredient into the nanocomposite.

Stabilization of metal nanoparticles by high-molecular compounds presents a major branch of modern polymer colloidal science. It studies generation regularities of dispersed systems with highly developed interfaces, their kinetic and aggregation stability, different surface phenomena arising at the interface and adsorption of macromolecules from liquids on solid surfaces [1]. The theory of improving stability of colloidal particles by polymers is described in [2–7]. This chapter will focus just on those questions that are connected with nanoparticles and nanocomposites on their base.

3.1 Stability of Nanoparticles in Solutions

Nanoparticles in solutions present typical colloidal systems consisting of a continuous phase which is a dispersed medium (solvent) and a dispersed phase (nanoparticle). Such systems having a solid dispersed phase of non-polymer and a liquid dispersed medium are called sols as contrasted to latexes where the dispersed phase is of a polymer origin. Colloidal systems in organic solvents are frequently called organosols, while analogous dispersions in water are called hydrosols.

A distinctive feature of colloidal solutions is their relatively low stability attributed to large particle size with perceptible free surface energy. Each nanoparticle appears to be an aggregate of atoms or more or less simple molecules. Any change in conditions might result in aggregate size variation and precipitate fallout. The stability of colloidal solutions is considered to be one of the key problems of colloidal chemistry. A system turns out to be stable when its dispersed phase can exist as separate individual particles for a long time (a few months and even years).

There are two types of stability of colloidal solutions. The kinetic one is stability of the systems relative to gravity forces. Crucial factors that determine kinetic stability of colloids are Brownian motion, dispersion, viscosity of the medium, etc. The stability of such systems drops abruptly as the particle size increases. Kinetically stable systems are where the velocity of dispersed particle precipitation under the action of gravity forces is so slow that it can be neglected. Stability is sometimes estimated by the presence of sediment upon centrifuging, e.g. centrifuging of the platinum–PVA system at 6000 g/20 min acceleration. By varying centrifuging acceleration and time one can control the size of colloidal particles in the Pt–PVA system [8]. Aggregation stability is the ability of the system to preserve the degree of particle dispersity. This type of stability is attributed to the ability of nanoparticles to create large aggregates to adsorb low-molecular ions on their surface from the solution leading to formation of an adsorption layer. This layer consists of potential determining ions and counter-ions and a diffusion layer containing residual counter-ions. Aggregation stability of these systems can be also explained by the formation of solvate shells of dispersed medium molecules round colloidal particles tightly bound with them. Investigations have shown [1] that due to their elasticity and elevated viscosity the shells act as wedges and hamper particles sticking together by drawing them apart. Aggregation stability factors differ from those of kinetic stability. For example, temperature rise hinders particle precipitation but assists their aggregation or intensified Brownian motion hampers particle precipitation but promotes particle sticking at collision.

Colloidal solutions are lyophobic and lyophilic types. The last ones are characterized by so weak an interaction of the dispersive medium with the dispersive phase (such systems are unstable in thermodynamic respects) that the attraction between particles results in their aggregation. Nevertheless, a noticeable repulsive barrier and remoteness of particles from one another can retard the aggregation process so that the system becomes stable almost without restrictions (kinetic stability). These colloids are often called thermodynamically metastable. The kinetic stability is commonly reached for lyophobic particles through electrostatic (charged) stabilization. Such systems are stable and provide repulsion potential surpasses the aggregation energy of colloidal particles [9], Fig. 3.1.

Lyophilic systems are thermodynamically stable. Their Gibbs shift energy is negative. The dispersed state can be characterized as the low energy state. Aggregation stability of these particles is attained through solvation.

Aggregation of colloidal systems is called coagulation in case densely packed formations are formed. When the particles are separated by the

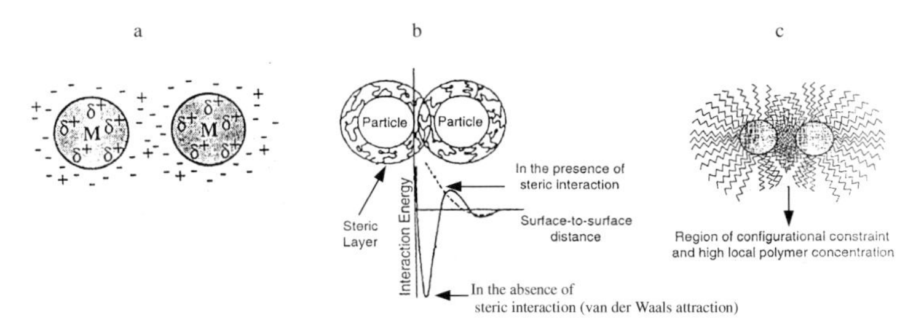

Fig. 3.1. Diagram of nanoparticle stabilization by polymers: (**a**) electrostatic stabilization; (**b**) steric stabilization; (**c**) region of high polymer concentration

dispersive medium, it is called flocculation. Coagulation of nanoparticles on polymer surfaces is considered as heteroaddagulation.

The state of colloidal systems changes in response to reducing dispersity of the solid phase in two stages: particle enlargement or coagulation and sedimentation of the solid phase. Coagulation velocity depends on such factors as the range of attraction forces, Brownian motion velocity, concentration of colloidal solution, presence of electrolytes. Uncharged particles coagulate quickly due to dispersive attraction. The coagulation mechanism is identical to that of the second order chemical reaction since two particles are to meet in both reactions. The difference is only in the fact that the reacted molecules might not participate in further reactions, whereas particle aggregates continue to be involved in coagulation processes thus forming still more intricate complexes. The process is described by velocity constants (k_c) of high-velocity coagulation by Smolukhovsky

$$k = 8kT/\eta \,, \tag{3.1}$$

where η is the viscosity of the dispersive medium.

The constants k_c characterize the long-range attraction between colloidal particles. Attraction forces arising from interaction between molecules and atoms of two neighboring particles are called van der Waalse forces. Three models reflecting interaction between particles of the dispersed phase are mostly used: Keesom's orientation forces (interaction between molecules with a constant dipole moment), Debye's induction forces (interaction between the constant dipole of one molecule and the induced dipole of the other) and London's dispersive forces (interaction between instantaneous dipoles). Ample description of the phenomenon is given by London's forces. The attraction between individual atoms or molecules is extended only to a distance of a few tenths of a nanometer. However, in the treated systems a multiatomic (multicentered) interaction takes place: each atom or molecule of a nanoparticle attracts each atom or molecule of another particle consisting of 10^6–10^{10} atoms. The strength of interaction of dispersed phase particles in coagulation structures is estimated on the average in 10^{-10} N per contact. There is strong attraction due to many interatomic interactions between nanoparticles.

The potential energy of attraction U_r between two spheres of particles of radius r and minimum distance l_0 between their surfaces can be given by the following equation

$$U_r \approx Ar/12l_0 \quad \text{when} \quad r \gg l_0 , \tag{3.2}$$

where A is the efficient Hamaker's constant with dimension of energy.

The constant stands close to the kT parameter for latex particles, although in metal dispersions it can be much higher. For example, for PMMC, PS, Ag and Al_2O_3 the constant A equals $6.3 \cdot 10^{-20}$, $6.15 \cdot 10^{-20}$, $40.0 \cdot 10^{-20}$, $15.5 \cdot 10^{-20}$ J, respectively [8]. Energy of attraction of two latex particles 120 nm in size is $\sim 10\,kT$ for each at a distance of 1 nm and 2 kT at 5 nm distance. This is enough for coagulation without the repulsion between particles as long as the energy of attraction exceeds heat energy of the particles equal to $\sim 3/2\ kT$ of three translation degrees of freedom.

Coagulation and flocculation processes of colloidal particles were studied by numerical methods [10, 11]. The aggregation process of colloidal particles at $N = 12$ has been reflected in molecular-dynamic models and dispersion evolution with time has been recorded [12]. It was found out that at a strong attraction between particles equal to $A = 10^{-20}$ J the depth of the potential minimum reflecting the dense contact between particles is $5.5\,kT$.

It can be presumed that the region of van der Waalse attraction in the systems considered acts to a distance of 5 nm. The attraction value and its range are the determining factors in choosing means of particle stabilization [2]. The range of attraction forces limits essentially stabilization potentialities. In order to stabilize nanoparticles, the repulsive forces between particles are to be effective within the range of attraction forces. Coagulation onsets in the isoelectric point when the particle charge becomes equal to zero.

Nanoparticles without stabilizers represent typical lyophobic colloids with low steadiness, so charge stabilization is insufficient, especially for nonpolar organic matter. High-molecular compounds and various substances are used to stabilize nanoparticles. As a result of such stabilization, called steric, nanoparticles turn out to be enveloped by a preserving barrier consisting of a continuous layer of long solvate polymer chains (Fig. 3.1c). The colloidal system becomes absolutely stable unless the protecting layer gets damaged. These stabilized systems can be treated as complex ones whose nucleus is lyophobic and outer layer is lyophilic.

3.2 Stabilizing Capability Characteristics of Polymers

Early studies in protecting lyophobic sols by high-molecular compounds presupposed a lyophilic adsorption layer formed on the surface of colloidal particles. A metal particle acquires all properties of absorbed polymer molecules. This is exhibited, in particular, in the elevated coagulation threshold (or the least electrolyte concentration, $\mathrm{mmol\,L^{-1}}$) resulting in coagulation within a certain time interval. What is more, in diluted solutions the dispersed phase concentration is directly proportional to the minimum amount of polymer stabilizer necessary for attaining roughly similar coagulation thresholds by the electrolyte [5] (Table 3.1).

Table 3.1. Coagulation threshold γ^{+}_{Na} of AgI and Sb_2S_3 sols by sodium chloride versus solid phase content with the addition of PEO

C_{DF}, g/L	C_{PEO} ($M = 1.7 \cdot 10^4$), mg/L	γ^{+}_{Na}, mmol/L	C_{PEO} ($M = 1.3 \cdot 10^6$), mg/L	γ^{+}_{Na}, mmol/L
		AgI sol		
0.5	3.2	140	1.7	240
1.0	6.4	145	3.4	250
2.0	9.6	140	5.1	250
		Sb_2S_3 sol		
0.2	12.8	60	1.7	350
0.4	25.6	65	3.4	325
0.6	36.4	65	5.1	340

However, contemporary experimental and theoretical findings on the structure of macromolecules and their solutions corrected the understanding of the protecting mechanism. These corrections were made from a comparison of nanoparticle size and that of protecting macromolecules which surpasses linear dimensions of colloidal particles of lyophobic sols by several fold. Particles of gold can be enlarged in the presence of quarternized poly-4-vinyl pyridine (P4VP). This is evidence that macrocations are chemisorbed on them and a soluble complex is formed [13]. The mean hydrodynamic radius of such stabilized particles (21 ± 0.5 nm) is larger than the corresponding radius of nanoparticles in water (13 ± 0.5 nm), Fig. 3.2. The mean hydrodynamic radius of PVA-stabilized platinum particles (molecular mass – 60 000) is 10.5 nm [8] and the metal particle size is small as compared to the protecting layer thickness.

Steric stabilization becomes probable because spatial dimensions of low-molecular compounds are commensurable with the range of London's forces of attraction or even exceed them. If we consider that the diameter of a macromolecule of a linear polymer coincides with the r.m.s. distance between its ends then the relation of the mean geometric radius of the particle $\langle r^2 \rangle^{0.5}$

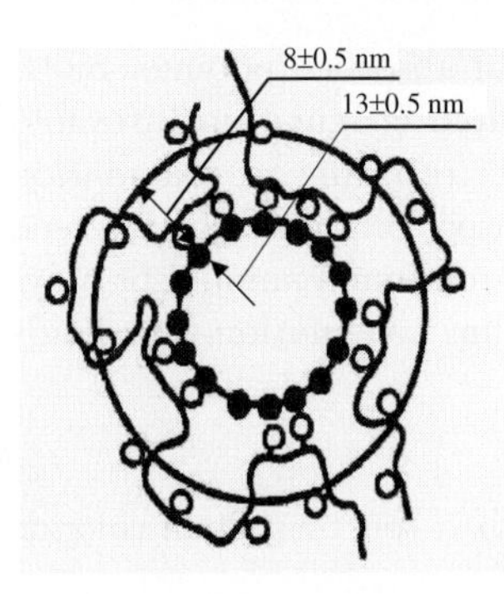

Fig. 3.2. Structure of a soluble polyelectrolyte complex formed by an aurum particle and macromolecules of a polycation

versus polymer molecular mass M can be expressed by the relation [14]:

$$\langle r^2 \rangle^{0.5} \approx 0.06 M^{0.5} \,. \tag{3.3}$$

The value reaches 20 nm for a polymer with $M = 10^4$, $\langle r^2 \rangle^{0.5} = 6$ nm and at $M = 10^5$. Macromolecules of polymers with $M > 10^4$ are of the sizes needed to stabilize colloidal particles (macromolecules are, of course, to induce repulsion of particles).

First attempts to quantitatively estimate stabilizing capabilities of polymers date back to Faraday's times [15]. He was the first to study the effect of gelatin on the variation of the color of gold from red to blue with the addition of NaCl and the capability of soluble polymers to inhibit gold coagulation by electrolytes.

To characterize various biopolymer stabilizers a notion of "golden number" was introduced. It corresponds to a polymer amount (mg) able to avert flocculation of 10 mg of gold when 1 mL of 10% solution of NaCl is added. The lower the golden number the more efficiently the polymer suppresses coagulation of the dispersed phase. Although there is a skeptical attitude to such empirically established numbers (more than ten are already known) [16], we believe they can bring certain benefits in comparative analysis. For example, golden numbers for gelatin, albumin, gum arabic and dextrin are 0.005–0.01; 0.1–0.2; 0.15–0.25 and 6–20 respectively.

The protective power of a polymer as its capability of stabilizing colloids is sometimes estimated by the gold mass (g) which is stabilized by 1 g of a polymer added to 1% water solution of NaCl. The protective power of gelatin, PVPr with a mean polymerization degree $P = 3250$, PVS ($P = 500$), PAA and PEI are 90, 50, 5, 0.07 and 0.04 respectively.

Stabilizing power of polymers can be estimated by critical concentration of KCl ($C_{\mathrm{KCl}} \cdot 10^3$) necessary to change the red coloring of auric sol on blue in the presence of a polymer of certain concentration (1.0 g/L). It is important that C_{KCl} depends upon the molecular mass of the polymer. For example, ($C_{\mathrm{KCl}} \cdot 10^3$) varies without polymers from 12.5 to 18, 22, 47 and to 480 in the presence of PEO with molecular mass 4000, 6000 and 9000, correspondingly.

A stabilizing macroligand during nucleation of nanoparticles undergoes considerable structural (including variations in macromolecular conformation, complexing of polymer functional groups) and chemical changes. The considered problems were so far scarcely elucidated in the literature. Three examples can be mentioned: 1. Oxidation of poly(2,6-dimethyl-1,4-phenylene oxide) (PPO) till quinone derivatives in the process of Ag salts reduction and formation of its nanoparticles in a chloroform–methanol solvent [17]; 2. Oxidation of Langmuir–Blodgett films by ruthenium complexes [18]; 3. Lessening of ability to stabilize small silver clusters by PAA at its decarboxylizing in the course of photoreducing salts [19,20].

3.3 Characteristics of Polymer Absorption on Metal Surfaces

Polymer stabilizing power, often considered as screening by a protecting colloid [21], is an after-effect of macromolecule absorption on nanoparticle surfaces [22]. It is of fundamental importance to understand the processes occurring in polymer interactions with nanoparticles since it will furnish information on the stabilization mechanism and the origin of adhesion on the forming interface too. It will become possible to create a reliable protection against corrosion and other undesirable factors [23–26]. The solution of these problems presumes examination of structure and properties of the surface and interfacial layers of the metal–polymer junction. The metal–polymer interaction mechanism is governed by many phenomena, the most significant among which are the origin of molecular forces, degree of perfection of nanoparticle surfaces, internal stress and electrical charge, solvent nature, temperature, and molecular mass of the polymer [1,27,28]. These factors are correlated with a number of energy and kinetic parameters in the metal–polymer boundary layer, including polymer cohesive energy, surface energy and segmental mobility of chains. Cohesive energy can be modeled by Gildebrant's parameter and segmental mobility – the glass transition temperature.

Interactions between polymer chains and nanoparticles differ by their nature and intensity, and might be displayed simultaneously. It is difficult to distinguish between the effects generated by various forces as a change of properties of high-molecular compounds at their contact with nanoparticles reflects to a certain degree the total effect of interaction on the surface of dispersed phase particles. Furthermore, adsorption of polymers proceeds with heat absorption, in contrast to that of low-molecular compounds. For example, $G = 0.0012\,\mathrm{g\,g^{-1}}$, $Q_1 = -25.9\,\mathrm{kJ\,mol^{-1}}$ in the system Fe–PVA–CCl_4; $G = 0.0025\,\mathrm{g\,g^{-1}}$, $Q_1 = 37.8\,\mathrm{kJ\,mol^{-1}}$ in Cu–PVA–benzene [29]. This can be attributed to processes where the total heat includes the heat Q_1 of polymer adsorption on the particle surface, the heat Q_2 of solvent molecule desorption from the surface and the heat Q_3 of polymer–solvent interaction. Thus, the negative value of the total polymer adsorption heat

$$-\Delta H = q = Q_1 - Q_2 - Q_3 \tag{3.4}$$

shows that the latter two members of (3.4) are larger than the first one.

The protecting polymer might interact with nanoparticles in two ways. The first presupposes that macromolecules fix on the particle surface through, e.g. physical adsorption. Physical adsorption presumes processes induced by van der Waalse forces, dipolar interactions and weak and easily broken hydrogen bonds. Noncovalent interaction of nanoparticles with a macromolecule is extremely weak – about 10^{-4} J·m^2. This is true for nanoparticle interaction with homopolymers, including PE PVS [30–32], PEO [33], PVPr [34, 35], polyethylene glycols [36] and others [37–41]. Homopolymers display high stabilizing power since are easily expelled from the interaction zone during Brownian movement of particles. Particularly, PEO interacts with metal surfaces, so it is a poor stabilizer for colloidal aurum [42].

Stabilization of nanoparticles is affected by the polymer nature, its functional groups and, in case of copolymers, by the composition and section branching in its chains (alternative, statistical, block and grafted). The comparison between the homopolymer P4VPy, the statistical copolymer of styrene and 4-VPy [43,44] has shown that the size of iron particles formed in P4VPy was less than that in the copolymer. Probably, excess fragments of the homopolymer aggregation of nanoparticles hamper as compared to the copolymer. The statistical copolymers are of low efficiency as stabilizers of nanoparticles [45,46].

The second procedure is based on chemisorption of macromolecules from the solution. The interaction of macromolecules (unfolded or globular) with nanoparticles depends upon the amount of polar groups of the adsorbed polymer per unit surface. A critical condition is that the polymer should contain polar groups that interact with surface atoms of the metal. The more electron-donator properties of polymer functional groups the stronger is their adhesion to dispersed phase particles. During chemisorption polymer chains can form covalent, ion or coordination links with the atoms of the metal surface layer which increases the total heat of polymer adsorption. A typical example is coordination interaction of the surface atoms of a zinc cluster with amine groups of methyl tetracyclododecene and 2,3-trans-bis(*tret*-butylamide)methylnorbornene-5 (Fig. 3.3) [42].

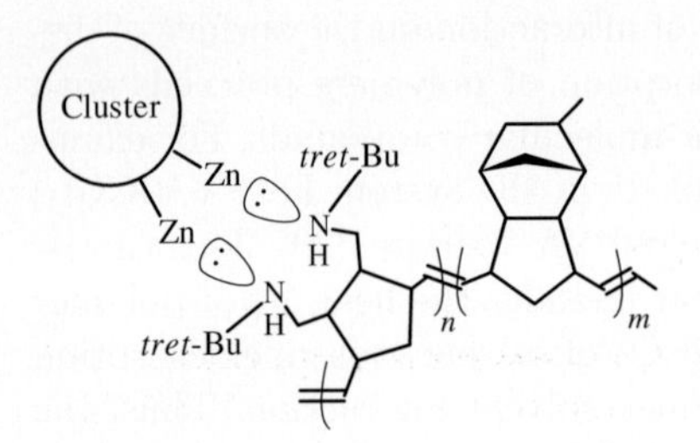

Fig. 3.3. Structure of a polycomplex zinc cluster with methyl tetracyclododecene and 2,3-trans-bis(*tret*-butylamido) methylnorbornen-5 copolymer

Introduction of various functional groups into the polymer results in a strengthened role of the acidic–base interactions in the polymer–metal systems and raised adherence of the polymer. Its adhesion to aluminum or copper rises greatly as the amount of oxygen or nitrogen-containing groups increases in styrene copolymer with acrylic acid or vinyl pyridine [47] (Fig. 3.4).

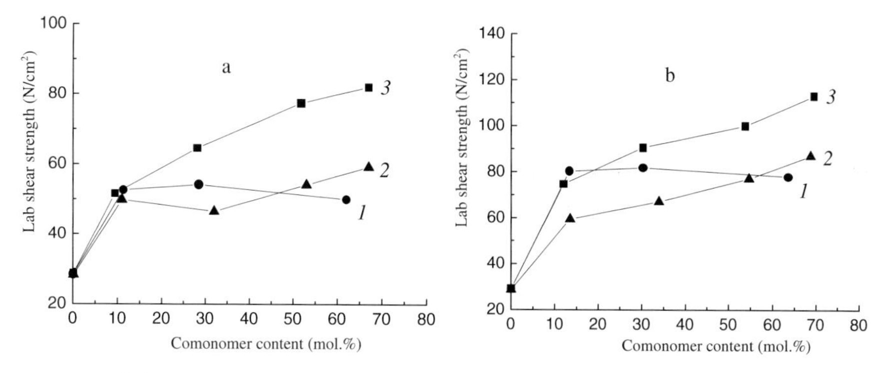

Fig. 3.4. (**a**) Shear strength between copper/poly(styrene-co-acrylic acid) copolymers (1), copper/poly(2-vinylpyridine) (2), copper/poly(4-vinylpyridine) (3) as a function of comonomer content. (**b**) Shear strength between aluminum/ poly(styrene-co-acrylic acid) copolymers (1), aluminum/poly(2-vinylpyridine) (2), aluminum/poly(4-vinylpyridine) (3) as a function of comonomer content

The efficiency of homopolymers and copolymers rises substantially in case some centers of specific interaction are formed either inside the polymers or on the surface of stabilizing particles similarly to centers with intricate in native state structure (secondary, tertiary and even quaternary) forming in polynucleotides. Such centers operate simultaneously analogous to charge and steric stabilization mechanism. Sometimes, the joint action is also observed when stabilizing clusters by synthetic polymers [48]. An example of specific interactions observed between nanoparticles and polymers is stabilization of $Fe(OH)_3$ particles having up to 120 iron atoms in the presence of PVA or polyvinyl ethers of various polymerization degrees (Fig. 3.5). The hydrolysis of Fe^{3+} in aqua solution leads to precipitation of $Fe_x(OH)_{3-x}nH_2O$ particles in which x and n values are dependent upon reaction conditions. Occasionally, α-FeO(OH) with 2 nm mean particle size is formed instead of $Fe(OH)_3$.

The mechanism of polymer interaction with nanoparticles is strongly influenced by the solution pH. At low pH values interaction between polyazocrown ester and red auric sol is connected with a reversible electrostatic interaction. At high pH values the interaction is conditioned by a strong coordination binding macromolecules and particles [49]. The effect of pH on PVP–30000 (polyvinyl pyrrolidone) adsorption on the surface of nanocrystalline particles of Ce_2 has been expounded elsewhere [50]. It was proved that the limiting adsorption reduced with rising pH of the solution (~3.7–11.5)

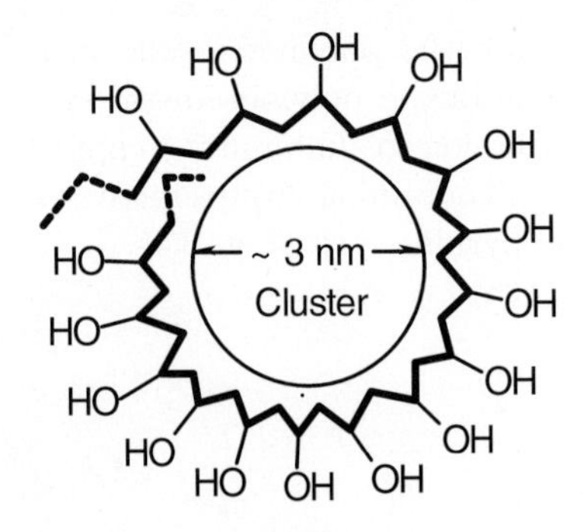

Fig. 3.5. Inclusion complexes consisting of clusters similar to PVA and $Fe(OH)_3$

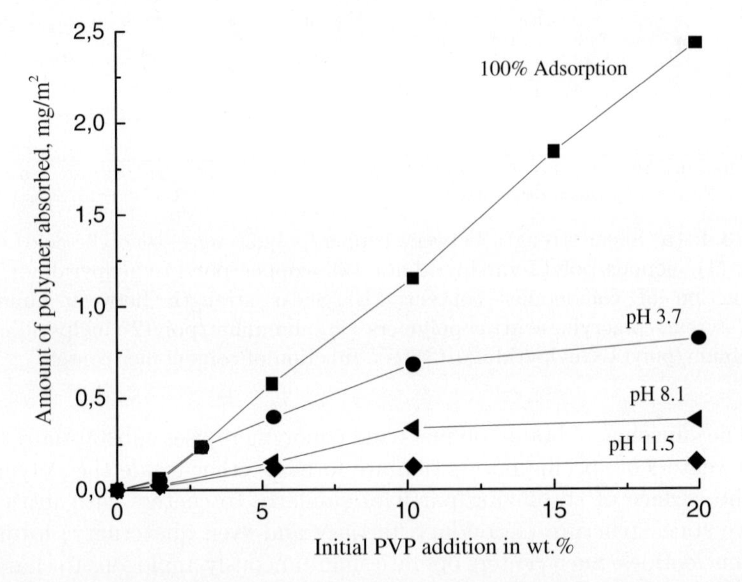

Fig. 3.6. Adsorption isotherms for polyvinylpyridine on nanocrystalline CeO_2 particles in aqueous suspensions at pH values of 3.7, 8.1 and 11.5

(Fig. 3.6). The specific character of local surroundings of macromolecule in water is connected with hydrophobic binding centers on the polymer chains and solvate shell of bonded water molecules. The incompatibility between the structures of water in polymer hydrated shell and hydrated metal cations was observed [51]. It can be supposed that increasing the reactivity of hydrated metal cations is determined by its particular dehydration near a polymer chain [52].

In spite of their low adsorptivity, linear homopolymers appear to be a convenient model for studying theoretical aspects of nanoparticle stabilization by polymers, particularly adsorption of macromolecules on metal surfaces.

Adsorption is a slow process that can be subdivided into two stages: diffusion of macromolecules to the surface of dispersed phase particles and the adsorption which is determined by the time of reaching an equilibrium state of the polymer in the adsorption layer. The time of polymer chain diffusion to the particle surface is a function of the dispersed phase concentration in the colloidal solution, molecular mass of the polymer, time and intensity of mixing and viscosity of the dispersive medium. The process runs quickly in diluted media even when mixing (during a few minutes or seconds). The time depends upon the contact between polymer chains and particles of the dispersed phase. Conformation of a macromolecule is usually stable and diffusion velocity depends on collisions between particles. The equilibrium in the adsorption layer is reached much more slowly. Sometimes it takes hours or even days. This stage depends on the chemical nature of the polymer, metal and adsorptive energy of elementary units on the surface, thermodynamic properties of the solvent and molecular-mass distribution of the polymer. Stability of nanoparticles depends on conformation of macromolecules adsorbed on the surface. That is why the stabilizing effect of high-molecular substances is determined by the time of contact between macromolecules and particles [5] (Fig. 3.7).

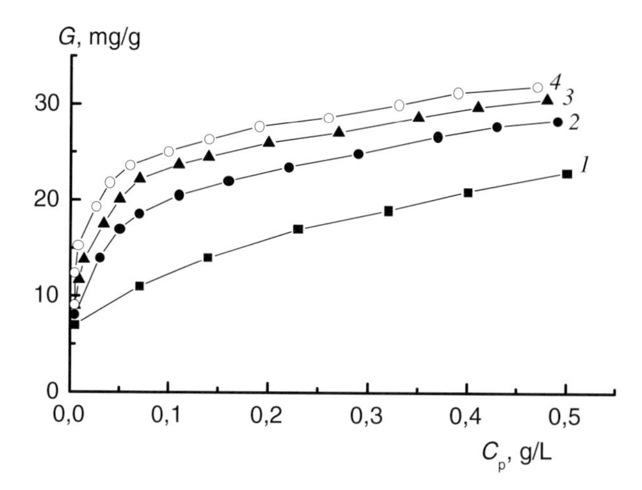

Fig. 3.7. Adsorption isotherms of PVA-18 (18% of unsaponified acetate groups in a molecule) with $M = 5.9 \cdot 10^4$ particles of silver iodide sol at a duration of macromolecule-to-particle contact: (1) 20 min, (2) 1 h, (3) 2 h and (4) 24 h

Adsorption of macromolecules on the surface of dispersed phase is, in the majority of cases, irreversible. All polymer links with a large number of surface links are strong enough not to let the adsorption run. In other words, irreversible behavior of adsorption of the high-molecular compound is more often conditioned by a relatively large amount of its contacts with the surface because of which a simultaneous breakage of all sections connected with it is statistically highly doubtful. Desorption velocity is so low that it is neglected in instantaneous contacts of nanoparticles in their collisions as a result of Brownian motion. The contact constituting portions of a second event in very viscous media can be calculated based on Stoke's law.

Adsorption is commonly described by Langmuir's isotherm

$$C_{\mathrm{p}}/G = 1/G_{\infty}b + C_{\mathrm{p}}/G_{\infty} \tag{3.5}$$

where G is the adsorption, C_{p} is the equilibrium concentration of polymer in solution, b is the constant, and G_{∞} is the limiting adsorption.

Isotherms of polymer adsorption on nanoparticle surfaces show a steep ascent characteristic of monomolecular Langmuir adsorption (Fig. 3.8). Relatively rigid macromolecules adsorbed on the surface are responsible for it. The second layer of adsorbed polymer chains easily desorbs into the solution. The experimental value of limiting adsorption G_{∞} corresponds to the formation of the monomolecular layer. Based on (3.5) the value of G_{∞} for the polymer can be obtained if the real adsorption saturation is not reached in given test conditions. Polymer adsorption isotherms are, as a rule, described by Langmuir's relation only in the case of diluted solutions. The resultant reduced equation fits the results of macromolecular adsorption on the surface of dispersed phase particles [29]. The adsorbed polymer amount depends on its molecular mass (Fig. 3.9).

Use of Langmuir's equation with this aim is, however, disputable since it presupposes an equilibrium between adsorption–desorption processes, whereas in high-molar substances adsorption is commonly irreversible. Besides, sometimes observed discrepancies between experimental and theoretical curves arise from violations of the main postulates of Langmuir's equation which says that only one active surface center and one molecule of the low-molecular substance are to participate in the reaction and all unreacted centers are equally reactive. It is obvious that these postulates are not met in adsorptive processes.

Polymer adsorption on the nanoparticle surface is characterized by diversified adsorption states. Most encountered states are three ones (Fig. 3.10)

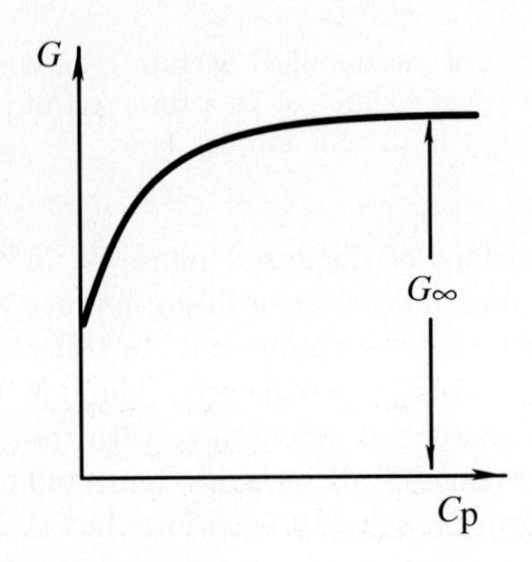

Fig. 3.8. Adsorption isotherm of polymers on metal nanoparticles (Langmuir's type isotherms) characterized by fast adsorption growth

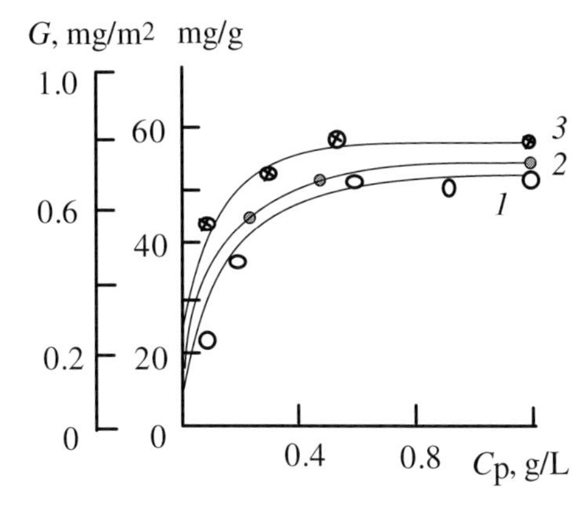

Fig. 3.9. Adsorption isotherm of PEO with $M = 2.3 \cdot 10^5$ (1), $1.3 \cdot 10^6$ (2) and $2.6 \cdot 10^6$ (3) on antimony sulfide particles ($\tau = 24\,\mathrm{h}$)

which correspond to horizontal and vertical positions of polymer molecules relative to the absorbing surface (if we neglect types such as brushes, umbrellas, etc.) and the loop type. The limiting adsorption is related to the polymer molecular mass through the relation

$$G_\infty = KM^\alpha , \tag{3.6}$$

where K and α are constants and α is the index of adsorptive state of the polymer (see Fig. 3.10).

Horizontal adsorption links, no doubt, the polymer and the dispersed phase surface most rigidly although the forming adsorption layer is very thin.

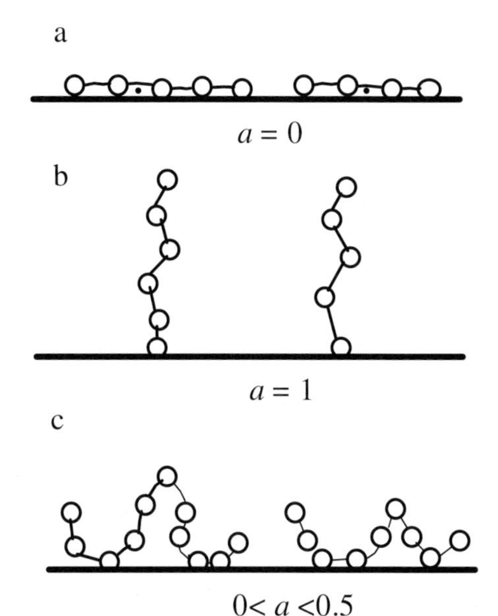

Fig. 3.10. The types of adsorption state of polymers into a surface. (**a**) Horizontal adsorption. (**b**) Vertical adsorption. (**c**) Adsorption of loop type

In contrast, during vertical absorption a volume layer is formed but adsorptive forces are so weak that polymer molecules can freely pass back into the solution. Loops have corresponding intermediate variants of adsorption interactions on the surface. Besides, when two nanoparticle surfaces especially in concentrated solutions are found close to each other, then bridges could arise when different ends of a molecule are adsorbed on two surfaces.

Major characteristics of the adsorption polymer layer are the share of elementary sections of the polymer chain being in direct contact with the surface (v), that of the surface occupied by adsorbed sections (θ) and the mean (efficient) thickness of the adsorption layer (δ). One can use various physical and physico-chemical methods to measure them. Different theoretical representations on polymer adsorption from solutions and on metal surfaces have been developed.

Adsorption of nonionic flexible-chain, water-soluble polymers on the surface of hydrophobic nanoparticle dispersions is the best theoretically interpreted. The adsorption layer is mostly described by the Boolas–Huve–Hesselink model according to which the polymer shell surface consists of a first dense layer and a second loose layer of polymer chains.

The first layer is characterized by a relatively large number of macromolecular contacts with the surface. The second one includes facing the solution unequal chains and loops whose scattering by size is described by an exponential dependence. The density of polymer chains on the boundary of the first and second layers changes jumpwise. The distribution of elementary chain density along the normal to the surface within the periphery of the adsorption layer bears, depending on conditions, either a linear, power or exponential dependence. The latter one is the most followed. Conformation of the adsorbed polymer chains on the metal surface varies constantly and can be described only by using statistical approaches. The structure of the adsorption layer and adsorption isotherms can be found from polymer molecules state in adsorptive state and in a solution by the methods of statistical thermodynamics based on the theories of polymer solutions.

The simplest model of the polymer molecule is the free conjugated chain where each link was equal length but is devoid of volume. This model can be studied based on Markov's statistics which considers variations in entropy due to limitations in conformation freedom of the polymer molecule.

The lattice model is another good model. The polymer unit can occupy a certain place in the lattice at once, then the model is reduced to the notion of a volume-free polymer chain. The formation of adsorption complexes of the polymer chain + particle type has been examined based on this [53–57] and Monte Carlo methods [58]. The nanoparticle radius R_j is considered to be much less than the mean length of free chains in the solution. An extended region is formed in the vicinity of the particle surface as a result of such interactions [53]. The boundary layer with thickness Δ is formed of the polymer chain portions having loop conformation with an average length $\sim\Delta$ (Fig. 3.11) [57]. Two alternative models of the adsorption complex structure are proposed – with long and short loops. In the first case the local density of polymer chains $\rho(r)$ inside the Δ layer is constant and greatly surpasses the mean density of links ρ in the nonadsorbed knot. So, $\Delta \gg \sigma$ of an individual polymer link ($\approx$1.0–2.0 nm corresponding to a few links of a real macromolecule) is roughly equal to the range of adsorption potential. It quickly (exponentially) drops to ρ in the radial direction from the particle surface at $r \gg \Delta\rho(r)$. For the case of the model with short loops, $\Delta \approx \sigma$ and $\rho(r)$ diminish more slowly at $r \geq \sigma$ than in the model with long loops. There are no limiting regions with a constant link density.

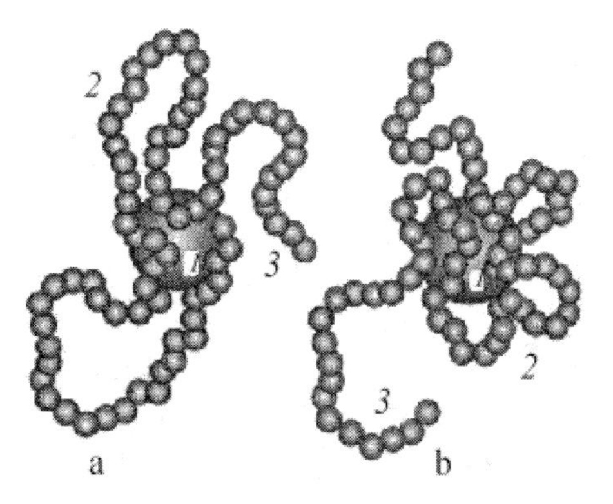

Fig. 3.11. Models of the interaction of polymer chain with a particle. (**a**) A "long loop". (**b**) A "short loop". 1 – the portion absorbed into surface of the particle; 2 – the loop-type portions; 3 – nonabsorbed "tails" of chains

As there is no reliable experimental data on the structure of the adsorption layer and conformation of the polymer chain, of more importance become calculations based on the Monte Carlo method [58] for different chain lengths, R_j values and energy parameters which characterize attraction of polymer links to nanoparticle surface. Calculations showed that polymer chain bonding with a small particle reflected by the model of short loops did not bring about elongated regions with a constant density of links.

As a result of polymer adsorption by a small particle, an energy $E = n\varepsilon$ is generated (n is the number of contacting particle links, and ε is the energy of link attraction to the particle surface in kT, which is independent of the total chain length).

When solving the problem of interaction between the polymer chain and a dispersed particle, it is important to know the relation of adsorption centers in the polymer versus those in a dispersed particle. It may exert an essential influence on interaction and spatial distribution of particles in the adsorption complex and on conformation of the macromolecule. More than one particle can be bound by the polymer chain into a structure of necklace type in a system with several adsorption centers. Such structures form a spatial adsorption polycomplex (Fig. 3.12) looking like a peculiar network or a web [14]. Polymer adsorption on particles could induce an efficient attraction of polymer links by the polymer knot compressing or polymer addition at strong adsorption could decrease aggregation stability of the dispersed phase [58]. Thus forming adsorption polycomplex shows intermediate (mesoscopic) inhomogeneities whose size is on the average $\sim 10\sigma$ for a model system and $\sim$10–100 nm for a real one. An excessive amount of adsorption centers on particles is exhibited in their mutual attraction and indirectly by the polymer (Fig. 3.13). Configuration of a chosen pair of particles is beneficial when the distance between particle surfaces equals the polymer link diameter and each link of the chain section found in this gap contacts both particles simultaneously (Fig. 3.13). When there is an excess of dispersed particles in the system relative to polymer content, then the dispersed particles of type 3 utilize some adsorption centers (Fig. 3.13) attracted by their free surfaces to the polymer shells enveloping particles of types 1 and 2. This results in formation of aggregates consisting of dispersed phase particles which violates the initial uniform distribution of particles in the volume.

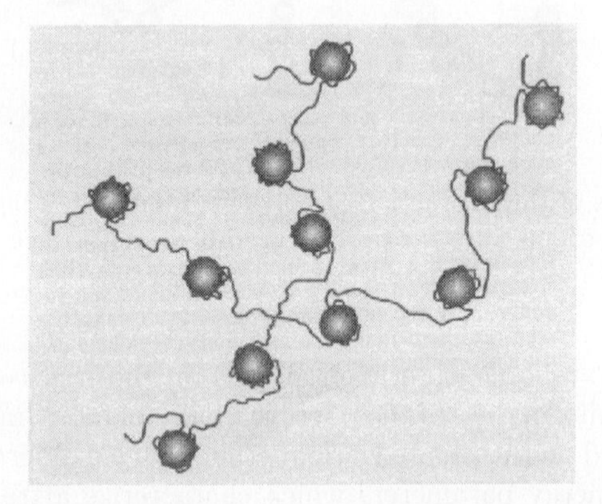

Fig. 3.12. Adsorption polycomplex of necklace type

Adsorption of water-soluble polymers by charged nanoparticles leading along with variation of van der Waalse forces of attraction to repulsion of the dual electric layers due to varied efficient Hamaker's constant is very interesting. In real dispersed systems these factors are manifested all together. Their stability is conditioned by a joint action of some of them.

The effect of temperature on polymer adsorption is exhibited in the changing mobility of macromolecules in the solution, thermodynamic quality of the solution, polymer solubility as well as a competing effect on polymer and solvent adsorption. However, the effect of temperature on the polymer–nanoparticle system is perceptibly less than that of the other factors.

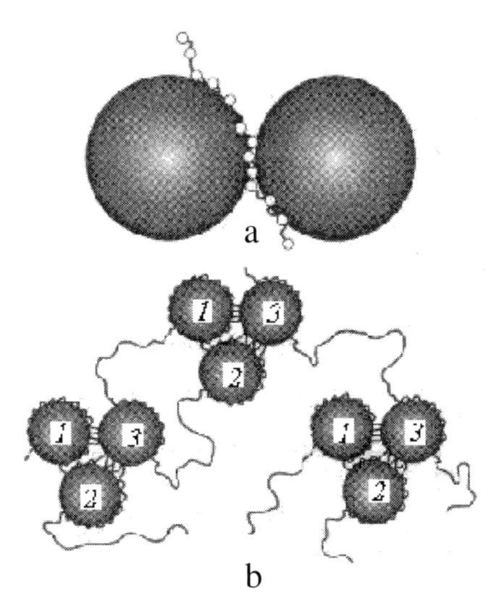

Fig. 3.13. The model polycomplexes. (**a**) The most favorable configuration for two disperse particles and polymer chain locating between them. (**b**) Aggregates consisting of a few disperse particles combined by polymer chain

3.4 Specifics of Polymer Surfactants as Stabilizers

Amphiphilic polymers including block and grafted copolymers, one fragment of which (anchor) is insoluble in the dispersive medium polymer and the other is soluble, are perfect nanoparticle stabilizers. The anchor forms a link with the surface [59], while the other component creates a steric barrier necessary for stabilization. The chemistry and physics of these systems, called surfactants [60], has been explained in detail in various textbooks [61–65]. In solutions surfactants separate into microphase structures, the simplest of which is a micelle forming in a diluted solution of certain solvents [66–68]. Self-assembly of surfactants into micelles is a well-known and studied phenomenon [69, 70]. This is true for micelle forming in block-copolymers [66, 67, 71–73].

Modifications of surfactant microstructures are water micelles, inverted micelles (e.g. water/sodium bis(2-ethylhexyl) sulfosuccinated/isooctane), microemulsions, vesicles, polymer vesicles, monolayers, Langmuir–Blodgett films (precipitated organized layers) and two-layered liquid membranes [74]. Surfactants fulfill several functions, namely size regulation, geometrical control and nanoparticle stabilization. They affect chemical properties of nanoparticles through hydrophobic and electrostatic phenomena. Aggregate behavior of surfactants depends on their nature, concentration, type of solvent and production method. The most typical surfactant microstructures are shown in Fig. 3.14. About 50–100 surfactant molecules are spontaneously joining

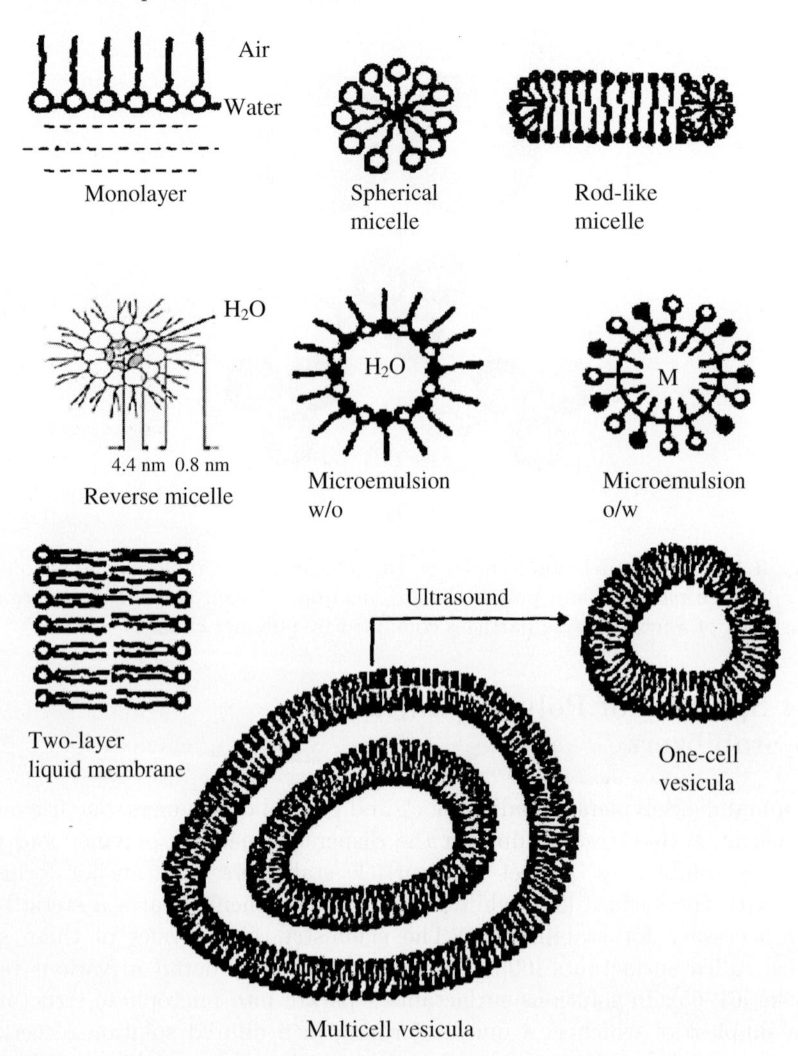

Fig. 3.14. Typical microstructures of surfactants

into aggregates known as micelles when the micelle-forming concentration in water is above critical (ACC).

The formation of water micelles is a cooperative process which is conditioned by the forces of interaction between polar head groups and association of the hydrophobic chains of surfactants. In a general case, the longer the hydrocarbon tail, the lower is ACC. With increasing surfactant concentration rod-like micelles and, consequently, liquid crystals start to form. Water-soluble surfactants transform in a hydrocarbon solvent into inverted micelles and increasing amount of water fixed by surfactant leads at a given surfactant concentration to formation of microemulsions of the water-in-oil type [75]. Such transformations are influenced by the hydrophilic–lipophilic ratio. PLA blocks in case of poly(ethylene glycol)-poly(lactic acid) diblock copolymers [76] is soluble neither in decane nor in water. Hydrophilic PEO blocks occur only in one phase as a mushroom (Fig. 3.15) or a brush (Fig. 3.15). Hydrodynamic diameters of these microemulsions depend on temperature. Inverted micelles capture large volumes of water and present unique systems. All water molecules at a relatively low water-to-surfactant ratio w (where w = 8–10) are strongly bound by the head groups and cannot be separated. A monolayer is formed at the water–air boundary when organic solution of a surfactant is sprayed on a water surface. When a dry plate or glass is dipped into the monolayer, the surfactant transfers onto the solid surface. If we immerse the plate with the monolayer sequentially into the solution, one can obtain Langmuir–Blodgett polylayered films. Depending on the type of immersion various polylayers (plate–tail–head–tail–head (X-type of polylayers), plate–tail–head–head–tail (Y-type) or plate–head–tail–head–tail (Z-type) are produced. Monolayers are an ideal medium for studying 2D processes, including those with nanoparticles.

Bound by their ends, closed two-layered aggregates of surfactants (vesicles) present the most beneficial model of biological membranes. Lamella vesicles with bulbiform 100–800 nm in size are formed as a result of liquid film swelling in water. Ultrasonic treatment of such vesicles under the temperature above which they transform from a gel into the liquid (phase transition temperature) yields 30–60 nm diameter extremely homogeneous simple two-layer vesicles (small unilamellar vesicles). Analogous systems can be obtained by ultracentrifuging and either by membrane or gel filtration.

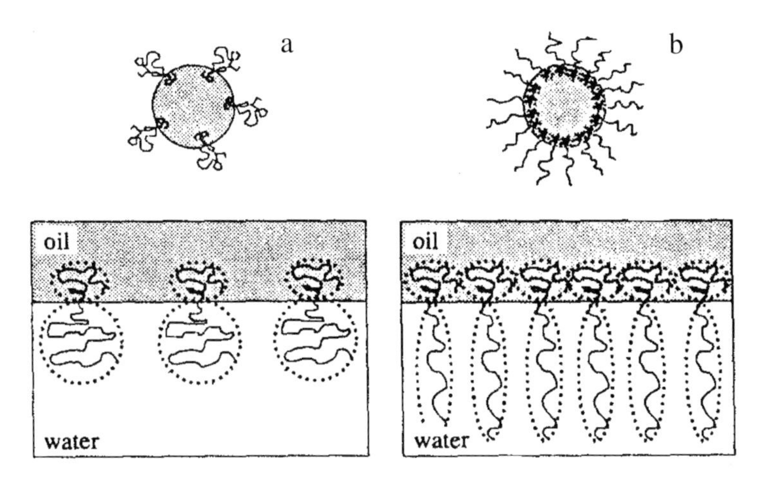

Fig. 3.15. The change of adsorption layer structure from the mushroom type (**a**) into a brush (**b**) with an increase in the concentration of absorbed surfactant

Both types of vesicles are subject to time–temperature phase transitions. A two-dimensional structure is formed at a temperature below that of the phase transition. Kinetic stability of vesicles is considerably higher than micelles. Particularly, there is no equivalent to ACC. Once formed, vesicles cannot be destroyed by dilution. They are usually stable for weeks and the stability can be elevated by polymerization. Vesicles have many places for solubilization. Hydrophobic molecules can scatter among hydrocarbon bilayers of vesicles, whereas polar ones are easily moving in vesicles captured by water. Ions are attracted to negatively charged outer ions or inert surfaces of vesicles. Particles with charges identical to vesicles fix on their surface thanks to long hydrocarbon tails. Block copolymers containing ion and non-ion water-soluble units form stoichiometric complexes together with low-molecular surfactants. The complexes generate small (120–85 nm) vesicles [77]. Modified β-cyclodextrins [78, 79], and heterofunctional PEO (to stabilize auric nanoparticles [80] and octylammonium bromide in combination with both high-molecular and low-molecular stabilizing agents, PVPr or cellulose in combination with alkyl thiols to stabilize ruthenium [81, 82]) are used for stabilizing nanoparticles (Pt, Pd, Au) and producing water-soluble materials.

Surfactants are capable of charged and steric stabilization [83]. For example, stabilized by a low-molecular surfactant (Fig. 3.16) (3-(methyldodecylammonium)propanesulfonate), 6 nm size platinum clusters are obtained as a 1M solution in water [84]. An analogous procedure is used to stabilize bimetallic clusters [85].

Low-molecular surfactants are less efficient than their polymer analogs. In contrast to low-molecular amphiphilic substances formed by amphiphilic polymer compounds, microphases are more stable in thermodynamic and kinetic respects. Besides, amphiphilic copolymers have expanded their range of varying molecular properties. It is possible to control the chemical structure

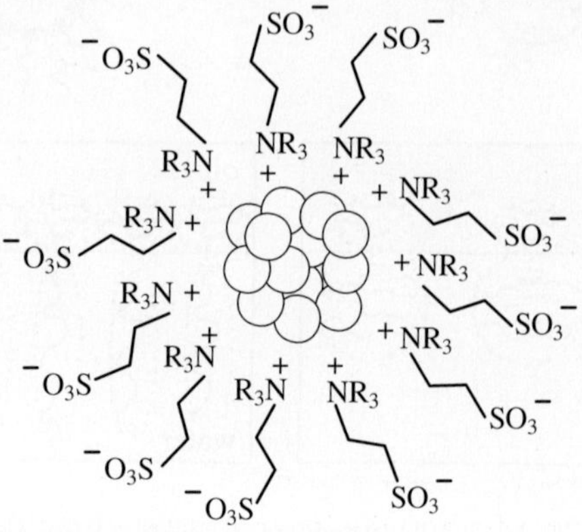

Fig. 3.16. Structure of aqueous-soluble metal cluster stabilized by betaine

of these polymers by the choice of a recurrent unit, copolymerization with functionalized comonomers and by the length and shape of individual building blocks.

The necessity of raising stability and control over reversible transitions and morphology has focused attention on polymer surfactant vesicles. Vesicle-forming surfactants are functionalized by vinyl, methacrylate, diacetylene, isocyanide and styrene groups in either hydrocarbon chains or head groups. Surfactant vesicles can be polymerized inside bilayers (Fig. 3.17) or head groups (Fig. 3.17 and Fig. 3.18) [86]. Of interest is the production of vesicles through polymerization of isodecyl acrylate in vesicles of sodium di-2-ethylhexyl phosphate [87, 88] and from a bipolar liquid monolayer [89–92]. Polymer monolayers and two-layer liquid membranes turned out to be more efficient stabilizers as compared to their monomer analogs.

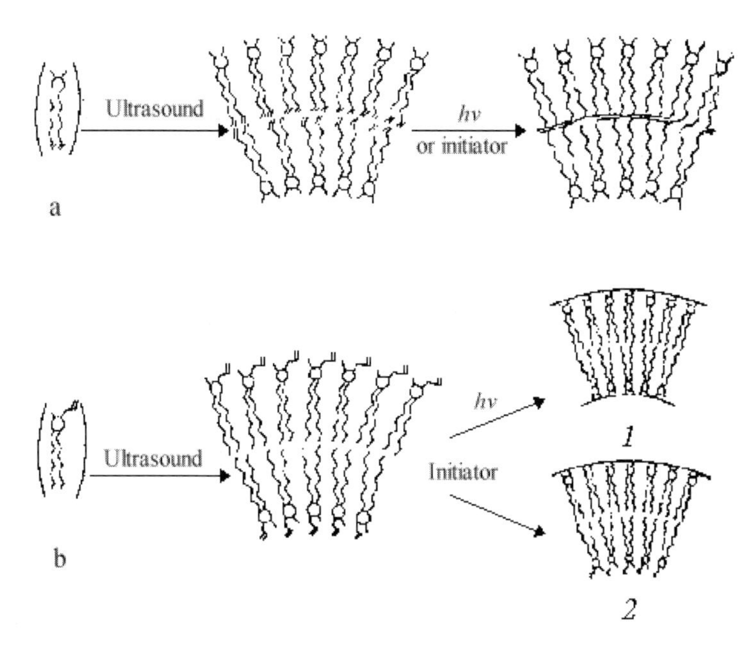

Fig. 3.17. A scheme of polymerization of surfactant vesicle

Anchor fragments of surfactants join nanoparticles as a result of physical adsorption due to van der Waalse forces, dipolar interaction and formation of weak easily broken hydrogen bonds, or as a result of chemisorption spurred by the formation of strong bonds of electrostatic or covalent origin or by the acidic–base interaction. Metal compounds usually link with functional groups of micelle nuclei thanks to formation of covalent and ionic bonds [93, 94]. As nanoparticle ligands it is possible to use surfactant groups, including Ag[bis(1-ethylhexyl)sulfosuccinate] (AgAOT) or a mixture of AgAOT–NaAOT. Sizes

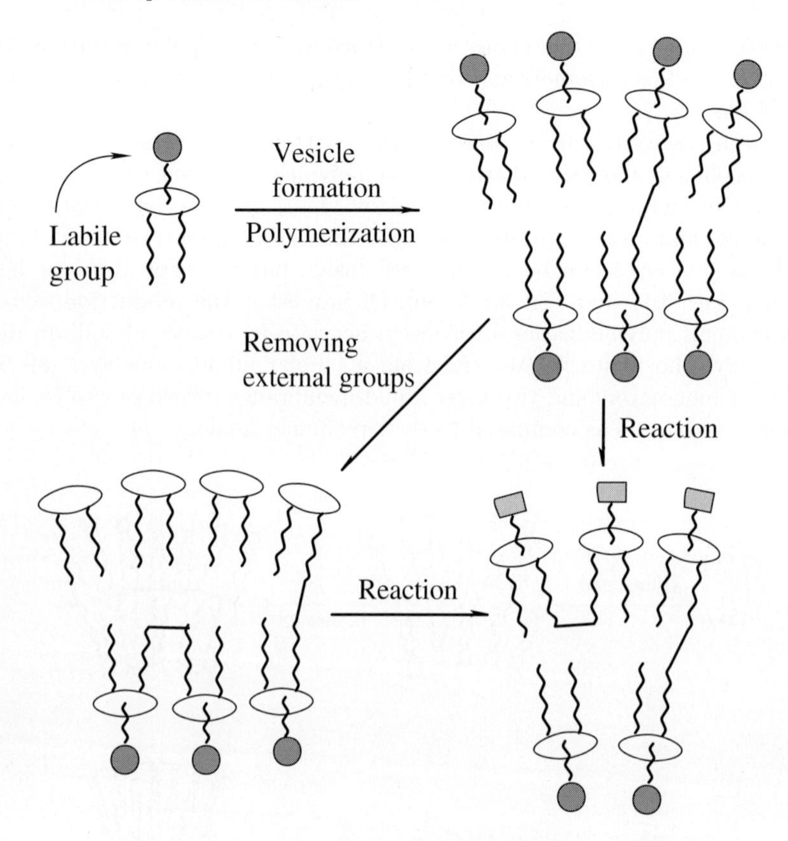

Fig. 3.18. Structure of asymmetrical polymer vesicles

of micelles are from 1.9 to 3.5 nm, that of Ag particles is from 2.7 to 7.5 nm [95]. Styrene and ethyl methacrylate copolymers bind nanoparticles of Au, Pd, In and Sn by a benzene ring of one monomer and ester group of another comonomer [95–98]. Surfactants present molecules with both hydrophobic and hydrophilic fragments. Depending on their chemical structure, they can be neutral (e.g., polyoxyethylene(9,5)octylphenol (Triton X-100)), cation, anion or zwitterion (e.g., block-polyelectrolytes, block-ionomers). Hydrophobic portions of surfactants have different lengths (usually 8–20 carbon atoms), contain double links and can include two and more hydrocarbon chains. The vesicular architecture also furnishes effective templates for the controlled construction of functional materials ranging from nano- to micrometer dimensions. The ability to control the composition, morphology and function of vesicles allows the formation of novel template materials with a diversity of function, including organic capsules [99, 100], inorganic capsules [101, 102], and biological models [103].

The relation between block-copolymer size (total molecular mass, composition) and micelle structure (ratio of nucleus to shell and shape) is very important. It is possible to use the micelle parameters (shape, size, and polydispersivity) as nanoreactors or templates [104–107]. Diblock copolymers join into films as soon as microdomains of lamellar, cylindrical or spherical shape with sizes 10–100 nm are generated [108–112]. The geometry and size of microdomains can be forecasted during variation by matching each block length against the total molecular mass [112–114] or by mixing with a homopolymer [115, 116]. It is important to know, for the block copolymers, the ratio of molecular masses of the first and second components (M_1 to M_2). As an approximation, it is convenient to take $M_1 \approx nM_2$, where n is the number of polymer chains of the second component linked to the polymer tail. M_1 lies within 10^4–10^5 and M_2 within 10^3–10^4. The higher the adsorption of an elementary unit (it is higher in water media), the lower the M_1 value needed to attain stabilization. M_2 magnitude should be high enough to make provision for a steric obstacle at a few tens of nanometers distance from the surface. A typical example is the epoxidation PS-block-PB further modified by various nucleophilic reagents [117]. Salts of valuable metals become soluble in toluene micelles of the above polymer that are not soluble in pure toluene. Formed polymer-stabilized nanoparticles have sizes ~10 nm and show low polydispersity. Described polymer micelles can serve as nanoreactors to obtain, e.g. CdS [118]. Out of many examples we shall refer to only few: styrene copolymers with 2-VPy [72, 119, 120], 4-VPy [75, 121], acrylonitrile [122], 4-methyl styrene [123], methyl methacrylate [121, 123], ethylene oxide [124], and so on [75,125–133]. More complex block copolymers have been elaborated lately, including carbazole-functionalized norbornene derivatives obtained by the method of active metathesis polymerization with cycle opening [134]. These works are reviewed in [135].

Nuclei of such block copolymer micelles serving as nanoreactors for nanoparticle synthesis stabilize, at the same time, nanoparticles by nonpolar blocks of these amphiphilic block-copolymers (by corona). Metal ions during the process generate a macrocomplex having units of a micelle nucleus at first [135–137]. Gold and silver nanoparticles were stabilized using these copolymers. The morphology of metal-containing polymers depends upon metal origin (Table 3.2). In particular, silver-containing polymer (Ag-1 sample) shows a lamellar morphology where the thickness of microdomains is about 9 nm and the gap between them is about 25 nm. Gold-containing polymer (sample Au-1) is characterized by a morphology with hexagonal location of metal-containing cylinders. The cylinder diameter is roughly 10 nm and the gap between microdomains is 30 nm. If we mix these polymers with a homopolymer (M_p = 9200, γ = 1.04), the structure of the microdomains changes in response to decreasing percentage of metal-containing blocks (see Table 3.2). Cylinders with a short-range order and disordered spheres were detected in the presence of metal-containing domains with diameter ~10 nm.

The structure of a microdomain depends also on the method of obtaining nanoparticles. This is true for application of the mentioned block copolymer

Table 3.2. Morphology of metal-containing homopolymers

Sample	Content of Metal-Containing Block, wt.%	Content of Metal, wt.%	Particle Size, nm	Morphology
Ag-1	28	5.9	2–10	Lamellar
Au-1	25	11	1.5–4	Cylindric
Ag-2[a]	20	3.7	–[c]	Cylindric
Au-2[b]	8.8	3.9	–[c]	Spherical

[a]Mixture of Ag-1 (70 wt.%) with homopolymer; [b]mixture of Au-1 (34 wt.%) with homopolymer; [c]not measured.

in the procedure during which samples with a spherical morphology of silver-containing copolymers are produced where a metal particle is found inside a microdomain (optimum control over clusterization is reached in spherical morphology) [136]. Analogous metal and metal-sulfide clusters with spherical microdomains inside micelles and vesicles have been studied elsewhere [133, 138–141]. Perfect correlation between the ionic nuclear radius of inverted micelles and the size of PS-block-PAA-Cd nanoparticles forming in the system was described in [131]. New block ionomers have been generated through copolymerization of n-butylacrylate and sodium styrene sulfonate [142]. The ionomers turned out to be good stabilizers of nanoparticles. Self-assembling organic and organometalic diblock copolymers, poly(isoprene-block-ferrocenylphosphine) and poly(ferrocenylsilane-block-dimethylsiloxane) produced by the active ionic sequential polymerization of corresponding monomers are also referred to the class of stabilizers [143, 144].

A promising approach has been proposed in [145] for producing and stabilizing silver nanoparticles using PS-block-PVA. The structure of this copolymer is divided into microphases where PVA plays the role of a nucleus located inside the PS shell. The general production scheme of silver nanoparticles is presented in Fig. 3.19. The maximum content of metal nanoparticles in the copolymer appeared to be 17%, the larger portion being inside and the less on the PVA nucleus surface. The forming nucleus in styrene block copolymer with 2-VPy undergoes additional cross-linking with the help of, say, 1,4-diiodbutane. As a result, 2-VPy links get quaternized [146–148]. Monodispersed microspheres are generated with a lamellar morphology and a "hairy ball" structure. Their PS shell does not swell in water while the cross-linked block phase of quaternized P2VPy does [149–151]. The thickness of the nucleus of P2VPy is 35 nm and the space between domains of the lamellar phase is 65 nm. The described structure of polymer stabilizers ensures localization of metal compounds and consequently, arising from them, nanoparticles with

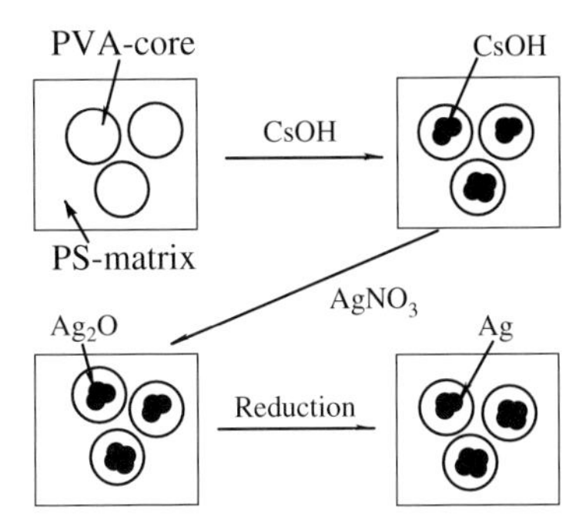

Fig. 3.19. Scheme of silver nanoparticles embedding into PS-block-PVA

diameter from 10 to 20 nm. This takes place only inside the nucleus of quaternized P2VP, whereas the PS phase does not contain metal. Nanoparticle dimensions are governed by concentration of the solution used for metal introduction and degree of P2VP cross-linking [152].

Analogous stabilizers were obtained from PS-block-PAA or PS-block-PAAm (polyacrylamide). PAA or PAAm chains were subjected to cross-linking which results in the copolymer outside block (Fig. 3.20) [153, 154]. These systems form spherical polymolecular micelles in water or in water mixture with tetrahydrofuran during which the diameter of micelles increases with cross-linking from 17 to 24 nm. The external shell can also be modified by various transformation reactions analogous to polymers [155].

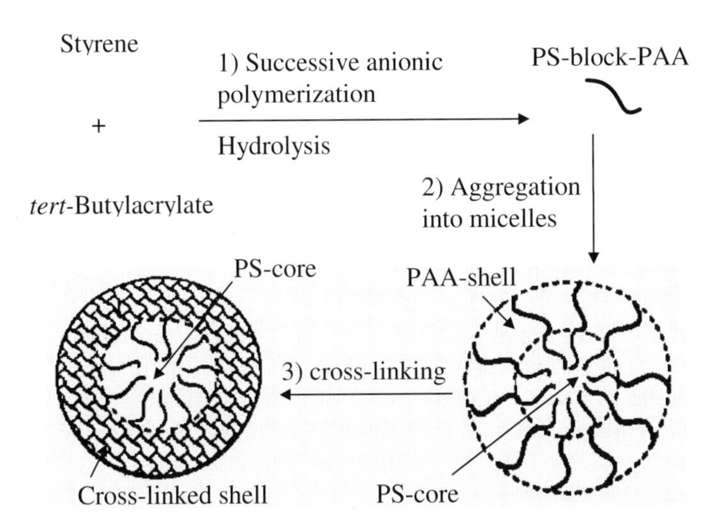

Fig. 3.20. Three-stage synthesis of block-copolymer with cross-linked shell. 1 – the formation of amphyphylic PS-block-PAA; 2 – self-assembling of block-copolymer into polymer micelle; 3 – cross-linking of hydrophylic segments of shell layer

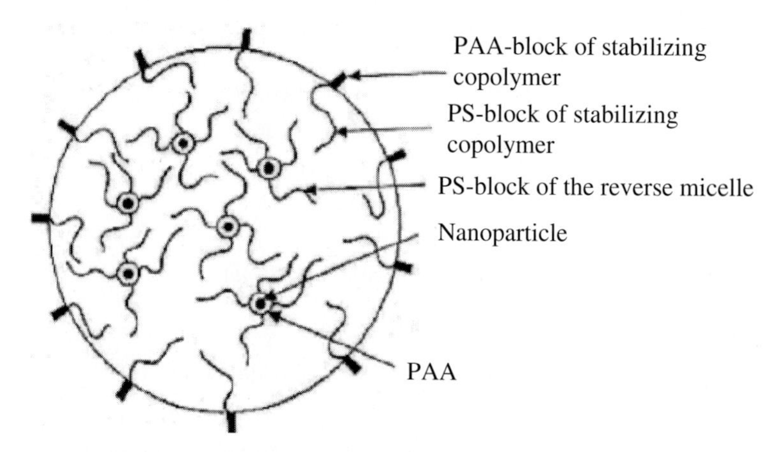

Fig. 3.21. Scheme of large micelles containing nanoparticles

Stable nanoparticles are generated [156] in a three-component system containing a metal phase, hydrophobic latex (e.g. PS or PVA) and a hydrophilic stabilizing polymer (e.g. PVPr). Some polymer stabilizers based on such surfactants as PS-block-PAA and PS-block-PEO are known as [157,158] micelles of large compounds. They present micro-size spheres of commonly high dispersity consisting of an ensemble of inverted micelles stabilized by a fine layer of hydrophilic chains. Metal nanoparticles (~3 nm) are inside the nucleus of such micelles (Fig. 3.21). When the stabilizer content is 12% of the total amount of the polymer, simple types of micelles of large compounds are formed ($D_{\text{av}} = 64\,\text{nm}$) containing on the average 58 inverted micelles and 87 stabilizing chains with mean area 148 nm^2 per stabilizing chain. At 10% stabilizer content two types of large compound micelles arise with $D_{\text{av}} = 52$ and 152 nm, correspondingly. If the stabilizer content exceeds 21%, a mixture of micelles of large compounds and inverted micelles is formed. The chemical behavior of block copolymer of butylene–PEO is similar to alkane-oxyethylene nonionic surfactants [159].

Interaction between the domain phase and the free interface is very important for cluster growth [126] and formation of the nanocomposite structure.

Vesicular architecture also furnishes effective templates for the controlled construction of functional material ranges from nano- to micrometer dimensions. Control of composition, morphology and function of vesicles allows for the formation of novel template materials with a diversity function, including organic capsules [160, 161], inorganic capsules [162, 163], and biological models [164].

3.5 Mechanism of Nanoparticle Stabilization by Polymers

The theory of "molecular solder" or so-called strong adhesive interactions between components [165, 166] can be used to describe the stabilization mechanism of nanoparticles by polymers. At the base of the theory lie representations on structural and mechanical factors of stability of dispersed systems and spatial nets of the coagulation structure type. A requisite condition of the stabilizing effect of a high-molecular compound is adequate activity (diophilic property) of a metal nanoparticle surface in relation to the polymer forming an extremely strengthened adsorption-solvate structurized film on the dispersed phase surface. Polymer protection generates strong structural networks on the nanoparticle surface within the dispersed phase bulk. The degree of steric stabilization varies in transitions from formation of structures with adsorption layers only till structurization of the whole volume of the dispersion medium. Adsorption layers arising at considerable concentrations of high-molecular substances are specifically stable. They represent peculiar film gels (lyogels) that are highly solvated by the dispersion medium. Under specific conditions a structural network appears in the solution where metal particles are grouped so that they form chains of different shape and length. The lyophobic colloidal particles serve to connect the links of different and similar macromolecules [14].

Several processes run simultaneously in the system and influence one another. These are processes of enlargement of the particles, and macromolecular adsorption on the surface of the original and forming particles. Aggregation stability depends on the relationship of the coagulation to adsorption velocity constants. The amount of sol coagulated on a unit area in certain time is calculated [167] by the equation

$$q = (C_0 - C_{\text{eq}})V_1 S_1 / m_2 S_2 \,, \tag{3.7}$$

where C_0 and C_{eq} are the initial and equilibrium mass concentrations of sol and metal, correspondingly; V_1 is its volume; m_2 is the polymer mass; and S_1 and S_2 are specific surfaces of metal sol and polymer, correspondingly.

Modeling of adsorption and coagulation processes presents a very complex problem. We shall confine ourselves to enumeration of the most prominent phenomena [2–7]. In the case of added macromolecules one can observe homosteric and intensified steric stabilizations, as well as flocculation (phase separations, substitution coagulation and crystallizing coagulation) and heterosteric stabilization (selective flocculation, heteroflocculation, phase separation). If free macromolecules are taken, substitution stabilization, flocculation and phase separation occur. The efficiency of sorption processes with coagulation of nanoparticles are determined by the overall balance of surface forces during interaction of different natural phases [168].

When two metal nanoparticles covered by a layer of adsorbed soluble polymer chains approach to a distance less the total thickness of adsorption layers, the polymer layers start to interact (Fig. 3.1). The interaction brings steric stabilization and leads, in the majority of cases, to repulsion between the colloidal particles. It was endeavored repeatedly to clarify its nature and determine the magnitude. Most frequently the problem is studied in terms of changing Gibbs' energy when two particles covered by an adsorbed polymer are approaching from infinite remoteness to finest distances.

The quantitative contribution of the polymer constituent to the interaction energy of particles is a function of polymer adsorption layer parameters on the surface (share of elementary links of a macromolecule contacting the surface, degree of its covering by the polymer, amount of the polymer in the first layer, layer thickness and so on) and those of macromolecules in the solution. The surface phenomenon in small systems play the considerable role surface phenomenon and plays the considerable role [169–171].

Polymer chains adsorbed on the surface of dispersed particles lessen the attraction energy for steric reasons (the minimum distance to which particles can approach increases) and change the efficient Hamaker's constant value. The attraction energy in expressions for U_r dependence on A is a function of not only the interaction constants of dispersed phase A_1, dispersion medium A_2 and the phase with the medium A_{12}, but of Hamaker's constant for adsorption layer A_3 too. The effect of polymer adsorption layers on molecular attraction of particles was considered theoretically in [2,3,5]. U_r can be defined by Lifshits macroscopic theory as:

$$U_r = -\frac{1}{12}\left(A_2^{1/2} - A_1^{1/2}\right)^2 (r/l_0) , \tag{3.8}$$

and an equation to consider unlagging London's forces in the microscopic approximation (see notation in Fig. 3.22)

$$\begin{aligned} U_r = -\frac{1}{12}\Big[&\left(A_2^{1/2} - A_3^{1/2}\right)^2 (r+\delta)/(r-2\delta) + \left(A_3^{1/2} - A_1^{1/2}\right)^2 (r/l_0) \\ &+4\left(A_2^{1/2} - A_3^{1/2}\right)^2 \left(A_3^{1/2} - A_1^{1/2}\right)^2 (r+\delta)/(l_0-\delta)(2+\delta/r)\Big] . \end{aligned} \tag{3.9}$$

Polymer layers adsorbed on nanoparticles can be considered as an unusual medium, a polymer solution specifying interaction forces between the dispersed phase particles. Evidently, the interaction between particles is negligibly weak when the adsorbed molecules are densely packed on the surface and macromolecules nearly lie on it. As soon as the particles approach and adsorbed polymer layers superimpose, the loops engage and an interaction with model (Fig. 3.22) occurs. This superimposing region serves as the medium (polymer solution). Interpenetration of two adsorption layers evokes the appearance of two effects related to a lessened amount of conformations in the macromolecular chain (leading to the loss of configuration entropy and, consequently, to growth of the system free energy) and varied concentration

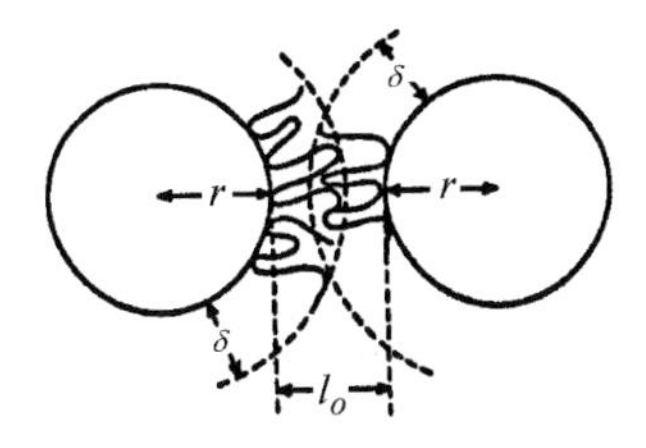

Fig. 3.22. The loops engage in the interaction of polymer layers absorbed into the surface of colloidal particles

of links in the zone of overlapping. This is accompanied by variation in the polymer–solvent interaction and generation of the local osmotic pressure. The first of the effects is called entropic and the second is the effect of osmotic pressure.

The total interaction energy of two colloidal particles with adsorbed polymer layers is described by the equation

$$U = U_1 + U_2 + U_r \,, \tag{3.10}$$

where U_1 is the effect of osmotic pressure, U_2 is the entropic effect, and U_r is the interaction energy of particles solved by (3.2).

The interaction energy for mixed polymer blends is given by the expression

$$U_1 = \frac{4}{3}\pi BkTC_a^2\,(\delta - l_0/2)^{\,2}\,(3r + 2\delta + l_0/2) \tag{3.11}$$

where π is the osmotic pressure, B is the second virial coefficient determining the interaction between the solvent and substance solved, and C_a is the polymer concentration in the adsorbed layer.

The entropic effect can be calculated by the methods of statistical mechanics. Calculation results show that the energy of interaction U_2 is on the order of U_1.

The dependencies of U_1, U_2 and U_r on l_0 are presented in Fig. 3.23. U_1 and U_2 tend to zero at distances twice as big as δ (2δ). The energy U_r is conditioned by the long-range London's forces, so the total energy of interaction U reaches its minimum value at a longer distance than 2δ. The transition through point $U_{\min}$ means coagulation of particles. For the process of particles interaction with the adsorption layer one can take that the condition $U_{\min} < 5\,kT$ corresponds to repulsion and $U_{\min} > 5\,kT$ to coagulation. The depth of the potential well $U_{\min}$ is determined by a number of factors, including polymer molecular mass, solvent–particle interaction, and particle size. With increasing polymer molecular mass due to the forming loops, the adsorption layer thickness increments too and the curve U_1 (Fig. 3.23) shifts to the right. $U_{\min}$ diminishes and the repulsion intensifies with growing thickness. The increasing total amount of the adsorbed polymer elevates C_a making U_1 go up and $U_{\min}$ go down. Once the solvent–particle interaction is intensified, the

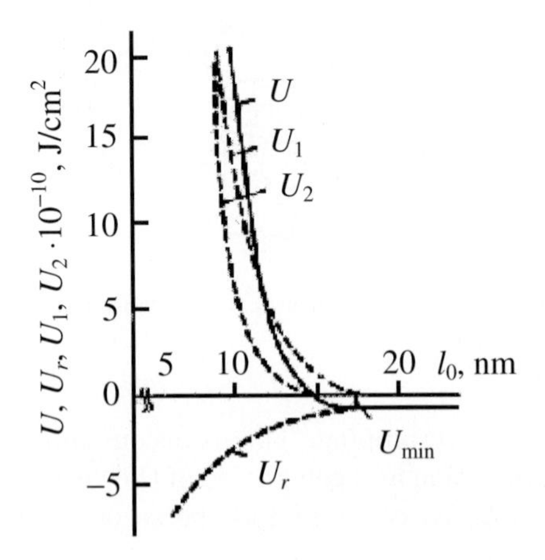

Fig. 3.23. The interaction energy of two particles absorbed into a surface polymer layer

thickness of the layer of adsorbed loops increments leading to the growth of the second virial coefficient B.

Two kinds of entropic effects are distinguished. The first one presupposes desorption of polymer chains as a result of the lowered entropy of the system. This augments the interphasal free energy and results in particle repulsion. The second kind is characteristic of nondesorbed polymer chains when a simultaneous breakage of a great amount of polymer links with the surface of colloidal particles is in fact improbable. It presumes that the adsorption equilibrium between the polymer and particles remains invariant during particle collisions. The free energy grows as a result of decreased entropy induced by redistribution or compression of adsorbed links in the interaction zone. Two limiting cases are to be mentioned in the model, called a constant adsorption model. They are: interpenetration of adsorption layers without constriction of the adsorbed polymer chains (effect of mixing) and compression of the layers without overlapping (effect of bound volume). Both cases are idealized since it is impossible to separate two phenomena in real systems. The effect of bound volume will dominate at high concentration of polymer units on the surface when polymers with low-molecular mass are adsorbed and the interacting surfaces converge most closely to each other. The effect of mixing prevails when the surface is only slightly covered by the polymer, high-molecular compounds are adsorbed and the distance between particles of the dispersed phase is substantial.

The notion of enthalpic (ΔH), entropic (ΔS) and mixed (enthalpy–entropic) stabilization at interpenetration of polymer chains is considered in [2]. The mixing energy is

$$U = \Delta H - T\Delta S. \tag{3.12}$$

The generalized Flory–Haggins parameter characterizing the thermodynamics of the polymer–solvent interaction can be presented as a sum of two constituents $\chi = \chi_h + \chi_s$, where χ_h is the parameter of mixing enthalpy, and χ_s is the additional entropic parameter specifying both volume effects of mixing and various structural transformations (e.g. rupture and formation of hydrogen bonds, conformation change, etc.).

Evidently, at endless dilution $\Delta H \to (-k)$ and $\Delta S \to (-\psi)$, where $k = \chi_h$ and $\psi = \frac{1}{2} - \chi_s$.

Proceeding from the above considerations, it is easy to define conditions under which one or another type of steric stabilization dominates (Table 3.3).

Table 3.3. Comparison of types of steric stabilization (by Napper [2])

K	ψ	ΔH	ΔS	$\Delta H/\Delta S$	U	Stabilization type
−	−	+	+	> 1	+	Enthalpic
+	+	−	−	< 1	+	Entropic
+	−	−	+	$> 1 <$	+	Mixed

Many polymers show negative k and ψ values in water solutions. This is connected with destruction of hydrogen bonds at growing polymer concentration in the system. Stabilization will be enthalpic in this case. Most polymer solutions in nonpolar media display positive k and ψ, so far the entropic stabilization can be expected. Systems with enthalpy type of stabilization show a tendency to lose stability at heating. Entropic systems tend to flocculate at cooling.

Adsorption of water-soluble polymers with numerous hydrophilic groups by nanoparticles imparts a considerable water-absorbing capacity to the dispersed phase surface. As a consequence, the role of solvation factor gains strength for stability. There are specific features of the microenvironment of the macromolecule in water: the presence of hydrophobic binding centers on the polymer chain and solvate shell of bound water molecules which differ from unbound molecules. The reactivity of hydrated silver cations increases and is determined by their partial dehydration near the polymer chain [172] as in the case with other metal cations [173]. The wedging pressure plays a critical role in the mechanism of stabilizing colloidal solutions using high-molecular compounds.

3.6 Stabilization of Nanoparticles by Electrolytes

Along with nonionic polymers, polyelectrolytes charged similarly with the solid phase particles can work as efficient stabilizers of metal-containing dispersions. As an example, one can cite stabilization of nanoparticles by quaternized P4VPy [13], PAA [33], PEI [30–32], polyvinyl phosphonic acid [35], cation polyelectrolytes, including poly(diallyldimethylammoniumchloride), poly(methacrylamidopropyltrimethylammonium chloride), poly (2-hydroxy-3-methacryloxypropyltrimethylammoniumchloride), poly(3-chlorine-2-hydroxypropyl-2-methacryloxyethyldimethylammoniumchloride) [35]. Optimum stabilizers turned out to be also polyphosphates [174–177]. Vistas in producing and stabilizing aurum salts with the help of a cross-linked poly(styrensulfonate) have been proved too [105]. There is great interest in procedures where along with being a stabilizer the polyelectrolyte serves as a reducing agent for generation of nanoparticles from corresponding salts. It was demonstrated on the example of $HAuCl_4$ reduced by a water-soluble oligomer-derived oligothiophen. The mean size of Au nanoparticles in these polyelectrolyte complexes is about 15 nm [178]. This procedure is rather often employed in practice.

Polyelectrolytes display an evident tendency to association of ionic groups which gather into ionic regions like microphases in a hydrophobic polymer. For example, in perfluoroalkyl acid (trade name "Nafion") some morphological formations have been detected [179] that contain ionic regions 3–4 nm in size consisting of sulfonate groups. These regions are interrelated through hydrophilic channels of ∼1 nm diameter and enclosed into a fluorocarbon matrix. Ionic centers concentrate in these regions that are given preference in metal cluster formation [180]. Matrices of this type show a confined diffusion of nanoparticles, like in solvate polymer networks in whose pores the confined diffusion coefficient of iron hydroxide particles (3–9 nm) is $2{\cdot}10^{-8}\,cm^2{\cdot}s^{-1}$ [181, 182].

Radiation-chemical reduction of Ag^+ ions in weakly alkaline water solutions is accompanied by their bright-blue coloring in the presence of PAA [183–186]. PAA molecules in the alkaline solution unfold into chains due to repulsion of COO groups. As a result, a provision is made for Ag^+ ions to interact actively with COO groups owing to which the resultant blue silver displays higher stability even in air and can be isolated by boiling down.

Polyelectrolyte adsorption is more complex in contrast to nonpolar polymers. Polyelectrolyte adsorption on the charged surface of a nanoparticle depends on factors typical for noionic polymers (e.g. adsorption energy of an elementary unit, polymer concentration, its molecular mass, thermodynamic quality of the solvent) and upon the charge density on the surface of a macroion, as well as the degree of screening of polyelectrolyte charges. The contribution of the electrostatic factor might dominate over others in many cases.

The lowering of specific adsorption of electrolytes is conditioned by intrinsic properties of macroions in the solution. Particularly, adsorption of proteins is apparently determined by conformation stability of their macromolecules and internal cohesion. The molecular structure of proteins is most stable at the isoelectric point. Macromolecules are less subject to deformation in this state which is the reason for their maximum adsorption at the isoelectric point. Coulomb forces arise not only in polyelectrolyte adsorption by uncharged nanoparticles but in adsorption of nonionic polymers on charged colloidal particles as well and sometimes in nonionic polymer interaction with uncharged nanoparticles. The reason is that polymer regions contacting metal particles may charge negatively due to injection of photoelectrons into the polymer that are further captured by the charge traps Ag^+ [187]. An opposite process is probable when electrons drain into the positively charged clusters from the neighboring to polymer regions. In both cases adjacent to nanoparticles or clusters, polymer regions acquire some charge (either positive or negative) relative to other regions of the polymer surface. That is why adsorption of nonionic polymer macromolecules on the charged surface of colloidal particles might lead to variation of the dual electric layer parameters. It will influence the electric constituent contribution into the total interaction energy of the dispersed phase particles.

Most typical changes are the following. First, the density of electric charges on the surface of colloidal particles changes due to increased amount of polymer chains in the adsorption layer. This happens because of variations in dielectric penetrability and in the dual layer capacity interconnected with adsorption leaps of potential (φ_d) in the oriented layer of dipolar links of the polymer surface. A competing adsorption of polymer chains and potential-determining ions on the nanoparticle surface yields an analogous result. By providing a permanent potential leap at the interface ($\varphi_\mathrm{o} = \varphi_\mathrm{i} + \varphi_\mathrm{d}$), the emergence of potential φ_d of a similar sign with ionic potential leap φ_i spurred by potential-determining ions and counterions will inevitably lead to lowered φ_i (if $\varphi_\mathrm{d} < \varphi_\mathrm{i}$) and even to a change of sign (if $\varphi_\mathrm{d} > \varphi_\mathrm{i}$). These phenomena were observed experimentally as recharging of colloidal particles in the presence of a nonionic polymer.

Second, polymer chains adsorbed on the nanoparticle surface are able to squeeze a portion of specifically adsorbed counterions out of the adsorption layer which will surely lead to augment of the potential φ_o. This effect is the cause of the initial growth of the kinetic potential of colloidal particles when a small amount of polymer is present.

Third, a highly dense adsorption polymer layer with homogeneous highly dense chains is formed on the surface of colloidal particles. The adsorption layer thickness increments and both potentials φ_o and ζ start to grow (electrokinetic zeta potentials generated at the interface of movable and steady phases of the colloidal micelle). This means that the system will be finally stabilized. The minimum stability and isoelectric point of dispersion do not usually coincide. Adsorption of polyelectrolytes leads mostly to growth of electrokinetic potential and stability.

This suggests that adsorption of high-molecular compounds on metal nanoparticles brings about various changes in the double electric layer structure. These changes are affected by the adsorption polymer layer characteristics (density, thickness, distribution of macromolecular units), chemical origin and molecular charge of the stabilizer. Similar charged polyelectrolytes show common regularities with those of nonionic polymers. In both cases their sol become stable under certain concentrations of high-molecular compounds and without restrictions. They do not tend to coagulate even in concentrated electrolyte solutions. In contrast to colloidal solutions stabilized by nonionic polymers, the stability of solutions stabilized by polyelectrolytes depends strongly on the coagulating ion charge. This is because the added multicharged counterions are bounded by ionogenic functional groups of the stabilizer molecule which leads to a seeming increment of the critical concentration of the electrolyte. The stabilizing effect of similarly charged polyelectrolytes augments with increasing charge of the macroion.

Polyelectrolytes admit both steric and charge stabilization [2]. To describe the stabilization mechanism of nanoparticles by polyelectrolytes a variety of theoretical models were proposed. For example, the effect of homopolymer-based polyelectrolytes has been analyzed at various concentration of low-molecular electrolytes in solution [188–193]. Stabilization of nanoparticles by block-polyelectrolytes was studied as a function of the nature of the solvent and the degree of low-molecular substance adsorption on the metal surface [194, 195] proceeding from models of nanoparticle stabilization by charged homopolymers and uncharged block-copolymers [196] in the presence of a selective solvent.

Israels et al. [197] consider adsorption of uncharged block-copolymers [198] and that of polyelectrolyte [199] aimed at non-selective solvents. A model for the case of a small amount of adsorbed uncharged block-copolymer was described in [200]. The obtained data show that for the low-charged and long unbound chains colloidal stability rises with increasing charge of the block. In the high-charged regime the picture is just the opposite – electrostatic repulsion enhances as the chain length or its charge lessens. This is attributed to the effect of electrostatic interactions on the amount of the polymer adsorbed. The stronger interactions lead to lower adsorption which leads to a weaker repulsion. The value of pH has an impact on the processes too. Electrostatic interactions are reversible under low pH whereas at high pH they are stipulated by strong coordination binding of macromolecules by particles and water repellency of the forming adsorption layer. When the pH is around its mean value, competition between the two kinds of interaction can be observed.

3.7 Surface Proofing as a Method of Stabilizing Nanoparticles by Polymers

The above considerations deal with volume stabilization in solutions, gels, and swollen polymers. Stabilization is often exercised on the surface, e.g. in thin films or in the fine subsurface layer of a polymer powder. In real polymers, including PE and PTFE with high, about 60–85 and 75–89%, crystallinity, natural voids are abundant. They arise from the packing of the polymer chains in transition from the amorphous to the crystalline phase. Amorphous regions are looser and contain many more hollows since links between macrochains are weaker than in crystalline blocks (Fig. 3.24). These voids might be the place where nanoparticles can localize and stabilize. Thus, ligand-free iron clusters (up to 35% by mass) stabilize in PE and PTFE matrices under a narrow enough particle distribution by size (the maximum of the distribution curve coincides with 1.48 and 1.81 nm size particles) [201].

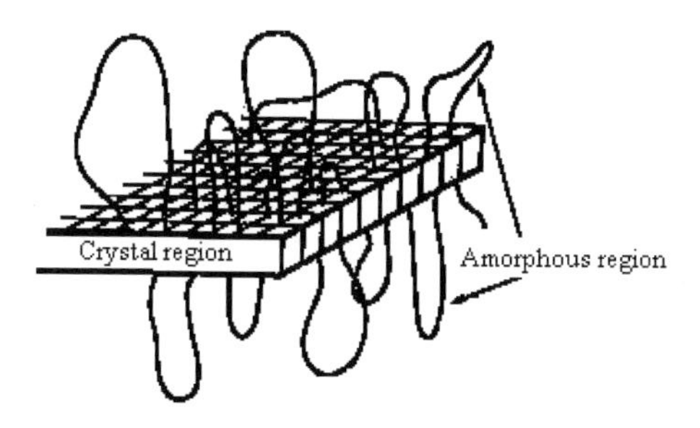

Fig. 3.24. Scheme of crystalline polymer

Pores play a significant part and determine the stabilizing capability of the matrix. The pores function as transporting arteries through which nanoparticles or their precursors penetrate into the polymer bulk. The intrusion of nanoparticles might distort the polymer structure as well. Porosity, estimated by specific surface S_{sp}, total volume V_o and pore radius r, also depends on a number of parameters. The most significant parameter is density of cross-links. The pores are subdivided according to their size into three groups: (1) Micropores ($r < 1.5$ nm, $<0.5\,cm^2/g$, S_{sp} is meaningless in this case). (2) Mesopores or transient pores ($r = 1.5$–30 nm, $V_o = 0.8\ cm^2/g$, $S_{sp} = 700$–$900\,m^2/g$). (3) Macropores ($r = 30$–6500 nm). The pores can also be closed-type (isolated from each other and without an outlet on the surface), blind (connected with the surface but isolated from each other) and through ones (the channels interconnected with the surface and with one another). Polymers contain pores of various types and sizes. Note that crystalline polymers (PE, PVA, PTFE, etc.) are considered to be nonporous (although the amorphous phase of crystalline polymers can be rather porous) and have $V_o = 0$,

$S_{sp} = 1$–$7\ m^2/g$. Sometimes pores are intentionally created in polymers for more efficient binding of nanoparticles.

Langmuir–Blodgett films appear to be good polymer stabilizers of nanoparticles too [202–209] (see Chap. 8).

Polyamide films were used to stabilize colloidal mono- (Pd, Ag) and bimetallic (Pd/Ag, Pd/Cu, Pd/Co, Pd/Pb) particles showing homogeneous distribution of 1–3 nm size nanoclusters. Notice that their size and distribution dependend strongly on production regimes, the nature of the solvent (N-methyl-2-pyrrolidone or tetrahydrofurane) and mixing time [210].

Matrices of grafted polymers turned out to be rather efficient in surface protection. The properties of these two-layer and most often macroporous materials depend on the type and amount of the grafted polymer and its distribution over the substrate polymer surface (optimum variants are PE, PP, PS and PTFE). They are mostly used as fine films or powders [211]. Surface proofing of nanoparticles employs grafting of small amounts of the monomer in contrast to widespread methods of grafted polymerization aimed at polymer modification. For example, the formation of a monolayer of grafted monomer with molecular mass 100 on a polymer with $S_{sp} = 10\ m^2/g$ is equivalent to its 1.5% grafting. The optimum thickness of the grafted layer does not usually exceed 10–30 nm [212].

High-molecular compounds do not stay indifferent during stabilization to the action of highly active nanoparticles. They undergo significant structural and sometimes chemical transformations. Generally speaking, for the case of glassy polymers there exists the possibility of creating 2D ordered lattices partially or fully submerged in the surface layer of the matrix [213]. The possibility is based on the known fact of a considerable reduction of the glass transition temperature on the polymer surface (T_g^s) as compared to the bulk. This can be attributed to the growing share of the free volume in the surface layer due to segregation of the end groups of macromolecules at the polymer–air interface (e.g. PS [214–216]). With this aim, copper or auric particles obtained by different methods were placed on a specially prepared film surface, e.g. PS [213,217] or polycarbonate [215]. The atomic-force microscope (AFM) is used to observe Au particles (18–20 nm) and X-ray photoelectron spectroscopy (XPS) to control the behavior of copper particles (1–4 nm). These particles are stabilized by the polymer surface layer and sink into the polymer. The motive force of the immersion is the excess of surface tension of nanoparticles (γ_m) over the sum of the interface tension polymer-nanoparticle and polymer $\gamma_m > \gamma_{mp} + \gamma_p$ [218]. For the case of PS, its surface layer deglassification (for block PS, $T_g = 375\,K$) occurs already at 350 K and Au particles start to immerse to a depth of 15–17 nm. Particles of Cu immerse into PS at a temperature lower than its T_g by 48° and lower than the glass transition temperature of polycarbonate by 20°. So, the formation of 2D nanocomposites through immersion of nanoparticles into the polymer can become a new method of studying dynamic properties of the polymer surface in the vicinity of its glass transition temperature. Besides, such investigations present considerable interest in view of polymer surface modification by metal nanoparticles, especially in case of plastic metallization. Modern physico-chemical methods allow us to control all stages of the process. Scanning probe microscopes, SPM and STM, AFM allow us not only to image the formed structures on the atomic and subatomic levels but to carry out the transfer and localization of the separate molecule into a solid surface [219–223].

3.8 On the Problem of Matrix Confinement

There exist two different means of stabilizing nanoparticles through matrix confinement. The commonest method consists of the addition of a suspension or solution of a polymer stabilizer into a ready particle dispersion (a variant of micro-encapsulation). There are modifications of the method presupposing mixing of metal powder with the latex. A latex is mixed with particles of the metal phase whose size is much smaller than those of the latex (100 and 700 nm, correspondingly). As soon as the mixture has stabilized (often coagulation takes place also) the dispersed medium is removed. After this, the coagulators are heated and compacted into metal–polymer samples.

Another approach includes preparation of a dispersion with addition of a stabilizer or isolation of the stabilizer from its precursors in a ready dispersion of nanoparticles. Of importance here is the origin of the polymer, metal and dispersion medium as well. A widespread modification of the method is the formation of nanoparticles in a polymer medium. This is one of the ways in which nanocomposites can be produced in which mixing precedes the formation of macrochains. The thus obtained materials differ not only in the extremely homogeneous distribution of metal particles in the polymer bulk, but by a stable chemical interaction of the metal and polymer as well. Very rarely both the polymer stabilizer and nanoparticles are prepared in situ (see Chap. 10).

Macromolecules are capable of affecting the formation of the new phase and predetermine the size of the forming nanoparticles depending on the polymer chain length. Noncovalent interaction between macromolecules and the surface of the growing particles followed by generation of protecting screens are considered to be the reason for ceased growth of the particles and stabilization of colloidal dispersions [41, 224]. As a result, distribution of small particles by size turns to be considerably narrow. Attempts to establish quantitative relations between conditions of the new phase formation in the presence of macromolecules the forming particles size were undertaken. The model is based on a principal affinity between named processes and matrix polymerization due to many reasons. The main one is noncovalent and cooperative molecular interactions between a macromolecular matrix A and growing daughter chain B:

$$-\text{A}-\text{A}-\text{A}-$$

$$\cdots$$

$$-\text{A}-\text{A}-\text{A}-$$

The role of the daughter chain is fulfilled by the whole surface of the growing nanoparticle or by the most reactive part of it (active centers). The stability of derived interaction products intensifies as the sequence of links A$\cdots$A increments in the generated complex. It is supposed that the appearance of such a complex can be characterized by equilibrium A + B $\leftrightarrow$ AB with equilibrium constant K. $K = K_1^P$, where P is the degree of polymerization of the growing chain, which in case of a nanoparticle is proportional to its surface area. The size of molecule-matrix A characterized by the root-mean-square distance between its ends is larger the growing object (i.e. at mutual saturation the macromolecule is able to bind more than one nanoparticle). The value of K_1 is determined from

$$K_1 = \exp(-\Delta G_i/RT) \tag{3.13}$$

where ΔG_i is the total energy of the (i+1)-th molecular link A$\cdots$B after the i-th one.

When a macromolecule interacts with a particle surface, the energy ΔG_i is expressed through variation of the energy of surface atoms during complexing. Absolute ΔG_i values for noncovalent interactions are usually rather small and, consequently, the equilibrium constant K at low P will also be small. The formation of any stable complex of the matrix with a growing particle is doubtful. Meanwhile, K quickly grows with enlarging particles and within some interval P (critical particle size) the matrix and the growing particle recognize each other. This results in the formation of a stable complex. Two variants of events become probable. Growth of the particles may either be stopped or prolonged. Upon recognition the macromolecule acquires an ability to control particle growth and the process becomes the master one. In case the macromolecule is incapable of controlling particle formation but can specify its size through the recognition, then the process is called a pseudo-matric.

The essence of the model for the formation of nanoparticles controlled by macromolecular pseudo-matrices can be reduced to the following. The fate of the growing particle is conditioned by two main competing processes, that is, the initial growth of the particle in the solution without any contact with macromolecules (see Chap. 2) and mutual recognition of the particle and macromolecule resulting in the formation of a complex where the particle ceases its growth due to its surface screening by the macromolecule (discontinued growth means nanoparticle stabilization). Within the framework of the model and supposing that the process is running in a solution with a great excess of macromolecules, a relation for the distribution $n(r)$ by size r

of the growing spherical particles of radius r has been derived. It is expressed through the probability of particle growth cessation $\varpi(r)$:

$$\varpi(r) = KF(C)/[1 + KF(C)]^{-1} \tag{3.14}$$

$$n\,(r) = \ln\,[1 + KF\,(C)]\exp\left\{-\int_{o}^{r}\ln\,[1 + KF\,(C)]\mathrm{d}r\right\} \tag{3.15}$$

where $K = K_i^S$, $S \sim r^2$; $F(C)$ is a function of the macromolecular concentration expressed either through a) concentration of macromolecular knots $F(C) = C_\mathrm{o}/P$, b) concentration of links $F(C) = C_\mathrm{o}$, or c) through the volume fraction of the knots $F(C) = C_\mathrm{o}/P^{3\chi-1}$; C_o is the basic–molar share of macromolecules; and χ is an exponent reflecting the dependence of the linear size of the knot on P ($\chi < 1$).

In Fig. 3.25 calculated dependencies $n(r)$ and $\varpi(r)$ for various stability constant values K_i are presented for $F(C) = C_\mathrm{o}$. The gain in free energy in formation of the macromolecule–nanoparticle complex is rather small. For $K_i = 1.5$ it is only $5{\cdot}10^{-4}$ J/cm^2, which is less than the specific surface energy at the metal–water interface for a polymer–metal composite (of the order of fractions of a percent). On conversion to a polymer unit this constitutes the energy of the portions of kT. However, as can be seen from Fig. 3.25, even such small energy parameters allow for the formation of a structurally homogeneous nanoparticle of ~1 nm diameter in the pseudo-matric process. The mean particle size for this model will be determined by the relation

$$r^2 = (\alpha \ln P - \ln C_\mathrm{o})/\ln K_1 \tag{3.16}$$

where $\alpha = 1.0$ or 1–3χ depending on the type of function $F(C)$.

The examples above describe different characters of particle-size dependence on the degree of polymerization. The average particle size may either grow with increasing P, (case a), remain constant (b) or diminish (c), which is supported experimentally. If the mean particle size is known, one can calculate ΔG_i, and from its temperature dependence define ΔH_i and ΔS_i. Furthermore, values of ΔG_i equal to, correspondingly $-2.8{\cdot}10^{-4}$ (290 K) and $-0.6{\cdot}10^{-4}$ (320 K) J/m^2 were obtained [191] by substituting the corresponding mean diameters of the particles forming during the pseudo-matric process at 290 and 320 K (5.5 and 12.5 nm). Subsequently the derived polymer–surface interaction enthalpy and entropy constituted $\Delta H_\mathrm{i} = -2.5{\cdot}10^{-3}$ J/m^2 and $\Delta S_\mathrm{i} = -7.4{\cdot}10^{-6}$ J/(m$^2{\cdot}$K). Gibbs energy values for interaction of the particle surface (e.g. nickel) with PVPr macromolecules are of the order of magnitude of the interaction energy between polymers and are by a few orders of magnitude less than the specific free surface energy characteristic of metals.

The critical size of the pseudo-matric is a function of the process temperature, providing it exerts some effect on the size of the forming nanoparticles.

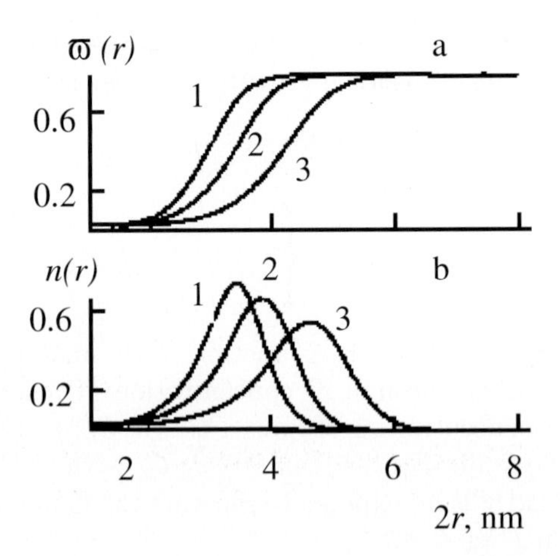

Fig. 3.25. The dependencies of the probability of particle growth termination on their size (**a**) and the distribution function of particles on size (**b**)

The size of copper particles and their stability depend essentially on the molecular mass of the stabilizing polymer and on the temperature of copper ion reduction in the presence of PVPr.

The process of recognition can be of the type $M_n + P_1 + P_2 \rightarrow (P_1{\cdot}M_n) + P_2$ in systems with nanoparticles and two types of macromolecules P_1 and P_2, each of which is capable of interacting with M_n. The mean copper nanoparticle size formed in PVPr solution is 4–7 nm and in aqua solutions of PEI it is 7–10 nm which is evidence of better stabilization of nanoparticles by the PVPr surface.

* * *

In conclusion, the following facts should be considered.

Although the chief theoretical postulates on the formation and stabilization of colloidal dispersions are mainly related to polymer emulsions (latex polymerization of vinyl and diene coatings, etc.), they can be certainly applied to nanoparticles too. The differences are more likely quantitative than qualitative. As a proof, we mention that forces of attraction between dispersion particles of organic polymers are much less than those between metal particles, their oxides or salts (compare to the above-cited Hamaker constants).

The effect of additions of high-molecular compounds on the stability of nanoparticles has been studied in many aspects, whereas the stabilization mechanism of polymers has not been completely clarified and the theoretical basis of their choice is just at its beginning. This is attributed to the great intricacy of polymer-containing dispersive systems and diversity of factors effecting stability of metal-containing nanoparticles. Based on experimental data a supposition can be put forward on a more promising stabilizer among nonionic polymers. They seem to be the stabilizers that form reasonably thick and dense adsorption layers on the particle surface. As for charged macromolecules, it is more expedient to use similarly charged polymers with high charge density. In the general case, the stabilizing effect of a high-molecular compound is determined by many supplementary factors, particularly, by polymerization level, type and amount of functional groups, character of their distribution along the chain and others. The strength and toughness of protecting layers, their spatial stretch and the way the polymer adjoins the particles are valuable characteristics of the stabilizing effect.

Two groups of problems can be singled out from the theory of the systems under study. The first one represents the subject of classical colloidal chemistry and physico-chemistry of surface phenomena. This, for example, includes the description of electric surface events in polymer-containing colloidal systems. The other group of problems is connected with traditional fields of physico-chemistry of polymers. In this domain a researcher usually digresses from treating details of interactions between colloidal particles but concentrates his attention on a pure contribution of the polymer interrelated with concentration of dissolved macromolecules, peculiarities of their interaction with particles, thermodynamic properties of the solvent and so on. In a real situation a combination of such approaches is needed for a unified general theory or perhaps for a multilevel scheme.

The quantitative theory of stabilization of colloidal particles proceeds from the analysis of diluted colloids whose important property is the kinetic (sedimentation) stability. In case of concentrated and highly concentrated systems (up to 80% of metal-containing powder of nanoparticles) participation of the dispersed phase in Brownian movement does not play a decisive role. Their distinctive features are a strongly developed interfacial surface and high dispersed phase content in the liquid dispersed medium. These properties give rise to the main outcome. As a result of particle cohesion inside the dispersion medium thermodynamically stable spatial structures appear in highly concentrated systems [165]. Nanoparticles are fixed in the spatial structure of the network thanks to either atomic or coagulation contacts between particles and their participation in Brownian motion is limited merely to oscillations about the equilibrium position. This is consistent with the heat propagation energy ($3/2\ kT$) that is considerably less than the total energy of interaction between neighboring particles. The latter is conditioned by the origin of elementary contacts, their number and distribution pattern in the bulk. So far the protecting high-molecular compounds are useful in producing concentrated metal hydrosols and dry preparations of their dispersed phase.

Thus screened sol acquire properties of reversible colloids. The dry dispersed phase which remained after evaporation of the dispersion medium is able to form once again a stable sol upon contact with the medium. Lyophobic nanoparticles in such systems are arranged as if in a cobweb of the protecting polymer. These materials are in strong demand since significant variation in their properties (elevated thermal and electric conductivity, heat capacity, ability to screen radiation, etc.) is reached only at a high content of nanoparticles (often more than 50% by volume or 90% by mass). However, the high viscosity of the polymer matrix melt inhibits the even distribution of nanoparticles thus impairing dielectric and mechanical characteristics of the material.

Synthesis and conservation processes frequently turn out to be separated. As will be shown further, wide potentialities are seen in integrating synthesis and conservation in a single production cycle of nanocomposites which is attained during generation of nanoparticles in the presence of polymers.

References

1. B.V. Deryagin: *The theory of stability of colloids and thin films. Surface forces* (Nauka, Moscow 1986)
2. D.H. Napper: *Polymeric Stabilization of Colloidal Disptrsions* (Academic Press, London 1983)
3. K.E.J. Barrett: *Dispersion polymerization in organic media* (Wiley, London 1975)
4. *Adsorption from Solution at the Solid–Liquid Interface* ed. by G.D. Parfitt, C.H. Rochester (Academ. Press, London 1983)
5. F. Baran: *Polymer-containing dispersed systems* (Navukova Dumka, Kiev 1986)
6. N.B. Uryev: *Highly concentrated dispersed systems* (Khimiya, Moscow 1980)
7. *Polymer Colloids* ed. by J.W. Goodwin, R. Buscall (Academic Press, New York 1995)
8. J. Visser: Adv. Colloid. Interface Sci **3**, 331 (1972)
9. J.S. Bradley: In: *Clusters and Colloids: from Theory to Applications* ed. by G. Smid (VCH Publ., Weinheim 1994) p. 469
10. E. Dickinson: Colloids Surfaces **39**, 143 (1989)
11. E. Dickinson: *The structure, dynamics and equlibrium properties of colloidal systems* (Kluwer Acad. Publ., 1990)
12. D.E. Ulberg, V.V. Ilyin, N.V. Churaev, Y.V. Nizhnik: Colloid J. **54** 151 (1992)
13. E.A. Nikologorskaya, N.V. Kasaikin, A.T. Savichev, V.A. Efremov: Colloidal J., **51**, 1136 (1989)
14. D. Cabane, R. Duplessix: J.Physique (France) **48**, 651 (1987)
15. M. Faraday: Philos. Trans. R. Soc. London **147**, 145–153 (1857)
16. A.M. Natanson, Z.P. Ulberg: *Colloidal metals and metal-polymers* (Navukova Dumka, Kiev (1971)
17. H.S. Kim, J.H. Rye, B. Jose, B.G. Lee, B.S. Ahn, Y.S. Kang: Langmuir, **17**, 5817 (2001)
18. M. Ferreira, K. Wohnrath, R.M. Torresi, C.J.L. Constantino, R.F. Aroka, O.N. Oliveira, Jr., J.A. Giacometti: Langmuir **18**, 540 (2002)
19. B.G. Ershov, A. Henglein: J. Phys. Chem. B **102**, 10663 (1998)
20. M.V. Kiryukhin, B.M. Sergeev, A.N. Prusov, V.G. Sergeev: Polym. Sci. Ser. B **42**, 2171 (2000)
21. J. Turkevich: Gold Bull. **18**, 86 (1985)

22. B.G. Ershov: Russ. Chem. Rev. **66**, 103 (1997)
23. *Metallization of Polymers* ed. by E. Sacher, J.J. Pireaux, S.P. Kowalczyk. (ACS Symp. Ser., Washington, D.C. 1990), **440**
24. *Metallized Plastics. Fundamental and Applied Aspects* ed. by K.L. Mittal, J.R. Susco (Plenum Press, New York 1989, 1991, 1992) Vol. 1–3
25. E. Sabatini, J.C. Boulakia, M. Bruening, I. Rubinstein: Langmuir **9**, 2974 (1989)
26. J. Beltan, S.I. Stupp: *Adhesion aspects of polymeric coating* (Plenum Press, New Year 1983)
27. A.A. Berlin, V.E. Basin: *Fundamentals of polymer adhesion* (Khimiya, Moscow 1974)
28. V.A. Vakula, L.M. Pritykin: *Physical chemistry of polymer adhesion* (Khimiya, Moscow 1984)
29. Yu.S. Lipatov: *Colloidal chemistry of polymers* (Elsevier, Amsterdam 1988)
30. H. Sonntag, B. Unterberger, S. Zimontkowski: Colloid and Polymer Sci. **257**, 286 (1979)
31. A.A. Baran, T.A. Polischuk: Ukr. Polym. J. **1**, 121 (1992)
32. L. Longenberger, G. Mills: J. Phys. Chem. **99**, 475 (1995)
33. H. Hirai, Y. Nakao, N. Toshima: J. Macromol. Sci.-Chem. A **13**, 727 (1979)
34. K.S. Suslick, M. Fang, T. Hyeon: J. Amer. Chem. Soc. **118**, 11960 (1996)
35. A.B.R. Mayer, J.E. Mark: In: *Nanotechnology of Molecularly Designed Materials* ed. by G.-M. Chow, K.E. Gonsalves, (ACS, Washington, DC 1996) p. 137
36. L. Longenberger, G. Mills: In: *Nanotechnology. Molecularly Designed Materials* ed. by G.-M. Chow, K.E. Gonsalves (ACS, Washington, DC 1996) p. 128
37. J.S. Bradley, J.M. Millar, E.W. Hill: J. Am. Chem. Soc. **113**, 4016 (1991)
38. F. Porta, F. Ragaini, S. Cenini, G. Scari: Gazz. Chim. Ital. **122**, 361 (1992)
39. M.S. El-Shall, W. Slack: Macromolecules **28**, 8456 (1995)
40. J.H. Golden, H. Deng, F.J. DiSalvo, J.M.J. Frechet, P.M. Thompson: Science **268**, 1463 (1995)
41. H. Hirai, N. Toshima: In: *Tailored metal Catalysts* ed. by Y. Iwasawa (D. Reidel Dordrecht 1986) p. 121
42. V. Sankaran, J. Yue, R.E. Cohen, R.R. Schrock, R.J. Silbey: Chem. Mater. **5**, 1133 (1993)
43. L. Chen, W.-J. Yang, C.-Z. Yang: J. Mater. Sci. **32**, 3571 (1997)
44. L. Chen, C.-Z. Yang, X.H. Yu: Chin. Chem. Lett. **5**, 443 (1994)
45. N.L. Holy, S.R. Shelton: Tetrahedron **39**, 25 (1981)
46. Y.P. Ning, M.Y. Tang, C.Y. Jiang, J.E. Mark: J. Appl. Polym. Sci. **29**, 3209 (1984)
47. D.H. Kim, W.H. Jo: *Polymer* **40**, 3989 (1999)
48. B.G. Ershov, N.I. Kartashev: Russ. Chem. Bull. 35 (1995)
49. L.A. Tsarkova: Colloid J. **58**, 849 (1996)
50. S. Lakhwani, M.N. Rahaman: J. Mater. Sci. **34**, 3909 (1999)
51. I.D. Robb: In: *The Chemistry and Technology of Water Soluble Polymers* (Plenum, London 1982) pp. 310–313
52. S. Kondo, M. Ozeki, N. Nakashima: Angew. Makromol. Chem. **163**, 139 (1988)
53. S. Alexander: J. Physiue (France) **38**, 977 (1977)
54. P.A. Pinkus, C.J. Sandroff, T.A. Witten: J. Physique (France) **45**, 725 (1984)
55. T.M. Birstein, O.V. Borisov: Polym. Sci. Ser. A **28**, 2265 (1986)

56. E.B. Zhulina: Polym. Sci. Ser. A **25**, 834 (1983)
57. D.K. Klimov, A.R. Khokhlov: Polymer **33** 2177 (1992)
58. E.M. Rozhkov, P.G. Khalatur: Colloid J. **58**, 823, 831 (1996)
59. T.F. Tadros: Polymer J. **23**, 683 (1991)
60. I. Piirma: *Polymeric Surfactants. Surfactants Science Series* **42** (Marcell Dekker, New York 1992)
61. Y. Moroi: *Micelles, Theoretical and Applied Aspects* (Plenum, New York 1992)
62. D.F. Evans, H. Wennerstrom: *The Colloidal Domain* (VCH, Weinheim 1994)
63. J.M. Seddon: Biochim. Biophys. Acta **1**, 1031 (1990)
64. *An introduction to polymer Colloids. NATO ASI Series* **303**. ed. by C F. Candau, R.H. Ottewill (Kluwer, Dodrecht 1990)
65. T. Sato, R. Ruch: *Stabilization of Colloidal Dispersions by Polymer Adsorption. Surfactant Science* Ser. 9. (Marcel Dekker, New York 1980)
66. B. Chu: Langmuir **11**, 414 (1995)
67. Z. Tuzar, P. Kratochvill: In: *Surface and Colloid Science* ed. by E. Matijevic (Plenum Press, New York 1993)
68. C. Price: In: *Developments in Block Copolymers* ed. by I. Goodman (Applied Science Publ., London 1982) **1**, p. 39
69. M. Antonietti, C. Goltner: Angew. Chem. Int. Ed. Engl. **36**, 910 (1997)
70. P. Mukedjy: In: *Micelle formation, solubilization and microemulasions* (Moscow, Mir 1980) p. 121
71. G. Riess, G. Hurtrez, P. Bahadur: In: *Encyclopedia of Polymer Science and Engineering* ed. by H.F. Mark, N.M. Bikales, C.S. Overberger, G. Menges (J. Wiley, New York 1985)
72. J. Selb, Y. Gallot: In: *Development in Block Copolymers* ed. by I. Goodman (Elsevier, London 1985)
73. A. Gast: NATO ASI Ser. **303**, 311 (1988)
74. J.H. Fendler: Chem. Rev. **87**, 877 (1987)
75. *Micelles, membranes, microemulsions and monolayers* ed. by W.M. Gelbart, B.S. Avinoam, R. Didier (Springer-Verlag, New York 1994)
76. V.G. Babak, R. Gref, E. Dellacherie: Mendeleev Commun. 105 (1998)
77. A.V. Kabanov, T.K. Bronich, V.A. Kabanov, K. Yu, A. Eisenberg: J. Am. Chem. Soc. **120**, 9941 (1998)
78. J. Liu, W. Ong, E. Roman, M.J. Lynn, A.E. Kaifer: Langmuir **16**, 3000 (2000)
79. J. Alvarez, J. Liu, E. Roman, A.E. Kaifer: Chem. Commun. 1151 (2000)
80. H. Otsuka, Y. Akiyama, Y. Nagasaki, K. Kataoka: J. Am. Chem. Soc. **123**, 8226 (2001)
81. M. Yamada, A. Kuzume, M. Kurihara, K. Kubo, H. Nishihara: Chem. Commun. 2476 (2001)
82. C. Pan, K. Pelzer, K. Phillippot, B. Chaudret, F. Dassenoy, P. Lecante, M.J. Casanove: J. Am. Chem. Soc. **123**, 7584 (2001)
83. J. Wiesner, A. Wokaun, H. Hoffmann: Progr. Colloid Polym. Sci. **76**, 271 (1988)
84. M.T. Reetz, A. Quaiser: Angew. Chem. Int. Ed. Engl. **34**, 2240 (1995)
85. M.T. Reetz, W. Helbig, S. Quaiser: Chem. Mater. **7**, 2227 (1995)
86. J.H. Fendler, P. Tundo: Acc. Chem. Res. **17**, 3 (1984)
87. N. Poulain, E. Nakache, A. Pina, G. Levesque: J. Polym. Sci. Polym. Chem. Ed. A **34**, 729 (1996)
88. N. Poulain, E. Nakache: J. Polym. Sci. Polym. Chem. Ed. A **36**, 3035 (1998)

89. Y. Okahata, T. Kunitake: J. Am. Chem. Soc. **101**, 5231 (1979)
90. S.C. Kushwaha, M. Kates, G.D. Sprott, I.C.P. Smith: Science, **211**, 1163 (1981)
91. T.A. Langworthy: Curr. Top. Membr. Transp. **17**, 45 (1982)
92. J.-H. Furhop, D. Fritsch: Acc. Chem. Res. **19**, 130 (1986)
93. M.P. Pileni: J. Phys. Chem. **97**, 6961 (1993)
94. *Structure and Reactivity in Reserse Micelles* ed. by M.P. Pileni (Elsevier, Amsterdam 1989)
95. C. Petit, P. Lixon, M.-P. Pileni: J. Phys. Chem. **97**, 12974 (1993)
96. G.T. Cardenas, E.C. Salgado, M.G. Gonsalves: Polym. Bull. (Berlin) **35**, 553 (1995)
97. G.T. Cardenas, J.E. Acuna, M.B. Rodrigues, H.H. Carbacho: Polym. Bull. (Berlin) **34**, 31 (1995)
98. G.T. Cardenas, E.C. Salgado, V.L. Vera, M.J. Yacaman: Macromol. Rapid Commun. **17**, 775 (1996)
99. F. Caruso: Chem. Eur. J. **6**, 413 (2000)
100. R.E. Holmlin, M. Schiavoni, C.Y. Chen, S.P. Smith, M.G. Prentiss, G.M. Whitesides: Angew. Chem. Int. Ed. **39**, 3503 (2000)
101. D.M. Antonelli: Microporous Mesoporous Mater. **33**, 209 (1999)
102. D.H. Hubert, M. Jung, A.L.German: Adv. Mater. **12**, 1291 (2000)
103. S. Oliver, A. Kuperman, N. Coombs, A. Lough, G.A. Ozin: Nature **378**, 47 (1995)
104. M. Antonietti, A. Thunemann, E. Wenz: Colloid Polym. Sci. **274**, 795 (1996)
105. M. Antonietti, F. Grohn, J. Hartmann, J. Bronstein: Angew. Chem. Int. Ed. Engl. **36**, 2080 (1997)
106. M. Antonietti, E. Wenz, L. Bronstein, M. Seregina: Adv. Mater. **7**, 1000 (1995)
107. O.A. Platonova, L.M. Bronstein, S.P. Solodovnikov, I.M. Yanovskaya, E.S. Obolokova, P.M. Valetsky, E. Wenz, M. Antonietti: Colloid. Polym. Sci. **275**, 426 (1997)
108. Y.Ng Cheong Chan, R.R. Schrock, R.E. Cohen: Chem. Mater. **4**, 24 (1992)
109. P.R. Quirk, D.J. Kinning, L.J. Fetters: In: *Comprehensive Polymer Science* ed. by G. Allen, J.C. Bevington, S.L. Aggarval: (Pergamon Press, New Year 1989) p. 7
110. F.S. Bates: Science **251**, 898 (1991)
111. L. Liebler: Macromolecules **13**, 1602 (1980)
112. E. Helfand, Z.R. Wasserman: Macromolecules **13**, 994 (1980)
113. G. Hadziioannou, A. Skoulios: Macromolecules **15**, 258 (1982)
114. T. Hashimoto, H. Tanaka, H. Hasegawa: Macromolecules **18**, 1864 (1985)
115. D.J. Kinning, E.L. Thomas, L.J. Fetters: J. Chem. Phys. **90**, 5806 (1989)
116. E.L. Thomas, K.I. Winey: Polym. Mater. Sci. Eng. **62**, 686 (1990)
117. M. Antonietti, S. Forster, J. Hartmann: Macromolecules **29**, 3800 (1996)
118. H. Zhao, E.P. Douglas, B.S. Harrison, K.S. Schanze: Langmuir **17**, 8428 (2001)
119. A. Roescher, M. Moller: Adv. Mat. **7**, 151 (1995)
120. J.P. Spatz, S.S. Mössner, C. Hartman, M. Möller, T. Herzog, M. Krieger, H.G. Boyen, P. Ziemann, B. Kabius: Langmuir, **16**, 407 (2000)
121. S. Forster, M. Zisenis, E. Wenz, M. Antonietti: J. Chem. Phys. **104**, 9956 (1996)
122. G.T. Cardenas, J.E. Acuna, M.B. Rodrigues, H.H. Carbacho: Int. J. Polym. Mater. **26**, 199 (1994)

123. G.T. Cardenas, C.C. Retamal, K.J. Klabunde: Polym. Bull. (Berlin) **27**, 983 (1992)
124. A. Roescher, M. Moller: In: *Nanotechnology. Molecularly Designed Materials* ed. by G.-M. Chow, K.E. Gonsalves (ACS, Washington, DC 1996) p. 116
125. Y.Ng Cheong Chan, G.S.W. Craig, R.R. Schrock, R.E. Cohen: Chem. Mater. **4**, 885 (1992)
126. B.H. Sohn, R.E. Cohen: Acta Polym. **47**, 340 (1996)
127. *Physics of Amphihiles: Micelles, Vesicles and Microemulsions.* ed. by V. Degiorgio, M. Corti (North Holland, Amsterdam 1985)
128. G. Riess, P. Bahadur, G. Hurtrez: In: *Encyclopedia of Polym. Sci. and Technology* **2** (J. Wiley Inc. Publ., New Year 1985) p. 324
129. R. Saito, S. Okamura, K. Ishizu: Polymer **33**, 1099 (1992)
130. J.P. Spatz, A. Roescher, M. Moller: Polym. Prepr. **37**, 409 (1996)
131. M. Moffit, L. McMahon, V. Pessel, A. Eisenberg: Chem. Mater. **7**, 1185 (1995)
132. M. Moffit, A. Eisenberg: Chem. Mater. **7**, 1178 (1995)
133. K. Kurihara, J.H. Fendler: J. Amer. Chem. Soc. **105**, 6152 (1983)
134. J. Gratt, R.E. Cohen: Macromolecules **30**, 3137 (1997)
135. R.E. Cohen, R.T. Clay, J.F. Ciebien, B.H. Sohn: In: *Polym. Mater. Encyclopedia* **6** ed. by J.C. Salamone (CRC Press 1996) p. 414
136. Y.Ng Cheon Chan, R.R. Schrock, R.E. Cohen: J. Am. Chem. Soc. **114** 7295 (1992)
137. R.T. Clay, R.E. Cohen: Supramolecul. Chem. **2**, 183 (1995); **4**, 113 (1997)
138. K. Kurihara, J. Kizling, P. Stenius, J.H. Fendler: J. Am. Chem. Soc. **105**, 2574 (1983)
139. M. Boutonnet, J. Kizling, P. Stenius, G. Maire: Colloids. Surf. **5**, 209 (1982)
140. M.L. Steigerwald, L.E. Brus: Annu. Rev. Mater. Sci. **19**, 471 (1989)
141. Y. Wang, A. Suna, W. Malher, R. Kasowski: J. Chem. Phys. **87**, 7315 (1987)
142. D.de Caro, T.O. Ely, A. Mari, B. Chaudret, E. Snoeck, M. Respaud, J.-M. Broto, A. Fert: Chem.Mater. **8**, 1987 (1996)
143. J.A. Massey, K. Temple, L. Cao, Y. Rharby, J. Raez, M.A. Winnik, I. Manners: J. Am. Chem. Soc. **122** 11577 (2000)
144. L. Cao, I. Manners, M.A. Winnik: Macromolecules, **34** 3353 (2001)
145. R. Saito, S. Okamura, K. Ishizu: Polymer **36**, 4515 (1995) **37**, 5255 (1996)
146. R. Saito, S. Okamura, K. Ishizu: Polymer **33**, 1099 (1992)
147. K. Ishizu, Y. Kashi, T. Fukutomi, T. Kakurai: Makromol. Chem. **183**, 3099 (1982)
148. K. Ishizu, K. Inagaki, T. Fukutomi: J. Polym. Sci. Polym. Chem. Ed. **23**, 1099 (1985)
149. R. Saito, H. Kotsubo, K. Ishizu: Polymer **33**, 1073 (1992)
150. K. Ishizu, T. Fukutomi: J. Polym. Sci. Polym. Lett. Ed. **26**, 28 (1988)
151. K. Ishizu: Polym. Commun. **30**, 209 (1989)
152. Y. Kurokawa, K. Ueno: J. Appl. Polym. Sci. **27**, 621 (1982)
153. H. Huang, T. Kowalewski, E.E. Remsen, R. Gertzmann, K.L. Wooley: J. Am. Chem. Soc. **119**, 11653 (1997)
154. B.K. Thurmond, T. Rjwalewski, K.L. Wooley: J. Am. Chem. Soc. **118**, 7239 (1996) **119**, 6656 (1997)
155. Q. Ma, E.E. Remsen, T. Kowalewski, K.L. Wooley: J. Am. Chem. Soc. **123**, 4627 (2001)
156. M. Moller: Synth. Met. **41-43,** 1159 (1991)

157. L. Zhang, A. Eisenberg: Science, **268**, 1728 (1995); J. Am. Chem. Soc. **118**, 3168 (1996)
158. K. Yu, A. Ensenberg: Langmuir **12**, 5980 (1996)
159. M.A. Hillmyer, F.S. Bates, K. Almdal, K. Mortensen, A.J. Ryan, J.P.A. Fairclough: Science **271**, 976 (1996)
160. F. Caruso: Chem. Europ. J. **6**, 413 (2000)
161. R.E. Holmin, M. Schiavoni, C.Y. Chen, S.P. Smith, M.G. Prentiss, G.M. Whitesides: Angew. Chem. Int. Ed. **39**, 3503 (2000)
162. D.M. Antonelli: Microporous Mesoporous Mater. **33**, 209 (1999)
163. D.H. Hubert, M. Jung, A.L. German: Adv. Mater. **12**, 1291 (2000)
164. S. Oliver, A. Kuperman, N. Coombs, A. Lough, G.A. Ozin: Nature, **378**, 47 (1995)
165. P.A. Rebinder: *Selected works. Surface phenomena in dispersed systems* ed. by G.I. Fux (Nauka, Moscow 1978)
166. E.D. Schukin: Colloid J. **59**, 270 (1997)
167. M.A. Lunina, M.G. Ivanova, A.A. Khachaturyan: Colloid J. **57**, 825 (1995)
168. *Adsorption from solutions of solid surfaces* ed. by G. Parphite, K. Rochester (Mir, Moscow 1986)
169. A.I. Rusanov: Russ. J. Gen. Chem. **72**, 532 (2002)
170. A.I. Rusanov, A.K. Shchekin: Colloids and Surfaces **192**, 357 (2001)
171. *Evolution of Size Effect in Chemical Dynamics* ed. by I. Prigogine, S. Rice (Wiley, New Year 1988) Part 2
172. I.D. Robb: In: *The chemistry and technology of water soluble polymers* (Plenum, London 1982) pp. 301–313
173. S. Kondo, M. Ozeki, N. Nakashima: Angew. Makromol. Chem. **163**, 139 (1988)
174. H. Welker: Angew. Chem. **105**, 43 (1993); Angew. Chem. Int. Ed. Engl. **32**, 41 (1993)
175. A. Henglein: Ber. Bunsenges Phys. Chem. **99**, 903 (1995)
176. G. Nimtz, P. Marquard, H. Gleiter: J. Cryst. Growth **86**, 66 (1988)
177. A. Fojtik, H. Weller, U. Koch, A. Henglein: Ber. Bunsenges. Phys. Chem. **88**, 969 (1984)
178. J.H. Youk, J. Locklin, C. Xia, M.-K. Park, R. Advincula: Langmuir **17**, 4681 (2001)
179. W.Y. Hsu, T.D. Gierke: Macromolecules **15**, 101 (1982)
180. D. Mattera Jr., D.M. Barnes, S.N. Chaudhuri, W.M. Risen Jr., R.D. Gonsalez, J. Phys. Chem. **90**, 4819 (1986)
181. A.S. Plachinda, V.E. Sedov, V.I. Khromov, I.P. Suzdalev, V.I. Goldanskii, G.U. Nienhaus, F. Parak: Phys. Rev. B Condens. Mater. **45**, 7716 (1992)
182. V.I. Khromov, A.S. Plachinda, S.I. Kamyshansky, I.P. Suzdalev: Russ. Chem. Bull. 886 (1996)
183. H. Yokoi, Y. Mori, Y. Fuijise: Bull. Chem. Soc. Jpn. **68**, 2061 (1995)
184. H. Yokoi, S. Kawata, M. Iwaizumi: J. Am. Chem. Soc. **108**, 3358 (1986)
185. H. Yokoi, Y. Mori, T. Mitani, S. Kawata: Bull. Chem. Soc. Jpn. **65**, 1989 (1992)
186. R.F. Ziolo, E.P. Gianneliss, B.A. Weinstein, M.P. O'Horo, B.N. Ganguly, V. Mehrotra, M.W. Russell, D.R. Huffman: Science **257**, 219 (1992)
187. H.M. Meyer, S.G. Anderson, K.J. Atanasoska, J.H. Weaver, J. Vac. Sci. Technol. A **6**, 30 (1988)
188. P. Pincus: Macromolecules **24**, 2912 (1991)

189. O.V. Borisov, T.M. Birshtein, E.B. Zhulina J. Phys. II **1**, 521 (1991)
190. E.B. Zhulina, O.V. Borisov, V.A. Priamitsin: J. Colloid. Interface Sci. **137**, 495 (1990)
191. E.B. Zhulina, O.V. Borisov, V.A. Priamitsin, T.M. Birshtein: Macromolecules **24**, 140 (1991)
192. E.B. Zhulina, T.M. Birshtein, O.V. Borisov: J. Phys. II **2**, 63 (1992)
193. R. Israels, F.A.M. Leermakers, G.J. Fleer, E.B. Zhulina: Macromolecules **27**, 3249 (1994)
194. J.F. Argilier, M. Tirrell: Theor. Chim. Acta **82**, 343 (1992)
195. J. Wittmer, J.-F. Joanny: Macromolecules **26**, 2691 (1993)
196. C.M. Marques, J.-F. Joanny, L. Leibler: Macromolecules **21**, 1051 (1988)
197. R. Israels, J.M.H.M. Scheutjens, G.J. Fleer: Macromolecules **26**, 5404 (1993)
198. O.A. Evers, J.M.H.M. Scheutjens, G.J. Fleer: J. Chem. Soc. Faraday Trans. **86**, 1333 (1990)
199. M.R. Bohmer, O.A. Evers, J.M.H.M. Scheutjens: Macromolecules **23**, 2288 (1990)
200. R. Israels, F.A.M. Leermakers, G.J. Fleer: Macromolecules, **28**, 1626 (1995)
201. S.P. Gubin, N.K. Eremenko: Mendeleev Chem. J. **36**, 718 (1991)
202. J. Schmitt, T. Grunewald, K. Kjaer, P. Pershan, G. Decher, M. Losche: Macromolecules **26**, 7058 (1993)
203. W. Chen, T.J. McCarthy: Macromolecules **30**, 78 (1997)
204. G. Decher: In: *The Polymeric Materials Encyclopedia. Synthesis, Properties, and Applications* ed. by J.C. Salamone (CRC Press, Boca Raton 1996) **6**, pp. 4540–4546
205. W. Knoll: Curr. Opinion in Coll. and Interface Sci. **1**, 137 (1996)
206. G. Decher: In: *Comprehensive Supramolecular Chemistry* ed. by J.-P. Sauvage, M.W. Hosseini (Pergamon Press, Oxford 1996) **9**, pp. 507–528
207. Y. Lvov, G. Decher, G. Sukhorukov: Macromolecules **26**, 5396 (1993)
208. J. Schmitt, G. Decher, W.J. Dressik, S.L. Brandow, R.E. Geer, R. Shashidhar, J.M. Calvert: Adv. Mater. **9**, 61 (1997)
209. J.H. Fendler, F.C. Meldrum: Adv. Mat. **7**, 607 (1995)
210. L. Troger, H. Hunnefeld, S. Nunes, M. Oehring, D. Fritsch: J. Phys. Chem. **101**, 1279 (1997)
211. D.A. Kritskaya, A.D. Pomogailo, A.N. Ponomarev, F.S. Diyachkovsky: Polym. Sci. USSR, Ser.A **21**, 1107 (1979); J. Appl. Polym. Sci. **25**, 349 (1980)
212. A.D. Pomogailo: *Polymer immobilized complex metal catalysts* (Nauka, Moscow 1988)
213. V.M. Rudoi, I.V. Yaminsky, O.V. Dementyev, V.A. Ogarev: Colloid. J. **61**, 861 (1999)
214. Y.C. Jean, R. Zhang, H. Cao, J.-P. Yuang, C.-M. Huang, B. Nielsen, P. Asoka-Kumar: Phys. Rev. B **56**, R8459 (1997)
215. T. Kajiyama, K. Tanaka, A. Takahara: Polymer **39**, 4665 (1998)
216. N. Satomi, A. Takahara, T. Kajiyama: Macromolecules **32**, 4474 (1999)
217. V. Zaporojtchenko, T. Strunskus, J. Erichen, F. Faupel: Macromolecules **34,** 1125 (2001)
218. V. Zaporojtchenko, K. Behnke, A. Thran, T. Strunskus, F. Faupel: Appl. Surf. Sci. 144, 355 (1999)
219. Y. Wada: Jpn. J. Appl. Phys. **39**, 3825 (200)
220. G.J. Fleer, M.A.C. Stuart, J.M.H. Scheutjiens, T. Cosgrove. B. Vicent: *Polymers at Interfaces* (Chapman and Hall, London 1993)

221. G.P. Lopinskii, D.D. Wayner, R.A. Wolkov: Nature **408**, 48 (200)
222. C.P. Collier: Science **285**, 391 (2000)
223. H. Sigenthaler: In: *Scanning tunneling Microscopy* (Springer, New York 1992), v.28, Chap. 2, p. 231
224. H. Hirai, N. Toshima: In: *Polymeric Materials Encyclopedia* ed. by J.C. Salamone (CPC Press, Boca Raton 1996) **2** p. 1310
225. I.M. Papisov, A.A. Litmanovich, K.I. Bolyachevskaya, Yu.S. Yablokov, A.I. Prokofiev, O.Ye. Litmanovich, S.V. Markov: Macromol. Symp. **106**, 287 (1996)
226. I.M. Papisov, Y.S. Yablokov, A.I. Prokofyev, A.A. Litmanovich: Polym. Sci. Ser. A **35**,515 (1993); **36**, 352 (1994)
227. A.A. Litmanovich, I.M. Papisov: Polym. Sci. Ser. A **39**, 323 (1997); **41**, 1824 (1999)
228. O.E. Litmanovich, A.G. Bogdanov, A.A. Litmanovich, I.M. Papisov: Polym. Sci. Ser. A **39**, 1875 (1997)
229. O.E. Litmanovich, A.G. Bogdanov, A.A. Litmanovich, I.M. Papisov: Polym. Sci. Ser. A **40**, 100 (1998); **39**, 1506 (1997); **43**, 135 (2001)
230. I.M. Papisov, A.A. Litmanovich: Colloids and Surfaces **151**, 399 (1999)

Part II

Synthetic Methods for Metallopolymer Nanocomposite Preparation

Nanoparticles are formed in two stages: molecular dispersion and grinding or reduction followed by atomic metal condensation. These reactions alter so quickly that they are in fact indivisible. They represent a single intricate process of nucleation and growth of the solid phase.

Composite materials consist of a polymer matrix inside which ultra-dispersed particles or clusters (an aggregate of adjoining discrete metal particles of indefinite form and value) are randomly distributed. Manufacturing method can be subdivided into three large groups: physical, chemical and physico-chemical. The division is very conventional and is based mostly on the method of nanoparticle formation and the character of their interactions with the matrix. The first group of methods includes the procedures seemingly devoid of any chemical interactions between the dispersed phase and the dispersion medium. Composites of this type are rare. More popular chemical methods of manufacturing nanocomposites are based on interactions between components anticipated or detected. The reduction reactions and precursors are well-known. Nevertheless, the new approaches have been elaborated from practical demands and many details of the chemical reduction are still unclear. The third group of methods are those when metal or their oxide nanoparticles are formed using high energy processes of atomic metal evaporation including low temperature discharge plasma, radiolysis and photolysis in the presence of monomer and polymer. Recently such methods have become widely used in vacuum metallization of polymer materials for electronics. These methods allow controlling the concentration of components and their size over a wide range. However, investigations devoted to synthesis of metallopolymer nanocomposites are only qualitative and limited by information about the dimensions of particles and the size distribution.

4 Physical Methods of Incorporating Nanoparticles into Polymers

Among physical methods are numerous modifications of nanoparticle mixing with either polymers or oligomers; their microencapsulation or conservation into polymer shells through solvent evaporation; setting, surface polymerization; heteroaddagulation of colloids based on polymer powders, films or fibers; co-extrusion, etc.

Co-solution of the polymer or its precursor is performed usually to produce systems without chemical binding between components. It is followed either by pouring of the film onto a suitable surface where the solvent is evaporated or the ready polymer film is saturated with the precursor solution. Membranes, including polydimethylsiloxane, are used for this purpose. These materials are sometimes additionally cross-linked.

Physico-chemical procedures include the formation of a new phase using various physico-chemical means of metal evaporation, thermolysis of precursors or the action of high-energy radiation on the *metal–(monomer)polymer* system.

Polymer materials with incorporated nanoparticles can be manufactured either by wet or dry methods. In the former case it is supposed that at least one of the components and at least during one stage of the multistage process is used in the form of a solution or dispersion in a solvent. Dry methods of manufacturing metal composites are used where it is possible. Poor solubility of the components (or chemical instability of some of them in definite solvents) and difficulties in solvent removal (including ecological reasons) limit the use of dry methods.

As has been noted in Chap. 2 the protecting polymer is brought into interaction with nanoparticles by two different methods: physical (processes induced by Van der Waalse forces, dipolar interactions or weak easily broken hydrogen bonds) or chemical adsorption. Noncovalent interactions between nanoparticles and macromolecules are very feeble (ca 10^{-4} J·m^{-2}) and their effectiveness is conditioned in case of chemisorption by the number of polar groups of the adsorbed polymer per unit surface independently of whether the macromolecules are unfolded or globular. So far it is critical for the polymer not only to contain certain functional groups but also to make them interact intensely with the surface atoms of nanoparticles when acting as, e.g. electron donors [1]. Polymer efficiency enhances significantly in case it includes centers of specific interactions adhering to charge and steric stabilization mechanisms simultaneously.

To gain further insight we will consider the main principles of producing metal–polymers in which nanoparticles are incorporated into polymers realizing that the processes are undoubtedly accompanied by chemical binding.

4.1 Mechanochemical Dispersion of Precursors Jointly with Polymers

Interest in chemical solid-phase processes can be explained in part by toughened requirements on the development of ecologically safe technologies by omitting liquid phase stages that employ toxic solvents or annealing furnaces. Mechanochemical synthesis runs at low temperatures, which clear the way to manufacturing materials in nanocrystalline and amorphous states. Mechanochemistry studies the effect of elastic and plastic strains on reactivity of solids [2]. Mechanochemical deformation and fracture of solids go together with mechano-emission phenomena that bring about oscillating and electron-excited bonds, metastable atomic structures, emission of electrons, static electrization, luminescence, and so on. The effect of stresses on the chemical reaction rate constant k in conditions of elastically strained bonds is estimated by the Eyring–Kozman equation

$$k = k_0 \exp[-(E_\mathrm{a} - \sigma \mathrm{V}^*)]/RT \, , \qquad (4.1)$$

where k_0 and E_a are the preexponential factor and activation energy of thermal decay, σV^* is the work of elastic stresses in the reaction step (σ is the stress, and V^* is the activation volume). The potential energy of the reactive center is expressed through the sum of the chemical interaction energy (U_chem) and strain energy (U_str). The energy barrier height of the reaction is equal to the system energy difference in the transient and original states. The reaction barrier of the solid phase is

$$E = (U^*_\mathrm{chem} + U^*_\mathrm{str}) - (U^0_\mathrm{chem} + U^0_\mathrm{str}) \, . \qquad (4.2)$$

The example is given further of the direct solid-phase synthesis of metal nanoparticles during introduction into the polymer matrix.

Coarse dispersed particles of original metals (>100 μm) or their salts are deformed together with the polymer matrix in different ways. High-speed planetary mills, disc integrators, various vibrators, including ultrasonic vibrators can be used to disperse magnetic particles (e.g. γ-Fe_2O_3) in a polymer a magnetodynamic disperser.

Typical examples [3] combine fragmentation (dispersion) of solid halogenides $TiCl_3$, VCl_3, etc. in PE and PP matrices (S_sp = 10–13 m^2/g) as well as PP and PE modified with L groups through grafting polymerization of 4-VPy or MMA (PE–L) [4]. The presence of PE initially affects the crystallite size slightly. Presumably, $TiCl_3$ agglomerate is encapsulated by PE, which does not however penetrate between crystallites and cannot hamper their reverse aggregation. Excess energy accumulated during mechanical operation in the reactive mixture can be carried by the residual elastic stresses, interfaces and intergrain boundaries, amorphous regions or those of oversaturated solid solutes and other metastable structures. Quantitative description of solid-phase reactions of the *polymer–inorganic precursor* kind is complicated by continuous variation of dispersity, phase composition and flaw structure during mechanical operation. The observed dispersion limit of MCl_3 can be attributed not to indivisibility of monocrystals but to the dynamic equilibrium established between disordering and spontaneous relaxation just like in other processes [5, 6]. This is what governs the dispersion limit, i.e. the least crystal block size preserved invariable irrespectively of the mechanical treatment time (45–20 nm) (Table 4.1).

Table 4.1. Variations of crystallite size during milling [3]

	CSA in Directions of 300 and 003, nm Milling Time, h					
Sample	0	0.3	1	3	5	10
$TiCl_3$/PE	47	14	13	14.5	12.5	13
	25	18	17	19	22	20
$TiCl_3$/PE-gr-P4VPy	47	32	20	10	~5	–
	25	10	8	~5	–	–
$TiCl_3$/PP	47	–	25	–	–	–
	25	–	5	–	–	–
$TiCl_3$/PP-gr-PMMA	47	–	25	–	–	–
	25	–	18	–	–	–
VCl_3 /PP	300	–	45	–	–	–
VCl_3/PP-gr-PAN	300	–	72	–	–	–

The combined milling of MCl_3 and polymers having functional groups follows another mechanism where particle size (coherent scattering area, CSA) depends upon milling time and L/M relation. This is connected not only with physical processes but with chemical interactions of MCl_3 with polymer functional groups as well and results in corresponding complexes formed simultaneously with size reduction of salt crystallites

$$\text{PE–L} + (\text{MCl}_3)_n \rightarrow [(\text{MCl}_3)_n \leftrightarrow (\text{MCl}_3)_m] + \text{MCl}_3\cdot\text{L–PE} \ (m < n).$$

More perceptible changes of chemical and physical character take place during micromechanical transformations in the substance being milled when part of the energy is spent on reorganizing the crystalline structure in MX_n. For example, during VCl_3 division in a ball mill a new, earlier unknown structure δ-VCl_3 is formed [7].

The formation of nanocomplexes has been observed in [8] in the course of mechanochemical milling of $AlCl_3 \cdot 6H_2O$ with styrene and divinylbenzene copolymers modified with aminodiacetate groups at a joint dispersion of salts with PVA and others.

Much energy is consumed in flaw formation in such systems. The quantitative characteristic of mechanochemical transformations here is an energy release G that depends upon the system susceptibility to mechanical effects and does not as a rule surpass 10^{-4}–10^{-1} mol MJ^{-1} [9].

Ultrasonic dispersion of metal suspensions with application of stabilizing effects of water-soluble polymers is one of the modern methods of mechanochemical milling. Surfactants are used to reduce adsorptive strength of the milled substances (Rehbinder's effect) and to stabilize the thus formed suspensions. It seems promising to use metal monomers as surfactants, in particular sodium nitrate, capable of polymerization under high shear stresses and contact loads in metal dispersion in the polymer–monomer solutions [10]. The influence of mechanical pulse methods on solid-phase mixtures [11] is highly efficient. These can be due to, e.g., Bridgman's anvil type effect on formation of cobalt nanoparticles in the presence of a polymer matrix [12]. Methods of synthesizing colloidal aurum are based on mechanical disintegration of aurum by, e.g. an electric arc in a liquid medium.

Chemical composition of solid surfaces differ significantly from volume phases. A newly formed surface is immediately involved in the interaction with the environment, gaseous or liquid medium, as a result of which impurities from the solid bulk are accumulated on the surface. Therefore, the surface always contains originally adsorbed substances, which are often difficult to remove.

Although mechanochemical means of forming metal–polymer composites have not yet gained popularity, they are the techniques of the near future.

4.2 Microencapsulation of Nanoparticles into Polymers

The term microencapsulation means having particles of some substance in protecting shells made of film-forming polymer materials [13] as well as the very principle of creating protecting systems and methods of targeted delivery of substances. The substance being encapsulated and forming the nucleus of microcapsules can be metal-complexes and metal nanoparticles. The polymer shell also functions as a stabilizing agent by separating nanoparticles from one another and from the environment. Various natural (proteins, polysaccharides) and synthetic polymers of either polymerization or polycondensation types (polymers or monomers to be polymerized) can shape the shell. Dimensions of the encapsulated substance can be tens of nanometers and those of the stabilizing shell can be from a few to several hundreds of micrometers.

Microencapsulation is known to involve various procedures of coacervation, physical adsorption, deposition by nonsolvent or evaporation. To generate a developed surface, the solvent is removed in vacuum by sublimation under low temperatures as the polymer remains in a glassy state. Methods of extrusion (squeezing of the encapsulated matter through a film-forming material during which the particles are encapsulated into a shell), spraying in a fluidized bed, vapor condensation, polymerization on the particle surface, etc. are widely used. Mixing of stabilized nanoparticles by polymer melts also belongs to this way of producing nanocomposites. The procedure is usually reduced to the introduction of well-dispersed ceramics or nanoparticles into polymer solutes (infrequently in melts). For instance, a solution-cast film has been prepared [14] by mixing silver nanoparticles stabilized by surfactants and PEHD in *p*-xylene at 403 K through solvent evaporation and heating to 453 K for 15 h. By further corotation melt-processing in a twin-screw mini-extruder a uniaxially oriented PE/Ag nanocomposite has been produced containing 2 mass % of nanoparticles with mean size 4.5 nm.

Configuration of microcapsules commonly coincides with that of the substance encapsulated whereas the shell thickness depends on encapsulation conditions and designation of the material obtained. The composition can be reprocessed directly into articles or be used for better matching nanoparticles with other polymers through the polymer shell.

Microencapsulation methods based on production of metal–polymer nanocomposites simultaneously with the polymer matrix are very interesting for specific applications. Already in the early stages of their evolution [15–17] it was reported that polymerization of vinyl monomers might be initiated in the course of intense mechanical dispersion (peculiar mechanochemical syntheses) of a series of inorganic substances, including metals Fe, Al, Mg, Cr and W. The transformation degree of the monomer (styrene, vinyl acetate, acrylonitrile, MMA) is a function of dispersion intensity. Freshly formed metal surfaces serve as polymerization catalysts or initiators. During the reaction electrons of the metal surface atoms are transferred onto the monomer thus giving rise to particles of the ion-radical type. Note that colloids of certain metals (Au, Tl and Pt) exert a strong effect on styrene polymerization in a block or solution [17].

To enhance stability nanoparticles can be covered with a latex film like some microencapsulating matter. For instance, hydrosols Fe_2O_3 with 3–5 nm particle size are adsorbed on latex grains resulted from terpolymerization of styrene, butadiene and acrylic acid and displaying negative charge on its surface (–COO^- – polymer groups, –SO_3 – emulsifier groups). Further comonomers are introduced into the system and are then polymerized on the encapsulated surface [18, 19]. As a result, a triple nanocomposite is obtained having TiO_2 or Fe_2O_3 as the middle layer (Fig. 4.1). The Fe_2O_3 nanoparticles (3–5 nm) were prepared by emulsion polymerization using PS–latex [20] to obtain polymer/Fe_2O_3/polymer heterostructured microspheres.

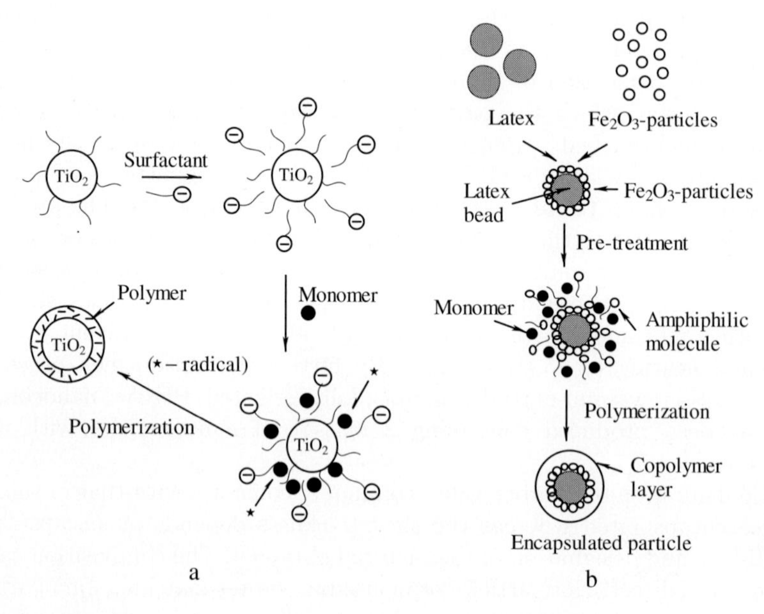

Fig. 4.1. Emulsion polymerization diagram on nanoparticle surfaces (**a**) and their encapsulation (**b**)

The size of nanoparticles during such polymerization increments, in contrast to their polydispersity index (PDI), remains invariable and the charge density on the particles diminishes (Table 4.2).

Table 4.2. Size,* polydispersity index and charge density

Sample	D_n, nm	D_w, nm	PDI	σ, μK/cm^2
Settling latex	52.39	56.23	1.07	7.5
Latex with fixed Fe_2O_3	62.2	68.4	1.08	45.2
Encapsulated particles	80.56	89.5	1.11	3.44

*D_n, D_w – number-average size and mass-average size of particles, PDI = D_w/D_n

Shells of nanoparticles, e.g. from TiO_2, are formed by treating with surfactants during which the polymerization initiator is generated and the needed monomer gets polymerized (Fig. 4.1a). The stabilization procedure also fits Pt and Pd nanoparticles that are formed in reduction of their salts [21]. As an alternative to this, $Pd(AcAc)_2$ can be dissolving in MMA followed by polymerization. However, the palladium content in the metal-composite is below 0.5%, which is limited by solubility of Pd salt in MMA [22].

The production procedure of nanocomposites by metal–monomer polymerization means polymerization-induced transformations in lyotropic (amphiphilic) liquid crystals. The principle of the approach is presented in [23–25]. A photoinitiator is introduced into the composition with Na, Co, Ni, Cd, Eu and Ce compounds in a water mixture with xylol solution of 2-hydroxy-2-methyl propiophenone. Photopolymerization is exercised leading to formation of a nanostructured polymer network. The nanocomposite is produced during heating of the photopolymerized metal–polymer blend with poly(n-phenylenevinylene) to 493 K in vacuum. Methods of acrylate, methacrylate, and metal sorbate polymerization, followed by their transformation into nanoparticles of respective metals, are described in [26, 27].

There exist diversified approaches to microencapsulation of nanoparticles in polymers. It is possible to use acetylcetonates or α- and β-hydroxy acids substituted afterwards by other ligands, including polymerizable ones (MMA, pyrrol, thyophene, etc.) as agents protecting against aggregation of nanoparticles. Nanocomposites displaying ferromagnetic properties are produced in turn by deriving γ-Fe_2O_3 and polyaniline and its copolymers in situ [28–30].

During this process the matrix is formed through polymerization or polycondensation. Epoxide or formaldehyde resins are most commonly employed as the organic binder. The composites (e.g. conducting ones) are set under the corresponding resin polymerization regimes. Polymer compositions are prepared [31] by introduction of dispersed metal powders (Al, Fe, Cr, Ta, W) into the oligomer, e.g. trioxiethylene dimethacrylate, and are set by exposing to a flow of accelerated electrons. Since these processes are comparatively time-consuming, there is a problem of ensuring sedimentation stability of the systems being cured, e.g. the epoxy–thiokol system with colloidal Pb when it forms under thermal treatment [32]. It is necessary also to deal with internal stresses on setting [33].

Many modifications of microencapsulating methods are known in the art. One typical example is described in Chap. 12. Efficient α-olefin polymerization catalysts [34] are obtained by the radical homo- or copolymerization of the corresponding monomers in solution or suspension of catalyst components MX_n ($TiCl_4$, VCl_4, $ZrCl_4$, $Ti(OR)_4$, $TiCl_3$, VCl_3, $VO(OR)_3$, etc.)

$$\text{Monomer (monomers)} + MX_n + \text{solvent} \xrightarrow{\text{initiation}}$$

$$\longrightarrow MX_n\text{(in polymer solution)} \xrightarrow{\text{reduction}} M_0 \text{ (in polymer)} \longrightarrow$$

$$\xrightarrow{\text{solvent removal}} M_n \text{ (in polymer)}$$

The thus forming macromolecules (e.g. PS) encapsulate MX_n into a microcapsule. In principle, MX_n can be a polymerization initiator too since a microporous (microreticular) shell structure incorporating MX_n is formed upon solvent removal.

There is wide use of inorganic pigments encapsulating iron oxides, those of aluminum, silicone in PVA matrix, titanium dioxide in PAA and its salts, isolating and waterproof shells for lithium $LiAlH_4$ from polyacrylates. Iron and cobalt nanoparticles are encapsulated in PMMA aimed at magnetic tapes. By irradiating dispersed AgBr particles in corresponding monomers photoemulsions are obtained with argentums bromide microcapsules. Photoconducting nanocomposites are formed similarly using CdS, microcapsules with aluminum particles and air in one polymer shell is added to detonating compositions. There are many other applications of microencapsulated particles. These particles are successfully used in chemical products of various purposes, among which are dyes for carbon-free paper, toners, fire-retardants, including liquid volatile ones, for polymer compositions and anaerobic sealants. In addition, they are used in foodstuffs, bioactive additions to fodder, pharmaceutical means of programmed and prolonged action, transdermal carriage systems, microencapsulated pesticides, and cosmetic components [35]. There are no evident constraints on microencapsulation of heterometallic nanoparticles (alloys).

4.3 Physical Deposition of Metal Particles on Polymers

The paramount sphere of using chemical deposition is the application of solid coatings on polymers, including usage of chemical metallization. Metallized plastics are used in the production of various instruments and photomasks. The conducting material is here most often copper that possesses high conductivity. Composite materials of this kind are used for creation of sensors and biosensors, and chemical monitoring systems [36]. Deposition of colloidal metals on polymer surfaces presents an actual problem of obtaining functional nanophase materials. Colloids deposited on polymer matrices are used in catalyst design, components of selective membranes, anticorrosive coatings, production of biological active medications, and compositions [37–40]. Special attention is paid to polymer films with incorporated metal nanoparticles [41]. The study of polymer adsorption from their solutions on nanoparticle surfaces is very important. The adsorption layer can be treated as a polymer shell of microcapsules as well.

We have reported in Chap. 3 that deposition (aggregation) of colloidal particles is called coagulation, during which densely packed formations appear. The coagulation of nanoparticles on polymer surfaces is sometimes termed *heteroaddagulation*. The simplest way of achieving polymer-immobilized nanoparticles could be heteroaddagulation on polymer powders, films or fibers [42]. Mostly inorganic materials along with heteroaddagulation [43] of unstable hydro- (8–12 nm) and propionolo- (4–1 nm) sols of Fe, Zn and Cu on lavsan and kapron fiber surfaces were investigated. The heteroaddagulation rate on the former surface is higher because of specific interactions and formation of M–O bonds, while the least heteroaddagulation rate is observed in iron sols. The heteroaddagulation process has a topochemical character since the reaction runs at the *product–reaction medium* interface. An inhomogeneous material is formed where nanoparticles are localized on polymer surfaces. In this respect the method is in affinity to chemical metal deposition on polymer from aqua solutions [44].

Deposition of layers is an intricate multistage process consisting of diffusion of the original either gaseous or liquid reagents, their interaction yielding intermediate products, adsorption, desorption, transformation of the intermediate into final products on the polymer surface, creation of a solid layer (film), and formation of complete structures. The actual process is much more complicated [44]. The film begins to nucleate from formation of isolated colloidal particles on separate surface regions giving birth to adsorbed clusters. Further the polymer gets filled by the new forming particles and films, and formations grow into agglomerates. Parallel to augmenting the number of particles on the polymer their inner structure undergoes changes. At the initial stages the prevailing amorphous particles are transformed into crystalline particles. Sometimes crystallization processes are long-term; moreover rigid matrices, including solid polymer layers of PVA, PEO, gelatin or quartz, are able to prolong the life of the intermediates. For instance, photo-generated ~1 nm diameter clusters of Ag_n in PVA stay stable for more than 2 days [45] (Chap. 5). Presumably, the surface functions here as a stabilizer of the growing particles. In the case with silica glass and photocarbon membranes the nanoparticle size is limited by that of the pores and channels. The films consist of spherical particles forming regular chain structures and their size might reach 50 nm (Fig. 4.2).

The processes of metal–polymer nanocomposite formation are usually connected with generation of either nanoparticles or polynuclear complexes. Another way of producing nanoparticles in polymers is metal or its oxide deposition in situ on the polymer matrix involving polymer or its precursor mixing with a soluble inorganic reagent. For example, metal alkoxides can be mixed with polymers or monomers and subjected to hydrolysis to yield nanoparticles of metal oxides in polymers. Nanoparticles and polymers can be

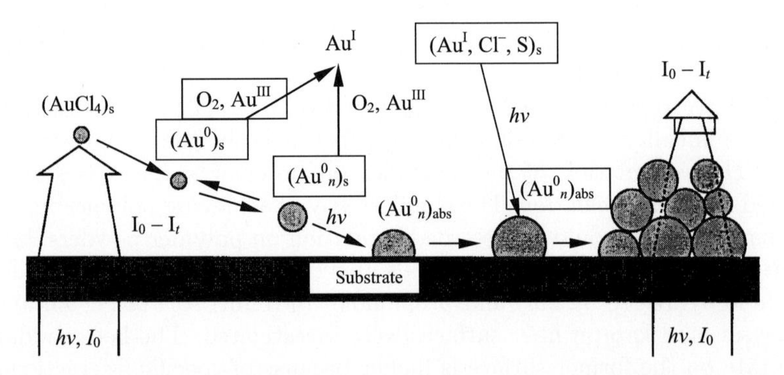

Fig. 4.2. Diagram of photochemical formation (using Au as an example) of colloidal metal film on quartz surface

formed simultaneously or in sequence. In a number of cases the polymer participates actively in deposition in which its phase state and diffusion effects govern the process. This is most peculiar to polynuclear hydrocomplexes of various metals, e.g. Al, Cr, Ti, Zr some of which are analyzed in Chap. 7. The processes of producing named nanocomposites are accompanied by the formation of hydrogen bonds between the polymer and polynuclear complex. Alkali hydrolysis of Al^{3+} salts in the presence of poly-1-vinylimidazole (PVIA) is accompanied by generation of a set of aluminum complexes of different nuclearity. Introduction of PVIA into the system shifts the reaction direction almost fully to the side of formation of $Al_{13}O_4(OH)_{24}^{7+}$ [46] hydrocomplexes. The latter form, together with PVA, a cooperative system of hydrogen bonds with hydrolyzed aluminum particles.

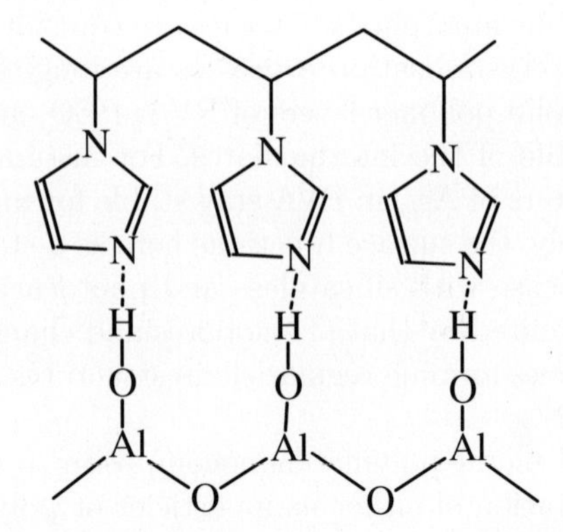

When the neutralization degree of Al^{3+} is rather high, polynuclear complexes Al_{13} start to transform under the action of PVIA into the polymer forms of aluminum hydroxide and crystallize into pseudoboehmite accompanied by breakage of hydrogen bonds with the polymer.

Thin films of semiconducting nanoparticles among which are those applied on polymers attract interest too. Use of colloidal solutions in practice is hampered by the necessity of isolating nanoparticles from the solution after, e.g. the catalytic process. So far, thin films of nanoparticle layers adsorbed on a substrate (carrier) can be obtained by precipitation of presynthesized nanoparticles from colloidal solutions on substrates through direct deposition of the forming nanoparticles on the substrates or oxidative hydrolysis of corresponding metal salts on them as demonstrated earlier. These approaches are especially convenient for application of semiconducting nanoparticles as films, including SnO_2, ZnO, TiO_2, WO_3, Fe_2O_3, etc. [47].

4.4 Formation of 2D Nanostructures on Polymers

Structure polymer surface layers are critical for improving adhesive and frictional properties of polymers, imparting biocompatibility, making two-dimensional metal–polymer nanocomposites by incorporating nanoparticle ensembles into polymer surface layers, so it would be helpful to give a special account. Processes on polymer surfaces on contact with nanoparticles is of fundamental importance for understanding the nature of interactions and adhesion on the interfaces [48].

Diffusion[1] of copper deposited into a polyamide film is known to be preceded by copper clusterization. The metal evaporation velocity on the polymer is significantly lower than in the case of a monolayer. This value constitutes 0.03 nm/min for Cu [50] which is attained by evaporation and thermal treatment in ultra-high vacuum ($\sim 8 \cdot 10^{-8}$ Pa). The post-deposition annealing near and above T_g of the film under high enough temperature (300–620 K) leads to metal immersion into the upper layers of the polymer. The process can be traced using the photoelectron spectroscopy method (ESCA) proceeding from the normalized intensity peak ratio of $Cu2p_{3/2}$ to $C1_s$ (Fig. 4.3).

Above T_g temperature of the polyamide film (693 K) copper almost fully dips into the subsurface layer of the polymer.

Many experiments showed that polymer surfaces at the interface with air are characterized by increased segmental mobility. This means that the T_g of the polymer subsurface layers (about 1 nm thick) is lower as compared to corresponding bulk values of T_g (378 K for PS with $M_w = 275\,000$).

The concept served as the basis for a novel substantiated experimental approach to the solution of two problems. One of them is the creation of 2D nanocomposites through incorporating multilayered nanoparticle ensembles into the surface layers of glassy polymers [51, 52]; the other is experimental evaluation of T_g. The annealing temperature under which Au nanoparticles of 18–20 nm in size begin to dip into the polymer is taken as T_g. According

[1] Diffusion of metals into polymers is conditioned by many factors. Its mechanism is discussed in detail in [49].

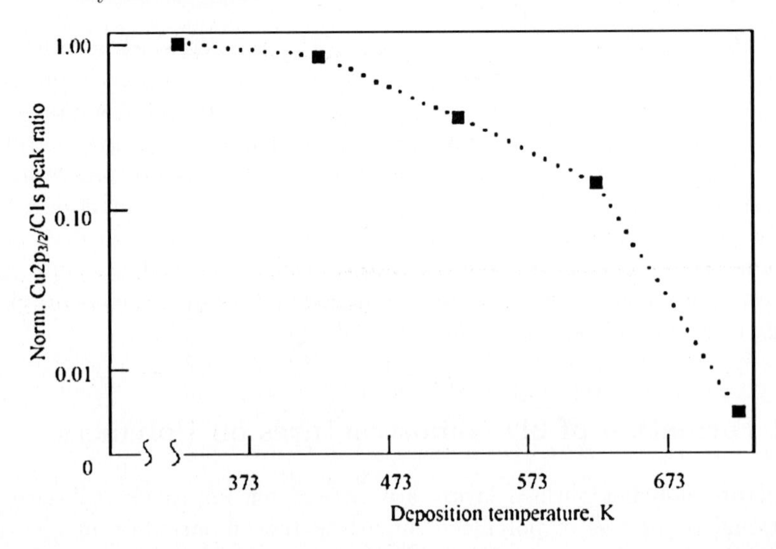

Fig. 4.3. $Cu2p_{3/2}/C1_s$ peak intensity ratio in ESCA as a function of deposition temperature after deposition of 1.6 nm Cu (nominal coverage rate 3s, deposition 0.16 nm/min) onto polyimide (T_g is about 693 K)

to AFM, the results of probing the PS surface with Au nanoparticles have proved noticeable immersion (~4 nm) to begin already after 8 h of polymer annealing at 313 K, and after 8 hour treatment at 338 K (40° lower than the bulk T_g) the nanoparticles get immersed to 12–14 nm; and at 353 K to 16 nm. Nanoparticles do not immerse fully into PS even after prolonged endurance (7 days) of the samples at room temperature. But they can be easily removed with a cantilever edge from the film surface (Fig. 4.4).

It is evident that T_g of PS is lower than its bulk values, being just 313 K.

Au nanoparticles aggregated on a PS surface did not immerse into the molten polymer matrix even after long-term annealing at above its T_g [53].

These results were used to solve the second problem of developing 2D metal–polymer nanocomposites by immersing nanoparticles into the surface layer of a glassy polymer.

In Chap. 6 nanocomposites of this type produced by deposition of thermally evaporated metal on the surface of a polymer or its melt will be analyzed in detail. The nanoparticles can be partially and even fully immersed into the polymer depending on the wettability of nanoparticles with the polymer. The first situation is realized when the wetting angle of nanoparticles with the polymer is $\Theta > 0$ which is true in case the free surface energy of metal nanoparticles exceeds that of the *nanoparticle–polymer* interfacial surface ($\gamma_m > \gamma_{mp}$). In case $\gamma_m > \gamma_p + \gamma_{mp}$ (γ_p is the free surface energy of the polymer), when the adhesion of the *polymer–nanoparticle* system surpasses the polymer cohesion, then full immersion of metal nanoparticles takes place. Two main forces affect the dipping particle in opposite directions [54]: the Van

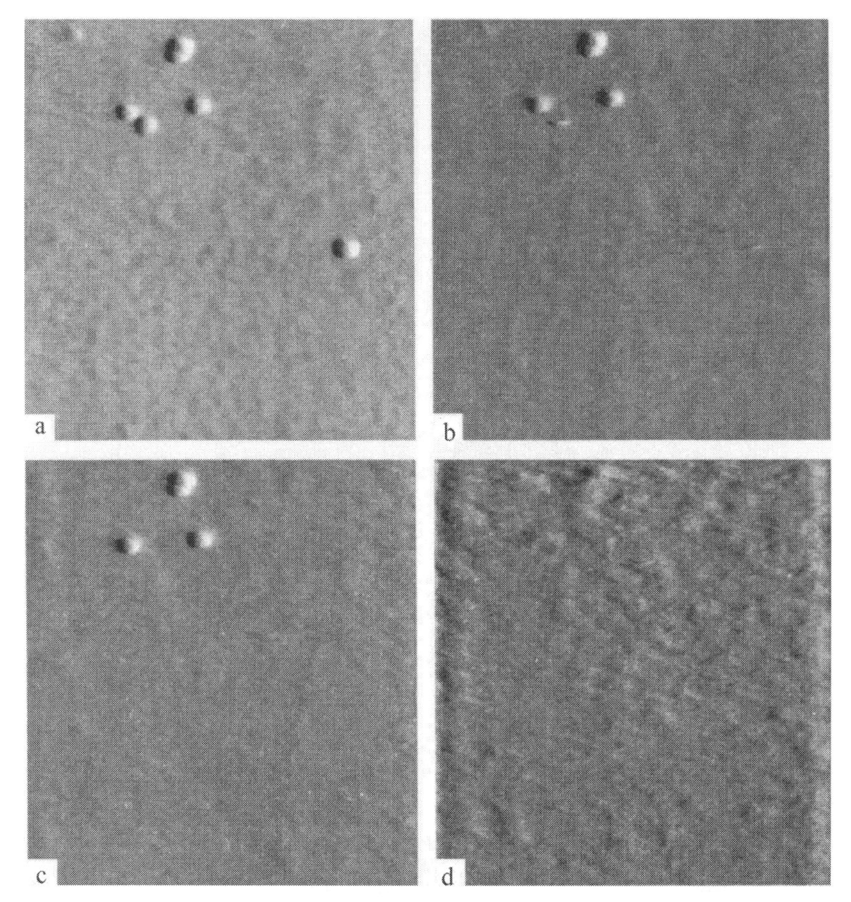

Fig. 4.4. AFM image of one and the same area of PS surface (1 μm^2) with deposited Au nanoparticles after four sequential scanning cycles

der Waalse force carries the particle into the polymer bulk and the entropic force induced by the polymer chain compression near the particle pushes it out onto the surface.

Calculations prove [52] that stabilization of a <50 nm radius monolayer of nanoparticles in the surface layer of the polymer melt is improbable since Brownian motion mixes nanoparticles sufficiently quickly within the polymer volume. Functionalized polymers present additional difficulties for designing 2D nanocomposites because their functional groups elevate the wettability of nanoparticles with the polymer and enable regulation of immersion into the polymer surface layer.

Formation of 2D nanocomposites will be explained further in the example of Langmuir–Blodgett films (Chap. 8).

4.5 Formation of Metal Nanoparticles in Polymer Matrix Voids (Pores)

It is possible to monitor the growth of metal nanoparticles in the polymer matrix by generating nanoparticles in polymer voids or pores. The size of nanoparticles is to be constrained by the void dimensions in which they are grown and should not exceed this critical size. In real systems this limitation is not always observed (see Sect. 5.1.5). Physical binding of nanoparticles in pores is exercised in case there are no functional groups capable of reacting with metal compounds and nanoparticle surfaces. An example is the formation of metal nanoparticles in the pores of a supercross-linked polystyrene (SCLPS) [55, 56]. This represents a stiff polymer network of unusual topology and its formal cross-linking degree is 200%. The polymer has acquired a large internal surface reaching 1000 m^2/g, narrow pore size distribution with the maximum about 2 nm, and the ability to swell in any liquid medium (Fig. 4.5).

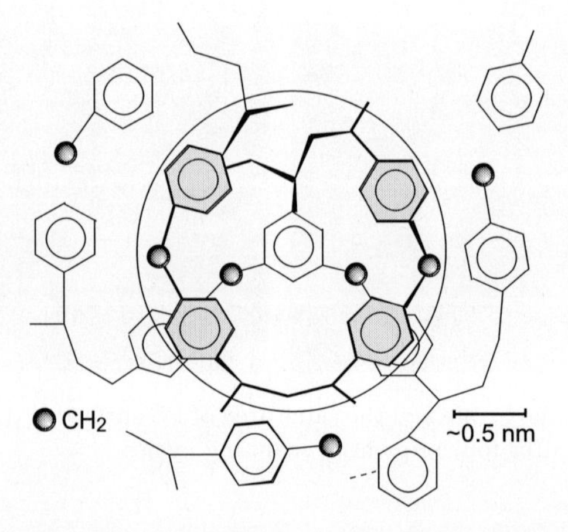

Fig. 4.5. Schematic diagram of SCLPS internal microstructure. Dark phenyl rings are on the foreground, light rings are on the background. An individual pore of the supercross-linked net is denoted by the circumference

SCLPS matrix is used for formation of Co and Pt nanoparticles [57, 58].

A cobalt compound from DMF solution ($Co_2(CO)_8$) is introduced into the SCLPS matrix or in *iso*-propanol. Cobalt nanoparticles are formed during thermolysis at 473 K. According to TEM the particle mean diameter is 2 ± 1 nm for samples containing 5 mass % of Co and 2.6 ± 1.5 nm for 10 wt.%.

The reason for the controlled growth of Co nanoparticles within nanosize voids of SCLPS matrix can be unspecific interaction between phenyl rings and metal surfaces leading to steric stabilization of nanoparticles. Coarse nanoparticles start to grow when the Co content reaches some saturation limit (above 8 wt.% concentration). Their size may be a reflection of the statistics of SCLPS pore distribution by size or recombination of SCLPS internal structure induced by the presence of metal which is because cobalt clusters can easily travel between SCLPS pores at 473 K and form cobalt nanoparticles.

Pt nanoparticles are formed in such pores through another mechanism [57]. They are produced by sorption of H_2PtCl_6 from THF solutions or in methanol followed by reduction with gaseous hydrogen under moderate temperatures, i.e. in conditions that limit platinum migration inside SCLPS. The mean size of Pt nanoparticles is about 1.3 ± 0.3 in contrast to those containing Co (ca. 2 nm). This difference can be explained by the formation mechanisms of Co and Pt nanoparticles. In the case with Pt the reduction occurs at room temperature and atoms or clusters and molecules of the precursor do not migrate, therefore nanoparticles coincide fully in size with those of the precursor filling the pores. When Co particles are formed, SCLPS matrix governs the size of metal nanoparticles by limiting their growth with the pore size (2 nm). As for Pt nanoparticles, the pores restrict the amount of the precursor filling an individual pore and the precursor volume diminishes (increase in particle density) in the course of reduction to a nanoparticle volume. This approach is a path to structurally organized synthesis of calibrated nanoparticles.

4.6 Physical Modification and Filling of Polymers with Metal

Polymers are often used just as stabilizers so in ingredient concentrations. In the latter case they speak of metal-polymer nanocomposites.

It allows one to form composites according to which the highest modifier concentrations (metal nanoparticles, oxides, salts, stabilized surfactants) are usually about 1–2 wt.%. It improves the physico-mechanical properties of nanocomposites due to interactions with the polymer matrix (adsorption, solubilization, encapsulation). Elastomer properties (natural, polyisoprene or polydiene rubbers) are regulated [59, 60] by introduction of $CaSO_4$ nanoparticles in the form of a stable sol in organic solvent into a rubber solution. Nanoparticles are synthesized in situ by mixing balanced micellar solutions of $CaCl_2$ and Na_2SO_4 with AOT micelles. They represent a polynuclear colloidal complex consisting of calcium salts AOT, – ($Ca\text{-}SO_4$)-segments and bound water 3–4 nm in size. The concluding stage is isolation of the modified rubber through water degassing and drying till a constant mass remains. Introduction of such colloidal clusters into the elastomer matrix assists in accelerating orientation and crystallization processes through adding mobility to macromolecules. The share of chains in the viscoelastic state drops to 5–10%; areas having strongly elevated ($\sim$4 times) electron density are 8–10%. This approach promotes targeted regulation of the physical structure formation in elastomers leading to raised elastic and strength characteristics. Some examples of incorporating metal nanoparticles in polymers can be found in Chaps. 8–10.

Incorporation of nanoparticles in polymers allows us to obtain nanocomposites with properties different from those observed in composites synthesized by other methods. Synthesis of such materials is poorly reproduced and colloids of base metals undergo fast oxidation in some cases. Synthesis of structurally homogeneous nanoparticles and materials with optimum electric, magnetic, optical and physico-mechanical properties on their base is improbable unless a specifically prepared matrix is present.

New vistas of nanotechnologies are related to molecular layer spraying. The layers are formed in conditions of self-assembling when structural organizations from nanoscale to micrometer level are governed by the structure of molecules being sprayed and their interactions. These molecular layers can arise from pliability of molecules to aggregation just as in precipitation in solutions, so in depositing on various substrates.

References

1. A.D. Pomogailo, A.S. Rozenberg, I.E. Uflyand: *Metal nanoparticles in polymers* (Khimiya, Moscow 2000)
2. P.Y. Butyagin: Russ. Chem. Rev. **63**, 1031 (1994)
3. S.L. Saratovskikh, A.D. Pomogailo, O.N. Babkina, F.S. Diyachkovsky: Kinet. Catal. **25**, 464 (1984)
4. D.A. Kritskaya, A.D. Pomogailo, A.N. Ponomarev, F.S. Diyachkovsky: Polym. Sci. USSR A **21**, 1107 (1979); J. Polym. Sci. Polym. Symp. **68**, 23 (1980); J. Appl. Polym. Sci. **25**, 349 (1980)
5. A.R. Yavari, P.J. Desre: Phys. Rev. Lett. **65**, 2571 (1990)
6. C.C. Koch: Nanostruct. Mater. **2**, 109 (1993)
7. D. Siew Hew Sam, P. Courtine, J.C. Jannel: Macromol. Chem. Rapid Commun. **6**, 631 (1986); Europ. Polym. J. **22**, 89 (1986)
8. N. Toshima, T. Teranishi, H. Asanuma, Y. Saito: Chem. Lett. 819 (1990)
9. P.Yu. Butyagin: Sov. Sci. Rev. B, Part 1 **14**, 1 (1989)
10. P.N. Logvinenko, T.M. Dmitrieva: Colloid. J. **52**, 1067 (1990)
11. N.S. Enikolopov: Russ. Chem. Rev. **60**, 586 (1991)
12. A.I. Alexandrov, A.I. Prokofiev, V.N. Lebedev, E.B. Balagurova, N.I. Bubnov, I.Y. Metlenkova, S.P. Solodovnikov, A.N. Ozerin: Russ. Chem. Bull. 2355 (1995)
13. V.D. Solodovnikov: *Microencapsulation* (Khimiya, Moscow 1980)
14. Y. Dirix, C. Bastiaansen, W. Caseri, P. Smith: J. Mater. Sci. **34**, 3859 (1999)
15. V.A. Kargin, N.A. Plate, I.A. Litvinov, V.P. Shibaev, E.G. Lurye: Polym. Sci. USSR **3** 1091 (1961)
16. N.A. Plate, V.V. Prokopenko, V.A. Kargin: Polym. Sci. USSR **1**, 1713 (1959)
17. A.D. Stepukhovich, A.L. Bortnichuk, E.A. Rafikov: Polym. Sci. USSR **4**, 85, 182, 516, 523 (1962)
18. A.B.R. Mayer, J.E. Mark: J. Polym. Sci. Polym. Phys. B **35**, 1207 (1997)
19. H. Du, P. Zhang, F. Liu, S. Kan, D. Wang, T. Li, X. Tang: Polymer Intern. **43**, 274 (1997)

20. H. Du, Y. Cao, Y. Bai, P. Zhang, X. Qian, D. Wang, T. Li, X. Tang: J. Phys. Chem. B **102**, 2329 (1998)
21. A.B.R. Mayer, J.E. Mark: J. Polym. Sci. Polym. Chem. A **35**, 3151 (1997)
22. Y. Nakao: J. Colloid Interface Sci. **171**, 386 (1995)
23. D.H. Gray, S. Hu, E. Huang, D.L. Gin: Adv. Mater. **9**, 731 (1997)
24. R.C. Smith, W.M. Fischer, D.L. Gin: J. Am. Chem. Soc. **119**, 4092 (1997)
25. H. Deng, D.L. Gin, R.C. Smith: J. Am. Chem. Soc. **120**, 3522 (1998)
26. D.D. Papakonstantinou, J. Huang, P. Lianos: J. Mater. Sci. Lett. **17**, 1571 (1998)
27. H.-J. Gläsel, E. Hartmann, R. Böttcher, C. Klimm, B. Milsch, D. Michel, H.-C. Semmelhack: J. Mater. Sci. **34** (1999)
28. M. Wan, M. Zhou, J. Li: Synth. Met. **78**, 27 (1996)
29. M. Wan, J. Fan: J. Polym. Sci. Polym. Chem. A **36**, 2749 (1998)
30. M. Wan, J. Li: J. Polym. Sci. Polym. Chem. A **36**, 2799 (1998)
31. N.V. Lomonosova, S.L. Dobretsov, V.K. Matveev, Y.A. Chikin, V.P. Osipov: Plastic Masses **3**, 26 (1995)
32. Z.P. Ulberg, V.A. Kompaniets, Z.T. Ilyina, N.V. Yavorskaya: Colloid. J. **32**, 278 (1970)
33. V.V. Vysotsky, V.I. Roldugin: Colloid. J. **58**, 312 (1996)
34. USSR Patent, 925965 (1982)
35. M.S. Vilesova, N.I. Aizenshtadt, M.S. Bosenko, A.D. Vilesov, E.P. Zhuravskii, A.T. Klimov, V.A. Marei, V.B. Moshkovskii, V.E. Mukhin, A.S. Radilov, L.A. Rubinchik, N.N. Saprykina, R.P. Stankevich, B.I. Tkachev, Yu.I. Trulev: Mendeleev Chem. J. **45** (5/6), 125 (2001)
36. J. Janata, M. Josowicz: Anal. Chem. **70** (12), R179 (1998)
37. S. Fennouh, S. Guyon, J. Livage, C. Roux: J. Sol–Gel Sci. Techn. **19**, 647 (2000)
38. C.-K. Chan, I.-M. Chu: Polymer, **42**, 6089 (2001)
39. S.K. Young, K.A. Mouritz: J. Polym. Sci. Polym. Phys. **39**, 1282 (2001)
40. M. Nagale, B.Y. Kim, M.L. Bruening: J. Am. Chem. Soc. **122**, 11670 (2000)
41. A. Heilmann: *Polymer Films with Embeded Metal Nanoparticles. Springer Series in Materials Science. Vol. 52* (Springer 2003)
42. K. Furusawa, C. Anzai: Colloid. Polym. Sci. **265**, 882 (1987)
43. M.A. Lunina, M.G. Ivanova, A.A. Khachaturyan: Colloid. J. **57**, 825 (1995)
44. V.V. Sviridov, T.N. Vorobyeva, T.V. Gaevskaya, L.I. Stepanova: *Chemical deposition of metals from aqua solutions* (Minsk, Univ. Publishers 1987)
45. T.B. Boitsova, V.V. Gorbunova, E.I. Volkova: Russ. J. Gen. Chem. **72**, 688 (2002)
46. V.V. Annenkov, E.A. Filina, E.N. Danilovtseva, S.V. Fedorov, L.N. Belonogova, A.I. Mikhaleva: Polym. Sci. Ser. A **44**, 1819 (2002)
47. R.H. Khairutdinov: Russ. Chem. Rev. **67**, 125 (1998)
48. *Metallization of Polymers. ACS Symp. Ser. 440* ed. by E. Sacher, J.J. Pireaux, S.P. Kowalczyk (ASC, Washington D.C. 1990)
49. F. Faupel, R. Willecke, A. Thran: Mater. Sci. Engn. Reports A, Rev. **R22**, 1 (1998)
50. M. Kiene, T. Strunskus, R. Peter, F. Faupel: Adv. Mater. **10**, 1357 (1998)
51. V.M. Rudoi, I.V. Yaminsky, O.V. Dementyeva, V.A. Ogarev: Colloid. J. **61**, 861 (1999)
52. V.M. Rudoi, O.V. Dementyeva, I.V. Yaminsky, V.M. Sukhov, M.E. Kartseva, V.A. Ogarev: Colloid. J. **64**, 823 (2002)

53. M.S. Kunz, K.R. Shull, A.J. Kellock: J. Colloid. Interface Sci. **156**, 240 (1993)
54. G.J. Kovacs, P.S. Vincett: J. Colloid. Interface Sci. **90**, 335 (1982)
55. V.A. Davankov, M.P. Tsyurupa: React. Polym. **13**, 27 (1990)
56. V.A. Davankov, M.P. Tsyurupa: J. Polym. Sci. Polym. Chem. Ed. **18**, 1399 (1980)
57. S.N. Sidorov, I.V. Volkov, V.A. Davankov, M.P. Tsyurupa, P.M. Valetsky, L.M. Bronstein, R. Karlinsey, J.W. Zwanziger, V.G. Matveeva, E.M. Sulman, N.V. Lakina, E.A. Wilder, R.J. Spontak: J. Am. Chem. Soc. **123**, 10502 (2001)
58. S.N. Sidorov, L.M. Bronstein, V.A. Davankov, M.P. Tsyurupa, S.P. Solodovnikov, P.M. Valetsky, E.A. Wilder, R.J. Spontak: Chem. Mater. **11**, 3210 (1999)
59. B.S. Grishin, T.I. Pisarenko, G.I. Esenkina, I.R. Markov: Kautch. Rezina 222 (1993)
60. B.S. Grishin, T.I. Pisarenko, G.I. Esenkina, V.P. Tarasova, A.K. Khitrin, V.L. Erofeev, I.R. Markov: Polym. Sci. A, **34**, 601 (1992)

5 Chemical Methods of Metal–Polymer Nanocomposite Production

Chemical methods are the most popular methods of manufacturing nanocomposites. They are characterized by narrow nanoparticles size distribution, relative simplicity of control over synthesis, and reliable stabilization of nanoparticles in the systems. The methods are based on various reduction procedures involving polymers or copolymers, dendrimers as well as thermal decomposition of metal-containing precursors.

A general scheme of metallopolymer nanocomposite synthesis by chemical immobilization into polymers can be presented as follows (Scheme 5.1).

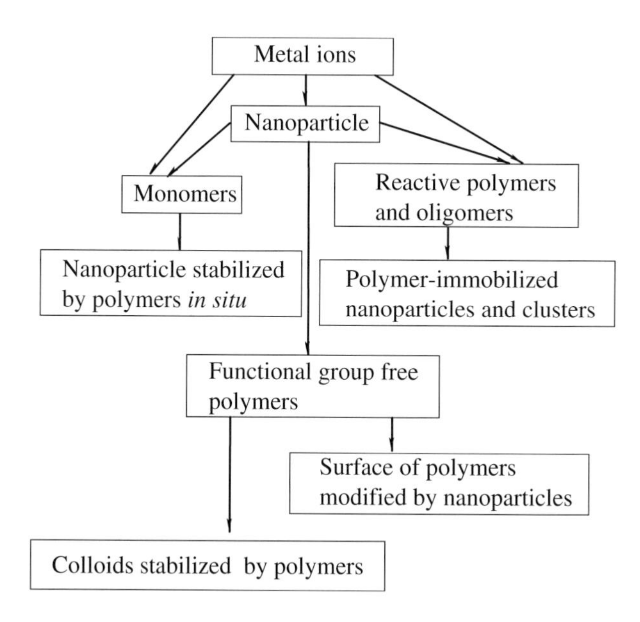

Scheme 5.1

As has been analyzed in Chap. 1, the term “embryo of solid phase” is ambiguous when referring to nanoparticles. Thus, already as many as 25 atoms of Pt in the cluster is enough to reveal the adsorption properties characteristic of bulk metal. The ion assemblies display the typical properties of metal in the reduction of gold chloride only when the aggregates accumulate as many as 300 or more gold atoms. The dimensions of some particles can be estimated by the change of color of their solution. For example, Faraday's red gold sols change from red to blue when the aggregation of particles occurs.

Sol colour	red	violet	lilac	blue
Size of Au particles, nm	23.7 ± 0.9	24.3 ± 0.7	32.5 ± 1.4	33.2 ± 2.0

Later we will address the color of nanoparticles of other metals too.

5.1 Reduction of Polymer-Bound Metal Complexes

The reduction methods are considered as the main procedures for obtaining metal–polymer nanocomposites with diversity of variants and resultant products. We shall review most typical modifications as an example.

5.1.1 Reducers and Reduction Processes

Similarly to reactions without polymers (see Sect. 2.5) the reducing agents are H_2, hydrogen-containing compounds NH_3, $NaBH_4$, NH_2NH_2, $NH_2NH_2{\cdot}H_2O$ (heterogeneous slowly reducing agent), hydrazine borane, phenylhydrazine, and photographic reducers (hydroquinone, n-phenylene diamine, pyrogallol and other). Highly effective reducers are sodium borane related to the class of strong reducers [1], superhydride $LiAlEt_3H$, triethylsilane Et_3SiH (homogeneous but slowly reducing agent), different alcohols (both monatomic and polyatomic) most often ethylene glycol (EG), diethylene glycol (DG), triethylene glycol (TG), etc., dextrins, aldehydes (formaldehyde, benzaldehyde), hydrogen peroxide,and tetraethylammonium chloride hydrate $(C_2H_5)_4NCl{\cdot}xH_2O$. The reduction is going on until metals can run just as in the polymer solution in its suspension. The chief regularities of metal ion reduction under the effect of chemical reducers Sect. 2.5 are also true for metal-polymer systems. Typical metals used to form nanoparticles by this method are found in the series of hydrogen. Metal ions having higher potential are reduced more easily. The reduction of metal ions with a positive electrochemical electrode potential by hydrogen is preferable in thermodynamic respects. Let us review the stoichiometry of some typical reactions, particularly the formation of zero-valent Pd and Au under the action of hydrazine aqua solution:

$$2H_2PdCl_4 + N_2H_5OH \rightarrow 2Pd^0 + 8HCl + N_2 + H_2O$$
$$4HAuCl_4 + 3N_2H_5OH \rightarrow 4Au^0 + 16HCl + 3N_2 + 3H_2O.$$

The reduction of Cu^{2+} ions by formaldehyde occurs only in alkali media through the reaction

$$Cu^{2+} + 2HCHO + 4OH^- \rightarrow Cu^0 + 2HCOO^- + H_2 + 2H_2O.$$

It is interesting that high-molecular reducing agents can function as metal ion reducers and stabilizers of the forming suspension of nanoparticles. The possibility of reducing chloroauric acid, PEO, PEI and PVPr [2] exists too. The kinetics of the process differ essentially from that of the traditional citrate reduction of $HAuCl_4$. Colloidal particles of Ag, Au and Pd are produced from aqua solutions of their salts using the reducers polyvinyl alcohol and polyethylene glycol functioning also as stabilizers [3–5]. As a result, low-dispersed metal nanoparticles close to spherical shape are formed. Silver reduces quickly and forms nanoparticles [6] when Ag salt is added to the chloroform solution poly(2,6-dimethyl-1,4-phenylene oxide). The most probable mechanism includes the oxidation reaction of polymer groups. This path is taken when Au particles are formed in the presence of amphiphilic derivatives of oligothiophenes [7] as a reducer-stabilizer. It results in formation of quasi-crystalline Au particles whose size is from $\sim$1.1 nm (40 atoms) to 1.9 nm ($\sim$200 atoms) [8].

Commonly the reducer is slowly added in an inert atmosphere at room or slightly higher temperature (in case of noble metals). This proceeds during boiling of alcohol, aqua, alcohol–aqua and other media where the molar ratio of the reducer versus metal ions is $\geq$1:1. The substance being reduced can be introduced directly with a polymer, e.g. $RuCl_3$ solution with PVP, in waterless acetone (in this case, the probability of performing macromolecular metal complexes in the system is higher) or a solution of the reducing agent with the polymer. Sometimes the average size of nanoparticles in polymers starts to gradually grow at moderate velocities of introduction of metal ions. In contrast, the reduction often ceases at the stage of formation of low-valent metal ion complexes with a macroligand at high introduction velocities.

Changes in formation process of nanocomposites are expressed in variations of the reacting solution color (its UV-vis-absorption spectra), e.g. in the course of Ru nanoparticle formation from $RuCl_3 \cdot nH_2O$ solution under the action of $Na_2B_4O_7$ in the presence of PVP (Scheme 5.2) [9]:

0–10 min		15 min		20 min		25 min		180 min
Dark red	→	Light yellow	→	Dark green	→	Dark brown	→	Dark brown
r.t. – 433 K		453 K		Reflux (471 K)		Reflux (471 K)		Reflux (471 K)
Starting solution								Ru colloids

Scheme 5.2

Chemical reduction is an intricate multistage process. The resultant low-valent metal forms are unstable and decompose spontaneously with metal isolation upon which intermediate particles become capable of catalyzing further reduction. There is also a probability of interaction between the metal being reduced and the reducer transformation products. Unfortunately, no attention is given to these aspects in many works. Nanocomposites are stable usually during storage in dry air. However, the reduction products of base metals get oxidized quickly during drying in air or contact with moisture. The composition of the forming composite (share of deposited substance) can be estimated by a chemical analysis or by overweight ($\Delta m = (m_b - m_p)/m_p$, where m_b is the mass of the original polymer) of the substance reduced. It might reach from fractions of percent to hundreds of a percent depending on the initial concentration of reagents, their ratio and origin of the reactive medium. Furthermore, phase composition, characteristics of the forming layer (its thickness and compactness), layer location within the polymer, and sizes of nanoparticles and crystallites are determined.

The reduction methods can be subdivided into two groups. The method of impregnation is brought to either chemical reduction of a metal from solutions or suspensions of its salts. Ammonia solutions such as Cu, Cr, Ag are often used in electrochemical or radiation-chemical reduction in the presence of macroligands. In this case in fact an unbounded amount of metal can be introduced into the polymer. Methods of the other group involve the reduction of preliminary obtained mononuclear metal complexes chemically combined with the polymer. As a rule, 1–3% of metal, sometimes 3–15%, is introduced into the polymer in one cycle of binding of metal ions – reduction. The concentration of metal ions bound by the polymer functional groups and dependent upon composition of forming complexes (sorptive capacity) is low. The degree of metal binding in multistage processes can reach tens and sometimes hundreds of percent relative to the polymer. This is because the complex-forming capacity of the majority of polymer functional groups (e.g. carboxylic in PAA) in reduction media is almost fully restored once the metal, its oxide or sulphide phase are formed. It is not excluded that a part of the polymer functional groups participate in the formation of chemical bonds with surface atoms of nanoparticles.

Both methods admit chemical interaction between the compound being reduced and the polymer matrix. For instance, a polyacetylene–MX_n macromolecule in polymer-based systems devoid of functional groups acts as a peculiar ligand, which hampers multicentered association of complexes and prevents enlargement of nanoparticles [10].

The first method gained popularity mainly thanks to the needs of catalysis, in particular, by the necessity to get niello and ultradispersed powders.

5.1.2 Reduction in Polymer Solutions

Kinetic rules of metal ion reduction in the polymer matrix are conditioned by many factors: diffusion rate of metal ions and reducer in the polymer, velocity of their motion in pores (in case of ion-exchange porous materials, suspensions, polymer extraction in adsorption active media) or in polymer balls (in the subsurface layer), the speed of the reduction process itself, and the crystallization rate of isolated metal or its oxide. The kinetic curve of ion silver reduction in a water–ethanol solution with addition of PVP [11] has an induction period and the effect of secondary self-catalytic acceleration of the reaction (Fig. 5.1). This can be attributed to the behavior of PVP as a polymer catalyzer that changes the solvate state of silver ions and transforms the reaction complex into an excited state $(PVP\cdots Ag^0\cdots Ag^+\cdots HOC_2H_5)^*$. The next stage is electron transfer to silver cation and proton to the alcohol molecule since the reduction of silver ions by ethanol does not take place in the absence of PVP. PVP can act as a polymer reagent also participating in the reduction of silver ions through its end aldehyde group. The latter is formed from depyrrolidonization of the end 2-hydroxypyrrolidone fragment of the PVP macromolecule during its water treatment [12]. Hence, the kinetic curve presented in Fig. 5.1 is a superposition of two parallel processes.

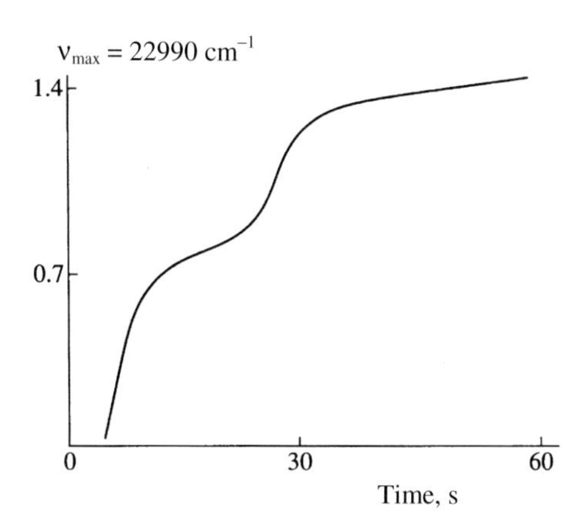

Fig. 5.1. The kinetic curve of the reduction of Ag ions in water–ethanol in the presence of PVP

The forming nanoparticle size is affected by the origin of the polymer (functional groups), molecular mass and, in the case of copolymers, their composition and even the type of their link distribution within the copolymer chain (alternative, statistical, block, grafted). The comparison of reduction products obtained in the presence of P4VPy to those of statistical styrene copolymer with 4-VPy shows that [13, 14] the Fe particle size formed in P4VPy is less in conditions of equal molar concentration of metal salts in the systems. Excess vinylpyridine fragments hamper aggregation of nanoparticles. On the contrary, elevated concentration of salts in the solution brings about more coarse nanoparticles. Although the size of nanoparticles is within 20–200 nm, these are almost always cluster particle aggregates. This tendency is common in production of nanoparticles by reduction methods [15]. The styrene monodispersed copolymer with 4-vinylpyridine containing nanoparticles of Ni and Co display ferromagnetic properties [16].

Polymetal-chelate films from aqua solutions of polyvinyl alcohol PVA and $AgNO_3$ are reduced by photographic reducing agents [17] yielding corresponding metal-composite materials. Monodispersed ultrafine gold [18, 19] and platinum powders [20–22] are produced during metal salt reduction in the presence of protecting colloids in the same way.

Let us consider $PdAc_2$ reduction in thin films of polyamidoimine from 10% solutions of THF or in N-methylpyrrolidone with application of $NaBH_4$ [23]. The mechanism of nanoparticle formation consists of three stages (Fig. 5.2). The first one includes mixing of components and uniform distribution of MX_n in the polymer solution (possibly with formation of chelate structures). During the second stage a film is formed from the solution by the pouring method and the solvent is gradually removed (10–15 h) until there is a residual content of 5-10%. This results in the formation of a metastable membrane film which is washed off initial salts with methanol or water. The third stage is a very fast reduction in which the formed nanoparticles are chemically combined with polymer chains. Different estimates give different Pd particle diameters depending on the synthesis conditions which constitute 3, 2 and 1 nm (correspondingly, ~960, ~290 and ~36 atoms) and Ag particles – 1.5–3.5 nm. Polymers of cation (most often PDAAC) or anion (PMMA) types as protective colloids [24], sometimes in combination with SAS [25], are used in synthesis of nanoparticles in solutions.

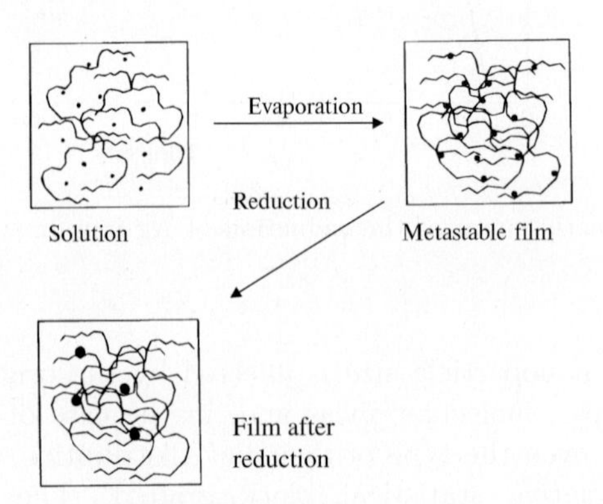

Fig. 5.2. Possible mechanism of nanoparticle formation

It seems important that there are favorable conditions for simultaneous emergence of a large number of nuclei of colloidal metal particles as a result of adsorption of reduced ions by protective polymers (see Chap. 3). The reduced metal is spent mostly on the formation of nuclei, whereas only a small portion is expended on their growth. This results in fine nanoparticles, which is in part spurred by high aggregation stability of sols.

The formation mechanism of nanoparticles in reducing systems is the subject of numerous discussions. Most likely metal condensation in a liquid phase occurs directly at the moment when its ions are reduced but not by reaching an oversaturated solution of metal atoms. Nuclei of nanoparticles are generated spontaneously mostly in places where the concentration of metal ions is the highest and their growth is induced by the reduction of ions adsorbed on the surface or from solution. This process is apparently affine to particle growth during cryochemical synthesis (see Chap. 6). It is critical for the chemical reduction rate to be limiting, in which case the size of nucleating nanoparticles remains in fact constant under these reaction conditions.

The probability of control over the shape of the forming nanoparticles must be specified separately. It can be attained by regulating the ratio of the protected polymer content against that of the forming nanoparticles. This fact has been studied in [26, 27] on Pt^{2+} reduction in 0.1 M water solutions of sodium polyacrylate by hydrogen at pH $= 7.5$. With varying γ–polymer relation: Pt $=$ 1:1, 2.5:1 and 5:1, the preferred form of nanoparticle plates changes (Fig. 5.3) from square cubic (89%) to triangular tetrahedral. Seven tetrahedral, polyhedral and irregular prismatic particles are formed when $\gamma = 2.5{:}1$. Mainly tetrahedral ($60 \pm 10\%$) are formed with slight addition of polyhedral and irregular prismatic particles at $\gamma = 5{:}1$. The reasons are still unclear but γ is seen to depend upon particle size and their distribution (Fig. 5.3c) and their mean sizes are correspondingly 11.0 ± 0.5, 8.0 ± 0.5 and 7.0 ± 0.5 nm. Polymer-stabilized nanoparticles of Rh acquired only a sponge-like shape [28], Ag are decahedral, Au are rod-like and plate like [29].

Choo et al. [30] studied how the reduction of H_2PdCl_4 ran in a single-stage procedure [31] in a boiling mixture of methanol (46% by volume) and water (54% by volume) with addition of different amounts of PVP. It was done to exclude the influence of the reducing agent on such fragile structural characteristics as the shape of the forming nanoparticles and to clarify the effect of the stabilizing polymer matrix on them. Concentration of the stabilizing polymer affects not only the size of the forming Pd nanoparticles, but also their morphological control.

The reduction mechanism of metal ions chemically combined with the polymer is comparatively complex and was studied only for formation of clusters of zero-valent Pd, Ru, Rh, Ag, Os, Ir, Pt, Au, Ni [32–35]. Salts of these metals during boiling in proton–donor solutions in the presence of polymers (particularly, $RhCl_3$ in methanol–water–PVA mixture) create the next chain of chemical transformation (Scheme 5.3).

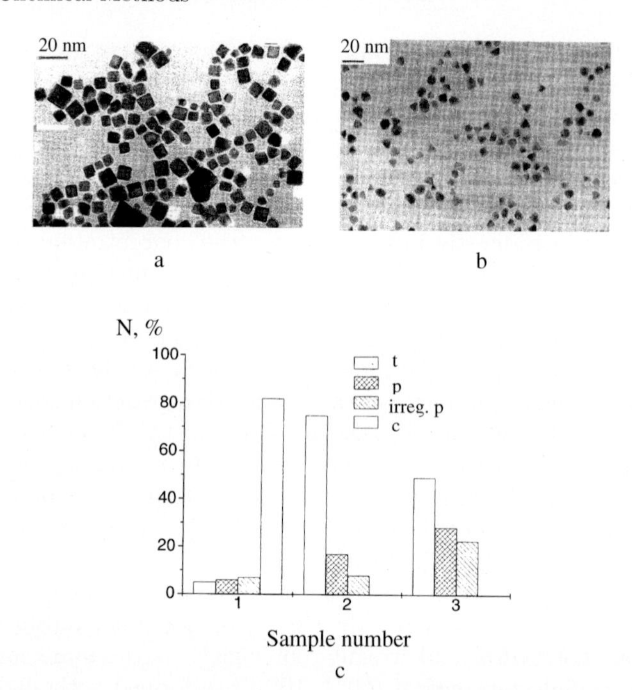

Fig. 5.3. Transmission electron micrographs of Pt nanoparticles with cubic (**a**) and tetrahedral (**b**) structures (N) and histograms of particles with different shapes (**c**)

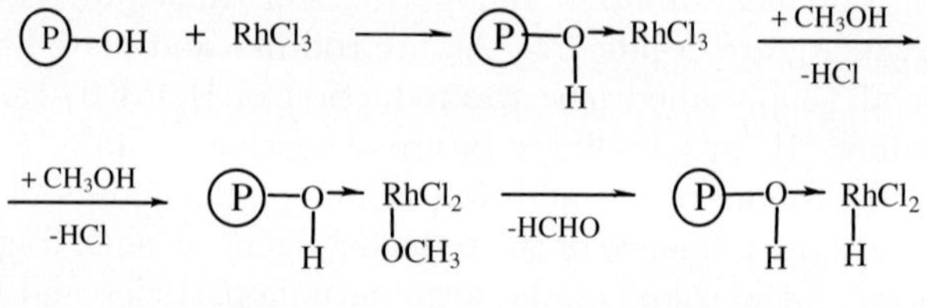

$$\text{(P)}\!-\!OH + RhCl_3 \longrightarrow \text{(P)}\!-\!\underset{\displaystyle H}{\underset{|}{O}}\!\rightarrow\! RhCl_3 \xrightarrow[-HCl]{+CH_3OH}$$

$$\xrightarrow[-HCl]{+CH_3OH} \text{(P)}\!-\!\underset{\displaystyle H}{\underset{|}{O}}\!\rightarrow\! \underset{\displaystyle OCH_3}{\underset{|}{RhCl_2}} \xrightarrow[-HCHO]{} \text{(P)}\!-\!\underset{\displaystyle H}{\underset{|}{O}}\!\rightarrow\! \underset{\displaystyle H}{\underset{|}{RhCl_2}}$$

Scheme 5.3

$RhCl_3$ is coordinated with hydroxylic PVA groups within the first stage, then an oxonium product is derived, further transformed through the alkoxide into a polymer-bound hydride complex. These reactions are critical for a homogeneous colloid arising from disproportioning rhodium hydrides followed by particle growth

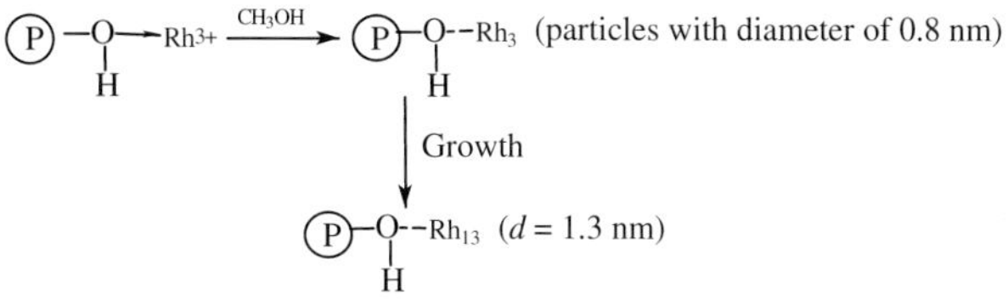

(P)—O(H)—Rh^{3+} $\xrightarrow{CH_3OH}$ (P)—O(H)--Rh_3 (particles with diameter of 0.8 nm)

↓ Growth

(P)—O(H)--Rh_{13} (d = 1.3 nm)

Scheme 5.4

The structure of the forming 13-nuclei clusters presents a face-centered cubic lattice with coordination number (NC) of rhodium atoms equal to 12 and 1.3 nm diameter (Scheme 5.4). They enter into interaction with the protective polymer via electrostatic attraction forces or physical adsorption, and may yield coordination links. It is possible to achieve metal-polymers with different cluster particle size by changing the reaction conditions, and the nature of the medium (ethanol, butanol, toluene/butanol), by introducing additional ligands (e.g. PPh_3), employing different polymers like PVP, MMA with N-vinyl-2-pyrrolidone, polyacrylamide gel, etc. [36–38]. For example, 0.9–4 nm nanoparticles for Rh^0, and 1.8 nm for Pd^0 which are stable in a protective PVP colloid, have a narrow particle size distribution range by size, show high stability and are not liberated from gel with pH variations from 2 to 12. Attempts to compare the reduction rate as a function of the metal ion being reduced turn out to be a success very rarely. Notice that, other conditions being equal, the reduction of Ru^{3+} in the presence of PVP ions $[PtCl_6]^2$ and $[PdCl_6]^{2-}$ [39,40] proceeds less readily [9]. This also relates to the formation of PVP-Ir.

Water-soluble polymers and alcohol work as a protective colloid and reducing agent, correspondingly, in these systems.

When studying the reduction of water solutions of $H_2PtCl_6{\cdot}6H_2O$ (4 MPa H_2, room temperature, 96 h), a supposition was made [41] on the participation of both the carbonyl group and nitrogen atom with an unshared pair of PVP electrons in binding $[PtCl_6]^{2-}$ ions. The Pt particle distribution and mean size depend negligibly on the nature of the solvent they were formed in and prepared. The nucleus–shell structure of Pt particles protected by PVP can be established by comparing the results of scanning tunnel microscopy (STM) and high-resolution transmission electron microscopy (TEM) [42]. The former detects the size of the enveloping polymer shell, the latter visualizes the inner metal nucleus size (3.5 and 2.6 nm, respectively) (Fig. 5.4).

A convenient method for studying the size of metal nanoparticles (including Pt stabilized in PVP matrices whose particle size is <1–7.5 nm) is spectroscopic investigation of CO adsorption on their surface [43]. The geometry of adsorbed CO binding depends upon particle size. For example, CO is chemisorbed on coarse platinum nanoparticles following only the bridge link mode ($\nu(CO) = 1944$ cm^{-1}), on minor cases it follows the terminal type ($\nu(CO) = 2037$ cm^{-1}), and on intermediate (1.8–2.3 nm) a combination of

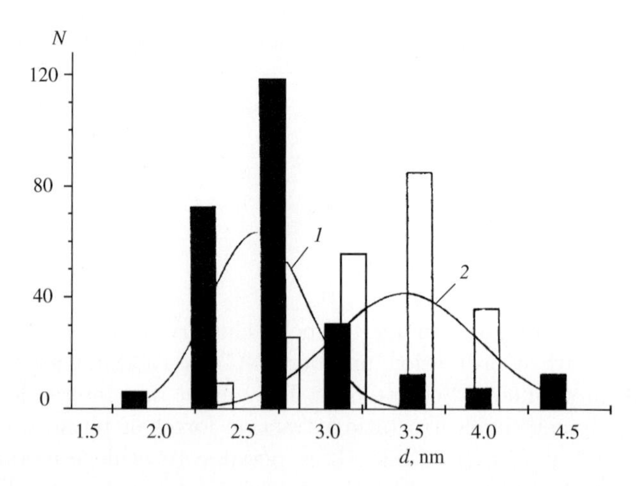

Fig. 5.4. Curves of particle distribution on the size with mean diameters of 2.6 (*1*) and 3.5 nm (*2*) and histograms of particle diameters (N is the number of particles)

both binding types. This is consistent with the NMR of ^{13}C and IR spectra of adsorbed $^{12}CO/^{13}CO$ mixtures [44]. Although this analysis can also establish the size of Ni nanoparticles [45] in nickel-boride sol (obtained by $NiCl_2$ reduction by $NaBH_4$ in ethyl alcohol in the presence of PVP), it is more appropriate for obtaining organic sols of noble metals.

One can observe [46] a gradual increment of the mean Ni nanoparticle size in PVP or PEO under moderate rates of introduction of $NiSO_4$ $(0.5–1.5)\cdot 10^{-4}$ mol/min^{-1}. The reduction runs not till a zero-valent state but ceases at the stage of Ni^+ complex formation with macroligands under high introduction rates.

The reduction products of Fe^{2+} are resistant to storage in dry air, but quickly oxidize on contact with moisture (polymers PVP, PEO, PAAM). The formed metal nanoparticles are often oxidized until corresponding oxides, e.g. Fe nanoparticles, are oxidized to Fe_3O_4 crystallites by NaOH at pH = 13–14 by intensive mixing for 2 h at 333 K followed by washing out until the pH is neutral. In the matrix of water-soluble polymer sols of Au, Fe and Cr were obtained by reduction of hydroxides of corresponding metals by hydrazine hydrate, in rubber films Pd (1–2 nm particle size) and Au (1–10 nm) were dispersed [47]. Reduction products of $H_2PtCl_6\cdot 6H_2O$ in water–alcohol media with addition of poly(N-isopropylacrylamide) possess specific properties [48]. This polymer displays unusual temperature-dependent properties, namely a reversible phase separation occurring above the lower critical solution temperature (LCST) in water (305 K). Pt nanoparticles 1–5 nm in size are formed in such systems (the mean particle size is 2.38 nm) [49].

Formation of nanoparticles in polymer solutions is similar to matrix polymerization due to its cooperative character of macromolecular interactions with the growing nanoparticle surface. This fact served as the basis for constructing a model of pseudo-matrix control over nanoparticle growth processes [50,51]. It is presumed that macromolecules are capable of capturing and stabilizing nuclei of the metal phase owing to cooperative interactions between chain segments of the type of matrix recognition by the growing chain during matrix polymerization [52]. A schematic diagram depicting complex formation between a macrochain and metal ions of the growing nanoparticle can be presented as follows (Scheme 5.5) [53]:

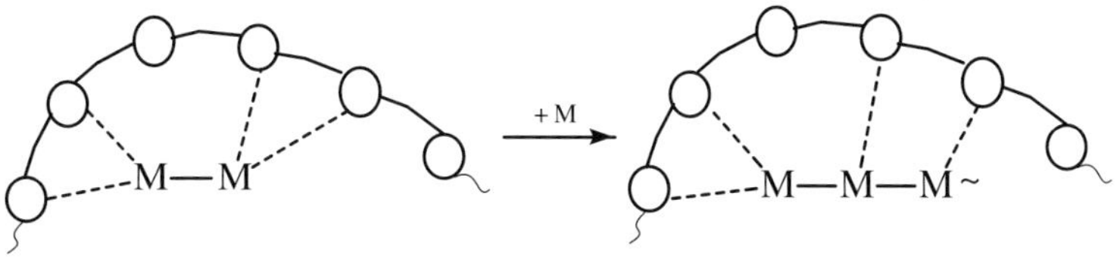

Scheme 5.5

The stability of this macrocomplex, characterized by the equilibrium constant K, rises as its size increments provided the macromolecular size is larger than the forming particle (see Chap. 3).

5.1.3 Reduction in Block-Copolymers in Solution

The reduction of metal ions in the presence of block-copolymers is very interesting. This is the way ligand-controlled synthesis of nanoparticles is realized. These systems are frequently called metal–supramolecular block copolymer micelles. It has been noted previously (Sect. 3.4) that block-copolymers, e.g. poly(styrene-4-vinylpyridine), poly(styrene-2-vinylpyridine), PS-PEO, and PS-PAA (block-ionomers) exist in the form of reverse micelles [54]. Block-copolymers segregate into microphases with spherical, cylindrical and lamellar morphology in organic solutions [55]. Metal ions combine with functional groups of micelle nuclei through generation of covalent or ion bonds. Salts of many metals insoluble in pure toluene can be dissolved in toluene micelles of similar copolymers. Their micelle structure is retained in, e.g. supramolecular micelles [56] of the diblock-copolymer PS_{20}-[Ru]-PEO_{70}. In the case of triblock-copolymers polystyrene-polybutadiene, Pt-, Pd- and Rh-containing complexes before reduction are combined through formation of π-olefine or π-allyl structures [57].

The situation with block-copolymers is analogous to synthesis of nanoparticles in situ in reverse micelles [58–61]. Almost calibrated particles of nanometer size are formed in these systems during reduction of metal salts [60] in which surface-active groups are acting as ligands, e.g. Ag[bis(2-ethylhexyl)sulphosuccinate] (AgAOT). The mixture AgAOT + NaAOT is used sometimes. The sizes of micelles are from 1.9 to 3.5 nm, that of Ag particles ranges from 2.7 to 7.5 nm depending on conditions (AgAOT-NaAOT-water-isooctane). Synthesis of nanoparticles of different metal salts can be exercised, including ZnS [62], $BaSO_4$ in mixing micelle emulsions of the reagents $BaCl_2$ and Na_2SO_4 in a toluene medium [63]. Surfactant groups can be simultaneously the reducing, stabilizing (e.g. tetra-*n*-octylammonium carboxylates $(n\text{-Oct})_4N^+(RCO_2)^-$ [64], and self-organizing agents [65]. More complex block-copolymers have been developed lately for production of nanocomposites. These block-copolymers are norbornen derivatives functionalized by carbazole groups through the living metathesis polymerization with cycle opening [66–68]. Nuclei of block-copolymer micelles may serve as peculiar nanoreactors for synthesizing nanoparticles that are simultaneously stabilized by nonpolar blocks of these amphiphilic block-copolymers (corona) [69]. This procedure for the production of metal nanoparticles in microsegregated blocks of polymer films was first reported in [70–74].

A general approach can be demonstrated on the example of formation of Ag, Au, Cu, Ni, Pb, Pd, Pt and some other clusters stabilized inside a segregated diblock block-copolymer $(\text{methyltetracyclododecene})_{400}$-$(\text{2-nonbornen-5,6-dicarboxylic acid})_{50}$ (Fig. 5.5). Ions of the reduced metal are commonly preliminary forming a macrocomplex with segments of micelle nuclei (loading can reach here high values, in optimum variants – above 1 g/g of polar comonomer in a block-copolymer, Fig. 5.6). Metal ions are reduced by traditional reducing agents, particularly H_2 at 388 K over 6 days and then nanoparticles are formed [68].

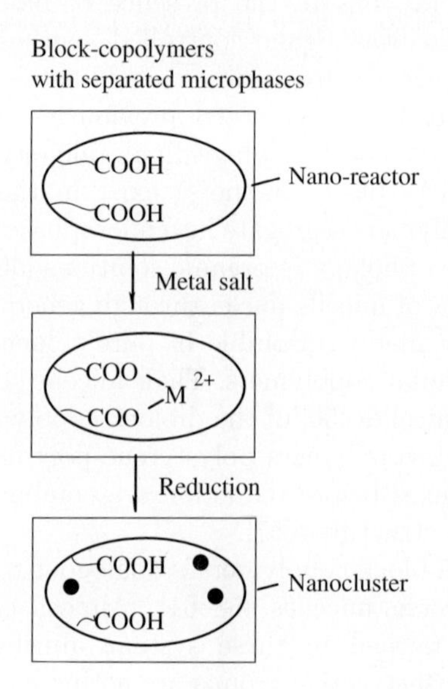

Fig. 5.5. Principal scheme of nanoparticle formation in block-copolymers

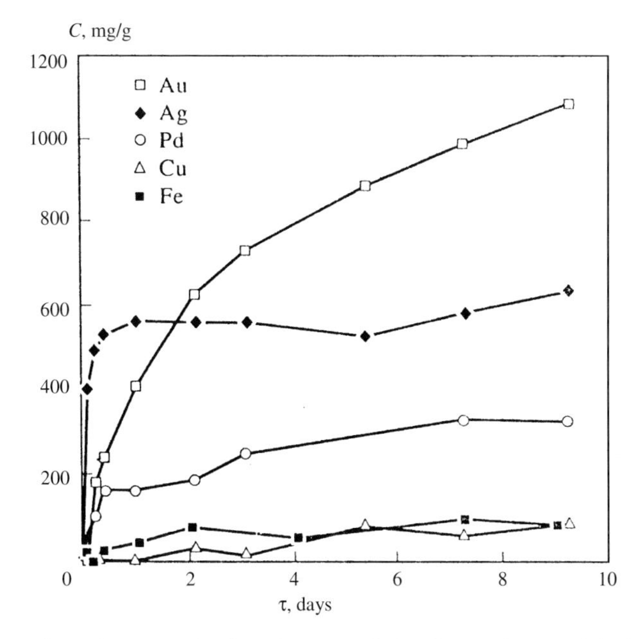

Fig. 5.6. The change of the degree of loading of the block-copolymer (methyltetracyclododecene)$_{400}$-(2-norbornene-5,6-dicarboxylic acid)$_{50}$ with time (C is the capacity, mg of metal/g of polymer)

In styrene block copolymers with vinylpyridines the forming nucleus can additionally undergo cross-linking, as, e.g. in the case of PS-block-P2VPy [76] with 1,4-diiodobutane accompanied by quarternization of 2-VPy segments [77]. As a result, monodispersed microspheres of a hairy ball structure are formed. The thickness of swelling in a water quarternized polyvinylpyridine nucleus is 35 nm and the microdomain space of the lamellar phase is 65 nm. This structure of the polymer phase is able to localize nanoparticles (10–20 nm in diameter) only in the block-copolymer nucleus. From the water–dioxane solution of Ag acetate (10% by volume) Ag^+ ions are introduced into the block-copolymers. The thus formed AgI, or the original argentic salt, is reduced by photochemical methods in the polymer film from 10 to 20 nm diameter nanoparticles. Unreacted components are washed out by $Na_2S_2O_3$. These polymers appeared to be effective for production of nanoclusters γ-Fe_2O_3, Cu and CuO [78]. Diblock copolymers as micelle-forming matter may include diphosphine-containing norbornen blocks [79, 80] to effectively bind Ag^+ ions followed by their transformation into nanoparticles. They can also include norbornen with *tret*-butylamine groups for binding Zn^{2+} ions [81]. In [82] crystallization of ZnO nanoparticles in PEO-block-PMAA diblock copolymers has been presented on models of biomineralization.

Each water-soluble micelle nucleus can form many cluster particles (e.g. 10–1000 [83]), can form in colloids of noble metals and are able to enlarge. Moreover, not only the size but also the shape of the arising nanoparticles might change during reduction. The example of Au stabilized by PS-block-P4VPy (Fig. 5.7) showed that at the beginning of reduction by hydrazine fine nanoparticles appeared. In 10 min particles turned into aubergine-like particles with mean dimensions 24×10 nm and after 24 h they become spheres 12 nm in diameter [84]. The reduction conditions and origin of the reducing agent have a significant impact on nanoparticle size. Tiny spherical particles are generated in the microgel of cross-linked polystyrenesulphonate micelles (Fig. 5.7a) during fast reduction of $AuCl_3$ in water with the addition of $NaBH_4$. A diminished reduction rate and changed pH of the medium (0.1 M of NaOH), all other conditions being equal, bring about fine wriggling threads of Au (Fig. 5.7b). Reducing by hydrazine gives "nanonuggets" looking like sea algae (Fig. 5.7c) [85]. If we subject $AuCl_4^-$ in PS-block-P2VPy micelles to pyrrole [86], the latter undergoes polymerization simultaneously with Au ion reduction and formation of nanoparticles $\sim$7 nm with dendrite-like shape (Fig. 5.7d).

When part of the salt being reduced will not embed into the micelle nucleus because of stoichiometric or other reasons and stays in the solution,

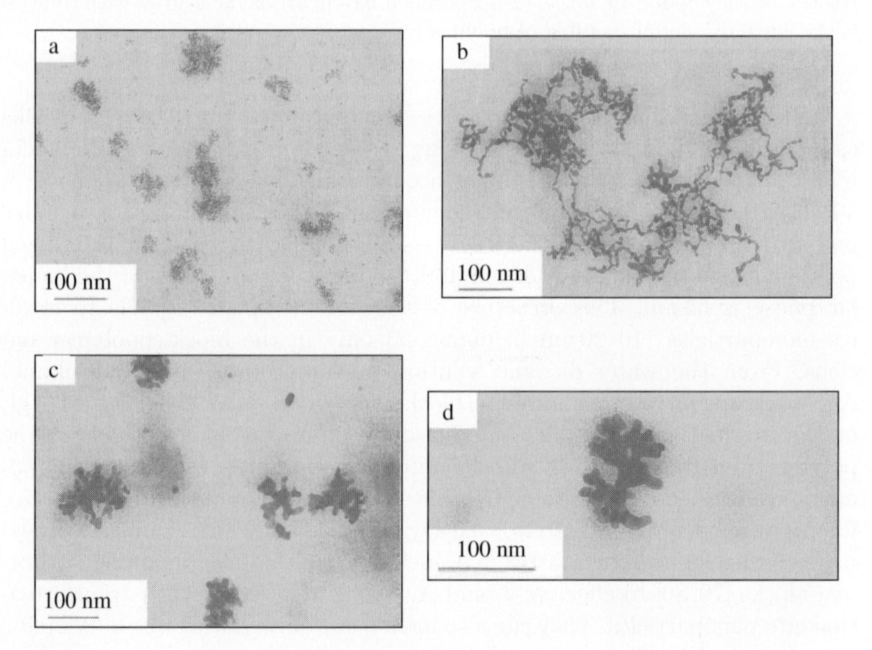

Fig. 5.7. Morphology of nanoparticles obtained under different conditions of reduction (description in the text)

it becomes a source of metal for further nanoparticle growth in micelles. Exchange of nanoparticles between micelle nuclei can be averted using cross-linking agents or by imparting better mobility via a selective solvent such as methanol. Micelle size after formation of nanoparticles in them, in particular Au, remains in fact constant. It is possible to realize the idea of nanoreactors under optimum regimes when the nanoparticle size is strictly observed in terms of micellar parameters. The particle diameter of Au is 3 nm and of Pd is 1.8 nm. Regeneration of protonolytic functional groups occurs along with formation of nanoparticles in micelles under the influence of the reducing agent (Fig. 5.5). The process of micelle loading by metal ions – reduction can be repeated more than once. It was found that [87] the second sequence is accompanied by an enlarging nanoparticle, e.g. in reduction of Ag^+ or $[PdCl_4]^{2-}$ by H_2 nanoparticles of 15–35 and 10–20 nm are formed after the first cycle; after the second cycle the nanoparticles reach 50–90 and 20–30 nm, respectively. Some variants are known when incorporating into a block-copolymer micelle metal ions are reduced without additional introduction of the reducing agent. In just this way Au nanoparticles were prepared in the triblock copolymer of polystyrene-oligothiophene-polystyrene [88].

The existence of micelles in conventional surfactants is possible at above critical concentration (CCM), usually making up 50–100 surfactant molecules. This equivalent is unavailable for polymer micelles. It is known that such micelles are formed in diblock PS-block-PMMA ionomers [89] if the number of insoluble replicate units (N_b) is less than the number of soluble replicate styrene (N_a) units, i.e. in case $N_b \ll N_a$. A compact nucleus and large corona are shaped. The relation of the aggregation number Z versus nucleus ion radius R is a function of block lengths $Z \sim N_b^{0.74\pm0.08}$ and $R \sim N_b^{0.58\pm0.03}$ which decrease within the metal series $Ni^{2+} > Cs^+ > Co^{2+} > Ba^{2+} > Cd^{2+} > Pb^{2+}$.

This original nanoreactor that appears to be the block-copolymer can be used for chemical synthesis of chalcogenide semiconducting nanocrystals CdS,[1] PbS, CuS, ZnS, CdSe, (CdSe)ZnS as well [91–99]. The interrelation between the size of CdS nanoparticles and that of the micelle nucleus is shown in Fig. 5.8 [98].

A typical scheme of their formation with participation of micellar functional groups used for template synthesis of metal–sulphide nanoclusters includes interaction with a metal–organic compound. The scheme is used with, e.g. diethyl zinc $Zn(C_2H_5)_2$, dimethyl cadmium $Cd(CH_3)_2$, tetraethyl plumbum $Pb(C_2H_5)_4$ as well as $ZnPh_2$, $Cd[3{,}5\text{-}(CF_3)_2C_6H_3]_2$, etc., followed by transformation of metal–carboxylates into metal–sulphide clusters. The micelles are often loaded by metal–organic compounds $CdCl_2$, $PbCl_2$, $CuCl_2$ and soluble metal salts, while treatment is exercised by H_2S, Na_2S,

[1] A clear-cut dependence [90] is observed between CdS particle size and their frequency in absorption spectra for particles of 2.1, 3.0, 3.9 and 5.7 nm; that of a massive CdS λ_{max} makes up 286, 367, 416, 461 and 490, correspondingly.

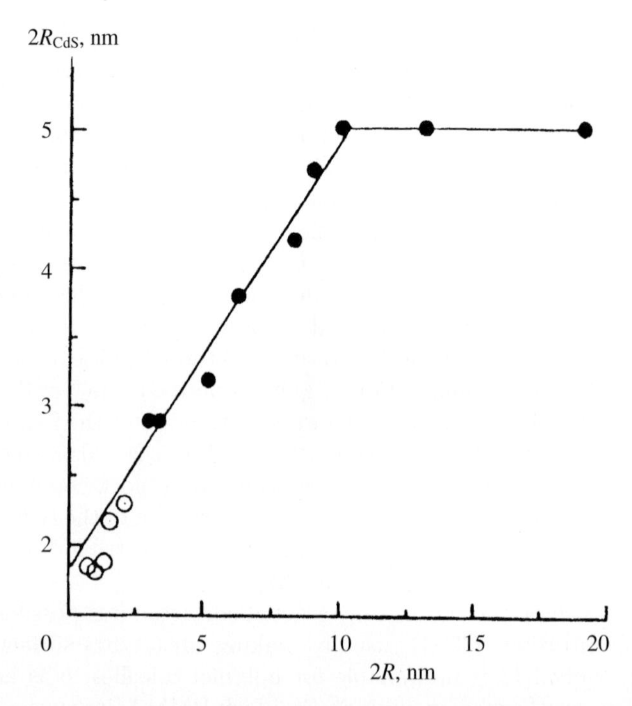

Fig. 5.8. Dependence of the CdS particle diameter on micelle core ionic radius

etc. [100–103]. Monodispersed nanoparticles with sizes 1.7–2 nm for PbS and 6 nm for ZnS are formed in these nanoreactors and their size is conditioned by a number of factors, including the nature and concentration of sulphide substance, temperature and treatment time, origin of coordination solvent, etc. The micelle size and copolymer morphology are critical parameters. It is very important that such reactions can be realized inside the micelles as binding of ZnF_2 followed by its conversion into ZnS through treating with H_2S [104]. Optical examinations have proved that the formed nanoparticles are not always monodispersed and the degree of homogeneity rises significantly if the following requirement is met – one micelle incorporates one particle. It is possible to conclude that nanocomposites on the base of metal chalcogenides can be produced by other means too, namely by incorporating into layered structures, Langmuir–Blodgett films and even during biomineralization.

5.1.4 Formation and Architecture of Metal Nanoparticles in Dendrimers

The size of dendrimer polymer chains is in the range of 1–15 nm. Precursor ions are accumulated within the dendrimer molecules due to electrostatic interaction, coordination to the amine groups or chemical reactions.

New types of nanocomposites with unusual architecture induced by the uncommon structure of strongly branched polymers have received a great deal of attention lately. Their place in macromolecular systems (Fig. 5.9) is assigned to dendrimers that represent a new class of regular spatially hyperbranched polymers characterized by an arborescent structure issuing from one center, a large number of branching centers and the absence of closed cycles. The star-burst dendrimer poly(amidoamine) (PAMAM) has been used as a stabilizer and template for inorganic nanoclusters in a solution [105–111]. It was shown that [112, 113] dendrimers were not simply a new modification of polymers but a new form of polymer matter organization. Their chief peculiarity is correlation of properties of a macromolecule and a particle. They are sometimes called cascade polymers having controlled molecular architecture [114]. Batch-produced polyamidomine or polypropylenamine possessing a diaminobutane nucleus are often employed. Nanomaterials are produced during metal ions sorbing inside the dendrimer whereupon the ions form a complex with amine groups. Chemical reduction generates metal nanoparticles encapsulated in dendrimers, where nanoparticles are composed of as many atoms as there were initially introduced into the dendrimer. The varying size of dendrimers (or their generation) and loading of metal compounds leads to a change in the size of metal nanoparticles. The range of particle distribution by size is very narrow and a fine control can be provided within the particle growth because of the absence of metal ion exchange between dendrimer macromolecules. The thus obtained nanocomposites are soluble in water and are stable over prolonged time (months). They form reverse micelles with a collapsed nucleus in toluene. Metal salt absorption gives rise to cylindrical multidendrimer structures with a swollen nucleus able to break up the metal salt [115]. Generation of dendrimers exerts a perceptible influence

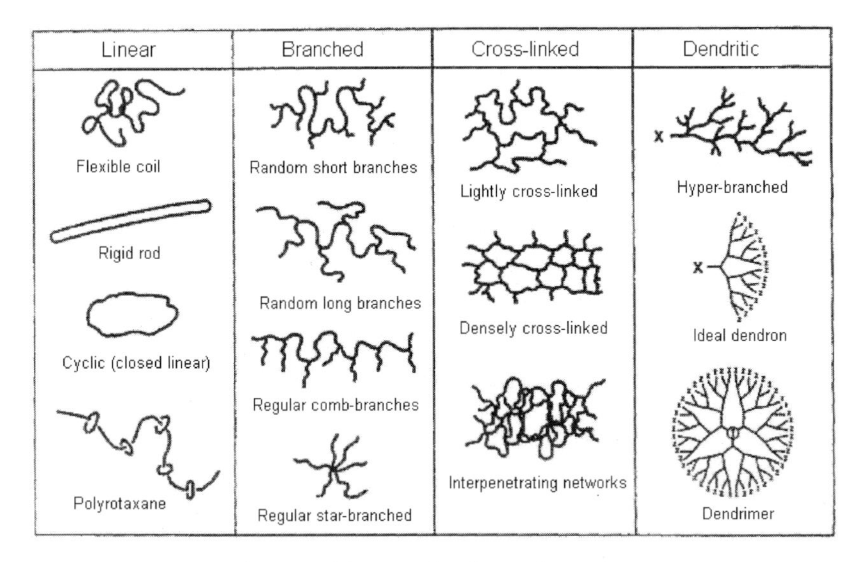

Fig. 5.9. Macromolecular architectures

on nanocomposite architecture as well. Low-generation dendrimers aggregate during nanoparticle stabilization [116–121] while dendrimers of high (to 10) generation stabilize just one or a few nanoparticles in a dendrimer.

End fragments in such formations play a critical role and their number grows exponentially with generation number. The spatial structure of dendrimers of third and fourth generations is close to spherical with a dense core and more friable periphery [122]. Metal–organic derivatives of dendrimers have been obtained [123,124] as well as their numerous macromolecular metal complexes, e.g. ruthenium, palladium, platinum, etc. [125]. It was found that large voids remained inside certain dendrimers which can be used as molecular containers. They served as the base for designing dendrimer-template nanocomposites [126, 127]. The formation of dendrimers following the divergence type (a) and copper-containing nanocomposite on its base (b) are presented in Fig. 5.10 for the example of polyamidoamine of the fourth generation (repetitive segment [$-CH_2CH_2C(O)N(H)CH_2CH_2N-$], I is the initiator, ($NH_3$) is the radial molecular direction from the nucleus to the surface, Z is the surface groups). Cu^{2+} forms a macrocomplex with polyamidoamine during the first stage and it forms Cu^0 nanoparticles of 4.5–6 nm size in the course of reduction by hydrazine hydrate. Copper nanoparticles synthesized in a template way in a dendrimer nanoreactor continue to be stable for more than three months at room temperature without oxygen. The formation dynamics of Cu^0 nanoparticles in polypropylenimine dendrimers (Am_n) with a diaminobutane nucleus (DAB) was studied as a function of generation using different physico-chemical methods [128–131], including EXAFS. It was confirmed that the complex DAB-Am_n-$Cu(II)_x$ (n = 4, 8, 16, 32, 64; $x = n/2$)

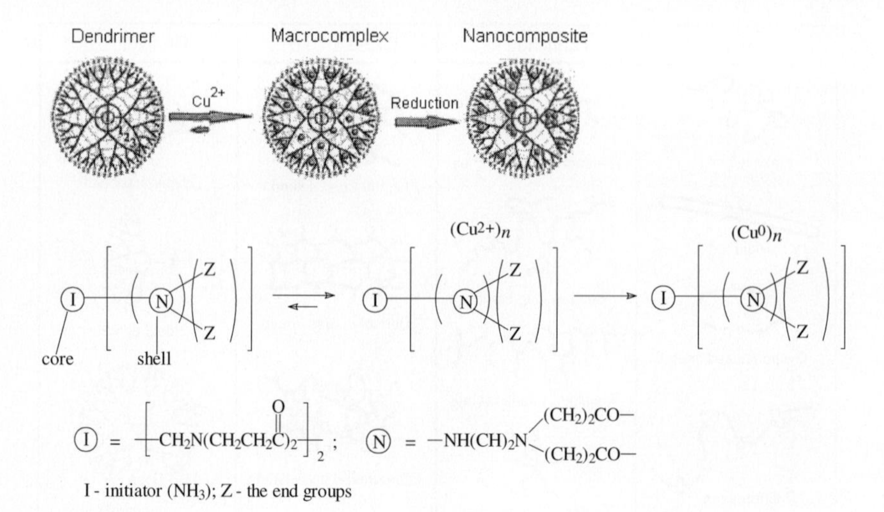

Fig. 5.10. Construction of dendrimer nanocomposites on the base of poly(amidoamine) and Cu^{2+}

is formed during the first stage. The geometry of tripropylenediamine complex Cu(II) with end groups of any generation of dendrimers is repeated as a square pyramid with three nitrogen atoms and two oxygen atoms (including axial). In the course of the macrocomplex reduction by $NaBH_4$ a system DAB-Am$_n$-Cu$^0_{cluster}$ is formed in which a systematic diminishing of the forming cluster size is observed as dendrimers generation rises since cluster size is a function of the n/x relation. Clusters of this type are monodispersed and small, e.g. at $n = 64$, $x = 16$ $r_{cluster} = 0.80 \pm 0.16$ nm.

It is necessary to use dendrimers functionalized by specific groups, e.g. by thiolic dendrimers for Au nanoparticles [132] to achieve dendron-stabilized highly stable nanoparticles with narrow distribution range by size. In this case the dendrimer has a dual role as stabilizing agent and as a permanent selective ligand intended to form a shell round Au nanoparticles. A diagram for a dendron-stabilized nanoparticle system (1–3 generations) is shown in Fig. 5.11. The distribution by size is very narrow in this case too, as it is with the dendrimers of the second generation, where nanoparticles acquire a monodispersed nucleus of 2.4 ± 0.2 nm. Organic–inorganic nanocomposites including metal sulphides in dendrimers can be obtained through sulfonation of macrocomplexes stabilized in them, e.g. copper [133]. Dispersed mixtures of dendrimers with polymer matrices can be used to create still new types of polymer–inorganic nanomaterials.

Higher generations of polyamidoamines in water are introduced into swollen polymer networks of poly(2-hydroxyethyl methacrylate). Besides,

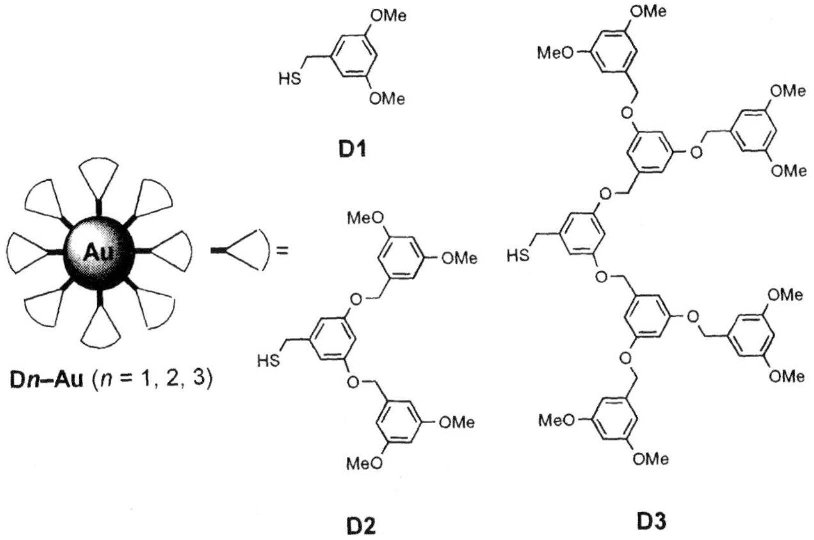

Fig. 5.11. Dendron stabilized Au nanoparticles

Cu^{2+}, Au^{3+} or Pt^{4+} ions are added which are complex-combined with the dendrimer [120]. Metal ion reduction in these compositions gives rise to a new type of organic–inorganic hybrid material.

Dendrimers are the objects of supramolecular chemistry. The microsturcture of nanocomposites on their base depends on Van der Waalse forces, weak hydrogen, hydrophilic or hydrophobic interactions. Their transformations are template directed and self-organized like in biological processes. It is worthwhile mentioning the strategy of assembly of coarse self-organizing nanoparticles with deposition of negatively charged macromolecules onto positively charged metal complexes [134]. Positively charged cetyltrimethyl ammonium-stabilized 73 ± 14 nm diameter PS has been used as a template. A nanocomposite whose layers contain 7600 metal ions each is produced by sequential adsorption of polystyrene sulfonate and Fe^{2+} complex with terpyridine (Fig. 5.12), where centrifuging alters with polyelectrolyte washing. The methods based on the layer by layer deposition technique will intensively develop in years to come.

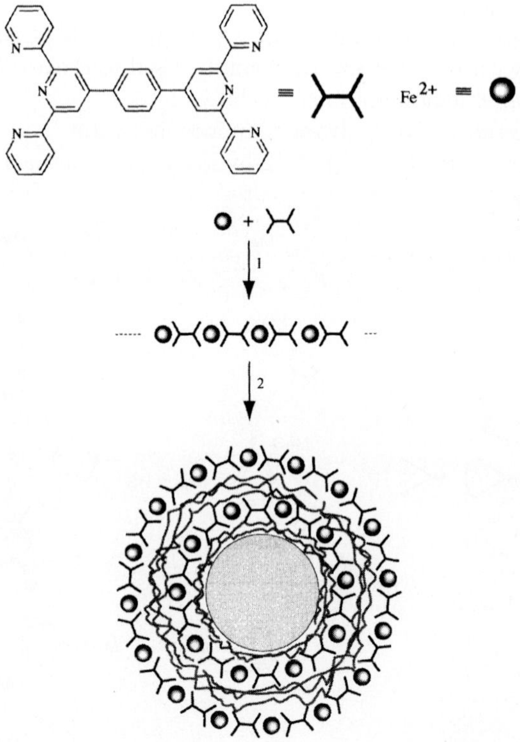

Fig. 5.12. Template assembling of nanocomposites

5.1.5 Formation of Nanoparticles in Heterogeneous Polymer Systems

Such materials are often called nanoheterogeneous composite materials. This term is most frequently applied to describe thin film structures and is referred to either metal or semiconducting nanoparticles incorporated in a dielectric matrix. These materials display unusual photo- and electrical properties imparted by specific types of processes during which minor charged particles interact and join into ensembles whose properties are dependent upon charge redistribution within the nanocomposite.

The formation of nanoparticles in the systems based on insoluble polymers is hampered by the necessity for the reduced ions to diffuse into the polymer matrix. This fact was studied in the example of Ni^{2+}, Cu^{2+}, Fe^{3+} metal ions by sodiumborane in water–methanol media in highly porous isotropic isotactic polypropylene films predeformed in adsorption-active liquid media. An original or annealed film of glassy or crystalline polymers is subjected to cold stretching in alcohol or aqueous-alcoholic media up to $\sim$200% of elongation. As a result, the polymer disperses to fine (1–100 nm) aggregates of oriented macromolecule-fibrils disunited in the space by microcavities (of the same size), i.e. it is in a high-dispersed oriented state [135]. It appeared possible to develop a porous structure with pore volume up to 45% and pore diameter 3–6 nm after the treatment following the mechanism of delocalized crazing in the matrix. This is a system permeable for reagents interpenetrative open pores separated by areas of a block unoriented polymer. Various substances dissolvable in the adsorptive active medium can be introduced. Metal salts start to crystallize forming highly ordered structures as soon as the liquid phase evaporates in the porous structure of the polymer matrix. Many authors [136] attribute this phenomenon to epitaxial crystallization on microcavity walls. The original reagents (metal salts, reducers) can be delivered into the polymer in different ways at the stage of polymer porous structure formation or counter-current diffusion into the preliminary prepared polymer matrix [137]. In such pores the reduced metal is localized in several stages, including metal ions and reducer permeation into the polymer matrix, diffusion of reagents inside it described by Fick's first law and, finally, the chemical reaction in the zone where the flows intermix.

Investigation of the kinetics of these processes consists of the analysis of diffusion rate and chemical reaction constants as well as their correlation. The size of the forming nanoparticles depends rather upon the interaction conditions and parameters of the polymer porous structure than on the nature of the metal [138, 139]. The amount of metal in the polymer is increased mainly through the growth of particle size but not their number. This fact makes it possible to obtain metal nanoparticles microencapsulated in pores 3–10 nm and agglomerated into polycrystalline formations of controllable size and narrow distribution range. In photography the generation of silver halides was used through exchange reaction of $AgNO_3$ and KI followed by their decomposition in a standard developer for photographic films and formation of colloidal silver in porous structure of PTFE. In the course of studying chemical processes underlying photography, amorphous PTFE film in a high-dispersed state stretched to 400% in KI solution (propanol–water) was used as the matrix. It was found that the process was accompanied by total disorientation of the crystalline structure of nanoparticles [140].

Table 5.1 shows the reduction of salts in a polymer stretched in an adsorptive active medium (aqua-alcoholic) to a high dispersed state (with known parameters of the porous structure) that allows one to achieve 5–30 nm nanoparticles. Their mass concentration is a function of production conditions (Fig. 5.13). Pores can have a discrete or layerwise distribution of the new phase particles depending on how metal salt is introduced into the polymer. Most likely, pore and fibril walls present a heterogeneity on which the new phase nucleates and grows. Nanocrystals grow until there exists the possibility of moving apart fibrils connecting craze walls, leaving them intact. Aggregates can form from metal nanocrystals or its oxides and polymer fibrils [141–143].

Table 5.1. Dependence of the mean size of crystallite of different metals on the nature of the matrix [138]

Matrix	Pore Diameter, nm	M_j, nm			
		Ag	Cu	Ni	Fe
PETF	3	12	12	12	–
PP	10	–	–	10	11
PE	16	30	24	13	–
Membrane of naphionic film	Swelling matrix			8	

Copper oxide is formed in swelling matrices (cellulose hydrate, Nafion, PVA, cellophane), while mostly copper is formed in porous matrices (PE, PETF). One distinguishing feature of such systems is the reversibility of the nanoparticle formation reaction in swelling polymers, at least that of copper. Due to the formation of the galvanic pair Cu^0/Cu^{2+} nanoparticles ≤ 3 oxidize with time (a few hours) and convert into mononuclear ones where copper ions migrate and characteristic metallic glitter of polymer films disappear. Although the metal layer is not dense and compact (Ni spherical particles of 30–40 nm diameter are threaded on polymer bars), intimate contact between nanoparticles are not excluded at comparatively high degree of polymer filling. Such nanoparticles can have an isolated structure unable to form percolation (fractal) aggregates important for conducting materials.

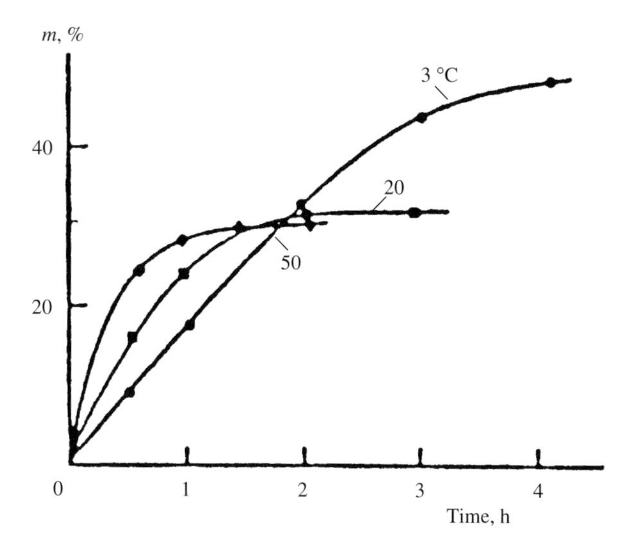

Fig. 5.13. Dependence of weight concentration of nanoparticles in a polymer on the duration and temperature of reduction

Dimensions of the nucleating crystallites are several times as big as the matrix pores (see Table 5.1). It is most likely that nanoparticles during their growth move fibrils apart thus increasing their size. Nanocomposites have low volume concentration of nanoparticles (1–2%) at which the percolation threshold of electric conductivity is reached [144]. At the same time the topochemistry of the metal layer along the polymer matrix cross-section depends on the reactive region width dependening on the diffusion factor D versus chemical reaction constant k ratio. For $D \ll k$ the rate of metal particle deposition is bounded by the diffusion rate and the reactive region width is minimal, whereas for $D \gg k$ the situation is the opposite, where the reaction region extends over the whole cross-section of the polymer film [145]. An optimum correlation between rates of the process stages and of metal layer parameters is reached in conditions of elevated viscosity of solutions and binding of Ni^{2+} ions in a complex with glycerin [146]. It follows that by adjusting relationships between named parameters it is possible to obtain nanocomposite materials possessing different model schemes (Fig. 5.14).

Since with temperature rise k grows much faster than D, the process transfers into the diffusion area and is accompanied by diminishing metal layer dimensions (from 20–25 to 10–12 and 5–7 nm at, correspondingly 3, 20 and 323 K, correspondingly). Though the formation of Ni_4B_3 and NiO admixtures is recorded in this case, Ni particles show strong stability (in PEO, PVP, acrylamide and acrylic acid copolymer, hydrogels of the urea-formaldehyde interpolymer complex and PAA [147]. Nickel obtained in the absence of a polymer oxidizes readily in water and air. The dispersion of reduced Ni mixed with the polymer solution does not result in stabilization

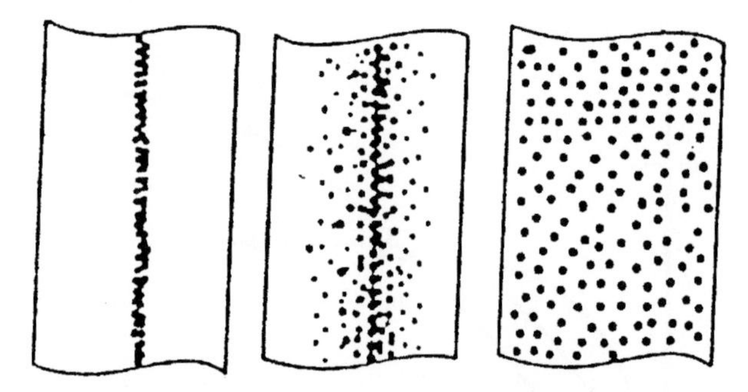

Fig. 5.14. Model of nanocomposite cross-section (dots are nanoparticles)

either, unless reduced in a polymer matrix Ni is oxidized during two weeks and more in water and is dried for half a year. Screening with the polymer evidently averts further growth of particle nuclei and imparts high resistance to oxidation.

The amount of introduced semiconducting filler (e.g. CuS with 10–20 nm average particle size) into swollen polymer films consisting of two polymers [148] (film-forming component PVA and PAA or PEI as complexing agents, and PA-6) may reach 150 mass % of the polymer mass, ∼50% by volume. The films are subjected to a high-cycle successive treatment with $CuSO_4$ and Na_2S. As a result, monocrystalline (∼15 nm) and polycrystalline (∼60–70 nm) nanoparticles are formed [149]. It is believed that along with concentration and diffusion factors nanoparticles and their distribution in the matrix are also effected by thermodynamic and kinetic processes. The most critical of them are the molecular interaction energy (energy of adhesion) and free surface energy at the polymer–particle interface (see Chap. 3).

One of the first attempts of using polymer films as matrix materials is related to CdS synthesis in phosphorus-containing PS [149]. The particle size increased up to 4.4–8.2 nm as the content of phosphorus-containing groups was raised. There are many options for introduction of metal chalcogenides in situ into insoluble Nafion-CdS films [150, 151], CdS and CdSe in PVA or PMMA [152].

The formation of nanoparticles in micelles of diblock copolymers has been extensively studied. Indeed, more homogeneous in structure, rather stable and with controlled size cluster particles are formed in these nanoreactors. Nevertheless, strong passivating and, in the first place, in catalysis may impose negative influence when diffusion limitations will turn essential too for reagents and reaction products. Most likely the methods involving grafted polymers as the matrix appears to be the most promising for production of nanoparticles. Though the method itself for grafted polymerization of monomers and ligand-carrying functional groups has been elaborated to impart new properties to modified materials, it can be equally employed to produce immobilized mononucleus metal complexes [153] or synthesize nanocomposites. The properties of such two-layered microporous materials are interrelated with the localization of the grafted layer on the substrate surface (optimum variants are PE, PP, PTFE, PS) being in the form of thin films or powders. General schemes for manufacturing materials of this type are described in Chap. 3. Just slight amounts of a monomer are necessary to be grafted for synthesis of polymer-immobilized nanoparticles. Particularly, formation of a monolayer of 100 molecular mass on a polymer with $S_{sp} = 10$ m^2/g is equivalent to grafting of 1.5 mass % and the optimum thickness of the grafted layer does not exceed 10–30 nm.

Grafted polymers have been created (in a sense block-copolymers) by this method. They were covered by a protonolytic functional sheath and fragments containing heteroatoms with unshared pair of electrons effectively combining different metal compounds and after reduction by standard methods possessing nanoparticles (Fig. 5.15) [153, 154]. Examples of metal ion reduction in polymer suspensions, including macrocomplexes, are numerous.

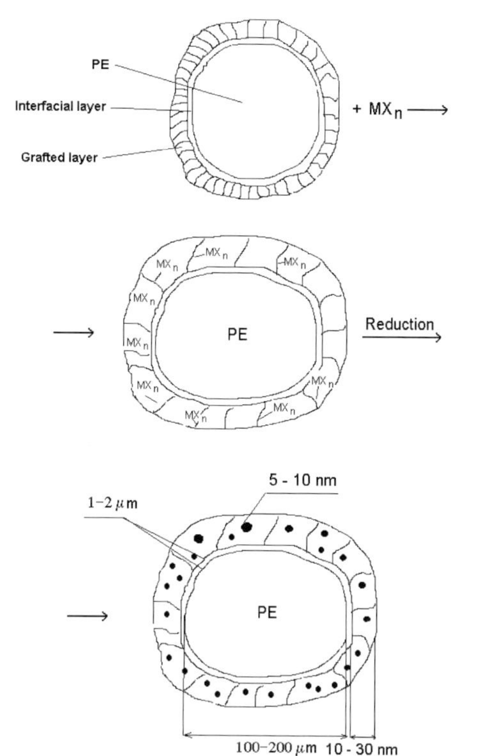

Fig. 5.15. The scheme of nanoparticle formation in a grafted layer

Fixed by the ion exchange method Rh, Ru and Pt compounds on fully fluorinated Nafion resin are reduced in a H_2 current at 473 K. As a result, metal clusters of 2.8, 3.3 and 3.4 nm size, respectively, are formed [155, 156].

Rhenium nickel is stabilized [157] by incorporating into the silicone rubber followed by polymer curing at room temperature.[2]

Natural polymers (gelatin, gum-arabic, agar-agar, proteins and their hydrolysis products, sodium prothalbuminate, starch, cellulose and its derivatives) and polymer substances soluble in organic media (rubber, linolein, naphthenic acids) have long and frequently been used as protective high-molecular compounds for producing metal sols, including silver, mercury, bismuth, copper, platinum group metals by reducing the corresponding ions (see Chap. 3). The stabilizing agent for colloidal Rh, Pt and Pd (with 1–100 nm particle size) can be β-cyclodextrin [159] while its thiolate derivatives (e.g. *per*-6-thio-β-cyclodextrin) stabilize water-soluble Pt (14.1 ± 2.2 nm particle size) and Pd (15.6 ± 1.3 nm) nanoparticles formed by reduction of their salts subjected to BH_4 [160, 161] as well as Au clusters [162]. Ultra-dispersed Ni particles (5–20 nm) are obtained by [163] diffusion of $Ni(NO_3)_2$ in gel membranes of cellulose acetate followed by H_2 induced reduction.

In this chapter we mention only the most used reduction mainly by polyatomic alcohols under elevated temperatures with addition of corresponding stabilizers and mixing of derived products with the polymer matrix.

5.1.6 Production of Nanocomposites by Reducing with Metals and Organometallic Compounds

Strong agents like aluminum hydride, its derivatives, tetrahydroaluminates of alkali and alkaline-earth metals, organometallic compounds are not often used as reducers in polymer systems. Out of a few examples we shall consider the generation of nanoparticles of Fe/Al alloys through reduction of Fe^{3+} salts using $LiAlH_4$ [164], and those of nickel through reduction of its salts in diethyl ether with addition of Et_2AlH [165]. Less attention was given in the literature to methods based on reduction of complexes, including the polymer matrix-bound, in an ester or hydrocarbon medium following the reaction with alkali metals (A):

$$MX_n + nA \rightarrow M^0 + nAX.$$

This variant is used in the production of Fe, Ni, Co and some other nanoparticles. Silver colloid (2–50 nm) that remains stable for a long time is formed at reduction of its salts by solvated electrons (metallic sodium dissolved in liquid ammonia) [166]. A special technique has been developed for polymer metallization by plating-out, whose essence is in the following. Metal–polymer chelate films are reduced by metals whose ion potential is higher than that connected with the polymer. For example, according to this method Cu^{2+} ions under the action of different metals [167] are reduced in the polymer till metallic copper is formed. Another way is to disperse powders

[2] Rubber can be cured by salts of unsaturated acids too, which yields cured metal-polymer structures [158].

of different metals showing low reduction potential in PVA water solution, dry and treat them with some other metal salt showing high reduction potential [168]. This reaction can run spontaneously due to the positive net electromotive force value (E^0_{net}), e.g.

$$Zn^0 + Cu^{2+} \rightarrow Zn^{2+} + Cu^0,\ E^0_{net} = +\ 1.100\ V$$

$$Cu^0 + 2Ag^- \rightarrow Cu^{2+} + 2Ag^0,\ E^0_{net} = +\ 0.462\ V$$

$$Pb^0 + Cu^{2+} \rightarrow Pb^{2+} + Cu^0,\ E^0_{net} = +\ 0.463\ V.$$

The reduced metal is strongly fixed to the metallized surface of a PVA film possessing low surface resistance ($\sim 10^0$–$10^1\ \Omega/cm^2$).

Polyatomic complexes chemically bound to the polymer may undergo various transformations that increase their nuclearity under the action of organometallic reducers. For example, the formation of polymer-immobilized metal clusters during a catalyzed reaction is prevailing, especially in applied onto polymers complex catalyzers [169, 170]. Important information about metamorphoses of immobilized Ni^{2+} during interactions with AlR_xCl_{3-x} [171] was obtained by measurements of magnetic susceptibility under high magnetic field intensities (up to 70 kE) and low temperature (4.2 K) [172]. The susceptibility of various spin states under such conditions depends nonlinearly on the magnetic field intensity. It allows obtaining the distribution of nickel states in terms of the magnetic moment values. The formation of ~1 in size particles contained from 10 to 30 Ni atoms that are not ferromagnetic (supermagnetic) but representing polymer-immobilized clusters with ferromagnetic exchange. Hydride nickel complexes being active in ethylene dimerization are known [173] to arise from nickel compounds at usual oxidation degrees (+2) through a series of transformations and participation of organo-aluminium compounds. Cycle regenerates active centers by involving the forming Ni^0 into the reduction. These centers are deactivated by formation of the colloidal nickel.

Similarly, the share of Co^{2+} is reduced till it is a metal in polymer-immobilized cobalt systems (butadiene polymerization catalyzers) and at least 90% of all ferromagnetic cobalt is accumulated in relatively coarse ~10 nm particles [174]. Increasing nuclearity characterizes other catalyzers as well. The reducing agents can be the very substrates, e.g. Ag, Cu, Rh, Ru, Pt and Ir cations combined with either perfluorethylenesulfacid, or partially sulfonated PS. They oxidize CO, NO, NH_3, N_2H_4, C_2H_2 at 373–473 K accompanied by formation of 2.5–4.0 nm colloidal particles [175].

5.1.7 Bi- and Polymetallic Nanoparticles in Polymers

Many problems can be solved by use of not monometallic but polymetallic nanoparticles (e.g. multicomponent catalysis, model investigations of alloy surfaces).

All methods of producing nanocomposite materials were treated previously as based on monometallic nanoparticle type. It is possible to produce such materials obtaining particles with two and more metals. Cryochemical syntheses can be used (see Chap. 6) [176]. Organosols of bimetallic particles of below 5 nm diameter have been produced by combining condensation of silver, tin and MMA with vacuum evaporation on the cooled surface of the glass reactor and subsequent melting and heating. This method, however, complicates the characteristics of the forming products, their composition, distribution and morphology.

To get a bimetallic colloid of the *nucleus–shell* structure it will be most efficient to deposit the other metal M′ on the preformed surface of either colloid or cluster of the first metal M. In this case the preliminary formed metal nucleus should not be protected by some ligand of triphenylphosphine type or a stabilizer:

$$M_x + yM'^{n+} + ne^- \rightarrow M_xM'_y.$$

This reaction is analogous to nanoparticle growth following the stage of nucleation if we ignore differences in velocities and metal nature. Preparation of bimetallic nanoparticles of the Pt/Pd or Cu/Au kind [177] stabilized in reverse micelles turned out to be a complex problem. Nevertheless, the conditions of generating Au/Pd nanoparticles in reverse AOT micelles by reduction of respective salts [178] using hydrazine hydrate have been optimized and bimetallic nanoparticles had sizes up 2–3 nm depending on their composition.

This reduction procedure appeared highly useful for synthesizing colloidal dispersions of polymer-protected bimetallic clusters Pd/Pt or Pd/Au [179–181]. It was made by combined reduction of $PdCl_2$ and H_2PtCl_6 or $HAuCl_4$ in the presence of PVPr (boiling in a water–ethanol mixture, in methanol or in exposure to visible light). The products stay stable during months at room temperature, and their size were, for Pd/Pt, about 1.5 nm, and 3.4 nm for Pd/Au clusters. The size of the latter depends essentially on the reduction scheme and correlation of the initial concentrations of components [182]. A mixture of metallic phases is generated following the scheme of simultaneous reduction or reduction of Au^{3+} on Pd^{2+}, whereas structures of the *nucleus–shell* type are formed only in case of sequential reduction of Au^{3+} on Pd^{2+}. The formation of Pd/Pt clusters in the presence of PVP was studied in [183] and proved that the corresponding alloy with particles of the same size is formed at molar ratio Pd/Pt = 4/1 (Fig. 5.16).

The coordination numbers of Pd and Pt atoms calculated on the basis of EXAFS spectra (Table 5.2) comply with a nanoparticle model incorporating 55 atoms: 42 atoms of Pd and 13 atoms of Pt.

The coordination numbers match up only the model where Pt atoms are not randomly arranged but constitute the nucleus of a cluster while Pd atoms are localized on the nucleus surface (an onion-shaped model).

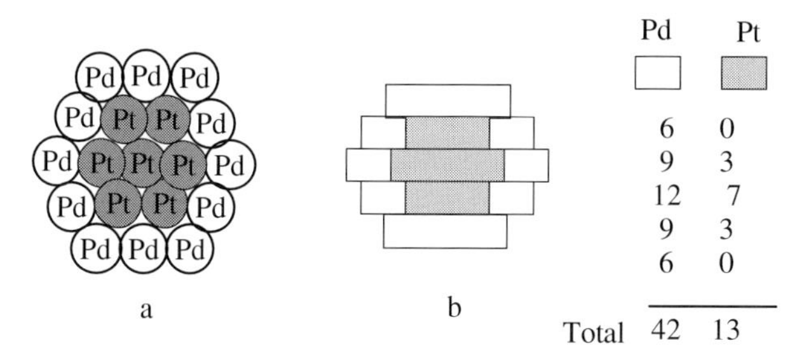

Fig. 5.16. Model of core–shell structure for a bimetallic cluster with a molar ratio Pd/Pt = 4/1 (**a**) and its projection (**b**)

Interactions between Pd and Pt bring about alloys [184]. Furthermore, electron spectroscopy has shown that with increasing Pd/Pt ratio absorption in the short-wave part of the spectrum shifts to the region of longer waves (to the side of Pt absorption) and reaches a shift maximum at Pd/Pt = 1/4. The long-wave part of the spectrum does not show any changes.

Table 5.2. Coordination numbers of Pd and Pt atoms in bimetallic cluster Pd/Pt = 1/4 (according to EXAFS spectroscopy) [183]

Adsorbed Metal	Metal in Solute (scattering)	Interatomic Distance r, nm	Coordination Numbers: Observed	Pt Nucleus	Statistical Distribution
Pt	Pd	0.274 ± 0.003	4.4 ± 0.4	4.6	6.0
Pd	Pt	0.273 ± 0.003	2.3 ± 0.2	2.0	1.9
Pt	Pt	0.273 ± 0.003	5.5 ± 0.05	5.5	1.9
Pt	Pd	0.272 ± 0.003	3.1 ± 0.7	6.5	6.0

The problem is simple only when ions of the deposited metal have lower redox potential in contrast to the metal forming the nucleus. Otherwise, some of the atoms making up the nucleus would transfer into the solute through the redox reaction

$$M_x + M'^{n+} \rightarrow M_{x-1}M' + M^{n+}.$$

To avert such reactions it is proposed [185] to use the protective function of hydrogen, which is based on the ability of H_2 to absorb on atoms of noble metals, form surface M-H bonds (α direction in Fig. 5.17) and break-up on them [186] (direction b in Fig. 5.17). The adsorbed hydrogen atom will furnish electrons for reduction (they leave the Pd surface as a proton) (direction c, Fig. 5.17) during interactions between metal ions with higher redox potential (e.g. Pt^{2+} and Pt^{4+}) and hydrogen-adsorbed Pd nuclei not palladium). The nucleus is surrounded by the stabilizing polymer PVP. It does not exert any essential effect on diffusion of the second metal to the nucleus (route d).

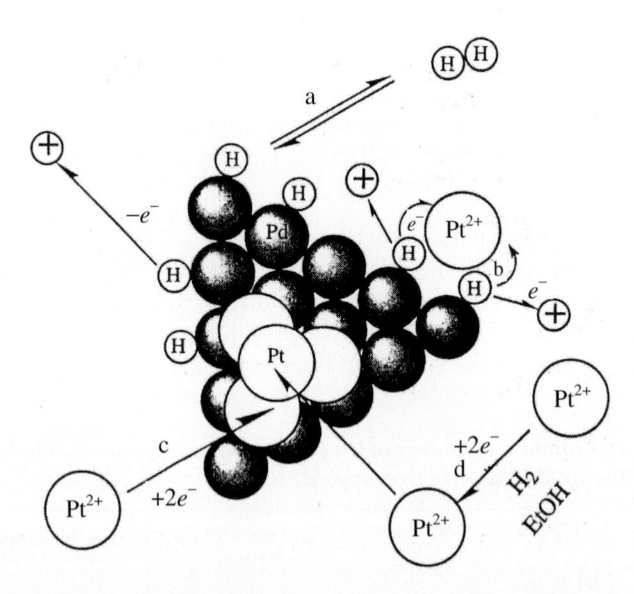

Fig. 5.17. The protective function of hydrogen in the synthesis of a bimetallic cluster with core–shell structure

The number of Pt atoms necessary to cover the Pd nucleus with a monometallic layer is calculated by the equation

$$R = \frac{\sum\limits_{f_i=f_{\min}}^{f_i} f_i\left[10(n_i+1)^2+2\right]}{\sum\limits_{f_i=f_{\min}}^{f_i} f_i[1/3(10n_i^3+15n_i^2+11n_i+3)]}, \tag{5.1}$$

where R is the atomic ratio Pt/Pd needed for covering with a monolayer; f_i is the particle share of size d_i; n_i is the number of layers of Pd atoms in a nucleus of size d_i; d_i: $n_i = (d_i - r)/2r$(r is the Pd–Pd distance).

Calculations prove that for this coating 90% of Pt atoms of Pd ones are required. This fact is supported experimentally by absorption of molecules checked by IR spectroscopy (see IR-CO analysis below).

Polymer-protected Pt/Co bimetallic colloids with Pt/Co ratio 3:1 and 1:1 have been derived by the reduction of respective precursors [187]. Pd/Cu alloys stabilized by polymer matrices (mainly PVP) are very interesting too.

They are produced [188–190] by boiling bivalent metal acetates in alcohol solutions in the presence of PVP ($M_w = 40{,}000$). Thus, nanoparticles are formed with a proportion of components Pd_xCu_{1-x} ($1 \geq x \geq 0.37$) and mean size ~4.5 nm. The model of these cluster particles is strongly debatable (in the part concerning the composition of the nucleus and its surface environs). These alloys [190–192] (of $Pd_{80}Cu_{20}$PVPr and $Pd_{90}Cu_{10}$PVPr composition) have enriched with Pd atoms surface that is consistent with IR spectroscopy data on CO adsorption.

Vistas in using this method for detecting nanoparticle size was discussed earlier. It can be employed to define their composition; in particular, the valent oscillation frequency is ν(Pd–CO), ν(Cu–CO) and ν(Pd–Cu–CO), and constitute, correspondingly 1940, 2090 and 2050 cm^{-1}.

The preparative method [190] ensured the formation of only bimetallic Pd/Cu particles since in conditions of synthesis in the alcohol solution only Pd^{2+} could reduce to Pd^0, whereas copper deposited on Pd^0 particle surfaces only in redox reactions. Relative redox potentials of Pd^0/Pd^{2+} and Cu^0/Cu^{2+} pairs as well as Cu^{2+}/Pd^0 hamper copper reduction till zero-valent state at the expense of Pd^0 and the most probable form of the reduced copper could be only Cu^+. Hence, when reducing in the alcohol solution (343–408 K) copper deposits on Pd^0 particles quickly covering their surface and then slowly dissolves in the subsurface layers of the colloidal alloy forming a solid solution in this concentrated zone [193] thus enriching it with palladium.

Almost all bimetallic clusters Pd/Ag, Pd/Cu, Pd/Co, Pd/Pb (Table 5.3) are 1–3 nm in diameter and Pd/Pb is the least of all (<1 nm). A more detailed description of structural parameters (EXAFS) of Cu/Pd alloys of different compositions stabilized by PVP is presented elsewhere [192].

In contrast to Cu-containing bimetallic nanoparticles, the production of Ni-containing nanoparticles is restricted because Ni possesses a lower redox

Table 5.3. Coordination numbers n and interatomic distances r between bimetallic nanoparticles incorporated in polymer membranes [23]

Pd/X nanocomposite	n (X–O)	n (X–X)	r (Pd–Pd)
Pd/Ag (3:1)		8.8	2.75
Pd/Pb (1:1)	2.0		2.75
Pd/Pb (9:1)	1.5		2.78
Pd/Cu (1:1)	4.1		2.73
Pd/Cu (3:1)	4.2		2.71
Pd/Cu (9:1)	3.6		2.76
Pd/Co (9:1)	6.7		2.75

potential value ($E^0_{Cu^{2+}/Cu^0} = 0.342$ and $E^0_{Ni^{2+}/Ni^0} = -0.257$ V, respectively). Therefore, Pd/Ni clusters (Pd/Ni = 3:2, 1:1 and 2:3) protected by PVP are obtained by [194] reduction of corresponding amounts of $NiSO_4{\cdot}7H_2O$ and $Pd(CH_3COO)_2$ (pH = 9–11) by glycol at 471 K over 2 h. Parallel to this colloidal dispersions of Ni and Pd in PVP are formed. Particles in compositions Pd/Ni = 1:1 are about 1.6–2.0 nm in size (means size are 1.89 nm) and they represent an alloy (Fig. 5.18).

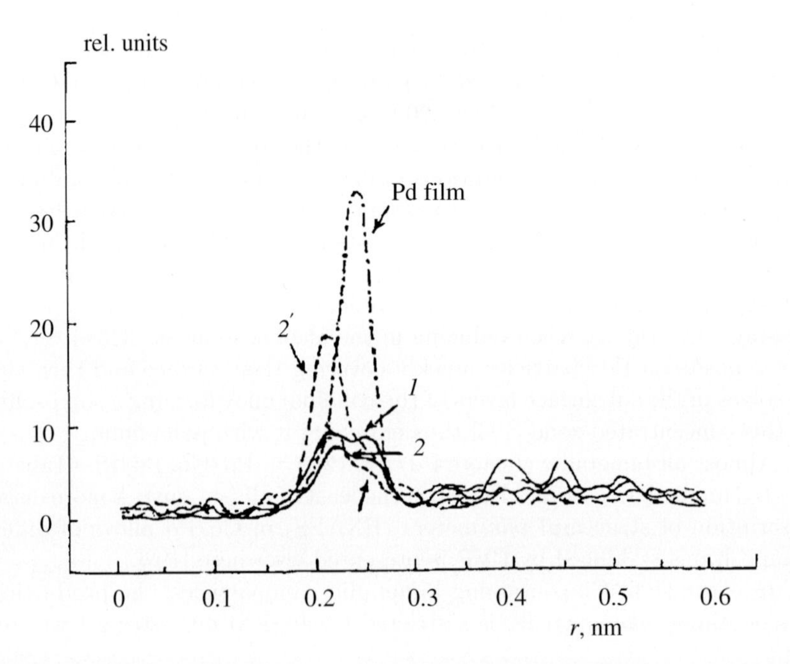

Fig. 5.18. Fourier transformation of the EXAFS spectra of Pd and its clusters with Ni in the ratio Pd/Ni 3:2 (*1*), 1:1 (*2*; *2′* – the same ratio for Pd/Ni film) and 2:3 (*3*)

Metal–polymer compositions based on low-melting multicomponent alloys (PE, PP and so on with Bi, Pb, Sn, Zn, Cd) are used [195] as electrode materials for batteries, PVP - Pt- MCl_n (MCl_n = NaCl, $ZnCl_2$, $NiCl_2$, $FeCl_3$) are used as selective catalysts [196].

It is possible to catalyze nanoparticles of Pd stabilized by aqua-alcoholic solutions of sodium polyacrylate (mean molecular mass is 2200). PVP (M_n= 40 000) are modified by Nd^{3+} ions through the introduction of a rated amount of $NdCl_3{\cdot}6H_2O$ or lanthanide ions into a ready suspension [197]. Nd-containing particles localize on the Pd nanoparticle surface.

Bimetallic Pd/Pt or Pd/Rh nanoparticles are used embedded in polymer films for electrocatalytic hydrogenation [198].

The treatment of different metal salts in polymer matrices with a common reagent could be effective for generation of other types of alloys and heterometallic materials, e.g. $(ZnCd)S_2$. H_2S of the triple co-polymer styrene-diacrylate zinc-diacrylate of cadmium ($M_n = 4.7 \cdot 10^4$, atomic ratio Zn/Cd = 3.3:1) [199] assists in the production of polymer composites having ZnS/CdS crystal with luminescence properties (see Chap. 11). Polymer polymetallic nanocomposites are anticipated to be promising for various protecting coatings, e.g. polyphenylensulphide with Ni-Al nanoparticles to inhibit corrosion and steel oxidation [200], and surface modification of $CrSi_2$ [201] nanocrystals.

Some acoustic-chemical methods of deriving nanoparticles of two metals were developed for homogeneous systems ($NiFe_2O_4$ from $Fe(CO)_5$ and $Ni(CO)_4$ [202], Fe/Co alloys from pentacarbonyl iron and tricarbonylnitrosyl cobalt [203]). There are no apparent principal restrictions for these reactions in polymer media. For example, polydispersed sols of plumbum chromate $PbCrO_4$ (20–30 – 70–80 nm particle size) [204] are obtained by precipitation of $Pb(NO_3)_2$ and K_2CrO_4 in the presence of PVA or gelatin functioning simultaneously as stabilizers of nanoparticles and film-forming agents. The concept of nanoreactors established in the generation of monometallic nanoparticles in block-copolymers (Sect. 5.1.3) is also applicable to the production of metal composites based on heterometallic nanoparticles, particularly of mixed metal oxide composition. With this aim, $FeCl_3$ and $CoCl_2$ were introduced into the block-copolymer formed by ring-opening metathesis polymerization of norbornene derivatives so that chloride ions were substituted and reduced at room temperature with addition of NaOH [205]. Emerging in the polymer matrix $CoFe_2O_4$ nanocrystals were of 4.8 ± 1.4 nm size with unimodal distribution and oval-shaped morphology. Inorganic synthesis of these particles proceeds as a rule under high enough (598 K) temperatures [206, 207].

PEO [208], PVA [209], PAA [210, 211], starch [212] and some other substances can be used to produce mixtures of metal oxides. The reaction follows the concept of rocket propellant chemistry where the reagents are treated as the propellant or oxidizer components and their ratio should correspond to the maximum combustion energy liberation [213]. It is possible to reach self-ignition of the polymer precursor yielding mixed metal oxides with PEO and PVA.

The situation is much more complicated if nanoparticles include three and more metals (multimetallic nanoparticles). γ-Radiation of corresponding salts results in formation of colloidal dispersions of trimetallic Au–Pb–Cd particles of 33 nm diameter and incorporating a Au nucleus acquiring a 18 nm size Pb mantle [214]. Surely, this complicates investigation of such particles structure still more. These metals are used in production of precursors in manufacturing superconducting ceramics, and specific multicomponent steels. In particular, the traditional methods of preparing melting stock for high-temperature superconducting (HTSC) ceramics (direct mixing of oxides, carbonates, oxalates, nitrates and other salts of respective metals) are strongly limited in reproducibility. This is attributed not only to the quality of ground matter and mixing of initial solid components (see Chap. 2) but also to intricate physico-chemical and mechano-chemical transformations occurring during preparation of samples. It results in the formation of microinhomogeneities, various, including nonconductivity, phases leading to low-quality HTSC ceramics with a fuzzy superconducting transition (Meissner's effect,T_s).

The initial components like $Y(NO_3)_3{\cdot}6H_2O$, $Ba(NO_3)_2$ and $Cu(NO_3)_2{\cdot}2H_2O$ are intermixed at the molecular level in a solution to eliminate the above phenomenon and to achieve structurally homogeneous superconducting ceramics of $YBa_3Cu_4O_8$ type. A homogeneous dispersion of HTSC precursor is obtained upon solvent evaporation (congruent evaporation). This problem applied to polymers can be solved in two ways. According to the first approach, ready HTSC ceramics is introduced into the polymer matrix, in the other case the ceramics are synthesized in the presence of the polymer matrix. In the first case, the dispersed HTSC phase $YBa_2Cu_3O_{7-\delta}$ (Y-123 ceramics) or $(Pb_xBi_{1-x})_2Ca_2Sr_2Cu_3O_y$ [215], as well as $Tl_2Ba_2Ca_2Cu_3O_y$ [216] ($T_s \sim 90$, 110 and 125 K, correspondingly) are introduced into the polymer matrix of polychlorotrifluoroethylene, PVC and rubber. The amount of ceramics makes up 50–75%. The advantages of HTSC in the composition of an optimum formulation (high T_s value are preserved, although the superconducting transition is somewhat widened along with magnetic properties and levitation) correlate with those of the polymers (mechanical strength, flexibility, machinability, resistance to atmospheric effects, hostile media, etc.). In addition, the procedure helps to eliminate the drawbacks of HTSC ceramics, including their high porosity, brittleness, and liability to degradation. However, this procedure necessitates rather coarse (of micron size) ceramic particles to be introduced into the polymer.

The second way is synthesis of nanoceramics through prior polymer synthesis that presupposes use of the polymer matrix with metal ions dispersed in it at the molecular level. This can also take the path of nanoparticles generation during polymerization or polycondensation. For instance, [217] citrates of corresponding metals are introduced into aqua solution of acrylamide and polymerized till a gel-like state is reached in the presence of radical initiators and chain conductors (N,N,N′N′-tetraethylenediamine). Metal cations turn out to be captured in the polyacrylamide gel traps. Upon calcinations ultrafine multicomponent oxide powders are produced, such as $YBa_2Cu_3O_{7-x}$ or $LaAlO_3$. $YBa_2Cu_3O_{7-x}$ ceramics [218] are derived with addition of either acrylic or methacrylic acids as chelating ligands following gel–sol synthesis. Polymer-salt solutions which are afterwards dried and sintered are often used to prepare Y-123 ceramics. Polymer alcohols and acids like PEO and PVP (see [219–221]) are suitable for this purpose. Compaction, agglomeration and oxygen annealing of these blends are exercised by the usual procedures.

The dimensions of all salt crystals in the presence of any polymer are considerably less than in crystallization of purely salt solutes. This is connected with complexing in solutions and macromolecular adsorption on a crystalline embryo that does not exclude crystallization processes because of the presence of a concentration gradient of the components across the film thickness. Sometimes, the films appear to be homogeneous since their formation occurs just as on the surface so in the bulk. The blocking capability of polymers depends upon their molecular mass as well. The final dimensions of ceramic crystallites obtained in powders in fact correspond to their dimensions in the films.

One more variant is when polymer complexes Y^{3+}, Ba^{2+} and Cu^{2+} with PMAA [222, 223] and YBC-chelates with polyamides [224, 225] are produced by polymer-analogous transformations using jellification with PVA [226]. The thus obtained HTSC ceramics show $T_s = 80$–$92\,K$ while the critical current density is 150–160 A/cm^2. Besides, epoxy polymers (YBC-epoxy composites) are known to be used with this aim [227]. Not only powders but also films and fibers can be produced from such materials. For example, it is possible to produce long fibers (>200 nm) of 1 mm diameter at $T_s = 92$–$94\,K$ and current density 300 A/cm^2 (at 77 K) [228] by dry molding using thermoplastic gels based on Y-123 ceramics and PVA.

A trend of polymer synthesis in the presence of HTSC components is known, i.e. polymerization of acrylic acid in a mixture with aqua solutes of Y^{3+} nitrate along with Ba^{2+} and Cu^{2+} acetates [218]. It has been established [229–231] that PVA fibers (PVAc precursors) of different saponification degree (DS) with incorporated YBC-123 ceramic components form a material with different current densities J_c. The minimal J_c value is observed at DS = 67% (mol), and the maximum ($J_c = 3.5 \cdot 10^4$ A/cm^2, 77 K) is at DS = 81% (mol). These parameters are affected by the preparation conditions of the samples (annealing and pyrolysis) that is attributed to the peculiarities of the distribution of ceramics on fibers.

One of the most promising trends in HTSC ceramic synthesis is copolymerization of corresponding metal-containing monomers [232]. A more homogeneous distribution of metal ions can be attained by introduction of a monomer in their molecules, i.e. prior to synthesis, after which copolymerization is carried out. These reactions run as follows. Either acrylates or acrylamide complexes Y^{3+}, Ba^{2+} and Cu^{2+} are mixed in molar proportions 1:2:3 in a minimal methanol quantity, dried and then copolymerized in a solid phase [233]. Similarly, bismuth ceramic components, acrylamide complexes Bi^{3+}, Pb^{2+}, Sr^{2+}, Ca^{2+} and Cu^{2+} taken in molar proportions 2:0.3:2:2:3 are spontaneously copolymerized in concentrated water solutions [234]. The resultant metal-containing copolymers are subjected to a standard thermal treatment aimed at burning out the organic phase. Figure 5.19 illustrates properties of HTSC ceramics produced by this method.

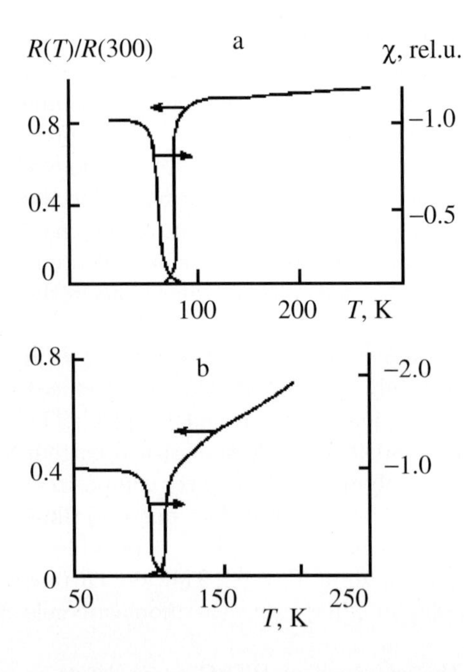

Fig. 5.19. Temperature dependence of the electrical resistance and the magnetic susceptibility of high-T_c superconducting ceramics: (**a**) ceramics from a copolymer of Y, Ba and Cu acrylates; (**b**) ceramics prepared by spontaneous copolymerization of acrylamide complexes of Bi, Ca, Sr, Pb and Cu

The analysis of the $R(T)$ dependence proves that at $T < T_s$ the ceramics behave like metals although their T_s is not so high as that of monocrystals. It is a matter of common experience that the quality of a polycrystalline sample is to a greater extent attributed to that of its intergrain contacts than to the quality of the grain structure itself. The superconducting transitions (87 and 110 K, correspondingly) are sufficiently abrupt and their width is within 0.1–0.9 of the full resistance drop. The behavior of the magnetic susceptibility dependencies on temperature $\chi(T)$ is also evidence of the abrupt transition and the presence of a single superconducting phase. The total volume of the diamagnetic screening of the phase estimated from the Pb bench-mark reaches 100%.

This method looks especially challenging for production of superconducting bismuth cuprates for which reproducible synthesis is highly problematic (phase admixtures are in fact always present at $T_s = 85$ K due to microinhomogeneity of the ceramics and sequential phase transition 2201→2212→2223). Products obtained from spontaneous polymerization are perfectly reproducible samples with $T_s = 110$ K and critical current density up to 240 A/cm^2. This approach will be useful for production of other single-phase materials in a nanocrystalline state too.

The use of nanostructured polymetallic materials is very important in multicomponent steels of M50 type in which size and homogeneity problems are much more important [235, 236]. M50 steel is used mostly in aircraft building for journal-bearings and turbine engines. It represents an iron-based alloy incorporating Cr – 4, Mo – 4.5, V – 1, C – 0.8 mass %, the rest being

Fe. The steel contains a large number of micron size carbon particles able to initiate fatigue cracks in bearing material. It was proposed to improve the mechanical characteristics of these structured materials by diminishing the grain size in flaw sites (microcrack healing). PVP was used in M50 steel as the polymer surfactant; the precursors were M50 – $Fe(CO)_5$, $Cr(Et_xC_6H_{6-x})_2$– bis-(ethylbenzyl)chrome, $x = 0 - 4$, $Mo(Et_xC_6H_{6-x})_2$– bis(ethylbenzyl)-molybdenum, $x = 0 - 4$ and $V(CO)_6$ in dry decalin under ultrasound exposure (acoustochemical synthesis). Upon removal of the solvent and gaseous phase, the thus formed colloidal particles were 7 nm in diameter and represented a homogeneous alloy. Chemical reduction of steel M50 precursors ($FeCl_3$, $MoCl_3$, $CrCl_3$ and VCl_3) by lithium triethyl-borohydride in THF is followed by solid LiCl removal with sublimation in vacuum at high temperature.

These approaches may turn out to be useful too for the creation of magnetic materials based on, e.g. polymetallic nanoparticles of $LaSrCr_xNi_{1-x}O_{4+\delta}$ type. A new method [237] has been proposed for preparation of $La_{1-x}Sr_xMnO_3$ ($x = 0$; 0.2; 0.4; 0.6) ceramics through presynthesis of the polymer (Fig. 5.20). Its essence is in binding metal nitrates in definite shares with a gel precursor PEO (M = 20 000), in which the molar ratio between polymer segments and the metal sum made up 1.25–10.0. Self-ignition was exercised at 573 K, oxidizers were $-NO_3$ groups and further thermal treatment of combustion products led to formation of a homogeneous phase. Its morphology and surface dimensions (1–7 m^2/g, crystallite size – 24–150 nm) were a function of concentration and production conditions.

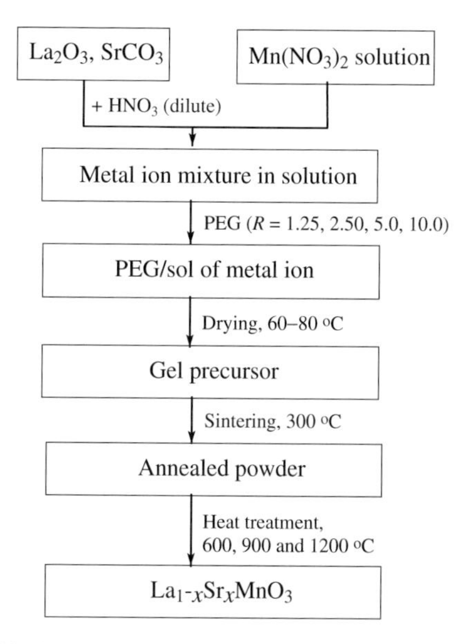

Fig. 5.20. Block scheme of $La_{1-x}Sr_xMnO_3$ synthesis using PEO

The analysis of the above phenomena showed that the reduction methods of metal ions immobilized in polymer matrices represented broad, unlimited vistas in designing polymer nanocomposites.

It does not present any difficulty for investigators to choose an optimum variant. This fact can be demonstrated by way of example of designing polymer–chalcogenide systems. It can be derived depending on the required nanocomposite properties through interactions between metal sulfides with addition of soluble polymers, in heterogeneous conditions using homo- and copolymers (with block-copolymers as well), even by incorporating into laminated structures and also by mineralization.

Many theoretical problems are awaiting their solution. They include physico-chemistry of nanoparticle nucleation and growth and the effect of composition and structure of the polymer matrix, formation of the metal-polymer phase, structural features and kinetic regularities of fractal cluster and fractal fiber growth in the polymer matrix, as well as the specifics and reasons that limit cluster–cluster aggregation in such systems.

These peculiarities have been extensively studied with respect to nanocomposites without polymers and quantitative relations have been derived (see Chap. 2). Investigations of metal–polymer systems will no doubt be progressing in this direction too. The level of quantitative investigations of named composites has immeasurably grown lately. A number of researchers have attempted to explain the reactive capability of atomic metal towards a polymer surface in terms of metal ionization potential and the lowest vacant molecular orbital of the polymer. Investigations have recently started in achieving bicomponent redox-active films through electrochemical reduction of C_{60} of $Pd_n(C_{60})$ type [238] or buckminsterfullerenes into unidimensional channels of microporous aluminum phosphates [239].

5.2 Nanocomposite Formation by Metal-Containing Precursor Thermolysis

A new and rapidly developing field in nanocomposite science is based on controlled pyrolysis of organic and nonorganic acid salts, metal complexes and metal-containing compounds in the presence of polymers or nanocomposite formation from thermolysis products. Such processes can be carried out into a gaseous phase (different variants of CVD synthesis), liquids (in high temperature solvents, in the suspension decomposition products from high-molecular liquids or polymer melts [240]) and solid state (in vacuum or in a self-generating atmosphere of gaseous transformation products). The processes temperature can be kept constant (isothermal conditions) or can be program-controlled (non-isothermal conditions, thermal analysis mode). The processes rate can be controlled by the amounts of evolving gaseous product, the change of the initial or final products or by the spectral and magnetic properties of the reacting system as a whole. Below we will consider the thermolysis of the most common precursors – the high-volatile metal-forming compounds, such as metal carbonyls and carbonitrosyls $M_a(CO)_b$, $M(CO)_x(NO)_y$, formiates, acetates and some π-allyl complexes. Naturally the metal-containing monomers are of special interest.

5.2.1 Metal Sol Formation in Polymers by Thermal Decomposition of Metal Carbonyls

The metal carbonyls are thermolysed according to Scheme 5.6:

$$Fe(CO)_5 \rightarrow Fe_2(CO)_9 \rightarrow Fe_3(CO)_{12} \rightarrow Fe_n^0$$

Scheme 5.6

forming nanoparticles.

Such reactions in the presence of polymers are the most widely used way to obtain polymer compositions having a large fraction (up to 90% in mass) of the colloidal metal particles. It is the main method to produce ferromagnetic nanoparticles incorporated into polymers due to macromolecule chemisorption on the nanoparticles directly at their moment of formation.

Metal carbonyl decomposition under special conditions can be accompanied by formation not only of metals but their compounds too. For example, one can obtain not only the α-ferrum but also Fe_3O_4 and FeOOH particles and ferric carbides using reactions (Scheme 5.7):

$$\begin{cases} Fe + CO \longrightarrow Fe_3O_4 + Fe_3C \\ Fe_3O_4 + CO \longrightarrow Fe_3C + CO_2 \\ CO \longrightarrow 0.5CO_2 + 0.5CO \end{cases}$$

Scheme 5.7

It was shown in [241, 242] that there were two methods to fabricate homogeneous polymer-immobilized dispersions of colloidal metal particles (Fe, Co, Cr, Mo, W, Mn, Re, Ni, Pd, Pt, Ru, Rh, Os, Ir) using the thermal decomposition of the precursors. In the former case an "active" polymer solution (containing amino-, amido-, imino-, nitrilo-, hydroxy- and other functional groups) is used. In an inert solvent a labile metal compound is gradually added to the solution (this operation creates the favorable conditions for the chemisorption interaction) followed by the thermal decomposition of the suspension at 370–440 K or by radiation.

A "passive" polymer in the second method is used. It can react with an initial metal complex if one ligand is lost (in our systems it is the CO group). Such polymers (e.g. PS, PB, styrene-butadiene copolymers, etc.) gradually added to the solution of the inert solvent of the initial complex at proper temperature lead to ligand separation, an anion complex binding with the passive polymer followed by its thermal decomposition.

Some stages of this multistep process (especially the particle growth) are similar to the mechanism of metal vapor condensation on the polymeric matrix, which we considered in Chap. 4. The thermolysis in the presence of polymers was studied to a great extent for carbonyls of cobalt (see for example [243, 244]) and iron [241, 244–246] (Table 5.4) when the process proceeds by direct $M_a(CO)_b$ vaporization over polymers or by preliminary adsorption on them.

In particular, $Fe(CO)_5$ thermolysis in a xylol solution (*cis*-PB or styrene-butadiene copolymer at 408 K, 24 hr, Ar atmosphere) has gone through the successive stages and led to the formation gave rise to the tricarbonyl iron(diene) chains $[C_8H_{12}Fe(CO)_3]_n$ (Scheme 5.8):

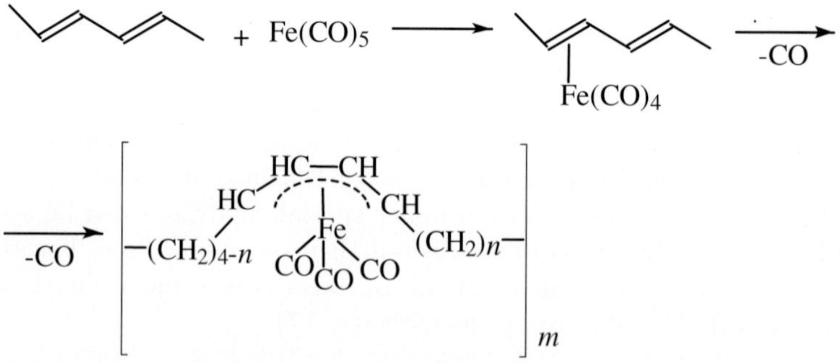

+ $Fe(CO)_5$ → $Fe(CO)_4$ →(-CO)

→(-CO) $[-(CH_2)_{4-n}$ HC HC—CH CH Fe CO CO CO $(CH_2)_n-]_m$

Scheme 5.8

At the initial stage of thermolysis a very active $Fe(CO)_4^-$ anion is formed reacting with the isolated double bonds after which chain double bond isomerization proceeds generating π-complexes with the tricarbonyl iron residues. The product is composed from the η^4-(butadienyl)tricarbonyl iron chains with *trans–trans* and *cis–trans* units. The tricarbonyl iron complexes, having two double nonconjugated bonds, are unstable and the intermolecular products in this situation can be formed. A distinctive feature of the process is that the same type of tricarbonyl iron complexes connected with allylic fragments were identified [254] in the $Co_2(CO)_8$ or $Fe_3(CO)_{12}$ interactions with polystyrene-polydiene block-copolymers. For polystyrene the $M(CO)_3$ fragments connected through the π-complexes with the phenyl ring are formed from $M(CO)_6$ (M = Cr, Mo, W) [255].

Such polymer-immobilized π-allylic complexes under thermal decomposition can form nanoparticles in polymer [256], during which the thermolysis can be carried out even without a solvent (for example by heating of the formed metallopolymer films using an infrared lamp). The nanoparticle sizes

Table 5.4. The thermal decomposition of metal carbonyls in polymer matrices

$M_a(CO)_b$	Polymer Matrix	Thermolysis Conditions	Polymer Content (mass %) and NPs Size (nm)	References
$Co_2(CO)_8$	(Without polymer)	Toluene	>100	[243]
$Co_2(CO)_8$	PP(atactic)	Toluene	75%, >100	[243]
$Co_2(CO)_8$	PS	Toluene	75%, 10–30	[243]
$Co_2(CO)_8$	Polyurethane	Toluene	75%, 5–30	[243]
$Co_2(CO)_8$	Polychloroprene	Toluene	93%, 30–60	[243]
$Co_2(CO)_8$	Polyesters	Toluene	75%, 6–20	
$Co_2(CO)_8$	Terpolymer MMA-ethylmethacrylate-vinilpyrrolidone (33:66:1)	Toluene	20–30	[244]
$Co_2(CO)_8$	Copolymer vinylchloride-vinylacetate alcohol (91:6:3)	Benzene chloride	75%, 7–47	[243]
$Co_2(CO)_8$	Copolymer MMA-vinilpyrrolidone (90:10)	Benzene chloride	75%, 6–25	[243]
$Co_2(CO)_8$	Styrene-acrylonitrile copolymer (88:12)	Toluene	75%, 6–13	[247]
$Fe(CO)_5$	PB	Decalin, 413–433 K	5–15	[241]
$Fe(CO)_5$	PB	Decalin, 433 K	~6	[246]
$Fe(CO)_5$	*Cys*-PB	Xylol, 408 K		[245, 246]
$Fe(CO)_5$	*Cys*-PB	1% solution dioxane-xylol, 408 K	37% Fe	[243]
$Fe(CO)_5$	PTFE	DMFA, 413 K	5–15	[246]
$Fe(CO)_5$	Natural rubber	8% solution dioxane-xylol, 398 K	17% Fe	[245]
$Fe(CO)_5$	Polystyrene -block-polybutadiene	Dichlorobenzene, 418 K	7–8	[241]

Table 5.4. (Cont.)

$M_a(CO)_b$	Polymer Matrix	Thermolysis Conditions	Polymer Content (mass %) and NPs Size (nm)	References
$Fe(CO)_5$	Styrene -butadiene copolymer (5.8:1)	Decalin, 433 K	~6	[248]
$Fe(CO)_5$	Styrene-4-vinilpyridine copolymer (1:0,05)	o-Dichlorobenzene	2% Fe; ~6	[249]
$Fe(CO)_5$	Styrene-4-vinilpyridine copolymer (1:0,1)	o-Dichlorobenzene	1.8 Fe, 16	[248]
$Fe_3(CO)_{12}$	*trans*-PB	Benzene, 10% ethanol, 353 K	23% Fe	[242]
$Fe_3(CO)_{12}$	Styrene-butadiene copolymer (25:75)	Benzene, 10% ethanol, 353 K	8% Fe	[250]
$Fe_3(CO)_{12}$	PB (81% 1.2–chains)	Benzene-dimethoxyethane 353 K	16% Fe	[241]
$Fe(CO)_5$	Isotactic PP (melt)		5% Fe	[251–253]
$Fe(CO)_5$	Atactic PP (melt)		5–30% Fe	[251–253]
$Fe(CO)_5$	PE (melt)		1–29% Fe, 1.5–7	[251–253]
$Cr(CO)_6$	PTFE (fluoroplast-40) (melt)		0.5–4% Cr, 1–5	[251–253]

depend on many factors such as the characteristics of the used dispersant polymer, its molecular mass (the optimal value is $M_w \approx 100\,000$), the nature of the functional groups (L) and solvent. The polymer crystallization proceeds in the solvent removing from the reacting systems. The formation of supermolecular structure depends on the cluster particle concentration and their size distribution. The nature of the solvent is very important because the $M_a(CO)_b$ complexes in the basic solvents (e.g. in DMFA) can disproportionate forming ionic type complexes where the metal formal charge equals +2. The product composition identified by IR-spectroscopy depends on the component ratio (Scheme 5.9):

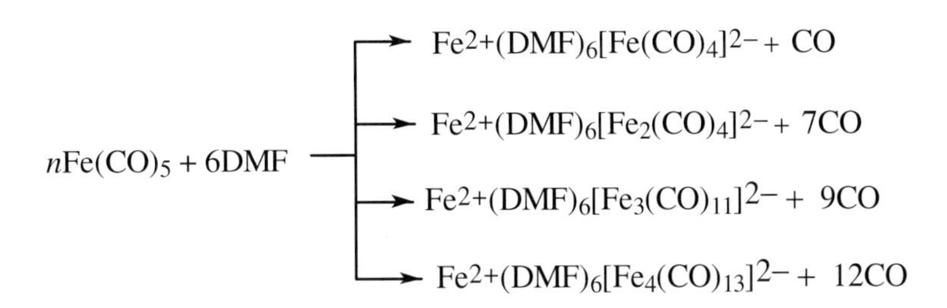

Scheme 5.9

In the general case one has:

$$Fe(CO)_5 + Nu \rightarrow [Fe^{2+}(Nu)_x]\ [Fe_n(CO)_m]^{2-}$$

where Nu is a nucleophyle (Py, N-methylpyrrolidone, etc. and $x = 2$–6), $n = 2$ or $m = 8$; $n = 3$, $m = 11$; or $n = 4$, $m = 13$.

The octacarbonyl–dicobalt transformation scheme in these conditions takes the form:

$$2Co_2(CO)_8 + 12DMFA \rightarrow 2[Co^{2+}(DMF)_6][Co(CO)_4^-]_2 + 8CO.$$

The basic solvents and high temperatures are favorable to metal carbonyls bonding by polymers and allow one to obtain stable colloidal dispersions (e.g. with iron particles 5–10 nm in size) by carbonyl thermolysis in dilute polymeric solutions. Such nascent nanoparticles are very reactive; in addition the small particles (<10 nm) are superparamagnetic and the larger ones (10–20 nm) exhibit magnetic hysteresis [248].

In solvents having a low dielectric permeability (e.g. in toluene) the obtained Co particle size (from 2 to 30 nm with a narrow size distribution) can be controlled by the reaction conditions (temperature, reagent concentration) or varying the polymer composition (tercopolymer of methylmethacrylate-ethylacrylate-vinylpyrrolidone [244]).

The nucleophilic fragments of macroligands (particularly in styrene-N-vinylpyrrolidone copolymer) bring about a polymer-catalyzed decomposition of the metal carbonyl [241] by Scheme 5.10:

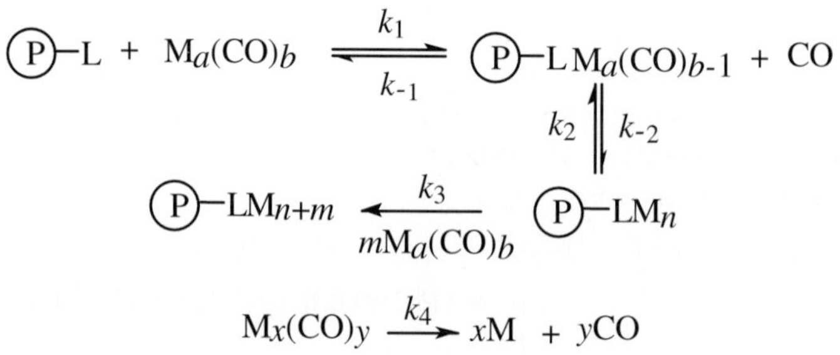

$$\text{(P)–L} + M_a(CO)_b \underset{k_{-1}}{\overset{k_1}{\rightleftharpoons}} \text{(P)–L}M_a(CO)_{b-1} + CO$$

$$\text{(P)–L}M_a(CO)_{b-1} \underset{k_{-2}}{\overset{k_2}{\rightleftharpoons}} \text{(P)–L}M_n$$

$$\text{(P)–L}M_n \xrightarrow[mM_a(CO)_b]{k_3} \text{(P)–L}M_{n+m}$$

$$M_x(CO)_y \xrightarrow{k_4} xM + yCO$$

Scheme 5.10

In the first stage a soluble macromolecular complex is formed with functional groups L. The complex transforms to a polymer-immobilized cluster particle which after the cleavage of CO groups undergoes a rise up to a nanoparticle. Actually we see the same development stages as in the above considered processes of nanoparticle formation in polymer, i.e. the initiation, particles growth and reaction termination (which can be presented as a simple disproportionation at the particle surface). The main condition for a proper matching of a functionalized polymer is connected with the greater reaction rate on the polymer surface to reduce the decomposition in the solution to a minimum, i.e. to provide

$$k_1 + k_2 - k_{-1} - k_{-2} > k_4,$$
$$k_3 > k_4.$$

The metal carbonyl decomposition in polar media as a rule proceeds by the main reaction routes. First is the main process, i.e. particle growth from the "hot" metal atoms and the formation of NP of 1–10 nm in size:

$$M_n + M \rightarrow M_{n+1},\; M_n + M_m \rightarrow M_{n+m}.$$

Coincidentally with the sizes of the forming nanoparticles the probability of termination of particle growth increases due to their noncovalent surface interactions with the macromolecule (it was mentioned in Chap. 3 that this interaction is very weak). For an interaction strength of order 10^{-4} J/m^2 even the particles of size 1–10 nm will be captured by the macromolecule and its growth will be stopped (as a result of a screening effect). The stronger the interaction the smaller the particle size. For small and modest-sized particles the reaction with the polymer is described by the scheme

$$\mathrm{M}_n + (\mathrm{P}) \longrightarrow (\mathrm{P})\!-\!\mathrm{M}_n$$

(in effect it is a chain termination) which usually leads to the formation of a non-magnetic material.

The disproportionate reaction products interact with the polymer chain, giving rise to some new side reactions (reticulation, chain destruction or isomerization and mononuclear carbonyl complex immobilization). As for instance in 1,4-*cis*-PB (containing 92% of 1,4-*cis*-, 4% 1,4-*trans*- and 4% 1,2-units with $M_w = 246{,}000$) after interacting with $Fe_3(CO)_{12}$ for 2 hrs at 350 K there proceeds a geometrical isomerization of the chain 1,4-units and the 1,4-*trans*-unit content rises up to 76% [245, 257].

Both processes run spontaneously and their competition (connected with the differences of the nature of the reacting particles and the various reaction conditions) is characterized by the ratio of the thermolysis and the M particle diffusion to the "hot" metal centers in a solid polymer matrix. If big particles (i.e. big in relation to the distances between the chains, crystalline blocks or polymer lattice points) are introduced in the system the polymer structure and its physical and mechanical properties degrade.

Carbonyl thermolysis in the absence of solvent can be accompanied by some peculiarities. In some systems the Fe, Cr, Co, Ni and Mn carbonyls can have a catalytic action on the polymer carbonization and graphitization processes, for example in petroleum mesogenic pitches (the specific matrices with polycondensed aromatic structures) being able to act as π-ligands stabilizing the metal clusters [258, 259]. Even the 1% content of metal carbonyl catalyzed the mentioned processes but the iron content rise up to 5% (at 423 K) leads to noticeable growth of the structures similar to the graphitic layer compounds.

Polymers having more polar groups promote the finest particle growth and the same can be said about the metal carbonyl concentration in the polymer. For a thermal $Fe(CO)_5$ decomposition in a PELD matrix a bimodal distribution of the iron particle size is observed (with the mean sizes of the distribution modes equal to 2–3 and 4–10 nm) and the particle content is ~20 and 80% (in volume) correspondingly as shown in Fig. 5.21 [260]. The majority of the oxide Fe_3O_4 particles have dimensions ~5 nm.

An original method of metallopolymer production by precursor thermal decomposition is to localize the particles being formed due to fast monomolecular decay of the solutions containing the metal compounds in polymer melts, i.e. in the natural voids of the polymer matrix (as PE, PP, PTFE, etc.). Such materials are called "cluspol" [251–253, 261] and for their production it is necessary to provide the highest possible melt temperature which must be considerably above the temperature of the initiation of carbonyl decay. For this purpose carbonyl dilute solutions are used under the conditions of the fast and complete removal of the split out ligand from the reaction system. This approach has many advantages because the temperature rise, on one hand,

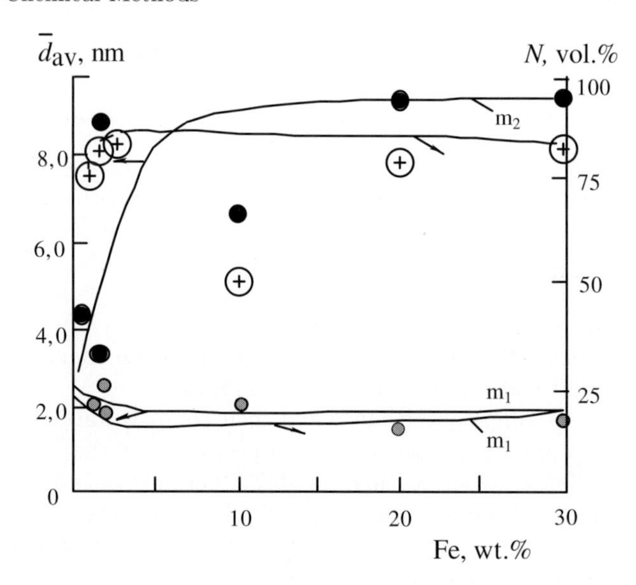

Fig. 5.21. The curves of Fe nanoparticle distribution on the size in a PELD matrix (m_1 and m_2 are the first and second modes of the distribution)

promotes the metal forming precursor decomposition and on the other hand decreases the byproduct yield. Besides in a melt (as distinct from a solution) the short-range order of the initial polymer structure is preserved and its voids become readily available for the localization of the forming nanoparticles. The obtained polymer-immobilized nanoparticles are characterized by a relatively high dispersion, a uniform distribution through the polymer volume and nonreversible macromolecular sorption on their surface. They are localized primarily in the disordered loosen interspherulite regions of the matrix, between the lamellas or in the spherolite centers. Such disposition hampers the segmental motion of the amorphous phase for a free volume decrease and possible interlacing. As a result the polymer thermostability increases (e.g. the T_m value for atactic PP rises by 50–80 degrees) and the products represent monolithic (pseudocrystalline) materials. The spatial periodicity of the ferrum clusters in such structures depends on concentration and averages between 9–12 nm for high (20–30% in mass) and 20–22 nm for low (2–3% in mass) concentrations. The particle size distribution is narrow (the half-width is about 1 nm) and moreover it is possible that the nanoparticles in the matrix intermolecular voids favor polymer crystalline phase destruction and its transformation to an amorphous state.

It is of interest that neither the physical nor chemical properties of such materials allow the detection of a metallic phase for the strong polymer-NP interactions to provide the polycrystalline matter formation. In other words such products at the some ingredients ratio present the monophase metallopolymers. The NP size and distribution depend on the chemical nature of the polymer, e.g. is it PELD, an alternating copolymer of ethylene and carboxide, polyacetonaphtalene, polycarbonate, etc. [260]. With a rise of the NP content in such polymers naturally their coherent scattering region (CSR) is widened and the crystallinity degrees of PELD and polycarbonates rise (Table 5.5). At the same time the polymer matrices cannot preserve the nanoparticles from oxidation (for example, the ferric nanoparticles in PELD with the mean size 50 nm after oxidation transform into Fe_3O_4 particles of the same size). It is of interest that metal carbonitriles as $Co_2(CO)_8$, form complexes with polyionic ligands (such as ethynylbensene-$Co_2(CO)_6$) can be decomposed explosively [262] at 513 K (generating CH_4 and H_2) and at 1070 K (forming multilayer nanotubes). It is possible that metabolized polygon pyrolysis can lead to moonwalk nanotube formation too [263] just as $Fe(CO)_5$ pyrolysis (at 1320 K) in the presence of other carbon sources [264]. The thermal treatment of platinum and cobalt containing poly(phenylene diacetylenes) leads to Pt (1–3 nm) or Co (5–100 nm) cluster formation in carboglasses [265]. The varied methods such as flame pyrolysis are described for synthesis of metal oxide nanoparticles [266].

Another method of production of metallocomposites by precursor thermolysis is metal carbonyl decomposition in haloid-containing matrices. Using the above mentioned classification such methods can be related to the "passive" ones. The PETF modifications can be performed by Fe^{3+} or Mn^{4+} oxides [267–269] using the sorption of the proper carbonyls and their following decomposition under the action of $KMnO_4$ or H_2O_2. After these reactions the oxides are introduced into the amorphous regions of PETF. The most common situation is that nanoparticles, derived in a polar solvent, form complicated complexes with metal carbonyls (see above) able to react with fluoropolymers. The main factors affecting the selective interaction of $Fe(CO)_5$ or $CO_2(CO)_8$ with PETF during the thermolysis are generalized in [270–272]. The IR-spectral method allows us to fix the cation–anion $[Fe(DMFA)_6]^{2+}[Fe_3(CO)_{11}]^{2-}$–PETF and $[Co(DMFA)_6]^{2+}[Co(CO)_4]_2^{2-}$–PETF complex formation during adsorption that allows performing kinetic studies of the process as a whole and thermal decomposition in particular. The thermolysis of intermediates (413 K, DMFA) is a first-order reaction but the rate constants and activation energies of these processes differ essentially (Table 5.6).

Attention is drawn to the fact that the rate constant of polymer-immobilized cobalt carbonyl is very high (its order of magnitude is comparable with the value for ferrum carbonyl) and correlates with a higher reactivity of the cobalt carbonyl anion. The very high value of E_a for neutral $Fe(CO)_5$ decomposition in PETF (as compared with the immobilized compounds) suggests that the reactions have different mechanisms. The thermolysis of Fe and Co carbonyls in situ leads to the formation of heterogeneous metal domains in a polymeric matrix. The polymer-connected carbonyls (both for Co and Fe) generate systems with uniformly distributed ferromagnetic particles of

Table 5.5. XRD data characteristics of Fe-composites materials [258]

Composite	CSRP, nm	Crystalline Phase Content, %	Particle Size, nm
An alternating copolymer of ethylene and carboxide	20	–	–
+ 10% Fe	21	–	–
+ 30% Fe	23	–	–
+ 50% Fe	23	–	α-Fe, 16.5 nm
			Fe_3O_4, 5 nm
Polyacenaphthylene	2.5	–	–
+ 0.1% Fe	2.6	–	–
+ 1% Fe	2.6	–	–
+ 10% Fe	2.6	–	Fe_3O_4, 3.5 nm
Polycarbonate	8.0	65	–
+ 0.5% Fe	9.0	60	–
+ 1.0% Fe	9.0	60	–
+ 5.0% Fe	10.0	50	–
+ 10% Fe	10.0	40	–
PEHP	21.5	30	–
+ 1.0% Fe	21.5	30	–
+ 5.0% Fe	26.5	20	–
+ 10% Fe	25.0	15	Fe_3O_4, 5.5 nm

Table 5.6. The characteristics of metal carbonyl thermolysis in PETF matrices

The Constants	PETF–$Fe(CO)_5$	PETF–$Fe_3(CO)_{11}^{2-}$	PETF–$Co(CO)_4^-$
$k{\cdot}10^{-5}$, s^{-1}	8.41	12.2	104
E_a, kJ/mol	142.6	51.62	45.80

5–10 nm in size and mean surface areas from 80 to 700 nm^2 per particle. Electron diffraction analysis of such systems [271] has revealed the formation of γ-Fe_2O_3 and FeF_2 particles for initial $Fe(CO)_5$ and Co, Co_2O_3 and CoF_2 for initial $Co_2(CO)_8$. The formation of metal fluorides is connected with the reactions involving the polymer chain and attended with the C–F-bond splitting (Scheme 5.11):

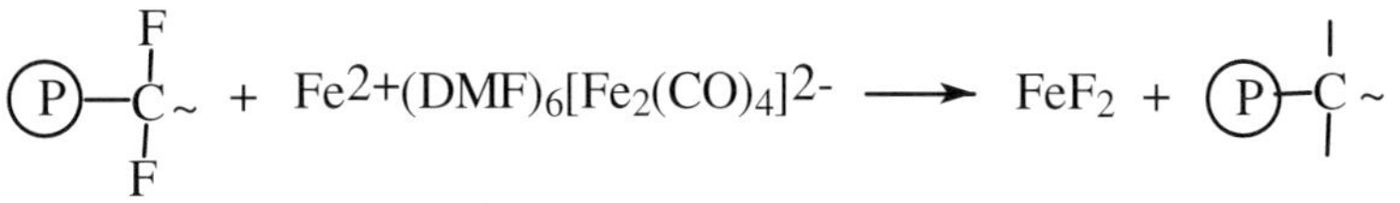

$$\text{(P)}\!-\!\underset{\text{F}}{\overset{\text{F}}{\text{C}}}\!\sim \; + \; Fe^{2+}(DMF)_6[Fe_2(CO)_4]^{2-} \longrightarrow FeF_2 + \text{(P)}\!-\!\underset{|}{\overset{|}{\text{C}}}\!\sim$$

Scheme 5.11

It follows that in such systems the two processes take place synchronously. The main process is nanoparticle formation due to thermolysis and the side process is the attack of the polymer chain by metal ions whereby the polymer goes through various transformations (destruction, cross-linking, binding with the metal complexes, etc). Moreover the secondary carbonium ions (formed from PETF) are very active and can be involved in many chemical reactions. As an illustration atomic zinc interacting with a PETF surface formed ZnF_2 [273,274]. The characteristics of the bonds between the formed clusters and nanoparticles with a polymer matrix and their topography have not been established conclusively but it is evident that the polymer–metal particle interactions are defined by the reactivity of metallocenters formed during the decomposition.

It is possible [275] that introducing $Co_2(CO)_8$ into the toluene solution of styrene block-copolymer with 4-VPy also gives rise to cation–anion complexes localized into micelle having 4-VPy-units (similar to DMF) with composition $[Co(4\text{-}VPy)_6]^{2+}[Co(CO)_4]^{2-}$. The topochemistry of such nanoparticles and their sizes or forms are defined by the molar ratio [4-VPy]/[Co] because the bounded (in micelles) cobalt forms small particles growing at the expense of dissolved $Co_2(CO)_8$ up to spheroids (~10 nm in size) or star-like particles (with nonregular forms and mean sizes 20–23 nm) built up from the initial anisotropic clusters. Only if $Co_2(CO)_8$ is present in large excess do the nanoparticles begin to originate and grow out of the micelle (cubes with mean size 21 nm). Such particles are vastly larger than the particles formed, for example, during $CoCl_2$ reduction by superhydride $Li(C_2H_5)_3BH$ [276].

In the last few years there are new trends in metal carbonyl decomposition in the presence of a polymer, namely the sonochemical reactions [277–279] connected with acoustical cavitation, including bubble initiation, growth and explosive "blow-in" into a high-boiling solvent. The process is characterized by a high local temperature (several thousand degrees), pressure (150–200 MPa) and acoustical emission of excited particles called sonoluminescence.

The acoustic field helps to generate the high-volatile metal compounds (particularly from metal carbonyls) with their following agglomeration in the presence of the polymer (e.g. PVP [280]) up to polymer-stabilized nanoparticles. The general scheme of such sonochemical synthesis of nanostructured materials can be present as follows (Scheme 5.12):

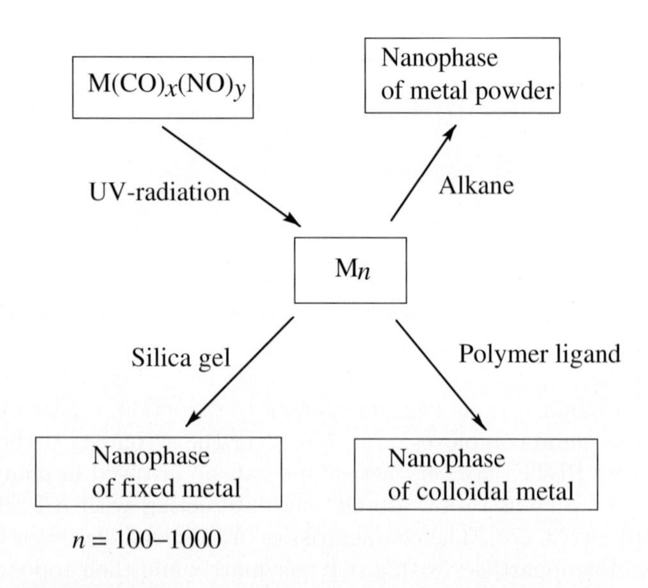

Scheme 5.12

Such reaction of $Fe(CO)_5$ (at 293–363 K, PVP) without ultrasonic radiation proceeds very slowly and only after a few days is material formed with very low Fe content (2%, isolated particles 2–5 nm in size). The sonochemical decomposition of $Fe(CO)_5$ does not proceed in the presence of PVP if THF is used as the solvent, but the reaction is very effective when anisole is used as the solvent and PFO as the polymer matrix [281]. As shown in Fig. 5.22, the black product formed contains up to 10% (in mass) of the spheric particles of non-oxidized Fe (mainly γ-Fe, with little content of α-Fe) with 1–12 nm size (the mean diameter is 3 nm). It is likely that the big particles present flocks of little ones ($\sim$2–2.5 nm). Sonochemical synthesis allows one to get functionalized amorphous nanoparticles of ferric oxide 5–16 nm in diameter [282–284]. The ultrasonic irradiation in the presence of PFO also allows the production of stabilized nanoparticles of copper and gold. In the literature the evidence is not about bimetallic particle formation in ultrasonic fields by carbonyl metal reduction in the presence of polymer matrices (as was made, for example, in the case of carbon-supported Pt/Ru from $PtRu_5C(CO)_{16}$ reduced clusters [285]).

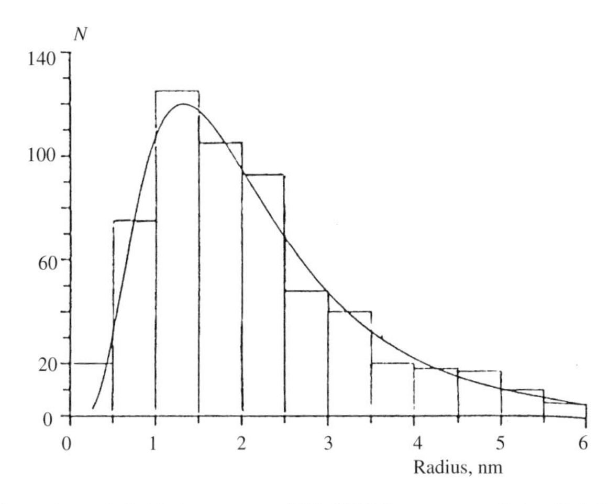

Fig. 5.22. Experimental histograms of Fe/PFO nanoparticle distribution on the size (N is the total number of particles)

5.2.2 The Features of Metal-Containing Polymer Thermolysis

This promising method of metal-polymer nanocomposite formation involves the synthesis of both nanoparticle and its stabilizing polymer matrix in one place and in one stage. This approach conceptually is unique and the systems under consideration are chemically self-regulating ones. They embodied the best solution of the problem (i.e. nanoparticle formation and stabilization in polymer systems). Although the method is realized on transition metals acrylates and maleates only, there are no principal limitations for its use with the other types of metallomonomers.

The kinetic studies of the thermal transformations were carried out in the self-generating atmosphere of many simple and clustered acrylates: $Co(CH_2{=}CHCOO)_2{\cdot}H_2O$ ($CoAcr_2$) [286], $Ni(CH_2{=}CHCOO)_2{\cdot}H_2O$ ($NiAcr_2$) [287], $Cu_2(CH_2{=}CHCOO)_4$ ($CuAcr_2$) [288], $Fe_3O(OH)(CH_2{=}CHCOO)_6{\cdot}3H_2O$ ($FeAcr_3$) [289], their cocrystallizates $[Fe_3O(OH)(CH_2{=}CHCOO)_6]{\cdot}[Co(CH_2{=}CHCOO)_2]_{2.4}$ (FeCoAcr) and $[Fe_3O(OH)(CH_2{=}CHCOO)_6]{\cdot}[Co(CH_2{=}CHCOO)_2]_{1.5}{\cdot}3H_2O$ (Fe_2CoAcr) [290, 291], maleate $Co(OCOCH{=}CHCOO){\cdot}2H_2O$ ($CoMal_2$) [292] and acid maleate $Fe_3O(OH)(OCOCH{=}CHCOOH)_6{\cdot}3H_2O$ ($FeMal_3$) [293]. Some common features can be presented as follows [294–297].

The thermal decomposition is attended with evolution of gas and mass loss of the samples. The processes have similar characteristics for the iron(III)-methacrylate thermal decomposition in the TA regime, studied by EPR- , NMR- and IR-spectroscopy methods [297]. It is due to the fact that the transformations go through three main macrostages with different temperatures [288–294, 298, 299]:

(1) Initial monomer dehydration (desolvation) at 303–473 K;
(2) The stage of solid-state homo- and copolymerization of the dehydrated monomer (at 473–573 K);
(3) The produced polymer decarboxylation to a metal-containing phase and oxygen-free polymer matrix at $T_{ex} > 523$ K (for copper acrylate $T_{ex} > 453$ K) with an intense gas emission.

A little weight loss of samples takes place during the polymerization and the monomer IR-spectra change due to absorption in the range of the symmetric and asymmetric valence vibrations of C–O and C=C bonds.

The gas emission kinetics with reference to a value $\eta(t)$, the degree of transformation for all compounds, can be approximated adequately by the expression

$$\eta(t) = \eta_{1f}[1 - \exp(-k_1\tau)] + (1 - \eta_{1f})[1 - \exp(-k_2\tau)],$$

where k_1, k_2 are effective rate constants; $\eta_{1f} = \eta(t)$ at $k_2 t \to 0$ and $k_1 t \to \infty$; $\tau = t - t_h$; and t_h is the sample warming up time.

The transformation kinetic parameters are presented at Table 5.7. The change of ratio m_0/V (m_0 is the initial mass of the sample; V is the reaction vessel volume) does not practically affect the studied carboxylate transformation rate. It is particularly remarkable that for $FeAcr_3$ thermolysis there are low- and high-temperature ranges of the gas emission characterized by the different kinetic parameters. It seems likely that this difference in the formal kinetic description of the transformation rates for the $FeCoAcr_5$, Fe_2CoAcr_8, $NiAcr_2$, $FeMal_3$ thermal decomposition (with other kinetic factors held constant, including η_{1f} too) is connected with these two parallel gas emission processes. The studied $MAcr_n$ compounds can be ordered according to their transformation initial rates (i.e. by a decrease of the ability of gas emission) in the following manner: Cu $\geq$ Fe $>$ Co $>$ Ni. The analysis of the thermal transformation of the gaseous products and the composition of the solid product (decarboxylated polymer, including metal or its oxide) allows determining a general scheme of metal acrylate thermal transformations (Scheme 5.13).

One of the main transformations is the origin of the acrylic $CH_2{=}CHCOO$ radical in the primary decomposition act that initiates metal-containing monomer polymerization with a consequent decarboxylation of metal-containing units. The process temperature has an impact on the product yield and their composition. It was shown [293, 300] that even at the dehydratation stage the surrounding metal ligand changes and during the process such reconstruction is growing. At high temperatures the decarboxylation is accompanied by practically complete removal of the oxygen-containing units from the polymer matrix.

Studies of the specific surface of nanocomposites and its topography [301, 302] (Table 5.8) show that they are powders where crystallinity does not exist on the distances correlating with transmitted light wavelength. The produced samples have high values of specific surface (15–30 m^2 g^{-1}) and

Table 5.7. Thermolysis kinetic parameters of transition metal unsaturated carboxylates

MR_n	T_{ex}, K	η_{1f}, $\Delta\alpha_{\Sigma,f}$	η_{1f}, $\Delta\alpha_{\Sigma,f}$ = $A\exp[-\Delta H/(RT)]$		k	$k = A\exp[-E_a/(RT)]$	
			A	ΔH, kJ/mol		A	E_a, kJ/mol
$CuAcr_2$	463–513	η_{1f}	$1.8 \cdot 10^4$	48.1	k_1	$9.5 \cdot 10^{11}$	154.7
		$\Delta\alpha_{\Sigma,f}$	3.6	12.5	k_2	$9.2 \cdot 10^{11}$	163.0
$CoAcr_2$	623–663	η_{1f}	1.0	0	k_1	$3.0 \cdot 10^{14}$	238.3
		$\Delta\alpha_{\Sigma,f}$	1.55	0	k_2	0	0
$FeAcr_3$	473–573	η_{1f}	1.0	0	k_1	$4.2 \cdot 10^{21}$	246.6
		$\Delta\alpha_{\Sigma,f}$	$1.6 \cdot 10^2$	25.5	k_2	0	0
	573–643	η_{1f}	1.0	0	k_1	$1.3 \cdot 10^6$	127.5
		$\Delta\alpha_{\Sigma,f}$	$1.7 \cdot 10^2$	26.3	k_2	0	0
$NiAcr_2$	573–633	$\eta 1f$	2.6	1.1	k_1	$1.7 \cdot 10^{17}$	242.4
		$\Delta\alpha_{\Sigma,f}$	$1.4 \cdot 10^{11}$	125.4 (<613 K)			
		$\Delta\alpha_{\Sigma,f}$	1.2	10.5 (>613 K)	k_2	$7.5 \cdot 10^8$	156.8
FeCoAcr	613–663	η_{1f}= 0.45(663 K) – 0.65(613 K)			k_1	$2.3 \cdot 10^{12}$	206.9
		$\Delta\alpha_{\Sigma,f}$	$5.25 \cdot 10^2$	7.5	k_2	$6.0 \cdot 10^8$	137.9
Fe_2CoAcr	613–663	η_{1f}= 0.35(663 K) – 0.50(613 K)			k_1	$2.6 \cdot 10^{12}$	204.8
		$\Delta\alpha_{\Sigma,f}$	$1.9 \cdot 10^2$	6.0	k_2	$6.6 \cdot 10^5$	125.4
$CoMal_2$	613–643	η_{1f}	1.0	0	k_1	$1.6 \cdot 10^6$	125.4
		$\Delta\alpha_{\Sigma,f}$	$1.3 \cdot 10^2$	23,4	k_2	0	0
$FeMal_3$	573–643	η_{1f}	$0.59 \cdot 10^2$	23,4	k_1	$3.3 \cdot 10^7$	133.8
		$\Delta\alpha_{\Sigma,f}$ = 4.78(573 K)–7.40 (643 K)			k_2	$1.0 \cdot 10^7$	110.8

a corresponding dispersity. The mean size of the metal-containing particles (using data on S^f_{sp} and assuming complete polymer decarboxylation) is ~20–30 nm. In some special cases ($CuAcr_2$, $CoAcr_2$ and partially $NiAcr_2$) the dispersion of the big aggregates, a decrease of particle mean size and a growth of S_{sp} [286,288] were observed. At deep stages of metal carboxylate pyrolysis the small (<1 μm) opaque particles sometimes generate fractal-type chain structures 50–70 μm in length constituted by the agglomerates from the 6–7 primary particles.

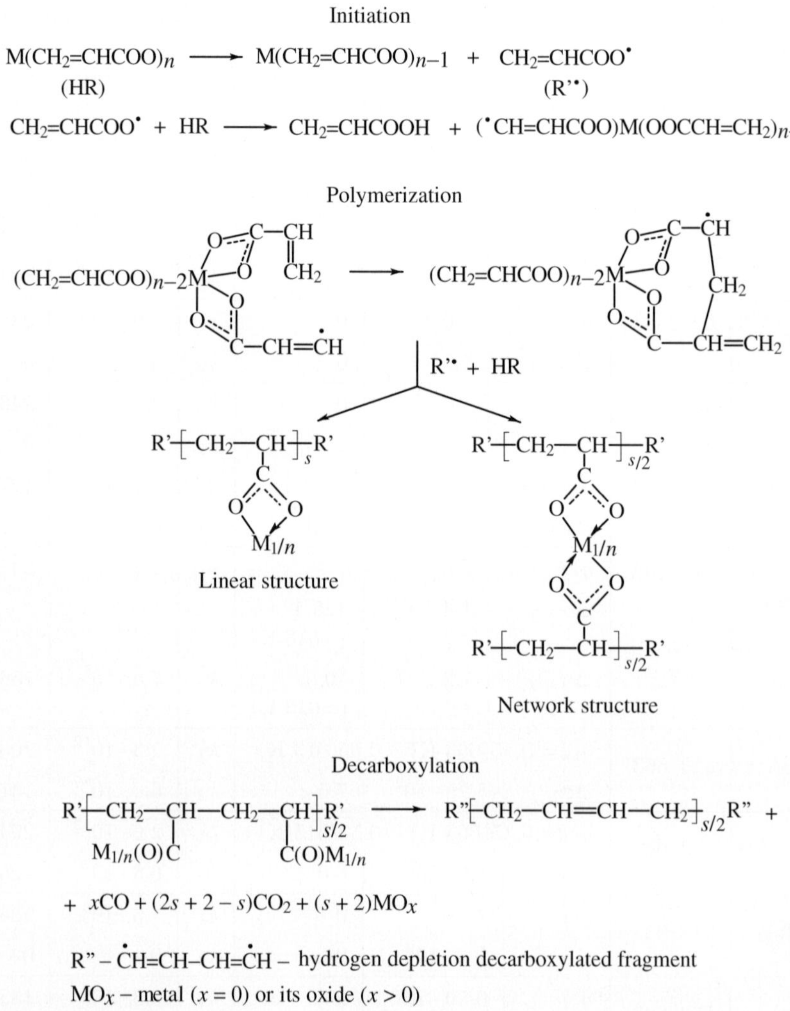

Initiation

$$\underset{(HR)}{M(CH_2{=}CHCOO)_n} \longrightarrow M(CH_2{=}CHCOO)_{n-1} + \underset{(R'^{\bullet})}{CH_2{=}CHCOO^{\bullet}}$$

$$CH_2{=}CHCOO^{\bullet} + HR \longrightarrow CH_2{=}CHCOOH + ({}^{\bullet}CH{=}CHCOO)M(OOCCH{=}CH_2)_{n-1}$$

Polymerization

$$(CH_2{=}CHCOO)_{n-2}M(OOC{-}CH{=}CH_2)(OOC{-}CH{=}\dot{C}H) \longrightarrow (CH_2{=}CHCOO)_{n-2}M(OOC{-}\dot{C}H{-}CH_2{-}CH{=}CH_2{-}COO)$$

$$\xrightarrow{R'^{\bullet} + HR}$$

$$R'{-}[CH_2{-}CH(COO M_{1/n})]_s{-}R'$$

Linear structure

$$R'{-}[CH_2{-}CH]_{s/2}{-}R' \;(\text{C(O)O}{\rightarrow}M_{1/n}{\leftarrow}\text{O(O)C})\; R'{-}[CH_2{-}CH]_{s/2}{-}R'$$

Network structure

Decarboxylation

$$R'{-}[CH_2{-}CH(C(O)M_{1/n}){-}CH_2{-}CH(C(O)M_{1/n})]_{s/2}{-}R' \longrightarrow R''{-}[CH_2{-}CH{=}CH{-}CH_2]_{s/2}{-}R'' +$$

$$+\ xCO + (2s + 2 - s)CO_2 + (s + 2)MO_x$$

$R'' - \dot{C}H{=}CH{-}CH{=}\dot{C}H$ – hydrogen depletion decarboxylated fragment

MO_x – metal ($x = 0$) or its oxide ($x > 0$)

Scheme 5.13

The topography and composition of the solid phase products were studied using electron microscopy and electron diffraction methods [289,291,292,298] and the results present for them a morphologically similar picture. There are the electron-dense metal-containing particles of oxides with a form near to spheric. They are presented as individual particles and aggregates (from 3–10 particles), being uniformly distributed through the matrix space with lower electronic density. The particles have a narrow size distribution with mean diameter $d = 4.0$–$9.0\,\text{nm}$ (Fig. 5.23) and the distance between them in the matrix is 8.0–10.0 nm. At the same time there are some big aggregates in

Table 5.8. Dispersity of starting metal carboxylate samples and the thermolysis products

Sample	$S_{0,sp}$, m^2/g	$S_{f,sp}$, m^2/g	$L_{OM,av}$, µm
$CuAcr_2$	14.7	48.0 (463 K) – 53.8(473 K) – 43,8(503 K)	5–50
$CoAcr_2$	20.2	24.1 (623 K)– 42.1(663 K)	100–150
$FeAcr_3$	15.0	15.0	1–5
$NiAcr_2$	16.0	55.0–60.5	60–100
FeCoAcr	9.0	13.6	5–10
Fe_2CoAcr	8.1	11.3	10–15
Fe_2NiAcr	8.5	13.5	100–200
CoMal	30.0	30.0	5–70
FeMal	24.0	26.0	30–50

the form of cubic crystals 10.0–20.0 nm in size [292]. The uniformity of the metal-containing particle space distribution and the narrow size distribution suggest that the decarboxylation and new phase formation processes are largely homogeneous. The estimates show [299] that the mean distance between forming nanoparticle centers is in the range 7.5–13.5 nm, i.e. near the above mentioned value 8.0–10.0 nm.

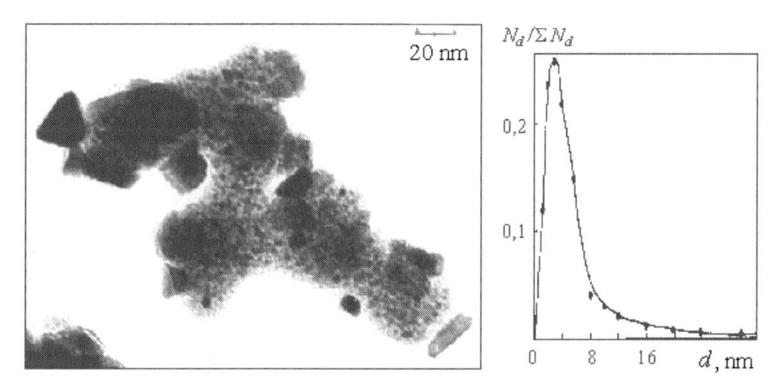

Fig. 5.23. Transmission electron micrographs of the product of thermolysis of $Co(OOCCH{=}CHCOO){\cdot}2H_2O$ (T_{ex} – 623 K) and the particle size distribution

Nanoparticle synthesis was combined with their synchronous stabilization by the forming decarboxylated matrix in the system of transition metal unsaturated carboxylates [298, 303].

5.2.3 Post-Thermolysis of Other Types of Precursors in Polymer Matrices

Besides carbonyls some other metal-forming precursors (usually formates, acetates, oxalates or organometallic compounds) are used in producing polymer-immobilized nanoparticles by thermolysis. For example, a metal–polymer composition can be prepared by thermolysis of copper format triethylene-diaminic complex $[Cu(EDA)_3](HCOO)_2$ into PS through a common solvent (DMF) [304]. The complex decays at 443 K and forms metallic copper in a highly dispersed state. The process temperature in PS is 20 degrees lower than in bulk copper formate. It may be associated with polymer catalytic action on the complex thermal decomposition.

The mechanism of the process is that the polymer reactive centers promote metal nucleation and aggregation, after which thermolysis occurs and the metal-containing substance is redistributed. The maximum amount of copper being introduced in PS through a common solvent is about 10%. At the same time the presence of the polymer increases the temperature of cadmium trihydrate-oxalate decomposition [305] and the decay products increase the initial temperature of the intensive destruction of PETF. The copper formate thermal decomposition in the highly dispersed PETF allows producing a metallopolymeric composition (20–34% of copper) where the nanoparticle size distribution is maximal at ~4 nm, without any chemical interaction between the components.

The metal clusters had formed in high temperature thermolysis (1270 K) of cobalt acetate in the presence of PS, PAA, PMVK catalyzing an oxygen electrolytic reduction [306]. As the result of thermal decomposition of silver trifluoro-acetyl-acetonate at 613 K polyimide film [307–309] becomes metallized (a structure called “film on film”) and a nanocomposite is formed with high surface conductivity and light reflection coefficient (above 80%).

There are some examples of nanoparticle formation in polymers by metal hydroxide decomposition. The thermal decay of silver hydroxide ammonia solution in the isotactic PP melt (temperature range 543–563 K in polymethylsiloxane) goes by the scheme:

$$4[Ag(NH_3)_2]OH \rightarrow 4Ag^0 + 8NH_3 + 2H_2O + O_2.$$

The silver content in PP varies from 3 to 15% (in mass), the particle sizes change in a wide range (2–25 nm), the small particles (<5 nm in diameter) are roentgen-amorphous but the bigger ones have a crystalline structure [310].

The Pd particles (with $S_{sp} = 26\,m^2/g$), produced by the decomposition of a palladium hydroxide fine dispersion in PVC formed the space structures with nodes where the high-dispersed Pd particles are located [311]. Such structure increases the temperature of complete dehydrochlorination of PVC.

An original method of copper containing composite production on porous PELP base is connected with using of Cu^{2+}-monoethanolamine complex [312]. The complex simultaneously combines the functions of an adsorption active medium, a complex-forming agent and a copper ion reductant. Under complex decay (at temperatures from 363 to 398 K) the copper nanoparticles form in the porous PE. They have a wide size distribution (from 60 to 320 nm) and the metal mass content is about 8%. Even a small amount of such filler (1–2 vol.%) improved considerably the mechanical properties of the polymer.

There are some examples with limited evidence of nanoparticle formation by organometallic compound decomposition in polymers. The cryochemically synthesized bisarene complexes $(C_6H_5CH_3)_2M$ (M = Ni,Co) were subjected to a fast decomposition in PELD at temperature higher than the complex decay one. The metal content in then produced material achieved 1.4–4.3% [313]. The cobalt nanoparticles (mean size ~1.6 nm, a narrow size distribution) were obtained by H_2 reduction of organometallic precursor $Co(\eta^3\text{–}C_8H_{13})(\eta^4\text{–}C_8H_{12})$ in the presence of PVP at 273–333 K [314]. As a precursor bis(cyclooctatetraen)-iron $Fe(C_8H_8)_2$ can be used too [315]. By contrast, the similar ruthenium complex $Ru(\eta^4\text{-}C_8H_{10})(\eta^6\text{-}C_8H_{12})$ decomposed slowly even at 0.3 MPa in the presence of PVP. The Mo, Cr and W hexacarbonyls did not photochemically decompose at high pressures (~35 MPa) of N_2 or H_2 and low temperatures (~30 K) in a polyethylene matrix but gave different substituents [316] as well as ferrum carbonyls in PE matrix at 190 K [317]. The nanoparticles of Ru (~1.2 nm [318]), Pd and Pt [319,320] can also be produced from the organometallic precursors in PVP matrix at 298 K. Nickel dicyclo-octadiene $Ni(COD)_2$ can spontaneously decay (at room temperature in a PVP dichloromethane solution), forming particles with mean size 2–3 nm [321]. The nanocomposites can be obtained in the same manner (by precursor decomposition) on the base of Nafion-type films of ionic polymer [322–324].

An original method is developed for orgametallic compound impregnation into the PS, PELD , PETF amorphous regions and polyacrylates with their consequent structural modification and clustering. The method is based on the usage of some liquids in a supercritical state (CO_2, 8–25 MPa, 303–313 K), as shown for cymanthrene $(\eta^5\text{-}C_5H_5)Mn(CO)_3)$ [325]. In another method the platinum dimethyl-cyclooctadiene(II) precursor was dissolved in CO_2 and the thin films of poly(4-methyl-1-pentene) or PETF were impregnated from the solution [322, 326]. The polymeric nanocomposites are formed with Pt particles (15–100 nm in size) after thermolysis or hydrogenolysis reduction. The kinetic data of impregnation for manganese cyclopentadienyl-tricarbonyl and copper hexafluoro-acetylacetonate, dissolved in the supercritical media (CO_2), show that the process is very intense in the polymer amorphous regions, very weak in the partially crystalline regions and practically absent in crystalline regions [323].

In summary of this section it must noted that, in spite of numerous studies, we still know very little about the thermolysis of carbonylhydrates and other substituted (mixed) carbonyls in polymeric systems, as well as in reactive plastics. For example, in some experiments the decomposing metal carbonyls were placed into an epoxide resin heated up to allow nanoparticle deposition on the forming polymer surface [219]. It is possible that the high reactive metal particles in such systems can initiate the epoxy cycle cleavages following by the formation of a three-dimensional space structure. Ferrum carbonyl being decomposed into polybenzimidazole suspension (in transformer oil at 473 K) forms ferrum nanoparticles (1–11 nm) capable of polymer thermostabization [327].

The simple picture of nanocomposite formation is observed only when the thermally unstable substances are used (e.g. the metal containing precursors) for the thermolysis. Otherwise the thermal destruction of the polymer matrix, the reactions connected with the chemical bond breakage and radical generation begin to play an important role along with nanoparticle formation processes. The thermodestruction products present low-molecular volatile compounds of complex type (including monomer) and nonvolatile residue. These products become a carbonized structure in the course of time. On the other hand, the highly dispersed metals of alternating valence or their lowest oxides (Fe, FeO, Ni, Cu, etc.) with a defective structure are often introduced in polymer specially for inhibiting its thermal or thermooxidative destruction processes [328,329]. For example, a small amount of iron particles in PELD (0.05–1.0% by mass) increases its thermal stability in comparison with the pure polymer.

The transformations are more complicated for thermolysed polymer composites with metal halogenides such as MX_n–PMMA (M = Cr, Mn, Zn; X = Cl, Br) [330]. For example, the $CrCl_3$–PMMA composite lost under pyrolysis at 523–773 K is about ~62% in mass. The volatile fraction contains monomer, carbon oxides, HCl, methane traces and non-identified organic products (in the condensed phase there are present, along with nonsaturated oligomers, highly dispersed solid chromium anhydride and oxides) [331–335]. A quantitatively similar picture of PMMA destruction was observed (at 373–873 K) for $MnCl_2$ [336], when manganese oxide formed during PMMA manganese ionomer thermolysis. The compound $CuBr_2$ (in concentrations 5–10%) has a thermostabilizing property for PEO [337].

Metallopolymeric systems under pyrolysis permit us to produce some other important and interesting products. For example, carbonic films with cobalt nanoparticles can be produced by macrocomplex pyrolysis ($CoCl_2$ with PAN at 1570 K) [338]. Thermotreated PAN complexes with copper halogenides have an electrical conductivity [339]. Macrocomplexes with two metals (e.g. Co, Ni or Rh) demonstrate a synergy in thermostabilization [340].

The Fe, Co and Ni nanoparticles can be used as catalysts for low-temperature graphitization of amorphous carbon [341]. Many such examples demonstrate that metal acetylacetonates in polymer matrices can perform a double function. They interact with the matrix forming the metal-containing products and initiating depolymerization and destruction [342].

The metal-containing precursors and nanoparticle interactions with the polymer matrix, and the product topography are not sufficiently well studied. But the role of metal centers and their activities are very important for metal-particle interaction with a polymer matrix.

5.3 Nanocomposite Formation in Monomer–Polymer Mixtures in Thermolysis

The separate components (monomer and nanoparticle precursor) are inserted into a polymerizing system by this method. The specific of this method is that post-thermolysis is used to get a nanoparticle-containing composite. A material produced by polymerization is heated and transformed into a metal–polymer composite with uniform distribution of the metallic component. This permits us to omit the tedious stage of ingredient mixing. But some other problems arise. The precursor affects the polymerization of matrix-forming monomer. It was shown that moderate amounts of potassium-organic compounds did not have an impact on the initial rate of MMA radical polymerization [343–345]. There are two ways to form nanoparticles in these conditions. The first one is based on the use of organometallic compounds, such as cadmium bis(triethylgermyl), with the decomposition temperature being near to the polymerization temperature. Another way is warming up the formed composition, containing components with the decomposition temperature higher then the polymerization temperature, e.g. MMA (cadmium alkyl compounds). The presence of such Cd-organic compounds decreases the forming PMMA molecular mass (e.g. from 800 000 to 60 000 at the content of only 2.5% mass). This can be explained by a chain transfer reaction to the organo-metallic compound. The nanoparticle formation scheme in such a monomer–polymer mixture can be as follows (Scheme 5.14):

$$\text{MMA} + [(C_2H_5)_3Ge]_2Cd \xrightarrow[t^\circ]{\dot{R}} Cd_j/\text{PMMA} + 2\dot{G}e(C_2H_5)_3$$

Formation of nanoparticles immobilized in PMMA

(a) initiation of MMA polymerization
(b) macroradical termination
(c) recombination

Scheme 5.14

The sizes of Cd nanoparticles formed during the polymerization stage are from 6.3 to 300 nm with the change of initial component content in monomeric mixture in the range 8–40% mass. The larger nanoparticles (up to 100 nm) with correct hexagonal form characteristic for Cd crystals are formed under heating.

The processes depend upon the rates of decomposition of organo-metallic components (or other precursors), the monomer polymerization, as well as the sedimentational stability of the formed metal dispersions in the monomer.

The same principle was used in [346] to get Rh, Pd, Pt, Ag or Au sols in PMMA at the stage of MMA polymerization. The process included dissolution of metal salt and initiator (benzoyl peroxide) in MMA with the consequent heating to 343 K for 0–60 min. The nanoparticles were formed by post-heating of the eliminated products at 410-430 K (Scheme 5.15):

$$\text{MMA} + \text{MX}_n \xrightarrow{\text{Solution}} \text{M}^{n+}/\text{MMA} \xrightarrow{\text{Initiation}} \underset{\text{Solid solution}}{\text{M}^{n+}/\text{PMMA}} \longrightarrow$$

$$\xrightarrow{\text{Heating}} \underset{\text{Solid sol}}{\text{M}_j/\text{PMMA}}$$

Scheme 5.15

The nanocomposite thermal decomposition temperature (Table 5.9) is well above the decomposition temperature of PMMA (373 K).

Table 5.9. The properties of nanoparticles produced by Scheme 5.15

Composite	Particle Sizes (nm) After Heating to 433 K	Concentration, μ mol	Decay Temperature, K
Rh-AA	5–10	50	>493-503
Pd-AA	1.5–2	0.5	>458
Pd-AA	1.0–2.0	50	
Pt-AA	1.5–2.0	0.5	>4685
Ag-CN-STMA	2.0–4.0	0.5	>230
Au-Cl-STMA	8.0–10.0	0.5	>503

Notes: AA is acetyl acetone residue; STMA is styryl-trimethyl-ammonia residue

MMA polymerization initiated by AIBN or BP in the presence of silver trifluoroacetate ($AgCF_3CO_2$) followed by thermal treatment (at 393 K) leads to Ag^0/PMMA–nanocomposite with the mean diameter of Ag particles 3–10 nm [347, 348]. The introduction of metal salt into the system decreased the polymerization rate (Fig. 5.24). Ag^+ reduction by the growing polymer radicals was observed even in the polymerization stage (333 K). The formed small silver clusters, dispersed in the polymer matrix, act as nucleation agents and give rise to a cluster aggregation and the formation of bigger nanoparticles under thermal treatment. Metal-composite materials based on the Au, Pt and Pd can be formed in the same conditions but with thermal treatment at 413 K [346,349].

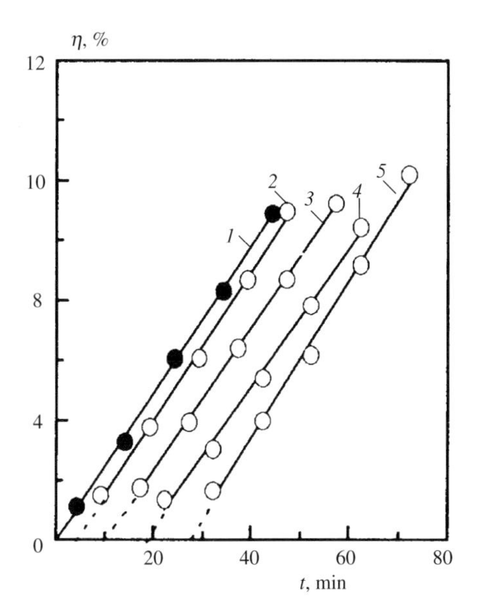

Fig. 5.24. The influence of $AgCF_3CO_2$ on the polymerization of MMA (333 K, initiator AIBN, $2.85 \cdot 10^{-2}$ mol/L) at the concentration $[AgCF_3CO_2] \cdot 10^3$ mol/L: 1 – 0, 2 – 1.3, 3 – 8.0, 4 – 14, 5 – 20)

Another technology of nanocomposite production is based on monomer polymerization and afterheating. This approach transforms liotropic (amphiphilic) liquid crystals into an inverse hexagonal phase. The main principle is that a photoinitiator is added into the composition, containing monomer (M = Na, Co, Ni, Cd, Eu, Ce), water and 2-hydroxi-2-methyl-propiophenone xylolic solution. It allows one to carry out the photopolymerization resulting in the formation of a nanostructured polymer network [350–352]. A mixture of the photopolymerized metallopolymer with poly(p-phenylene-vinylene) forms a composite after heating to 493 K in vacuum. There are other methods of polymerization of metal acrylates, methacrylates [353] and sorbates [354] with their subsequent transformation into nanoparticles of corresponding metals.

5.4 Nanocomposites on the Base of Polymer-Immobilized Metalloclusters

In many cases it is necessary to introduce a cluster-containing group into a polymer chain. For example, polymer-immobilized metal catalysts are produced for the various organic reactions. Such systems are interesting theoretically as they allow one to study the dimensional effects in catalytic activity, to consider and illustrate many general notions of catalysis theory and to "construct bridges" between homogeneous and heterogeneous catalysis. The cluster formations are very useful objects for metal–polymer interface modelling, the study of metal-polymer adhesion, atomic metal vaporization on the polymers or polymer surface decoration. The clusters are widely used to improve the tribotechnical characteristics of the material or antifrictional coatings and to get new components of greases. In general metalloclusters are the most promising for producing hybrid polymer-inorganic nanocomposites [355]. In recent years methods of cluster immobilization with known composition M_mL_n on polymers were widely recognized.

5.4.1 Monometal-Type Cluster Immobilization

Many problems of cluster bonding with polymers (e.g. a justified macroligand choice, the reaction conditions, the product characteristics and composition) still remain unsolved. The functionalized macroligands, especially the phosphorylated styrene-divinylbenzene copolymers ($SDVBC–PPh_2$) or their species are the most widely used. Styrene-divinylbenzene-vinyl(styryl)diphenylphosphine triple block-copolymers ($\mathbf{P}–PPh_2$) are used as powders or grains (with particles 0.1–0.05 mm, high S_{sp} and porosity) and sometimes as thin membranes. The preparation of the functionalized phosphorus containing polymers and films with thickness 11 μm is described in [356]. The metallocluster bonding is carried out by the methods of ligand or ionic exchange, oxidative addition, decarbonylation, and ligand addition. The preparative methods of the phosphorylated PS and CSDVB synthesis and cluster addition to them can be presented as the simplified Scheme 5.16. The first attempts to prepare the polymer-bonded clusters relate to $Rh_6(CO)_{16}$ and $Rh_4(CO)_{12}$ with $CSDVB–PPh_2$ [357]. Sometimes at cluster bonding by polymer the cluster is decomposed to a mononuclear complex, as for photogeneration of polymer-bonded ferrum carbonyls [358]. The synthetic methods of nondestructed carbonyl clusters on such surfaces were reviewed in [359], the data on bonded triosmium clusters are presented in [360] and tetrairidium clusters are in [361]. The tetrairidium carbonyl cluster was composed in situ on a polymer with mononuclear complex [359].

The main types of clusters chemically bonded with polymers and produced by polymer-analogue transformations are listed in Scheme 5.17. Comparison of IR and NMR spectra of the initial or model compounds and the

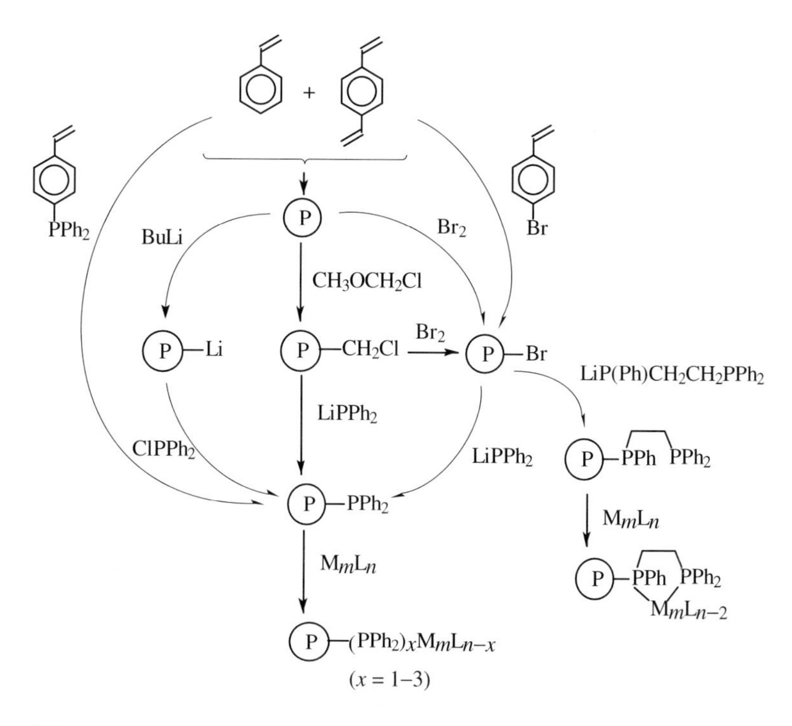

Scheme 5.16

formed products define the structure of such formation. This is especially true for three-nuclear Ru and Os clusters because the absence of bridge ligands and high symmetry in such clusters provide important spectroscopic information. These methods are often used for clusters with higher nuclearity too, e.g. Ir_4 (Fig. 5.25); at the same time they serve as initial structurally characterized subjects. Sometimes other methods are used, such as EXAFS, UV-, XP- or γ-resonance spectroscopy.

The most common and characteristic peculiarities of polymer-immobilized cluster formation are listed in [362–367].

The metalloclusters are connected with macroligands through the only one PPh_2-group at low concentrations of phosphine groups (i.e. when the phosphorylated benzene ring ratio is less than 2%, see Scheme 5.16) or when PPh_2-groups are statistically distributed in the polymer (this is particularly true for iridium metalloclusters). If the PPh_2-group concentration in ligands is sufficiently high (i.e. more than 3% of benzene rings are functionalized), a mixture of cluster units bonded with one or two groups in the polymer occurs. The clusters connected with two groups only are formed if styrene-divinyl-benzene-vinyldiphenyl-phosphine block copolymer (having 8–15% PPh_2-groups) is used.

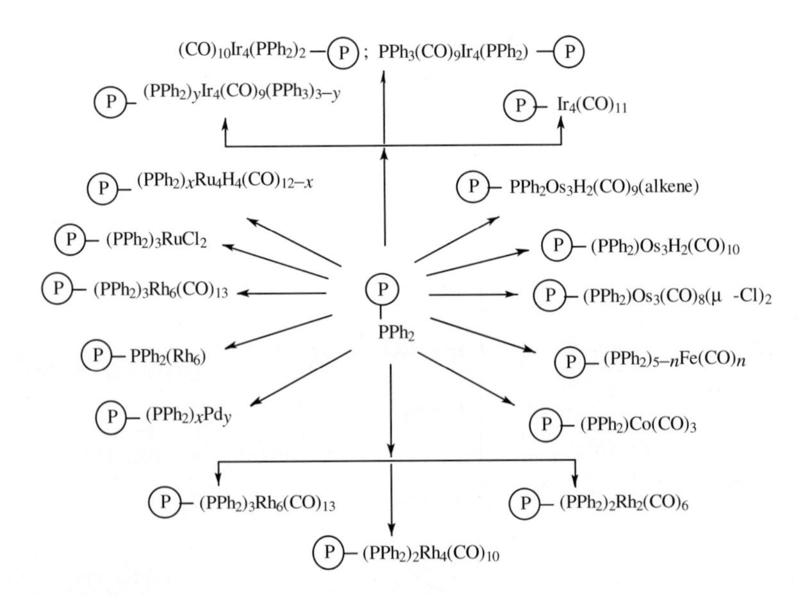

Scheme 5.17

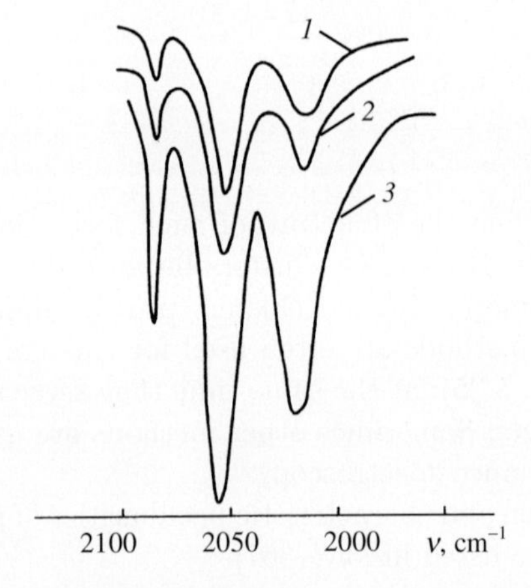

Fig. 5.25. IR spectra of **P**–$PPh_2Ir_4(CO)_{11}$ (*1*), **P**–DIOP$[Ir_4(CO)_{11}]_2$ (*2*) and $Ir_4(CO)_{11}PPh_3$ (*3*) in methylenechloride

The ligand exchange of the equivalent (by electron number) functional groups is the standard method of cluster immobilization if the latter are subjected to a fast dissociation that is most pronounced for $H_4Ru_4(CO)_{12}$ type clusters. The phosphine ligand is changed to the polymer phosphorus-containing group at milder conditions such with the exchange of the carbonyl cluster ligand on a phosphorus-containing one in the polymer. Therefore attempts to bond $Ir_4(CO)_{12}$ on a phosphorus-containing polymer have failed. For this reason the precursors of the polymer-immobilized clusters are often constructed as mixed (i.e. phosphorus and carbonyl containing) ones, as shown in Scheme 5.17.

Sometimes the immobilization processes are accompanied by indefinite structure aggregations, e.g. more than hexarodium compound formation for $Rh_6(CO)_{16}$. Such polymer-bonded particles can be reversibly carbonylated or decarbonylated to the given structure clusters in accordance with the reaction conditions. Thus, if $Rh_4(CO)_{12}$ contacts with **P**–PPh_2 in hexane at 298 K (instead of 323 K), the two-nuclear clusters were formed [365], but when this operation is proceeded in benzene with $Rh_4(CO)_{16}$ [366], a deposition of metallic rhodium (in the form of small black spots 2.5–4.0 nm in size, i.e. about 100 Rh atoms in every spot) was noted at the reactor walls. A lower temperature and use of CO in the reaction led to cluster immobilization. The IR spectrum corresponded to $Rh_6(CO)_{13}(PPh_3)_3$ (absorption bands with maxima at 2078 and 2008 cm^{-1}) (Table 5.10).

Temperature rise affects the genesis of immobilized Ru clusters that are decomposed at 373–423 K, forming ruthenium clusters in the polymer. The cluster anion $[Ru_5Cl_{12}]^{2-}$ bonded on **P**–PPh_2 has some steric hindrances, preventing this penetration of the cluster anion into the polymer matrix and as a result the cluster is split into two immobilized complexes (mono- and binuclear). Decarbonylation parallel with the bonding reaction can takes place. It is most pronounced for iron and cobalt clusters, probably because the values of their M–M and M–CO bond energies are close (about 113 kJ/mol) [368, 369]. For example, addition of $Fe_3(CO)_{12}$ to **P**–PPh_2 is accompanied by fragmentation and formation of a monosubstituted **P**–$PPh_2Fe(CO)_4$ complex with traces of **P**–$PPh_2Fe(CO)_3$ [370]. In the homogeneous variant $Fe_3(CO)_{12}$ reacts with an excess of PPh_3 and the mono- and disubstituted product yields are practically equal. Thus the polymer ligand can carry out some steric control of the reaction. The following thermal or photochemical action leads to mononuclear carbonyl complex decomposition and forming of "ligandless" clusters being dispersed in the polymer.

Other P-containing macroligands, modified by optically active DIOP groups, are used for immobilizing $Ir_4(CO)_{12}$ cluster fragments [371] (see Table 5.10).

The uniformity of the functional groups and forming cluster distributions over particle volumes or surfaces is very important. The dispersion uniformity is the main factor of the uniformity of the polymer characteristics. Distributions of Pd and P along the diameter of spherical granules of the immobilized trinuclear Pd clusters were studied in [379]. The reaction can proceed in such a way that the metalloclusters will be either distributed along the granule diameters uniformly (cross-linking degree – 2%) or localized in a thin near-surface layer (cross-linking degree – 20%) with thickness up to 0.2 mm (for

Table 5.10. Spectral characteristics of molecular and polymer clusters

Cluster	IR-Spectrum (ν(CO), cm^{-1})	NMR-Spectrum (δ, ppm)	References
$Ir_4(CO)_{11}PPh_3$	2088, 2056, 2020, 1887, 1847, 1825	–	[372]
P–$Ir_4(CO)_{11}$	2087, 2055, 2017, 1848 (membrane)	–	[372]
$Ir_4(CO)_{10}$DIOP	2070, 2040, 2011, 1986, 1870, 1833, 1791 (CH_2Cl_2)	–	[372]
P–DIOP$[Ir_4(CO)_{11}]_2$	2088, 2057, 2020, 1845	–	[372]
P<DIOP$[Ir_4(CO)_{10}]$ / DIOP$[Ir_4(CO)_{11}]_2$	2068, 2040, 2012, 1991, 2088, 2054, 2021 (membrane)	–	[373]
$Os_3(CO)_8Cl_2(PPh_3)_2$	2079, 2014, 2004, 1967, 1943	–	[373]
$Os_3(CO)_8(PPh_2$–**P**$)_2$	2096, 2073, 2036, 2007, 1997, 1959, 1939	–	[373]
$Os_3(CO)_{10}Cl_2(PPh_2$ – **P**$)$	2096, 2073, 2060, 2029, 2007, 1994, 1955, 1940	–	[373]
$(PPh_3Cu$–$\mu_3)FeCo_3(CO)_{12}$	2074, 2012, 1978, 1972, 1850	–	[374]
(**P**–PPh_2Cu–$\mu_3)FeCo_3(CO)_{12}$	2065, 2000, 1969, 1956, 1840	–	[374]
$(PPh_3Au$-$\mu_3)FeCo_3(CO)_{12}$	2075, 2003, 1969, 1930, 1823	–	[374]
(**P**–PPh_2Au-$\mu_3)FeCo_3(CO)_{12}$	2066, 1993, 1960, 1930, 1813	–	[374]
$(\mu$–H$)Os_3(\mu$–4-VPy$)(CO)_{10}$	2101, 2061, 2051, 2020, 2008, 2000, 1988, 1973	8.01; 7.25; 6.52; 5.72; −14.85 $[Os_2(\mu$-H$)]$; ($CDCl_3$)	[375]

Table 5.10. (Cont.)

Cluster	IR-Spectrum (ν(CO), cm^{-1})	NMR-Spectrum (δ, ppm)	References
$(\mu$-H$)Os_3(\mu$-4-VPy$)(PPh_3)(CO)_9$	2087, 2047, 2012, 2004, 1997, 1984, 1975, 1945	7.32; 6.51; 5.71; -14.13 $[Os_2(\mu$-H$)]$; (CD_2Cl_2)	[375]
$Os_3H(CO)_9Py$	1940, 2040	10.5; 12.0; 13.75 $(CDCl_3)$	[376]
$(\mu$-H$)Ru_3(\mu$-$SCH_2CH{=}CH_2)(CO)_{10}$	2108, 2069, 2059, 2022, 2020, 2000, 1988, 1955, 1636	5.87; 5.28; 2.82; –15.38 $[Ru_2(\mu$-H$)]$ $(CDCl_3)$	[375]
$Rh_6(CO)_{15}$(4-VPy)	2104, 2068, 2038, 2010, 1788	7.39; 8.88; 6.68; 6.10; 5.68 $(CDCl_3)$	[377]
$Rh_6(CO)_{14}$(4-VPy$)_2$	2090, 2056, 2028, 1760	7.35; 8.81; 6.65; 6.06; 5.63	[377]
P–PPh_2	2060, 1975, 1940,	–	[378]
$Fe(CO)_4PPh_3$	1897, 1875 1875	–	[378]

Note: DIOP-2,3-isopropylidenedioxy-1,4-bis(diphenylphosphino)butane groups.

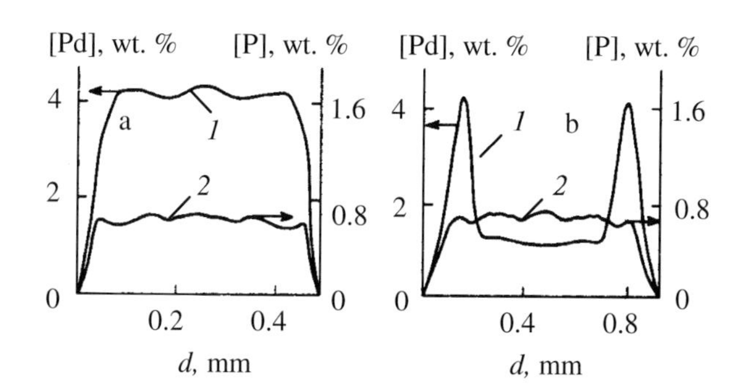

Fig. 5.26. Distributions of Pd (*1*) and P (*2*) on the granule diameter in palladium-immobilized phosphorylated polystyrene with cross-linking degrees of 1 (**a**) and 20% (**b**)

granules having 1 mm in diameter), though the phosphorus-containing group distribution over the granules volume is uniform in both cases (Fig. 5.26).

The polymer-immobilized cluster structure depends on the [P]:[Pd] ratio and for polymers with 2% of cross-linking (at [P]:[Pd] = 0.5) can be presented as

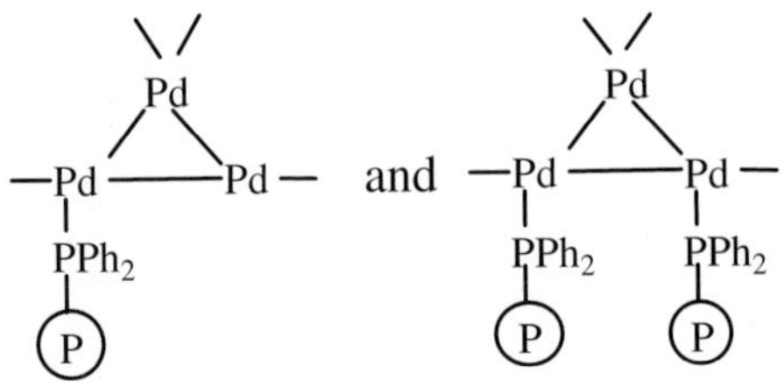

It is difficult to imagine the hypothetical structures formed at [P]:[Pd] = 1:4 (if for no other reasons than steric), but they can have the following forms:

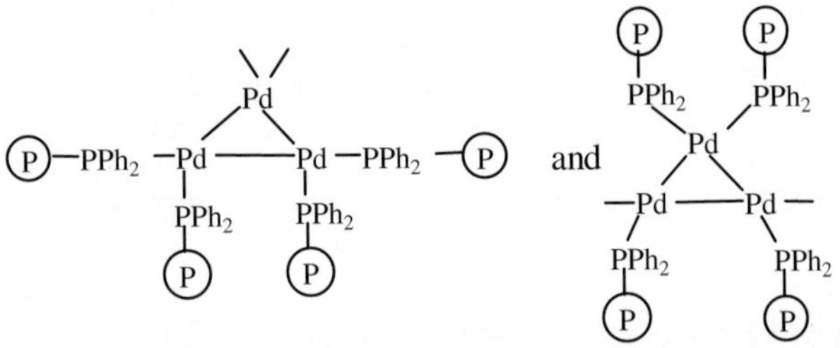

Phosphorylated SDVBC (with cross-linking from 2 to 20%) readily bonds $Mo_6Cl_{12}L_2$ type clusters forming a covalent bonded cluster core Mo_6Cl_{12} [380].

The cluster immobilization by polymers with other functional groups is poorly known. Some works were devoted to nitrogen-containing polymers, such as aminated PS, P4VPy, ion-exchange amine resins (Amberlyst A-21, etc.). Thus the interactions of $Rh_6(CO)_{16}$, $Rh_4(CO)_{12}$ and $Rh_2(CO)_4Cl_2$ with aminated PS proceed by Scheme 5.18 [381]:

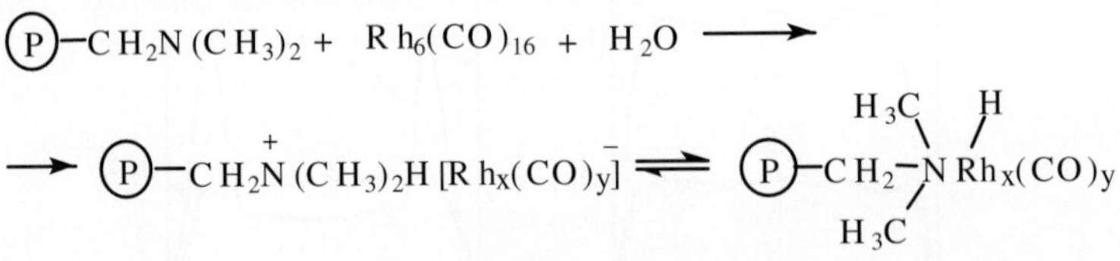

$$\text{(P)}-CH_2N(CH_3)_2 + Rh_6(CO)_{16} + H_2O \longrightarrow$$

$$\longrightarrow \text{(P)}-CH_2\overset{+}{N}(CH_3)_2H\,[Rh_x(CO)_y]^{-} \rightleftharpoons \text{(P)}-CH_2-N(CH_3)_2(H)Rh_x(CO)_y$$

Scheme 5.18

The product of Os_3 cluster bonding with this ligand [382] is $HOs(CO)_{11}\overset{+}{N}Et_3CH_2$–**P**.

In a similar way $Ru_3(CO)_{12}$ is fixed on macroporous chelated polymers, functionalized by dipyridylic, 2-aminopropylic, 2-aminophenolic and other type groups. An anionic triferrum cluster is bonded to an ion-exchange resin (in the form of a membrane with thickness 10 μm) by $^{+}NR_4\overset{+}{N}R_4$-groups. The $[Pt_{15}(CO)_{30}]^{2-}$ cluster can be bonded with anion-exchange resin, containing diethyl(2-hydroxipropyl)-aminoethyl groups (the trade mark of the polymer is QAE-SEPHADEX) [383].

The reactions of nitrogen-containing macroligands, contrary to the reactions of phosphorus-containing ones, proceed by forming many by-products (especially with the polymer having primary and secondary aminogroups). Thus in ruthenium and osmium carbonyl cluster reactions an activation of α- and β-C–H-bonds, a transalkylation (at 298–315 K) and even a C–N-bond cleavage (for the tertiary amine clusters with Os_3 at 398–416 K) are observed that clearly are complicated by the polymeric nature of the reagents.

The mechanism of $Rh_4(CO)_{12}$ bonding by poly-4-vinylpyridine has not been adequately investigated. At the same time the interaction of $Os_3(CO)_{12}$ with P4VPr [364] is a very peculiar one, because a cluster is bonded in DMF (at 383 K, CO atmosphere). Its structure is very complicated to interpret (there are two intense absorption bands with maximum at 2040 and 1940 cm^{-1}). Using the NMR data (see Table 5.10) one can propose that the cluster is bonding by the polymer with two pyridine rings, one of them being chelated. Such bonding is provided by the polymeric chain flexibility:

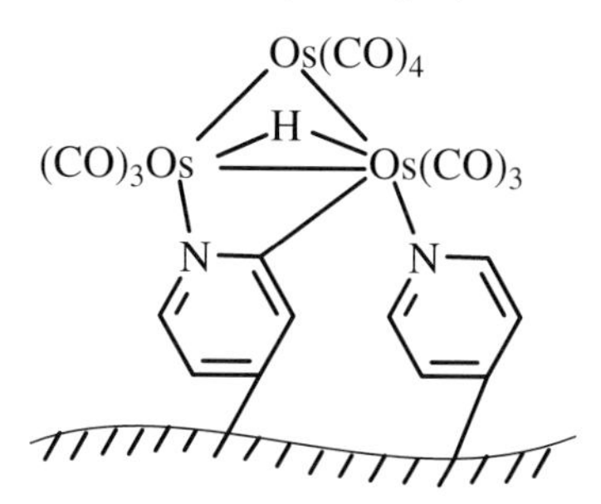

There are some interesting examples of these systems. For example, the Mo_6Cl_{12} cluster (immobilized on P4VPr, cross-linked by 2% of divinylbenzene) has unusual physical properties and singlet oxygen chemistry [384], because in the cluster every nucleus is connected with polymer chain by two pyridine fragments. Polypyrrole with a covalent bonded trinuclear ruthenium is described in [385], and $[Fe_4S_4(SPh)_4]^{2-}$ was introduced into ion-exchange polymer (based on N-substituted polypyrrole) by an electrode process [386].

The conservation (or decay) of the polynuclear structure depends on the specific nature of the cluster ligand groups. Usually the surrounding ligand gives its own contribution to the cluster total energy, which is sufficiently large and is decreasing when the nuclearity rises [387]. The binuclear structure and oxidation degree of Rh^{2+} are conserved in systems of the acetate binuclear complex $Rh_2(OCCH_3)_4$ with polymers (containing 3(5)-methyl-pyrasol or imidasol groups) [388]. But the Rh–Rh bond breaks and the rhodium oxidation degree grows for the binuclear complexes (such as sulfate, acetonitrile and hexa-fluoroacety-acetonate ones). The main factor preserving the cluster structure is initial cluster equatorial ligand inertness, because the labile ligands substituted to aminogroups would favor Rh–Rh bond breaking. The bridge acetate groups for the binuclear structure of Rh^{2+} have a tightening effect.

Sometimes for cluster immobilization the O and S containing matrices are used (specifically, polymeric alcohols and ketones). For example, a triosmium cluster can join to polymeric alcohols (just as to inorganic oxides) by an oxidative addition mechanism [389]:

$$\text{(P)}{-}CH_2CH_2OH + Os_3(CO)_{12} \longrightarrow \text{(P)}{-}CH_2CH_2OOs_3H(CO)_{10}.$$

Scheme 5.19

The very stable di- and trinuclear ruthenium complexes (Ru^{3+}–O–Ru^{4+}–O–Ru^{3+}-type) are easily bonded by ion-exchange resin Diaion CR-10 containing iminodiacetate groups [390]. This is typical for the Mo_2 clusters in PEO or PPG [391].

A generic connection between synthetic and inorganic polymers can be demonstrated by immobilization of the formally polynuclear keggin type 12-heteropolyacids, as $H_n[XM_{12}O_{40}]$ (X – heteroatom, M – Mo or W) [392] with polymers. Under mixing with P4VP (303–308 K) in such systems a product is formed [393] where (as a chemical analysis and IR spectroscopy show) four protonated vinyl-pyridine units are bonded with one $H_4[SiMo_{12}O_{40}]{\cdot}13H_2O$ molecule (or three units are bonded with $H_3[PMo_{12}O_{40}]$), i.e. the repeating units of such structures can be expressed in the form $(H{-}PyCHCH_2)_4[SiMo_{12}O_{40}]$ or $(H{-}PyCHCH_2)_3[PMo_{12}O_{40}]$. The same type of bonding was found in 12-heteropolyacids immobilization by some water-soluble polymeric eletrolytes, such as alkylated poly(2-methyl-5-vinylpyridine) or poly(dimethyl-diallyl-ammonia) chloride [379]. The oxidized forms of heteropolyacids react due to heteropolyanion negative charge compensation by the positive charged groups of the polymer chain.

This type of metallocluster can be used to produce hybrid organic–inorganic nanomaterials. For example, dendrimers $Mo_6(\mu_3\text{-}Cl)_8(OR)_6$ where R are the focally substituted phenol dendrimers can be received by substituting the outer spheres ligands [394]. Removing the outer spheres chlorine ions in $(Mo_6Cl_8)Cl_4$ at interaction with Na^+-montmorillonite a molybdenum pillared montmorillonite was obtained [395]. The surface hydroxyl groups reaction with the silica gel supported clusters are a potentially useful medium for photochemical transformations.

The same is true for self-assembling inorganic–organic networks on the base of molybdenum clusters and SDVBC-PPh_2 [396].

5.4.2 Heterometallic Clusters, Chemically Bonded with Polymers

The heterometallic polynuclear centers are favored over monometallic ones in catalysis, alloy surface models and many other situations. Such immobilized particles can be considered as a model of bimetallic catalysts (similarly the contact surface crystallites) widely used in industry of the oils refineries. In contrast to alloy surfaces, the bimetallic clusters are structurally uniform.

The same methods of heterometallic cluster bonding and identification are usually applied to monometallic ones. The most frequently used method is assembling from the monometallic complexes or clusters (for example, by a synchronous decomposition of H_2PtCl_6 – $Fe(NO_3)_2$ or $Rh_4(CO)_{12}$– $CoCl_2{\cdot}6H_2O$ mixtures in polymer systems). It was proved in Sect. 5.1 that the bimetallic clusters can be formed [179–181, 393] into PVP stabilized dispersions of Pt and Pd at initial molar ratio [Pd]:[Pt] = 4:1. Sometimes the cluster-type (but not monomeric) compounds of different metals are used in such systems. For example, a mixture of $Rh_4(CO)_{12}$ and $Co_4(CO)_{12}$ in molar proportions 3:1–1:1 was reacting with ion-exchange resins [371].

It looks favorable to use heterometallic clusters of the identified structure (just as for monometallic type cluster bonding).

It should be considered that the general number of molecules with more than one type of metal atoms is over a hundred. Generally they are synthesized by changing one metal atom to another into high-symmetric metallopolyhedrons (on retention of the total cluster valent electron number and ligand type).

The most frequently used matrix is **P**–PPh_2, and the immobilization methods are the ligand and ionic exchanges. The most reliable method of polymer-bonded products identification is based on the comparison of their carbonyl group IR spectra with model molecular spectra. On Scheme 5.20 typical examples of the immobilized heterometallic clusters on **P**–PPh_2 are presented [239, 363, 373, 397, 398].

There are a few other interesting examples: bonding of $H_2FeOs_3(CO)_{13}$ by cross-linked macroporous chelating type polymers [399]; fixation of $Co_2Rh_2(CO)_{12}$ on Dauex-1 type ion-exchange resin [400]; $H_2FeRu_3(CO)_{13}$, $[FeRh_4(CO)_{15}](NMe_4)_2$ and $Rh_{4-x}Co_x(CO)_{12}$ ($x = 2$; 3) fixation on various macroligands. The mechanisms of such reactions are not properly explained.

The mechanisms of bimetallic keggin hetepolyacids bonding to P4VP are identical to that of monometallic ones, e.g. $H_6[CoW_{12}O_{40}]$ with repeating units, having a composition $(H\text{-}4\text{-}VPy)_4H_2[CoW_{12}O_{40}]$ [393]. These processes can also be realized with polyoxoanions of niobium-containing polytungstates such as $[(CO)_2Rh]_5[Nb_2W_4O_{19}]_2^{3-}$, $[(CO)_2Ir]_2H[Nb_2W_4O_{19}]_2^{5-}$ [401].

The above listed compounds are formally trimetallic type clusters. The trimetallic clusters bonded to **P**–PPh_2 can be produced by the assembling method [374], i.e. through the polymer interaction (L is a functional group) with $MPPh_3Cl$ (M = Cu or Au) and a consequent contact of the product with tetranuclear bimetallic cluster $FeCo_3(CO)_{12}$. The general strategy of heterometallic polynuclear clusters production can be expressed by Scheme 5.21:

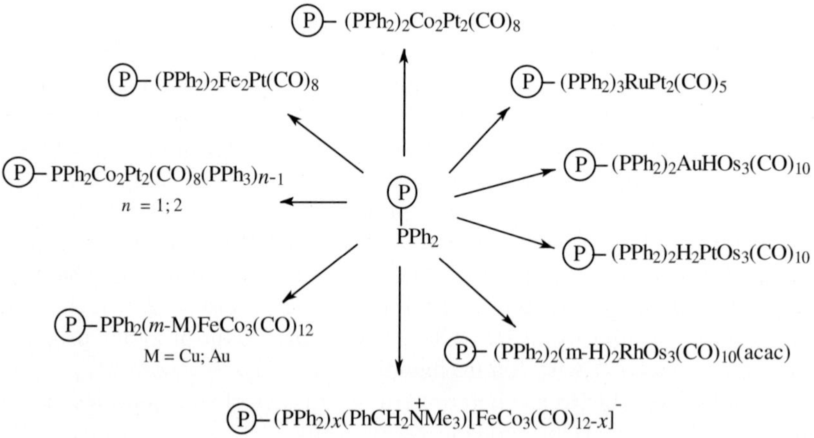
(P)– (PPh2)2Co2Pt2(CO)8
(P)–(PPh2)2Fe2Pt(CO)8
(P)– (PPh2)3RuPt2(CO)5
(P)– PPh2Co2Pt2(CO)8(PPh3)n-1
n = 1; 2
(P)–(PPh2)2AuHOs3(CO)10
(P)
PPh2
(P)– (PPh2)2H2PtOs3(CO)10
(P)–PPh2(m-M)FeCo3(CO)12
M = Cu; Au
(P)– (PPh2)2(m-H)2RhOs3(CO)10(acac)
(P)– (PPh2)x(PhCH2N+Me3)[FeCo3(CO)12-x]-

Scheme 5.20

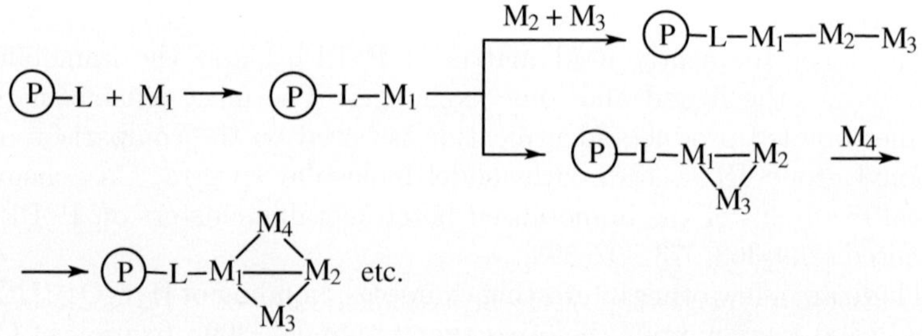
M2 + M3
(P)–L–M1—M2—M3
(P)–L + M1
(P)–L–M1
(P)–L–M1—M2
M3
M4
M4
(P)–L–M1—M2 etc.
M3

Scheme 5.21

The reaction most probably proceeds through heterometallic cycles formation and this type of trimetallic unit are the "structural blocks" of cluster chemistry [402].

A pentametallic cluster bonded with polymer can have the following structure [374]:

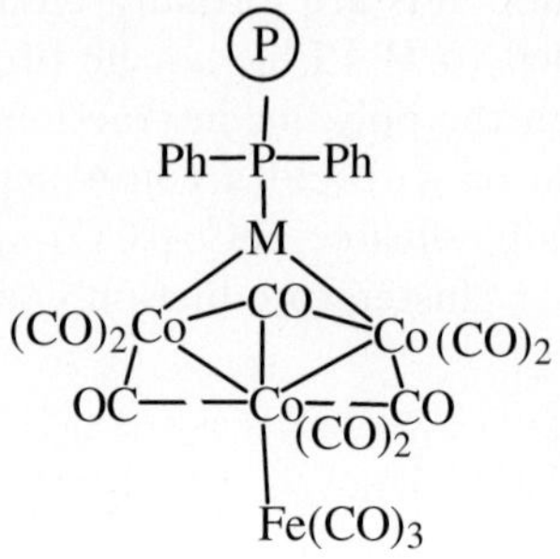
(P)
Ph—P—Ph
M
(CO)2Co
CO
Co (CO)2
OC
Co
CO
(CO)2
Fe(CO)3

The polymer-immobilized clusters of polymetallic type are more stable than the initial ones. For example, the decay temperature of the above mentioned initial cluster (M = Cu) is 443–451 K, but its polymer analog is decomposed at 459–466 K. In all instances the Cu-containing clusters were more stable than Au-containing ones (436–438 and 444–449 K correspondingly). In general, the immobilization increases the thermal stability of both the clusters and bonding polymer.

The "layering-up" method was used [403] to bond Berlin blue and its analogs with polyvinylamine hydrochloride. The MCl_2 (M = Fe, Co and Cu) compounds were fixed on the polymer in water solutions and to the ions $[Fe(CN)_6]^{n-}$ (n= 3; 4).

In general, nanocomposite production on the basis of both mono- and heterometallic clusters is being intensively developed, although there are many unsolved problems, related mainly to the interaction mechanisms (e.g. between the cluster and polymer functional groups), the transformations of the immobilized reactants (including the change of nuclearity) and the structure of the products.

5.4.3 Cluster-Containing Monomer Synthesis and Copolymerization as a Way to Produce Metallopolymeric Composites

Nanocomposite formation in the polymer-analogous transformation was considered above and it was shown that a synchronous forming of matrix and nanoparticles is possible (see Sect. 5.1). By analogy, cluster-containing monomer copolymerization might be a simple, one-step process directly leading to cluster-containing polymer production. Thus we suppose that the problems may be connected with the corresponding monomer synthesis only.

Cluster-Containing Monomers

In recent years the two promising methods of cluster-containing monomer synthesis were developed [404]. A polymerizable group was introduced as a ligand surrounding typical clusters L_nM_m (the own specific ligand substitution to their analogs with multiple bonds, their oxidative addition, an addition by M–M bonds at mild conditions). Such reactions can proceed with increase or decrease of nuclearity. For example, the different complexes with various nuclearity are formed in the interaction of $Ru_3(CO)_{12}$ with dimethylfumarate in boiling heptane and their relative yield depends on the reagent ratio and process duration [405]. But the same cluster with dienyl derivatives can form μ_3, η^2-type structures [406].

The mechanism of $Ru_3(CO)_{12}$ interaction with styrene and its derivatives (4-methylstyrene and trifluorostyrene) was studied in [407]. It was shown that the polymerizable double bonds during the reaction took part in a π-bonding, resulting in cluster aggregation and fragmentation (Scheme 5.22).

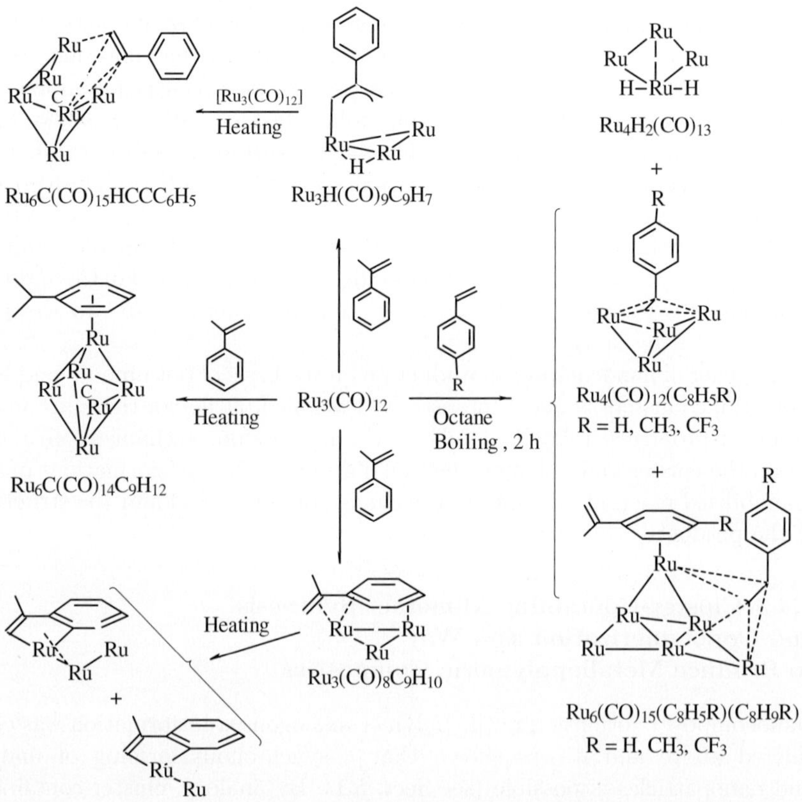

Scheme 5.22

Below a representative example when traditional type monomers are added on by clusters is given: the cluster monomer formation on the base of *n*-vinyldithiobenzoic acid methyl ester [408]:

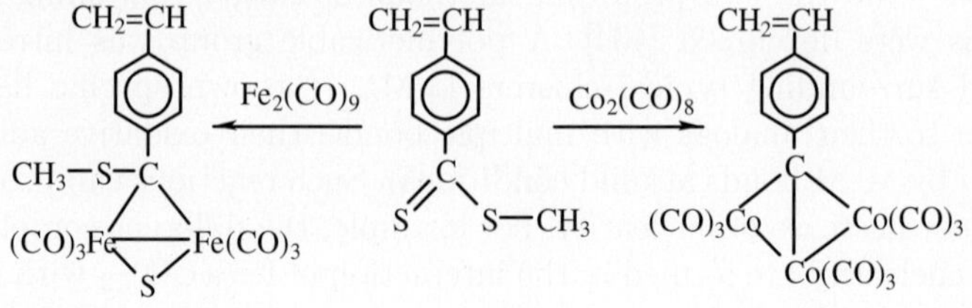

Scheme 5.23

Usually the target product yield of such reactions is low (<40%) because of side reactions (desulfuration, metal sulfides and organometallic product formation, nanoparticles generation). It is important that during the process the cluster skeletons are reserved and the forming monomers are able to go to the different polymerization transformations. The μ-ethynyl groups are split off even under mild conditions in various clusters on the Fe_2 base, including the thiolate-bridged $[Fe_2(CO)_6(\mu\text{-}\eta^2\text{-}HC{=}CH_2)(\mu\text{-}SR)]$ [409]. The tricobalt-alkylidene complexes $Co_3(\mu^3\text{-}CR)(CO)_9$ react with vinyl-phosphine ligands in heptane 308 K. The compounds $Co_3(\mu_3\text{-}CR)(CO)_{9-x}(PPh_2CH{=}CH_2)_x$ (x = 1; 2; 3) were derived and characterized by this method [410]. The product (with $x = 2$) on heating to 343 K is a split CO group (Scheme 5.24) with additional coordination between the vinyl group and cobalt atom.

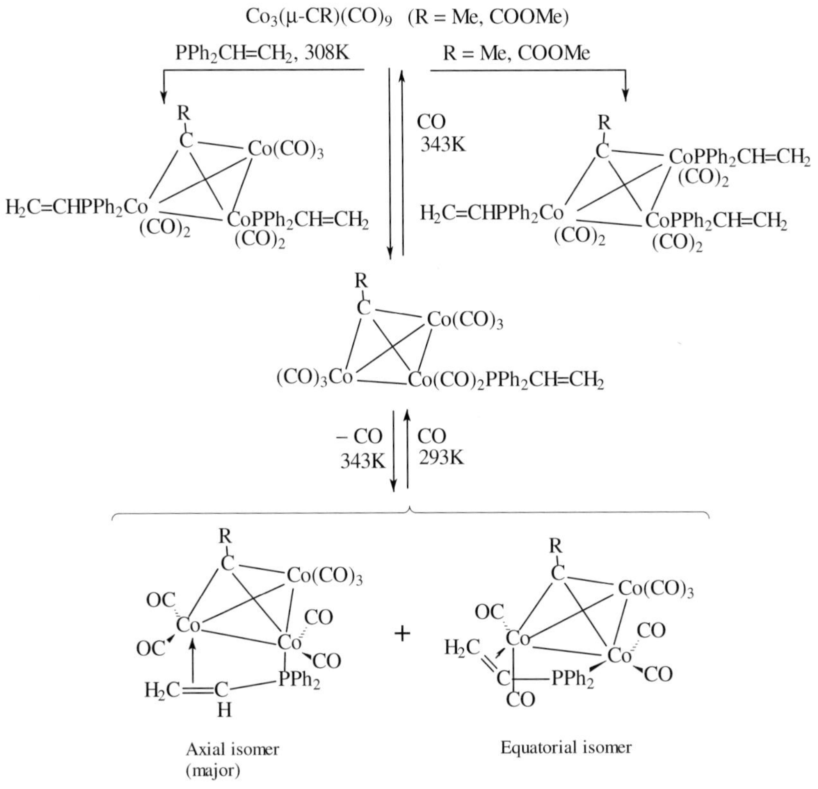

Scheme 5.24

The diphenyl-vinylphosphine ligand is very versatile in carbonyl cluster reactions and form both the terminal type clusters with a phosphorus atom and the μ_2- or μ_3-bridges including vinyl group π-coordination by Scheme 5.25 [411]. In Table 5.11, some characteristics of the products are given.

The high product yield can be obtained on the base of trinuclear clusters $M_3(CO)_{12}$ (M = Os, Ru) or $Os_3(CO)_{11}(CH_3CN)$, $Os_3(CO)_{10}(CH_3CN)_2$,

$(\mu\text{-H})Os_3(CO)_{10}(\mu\text{-OR})$ (R = H, Ph) and the traditional type monomers: 4-VPy, acrylic acid (AA), allylamine (AllA), allylsulfide (AllS), etc. [375] (Scheme 5.26).

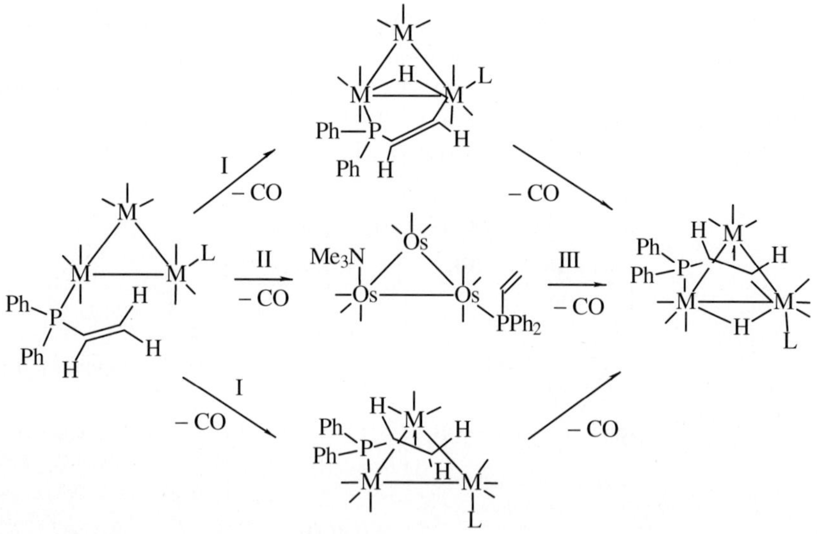

M = Os, Ru;
L = CO, $Ph_2PCH{=}CH_2$;
I: M = Os, boiling in dibutyl ether; M = Ru, boiling or microwave heating;
II: M = Os, Me_3ONO, CH_2Cl_2; III: M = Os, boiling in CH_2Cl_2

Scheme 5.25

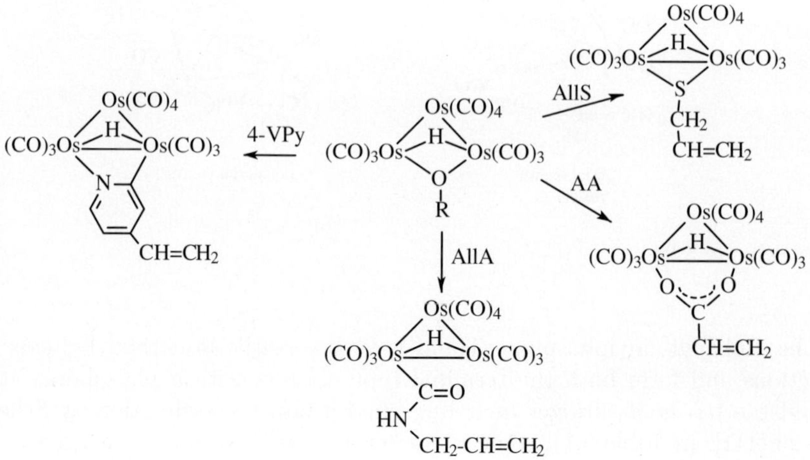

Scheme 5.26

Table 5.11. Spectral characteristics of vinyldiphenylphosphine based mono- and heteronuclear clusters

Cluster	IR-Spectra (ν_{CO}, cm^{-1})	^1H NMR (δ, ppm)	^{31}P NMR (δ, ppm)	Mass Spectra (m/z)	References
$Os_3(CO)_{11}$ $(Ph_2PCH{=}CH_2)$	2104, 2064, 2050, 2026, 2016, 2000	7.71–7.42	–150.88	–	[412]
$Os_3(CO)_{10}$ $(Ph_2PCH{=}CH_2)_2$	2084, 2058,1999, 1965, 1986	7.70–7.15	–156.7	1280	[411]
Ru_3 $(CO)_{11}$ $(Ph_2PCH{=}CH_2)$	2095, 2043, 2025, 2010	7,70–7,15	–115.0	823	[411]
Ru_3 $(CO)_{10}$ $(Ph_2PCH{=}CH_2)_2$	2069, 2020, 1994	7.70–7.15	–115.0	1007	[411]
$Os_3(\mu\text{-}H)(CO)_9(\mu_3\text{-}Ph_2PCH{=}CH_2)$	2089, 2061, 2032, 2013, 1990, 1975	7.70–7.15	–150.1	1040	[411]
Ru_3 $(\mu\text{-}H)(CO)_8$– $(Ph_2PCH{=}CH_2)$ $(\mu_3\text{-}Ph_2PCH{=}CH_2)$	2070, 2032, 2017, 1998, 1970	7.93–7.07	–103.4	–	[412]
$MoMn(\mu\text{-}C_5H_4PPh_2)_2(CO)_6$– $(PPh_2CH{=}CH_2)$	2042, 1978, 1967, 1925, 1897, 1847	8.0–7.4	–114.3	–	[413]
$MoMn(\mu\text{-}C_5H_4PPh_2)_2(CO)_5$– $(PPh_2CH{=}CH_2)_2$	1931, 1920, 1906, 1883, 1817	8.1–7.5	–	–	[413]

The noncoordinated allyl fragment is transformed with a terminal double bond migration to the α-position and a new $(\mu\text{-}H)Os_3(CO)_{10}(\mu\text{-}OCNHCH_2CH{=}CH_2)$ cluster formation even at room temperatures [414]. The pyridine ring for a monomer based on the 4-VPy is chelated (just as for the above analyzed complex $Os_3(CO)_{12}$ immobilized on P4VPy). The $Os_3(CO)_{10}(CH_3CN)_2$ complex readily reacts with 2-VPy, forming a trinuclear cluster $HOs_3(CO)_9L$ $(NC_5H_4\ CH{=}CH_2)$ (L = CO or PMe_2Ph). The triosmium alkylidene cluster $H_2Os_3(CO)_9(\mu\text{-}CNC_5H_4\text{-}CH{=}CH_2)$ can be used as an initial compound to obtain the planar hexaosmium carbonyl cluster with an open structure [415] (Scheme 5.27).

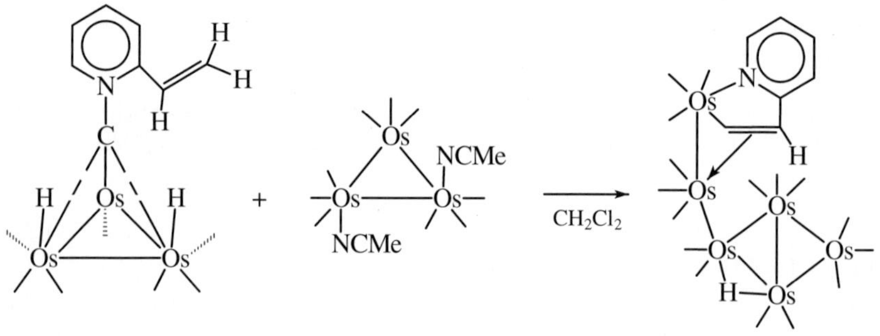

Scheme 5.27

4-VPy interaction with $Rh_6(CO)_{16}$ in the presence of acetonitrile or trimethylamine N-oxide proceeds in mild conditions and gives a main product – monosubstituted derivative of $Rh_6(CO)_{15}$(4-VPy) with a small amount of disubstituted compound $Rh_6(CO)_{14}$(4-VPy)$_2$ [377], which is easily separated by chromatographic methods in the individual states. Contrary to the Os_3-derivative, the $Rh_6(CO)_{15}$(4-VPy) monomer is an octahedral cluster with 11 terminal and four μ_3-bridged carbonyl ligands. The vinylpyridine group is connected with the rhodium atom by a nitrogen atom only and its coordination position is the twelfth terminal group (the mean length of the Rh–Rh bond is 0.2762 nm, which is near to the value for other Rh clusters). The $Rh_6(CO)_{14}L_2$ composition clusters are identified also in solutions with diene (cyclooctene, dimethylbutadiene) [416] and 1,3,5-hexadiene ligands [417]. The rhodium clusters interact with the allyl ligand $Ph_2PCHCH{=}CH_2$ probably by the same mechanism, because X-ray analysis shows that the allyl group forms a π-complex with the Rh atom and the unshared electron pair of the phosphorus atom forms a donor bond [418–420].

These metallomonomers include the carbonyl type clusters only, although the other clusters containing monomers and polymers are of interest (e.g. halogen-containing ones). The halogenides derivatives Mo(II) are forming polynuclear complexes such as Mo_6Cl_{12} with a strong associating tendency of bivalent molybdenum to an association [421] including the stable group $[Mo_6Cl_8]^{4+}$ (Fig. 5.27). They present a weakly distorted octahedron, composed from six molybdenum atoms, inscribed into a cube, having in chlorine atoms at the eight vertexes. The surrounding of every Mo atom is near to a plane-square one (when every Cl atom is coordinated at the same time with three Mo atoms). Such a cluster has diameter 1 nm (the Cl–Cl cube diagonal is 0.6 nm plus two Cl ions, having a radius about 0.18 nm). The cluster grouping can add up to six axial ligands in the form of negative ions or polar groups (including the groups having multiple bonds) able to the polymeric transformations. The possibility was first noted in [422], when N-vinylimidazole (NVI) polymerization was studied in the presence of a molybdenum cluster, but the

MCM itself was not separated and characterized. The main problem connected with such reactions is the task of all out-sphere chlorine atom substitution by the labile ligands and this was solved only recently. Use of the intermediate trifluoric groups CF_3COO^- allowed substitution to the acrylate groups (in mild conditions). The outer-sphere substitution

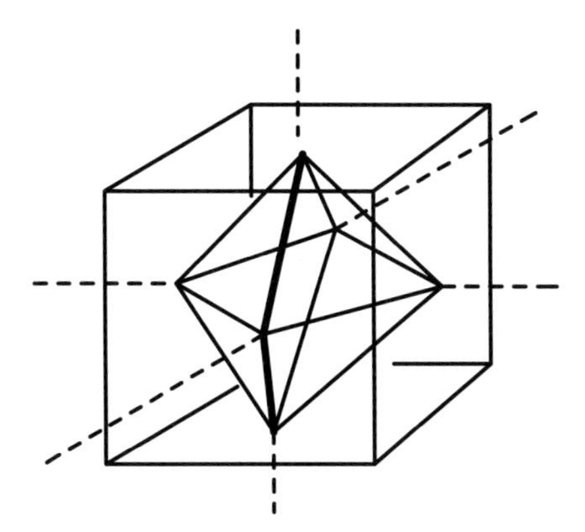

Fig. 5.27. Staffilonuclear structure $[Mo_6Cl_8]^{4+}$

principle is a base of our own synthesis of many cluster-containing monomers (Scheme 5.28) [423, 424]:

$$[(Mo_6Cl_8)Cl_6]^{2-} \xrightarrow{CF_3COO^-} [(Mo_6Cl_8)(CF_3COO)_6]^{2-} \xrightarrow{CH_2{=}CHCOO^-}$$

$$\longrightarrow [(Mo_6Cl_8)(CF_3COO)_{6-n}(CH_2{=}CHCOO)_n]^{2-}$$

Scheme 5.28

The stable staffilonuclear group is the central one for the complex reactions [425]. It is able to add up to six axial ligands L, being negative ions or polymerizable polar molecules, including the multiple bonds (many traditional vinyl type monomers, 4-VPy, acrylonitrile, acrylamide). Unfortunately many of the required monomers are still not synthesized.

In general, this direction of synthesis and polymerization of the cluster-containing monomers is new and very promising. There are no principal limitations for the use of these methods (see Sect. 5.1) of atomic metal vaporization in the following way (Scheme 5.29):

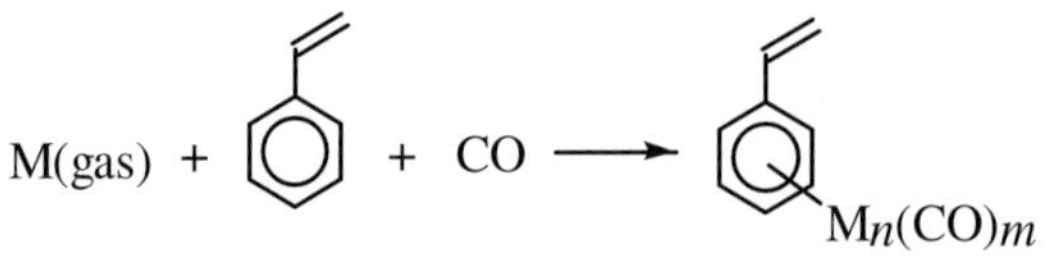

$$M(gas) + C_6H_5CH{=}CH_2 + CO \longrightarrow (\eta^6\text{-}C_6H_5CH{=}CH_2)M_n(CO)_m$$

Scheme 5.29

The case in point is to use in the polymerization processes not only the polynuclear and cluster MCM, but to produce higher nuclearity compounds (for example, nanoparticles 1–100 nm in size). A hypothetical structure of such a metallomonomer (it can be called a nanomer) can be presented (with MMA, for example) in the following form:

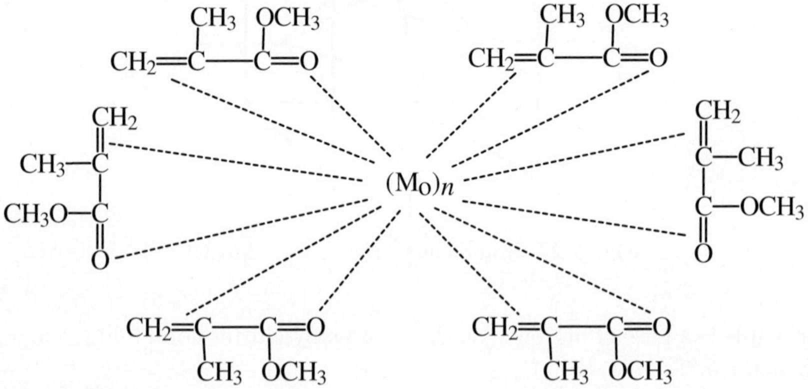

$$(M_0)_n \cdots [CH_2{=}C(CH_3){-}C(OCH_3){=}O]_6$$

Scheme 5.30

The monomeric shell can stabilize, in this structure, the energetically filled nanoparticles. But at the same time it can be involved in a copolymerization with some traditional monomers. The way directly leads to forming and production of metallopolymeric nanocomposites

The analyzed methods refer equally to monomer construction on heteropolynuclear clusters. One of the methods is to introduce multiple bond formations into the heterometallic complexes. Treatment of the iron-ruthenium complex $[FeRu(CO)_4(\eta^5\text{-}C_5H_5)_2]$ by olefins gives the μ-vinylic complexes (R = H, Ph) with the Fe–Ru bond [426]. The double bond in the μ-vinylidene Mn–Pt-complexes is not coordinated [427], because the σ-bond is with Fe, but the π-bond is with Ru:

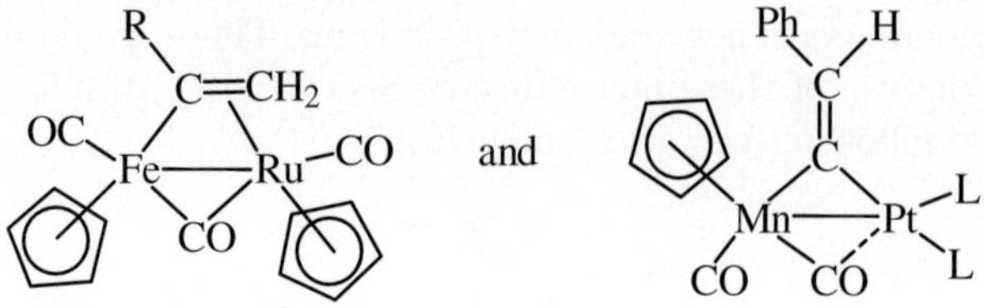

$$(OC)(\eta^5\text{-}C_5H_5)Fe(\mu\text{-}CO)(\mu\text{-}C(R){=}CH_2)Ru(CO)(\eta^5\text{-}C_5H_5) \quad \text{and} \quad (\eta^5\text{-}C_5H_5)(CO)Mn(\mu\text{-}CO)(\mu\text{-}C{=}C(Ph)H)PtL_2$$

The reactivity of the polymetal complexes (in particular, relative unsaturated hydrocarbons) is similar to that of homonuclear ones, but different by a sequence: FeRu > Fe_2 > Ru_2. To the same group of potentially heterometallic monomers can be related $FeCo(CO)_7(\mu\text{-}CH{=}CHR)$ [428] and some others. As for monometallic monomer vinylphenylphosphine coordination can be carried out by different mechanisms using both the unshared electron pair (of phosphorus atom) and the double bond π-electrons [413] (Scheme 5.31).

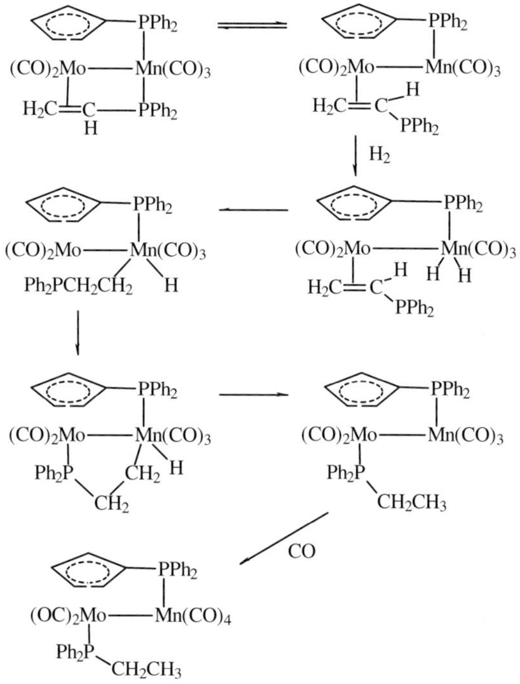

Scheme 5.31

Polymeric Transformation of the Cluster-Containing Monomers

There were attempts to (co)polymerize the cluster monomers. But the presence of the cluster fragments with specific electronic structure and radicals or ionic particle interaction with cluster-containing monomers often results in their decay. Therefore such exotic monomer homopolymerization is practically precluded. The most successful was the described copolymerization of monomers, produced from dicobalt octacarbonyls or diferrum nanocarbonyls and dithioethers or thioamides (see Scheme 5.23) with styrene or N-[tris(hydroximethyl)methyl]acrylamide. The products included up to 1–2%

(mol.) units with nondestructed cluster skeletons (Table 5.12). The copolymer molecular mass was sufficiently high.

The homopolymerization of (μ-H)Os_3(μ-4-VPy)$(CO)_{10}$ is connected with difficulties: it proceeds in benzene at 343 K; an oligomer was produced (by an AIBN initiator) with yield 1–2%, containing 5–6 units in the chain. The Os_3-containing monomer copolymerization with styrene or acrylonitrile leads at these conditions to the content of cluster units in the chain (up to 3%), but does not change the cluster composition and structure [429]. The cluster skeleton carbonyl IR spectra correspond to the initial monomers and a small intensity change and some absorption line widening are observed. A conjugation with the double bond in (μ-H)Os_3(μ-4-VPy)$(CO)_{10}$ leads presumably to an electronic density redistribution on the Os–Os bonds and their shortening. This shortening in the polymer disappears and the skeleton photoelectronic spectra are near to the spectra of nonsubstituted $Os_3(CO)_{12}$ clusters. It allows us to suppose that the Os_3 monomer is attached to a growing macroradical and limited the chain growth, but this process does not decrease the active particle concentration due to fast chain transfer (Scheme 5.32). The $Rh_6(CO)_{15}$(4-VPy) copolymerization with styrene in toluene [430] proceeds sufficiently easier and the cluster monomer content can achieve 10%. This can be explained by the fact that 4-VPy is connected with the cluster skeleton by a donor0-acceptor bond only. It can be supposed that the cluster metal particles are only slightly bigger than the initial clusters, lower in their content in copolymers, because a forming polymer segment (from the copolymer) is a stabilizing factor, preventing an associative process. There is direct evidence of that based on radical block-copolymerization of $Mo_6Cl_6(VIA)_6(CF_3COO)_4$ (VIA – N-vinylimidazole) in solubilizing monomer media [422]. In this case VIA was used coincidentally as a cluster monomer solvent (0.1 and 0.5 mol/L) and comonomer. The fast homogeneous copolymerization is observed at 343 K. It was followed by cross-linking and reacting media thickening (including the chemically bonded monodispersed clusters of comparatively high concentration). The polymer-immobilized cluster dimensions are near to the initial ones and are about 1 nm (Fig. 5.28). Therefore it is possible that the "monomer as solvent" method will become the main method of polymer-immobilized monodispersed cluster fabrication.

The introduction of cluster monomers into a copolymeric chain (even their number is small) improved the characteristics both of the metalloclusters and polymer. For example, this happened in Os_3-copolymers, where the mutual stabilization of PS and connected clusters were observed [431]. Cluster decarbonylation begins at about 470 K, but in copolymers it increases up to 620 K. The thermal stability of the cluster increases only if they were connected with the polymer chain and no such effect was found in the mechanical mixtures. One of the main reasons for this may be an energy transfer at high temperatures from the rotational–vibrational degrees of freedom to the translational

Table 5.12. Some characteristics of cluster-containing copolymers

Monomers		Content, % (mol.)		Molecular-mass Characteristics		Notes
M_1	M_2	Monomer Mixture M_2:M_1	of Copolymer m_2	$\overline{M}_w$	$\overline{M}_n$	
CH_2=CH C $(CO)_3Co$ $Co(CO)_3$ $Co(CO)_3$	Styrene	99:1	3.7	14720	6400	
	MMA	99:1	1.0	44700	21300	–
	N-[tris(hydroximethyl)-methyl]acrylamide	99:1	1.6	–		Water-soluble copolymer
CH_2=CH C $(CO)_3Co$ $Co(CO)_3$ $Co(CO)_3$	Styrene	99:1	1.3	16900	7200	–
	MMA	99:1	1.8	41300	24600	–
	N-[tris(hydroximethyl)-methyl]acrylamide	99:1	1.6	–		Water-soluble copolymer

Table 5.12. (Cont.)

	Styrene	98.8:1.2	1.4	3320	7000–10060	
	MMA	99:1	1.1	18500	2250–5500	
	N-[tris-(hydroximethyl)-methyl]acrylamide	98.5:1.5	1.5			
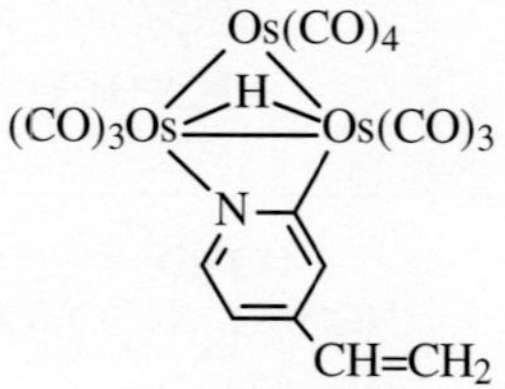	Styrene	(99-99.5):(0.5-1.0)	0.2-1.0	–	5000-36500	Polymerization in bulk
	AN	(99–99.5):(0.5–1.0)	0.1–0.2	–	15000	in benzene
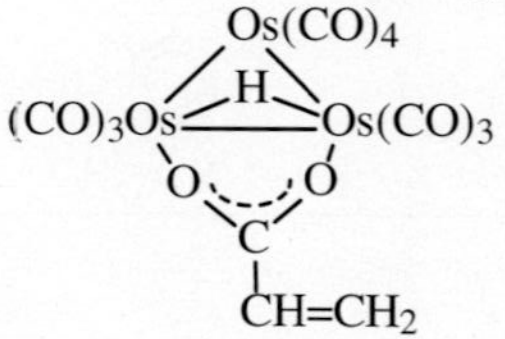	Styrene	(99–99.5):(0.5–1.0)	0.1–1.0	–	12000–25500	Polymerization in bulk
	AN	(99–99,9):(0,1–1,0)	0.1–1.0	–	8000–26000	in benzene

Table 5.12. (Cont.)

$Os(CO)_4$, H, $(CO)_3Os$—$Os(CO)_3$, C—O, HN, CH_2-CH=CH_2	Styrene	(99–99.5):(0.5–1.0)	0.1–1.0	–	21000–26000	Polymerization in bulk
	AN	(99–99.5):(0.1–1.0)	0.1–1.0	–	15000–17000	in benzene
Mo_6Cl_6(NVI)]$(CF_3COO)_4$	N-vinylimidazole	(99–99.5):(0.5–1.0)	0.1–0.5	–		Comonomer as solvent

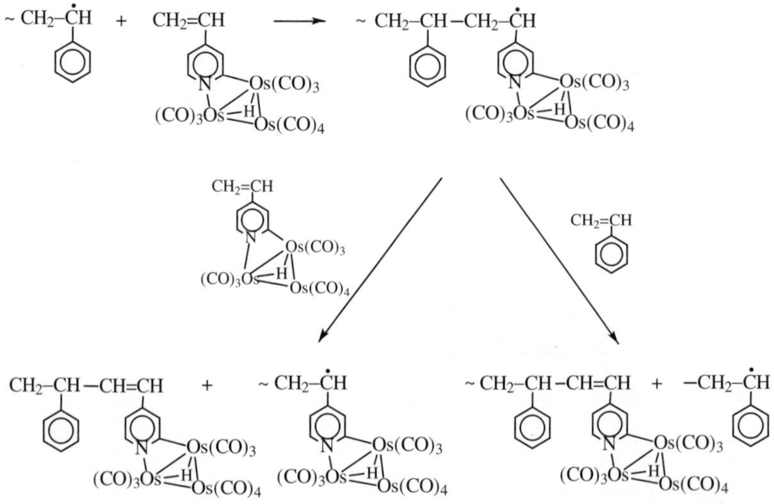

Scheme 5.32

ones of the polymer chain segments. The temperature of the polymer decay begins to rise at 323–333 K, the effective thermal destruction activation energy rises at 50–60 kJ/mol which can be explained by the stabilizing action of the metalloclusters. It was found that the glass transition temperatures for polyvinylimidazole and its copolymers with imidazole clusters (0.1 and 0.5 mol/L) are equal to 437, 493 and 530 K, correspondingly [422].

At the present time we have only scattered data on polymer-immobilized cluster formation in situ. In principle it is possible to create nanophase materials by a controlled deposition of the clusters with high kinetic energy on the polymer surface, similar to the ordered Cu_{147} cluster deposition on the cooled substrates [432], followed by the implantation, disordering and cluster spreading over the surface.

All these clusters were produced in the various polymer systems by the polymerization methods. There are no restrictions for the polycondensation reactions use for this purpose. But there are problems with the synthesis of the metallomonomers, containing the required bi- and polyfunctional ligands just as in the polymerization methods.

Such pyrolysis (Sect. 5.2) will allow us to produce high dispersed metal powders. Heteronuclear cluster polymer pyrolysis is a potential method for nanosize powder alloys (similarly to the pyrolysis of initial compounds $HFeCo_3(CO)_{12}$, $[Co(MeCOMe)_6][FeCo_3(CO)_{12}]$ [433]).

The metal nanoparticles capsulated into the carbon flakes [434] are produced by the thermolysis of chromium, palladium or platinum metallocomplexes, immobilized on hexo-2,4-diin-1,6-diol polymer (at 670–770 K, in

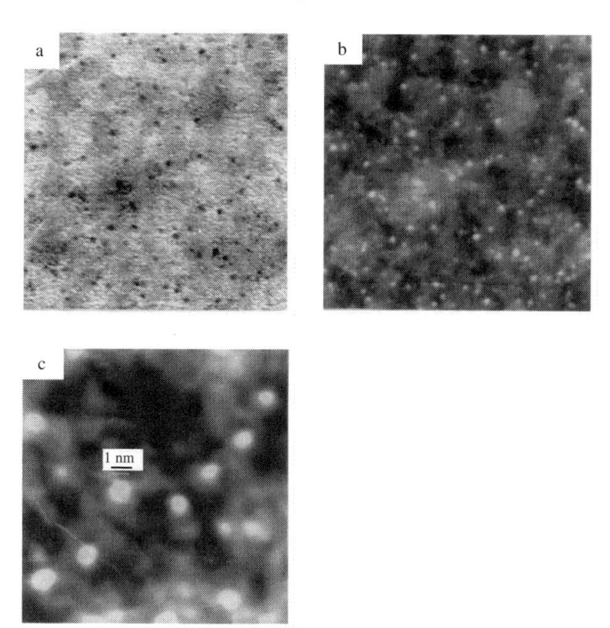

Fig. 5.28. Micrographs of Mo_6-polymer clusters in transmitted beam: **a** – phase-contrasting image; **b** – Z-contrasting image; **c** – "star formation" of monodispersed clusters with diameter ~1 nm at 10^6 magnification

argon). The incorporation of platinum clusters into a glass-carbon precursor was carried out [435] by mononuclear ethylene–bis(triphenylphosphino)-platinum(0) immobilization, using the triple bond of the poly(phenyldiacetylene) oligomer. A toluene-soluble complex (of 1:4 or 1:2 composition) heated to 870 K produces a platinum doped glass-carbon in the form of thin films with narrow Pt cluster distribution (mean diameter 1.6 nm). The big Pt particles (600 nm) were formed when the separated complexes (not bonded to a polymer) were used. The cluster-containing diacetylene olygomers pyrolysis (at 1620 K) gives ceramics with composition $M_xSi_yC_z$ (M – Fe, Co) [436].

The metallocluster polymeric transformation is now one of the most promising directions in nanopolymeric material science. There are two distinct ways of development: the synthesis and description of convenient monomers and their polymerization. The first favored one includes the new developed synthesis methods, optimized reactions and product preserving methods (including the remaining reactive multiple bonds, their activation on the bimetallic centers). Some exotic products such as buckminsterfullerene containing clusters of $\{[(C_2H_5)_3P]_2Pt\}_6C_{60}$ type are produced and described in [437]. The second less popular way includes studies of the cluster containing monomers involving in the polymerization processes. The theoretical notions of the role of metal cluster formations in the reactions and the problems of such monomer copolymerization with traditional polymers are under development.

* * *

It is clear that the reduction methods of nanoparticle synthesis in polymer matrices give great possibilities for the creation and production of polymer nanocomposites. The main problem now is to find an optimal way amongst the many others. This can be clearly demonstrated by the example of a polymer–chalcogenide system, when a nanocomposite can be produced by metal sulfides (including synthesized in situ) with soluble polymers, by a polymerization in heterogeneous conditions (with homopolymers, copolymers or even block-copolymers), by introducing the compounds into layered structures (see Chap. 8) or even by biomineralization (see Chap. 9).

There are many theoretical problems connected primarily with the physical and chemical aspects of the reduced generation and growth of nanoparticles, as well as the role of polymer matrix structure and composition, a metal-polymer phase forming, the fractal clusters or filaments growth peculiarities in the given polymer matrices, and cluster–cluster aggregation characteristics. These problems for the nanostructure in nonpolymeric systems are well studied and are described by quantitative relations (as was shown in Chap. 2) and there is no question that studies of metal–polymer systems will be developed in this direction, because a quantitative description of nanocomposites is growing and now there are attempts to describe atomic metal reactivity (in relation to a polymer surface) on the basis of the metal ionization potential and the lowest unoccupied molecular orbital of the given polymer.

References

1. N.N. Maltseva, V.S. Khain: *Sodium borohydride. Properties and application* (Khimiya, Moscow 1985)
2. L.A. Dykman, A.A. Lyakhov, V.A. Bogatyrev, S.Y. Schegolev: Colloid J. **60**, 757 (1998)
3. J.S. Bradley, E.W. Hill, S. Behal, C. Klein, B.M. Chaudret, A. Duteil: Chem. Mater. **4**, 1234 (1992)
4. L. Loudenberger, G.J. Mills: Phys. Chem. **99**, 475 (1995)
5. K. Ohitsu, Y. Vizukoshi, H. Brandow, T.A. Yamamoto, Y. Naguta, Y. Maeda: J. Phys. Chem. B **101**, 5470 (1997)
6. H.S. Kim, J.H. Ryu, B. Jose, B.G. Lee, B.S. Ahn, Y.S. Kang: Langmuir **17**, 5817 (2001)
7. J.H. Youk, J. Locklin, C. Xia, M.-K. Park, R. Advincula: Langmuir **17**, 4681 (2001)
8. T.G. Schaaff, M.N. Shafigullin, J.T. Khoury, I. Vezmar, R.I. Whetten, W.G. Cullen, P.N. First, C.J. Gutierrez-Wing, M.J. Ascensio, Jose-Yacaman: J. Phys. Chem. B **101**, 7885 (1997)
9. X. Yan, H. Liu, K.Y. Liew: J. Mater. Chem. **11**, 3387 (2001)

10. D.V. Sokolsky, N.F. Noskova: *Zigler-Natta's catalysts in hydrogenation reactions* (Nauka, Alma-Ata 1977)
11. V.V. Kopeikin, E.F. Panarin: Dokl. Phys. Chem. **380**, 497 (2001)
12. V.V. Kopeikin, Y.G. Saturyan, M.Y. Danilova: Chem-Pharm. J. 1253 (1989)
13. L. Chen, W.-J. Yang, C.-Z. Yang: J. Mater. Sci. **32**, 3571 (1997)
14. L. Chen, C.-Z. Yang, X.H. Yu: Chin. Chem. Lett. **5**, 443 (1994)
15. *Coulombic interactions in macromolecular systems* ed. by A. Eisenberg, E. Bailey (ACS, Washington, DC 1986)
16. Y. Wang, L. Feng, C. Pan: J. Appl. Polym. Sci. **70**, 2307 (1998)
17. C.-C. Yen: J. Appl. Polym. Sci. **60**, 605 (1996)
18. K. Kurihara, J. Kizling, P. Stenius, J.H. Fendler: J. Amer. Chem. Soc. **105**, 2574 (1983)
19. K. Torigoe, K. Eaumi: Langmuir **8**, 59 (1992)
20. M. Boutanner, J. Kizling, P. Stenius, G. Marie: J. Appl. Catal. **20**, 163 (1986)
21. N. Toshima, T. Takashi: Bull. Chem. Soc. Jpn. **65** 400 (1992)
22. J.H. Clint, I.R. Collins, J.A. Williams: Chem. Soc. Faraday Discuss. **95**, 219 (1993)
23. L. Tröger, H. Hünnefeld, S. Nunes, M. Oehring, D. Fritsch: J. Phys. Chem. B **101**, 1279 (1997)
24. A.B.R. Mayer, J.E. Mark, S.H. Hausner: J. Appl. Polym. Sci. **70**, 1209 (1998)
25. D.I. Svergun, E.V. Shtykova, A.T. Dembo, L.M. Bronstein, O.A. Platonova, A.N. Yakunin, P.M. Valetsky, A.R. Khokhlov: J. Chem. Phys. **109**, 11109 (1998)
26. T.S. Ahmadi, Z.L. Wang, T.C. Green, A. Henglein, M.A. El-Sayed: Science **272**, 1924 (1996)
27. T.S. Ahmadi, Z.L. Wang, A. Heglein, M.A. El-Saed: Chem. Mater. **8**, 1161 (1996)
28. O. Vidoni, K. Philippot, C. Amines, B. Chaudret, O. Balmes, J. Malm, J. Bovin, F. Senocq, M. Casanove: Angew. Chem., Int. Ed. **38**, 3736 (1999)
29. Y. Zhou, C.Y. Wang, Y.R. Zhu, Z.Y. Chen: Chem. Mater. **11**, 2310 (1999)
30. H.P. Choo, K.Y. Liew, W.A.K. Mahmood, H. Liu: J. Mater. Chem. **11**, 2906 (2001)
31. W. Yu, M. Liu, H. Liu, J. Zheng: J. Colloid Interface Sci. **210**, 218 (1999)
32. H. Hirai: J. Macromol. Sci. A **13**, 633 (1979)
33. H. Hirai, Y. Nakao, N. Toshima: Chem. Lett. 545 (1978), 905 (1976); J. Macromol. Sci. A **12**, 1117 (1978), **13**, 727 (1979)
34. H. Hirai, H. Chawanya, N. Toshima: J. Chem. Soc. Jpn., Chem. Ind. Chem. 1027 (1984); Kobunshu Ronbunshu **43**, 161 (1986)
35. T. Teranishi, M. Hosoe, M. Miyake: Adv. Mater. **9**, 65 (1997)
36. H. Hirai, M. Ohtaki, M. Komiyama: Chem. Lett. 149 (1987)
37. H. Hirai, N. Yakura, Y. Seta, S. Hodoshima: React. Funct. Polym. **37**, 121 (1988)
38. W. Yu, M. Liu, H. Liu, X. An, Z. Liu, X. Ma: J. Mol. Catal. A **142**, 201 (1999)
39. T. Teranishi, M. House, T. Tanaka, M. Miyake: J. Phys. Chem. B **103**, 3818 (1999)
40. T. Teranishi, M. Miyake: Chem. Mater. **10**, 594 (1998)
41. W. Yu, H. Liu: Chem. Mater. **10**, 1205 (1998)
42. M. Su, C. Bai, C. Wang: Solid State Commun. **106**, 643 (1998)
43. J.S. Bradley, J.M. Millar, E.W. Hill, S. Behal, B. Chaudret, A. Duteil: J. Chem. Soc. Faraday Discuss. **92**, 255 (1991)

44. N. Toshima, T. Takashi: Bull. Chem. Soc. Jpn. **65,** 400 (1992)
45. D. Caro, J.S. Bradley: Langmuir **14**, 245 (1998)
46. Y.S. Yablokov, A.I. Prokofyev: Chem. Phys. **15**, 114 (1996)
47. R. Payne, G. Fritz, H. Narmann: Angew. Makromol. Chem. **144**, 51 (1986)
48. C.-W. Chen, M. Akashi: J. Polym. Sci., Polym. Chem. A **35**, 1329 (1997)
49. H. Liu, N. Toshima: Chem. Comm. 1095 (1992)
50. A.A. Litmanovich, I.M. Papisov: J. Polym. Sci., Ser. B **39**, 323 (1997)
51. O.E. Litmanovich, A.G. Bogdanov, F.F. Litmanovich, I.M. Papisov: J. Polym. Sci. USSR, Ser. B **40**, 100 (1988)
52. V.A. Kabanov, I.M. Papisov: J. Polym. Sci. USSR, Ser. A **21**, 243 (1979)
53. A.D. Pomogailo, I.E. Uflyand, E.F. Vainstei: Russ. Chem. Rev. **64**, 913 (1995)
54. I.W. Hamley: *The Physics of Block Copolymers* (Oxford University Press, Oxford 1998)
55. E. Helfand, Z.R. Wasserman: Macromolecules **9**, 879 (1976), **11,** 960 (1978), **13** , 994 (1980)
56. J.-F. Gohy, B.G.G. Lohmeijer, U.S. Schubert: Macromolecules **35**, 4560 (2002)
57. L.M. Bronstein, M.V. Seregina, O.A. Platonova, Y.A. Kabachii, D.N. Chernyshov, M.G. Ezernitskaya, L.V. Dubrovina, T.P. Bragina, P.M. Valetsky: Macromol. Chem. Phys. **199**, 1357 (1998)
58. J. Fendler: Chem. Rev. **87**, 877 (1987)
59. *Structure and Reactivity in Reverse Micelles* ed. by M.P. Pileni (Elsevier, Amsterdam 1989)
60. C. Petit, P. Lixon, M.-P. Pileni: J. Phys. Chem. **97**, 12974 (1993)
61. L. Motte, M.-P. Pileni: J. Phys. Chem. B **102**, 4104 (1998)
62. T. Hanaoka, T. Tago, M. Kishida, K. Wakabayashi: Bull. Chem. Soc. **74**, 1349 (2001)
63. L.N. Erofeev, A.V. Raevskii, O.I. Kolesova, B.S. Grishin, T.I. Pisarenko: Chem. Phys. Reports **15**, 427 (1996)
64. J.S. Bradley, B. Tesche, W. Busser, M. Maase, M.T. Reetz: J. Am. Chem. Soc. **122**, 4631 (2000)
65. M. Rong, M. Zhang, H. Liu, H. Zeng: Polymer **40**, 6169 (1999)
66. J. Gratt, R.E. Cohen: Macromolecules **30**, 3137 (1997)
67. R.E. Cohen, R.T. Clay, J.F. Ciebien, B.H. Sohn;. In: *Polymeric Materials Encyclopedia*ed. by J.C. Salamone (CRC Press, Boca Raton 1996) V. 6, p. 4143
68. R.T. Clay, R.E. Cohen: Supramolecular Science **4**, 113 (1997)
69. M. Moffitt, H. Vali, A. Eisenberg: Chem. Mater. **10**, 102 (1998)
70. R. Saito, S. Okamura, K. Ishizu: Polymer **33**, 1099 (1992), **34**, 1189 (1993)
71. M. Antonietti, S. Henz: Nachr. Chem. Tech. Lab. **40**, 308 (1992)
72. Y.N.C. Chan, R.R. Schrock, R.E. Cohen: Chem. Mater. **4**, 24, 885 (1992)
73. Y.N.C. Chan, G.S.W. Craig, R.R. Schrock, R.E. Cohen: Chem. Mater. **4**, 88 (1992)
74. M. Möller, H. Kunstle, M. Kunz: Synth. Met. **41–43**, 1159 (1991)
75. R.T. Clay, R.E. Cohen: Supramolecular Science **2**, 183 (1995)
76. S.T. Selvan, T. Hyakawa, M. Nogami, M. Möller: J. Phys. Chem. B **103**, 7441 (1999)
77. R. Saito, H. Kotsubo, K. Ishizu: Polymer **33**, 1073 (1992)
78. B.H. Sohn, R.E. Cohen: Chem. Mater. **9**, 264 (1997)
79. Y.N.C. Chan, R.R. Schrock, R.E. Cohen: J. Am. Chem. Soc. **114**, 7295 (1992)
80. B.H. Sohn, R.E. Cohen: Acta Polym. **47,** 340 (1996)

81. V. Sankaran, J. Yue, R.E. Cohen, R.R. Schrock, R.J. Silbey: Chem. Mater. **5**, 1133 (1993)
82. M. Öner, J. Norwig, W.H. Meyer, G. Wegner: Chem. Mater. **10**, 460 (1998)
83. M. Antonietti, E. Wenz, L. Bronstein, M. Seregina: Adv. Mater. **7**, 1000 (1995)
84. M. Antonietti, A. Thüneman, E. Wenz: Colloid Polym. Sci. **274**, 795 (1996)
85. M. Antonietti, F. Gröhn, J. Hartmann, L. Bronstein: Angew. Chem. Int. Ed. Engl. **36**, 2080 (1997)
86. S. Tamil Selvan: Chem. Commun. 351 (1998)
87. K. Misawa, H. Yao, T. Hayashi, T. Kobayashi: J. Chem. Phys. **94**, 4131 (1991)
88. M.A. Hempenius, B.M.W. Langeveld, J.A. van Haare, R.A. Janssen, S.S. Sheiko, J.P. Spatz, M. Möller, E.W. Meijer: J. Am. Chem. Soc. **120**, 2798 (1998)
89. M. Moffitt, A. Eisenberg: Macromolecules **30**, 4363 (1997)
90. R. Rosetti, J.L. Ellison, J.M. Gibson, L.E. Brus: J. Chem. Phys. **80**, 4464 (1984)
91. A. Röscher, M. Möller: Adv. Mater. **7**, 151 (1995)
92. S.S. Im, H.S. Im, E.Y. Kang: J. Appl. Polym. Sci. **41**, 1517 (1990)
93. M. Möller, J.P. Spatz, A. Röscher, S. Mössmer, S. Tamil Selvan, H.-A. Klok: Macromol. Symp. **117**, 207 (1997)
94. M. Moffitt, L. McMahon, V. Pessel, A. Eisenberg: Chem. Mater. **7**, 1185 (1995)
95. J.P. Spatz, A. Roescher, M. Mö ller: Adv. Mater. **8**, 337 (1996); Chem. Eur. J. **2**, 1552 (1996)
96. D.E. Fogg, L.H. Radzilowski, R. Blanski, R.R. Schrock, E.L. Thomas: Macromolecules **30**, 417 (1997)
97. D.E. Fogg, L.H. Radzilowski, B.O. Dabbousi, R.R. Schrock, E.L. Thomas, M.G. Bawendi: Macromolecules **30**, 8433 (1997)
98. M. Moffitt, K. Khougaz, A. Eisenberg: Acc. Chem. Res. **29**, 95 (1996)
99. H. Zhao, E.P. Douglas, B.S. Harrison, K.S. Schanze: Langmuir **17**, 8428 (2001)
100. V. Sankaran, C.C. Cummins, R.R. Schrock, R.E. Cohen, R.J. Silbey: J. Am. Chem. Soc. **112**, 6858 (1990)
101. C.C. Cummins, R.R. Schrock, R.E. Cohen: Chem. Mater. **4**, 27 (1992)
102. J. Yue, R.E. Cohen: Supramolecular Science **1**, 117 (1994)
103. R.S. Kane, R.E. Cohen, R. Silbey: Chem. Mater. **8**, 1919 (1996)
104. J. Yue, V. Sankaran, R.E. Cohen, R.R. Schrock: J. Am. Chem. Soc. **115**, 4409 (1993)
105. J.R. Lakowicz, I. Gryczynski, Z. Gryczynski, C. Murphy: J. Phys. Chem. **103**, 7613 (1999)
106. K. Sooklal, L.H. Hanus, H.J. Ploehn, C. Murphy: J. Adv. Mater. **10**, 1083 (1998)
107. F. Grohn, G. Kim, B.J. Bauer, E.J. Amis: Macromolecules **34**, 2179 (2001)
108. F. Grohn, B.J. Bauer, Y.A. Akpalu, C.L. Jackson, E.J. Amis: Macromolecules **33**, 6042 (2000)
109. V. Chechik, R.M. Crooks: J. Am. Chem. Soc. **122**, 1243 (2000)
110. R.G. Ispasoiu, L. Balogh, O.P. Varnavski, D.A. Tomalia, T. Goodson: J. Am. Chem. Soc. **122**, 11005 (2000)
111. D.A. Tomalia, P.R. Dvornic: In: *The Polymeric Materials Encyclopedia* ed. by J. Salamone (CRC Press, Boca Raton 1996) V. 3, p. 1814)
112. A.M. Muzafarov, E.A. Rebrov: J. Polym. Sci. Ser. B **42**, 2015 (2000)
113. F. Aulentaa, W. Hayes, S. Rannardb: Europ. Polym. J. **39**, 1741 (2003)

114. K.L. Wooley, C.J. Hawker, J.M. Frechet: J. Chem. Soc. Perkin Trans. I, 1059 (1991)
115. F. Gröhn, B.J. Bauer, Y.A. Akpalu, C.L. Jackson, E.J. Amis: Macromolecules **33**, 6042 (2000)
116. M. Zhao, R.M. Crooks: Adv. Mater. **11**, 217 (1999)
117. M. Zhao, R.M. Crooks: Angew. Chem., Int. Ed. **38**, 364 (1999)
118. K. Esumi, A. Suzuki, N. Aihara, K. Usui, K. Torigoe: Langmuir **14**, 3157 (1998)
119. F. Gröhn, B.J. Bauer, Y.A. Akpalu, C.L. Jackson, E.J. Amis: Macromolecules **33**, 6042 (2000)
120. F. Gröhn, G. Kim, B.J. Bauer, E.J. Amis: Macromolecules **34**, 2179 (2001)
121. F. Gröhn, B.J. Bauer, E.J. Amis: Macromolecules **34**, 6701 (2001)
122. M.A. Mazo, E.B. Gusarov, N.K. Balabaev: J. Phys. Chem. **74**, 1985 (2000)
123. M.A. Hearshaw, J.R. Moss: Chem. Commun. 1 (1999)
124. M. Urbanczyk-Lipowska, J. Janiszeska: Polimery **48**, 54 (2003)
125. F. Zeng, S.C. Zimmerman: Chem. Rev. **97**, 1681 (1997)
126. L. Balogh, D.A. Tomalia: J. Am. Chem. Soc. **120**, 7355 (1998)
127. M. Zhao, L. Sun, R.M. Crooks: J. Am. Chem. Soc. **120**, 4877 (1998)
128. P.N. Floriano, C.O. Noble, J.M. Schoonmaker, E.D. Poliakoff, R.L. McCarley: J. Am. Chem. Soc. **123**, 10545 (2001)
129. L. Zhou, D.H. Russell, M. Zhao, R.M. Crooks: Macromolecules **34**, 3567 (2001) 1
130. B.I. Lemon, R.M. Crooks J. Am. Chem. Soc. **122**, 12886 (2000)
131. J.M. Koonnen, W.K. Russell, J.M. Hettick, D.H. Russell, Anal. Chem. **72**, 3860 (2000)
132. M.-K. Kim, Y.-M. Jeon, W.S. Jeon, H.-J. Kim., S.G. Hong, C.G. Park, K. Kim: Chem. Commun. 667 (2001)
133. N.C. Beck Tan, L. Balogh, S.F. Trevino, D.A. Tomalia, J.S. Lin: Polymer, **40**, 2537 (1999)
134. D.G. Kurth, F. Caruso, G. Schüler: Chem. Commun. 1579 (1999)
135. A.L. Volynsky, N.F. Bakeev: *High-dispersed oriented state of polymers* (Khimiya, Moscow 1984)
136. A.L. Volynsky, T.E. Gorokhovskaya, N.S. Shitov, N.F. Bakeev: J. Polym. Sci. USSR, Ser. A **24**, 1266 (1982)
137. S.V. Stakhanova, N.I. Nikonorova, V.D. Zanegin, G.M. Lukovkin, A.L. Volynsky, N.F. Bakeev: J. Polym. Sci., Ser. A **34**, 133 (1992)
138. N.I. Nikonorova, E.V. Semenova, V.D. Zanegin, G.M. Lukovkin, A.L. Volynsky, N.F. Bakeev: J. Polym. Sci., Ser. A **34,** 123 (1992)
139. V.S. Stakhanova, N.I. Nikonorova, G.M. Lukovkin, A.L. Volynsky, N.F. Bakeev: J. Polym. Sci., Ser. B **34**, 28 (1992)
140. N.A. Shitov, A.L. Volynsky, A.S. Chegolya, N.F. Bakeev: J. Polym. Sci. USSR, Ser. B **25**, 393 (1983)
141. N.I. Nikonorova, E.S. Trofimchuk, P.G. Elkin, N.E. Belova, S.S. Fanchenko, A.L. Volynsky, N.F. Bekeev: J. Polym. Sci., Ser. A **44**, 1185 (2002)
142. A.L. Volynsky, E.S. Trofimchuk, N.I. Nikonorova, N.F. Bakeev: Russ. J. Gen. Chem. **72**, 575 (2002)
143. N.I. Nikonorova, E.S. Trofimhuk, E.V. Semenova, A.L. Volynsky, N.F. Bakeev: J. Polym. Sci., Ser. A **42**, 1298 (2000)

144. N.I. Nikonorova, S.V. Stakhanova, I.A. Chmutin, E.S. Trofimchuk, P.A. Chernavsky, A.L. Volynsky, A.T. Ponomarenko, N.F. Bakeev: J. Polym. Sci., Ser. B **40**, 487 (1998)
145. S.V. Stakhanova, N.I. Nikonorova, A.L. Volynsky, N.F. Bakeev: J. Polym. Sci., Ser. A **39**, 312 (1997)
146. S.V. Stakhanova, E.S. Trofimchuk, N.I. Nikonorova, A.V. Rebrov, A.N. Ozerin, A.L. Volynsky, N.F. Bakeev: J. Polym. Sci., Ser. A **39,** 318 (1997)
147. I.M. Papisov, Y.S. Yablokov, A.I. Prokofyev, A.A. Litmanovich: J. Polym. Sci., Ser. A **35**, 515 (1993), **36**, 352 (1994)
148. A.V. Volkov, M.A. Moskvina, I.V. Karachevtsev, A.V. Rebrov, A.L. Volynsky, N.F. Bakeev: J. Polym. Sci., Ser. A **40**, 45 (1998)
149. Y. Yuan, J. Fendler, I. Cabasso: Macromolecules **23**, 3198 (1990); Chem. Mater **4**, 312 (1992)
150. E.F. Hillinski, P.A. Lucas, Y. Wang: J. Chem. Phys. **90**, 3435 (1988), **92**, 6927 (1990)
151. M. Krishnan, J.R. White, M.A. Fox, M.J. Bard: J. Amer. Chem. Soc. **105**, 7002 (1983)
152. S.A. Majetich, A.C. Carter: J. Phys. Chem. **97**, 8727 (1993)
153. A.D. Pomogailo: *Polymer immobilized metal-complex catalysts* (Nauka, Moscow 1988)
154. A.D. Pomogailo: Dokl. Phys. Chem. 2004 (in print)
155. G.D. Chryssicos, V.D. Mattera, A.T. Tsatsas, W.M. Risen: J. Catal. **93**, 430 (1985)
156. V.D. Mattera, D.M. Barnes, S.N. Chaudhri, W.M. Risen, R.D. Gonzalez: J. Phys. Chem. **90**, 4819 (1986)
157. A. Tai, Y. Imachi, T. Harada, Y. Izumi: Chem. Lett. 1651 (1981)
158. A.A. Dontsov, V.F. Soldatov, A.N. Kamesky, B.A. Dogadkin: Colloid. J. USSR, **31**, 370 (1969)
159. N. Thoshima: Catalyst. **27**, 488 (1985); High. Polym. Jap. **36**, 670 (1987); Chem. Lett. 1769 (1989)
160. C. Yee, M. Scotti, A. Ulmann, H. White, R. Rafailovich, J. Sokolov: Langmuir **15**, 4314 (1999)
161. J. Alvarez, J. Liu, E. Roman, A.E. Kaifer: Chem. Commun. 1151 (2000)
162. J. Liu, W. Ong, E. Roman, M.J. Linn, A.E. Kaifer: Langmuir **16**, 3000 (2000)
163. J. Shiyama, T. Shirakawa, Y. Kurokawa, S. Imaizumi: Angew. Makromol. Chem. **156**, 179 (1988)
164. H. Ishizuka, T. Tano, K. Torigoe: Colloids Surf. **63**, 337 (1992)
165. A. Duteil, G. Schmidt, W. Meyer-Zaika: Chem. Commun. 31 (1995)
166. N.E. Bogdanchikova, V.V. Tretiyakov, V.I. Zaikovsky: Kinet. Catal. **30**, 1468 (1989)
167. C.J. Huang, C.C. Jen, T.C. Chang: J. Appl. Polym. Sci. **42**, 2237 (1991)
168. C.-C. Yen: J. Appl. Polym. Sci. **71**, 1361 (1999)
169. A.D. Pomogailo: *Polymer-immobilized metal-complex catalysts* (Nauka, Moscow 1988)
170. A.D. Pomogailo: Russ. Chem. Rev. **61**, 257 (1992)
171. A.D. Pomogailo, F.A. Khrisostomov, F.S. Diyachkovsky: Kinet. Catal. **26**, 1104 (1985)
172. S.B. Echmaev, I.N. Ivleva, N.D. Golubeva, A.D. Pomogailo, Y.G. Borodyko: Kinet. Catal. **27**, 394 (1986)

173. F.K. Shmidt: *Catalysis of hydration reactions and dimerization by metal complexes of the first transient series* (Irkutsk Univ. Publ., Irkutsk 1986)
174. I.N. Ivleva, A.D. Pomogailo, S.B. Echmaev, M.S. Ioffe, N.D. Golubeva, Y.G. Borodyko: Kinet. Catal. **20**, 1282 (1979)
175. W.M.Risen: In: *Proc. NATO Adv. Res. Workshop Struct. and Prop. Ionomers, 1986* (Villard, Dodrecht 1987) p. 331
176. B.M. Sergeev, G.B. Sergeev, A.N. Prusov: Mendeleev Commun. 1 (1998)
177. C. Sangregorio, M. Galeotti, U. Bardi, P. Baglioni: Langmuir **12**, 5800 (1996)
178. M.-L. Wu, D.-H. Chen, T.-C. Huang: Langmuir **17**, 3877 (2001)
179. H. Hirai, N. Toshima: In: *Tailored Metal Catalysts* ed. by Y. Iwasawa (Reidel, Dodrecht 1986) p. 87
180. N. Toshima, H. Liu: Chem. Lett. 1925 (1992)
181. N. Toshima: J. Macromol. Sci. A **27**, 1125 (1990)
182. M. Harada, K. Asakura, N. Toshima: J. Phys. Chem. **97**, 5103 (1993)
183. N. Toshima, T. Yonezawa, M. Harada, K. Asakura, Y. Iwasawa: Chem. Lett. 815 (1990)
184. N. Toshima, K. Kushihashi, T. Yonezawa, H. Hirai: Chem. Lett. 1769 (1989)
185. Y. Wang, N. Toshima: J. Phys. Chem. B **101**, 5301 (1997)
186. M.T. Miller, A.N. Izumi, Y.-S. Shih, G.M. Whitesides: J. Am. Chem. Soc. **110,** 3146 (1988)
187. W. Yu, Y. Wang, H. Liu, W. Zheng: J. Catal. A **112**, 105 (1996)
188. J.S. Bradley: In:*Clusters and Colloids* ed. by G. Schmid (VCH, Weinheim 1993) p. 459
189. J.S. Bradley, E.W. Hill, C. Klein, B. Chaudret, A. Duteil: Chem. Mater. **5**, 254 (1993)
190. J.S. Bradley, E.W. Hill, B. Chaudret, A. Duteil: Langmuir **11**, 693 (1995)
191. N. Toshima, Y. Wang: Langmuir **10**, 4574 (1994); Chem. Lett. 1611 (1993)
192. J.S. Bradley, G.H. Via, L. Bonneviot, E.W. Hill: Chem. Mater. **8**, 1895 (1996)
193. *Binary Alloy Phase Diagrams* ed. by T.B. Massalski, H. Okamoto, P.R. Subramanian, L. Kacprzak (ASM International, Metals Park. OH 1990) v. 1
194. N. Toshima, P. Lu: Chem. Lett. 729 (1996)
195. A. Bukowski: Polymery **41**, 139 (1996)
196. W.-Y. Yu, H.-F. Liu, Q. Tao: Chem. Commun. 1773 (1996)
197. T. Teranishi, K. Nakata, M. Miyake, N. Toshima: Chem. Lett. 277 (1996); React. Funct. Polym. **37**, 111 (1998)
198. J.-C. Moutet, Y. Ouennoughi, A. Ourari, S. Hamar-Thibault: Electrochim. Acta **40**, 1827 (1995)
199. Y. Yang, J. Huang, B. Yang, S. Liu, J. Shen: Synth. Met. **91**, 347 (1997)
200. T. Sugama: J. Mater. Sci. **33**, 3791 (1998)
201. J. Lu, B. Liu, H. Yang, W. Luo, G. Zou: J. Mater. Sci. Lett. **17**, 1605 (1998)
202. K.V.P.M. Shafi, A. Gedanken, R. Prozorov, J. Balgoh, J. Lendrai, I. Felner: J. Phys. Chem. B **101**,6409 (1997)
203. K.S. Suslick, T. Hyeon, M. Fang, A.A. Cichowlas: Mater. Sci. Eng. A **204**, 186 (1996); Chem. Mater. **8**, 2172 (1996)
204. V.V. Sviridov, G.P. Shevchenko, Z.M. Afanasyeva, A.N. Ponyavina, N.V. Loginova: Colloid. J. **58**, 390 (1996)
205. S.R. Ahmed, P. Kofinas: Macromolecules, **35**, 3338 (2002)
206. N. Moumen, P. Veillet, M.P. Pileni: J. Magn. Magn. Mater. **149**, 67 (1995)

207. M. Grigorova, H.J. Blythe, V. Blaskov, V. Rusanov, V. Petkov, V. Masheva, D. Nihtianova, L.M. Martinez, J.S. Munoz, M. Mikhov: J. Magn. Magn. Mater. **183**, 163 (1998)
208. X. Li, H. Zhang, F. Chi, S. Li, B. Xu, M. Zhao: Mater. Sci. Eng. B **18**, 209 (1993)
209. S.K. Saha, A. Pathac, P. Pramanik: J. Mater. Sci. Lett. **14**, 35 (1995)
210. P.A. Lessing: Amer. Ceram. Soc. Bull. **68**, 1002 (1989)
211. H. Taguchi, D. Matsuda, M. Nagao: Amer. Ceram. Soc. Bull. **75**, 201 (1992); J. Mater. Sci. Lett. **12**, 891 (1993); J. Solid State Chem. **104**, 460 (1993)
212. L.-W. Tai, H.U. Anderson, P.A. Lessing: Amer. Ceram. Soc. Bull. **75**, 3490 (1992)
213. K.C. Jain, V.R. Adiga, P. Verneker: Combustion and Flame **40**, 71 (1981)
214. P. Mulvaney, M. Giersig, A. Henglein: J. Phys. Chem. **96**, 10419 (1992)
215. P.A. Vozny, L.V. Galushko, P.P. Gorbik, V.V. Dyakin, A.A. Levchenko, V.V. Levandobsky, V.N. Lysenko, V.M. Ogenko, L.K. Yanchevsky: Superconductivity: physics, chemistry, engineering **5**, 1478 (1992)
216. L.K. Yanchevsky, V.V. Levandovsky, N.V. Abramov, P.P. Gorbik, P.A. Vozny, I.V. Dubrovin, M.V. Bakuntseva: Plast. massy (9), 18 (1997)
217. A. Douy, P. Odier: Mat. Res. Bull. **24**, 1119 (1989)
218. I. Valente, C. Sanchez, M. Henri, J. Livage: Industrie Ceramique **836**, 193 (1989)
219. A.A. Ostroushko, L.I. Zhuravlev, S.M. Portnova, Y.I. Krasilov: J. Inorg. Chem. **36**, 3, 1099 (1991)
220. A.A. Ostroushko, S.M. Portnova, Y.I. Krasilov, I.P. Ostroushko: J. Inorg. Chem. **36**, 823 (1991)
221. A.A. Ostroushko, N.V. Mironova, I.P. Ostroushko, A.N. Petrov: J. Inorg. Chem. **37,** 2627 (1992)
222. J.C.W. Chien, B.M. Gong, J.M. Madsen, R.B. Hallock: Phys. Rev. B **38**, 11853 (1988)
223. J.C.W. Chien, B.M. Gong, X. Mu, Y. Yang: J. Polym. Sci., Polym. Chem. Ed. **28**, 1999 (1990)
224. S. Maeda, Y. Tsurusaki, Y. Tachiyama, K. Naka, A. Ohki, T. Ohgushi, T. Takeshita: J. Polym. Sci., Polym. Chem. A **32**, 1729 (1994)
225. K. Naka, Y. Tachiyama, A. Ohki, S. Maeda: J. Polym. Sci., Polym. Chem. A **34**, 1003 (1996)
226. H. Tamura, H. Hineta, M. Tatsumi, J. Tanishita, S. Yamamoto: Chem. Lett. 1147 (1994)
227. G. Mohazzab, I.M. Low: J. Appl. Polym. Sci. **56**, 1679 (1995)
228. P. Catania, W. Hovnanian, L. Cot: Mat. Res. Bull. **25**, 1477 (1990)
229. T. Goto, T. Takahashi: J. Mater. Res. **9**, 852 (1994)
230. H. Tomita, T. Goto, K. Takahashi: Supercond. Sci. Technol. **9**, 363, 1099 (1996)
231. H. Tomita, T. Omori, T. Goto, K. Takahashi: J. Mater. Sci. **33**, 247 (1998).
232. A.D. Pomogailo, V.S. Savostiyanov, G.I. Dzhardimalieva, A.V. Dubovitsky, A.N. Ponomarev: Russ. Chem. Bull. **44**, 1056 (1995)
233. V.S. Savostiyanov, V.A. Zhorin, G.I. Dzhardimalieva, A.D. Pomogailo, A.V. Dubovitsky, V.N. Topnikov, M.K. Makova, A.N. Ponomarev: Dokl. Phys. Chem. **318**, 378 (1991)
234. V.S. Savostiyanov, V.N. Vasilets, O.V. Ermakov, E.A. Sokolov, A.D. Pomogailo, D.A. Kritskaya: Bull. Russ. Acad. Sci., Div. Chem. Sci. 2073 (1992)

235. K.E. Gonsalves, S.P. Rangarajan, C.C. Law, C.R. Feng, G.-M. Chow, A. Garcia-Ruiz: In: *Nanotechnology. Molecular Designed Materials. ACS Symposium Book Series, N622* ed. by G.-M. Chow, K.E. Gonsalves (ACS, Washington, DC 1996) p. 220
236. K.E. Gonsalves, S.P. Rangarajan: J. Appl. Polym. Sci. **64,** 2667 (1997)
237. H.-B. Park, H.-J. Kweon, Y.-Sic Hong, S.-J. Kim: J. Mater. Sci. **32**, 57 (1997)
238. A.L. Balch, D.A. Costa, K. Winkler: J. Am. Chem. Soc. **120**, 9614 (1998)
239. M.W. Anderson, J. Shi, D.A. Leigh, A.E. Moody, F.A. Wade, B. Hamilton, S.W. Carr: Chem. Commun. 533 (1993)
240. M.J. Hampden-Smith, T.T. Kodas: Chem. Vap. Deposition **1**, 8 (1995)
241. T.W. Smith, D. Wychick: J. Phys. Chem. **84**, 1621 (1980)
242. Pat. 4252671–4252678, USA
243. P.H. Hess, P.H. Parker, Jr.: J. Appl. Polym. Sci **10**, 1915 (1966)
244. J.R. Thomas: J. Appl. Phys. **37**, 2914 (1966)
245. M. Berger, T.A. Manuel: J. Polym. Sci. A-1 **4,** 1509 (1966)
246. T.W. Smith, D.J. Luca: In: *Proc. Symp. Modif. Polym., Las Vegas (Nevada) 1982* (New York, London, 1983) p. 85
247. A.B. Gilman, V.M. Kolotyrkin: High Energy Chem. **12**, 450 (1978)
248. C.H. Griffiths, M.P. O'Horo, T.W. Smith: J. Appl. Phys. **50**, 7108 (1979)
249. H. Biederman: Vacuum **34**, 405 (1984)
250. C.E. Kerr, B.E. Eaton, J.A. Kadue: Organometallics **14**, 269 (1995)
251. S.P. Gubin, I.D. Kosobudskii, G.I. Petrakovskii: Dokl. Phys. Chem. **260**, 655 (1981)
252. S.P. Gubin, I.D. Kosobudskii: Dokl. Phys. Chem. **272**, 1155 (1983)
253. S.P. Gubin, I.D. Kosobudskii: Russ. Chem. Rev. **52**, 1350 (1983)
254. L.M. Bronstein, P.M. Valetskii, S.V. Vinogradova, A.I. Kuzaev, V.V. Korshak: Polym. Sci. USSR, Ser. A. **29**, 1694 (1987)
255. C.U. Pittman Jr., P.L. Grube, O.E. Ayers, S.P. McManus, M.D. Rausch, G.A. Moser: J. Polym. Sci. A-1 **10**, 379 (1972)
256. L.M. Bronstein, S.P. Solodovnikov, E.Sh. Mirzoeva, E.Yu. Baukova, P.M. Valetsky: In: *Proceed. ACS Div. Polym. Mater.Sci. and Engin. 1994* v. 71, p. 397
257. M. Berger, D.J. Buckley: J. Polym. Sci. A-1 **1**, 2945 (1963)
258. R.A. Arents, Yu.V. Maksimov, I.P. Suzdalev, Yu.B. Amerik: Hyperfine Interactions, **56**, 167 (1990)
259. Yu.B. Amerik, Yu.M. Korolev, V.N. Rogovoi: Petroleum Chemistry **36**, 304 (1996)
260. Yu.M. Korolev, A.L. Bykova, Yu.B. Amerik: Polym. Sci. Ser. B **39**, 1856 (1997)
261. I.D. Kosobudskii, S.P. Gubin, V.P. Piskorskii, G.A. Petrakovskii, L.V. Kashkina, V.N. Kolomiichuk, N.M. Svirskya: In: *Electronics of organic materials* ed. by A.A. Ovchinnikov (Nauka, Moscow 1985) p.62
262. P.I. Dosa, C. Erben, V.S. Iyer, K.P.C. Vollhard, I.M. Vasser: J. Am. Chem. Soc. **121**, 10430 (1999)
263. C.-H. Kiang, W.A. Goddard; Phys. Rev. Lett. **76**, 2515 (1996)
264. R. Sen, A. Govindaraj, C.N.R. Rao: Chem. Mater. **9**, 2078 (1997)
265. T.X. Neenan, M.R. Callstrom, O.J. Schueller: Macromol. Symp. **80**, 315 (1994)
266. S. Grimm, M. Schultz, S. Barth, R. Muller: J. Mater. Sci. **32**, 1083 (1997)
267. F. Galembeck, C.C. Chironi, C.A. Ribeiro: J. Appl. Polym. Sci. **25**, 1427 (1980)

268. R. Baumhardt-Neto, S.E. Galembeck, I. Joekes, F. Galembeck: J. Polym. Sci., Polym. Chem. Ed. **19**, 819 (1981)
269. F. Galembeck: J. Polym. Sci., Polym. Lett. Ed. **15**, 107 (1977); J. Polym. Sci., Polym. Chem. Ed. **16**, 3015 (1978)
270. R. Tannenbaum, C.L. Flenniken, E.P. Goldberg: In: *Metal-Containing Polymeric Systems* ed. by C. Carraher, C. Pittman, J. Sheats (Plenum, N.Y. 1985) p. 303
271. R. Tannenbaum, C.L. Flenniken, E.P. Goldberg: J. Polym. Sci., Polym. Phys. B **28**, 2421 1990
272. S. Reich, E.P. Goldberg: J. Polym. Sci., Polym. Phys. B **21**, 869 (1983)
273. A.I. Pertsin, Yu.M. Poshutin: Polym. Sci., Ser. B **38**, 919 (1996)
274. A.I. Pertsin, I.O. Volkov: Polym. Sci., Ser. B **38**, 1249 (1996)
275. L.M. Bronshtein, P.M. Valetskii, M. Antonietti: Polym. Sci., Ser. A **39**, 1847 (1997)
276. O.A. Platonova, L.M. Bronstein, S.P. Solodovnikov, I.M. Yanovskaya, E.S. Obolonkova, P.M. Valetsky, E. Wenz, M. Antonietti: Colloid Polym. Sci. **275**, 426 (1997)
277. *Ultrasound: Its Chemical, Physical, and Biological Effects* ed. by K.S. Suslick (VCH Press, N.Y. 1988)
278. K.S. Suslick: Science **247**, 1439 (1990)
279. K.S. Suslick, S.B. Choe, A.A. Cichowlas, M.W. Grinstaff: Nature **353**, 414 (1991)
280. K.S. Suslick, T. Hyeon, M. Fang, A.A. Cichowlas: Mater. Sci. Eng. **A204**, 186 (1996); Chem. Mater. **8**, 2172 (1996)
281. D. Caro, T.O. Ely, A. Mari, B. Chaudret, E. Snoeck, M. Respaud, J.-M. Broto, A. Fert: Chem. Mater. **8**, 1987 (1996)
282. G. Kataby, T. Prozorov, Yu. Koltyp, C.N. Sukenik, A. Ulman, A. Gedanken: Langmuir, **13**,615 (1997); **15**, 1702 (1999)
283. K.V.P.M. Shafi, A. Gedanken, R. Prozorov, J. Balogh: Chem. Mater. **10**, 3445 (1998)
284. V.P.M. Shafi, A. Ulman, X. Yan, N.L. Yang, C. Estournes, H. White, M. Rafailovich: Langmuir **17**, 5093 (2001)
285. M.S. Nashner, A.I. Frenkel, D. Somerville, C.W. Hills, J.R. Shapley, R.G. Nuzzo: J. Am. Chem. Soc. **120**, 8093 (1998)
286. E.I. Aleksandrova, G.I. Dzhardimalieva, A.S. Rozenberg, A.D. Pomogailo: Russ. Chem. Bull. **42**, 264 (1993)
287. A.N. Timofeev, I.Yu. Filatov, V.G. Savostyanov, K.I. Marushkin: High pure compounds (5) 45 (1994)
288. E.I. Aleksandrova, G.I. Dzhardimalieva, A.S. Rozenberg, A.D. Pomogailo: Russ. Chem. Bull. **42**, 259 (1993)
289. A.S. Rozenberg, E.I. Aleksandrova, G.I. Dzhardimalieva, A.N. Titkov, A.D. Pomogailo: Russ. Chem. Bull. **42**, 1666 (1993)
290. A.D. Pomogailo, A.S. Rozenberg, G.I. Dzhardimalieva: In: *Metal-containing Polymers Materials* ed. by C.U. Pittman, Jr., C.E. Carraher, Jr., M. Zeldin, B. Culberston (Plenum Press, N.Y. 1996) p. 313.
291. A.S. Rozenberg, E.I. Aleksandrova, G.I. Dzhardimalieva, N.V. Kiryakov, P.E. Chizhov, V.I. Petinov, A.D. Pomogailo: Russ. Chem. Bull. **44**, 858 (1995)
292. A.S. Rozenberg, E.I. Aleksandrova, N.P. Ivleva, G.I. Dzhardimalieva, A.V. Raevskii, O.I. Kolesova, I.E. Uflyand, A.D. Pomogailo: Russ. Chem. Bull. **47**, 259 (1998)

293. A.T. Shuvaev, A.S. Rozenberg, G.I. Dzhardimalieva, I.N. Ivleva, V.G. Vlasenko, T.I. Nedoseikina, T.A. Lyubeznova, I.E. Uflyand, A.D. Pomogailo: Russ. Chem. Bull. **47**, 1460 (1998)
294. I. Skupinska, H. Wilezura, H. Bonink: J. Therm. Anal. **31**, 1017 (1986)
295. A.S. Rosenberg, G.I. Dzhardimalieva, A.D. Pomogailo: Polym. Adv. Technol. **9**, 52 (1998)
296. A. Gronowski, Z. Wojtezak: J. Therm. Anal. **36**, 2357 (1990)
297. P.A. Vasil'ev, A.L. Ivanov, A.N. Glebov: Russ. J. Gen. Chem. **68**, 535 (1998)
298. A.S. Rozenberg, G.I. Dzhardimalieva, A.D. Pomogailo: Dokl. Phys. Chem. **356**, 294 (1997)
299. A.S. Rosenberg, G.I. Dzhardimalieva, A.D. Pomogailo: Polym. Adv. Technol. **9**, 52 (1998)
300. A.D. Pomogailo, V.G. Vlasenko, A.T. Schuvaev, A.S. Rozenberg, G.I. Dzhardimalieva: Colloid J. **64**, 524 (2002)
301. A.D. Pomogailo, A.S. Rozenberg, G.I. Dzhardimalieva, M. Leonowicz: Adv. Mater. Sci. **1**, 19 (2001)
302. A.D. Pomogailo, G.I. Dzhardimalieva, A.S. Rosenberg: Acta Physica Polonica A **102**, 135 (2002)
303. A.D. Pomogailo: Russ. Chem. Rev. **66**, 750 (1997)
304. T.Yu. Ryabova, A.S.'Chirkov, L.S. Radkevich, N.V. Evtushok: Ukr. Chem. J. **59**, 1329 (1993)
305. A.I. Savitskii, Sh.Ya. Korovskii, V.I. Prosvirin: Colloid J. USSR, **41**, 88 (1979); **42**, 998 (1980)
306. M.C. Alves, G. Tourillon: J. Phys. Chem. **100**, 7566 (1996).
307. J.E. Millburn, M.J. Rosseinsky: Chem. Mater. **9**, 511 (1997)
308. R.E. Southward, D.W. Thompson, A.K.St. Clair: Chem. Mater. **9**, 501 (1997)
309. R.E. Southward, D.S. Thompson, D.W. Thompson, A.K.St. Clair: Chem. Mater. **9**, 1691 (1997)
310. S.P. Pogorelova, G.M. Plavnik, A.P. Tikhonov, T.P. Puryaeva: Collooid J. **61**,100 (1999)
311. Yu.M. Khimchenko, L.S. Radkevich: Plast. Massy (1), 53 (1975)
312. N.I. Nikonorova, E.S. Trofimchuk, E.V. Semenova, A.L. Volynskii, N.F. Bakeev: Polym. Sci., Ser. A **42**, 1298 (2000)
313. A.Yu. Vasilkov, P.V. Pribytko, E.A. Fedorovskaya, A.A. Slinkin, A.S. Kogan, V.A. Sergeev: Dokl. Phys. Chem. **331**, 179 (1993)
314. J. Osuna, D. Caro, C. Amiens, B. Chaudret, E. Snoeck, M. Respaund, J.-M. Broto, A. Fest: J.Phys.Chem. **100**, 14571 (1996)
315. A. Carbonaro, A. Greco, G. Dall/asta J. Organomet. Chem. **20**, 177 (1969)
316. S.E.J. Goff, T.F. Nolan, M.W. George, M. Poliakoff: Organometallics **17**, 2730 (1998)
317. A.I. Cooper, M. Poliakoff: Chem. Phys. Lett. **212**, 611 (1993)
318. J.S. Bradley, J.M. Millar, E.W. Hill, S. Behal, B. Chaudret, A. Duteil: Faraday Discuss. Chem. Soc. **92**, 255 (1991)
319. A. Duteil, R. Queau, B.M. Chaudret, C. Roucau, J.S. Bradley: Chem. Mater. **5**, 341 (1993)
320. J.S. Bradley, E.W. Hill, S. Behal, C. Klein., B.M. Chaudret, A. Duteil: Chem. Mater. **4**, 1234 (1992)
321. D. Caro, J.S. Bradley: Langmuir **13**, 3067 (1997)
322. F. Waller: J .Catal. Rev.-Sci. Eng. **28**, 1 (1986)

323. A.M. Hodges, M. Linton, A.W.-H. Mau, K.J. Cavell, J.A. Hey, A. Seen: J. Appl. Organomet. Chem. **4**, 465 (1990)
324. M. Pehnt, D.L. Schulz, C.J. Curtis, K.M. Jones, D.S. Ginley: Appl. Phys. Lett. **67**, 2176 (1995)
325. E.E. Said-Galiev, L.N. Nikitin, Yu.P. Kudryavtsev, A.L. Rusanov, O.L. Lependina, V.K. Popov, M. Polyakoff, S.M. Khould: Chem. Phys. **14**, 190 (1995)
326. J.J. Watkins, T.J. McCarthy: Chem. Mater. **7**, 1991 (1995)
327. D.M. Mognonov, V.G. Samsonova, V.V. Khakinov, I.N. Pinchuk: Polym. Sci., Ser. B **39**, 1250 (1997)
328. G.P. Gladyshev, O.A. Vasnetsova, N.I. Mashukov: Mendeleev Chem. J. **34**, 575 (1990)
329. G.P. Gladyshev, N.I. Mashukov, A.K. Mikitaev, S.A. Eltsin: Polym. Sci. USSR, Ser. B **28**, 62 (1986)
330. L.V. Ruban, G.E. Zaikov: Russ. Chem. Rev. **63**, 373 (1994)
331. R.S. Beer, C.A. Wilkie, M.L. Mittleman: J. Appl. Polym. Sci. **46**, 1095 (1992)
332. I.C. McNeill, R.C. McGuiness: Polym. Degrad. Stabil. **9**, 1 (1984)
333. C.A. Wilkie, J.W. Pettergreww, C.E. Brown: J. Polym. Sci., Polym. Lett. **19**, 409 (1981)
334. L.E. Manring: Macromolecules **21**, 528 (1988); **22**, 2673 (1989); **24**, 3304 (1991)
335. L.E. Manring, D.Y. Sogah, G.M. Cohen: Macromolecules **22**, 4652 (1989)
336. C.A. Wilkie, J.T. Leone, M.L. Mittleman: J. Appl. Polym. Sci. **42**, 1133 (1991)
337. K. Gjurova, C. Uzov, M. Zagortcheva, G. Gavrilova: J. Appl. Polym. Sci. **74**, 3324 (1999)
338. T.-H. Ko, C.-Y. Chen: J. Appl. Polym. Sci. **71**, 2219 (1999)
339. A.M. Summan: J. Polym. Sci., Polym. Chem. A **37**, 3057 (1999)
340. L. Belfiore, M.P. McCurdie, E. Ueda: Macromolecules **26**, 6908 (1993)
341. O.P. Krivoruchko, V.I. Zaikovskii: Mendeleev Commun. 97 (1998)
342. C. McNeill, J.J. Liggat: Polym. Degrad. Stabil. **29**, 93 (1990); **37**, 25 (1992)
343. A.s. 707929 USSR; Bull. Inv. (1) 98 (1980)
344. V.G. Elson, Yu.D. Semchikov, D.N. Emelyanov, N.L. Khvatova: Polym. Sci. USSR, Ser. B **21**, 609 (1979)
345. Yu.D. Semchikov, N.L. Khvatova, V.G. Elson, R.F. Galliulina: Polym. Sci. USSR, Ser. A **29**, 503 (1987)
346. Y. Nakao: Chem. Commun. 826 (1993); J. Coll. Interf. Sci. **171**, 386 (1995)
347. N. Yanagihara, Y. Ishii, T. Kawase, T. Kaneko, H. Horie, T. Hara: Mat. Res. Soc. Symp. Proc. **457**, 469 (1997)
348. N. Yanagihara: Chem. Lett. 305 (1998)
349. Y. Nakao: Kobunshi **43**, 852 (1994); Zairyou Kagaku **31**, 28 (1994)
350. D.H. Gray, S. Hu, E. Huang, D.L. Gin: Adv. Mater. **9**, 731 (1997)
351. R.C. Smith, W.M. Fischer, D.L. Gin: J. Am. Chem. Soc. **119**, 4092 (1997)
352. H. Deng, D.L. Gin, R.C. Smith: J. Am. Chem. Soc. **120**, 3522 (1998)
353. D.D. Papakonstantinou, J. Huang, P. Lianos: J. Mater. Sci. Lett. **17**, 1571 (1998)
354. H.-J. Gläsel, E. Hartmann, R. Böttcher, C. Klimm, B. Milsch, D. Michel, H.-C. Semmelhack: J. Mater. Sci. **34** (1999)
355. A.D. Pomogailo: Russ. Chem. Rev. **67**, 60 (2000)
356. J.V. Minkiewicz, D. Milstein, J. Lieto, B.C. Gates, R.L. Albright: ACS Symp. Ser. **192**, 9 (1982)

357. J.P. Collman, L.S. Hegedus, M.P. Cooke, J.P. Norton, G. Dolcett, D.N. Marquardt: J. Am. Chem. Soc. **94**, 1789 (1972)
358. C.U. Pittman, Jr., W.D. Honnick, M.S. Wrighton, R.D. Sanner, R.G. Austin: In: *Fundamental Research in Homogeneous Catalysis. V. 3.* ed. by M. Tsutsui (Plenum Press, N.Y., London 1979)
359. D. Milstein, B.C. Gates: In: *179th ACS Nat.Meet. Houston Tex.* (ACS, Washington D.C. 1980)
360. S.C. Brown, J.J. Evans: Chem. Commun. 1063 (1978)
361. J.J. Rafalko, J. Lieto, B. Gates, G.L. Schrader Jr.: Chem. Commun. 540 (1978)
362. J. Lieto, J.J. Rafalko, B.C. Gates: J. Catal. **62**, 149 (1980)
363. M.B. Freeman, M.A. Patrickk, B.C. Gates: J. Catal. **73**, 82 (1982)
364. J.B. N'Guini-Effa, J. Lieto, J.P. Aune: J. Mol. Catal. **15**, 367 (1982)
365. R. Whymon: Chem. Commun. 230 (1970)
366. M.S. Jarrell, B.C. Gates, E.D. Nicholson: J. Am. Chem. Soc. **100**, 5727 (1978)
367. B.C. Gates, J. Lieto: Chem. Tech. **10**, 248 (1980)
368. L.N. Arsamaskova, Yu.I. Ermakov: Mendeleev Chem. J. **32**, 75 (1987)
369. C. Gates, H.H. Lamb: J. Mol. Catal. **52**, 1 (1989)
370. A.W. Olsen, Z.H. Kafafi: In: *Proc. Symp. on Clusters and Cluster-Assembled Materials* (Materials Research Society, Pittsburg 1991) v. 206, p. 175
371. Pat. 71.927 Finland (1986).
372. J. Lieto, J.J. Rafalko, J.V. Minkiewicz, P.W. Rafalko, B.C. Gates: In: *Fundamental Research in Homogeneous Catalysis. V. 3.* ed. by M.Tsutsui (Plenum Press, N.Y., London 1979)
373. Wolf, J. Lieto, B.A. Matrana, D.B. Arnold, B.C. Gates, H. Knosinger: J. Catal., **89**, 100 (1984)
374. C.-G. Jia, Y.-P. Wang, H.-Y. Feng: React. Polym. **18**, 203 (1992)
375. V.A. Maksakov, V.P. Kirin, S.N. Konchenko, N.M. Bravaya, A.D. Pomogailo, A.V. Virovets, N.V. Podberezskaya, I.G. Baranovskaya, S.V. Tkachev: Russ. Chem. Bull., 1293 (1993).
376. S. Bhaduri, H. Khwaja, B.A. Narayaman: J. Chem. Soc., Dalton Trans. 2327 (1984)
377. S.P. Tunik, S.I. Pomogailo, G.I. Dzhardimalieva, A.D. Pomogailo, I.I. Chuev, S.M. Aldoshin, A.B. Nikolskii: Russ. Chem. Bull. **42**, 937 (1993)
378. G.Strathdee, R. Given: Can. J. Chem. **52**, 3000 (1974)
379. J.B. Sobczak, J. Wernisch: Z. Phys. Chem. (BRD) **137**, 119 (1983) .
380. J.A. Jackson, M.D. Newsham, C. Worsham, D.G. Nocera: Chem. Mater. **8**, 558 (1996)
381. K. Kaneda, M. Kobayashi, T. Imanaka, S. Teranishi: Chem. Lett. 1483 (1984); Shokubai (Catalyst) **7**, 419 (1985)
382. H. Marrakchi, J.-B. N'Guini-Effa, M. Heimeur, J. Lieto, P. Aune: J. Mol. Catal. **30**, 101 (1985)
383. S. Bhaduri, K. Sharma: Chem. Commun. 207 (1996)
384. L.M. Robinson, D.F. Shriver: Coord. Chem. Rev., **37**, 119 (1996)
385. S. Cosnier, A. Deronzier, A. Llobet: J. Electroanal. Chem. **280**, 213 (1990)
386. C. Moutet, C.J. Pickett: Chem. Commun. 188 (1989)
387. S.P. Gubin: *Chemistry of clusters* (Nauka, Moscow 1987)
388. V.I. Isaeva, V.Z. Sharf, A.N. Zhilyaev: Bull. Russ. Acad. Sci., Div. Chem. Sci. 311 (1992)
389. J.Lieto: J. Mol. Catal. **31**, 89; 147 (1985)

390. R. Ramaraj, A. Kia, M. Kaneko: J. Chem. Soc., Faraday Trans., Part I **82**, 3515 (1986); **83**, 1539 (1987)
391. S. Takamaizawa, M. Furihata, S. Takeda, K. Yamaguchi, W. Mori: Macromolecules 33, 6222 (2000)
392. *Polyoxometalates; From Platonic Solds to Antiretrivoral Activity* ed. by M.T. Pope, A. Müller (Kluwer Acad. Publ., Dodrecht 1994) V. 10
393. K. Nomiy, H. Murasaki, M. Miwa: Polyhedron **5**,1031 (1986)
394. C.B. Gorman, W.Y. Su, H. Jiang, C.M. Watson, P. Boyle: Chem. Commun. 877(1999)
395. S.P. Christiano, T.J. Pinnavaia: J. Solid State Chem. **64**, 232 (1986)
396. N. Prokopuk, C.S. Weinert, D.P. Siska, C.L. Stern, D.F. Shriver: Angew. Chem. Int. Ed. **39**, 3312 (2000)
397. *Metal clusters in catalysis* ed. by B.C. Gates, L. Guczi, H. Knosinger (Elsevier, Amsterdam 1986)
398. *Clusters and Colloids* ed. by G. Schmid (VCH, Weinheim 1994)
399. Pat. 4.596.831 USA (1986)
400. Pat. 4.144.191 USA (1979)
401. M. Pohl, R.G. Finke: Organometallics **12**, 1453 (1993)
402. D. Imhof, L.M. Venanzi: Chem. Soc. Rev. **23**, 185 (1994)
403. T. Koyama, R. Hayashi, T. Yamane, E. Masuda, A. Kurose, K. Hanabusa, H. Sirai, T. Hayakawa, N. Hojo: Colloid and Polym. Sci. **265**, 786 (1987)
404. A.D. Pomogailo, V.S. Savost'yanov: *Synthesis and Polymerization of metal-containing monomers* (CRC Press, Boca Raton, Ann Arbor, London, Tokyo 1994)
405. N.A. Shteltser, L.V. Bybin, E.A. Petrovskaya, A.S. Batsanov, M.Kh. Dzhafarov, Yu.T. Struchkov, M.I. Rybinskaya, P.V. Petrovskii: Organomet. Chem. USSR **5,** 1009 (1992)
406. S.P. Tunik, E.V. Grachova, V.R. Denisov, G.L. Starova, A.B. Nikol'skii, F.M. Dolgushin, A.I. Yanovsky, Yu.T. Struchkov: J. Organomet. Chem. **536–537**, 339 (1997)
407. B.F.G. Johnson, J.M. Matters, P.E. Gaede, S.L. Ingham, N. Choi, M. McPartlin, M.-A. Persall: J. Chem. Soc., Dalton Trans. 3251 (1997)
408. J.C. Gressier: In book: *Metal-containing Polymer Systems* ed. by J.E. Sheats, C.E. Carraher, Jr., C.U. Pittman, Jr. (Plenum Press, New York 1985) p. 291
409. D. Glushove, G. Hogarth, M.H. Lavender: J. Organomet. Chem. **528**, 3 (1997)
410. G.A. Acum, M.J. Mays, P.R. Raithby, G.A. Solan: J. Organomet. Chem. **508**, 137 (1996)
411. B.F.G. Johnson, J. Lewis, E. Nordlander, P.R. Raithby: J. Chem. Soc., Dalton Trans. 3825 (1996)
412. R. Gobetto, E. Sappa, A. Tiripicchio, M.T. Cfmtllini, M.J. Mays: J. Chem. Soc., Dalton Trans. 807 (1990)
413. M.J. Doyle, T.J. Duckwortth, L. Manojlovic-Muir, M.J. Mays, P.R. Raithby, F.J. Robertson: J. Chem. Soc., Dalton Trans. 2703 (1992)
414. V.A. Maksakov, V.A. Ershova, V.P. Kirin, A.V. Golovin: J. Organomet. Chem. **532**, 11 (1997)
415. Y. Wong, W.-T. Wong: J. Organomet. Chem. **513**, 27 (1996)
416. S.P. Tunik, I.S. Podkorytov, B.T. Heaton, J.A. Iggo, J. Sampanthar: J. Organomet. Chem. **550**, 221 (1998)
417. R.D. Adams, W. Wu: Organometallics **12**, 1243 (1993)

418. S.I. Pomogailo, I.I. Chuev, G.I. Dzhardimalieva, A.V. Yarmolenko, V.D. Makhaev, S.M. Aldoshin, A.D. Pomogailo: Russ. Chem. Bull. **48**, 1174 (1999)
419. S.I. Pomogailo, G.I. Dzhardimalieva, V.A. Ershova, S.M. Aldoshin, A.D. Pomogailo: Macromol. Symp. **186**, 155 (2002)
420. S. Pomogailo, G. Dzhardimalieva, A. Pomogailo: Solid State Phenomena, **94**, 319 (2003)
421. A.A. Opalovskii, I.I. Tychynskaya, Z.M. Kuznetsova, P.P. Samoilov: *Molybdenum halogenides* (Nauka Sibirian branch AN USSR, Novosibirsk 1972)
422. J.H. Golden, H. Deng, F.J. Di Salvo, J.M.J. Frechet, P.M. Thompson: Science **268**, 1463 (1995)
423. N.D. Golubeva, O.A. Adamenko, G.N. Boiko, A.D. Pomogailo: Inorg. Mater. **40**, 363 (2004)
424. O.A. Adamenko, G.V. Lukova, N.D. Golubeva, V.A. Smirnov, G.N. Boiko, A.D. Pomogailo, I.E. Uflyand: Dokl. Phys. Chem. **381**, 360 (2001)
425. D.H. Johnston, D.C. Gaswick, M.C. Lonergan, C.L. Stem, D.F. Shriver: Inorg. Chem. **31**, 1869 (1992)
426. B.P. Gracey, S.A.R. Knox, K.A. Macpherson, A.G. Orpen, S.R. Stobart.: J. Chem. Soc., Dalton Trans. 1935 (1985)
427. O.G. Senotrusov, A.I. Rubailo, N.I. Pavlenko, A.B. Antonova, S.V. Kovalenko, A.G. Ginzburg: Organomet. Chem. USSR, **4**, 464 (1991)
428. I. Moldes, J. Ros, Z. Yonez: J. Organomet. Chem. **315**, C22 (1986)
429. N.M. Bravaya, A.D. Pomogailo, V.A. Maksakov, V.P. Kirin, V.P. Grachev, A.I. Kuzaev: Russ. Chem. Bull. 1102 (1995)
430. N.M. Bravaja, A.D. Pomogailo: In book: *Metal-containing Polymeric Materials* ed. by C.U. Pittman, Jr., C.E. Carraher, Jr., M. Zeldin, B. Culberston (Plenum Publ.Corp., New York 1996) p. 51.
431. N.M. Bravaya, A.D. Pomogailo, V.A. Maksakov, V.P. Kirin, G.P. Belov, T.I. Solov'eva: Russ. Chem. Bull., 423 (1994)
432. H.-P. Cheng, U. Landman: J. Phys. Chem. **98**, 3527 (1994)
433. K.E. Gonsalves, K.T. Kembaiyan: J. Mater. Sci., Lett. **9**, 59 (1990)
434. P. Gerbier, C. Guérin, B. Henner: Chem. Mater. **10**, 2304 (1998)
435. N.L. Pocard, D.C. Pocard, D.C. Alsmeyer, R.L. McCreery, T.X. Neenan, M.R. Callstrom: J. Am. Chem. Soc. **114**, 769 (1992)
436. R.J.P. Corrin, N. Devylder, C. Guerin, B. Henner, A. Jean: J. Organomet. Chem. **509**, 249 (1996)
437. P.J. Fagan, J.C. Calabrese, B. Malone: J. Am. Chem. Soc. **113**, 9408 (1991)

6 Physico-Chemical Methods of Metal-Polymer Nanocomposite Production

There is no distinct boundary between physical, chemical and physico-chemical methods of producing nanocomposites. One may refer to the procedures during which metal nanoparticles are deposited on polymers, especially on thin polymer films, as physico-chemical methods. They differ in means of evaporation of the atomic metal on the polymers heated to different temperatures. The methods should meet requirements such as reliable generation of nanoparticles of a needed size, high homogeneity degree, stability, good reproducibility of syntheses, etc.

Investigations of the processes on polymer surfaces during atomic metal deposition are very important as they help us to understand the origin of interactions and adhesion at the forming interface [1–3]. Certain polymer areas contacting nanoparticles can be negatively charged due to injection of photoelectrons from metal clusters into the polymer and their capture by the charge traps [4]. The opposite process of draining of electrons into positively charged clusters from the neighboring areas of the polymer is possible too. In both cases such areas of the polymer are to acquire some charge either positive or negative with regard to the rest of the polymer.

6.1 Cryochemical Methods of Atomic Metal Deposition on Polymers

Cryochemical methods are based on atomic metal (gasiform) deposition in vacuum (10^{-1}–10^{-4} Pa) on low-temperature fine polymer materials ($\sim$1000 nm). High-molecular paraffins, polyesters, oligo- and polyolefines, polydienes, vinyl- and phenyl siloxane polymers are usually used for these purposes. Metal atoms are deposited at below 77 K (the so-called technique of "naked" atoms) [5–12] upon which fine clusters are stabilized. A schematic diagram of the reactor is presented in Fig. 6.1.

At an evaporation rate of 10^{-8} mol·min^{-1} both homo- and heterometallic cluster particles are generated and then stabilized.

Scheme 6.1 illustrates the formation of a metal atom linked with functional groups of the polymer (I), further growth of the metal–polymer chains (II) and stabilization of the full-grown clusters through closure (III) or their

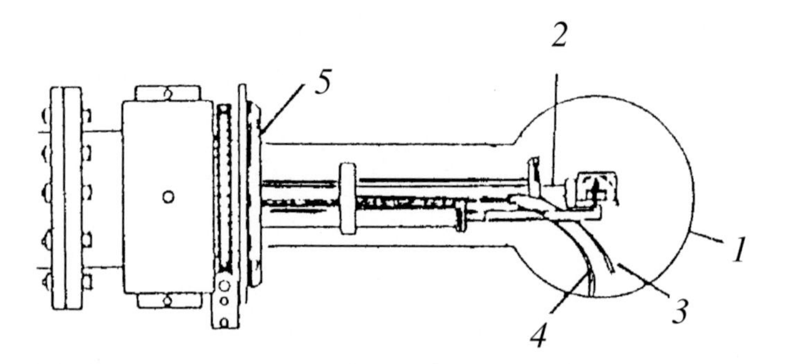

Fig. 6.1. Scheme of the reactor – electron gun for preparation of nanoparticles in polymers by vaporization: *1* – reaction volume, *2* – metal source, *3* – cooled carrier, *4,5* – fluid inlet and outlet, *6* – rotary vacuum vessel with gland

inclusion in a highly viscous metal–polymer network (IV), and finally enlargement of cluster particles (V) (∼O–O ∼ – functional groups or heteroatoms of the polymer chain maintaining the interaction with nanoparticles – may be ionic, electrostatic, covalent, hydrogen bonding, and so on).

This results usually in formation of mini-clusters whose dimensions depend upon process conditions and the nature of the metal (most often transient). In the case of chromium they consist of 2–3, Mo has 2–5 metal atoms, and in other metals they do not exceed 10.

Spectral methods are used to identify polymer-immobilized clusters. It is known that molybdenum mono-, di and trimers fixed on polydimethylphenylsiloxane (PMPhSi) absorb at $\lambda_{max} = 318$, 418 and 502 nm, respectively, while tetra- and pentanuclear absorb at $\lambda_{max} = 598$ and 640 nm. It helps to trace the cluster particle growth velocity.[1]

Condensation kinetics of metal vapors are analogous to coagulation processes of an ideal metal, found in an unstable state suspension that can be quantitatively described by Smolukhovsky's equation for rapid coagulation (see Chap. 2).

In accordance with Scheme 6.1, the process kinetics are described by a sequence of irreversible bimolecular reactions

$$
\begin{array}{l}
\mathrm{M} + \mathrm{L} \xrightarrow{k_1} \mathrm{ML} \\
\mathrm{ML} + \mathrm{L} \xrightarrow{k_2} \mathrm{ML_2} \\
\mathrm{ML_2} + \mathrm{M} \xrightarrow{k_3} \mathrm{M_2L_2} \\
\ldots\ldots\ldots\ldots\ldots\ldots\ldots\ldots \\
\mathrm{M}_{j-1} + \mathrm{M} \xrightarrow{k_n} \mathrm{M}_j\mathrm{L_2}
\end{array}
$$

[1] These formations become stable only below 290 K upon which they are rapidly decomposed by oxygen (even at 133 Pa). For comparison, the bonding energy of Cr–Cr is 150, and Mo–Mo is 400 kJ·mol^{-1}. Chromium is referred to reactive metals vigorously interacting with polymers [13].

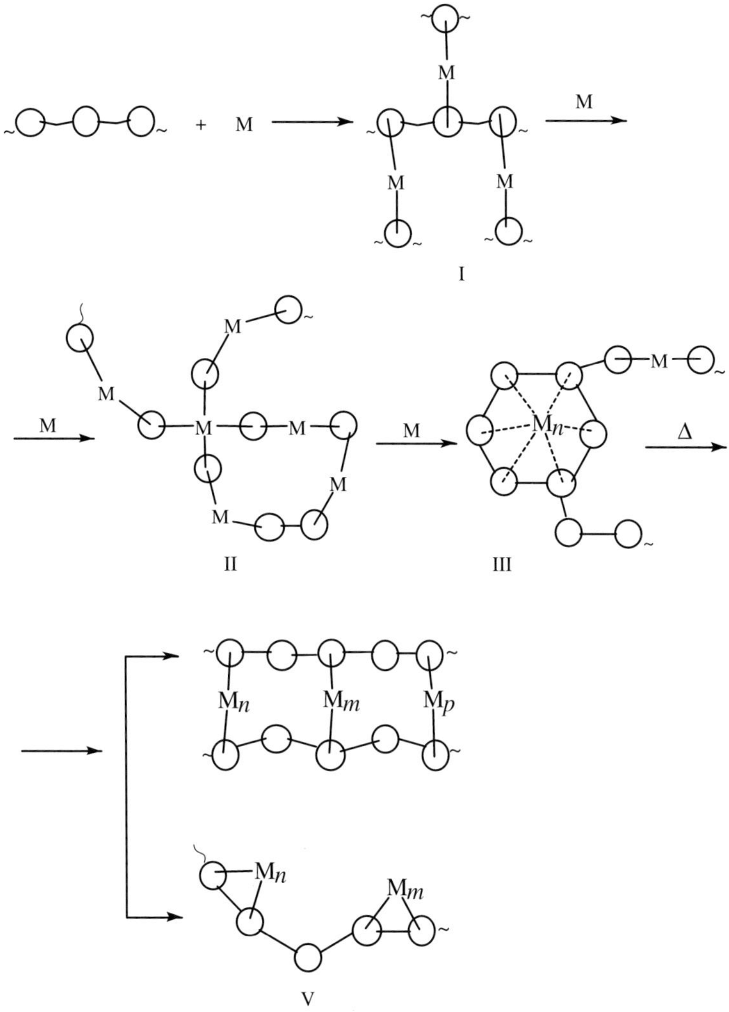

Scheme 6.1

and by a series of linear differential equations of the kind

$$\frac{d(ML_2)}{d(L)} = \frac{k_3}{2k_1}\frac{[ML_2]}{[L]}, \tag{6.1}$$

$$\frac{d(M_2L_2)}{d(L)} = -\frac{k_3}{2k_1}\frac{[ML_2]}{[L]}, \tag{6.2}$$

$$\frac{d(ML_2)}{d(M_2L_2)} = \frac{k_1}{k_3}\frac{[L]}{[ML_2]} - 1. \tag{6.3}$$

The velocity of cluster formation can be presented as a continuous deposition of metal atoms on the polymer film in time t at velocity V:

$$\frac{\mathrm{d[M]}}{\mathrm{d}t} = \mathrm{V} - \sum_{j=0}^{\mathrm{p}} k_m [\mathrm{M}][\mathrm{M}_j\mathrm{L}_2] \, , \tag{6.4}$$

$$\frac{\mathrm{d}[\mathrm{M}_n\mathrm{L}_2]}{\mathrm{d}t} = k_m [\mathrm{M}]\{[\mathrm{M}_{n-1}\mathrm{L}_2] - [\mathrm{M}_n\mathrm{L}_2]\}. \tag{6.5}$$

The concentration of cluster centers of different nuclearity $[M_nL_2]$ on the polymer is expressed by the relation

$$\frac{[\mathrm{M}_n\mathrm{L}_2]}{[\mathrm{ML}_2]} = [\mathrm{M_o}]^{n-1} \frac{2^{n-1}}{n![\mathrm{L_o}]^{n-1}} = k[\mathrm{M_o}]^{n-1} \, , \tag{6.6}$$

where M_o is the total mass of metal atoms deposited on polymer films in time t.

The slope of the curves $\lg[M_nL_2]/[ML_2] - \lg[M_o]$ gives a value $n-1$, in agreement with experimentally derived results for systems of a polymer and clusters V_n, Cr_n and Mo_n (Fig. 6.2). This approach presumes cooling of the matrix as a modification of the cluster series. It means [14] there is a sequence of clusters with monotonically growing nuclearity, and one and the same pattern of ligands (M_nL_m, where $n = qx$; x is the nuclearity of an elementary unit of the pattern; and q is a whole number). The major structural elements (elementary building blocks) in these patterns are individual metal atoms serving as models of the atom-by-atom growth of metal particles. It is possible to visualize their formation by enlarging these "bricks" as the elements of nucleation.

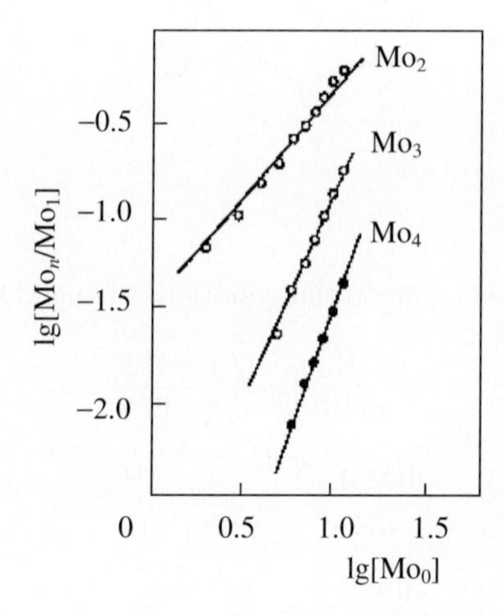

Fig. 6.2. Linear kinetic curves for determining nuclearity of the clusters formed from mononuclear Ar_2Mo centers

It is possible to produce not only homometallic but heterometallic polymer-immobilized cluster particles by the cryochemical method (see Chap. 5). For example, by sequential or simultaneous vapor deposition of different metals (e.g. Mo/Ti or Ti/Cr), especially at their high concentration, it is possible to obtain non-solvated bimetallic clusters $M_x^1M_y^2$ approaching colloidal particles in size (Scheme 6.2).

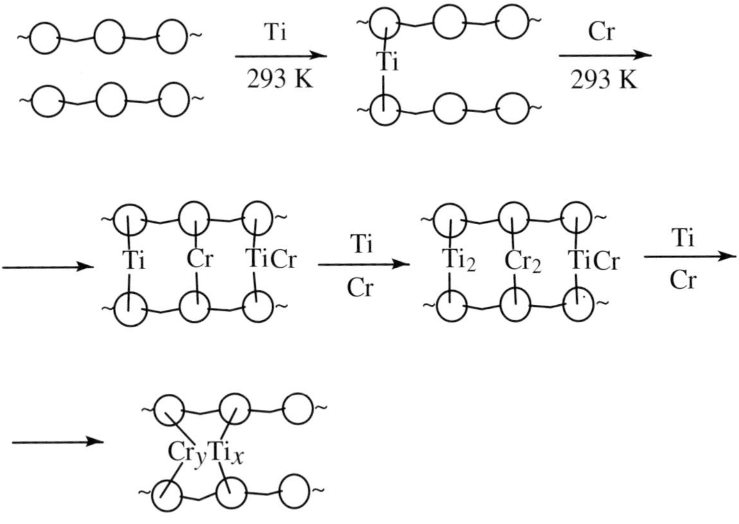

Ti, 293 K; Cr, 293 K; Ti, Cr; Ti, Cr

Ti Cr $TiCr$ → Ti_2 Cr_2 $TiCr$ → Cr_yTi_x

Scheme 6.2

The resultant fine particles are chemically bonded to the polymer and are, probably, cross-linking its chains.

One can obtain polymer-immobilized nanoparticles of other metals, e.g. of colloidal Ag 1.0–20 nm in size [9] and 2 nm mean size by selecting evaporation regimes.

Absorption of colloidal Ag particles <40 nm in diameter is observed within the ~400 nm region [15] (see the λ_{max} dependence on cluster size below). Aurum vapors of an organic solvent [16] condensed on poly(diphenylbutadiene) at 77 K are of about 2 nm size. Ferromagnetic nanoparticles can be generated similarly [17, 18].

The dimensions of ultra-dispersed particles produced by vapor deposition on a frozen organic solvent with polymer additions depend linearly on the gas-carrier (helium) pressure. Ag microclusters are 1.0 nm in diameter under 100 Pa [19]. Just as with the solvent [12], the origin of the polymer affects the reaction direction in cryochemical synthesis and the stability of the products formed [20–22]. For example, macromolecular sandwich complexes and polymer-immobilized cluster particles were obtained by low-temperature interaction between chromium vapor and oligoarylen matrices in different solvents [23]. To generate structurally homogeneous products sometimes it is necessary to avert the direct interaction of the atomic metal with the matrix through a co-condensation of the metal and solvent vapors (e.g. chromium and diglyme [24]). This unusual procedure replaces the direct metal–matrix interaction by the process of exchanging one stabilizing ligand by another high-molecular one (called resolvation).

Sometimes non-ionogenic surfactants are added to substitute a colloidal dispersion medium. Particularly, organic silver sols cryochemically produced in acetone (3–12 nm) are stabilized by 0.2% aqua solution of Triton X-100 and are embedded into a cross-linked polyacrylamide gel [25]. Resolvation may precede the formation of polymer-immobilized nanoparticles. In particular, cryochemical synthesis of colloidal Ag particles is conducted in acetone substituted afterwards for formamide. Such dispersion is stable for several days. Another approach is feasible too. An Ag (2–5 nm)–formamide system is employed as a solvent for obtaining cross-linked polyacrylamide gel [26,27]. Three resolvations were tried out for synthesizing metal–polymer nanocomposites [28]:

1. Cryochemically produced Co-nonadecane composite is mixed with the nonadecane solution of the polymer and is subjected then to ultrasonic treatment;

2. Co or Ni nanoparticles in toluene are introduced in PE solution in toluene at 363 K;

3. The polymer solution in toluene (383 K) is cooled down to 185 K and the cryochemically obtained nanoparticles are introduced into this jelly-like medium and the whole system is exposed to an ultrasonic effect.

These options assist in avoiding particle aggregation at all intermediate stages, although even more complex interactions between nanoparticles and the solvent can be observed along with peptization and resolvation.

A method based on the solid-phase cryochemical synthesis of nanocomposites was realized [29–31] in the variant of concurrent formation of the matrix and nanoparticles. Condensation of metal vapors (Cr, Ag, Pb, Cd or Mg) and monomers was arranged on a substrate (most often quartz) cooled by liquid nitrogen. These monomers are capable of polymerizing at rather low temperatures with, e.g. *p*-xylylene synthesized through pyrolysis of strained di-*p*-xylylene (*p*-cyclophane). The sublimation temperature at 10^{-1} Pa pressure is Si – 1405 K, Ni – 1653 K and PbS – 887 K. Poly-*p*-xylylene films with metal nanoparticle distribution of narrow size range were produced (Scheme 6.3)

Tiny metal clusters are also formed along with macromolecular bisyringic complexes (except for Fe). The process [33] includes the zones of sublimation and pyrolysis, a polymerization reactor, liquid nitrogen-cooled substrate and a metal source (Fig. 6.3). Monomer-*p*-xylylene vapors get condensed on

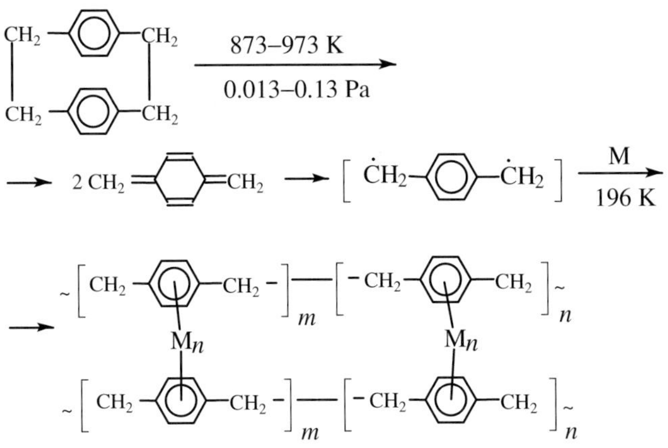

873–973 K

0.013–0.13 Pa

M

196 K

Scheme 6.3

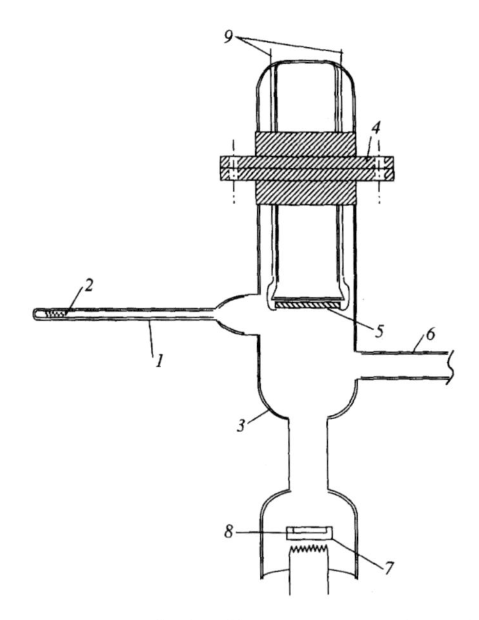

Fig. 6.3. Setup for synthesis of thin-film nanocomposites: *1* – quartz tube for pyrolysis, *2* – glass ampula with *p*-cyclophane, *3* – reactor, *4* – metal flange, *5* – quartz substrate layer, *6* – outlet to vacuum pump, *7* – tantalum trough, *8* – evaporated metal, *9* – molybdenum wires

the cooled substrate which is simultaneously hit by a flow of metal atoms. The last ones come from the block metal heating in a tantalum container up to the melting point of the metal. The concentration of metal atoms in the flow and in the nanocomposite can be changed by varying the metal heating conditions. The composite (aggregation of atoms) is formed as a result of *p*-xylylene polymerization under UV exposure (at 80 K) or during heating to 170–190 K, which is just 50 degrees lower than the monomer melting point. These films present a partially crystallized poly-*p*-xylylene matrix with incorporated metal clusters having less than 20 atoms [34]. The clusters and nanoparticles organized in planes between stacks of crystalline lamellas are formed at room temperature in free spaces. The probability of tiny cluster liberation from such traps and their movement towards nuclei depends upon the matrix structural relaxation barriers, defrosting of chain mobility and recombination of the matrix in the vicinity of the cluster. The nanoparticle size, e.g., of Ag is 12 nm [33]. It is possible to produce nanocomposites incorporating Cu, Ag, Fe, Pd, Pb and other some clusters [35].

Polymers with incorporated ZnS and PbS can be produced too. Polymerization starts 10–20 degrees lower as compared to the original monomers with 0.1–6.5% of Pb and the particle size is 3.4–6.7 nm. Metal–organic polymers are produced by cryohemical polymerization [36] with *p*-cyclophanes as monomers and Ge and Sn as organic bridges (their synthesis has been described in [37]) during which Ge and SnO_2 nanocrystals are formed in polymer matrices.

Thin-film metal–polymer composites are also produced [38] by palladium and paraxylylene co-deposition in vacuum on a substrate cooled to 77 K. The nanocomposite structure is defined by the proportion of initial ingredients. Analogous regularities have been observed in part in cryochemical synthesis and polymerization of chloro-*p*-xylylene complex with Ag [39] and Mg [40,41].

Polymer–metal cluster composites were also generated [42] during metal vapor deposition on liquid styrene or MMA followed by polymerization. Like in the case with Cd organic compounds polymerization of styrene in the presence of silver, palladium or auric salts are accompanied by the decreasing molecular masses of the forming polymers (Table 6.1).

The kinetic regularities (particularly, sequence of reactions) are similar to those of styrene homogenization. Particle sizes in a liquid monomer and in the arising polymer are identical, so long as no particle agglomeration takes place at polymerization. Similarly, Ag organosols with particle size below 15 nm are produced [27] by joint condensation of Ag and MMA vapors on the surface cooled down by liquid nitrogen followed by heating to room temperature.

The matrix is not usually indifferent to metal fixing in almost all production methods of polymer-immobilized nanoparticles. This can bring about abrupt reduction of the polymer molecular mass (e.g. oligoarylenes [24]), structurization of the matrix leading to solubility loss and sometimes fission of ether and other bonds. Besides, interaction products are formed out of the polymer functional groups with the metal (e.g. $CrCl_3$ due to chlorine end atoms splitting off the oligobenzyl matrix synthesized without benzylchloride). There is still no answer to the question of how the cryochemical chain reactions and auto-wave regimes of transformations in

Table 6.1. Polymerization characteristics of styrene and Au-styrene systems

System	AIBN, % (mol)	Yield of PS, %	PS Molecular Mass	Color
Styrene	0,1	2,0	128000	white
Styrene	0.2	2.3	83000	white
Styrene	0.5	4.2	65000	white
Styrene	1.0	6.2	47000	white
Au-Styrene[a]	0.1	2.1	53000	light pink
Au-Styrene[a]	0.2	3.2	50000	light pink
Au-Styrene[a]	0.5	6.2	24000	white
Au-Styrene[a]	1.0	7.3	21000	white
Au-Styrene[b]	0.1	2.2	72000	pink
Au-Styrene[b]	0.2	6.6	41000	pink
Au-styrene[b]	0.5	11.0	32000	pink
Au-styrene[b]	1.0	16.6	22000	pink

Note: in all tests styrene – 10 mL, [a][Au] = $1.09 \cdot 10^{-3}$ mol·L^{-1}, [b][Au] = $1.5 \cdot 10^{-3}$ mol·L^{-1}

systems subjected to radiolysis and especially at matrix devitrification influence processes [43]. The interaction of atomic Zn with polymer surfaces is accompanied by a deep going destruction of its subsurface layers after which Zn interacts with the destruction products [43, 45]. When strong ferromagnetics (Fe, Co, Ni) are used, alkane is destroyed even in inert matrices like nanodecane and PE [46] on the highly active cluster surfaces forming weakly magnetic structures.

There are many other types of cryochemical synthesis with application of polymers, e.g. using an oscillatory circuit in electric condensation-induced dispersion of metals [47]. Dispersions such as Ag, Al, Bi, Cd, Cu, Co, Mo, Fe, Ni, Zr, Pt, Sn, W with 1–10 nm particle size are unstable in nonpolar media unless stabilizing polymer rubbers capable of forming bulk structures are present. They immediately coagulate without the rubber.

Cryochemical modification, the so-called relaxation self-dispersion method, has special application [48]. The method is based on a thin film (a few micrometers thick of wax-like consistency) thermal deposition in vacuum. This can be, in particular, nylon-11 film deposition on a glass plate cooled to liquid nitrogen temperature which is subsequently overlaid with ~600 nm thick Au, Ag or Cu layer. This method is distinguished by the formation of a thermodynamic metastable system that upon heating for 10 min to 393 K (above polymer T_g, ~343 K) displays an effective thermal relaxation and creation of stable nanocomposite with ultrafine (1–10 nm) metal particles homogeneously distributed over the matrix. CeO_2 nanoparticles (~4 nm) in nylon-11 matrix (150 nm film thickness) were made by this method too [49, 50]. Composite films of cerium were produced and then they were oxidized in air into CeO_2. The relaxation self-dispersion method is hundreds of times more efficient than all other methods of manufacturing nanocomposites by co-evaporation or co-fission.

6.2 Metal Evaporation Methods on Polymers Localized at Room Temperature

Thermal procedures (see Chap. 2) presuppose that a stream of atomic metal deposited on the polymer kept at room temperature is spurred by either sublimation or thermal evaporation of a heated block metal, ion sputtering, evaporation induced by accelerated electrons, or by laser pyrolysis, etc. These methods are grouped into so-called vapor phase deposition of polymers (VPDP).

They are used mostly in vacuum metal-plating techniques of polymer materials where they have replaced galvanic technologies, in electronics in production of switching boards, ribbon cables and many other devices for microelectronics [52, 53]. In recent years this trend was explored for design of nanocomposites with emphasis on the details of metal evaporation processes and their optimization.

Bombardment of the polymer surface by hot metal atoms is accompanied by deeper penetration as compared to cryosynthesis destruction processes in the subsurface layers and chemical interactions between metal atoms and polymer functional groups. This process may not affect the adhesive strength of the metal–polymer pair. No destruction of the polymer layer during thermal deposition of chemically inert aurum on the polyimide surface [4] but only injection of photoelectrons emitted by metal clusters into the polymer were detected. The first stage of this process is condensation of metal atoms on the polymer surface after which it gets aggregated to nanoclusters and metal atoms start to diffuse either into the polymer subsurface layer or into the bulk (see Chap. 4). The technique of radioactive tracers (^{110}Ag) allowed us to define condensing coefficients of silver on polymers [54], which appeared to be a function of the polymer surface energy and diminishing abruptly at temperatures approaching the polymer T_g.

There are strict requirements to metal finish, e.g. metals with 99.997% [47] finish are needed in Cr evaporation. As Cr is deposited (evaporation rate is ~0.1 nm/min) on poly(bis-phenol-A-carbonate) films the metal interacts with the π-electron system of benzene rings leading to formation of semi-sandwich structures, and afterwards with carbonyl groups. The latter direction dominates in case with Al. Evaporation of bis(acetylacetonate)palladium(II) on thin PA-6 (15 μm), PETF or PMMA films [55] gives perfect results, in which case the coatings consist of palladium nanoparticles 3.9–4.3 nm in size. This approach was also realized for epoxy resin films (500 μm). The surface layer of such metal–composite films with Pd nanoparticles reaches 500 nm thickness, although strongly cross-linked polymers are limited with respect to diffusion of inorganic salts. Probably, it is worthwhile using acetylacetonates of Pt and Cu for these purposes.

Thermal deposition of metals in polymer film metal plating, especially polyimide ones, is widely used in food packaging, integrated circuits in microelectronics, etc. Fundamental problems of polymer metallization have been summarized recently in [56, 57]. Certain efforts are spent on provision of control over the microstructure and thermal stability of interfacial metal–polymer layers, improvement of adhesion and elimination of degradation.

It is very important that thermal evaporation on polymers employs not only metal nanoparticles but metalloid ones as well. For instance, selenium is introduced into thin carbon films through evaporation of selenium and acrylic acid at 1170 K [58] or Si crystals on silicone-organic polymers [59] accompanied by formation of polymer–semiconductor composites.

Evaporation of solid polymer particles is rarely used (at considerable pyrolysis) and co-deposition of volatile products (mainly radical ones) with metal vapors. As a result, it is possible to obtain copper islets 10 nm in diameter in a PE matrix [60]. A modification of the method is flash evaporation of a polymer-coated metal wire (metal evaporates instantaneously and polymer breaks into fragments as a result of short circuit) [61]. Another example is the wire electric explosion technique used to produce selective catalyzers and for silver nanoparticle deposition on inorganic powders [62].

The method of microwave heating was reported to be often used for synthesis of chemical compounds [63]. It consists of heating the substance through interactions of the dipole moment and ions in the electric field that accelerates their chemical reaction. Nickel nanoparticles (4–12 nm in size and 7 nm in diameter) are generated in $Ni(OH)_2$ reduction in ethylene glycol containing PVA under microwave irradiation (2.45 GHz, 200 W, 423 K, 5 min) [64]. In addition, 0.5 mol% of H_2PtCl_6 is introduced as the nucleation catalyzer [65].

Metal is often evaporated under the action of accelerated electrons. Particularly, an adhesive interfacial Ti layer (5 nm) is deposited on a PMPhSi surface at a 0.1–0.3 nm/s rate [66]. Local heating of the PMPhSi surface is insignificant, although new cross-links can appear besides the formation of Ti–C and Ti–O links. Afterwards, a thick silver layer (20–25 μm) [67] is deposited on this modified PMPhSi surface, and the thus formed material can be used in lithography for production of surface nanolithographic templates from a previously prepared mono- and bimetallic clusters [68, 69], in optical sensors. The methods of laser pyrolysis of metal-organic compounds aimed at production of nanostructure materials are not so widely used [68, 69].

Laser-induced CO_2 pyrolysis of $Fe(CO)_5$ [70, 71] is known to result in the formation of a pure γ-Fe with particles from 30 to 85 nm. The formation of carbide titanium clusters was recorded [72] from $Ti_3C_8^-$ to $Ti_9C_{15}^-$ during laser evaporation of Ti (CH_4 + He mixture)

Ion sputtering of metals on polymer surfaces consists of bombardment of a metal target by ions of noble gases with application of standard ion guns (see Chap. 2). Their energy reaches a few keV (for $Ar^+ = 2$ keV) and the average energy of the metal ions being sputtered makes up several eV, which corresponds to 10^4 K temperatures [73]. The high kinetic energy of the implanted ions is transmitted to the polymer matrix and increases the probability of polymer destruction and promotes chemical reactions with the polymer surface functional groups in which the usual metal atoms do not participate.

Modification of the polymer surface by ion beams includes physical and chemical processes during which interactions of the colliding ions with the multiatomic target are accompanied by the radiation and pyrolytic processes of the transformation of the polymer matrix. Ion sputtering of high kinetic energy has an impact on the surface layers (~10 nm long) in contrast to thermal treatment. For example, Au sputtering evokes destruction of PE, PTFE and PFO surface layers. In the case with PTFE, there is defluorination that follows the nucleation mechanism, as for PFO there is breakage of benzol rings and loss of oxygen atoms. Ion sputtering of more reactive Zn on PE, PTFE or PI surfaces is accompanied with much deeper processes [45]. At the initial sputtering stages Zn is first combined with oxygen on the PE surface and then deposition occurs. This is accompanied by dehydration of PE surface layers, partial destruction of its chains and cross-linking. When PTFE is bombarded by Zn atoms, the polymer undergoes defluorination and about half the fluorine atoms are removed from the surface, while Zn becomes fully bound (ZnF_2). In case PI is used, its surface layer undergoes destruction during the first stage and then Zn oxidizes to ZnO.

Analogous regularities were observed in Cu sputtering on PI and PTFE surfaces that are widely used in electronics thanks to their unique dielectric properties [74]. Ion sputtering is characterized by a deep destruction of the polymer in contrast to thermal deposition of Cu on the PI surface, which is accompanied only by changes in its chemical structure in the vicinity of carbonyl groups and formation of univalent copper compounds $\sim$O–Cu^+. Defluorination of the polymer surface layers in the Cu–PTFE system did not promote the formation of any copper compounds with the polymer decay products. Metal permeation inside the polymer bulk impaired its dielectric properties. Properties of polymer surface layers modified by ion implantation and diffusive alloying aimed at imparting conducting properties are reviewed in [75]. Ion implantation of Cu on PI [76, 77] is connected with more perceptible structural changes in PI than in the case of its thermal modification [78] but it is affined to changes occurring at thermal destruction in vacuum at 800 K. Deposition of thermal evaporated metal into the surface of polymer melt is described in [79, 80].

Less applicable methods are pulsed laser sequential deposition [81,82] and the co-deposition method of the polymer and metal [83].

6.3 Synthesis of Nanocomposites in a Plasma-Chemical Process

Polymerization in glowing discharge (low-temperature plasma) is initiated by ions, exited molecules and photons of rather high energy. The functional diagram of the setup and reactor for plasma synthesis of composite films is shown in Fig. 6.4 [84]. The pressure of vapors (monomer concentration) during polymerization in plasma is too low (10^{-1}–1000 Pa) for production of polymers by traditional methods in a plasmatron, but is quite sufficient for formation of thin films.

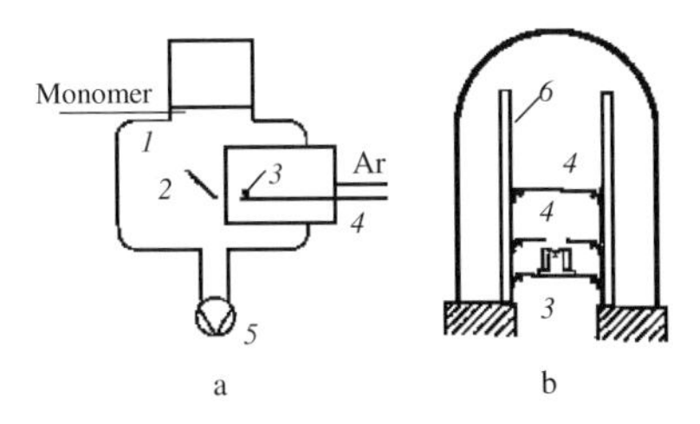

Fig. 6.4. Scheme of the setup (**a**) and reactor (**b**) for the deposition of nanoparticles into a polymer film: *1* – plasma camera, *2* – substrate, *3* – metal source, *4* – electrodes, *5* – vacuum pump, *6* – stand for electrodes and vaporizer

Compounds like CH_4, CF_4, benzol, chlorobenzol, perfluoropropane, hexamethyl disalazan and carbon disulfide are used as the film-forming substances. Complex compounds like *p*-xylylene dimers (N parylene) and traditional monomers are used rarely. In particular, homogeneous beryllium-enriched coatings are produced by simultaneous thermal evaporation of beryllium and deposition of the plasma polymer from *trans*-butene-2 [85]. Gaseous mixtures of monomers and argon are added during the process and the discharge parameters (current, pressure, treatment time, substrate temperature, metal source to substrate distance) and the nature of the film-forming substance are varied. The reactions between active particles dominate in the macrochain growth mechanism, which is sufficiently rapid and stepwise. These reactions are called polymer-combination reactions. The chain mechanism does not affect essentially the polymer vapor deposition. The resultant products are referred to as polymers, although having little in common with the polymers in their structural specifics. They are usually 3D highly cross-linked products mostly in case of prolonged plasma treatment (5–10 min). The plasma-activated thermal decomposition of PP on metal substrates leads to formation of coatings (50 nm thick film) with a diamond-like structure [86–88].

During polymerization in plasma two competing processes take place: polymer deposition from the gaseous phase and etching (destruction), which lead to removal of the formed polymer. The ratio of the processes at individual portions is conditioned by the energy of excited particles, their concentration, surface temperature and the origin of the monomer. For example, polymerization dominates in the case with C_3F_8 [89]. A method of generating ultra-thin polymer films by activating the electron beam destruction products, e.g. PTFE, became popular recently. These films grow mostly at the expense of the free radical particles CF_3 and C_3F_5 [90].

The main difference of plasma-induced polymerization from traditional polymer production procedures is in the formation of defectless ultra-thin (50–100 nm) films on various substrates. Polymerization in plasma as a method of surface modification of polymers is not limited in the choice of materials and substrate shape. It allows obtaining microencapsulated nanoparticles – so-called metal-doped polymer films [91].

Thin composite films may also be formed in the course of simultaneous polymerization of respective compounds and metal evaporation in vacuum. The plasma polymerization technique and metal deposition on such films was first reported in [92] and modified in [93–96]. The following variants of synthesis were developed based on manufacturing metal-filled fluorine polymer thin films [50, 97–100]:

(i) sputtering of a metal target in the glowing discharge plasma of a fluorine-containing monomer;
(ii) plasma polymerization of a fluorine-containing monomer simultaneously with metal evaporation;
(iii) combined sputtering and condensation in the glowing discharge plasma of the metal target and PTFE.

Metal films with 2D and 3D distribution of metal particles in the polymer matrix can be attained depending on the reaction conditions (Fig. 6.5).

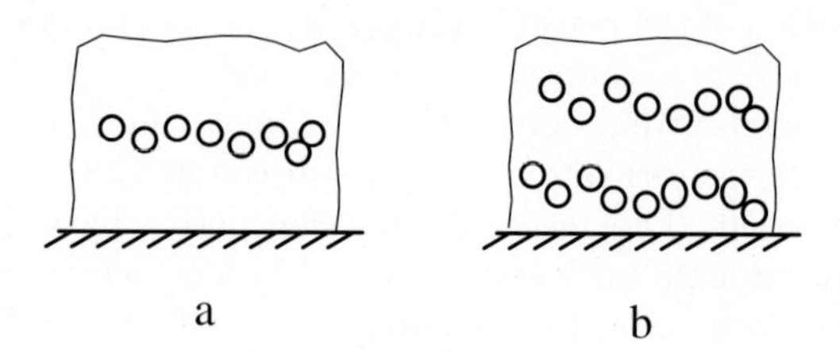

Fig. 6.5. Two- (**a**) and three- (**b**) dimensional distribution of nanoparticles in a polymer matrix

Two-dimensional films are produced by either sequential plasma-polymerization and metal deposition or by simultaneous short-term polymerization and deposition. The 3D structures are formed as a result of prolonged simultaneous plasma-polymerization and metal deposition with multiple repetitions of these cycles at high velocities of monomer feeding.

All modifications differ in the hardware used, and in the means of introduction and formation of nanoparticles.[2] According to one of them [60] the polymer and metal are evaporated simultaneously from different sources (plasmatrons); another more widespread procedure includes plasma-polymerization and metal evaporation simultaneously [91, 94]. An original new method uses simultaneous plasma-induced grafted polymerization of traditional monomers (e.g. vinyl imidazole on PI films [107]) and metal evaporation, e.g. Cu. Such treatment improves metal adhesion to the polymer like in the case with Cu and polyimides containing imidazole segments of the main chain [108].

Decorative coatings are applied on polymer surfaces (PE, PVC, PTFE, polyester) in glowing discharge conditions. Heated vapors of phthalocyanine or copper acetylacetonate $Cu(AcAc)_2$ [109] are supplied on cold substrates through the discharge zone and a hole in the anode. Molecules of copper phthalocyanine do not experience any essential changes on transfer on the polymer $Cu(AcAc)_2$ decomposes partially forming copper oxide. This phenomenon may occur both in the gaseous phase and on the polymer surface as well. Thin-film composite materials from copper phthalocyanine vapors (having several α, β and γ polymorphous forms) and benzene were synthesized in a high-frequency plasma of glowing discharge at a plasma-forming gas pressure $\sim$50 Pa and deposited on different substrates (100–600 nm thick) [84]. Gas-flame synthesis allows receiving composites of high mechanical strength. The films include microcrystals of α-form and the particles of the initial copper phthalocyanine. The latter dispersed in polyimide and polycarbamide have the same structure [110].

It is important that the plasma-polymerization of metal–organic compounds [111] distinguished by simplicity of hardware is used for plasma-polymerization of diethylberillium.

Limited information is available about the plasma-polymerization of two or more monomers. For example, there are only descriptions of the production of films from a mixture of organic and silicon-containing compounds [103,104], tetraethyl and tetrafluoroethylene in the electric discharge plasma [112]. In the latter case amorphous copolymer films are formed and tin is chemically bound and the C/Sn ratio changes within 3:6 limits. Tin isolates in tetraethyl into a self-maintained phase in the form of coarse (to 300 nm) almost spherical particles under high temperatures of the substrate. Nanoparticles of other metals can be introduced into the polymer, e.g. Te and Se by $Te(C_2H_5)_2$

[2] The conditions of low-temperature plasma formation and plasmo-chemical process technology have been described in [101–106].

evaporation or $Se(C_2H_5)_2$ [113], or tellurium-containing alloy of the type $As_{20}Te_{60}Se_{20}$. As the precursor concentration raises from 40 to 80% by volume, the size of tellurium nanoparticles increments from 2–20 to ∼40 nm. GaAs particles were obtained on fine polyimide films under the effect of plasma generated in conditions of electronic cyclotron resonance [114].

Especially often, combined targets are employed that contain a metal and a polymer [50]. In this case the target materials are sputtered with the help of plasma and the sputtered product deposits usually outside its area. This is convenient when the metal source is the cathode itself (Mo, Cu, Au). Thus, in the *trifluoropropanemetal* systems the amount of metal in the forming thin-film material reaches 18–26 mass%. Metal–polymers with the formal composition $(C_3F_4O_{0.6}Mo_{0.3})_n$ and $(C_3F_{3.9}O_{0.3}Cu_{0.3})_m$ [101] were isolated and characterized. In the same manner Al-containing material with improved adhesive properties was produced in a magnetron system of the glowing discharge by plasma polymerization of CH_4 having Al as the electrode material. Even under comparatively low frequency (10 kHz) the electrode surface undergoes such a strong ion bombardment that Al evaporates and gets included into the growing macromolecule and the metal is evenly distributed over the deposited material thickness.

The volume share (filling factor) φ of the metal included in metal–polymer materials is found from the equation [115]

$$\varphi = [S(\varphi) - C_{\mathrm{f}}]/[S_{\mathrm{M}} - S_{\mathrm{f}}] , \tag{6.7}$$

where $S(\varphi)$ is the forming metal film density; and C_{f} and S_{M} are the density of the metal-free film and block metal, correspondingly.

The value of φ varies within 0.1–1.0 and affects the structure of polymer-immobilized nanoparticles. 2D particles are derived from silver and benzene plasma-polymers formed under low φ values within a short period of plasma treatment [116]. It was shown that [117] three different structural zones with different φ values were observed in composite plasma-polymersilver films. Figure 6.6 has a metal zone consisting of predominantly metal particles with polymer inclusions, (c) a percolation zone devoid of any metal particles, but where cluster metal particles start to be formed (d-h) the zone consisting of fully isolated by the polymer cluster particles – microencapsulated polycrystalline silver particles.

Cu, Au and Ag particles of colloidal size encapsulated into thin films [118, 119] have acquired a similar structure (Fig. 6.6 d–h). They can be encapsulated in fluorine polymers as well – Au [89, 91, 120–122], Co, Al, Mo [123] and Cu in PE [60], Ag in chlorobenzene [94] and benzene [116] polymers. Distribution of Au nanoparticles in a polymer formed in plasma treatment of a C_3F_8–Ar blend was examined. The mean Au particle size was 5.5 nm ($\varphi =$ 0.15). The dimensions of silver crystallites shaping in thin composite films at chlorobenzene polymerization and vacuum evaporation of Ag [94] are 5 to 10 nm, and at high Ag content they are 100 to 200 nm. In this case AgCl and silver carbide were not isolated.

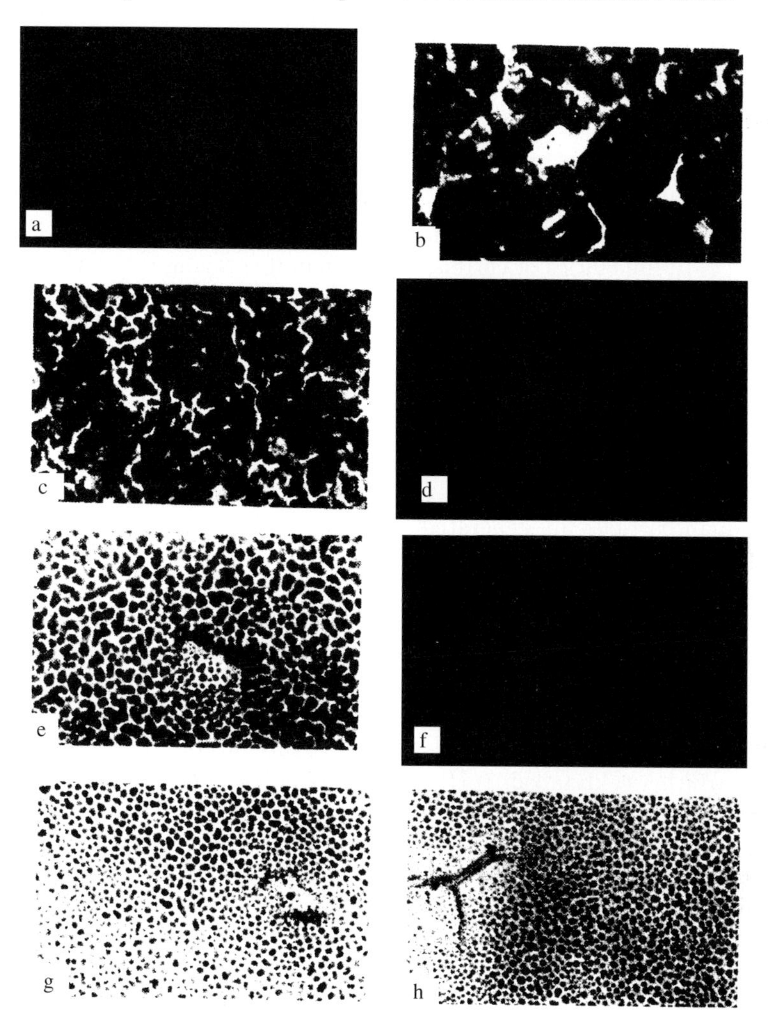

Fig. 6.6. Electron microscopic micrographs of plasma–Ag composite films (filling degree decreases from **a** to **h**)

Vinyltrimethylsilan and tetraethoxysilan were used as polymer-forming precursors for production of polymer cluster films [124–127]. Silicone-organic films as a result of thermal treatment are inert, insoluble, stable in mechanical and thermal respects, and can be used as membranes or protecting coatings [128]. The mean Au particle size depending on the reaction conditions constitutes (mean atomic numbers are in brackets): 1.4 (85); 2.0 (248); 2.5 (484); 3.7 (1571) and 5.0 (3876) nm. Absorption spectra depend on particle size from 449 nm (2 nm) to 545.5 nm (5.0 nm). Similarly, palladium/plasma polymer composites using vinyltrimethylsilan monomer are obtained with particle size 1.4–5 nm [129]. Thus, (hexamethyldisiloxane monomer) copper particles forming in plasma treatment are from 2.9 to 5.4 nm and depend on treatment conditions and can exist, just as in the form of Cu^0, so CuO and even Cu_2O [130].

Hence, a gas-discharge plasma is more efficient than conventional procedures in situ formation of thin metal–polymer films of controlled size on nanoparticle surfaces in conditions of low temperature. A drawback of the method is in difficulties of correlating the metal spraying rate and its incorporation into the forming polymer.

6.4 Radiolysis in Polymer Solutions

Radiation-chemical (γ-radiolysis, exposure to a flow of electrons, etc.) and photoreduction (photolysis) are preferable to hydrogen-induced reduction when small particles are generated with a narrow distribution range by size. This is because of diminished starting concentration of initiating particles due to diffusion of gaseous hydrogen through the aqua solution [131].

The radiation-chemical reduction is based on generation of strong reducers within the reactive system, including electrons, hydrogen atoms, organic and inorganic radicals. The process is terminated by precipitation of a sediment unless stabilizers like polyphosphates, PAA, PVA and others participate in the process. For example, there is the procedure of obtaining "blue silver", during which weakly alkali (pH = 8–10) aqua solutions of Ag^+ containing isopropyl alcohol are exposed to γ-radiation accompanied by coloring into a bright blue color (absorption bands λ_{max} at 290 and 700 nm) [132]. The value of λ_{max} is a criterion of the mean nanoparticle size [133,134]. It is also believed that [135] the blue silver presents a PAA-stabilized linear silver cluster. Its formation mechanism consists of Ag^+ binding by carboxylic groups of PAA macromolecules unfolded in the alkali medium and further reduction by γ-radiolysis.

During further radiation–chemical reduction it is transformed through a series of sequential stages [the λ_{max} band shifts from 320 to 270 corresponding to clusters Ag_2^+, Ag_4^{2+} ($\lambda_{max} = 277$ nm)]. In this case, the hypsochromic band shift responding to dimerization and formation of more coarse atomic aggregates (λ_{max} at 365 and 460 nm) [136] is evidence of the formation of coarse atomic clusters and their shape variation from spherical to ellipsoidal.[3] The low-intensity band with $\lambda_{max} = 325$ nm is related to the cluster Ag_9^+ [140], while absorption in the region of 360–380 nm is related to the two-charge cluster Ag_{14}^{2+} [141] closed by coalescence into immobilized colloidal silver.

[3] Formation of silver aggregates in inverse micelles under the effect of γ-radiation [137–139] brings about the usual spherical silver nanoparticles of $\lambda_{max} = 430$ nm along with light-absorbing aggregates at 380 and 530 nm identified as particles of irregular shape able to form complexes with SAS.

The thus formed colloidal metal phase catalyzes further reduction of metal ions.

Evolution of silver clusters is accompanied by changes in coloring from blue through green, red and brown to intensely yellow – the color of the subcolloidal silver (not yet exhibiting metallic properties since it has only 8–16 metal atoms).

Coagulation of these particles results in formation of colloidal metal of a narrow plasmon band at $\lambda_{\max} = 380$–$394\,\text{nm}$ with particle diameter from 1–10 to 5–20 nm [142]. The extinction coefficient of the blue silver linear cluster is estimated as $\sim 2 \cdot 10^4$ L·mol^{-1}·cm^{-1} which is very close to that of Ag_2^+, Ag_4^{2+}, Ag_8^{2+} [143]. Blue silver in oxygen over several days has high stability, and it can be isolated in a pure form [135, 144, 145]. Such silver clusters represent long-living oligomers stabilized by the polymer chain. Ag_n clusters of different size are generated [146] with absorption bands having $\lambda_{\max} = 280$, 295, 335, 370 and 480 nm (subcolloidal particles) depending on the reactions (pH, correlation of Ag_2SO_4 aqua solutions and polyacrylic acid or polyacrylate anion content, γ-radiation dose).

Nanoparticles are very effective in preparation of homogeneous and evenly distributed across the polymer matrix volume. It was used in the method of combined γ-radiation of $AgNO_3$ aqua solutions and acrylamide at room temperature [147]. In the course of irradiation (0.05 mol of $AgNO_3$ and 3.52 mol of acrylamide, at $2.7 \cdot 10^4$ Gy radiation dose) Ag^+ ions get reduced under the effect of hydrated electrons yielding the particles

$$Ag^+ + e^-_{eq} \quad \rightarrow \quad Ag^0,$$

$$n\,Ag^0 \quad \rightarrow \quad Ag_n.$$

This results in a brown-red transparent jelly-like product which can be dried. The dimensions of its quasi-spherical particles are from 2 to 20 nm with the mean size 10.8 nm. It is believed that this approach can be used for manufacturing nanocomposites of the polymer–metal oxide or polymer–metal sulfide type, where other precursors can also be used as the polymer matrix, e.g. mixtures of butyl acrylate, styrene and 2-hydroxy-α-methacrylate [148].

Radiolysis-induced formation of islet films of other metal nanoparticles from their salts has been studied less than that of silver. Interaction of H_2 with colloidal Pd obtained by radiation-chemical reduction of Pd^{2+} in aqua solution of PAA does not affect the size of Pd nanoparticles (2–4 nm) but is accompanied by hydrogen sorption on the nanoparticle surface [149, 150]. The reduction processes of Pt ions under the effect of various stabilizers have been compared: radiolysis, H_2, and citrates in the presence of polymer stabilizers [151]. The radiolytic reduction of $PtCl_4^{2-}$ with addition of sodium polyphosphate and methanol leads to formation of pointed particles, irregular in shape, 1–3 nm in size (35–900 Pt atoms in a particle). The reduction follows at least partially the autocatalytic mechanism. Autocomplexes of the type $PtCl_2(H_2O)_2$, $PtCl_3(H_2O)^-$ are reduced by H_2 in the presence of polyphosphate or sodium polyacrylate and isolate particles of regular shape (5–8 nm), while citrate platinum particles are of about 2.5 nm size. Spectra of the first two nanoparticle types show an absorption band with $\lambda_{max} = 215$ nm, whereas unstabilized Pt clusters are absorbed at 260 nm. Differences in spectra and in reactivity towards O_2, H_2, etc. are attributed to dimensional effects and different oxidative states of nanoparticle surfaces.

6.5 Photolysis of Metal-Polymer Systems as Means of Obtaining Nanocomposites

Optical excitation of a transient metal promotes a single-electron reduction of the metal ion (S – solvent) complexes within the regions of the ligand→metal charge transfer

$$\begin{array}{lcl} (MX_n)_s & \overset{h\nu}{\longleftrightarrow} & (M^+X_{n-1}, X^{\cdot})_s \\ (M^+X_{n-1}, X^{\cdot})_s & \longrightarrow & (M^+_{n-1})_s + X + S^{\cdot}. \end{array}$$

Under definite conditions the process might terminate in the formation of zero-valent forms and metal nanoparticles

$$M^+X_{n-1})_s \overset{(h\nu, S^{\cdot})}{\longleftrightarrow} (M^o)_s \longleftrightarrow (M_n).$$

The analysis of the phenomenon proves that only highly light-sensitive compounds whose quantum output of the reaction is above 0.1 can be used for these purposes. Complexes with ligands such as carbonic or amino acid anions, tetraphenyl borate, etc. in the inner or outer coordination sphere belong to such compounds. Besides, the accepting media (solvents, polymers) should behave as secondary reducers of intermediates that contribute to the reduction process.

Investigations in the photolysis of silver salts derived in a microemulsion medium (polymer solutions, e.g. gelatin, are introduced in the micelle solution [152]) are important for the nanotechnology of halogen-silver recording materials [153]. A band with $\lambda_{max} = 1100$ nm strongly shifted to the red region was registered for coarse (15–20 nm) ellipsoidal silver particles in gelatin photo emulsions [154].

Nanoparticles in situ formed by simultaneous photopolymerization (acrylamide, 0.08% of N,N-methylenbisacrylamide, exposure to a full spectrum of a mercury lamp) and photoreduction of Ag^+ ions show high stability (nascent samples in the presence of oxygen for a few weeks, and already formed films – during several months) [52].

Ag particles with sizes less than 20 nm are commensurable with those obtained by low-temperature co-condensation of metal vapors. These composites can be obtained by polymerization in the presence of MMA or acrylamide with addition of $AgNO_3$ in aqua media (or by the latter dissolving in

MMA) in UV radiation during 48 h at room temperature [155]. Silver particles and the stabilizing polymer matrix are formed similarly to the case with γ-radiation. Silver particles in Ag/PMMA composite are 10–20 nm in size. Polymerization of PMMA or acrylamide does not occur at UV radiation in room temperature probably because Ag^+ ions diminish the electron density of C=C bonds by coordination with the monomer molecules thus promoting the formation of radicals at exposure and initiating polymerization.

Photolysis of silver salts consists of intermediate stages during which low-atomic charged silver clusters are formed and catalyze further synthesis of nanoparticles incorporating more than 10^3 atoms. They are formed at the stage when photoelectrons enter into interaction with interstitial silver ions $e + Ag_o^+ \rightarrow Ag_o$ or as a result of interactions between silver ions and atoms: $Ag + Ag_o^+ \rightarrow Ag_2^+$. The final stage of the process is $Ag_n^+ + e \rightarrow Ag_n$. Study of polymer-stabilized silver nanoparticles is very important as knowledge of specific properties of its clusters and colloidal particles helps to correlate the position and form of bands in absorption spectra with the size, concentration, aggregation level and shape of particles. For instance [156], the band $\lambda_{max} = 277$ nm belonging to the doubly-charged silver cluster Ag_4^{2+} has been recorded during both chemical and photochemical reduction of silver ions in polymers.

Islet silver clusters were detected [157] in photochemical reduction of $AgNO_3$ or $[Ag(NH_3)_2]BF_4$ in DMPA in PVA, PEO and gelatin matrices exposed to monochromic light of $\lambda_{exc} = 254$ at light flow intensity $2{\cdot}10^{16}$ quantum$\cdot$cm$^{-2}\cdot$s^{-1}. The intensity threshold is attributed to simultaneously running processes of dark re-oxidation of the forming atoms and tiny nuclei, at higher intensities the colloidal particle breakdown. Photochemical formation of colloidal silver in polymer matrices is preceded by photo-generation of low-atomic clusters agglomerating afterwards into subcolloidal particles (growing of nuclei). Bands in the absorption spectra of photolysis products at various expositions have stable position, probably because clusters increment in mass as a result of their concentration growth but not the size. Initially, concentration of silver clusters is proportional to the degree of lighting, while the distribution by size corresponds to the logarithmic variance value. It was presumed [158] that nanoparticles start to grow due to photoreduction of the initial complexes on their surfaces in the photolyte proceeding from high enough viscosity of the polymer and comparatively low cluster mobility at the photolysis stage. Diminishing of their starting content should noticeably decrease the nanoparticle size, which is in fact observed during the experiment. Photoresponse (0.15–0.28) augments the complex concentration growth in the photolyte too. The kinetic dependencies of silver photoreduction are significantly affected by the polymer nature as well. The rate of silver colloid formation in PEO is 10 times as low as in PVA. Optical density of silver plasmons in PVA continues to grow even after photoradiation [158] since PVA is a rather strong secondary reducer of silver ions. However, the dark process flows far too slowly ($\sim$50 times) compared to the photochemical process. Therefore the particles arising from catalytic reduction of the initial silver ions on their surfaces continue to grow further. Despite the fact that the rated values of the submicrocrystalline lattice constant are identical to the block metal, the 3d-electron bonding energy of colloidal particles is, according to ESCA, (373 eV) considerably higher than the massive crystals have (368.1 eV) and is a function of the particle size [159]. It was established that coarse rod-like silver nanoparticles of 15–20 nm size and 250–300 nm length stabilized by a deprotonized form of PVA were formed during the process. The particles can be presented as cylinders 20 nm in diameter and 300 nm height. The light-sensitive gelatin layers based on AgBr nanocrystals and a physical macromolecular network on model layer surfaces of a PTFE film [161] are produced by in situ synthesis of Ag nanoparticles.

Photolysis of MX_n–macroligand systems may be accompanied by the photolysis of the polymer matrix. This results, e.g. for PAA, in its decarboxylation reaching 68% in 90 min [162] and even destruction [163, 164], which happens, probably, because of the forming of radicals in UV-induced photolysis and their interaction. The photo-oxidative destruction at UV radiation was observed in PS [165] during which the PS surface layer transformed gradually with increasing radiation dose even at room temperature from the glassy into a highly elastic state and then into the viscous flow. The glass temperature in the surface layer of PS is lower than that for bulk polymer. Especially, it is decreased after UV modification [166]. In contrast, when a thin Au layer (10 nm) is deposited in vacuum on PMPhSi films followed by UV exposure [167] the films get oxidized as a result of catalysis and the Si–Si and Si–O–Si bonds get broken. The phenomenon was detected in polysilane-Au systems only.

Metal–polymer nanocomposites were produced on the base of cobalt salts and styrene copolymer with acrylic acid under UV radiation [168]. The polymer in the form of spherical particles (300 nm in diameter) obtained by emulsion copolymerization is brought into interaction with $Co(CH_3COO)_2$ and is exposed to the radiation of a mercury lamp for 50 h at room temperature. The reduction is induced by hydrated electrons whose redox potential is 2.27 V, whereas that of cobalt is 0.277 V

$$Co^{2+} + 2e^-_{eq} \rightarrow Co^0 \text{ (reduction)}$$

$$nCo^0 \rightarrow Co_n \text{ (aggregation).}$$

Thus, spherical polymer particles of nanometer size get overgrown with cobalt nanoparticles of 15 nm size.

Photosyntheses was compared of silver, copper (reduction of $(Cuen_2)(BPh_4)_2$) and gold (reduction of $(Auen_2)Cl_3$) in the presence of PEO, PVA or in nafionic membranes [158]. It is proved that the particles formed are spherical in shape and they don't agglomerate during photolysis. The presence of intensive narrow bands of copper, silver and gold with λ_{max} 580, 430 and 580 nm, correspondingly can be seen in Fig. 6.7. Photometry is very useful

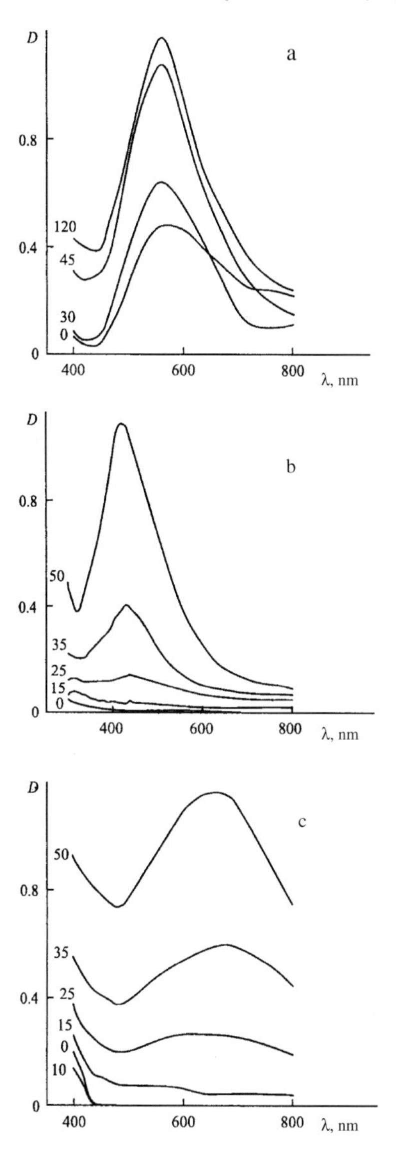

Fig. 6.7. The change of adsorption spectra of copper (**a**), silver (**b**) and gold (**c**) colloids in nafionic membranes obtained in photoreduction of $[Cuen_2](BPh_4)_2$, $[Ag(NH_3)_2]BF_4$, and $HAuCl_4$ (0.05 mol/L) in DMF (λ_{exc} 254 nm, intensity (I) is $7.0 \cdot 10^{15}$ quantum/cm$^{-2} \cdot$ s^{-1})

because metal nanoparticles have the specific plasmon absorption in UV and visible spectra. The maximum, intensity and form of plasmon bands are a qualitative criterion of the structure of nanophase materials. Rated extinction coefficients of colloidal metals are (L·mol^{-1}·cm^{-1}): $5.2 \cdot 10^3$(Au), $5.8 \cdot 10^3$(Ag),

$6.4{\cdot}10^3$(Cu), $4.8{\cdot}10^3$(Ni) $4.6{\cdot}10^3$(Pd) [169]. As expected, the transition from liquid to solid photolytes (reduction of Ag^+, Cu^{2+} and Au^{3+} ions in polymer membranes) is hampered due to difficulties in mass exchange. According to the data of XPS, the energy of the 4f-electron bond (90.2 eV) for colloid particles of Au is substantially higher than that for crystal bulk Au (84.0 eV).

Some other metal complexes can be used for photolytes as well, e.g. $Ni(en)_2(BPh_4)_2$, $Pd(en)_2(BPh_4)_2$, $Pd(ClO_4)_2$ and even Al (polymer–poly (styrene)-co-(4-chloromethylstyrene) complexes [170].

Nanoparticles grow during photosynthesis to definite photo-stable size in PVA and constitute for Cu 30–40 nm, for palladium nanoparticles in the film ~50 nm. Ag, Au, Pd colloids in PVA films preserve their spectral properties for up to two years during storage in darkness in air; those of Cu – for 2 to 3 months. Most probably, the formation of intra- and intermolecular cross-links, isolation of a microgel from PVA and their photolysis products in the vicinity of nanoparticles ensure their electrostatic and steric stabilization, high aggregative stability and resistance to oxidation of nanoparticles in aqua solutions. The diminishing capability of stabilizing silver clusters was observed in acrylic acid copolymer with acrylamide in the course of decarboxylation [171].

Approaches used in practice are much wider than we've just considered. For instance, the template technique of optical lithography employs special masks covered with a thin layer of a light-sensitive polymer. To bind Ag^+, a process of chemical or photochemical reduction [172] is used aimed at selective metallizing of the thin polymer film. Another way [173,174] of producing thin photoactive 2D films consists of UV exposure of Au nanoparticles stabilized by ammonia salts with addition of paraffin derivatives from which a polymer lattice is shaped during irradiation.

As can be seen, among the advantages of the physico-chemical methods is simplicity of the equipment employed, low operation cost, high purity of reducers subjected to a flow of electrons, light quanta, bombardment with ions, high-energy excited molecules, atomic metal flows deposited in deep vacuum on a polymer obtained by sublimation or evaporation of a block metal, etc. Besides, among merits of the methods under study are the reduction rapidity and efficiency of numerous metal ions, and perfect reproducibility of obtaining nanoparticles within a broad spectrum of temperatures and polymer matrices.

The method of joint vapor deposition of a monomer and a metal is very interesting too. During this procedure it is possible to stabilize a wide range of nanoparticles of different metals and monitor their concentration. It displays many advantages over other methods of producing nanocomposites: very neat substances are achieved since deposition proceeds from the gaseous phase and at deep vacuum; separate control over gaseous flows allows regulating the concentration of components in the material; the method does not impose any limitations on chemical or temperature parameters, on metal type, etc.

6.6 Electrochemical Methods of Nanocomposite Formation

The electrochemical and electroflotation methods are well known in production in situ of macromolecules with chemisorbed metal colloidal particles. The electrochemical polymerization reactions are heterogeneous, i.e. the initiation occurs at the electrode surfaces but the next stages (chain growth, limitations or termination) proceed usually in the surrounding of liquid media. The physical and chemical nature of the electrode and its special surface properties (such as the electrode overvoltage potentials, current density, etc.) affect these reactions. The nature of the electrode material (metals, alloys, thin coatings and so on) can have an impact on the characteristics of the electron transfer (i.e. the initiation) and polymerization reaction. The optimal processes of electrochemical polymerization are based on the direct electronic transitions (between the electrode and monomer), cathodic decomposition and anodic metal dissolution, when metal salts (an precursors of nanoparticles) can in principle serve as a peculiar electrochemical activator carrying out a direct chemical addition to the monomer (by a transfer, decomposition or catalysis). In general the electrochemical polymerization has limited application in the nanocomposite production.

The effective method of metal organosols production is based on use of double-layer electrolysis bath methods [175] (Fig. 6.8) with moving electrodes (rotation rate $\sim$60 rpm) wetted by a dispersed medium. The nanoparticles are giving off from the water solutions as highly dispersed cathodic depositions at the upper organic layer of the electrolytic bath. The layer presents a low-concentrated polymeric solution in the organic solvent (sometimes with SAS additives). The colloidal metallic particles interact with the solvent on the phase interface just in the moment of formation in situ. Nanoparticle formation rarely proceeds on polymers mainly by electrophoretic or electrochemical metal deposition on a polymeric suspension used as a dopant in electrolysis [176]. In this case the metallopolymer formation stage is defined both by a polarization interaction between polymeric and metal particles during the coprecipitation at the electrode and a macromolecule chemisorption interaction with the metal surface in the escaping moment.

The process is obviously multistage (M^{n+} ion discharge, metallic deposition crystallization, parallel electrochemical reactions, etc.). The transition from polymer-immobilized nanoparticles to metallic dense electrochemical coating depends on the dispersed phase and electrolyte concentrations and the particle charge. At the same time the screening characteristics of the polymeric particles result in a blocking effect for the metal particles or electrode surface that gives rise a specific type of interaction, namely electrophoretic polarization which can markedly influence the metal electrocrystallization. The polymer deposits usually as a separate phase (solid or liquid) on the electrode surface. As a result, the transfer of ions, electrons or neutral particles proceeds through the new phase layer and is accompanied by the serious

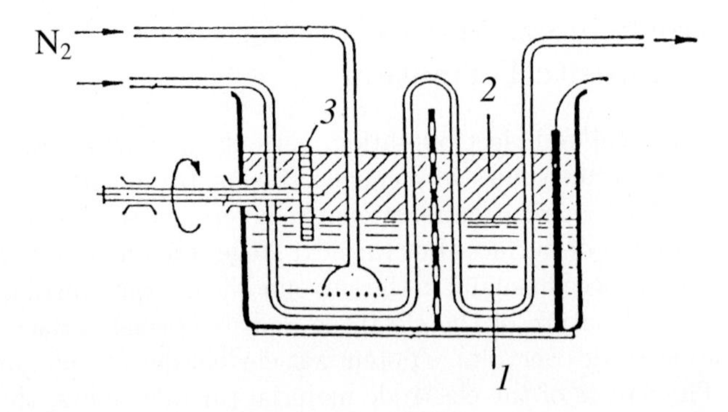

Fig. 6.8. A scheme of double-layer electrolytic bath for synthesis of metallopolymer: *1* – water solution of electrolyte, *2* – diluted solution of polymer and SAS in hydrocarbon, *3* – cathode

kinetic limitations connected with the polymeric coating permeability. The deposited polymer layer decreases the electrode general activity and the current but this effect can be partly reduced using liquid electrodes such as mercury-dropping electrodes. Some examples are given below.

The colloidal Pd particles in a nascent state were obtained [177] by the electrolysis of palladium chloride solutions at pH = 1 into a double-layer bath in the presence of a hydrocarbon solvent and epoxy-dianic resin or SAS. The particles arise because the polymer chemisorption bonds stabilize the formed colloidal Pd organosols. The products (after removing the solvent and electrolyte remains) were defined as metallopolymers, containing up to 90–95% of Pd. Their formation proceeded at the high polarization of the cathode and was accompanied by abundant release of hydrogen adsorbing on nanoparticles (5.5–7.8 nm in size).

The coprecipitation of Ni nanoparticles with PTFE suspension from the nickel sulfaminoacid electrolyte at room temperature leads to polymer incorporation into the Ni deposit (up to 20% by mass). The process is controlled by the cathodic current density and suspension concentration [178]. The relation between the homo- and heterocoagulation processes (i.e. between the metal particles and polymer) defines the rates of electrochemical deposition and metal release in Ni–PTFE systems [179]. The electrode material can be a precursor for noble metal nanoaprticles introduced for example in the polypyrrolic films being functionalized by 4,4′-dipyridyl, ammonium alkyl or other groups [180–184]. The electrochemically formed fine ferrous dispersions (~200 nm) can be used as the modification coatings of many materials [185]. Such metallopolymeric systems can be obtained electrochemically using porous films deformed into the adsorption active [186]. With this purpose, a cathode is covered by a thin porous film (Fig. 6.9) in the pores of which the metals (Cu, Ni, Co, Fe, Ag, etc.) can be released. The amount of deposition metal is controlled by the electrolysis process time and matrix

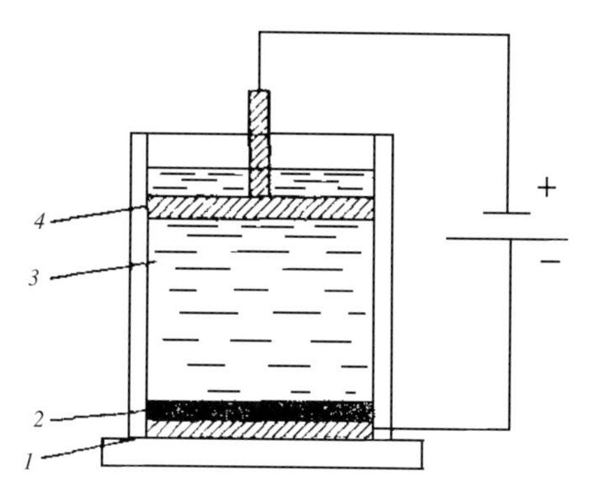

Fig. 6.9. Scheme of the electrochemical cell: *1* – cathode, *2* – film of porous polymer, *3* – metal salt solution, *4* – anode

porosity. All the craze volume in the ideal case must be filled by metal (for example, by Cu crystals 12–20 nm in size). Co nanoparticles 1–2 nm in size and with narrow size distribution can be generated electrochemically [187]. Dissolution of anode and formation of nanoparticles in a near cathode layer occur under current in a standard electrochemical cell containing tetraalkylammonium halide solution in ethanol.

There is little evidence about semiconductor type nanocomposites forming by electrochemical methods. For example, the formation of colloidal CdS (with particles 80–150 nm in size) in polypyrrole or polypyrrole/polystyrenesulfonate system on the electrodes can be marked [188] where two convenient approaches were found. In the former case cadmium ions were bound with a composition (obtained by pyrrole electrochemical polymerization) and treated by disulfide ions (HS^-). In the latter case the HS^- ions were electrochemically oxidized to elemental sulfur in a matrix from a polypyrrole thin film using the necessary cadmium ions in the electrode system.

The electrophoretic deposition of polymer-immobilized metal particles on the electrodes is very interesting too. For example, the $H_2PtCl_6 \cdot 6\,H_2O$ (stabilized by SAS) reduction on cuprous electrodes in spirit-water media allows us to form 2D coatings having the mean size of Pt particles 1.9–3.3 nm [189]. This method is suitable for the production of colloidal silver (on PAA) or gold stabilized by cationic PDDA polymer [190]. The polymeric films can be used for an effective charge separation between metal NP and solution in the different electrochemical devices [191]. There is little evidence about electrochemical methods of multimetallic nanoparticle formation in the polymers. The problem is very complicated because of differences in the electrochemical potentials during the various metals deposition. For example, the reduction potential of $AuCl_4^-$ is superior to $PdCl_4^-$ and therefore $AuCl_4^-$ in mixtures is reduced more rapidly than $PdCl_4^-$ [192]:

$$AuCl_4^- + 3e^- \quad \rightarrow \quad Au + 4Cl^-$$

$$PdCl_4^- + 2e^- \rightarrow Pd + 4Cl_4^-.$$

The deposition of electrochemical nanoparticles on polymers can be used for the production of materials having special properties. The thermal methods have the essential advantages over electrochemical ones, because an intense decomposition of the dispersed media, a strong oxidation of the forming colloidal particles, etc.

References

1. *Metallization of Polymers, Symp. Ser.* ed. by E. Sacher, J.J. Pireaux, S.P. Kowalczyk (ACS, Washington D.C. 1990) N440
2. D.M. Kolb, M. Andrews: Chem. Int. Ed. **40**, 1162 (2001)
3. A.N. Shipway, E. Katz, I. Willner: Chem. Phys., Phys. Chem. **1**, 18 (2000).
4. H.M. Meyer, S.G. Anderson, L.J. Atanasoska, J.H. Weaver: J. Vac. Sci. Technol. A **6**, 30 (1988)
5. G.A. Ozin: *Cryochemistry* (Mir, Moscow 1979)
6. C.G. Francis, H. Hubert, G.A. Ozin: Inorg. Chem. **19**, 219 (1980)
7. G.A. Ozin: J. Macromol. Sci. A **16**, 167 (1981)
8. G.A. Ozin, M.P. Andrews: In: [3], pp. 265-356
9. G.A. Ozin, C.G. Francis, H.X. Hubert, M. Andrews, L.S. Nazar: Coord. Chem. Rev. **48**, 203 (1983)
10. M.P. Andrews, G.A. Ozin: Chem. Mater. **1**, 174 (1989)
11. G.A. Ozin, M.P. Andrews, C.G. Francis, H.X. Huber, K. Molnar: Inorg. Chem. **29**, 1068 (1990)
12. K.W. Devenish, T. Goulding, B.T. Heaton, R. Whyman: J. Chem. Soc., Dalton Trans. 673 (1996)
13. T. Strunskus, M. Grunze, G. Kochendoerfer, Ch. Wöll: Langmuir **12**, 2712 (1996)
14. S.P. Gubin, N.K. Eremenko: Mendeleev Chem. J. **36**, 718 (1991)
15. M. Kerker: Colloid. Interface Sci. **105**, 297 (1985)
16. A.W. Olsen, Z.H. Kafafi: J. Am. Chem. Soc. **113**, 7758 (1991)
17. S.P. Solodovnikov, A.Y. Vasilkov, A.Y. Olenin, V.A. Sergeev: J. Magnetism & Magn. Mater. **129**, 317 (1994)
18. G.B. Sergeev: *Nanosize Metal particles in polymer films. Modern trends in low temperature chemistry* (Moscow University, Moscow 1994)
19. T. Susumu, I. Masayuki, T. Kanji: Bull. Inst. Atom. Energy Kyoto **83**, 85 (1993)
20. M.S. El-Shall, W. Slack: Macromolecules **28**, 8456 (1995)
21. G. Sergeev, V. Zagorsky, M. Petrukhina: J. Mater. Chem. **5**, 31 (1995)
22. A.Y. Vasilkov, A.Y. Olenin, V.A. Sergeev, A.N. Karanov, E.G. Olenina, V.M. Gryaznov: J. Cluster Sci. **2**, 117 (1991)
23. V.A. Sergeev, L.I. Vdovina, Y.Ya. Smetannikov: Polym. Sci. USSR B **29**, 431 (1987)
24. V.A. Sergeev, L.I. Vdovina, Y.Ya. Smetannikov, E.M. Belavtsev, A.Y. Vasilkov: Mendeleev Chem. J. **34**, 427 (1989); **36**, 255 (1991)

25. B.M. Sergeev, I.A. Gromchenko, A.N. Prusov, G.B. Sergeev: Vestn. Mosk. Univ. Ser. Khim. **36** 365 (1995)
26. B.M. Sergeev, I.A. Gromchenko, G.B. Sergeev: Vestn. Mosk. Univ. Ser. Khim. **35**,331 (1994); **40**, 129 (1999)
27. B.M. Sergeev, G.B. Sergeev, Y.J. Lee, A.N. Prusov, V.A. Polyakov: Mendeleev Commun. 151 (1997)
28. A.Y. Vasilkov, P.V. Pribytko, E.A. Fedorovskaya, A.A. Slinkin, A.S. Kogan, V.A. Sergeev: Dokl. Phys. Chem. **331**, 179 (1993)
29. V.A. Sergeev, L.I. Vdovina, Y.V. Smetannikov, A.Y. Vasilkov, E.M. Belavtseva, L.G. Radchenko, V.N. Guryshev: Organomet. Chem. USSR **3**, 919 (1990)
30. G.B. Sergeev: In: *Chemical physics at the threshold of the XXI-st century. To the centenary of Academician N.N. Semenov* ed. by G.B. Sergeev, A.E. Shilov (Nauka, Moscow 1996) 149 pp.
31. V.V. Zagorsky, E.A. Nasonova, M.A. Petrukhina, G.B. Sergeev: Vestn. Mosk. Univ. Ser. Khim. **36**, 159 (1995)
32. D.R. Linde, H.P.R. Frederikse (Eds.): *Handbook of Chemistry and Physics*, 78th edn. (CRC, Boca Raton 1997)
33. S.A. Ozerin, S.A. Zaviyalov, S.N. Chvalun: Polym. Sci. Ser. A **43**, 1993 (2001)
34. G.N. Gerasimov, E.V. Nikolaeva, E.I. Smirnova, V.A. Sochilin, L.I. Trakhtenberg: Dokl. Phys. Chem. **380**, 213 (2001)
35. B.Sh. Galyamov, S.A. Zaviyalov, L.Y. Kupriyanov: Russ. J. Phys. Chem. **74**, 459 (2000)
36. G.N. Gerasimov, E.L. Popova, E.V. Nikolaeva, S.N. Chvalun, E.I. Grigoriev, L.I. Trakhtenberg, V.I. Rozenberg, H. Hopf: Macromol. Chem. Phys. **199**, 2179 (1998)
37. E.L. Popova, D.Yu. Antonov, E.V. Sergeeva, V.I. Rosenberg, H. Hopf: Eur. J. Inorg. Chem. **2**, 1733 (1998)
38. P.S. Vorontsov, G.N. Gerasimov, E.N. Golubeva, E.I. Grigoriev, S.A. Zaviyalov, L.M. Zaviyalova, L.I. Trakhtenberg: Russ. J. Phys. Chem. **72**, 1912 (1998)
39. G.N. Gerasimov, V.A. Sochilin, S.N. Chvalun, L.V. Volkova, I.Ye. Kardash: Macromol. Chem. Phys. **197**, 1387 (1996)
40. L.N. Alexandrova, V.A. Sochilin, G.N. Gerasimov, I.E. Kardash: Polymer **38**, 721 (1997)
41. V.A. Sochilin, I.E. Kardash, G.N. Gerasimov: Polym. Sci. Ser. A **37** 1938 (1995)
42. K.J. Klabunde, J. Habdas, G. Cardenas-Trivino: Chem. Mater. **1**, 481 (1989)
43. I.M. Barkakov, D.P. Kirukhin: Intern. Rev. Phys. Chem. **11**,263 (1992)
44. A.I. Pertsin, Y.M. Poshutin: Polym. Sci. Ser. B **38**, 919 (1996)
45. A.I. Pertsin, I.O. Volkov: Polym. Sci. Ser. B **38**, 1249 (1996)
46. A.Y. Vasilkov: Proceed. I Russian Conference on Cluster Chem., June 27–July 1, 1994, (St. Petersburg 1996) p. 20
47. M.A. Lunina, Y.A. Novozhilov: Colloid. J. **31**, 467 (1969)
48. T. Noguchi, K. Gotoh, Y. Yamaguchi, S. Deki: J. Mater. Sci. Lett. **10**, 477 (1991); **11**, 648, 797 (1992)
49. T. Masui, K. Machida, T. Sakata, H. Mori, G. Adachi: Chem. Lett. 75 (1996)
50. R.A. Roy, R. Messier, S.V. Krishnaswamy: Thin Solid Films **109**, 27 (1983)
51. K. Kashiwagi, Y. Yoshida, Y. Murayama: J. Vacuum Sci. Technol. A **5**, 1828 (1987)

52. *Metallized plastics. Fundamental and applied aspects* ed. by K.L. Mittal, J.R. Susko (Plenum Press, N.Y. **1** 1989, **2** 1991, **3** 1993, **4** 1995)
53. *Metallized plastics. Fundamentals and Applications* ed. by K.L. Mittal (Marcel Dekker, N.Y. 1998)
54. A. Thran, M. Kiene, V. Zaporojtchenko, F. Faupel: Phys. Rev. Lett. **82**, 1903 (1999)
55. Y. Nakao: Chem. Lett. 766 (2000)
56. F. Faupel, V. Zaporojtchenko, T. Strunskus, J. Erichsen, K. Dolgner, A. Thran, M. Kiene: In: *Metallization of polymers. ACS Symposium Ser. 2* (Kluwer Academic/Plenum Publ. N.Y. 2001)
57. F. Faupel, T. Strunskus, M. Kiene, A. Thran, C. Bechtolsheim, V. Zaporojtchenko: Mat. Res. Soc. Symp. Proc., Materials Research Society, **511**, 15 (1998)
58. D. Mendoza, S. López, S. Granandos, F. Morales, R. Escudero: Synthetic Metals **89**, 71 (1997)
59. V.I. Fistul: In: *Fullerenes and Atom Clusters, International Workshoop at St.Petersburg, Russia, October 4–October 9, 1993* (St.Peterburg 1993) p. 21
60. S. Boonthanum, M. White: Thin Solid Films **24**, 295 (1974)
61. A.M. Krasovsky, V.A. Rogachev: Polym. Sci. USSR, Ser. B **22**, 610 (1980)
62. B.S. Balzhinimaev, V.I. Zaikovskii, L.G. Pinaeva, A.V. Romanenko, G.V. Ivanov: Mendeleev Commun. 100 (1998)
63. *Microwave-enhanced chemistry* ed. by H.M. Kingsytom, S.J. Haswell (Am. Chem. Soc., Washington 1997)
64. Y. Wada, H. Kuramoto, T. Sakata, H. Mori, T. Sumida, T. Kitamura, S. Yanagida: Chem. Lett. 607 (1999)
65. M.S. Hegde, D. Larcher, L. Dupont, B. Beaudoin, K. Tekaia-Elhsissen, J. Tarascon: Solid State Ionics **93**, 33 (1997)
66. P. Bodö, J.-E. Sundgren: Thin Solid Films **136**, 147 (1986)
67. N. Bowden, S. Brittain, A.G. Evans, J.W. Hutchinson, G.M. Whitesides: Nature **393**, 146 (1998)
68. M.T. Reetz, M. Winter, G. Dumpich, J. Lohau, S. Friedrichowski: J. Am. Chem. Soc. **119**, 4539 (1997)
69. J. Lohau, S. Friedrichowski, G. Dumpich, E.F. Wassermann, M. Winter, M.T. Reetz: J. Vac. Sci. Technol. B **16**, 77 (1998)
70. X.Q. Zhao, F. Zheng, Y. Liang, Z.Q. Hu, Y.B. Xu: Mater. Lett. **21**, 285 (1994)
71. *The Chemistry of Metal CDV* ed. by T. Kodas, M.J. Hampden-Smith (VCH, Weinheim 1994)
72. L.-S. Wang, X.-B. Wang, H. Wu, H. Cheng: J. Am. Chem. Soc. **120**, 6556 (1998)
73. L. Friemens, J. Vannik, V. Decaser: *Electronic and ionic spectroscopy of solids* (Mir, Mocsow 1984)
74. Y.M. Pashutin, A.I. Pertsyn: Polym. Sci. Ser. A **32**, 1983 (1990)
75. V.N. Popok: Surface **6**, 103 (1998)
76. Y.M. Pashutin, A.I. Pertsyn: Surface **3**, 130 (1990)
77. Y. Takeda, T. Hioki, T. Motohiro, S. Noda: Appl. Phys. Lett. **63**, 3420 (1993)
78. S.G. Anderson, H.M. Meyer, L.J. Atanasoska, J.H. Weaver: J. Vac. Sci. Technol. A **6**, 38 (1988)
79. V. Zaporojtchenko, T. Strunskus, K. Behnke et al.: J. Adhesion Sci. Technol. **14**, 467 (2000)

80. A.L. Stepanov, S.N. Abdulin, I.B. Khaibullin: J. Non-Cryst. Solids **223**, 250 (1998)
81. W.De Cruz, L.C. Araiza: Phys. Status Solidi B **220**, 569 (2000)
82. N.M. Dempsey, L. Ranno, D. Givord, J. Serna, G.T. Fei, A.K. Petford-Long, R.C. Doole, D.E. Hole: J. Appl. Phys. **90**, 6268 (2001)
83. A.E. Berkowitz: Phys. Rev. Lett. **68**, 3745 (1992)
84. A.E. Pachteny, D.I. Sagaidak, G.G. Fedoruk: Polym. Sci. Ser. A **39**, 1199 (1997)
85. N.C. Morosoff, N.E. Barr, W.J. James, R.B. Stephens: In: *12-th International Symposium on Plasma Chem. at Minnesota, August 21–August 25, 1995*, ed. by J.V. Hebberlleing, D.W. Ernie, J.T. Roberts (Univ. Minnesota 1995) textbf1, p. 147
86. G.N. Savenkov, Yu.P. Baidarovtsev, K.N. Yanchivenko, A.N. Ponomarev: Chem. High Energ. **30**, 65 (1996)
87. Yu.P. Baidarovtsev, G.N. Savenkov, V.V. Shevchenko, Yu.M. Shulga: Polym. Sci. A **42**, 287 (2000)
88. Yu.P. Baidarovtsev, G.N. Savenkov, E.D. Obraztsova, V.V. Shevchenko, Yu.M. Shulga: Polym. Sci. A, **42**, 1006 (2000)
89. E. Kay, M. Hecq: J. Appl. Phys. **55**, 370 (1984)
90. K.P. Gritsenko: Ukr. Chem. J. **57**, 782 (1991)
91. L. Martinu, H. Biederman, J. Zemek: Vacuum **35,** 171 (1985)
92. E.Z. Kay: Z. Phys. D **3,** 251 (1986)
93. H. Biederman, L. Martinu, D. Slavinska, I. Chudacek: Pure Appl. Chem. **60,** 607 (1988)
94. G. Kampfrath, A. Heilmann, C. Hamann: Vacuum **38** 1 (1988)
95. A. Heilmann, C. Hamann: Progr. Colloid Polymer Sci. **85**, 102 (1991)
96. A. Heilmann, J. Werner, V. Hopfe: Supplementto Z. Phys. D **39** 1993
97. N. Yasuda, M. Mashita, T. Yoneyama: Res. Develop. Jpn. 57 (1984)
98. M. Hecq, P. Zieman, E. Kay: J. Vac. Sci. Techn. A **1** 364 (1983)
99. L. Martinu: Thin Solid Films **140**, 307 (1986)
100. V.V. Petrov, K.P. Gritsenko, A.A. Kryuchin: Ukr. Dokl. Phys.-Math.&Techn. Sci. A **12**, 64 (1989)
101. H. Yasuda: *Plasma Polymerization* (Academic Press, Orlando 1985)
102. S.A. Krapivina: *Plasma-Chemical Technological Processes* (Khimiya, Leningrad 1981)
103. A.B. Gilman, V.M. Kolotyrkin: High Energy Chem. **12**, 450 (1978)
104. J. Sakata, M. Yamamoto, J. Tajima: J. Polym. Sci. A Polym. Chem. Ed. **26**, 1723 (1988)
105. *Plasma Polymerization. ACS Symposium Series* ed. by M. Shen, A.T. Bell (ACS, Washington, D.C. 1979) **108**
106. B.S. Danilin: *Low-Temperature Application for Deposition of Thin Films* (Energoatomizdat, Moscow 1989)
107. N. Inagaki, S. Tasuka, M. Masumoto: Macromolecules **29**, 1642 (1996)
108. H.L. Chen, S.H. Ho, T.H. Wang, K.M. Chen, J.P. Pan, S.M. Liang, A. Hung: J. Appl. Polym. Sci. **51**, 1647 (1994)
109. V.F. Sokolov, Y.A. Sokolov, D.V. Tremasova: Trans. of Higher Educ. Est. Chem. & Techn. **38**, 148 (1995)
110. K. Tsukagoshi, Y. Sakakibara, M. Ijima, Y. Takahashi: Jpn. J. Appl. Phys. Pt. 2 **33**, L463 (1994)

111. K. Sadir, H.E. Saunders: J. Vac. Sci. Technol. A **3**, 2093 (1985)
112. A.M. Krasovsky, A.I. Kuzavkov, V.A. Shelestova, N.P. Glazyrin: Belarussian Dokl. Phys.-Techn. Sci. 92 (1993)
113. K.P. Gritsenko: Thin Solid Films **227**, 1 (1993)
114. R.J. Shul, M.L. Lovejoy, J.C. Word, A.J. Howard, D.J. Rieger, S.H. Kravitz: J. Vac. Sci. Technol. B **15** 657 (1997)
115. J. Perrin, B. Despax, E. Key: Phys. Rev. B **32**, 719 (1985)
116. A. Heilmann, G. Kampfrath, V. Hopfe: J. Phys. D Appl. Phys. **21**, 986 (1988)
117. B. Abeles, P. Sheng, M.D. Coutts, Y. Arie: Adv. Phys. **24** 407 (1975)
118. U. Kreibig, A. Althoff, H. Pressmann: Surf. Sci. **106**, 308 (1981)
119. A. Liebsch, B.N.J. Persson: J. Phys. C **16**, 5375 (1983)
120. J. Perrin, B. Despax, V. Hanchett, E. Kay: J. Vac. Sci. Technol. A **4**, 46 (1986)
121. H. Biederman: Vacuum **34**, 405 (1984)
122. L. Martinu, H. Biederman: J. Vac. Sci. Technol. A **3**, 2639 (1985)
123. E. Kay, A. Dilks, D. Seybold: J. Appl. Phys. **51**, 5678 (1980)
124. R. Lamber, A. Baalmann, N.I. Jaeger, G. Schulz-Ekloff: Adv. Mater. **6**, 223 (1994)
125. R. Lamber, S. Wetjen, G. Schulz-Ekloff: J. Phys. Chem. **99**, 13834 (1995)
126. D. Salz, A. Baalmann, U. Simon, M. Wark, A. Baalmann, U. Simon, N. Jaeger: Chem. Phys. **4**, 2438 (2002)
127. D. Salz, R. Lamber, M. Wark, A. Baalmann, N. Jaeger: Chem. Phys. **1**, 4447 (1999)
128. C. Laurent, E. Kay: Z. Phys. D, **12**, 465 (1989)
129. R. Lamber, S. Wetjen, N.I. Jaeger: Phys. Chem. Rev. B **51**, 10968 (1995)
130. D. Salz, B. Mahltig, A. Baalmann, M. Wark, N. Jaeger: Phys. Chem. Chem. Phys. **2**, 3105 (2000)
131. K. Meguro, T. Adachi, R. Fukunishi, K. Esumi: Langmuir **4**, 1160 (1988)
132. M. Mostafavi, N. Keghouche, M.-O. Delcourt: Chem. Phys. Lett. **169,** 81 (1990)
133. *Absorption and scattering of light by small particles* ed. by C.F. Bohren, D.R. Huffman (Willey, N.Y. 1983)
134. U. Kreibig, M. Vollmer: *Optical Properties of Metal Clusters* (Springer, Berlin 1995)
135. B.G. Ershov, N.I. Kartashov: Russ. Chem. Bull. 35 (1995)
136. B.G. Ershov, E. Janata, A. Henglein: J. Chem. Phys. **97**, 339 (1993)
137. A.G. Dokuchaev, T.G. Myasoedova, A.A. Revina: High Energy Chem. **31**, 353 (1997)
138. A.A. Revina, E.M. Egorova, A.D. Karataeva: Russ. J. Phys. Chem. **73**, 1708 (1999)
139. E.M. Egorova, A.A. Revina: Colloid and Solid Surfaces A **168**, 87 (2000)
140. A. Henglein: J. Phys. Chem. **97**, 5457 (1993)
141. B.G. Ershov: Russ. Chem. Bull. 1 (1991)
142. M. Gutierrez, A. Henglein: J. Phys. Chem. **97**, 11368 (1993)
143. B.G. Ershov, E. Jananta, A. Henglein, A. Fojtic: J. Phys. Chem. **97**, 339, 4589 (1993)
144. A. Henglein, T. Linnert, P. Mulvaney: Ber. Bunsenges. Phys. Chem. **94**, 1449 (1990)
145. M. Mostafavi, N. Keghouche, M.-O. Delcourt, J. Belloni: Chem. Phys. Lett. **167**, 1933 (1990); **169**, 81 (1990)

146. M. Mostafavi, M.O. Delcourt, G. Picq: Radiat. Phys. and Chem. **41**, 453 (1993)
147. Y. Zhu, Y. Qian, X. Li, M. Zhang: Chem. Commun. 1081 (1997)
148. Y. Yin, X. Xu, C. Xia, X. Ge, Z. Zhang: Chem. Commun. 941 (1998)
149. B.G. Ershov, D.A. Troitsky: J. Phys. Chem. **69**, 2179 (1995)
150. B.G. Ershov: Russ. Chem. Bull. 600 (2001)
151. A. Henglein, B.G. Ershov, M. Malow: J. Phys. Chem. **99**, 14129 (1995)
152. A.N. Krasovsky, A.I. Andreeva: J. Appl. Chem. **72**, 1875 (1999)
153. H. Bekhter, I. Epperlien, A.V. Eltsov: *Modern Systems of Information Recording: Main Principles, Processes, Materials* (Synthesis, St. Petersburg 1992)
154. T. Sato, T. Ichikawa, T. Ito, Y. Yonezawa, K. Kodono, T. Sakaguchi, M. Miya: Chem. Phys. Lett. **242**, 310 (1995)
155. Y. Zhou, S. Yu, C. Wang, Y. Zhu, Z. Chen: Chem. Lett. 677 (1999)
156. S. Fredrigo, W. Harbick, J. Buttet: Phys. Rev. **47**, 10706 (1993)
157. T.B. Boitsova, V.V. Gorbunova, A.V. Loginov: Russ. J. Gen. Chem. **67**, 1741 (1997)
158. T.B. Boitsova, V.V. Gorbunova, A.V. Loginov: Russ. J. Gen. Chem. **69**, 1937 (1999)
159. V.V. Gorbunova, Y.M. Voronin, T.B. Boitsova: Optic. J. **68**, 1746 (2001)
160. B.M. Sergeev, M.V. Kiryakhin, A.N. Prusov, V.G. Sergeev: Vestn. Moscow Univ. Ser. Khim. **40**, 129 (1999)
161. A.I. Andreeva, A.N. Krasovsky, D.V. Novikov: J. Appl. Chem. **72**, 156 (1999)
162. M.V. Kiryukhin, B.M. Sergeev, A.N. Prusov, V.G. Serheev: Polym. Sci. Ser. B **42**, 2171 (2000)
163. P. Ulanski, E. Bothe, K. Hildenbrand, J.M. Rostiak, C. Sonntag: J. Chem. Soc. Perkin Trans. **2**, 5 (1996)
164. P. Ulanski, E. Bothe, J.M. Rostiak, C. Sonntag: J. Chem. Soc. Perkin Trans. **2**, 13, 23 (1996)
165. V.M. Rudoi, I.V. Yaminsky, V.A. Ogarev: Polym. Sci. Ser. B **41**, 1671 (1999)
166. V.M. Rudoi, O.V. Dementeva, I.V. Yaminskii, V.M. Sukhov, M.E. Kartseva, V.A. Ogarev: Russ. Colloid J. **64**, 823 (2002)
167. N. Nagayama, T. Maeda, M. Yokoyama: Chem. Lett. 397 (1997)
168. B. Cheng, Y.R. Zhu, Z.Y. Chen, W.Q. Jiang: J. Mater. Sci. Lett. **18**, 1859 (1999)
169. T.B. Boitsova, V.V. Gorbunova, E.I. Volkova: Russ. J. Gen. Chem. **72**, 688 (2002)
170. M. Suzuke, Y. Ohta, H. Nagal, T. Ichinohe, M. Kimura, K. Hanabusa, H. Shirai, D. Wohrle: Chem. Commun. 213 (2000)
171. B.G. Ershov, A. Henglein: J. Phys. Chem. B **102**, 10663 (1998)
172. Ch. Loppacher, S. Trogisch, F. Braun, A. Zerebow, S. Grafström: Macromolecules **35**, 1936 (2002)
173. M. Yamada, A. Kuzume, M. Kurichara, K. Kubo, H. Nishihara: Chem. Commun. 2476 (2001)
174. M. Yamada, T. Tadera, K. Kubo, H. Nishihara: Langmuir **17**, 2363 (2001)
175. A.L. Balch, D.A. Costa, K. Winkler: J. Am. Chem. Soc. **120**, 9614 (1998)
176. M.W. Anderson, J. Shi, D.A. Leigh, A.E. Moody, F.A. Wade, B. Hamilton, S.W. Carr: Chem. Commun. 533 (1993)
177. V.M. Varushchenko, S.A. Mikhalyuk, E.M. Natanson, B.D. Polkovnikov: Izv. AN SSSR, Ser. Khim. 1346 (1971)

178. E.V. Kuznetsova: Russ. J. Appl. Chem. **66**, 1155 (1993)
179. N.M. Teterina, G.V. Khaldeev: Russ. J. Appl. Chem. **67**, 1528 (1994)
180. L. Coche, J.-M. Moutet: J. Am. Chem. Soc. **109**, 6887 (1987)
181. I.M.F. Oliveira, J.-M. Moutet, S. Hamar-Thibault: J. Mater. Chem. **2**, 167 (1992)
182. L. Coche, B. Ehui, D. Liimosin, J.-M. Moutet: J. Org. Chem. **55**, 5905 (1990)
183. J.-M. Moutet, A. Ourari, A. Zouaoui: Electrochim. Acta **37**, 1261 (1992)
184. J.-C. Moutet, Y. Ouennoughi, A. Ourari, S. Hamar-Thibault: Electrochim. Acta **40**, 1827 (1995)
185. O.M. Mikhailik, V.I. Povstugar, S.S. Mikhailova, A.M. Lyakhovich, O.M. Tedorenko, G.T. Kurbatova, N.I. Shklouskoya, A.A. Chuiko: Colloids and Surfaces **52**, 315; 325; 331 (1991)
186. A.L. Volynskii, L.M. Yarysheva, G.M. Lukovkin, N.F. Bakeev, Polym. Sci. Ser. A **34**, 24 (1992)
187. J.A. Becker, R. Schagter, J.R. Festag et al.: Surf. Rev. Lett. **3**, 1121 (1996)
188. M. Hepel, E. Seymour, D. Yogev, J.H. Fendler: Chem. Mater. **4**, 211 (1992)
189. T. Teranishi, M. Hosoe, T. Tanaka, M. Miyake: J. Phys. Chem. B **103**, 3818 (1999)
190. R. Chapman, P. Mulvaney: Chem. Phys. Lett. **349**, 358 (2001)
191. B.I. Lemon, L.A. Lyon, J.T. Hupp: In: *Nanoparticles and Nanostructured Films* ed. by J. Fendler (Wiley-VCH, Weinheim 1998)
192. N. Toshima, T. Yonezawa: New J. Chem. 1179 (1998)

Part III

Hybrid Polymer–Inorganic Nanocomposites

These nanocomposites have in common the nanometer parameters of their structure, distances between lattices and layers formed by the polymer and inorganic ingredients, and the nanometer size of the forming particles, including metal-containing ones.

This class of advanced materials displays synergism of properties of the original components. Their organic phase may entrap metal particles inside peculiar traps of optimum parameters of the oxide network or a polymer grain. Silicone, titanium, aluminum, vanadium and molybdenum oxides along with glass, clay, lamellar silicates, zeolites, phosphates, metal chalcogenides, ferric oxychlorides, and graphite are used as inorganic precursors. The use of natural silicates and zeolites with methods of regulating pore size looks very interesting too.

In the past decade a lot of attention was given to the issue of developing sol–gel analysis in production of hybrid materials and to intercalation of polymers and nanoparticles in layered systems using the chemistry of the intracrystalline guest–host structures. The combinations of the mineral and organic components in their formation processes were tested with sol–gel synthesis and control over the formation of inorganic and polymer networks. Synthesis procedures and properties of metal–polymer Langmuir–Blodgett films were studied too. Many natural composite materials have similar structure (e.g. bone, borosilicate glass). The major synthetic methods of hybrid nanocomposite production are allied to the living nature and are observed in various bioprocesses, biosorption, and biomineralization.

7 The Sol–Gel Process in Formation of Hybrid Nanocomposites

Polymer–inorganic materials are distinguished by elevated mechanical strength and thermal stability in combination with optimum heat transfer features [1–3]. These materials are widely used in components of metal matrix composites, where they improve the strength and hardness of soft materials, and their thermochemical, rheological, electrical and optical properties [4–6]. These composites are used as chromatographic carriers, membrane materials [7], and as new classes of plastics for aerospace applications. Control of surface properties of colloidal materials is one of the major problems in pharmaceutical, cosmetic, semiconductor, catalysis, printing, food, biological, and medical fields. These problems have been generalized in a series of monographs and reviews [8–28], and special issues of journals were devoted to these aspects [29–32].

Optimum in the ecological respect is zero-discharge methods of composite material production, to which belong the sol–gel or spin-on-glass processes. They employ materials such as ceramics, luminophores, carriers, catalyzers, plastics and rubber reinforcing agents, binders, adsorbents in pharmaceutical and cosmetic industries [33–40], and photocatalysts for cleaning water and air. According to [41], sol–gel processes were used since 1980s for studying organic/inorganic hybrid materials.[1] In 1996 the review [42] on the use of sol–gel materials in electrochemistry only contained more than 300 references and the peak of investigations in this domain hasn't been yet reached. The sol–gel process is a convenient means of producing dispersed materials often called *ceramers* through growing metal-oxide polymers in solutions based on inorganic polymerization reactions. Organotitanium acrylic polymers, like titanium alkoxides, appear to be self-polishing free of toxicity coatings (release of biocides) and are completely nontoxic for the marine environment [44].

Hybrid nanocomposites depend on the prehistory of the organic and inorganic phase introduced into the material. The organic component can be introduced as: (a) a precursor, which can be a monomer or an oligomer [45], (b) a preformed linear polymer (in solution, molten, emulsion states) [46–50],

---

[1] Generally speaking, this statement isn't strictly accurate. The production process of low-density silica by tetraethyl ortosilicate hydrolysis in the presence of cation-ion surfactants was patented in 1971 (USA patent 3,556,725, Jan. 19, 1971, V. Chiola, J.E. Ritsko, C.D. Vanderpool).

or (c) a preformed polymer network, physically or chemically cross-linked [12, 34, 39, 51]. The mineral ingredient can be included into a hybrid material in three different ways: (a) as a metal-oxide monomer (see [39, 45–49, 53, 54]), (b) as preformed nanoparticles [55, 56], or (c) as an existed nanoporous structure [10, 20, 57, 58].

7.1 General Characteristics of Sol–Gel Reactions

The following methods of manufacturing ceramics with required properties are the most well known: coprecipitation, sol–gel processing of colloids or organometallic compounds, hydrothermal processing, spray pyrolysis processes, gel processing of organic polymers or polymerizable media in the presence of metal complexes.

The sol–gel process is a two-step network-forming process and consists of the following main stages: preparation of the solute → gel formation → drying → thermal treatment. The initial agents of the process are metal alkoxides which are brought into the reaction in the organic solvent medium where water serves as the reagent evoking hydrolysis $M(OR)_n$ (M = Si, Ti, Zr, VO, Zn, Al, Sn, Ce, Mo, W, and lanthanides). Further condensation of the forming compounds leads to formation of a gel. The acidic hydrolysis (for alkoxides with $n = 4$) and subsequent condensation can be presented by

$$M(OR)_4 + 4H_2O \rightarrow M(OH)_4 + 4ROH;$$
$$mM(OH)_4 \rightarrow (MO_2)_m + 2mH_2O.$$

The real process is naturally more intricate and follows a multiroute mechanism. Hydrolysis of alcoholates can result in oxoalcoholates of metals as intermediate forms which may include polynuclear forms of the $Ti_xO_y(OR)_{4x-2y}$ type many of which have already been discovered and characterized [59]. In the course of formation of the solid phase of a monodisperse TiO_2 powder, the sequential stages can be observed: hydrolysis → condensation → nucleation → particle growth [60].

During a traditional sol–gel process oxooligomers or polymers with rather high molecular mass as well as network polymers are formed:

Alcoxolation:

$$-\overset{|}{\underset{|}{Si}}-OH + EtO-\overset{|}{\underset{|}{Si}}- \xrightarrow[OH^-]{H^+} -\overset{|}{\underset{|}{Si}}-O-\overset{|}{\underset{|}{Si}}- + EtOH$$

Oxolation:

$$-\overset{|}{\underset{|}{Si}}-OH + HO-\overset{|}{\underset{|}{Si}}- \xrightarrow[OH^-]{H^+} -\overset{|}{\underset{|}{Si}}-O-\overset{|}{\underset{|}{Si}}- + H_2O$$

The particles are supposed to grow by the diffusion mechanism [61] and the diffusion constant is about 10^{-9} $cm^2 \cdot s^{-1}$. Nevertheless, by a combination of Fourier transform infrared spectroscopy (FTIR) and small-angle X-ray scattering (SAXS) it was proved that the major stage of $Zr(OBu)_4$ hydrolysis started in the very early moments of the process (in 1 s) and condensation occurred even at that so instantaneous period [62]. Heat effects during the $Ti(OBu)_4$ hydrolysis reaction in ROH (R= Et, Pr^i, Bu) at different ratios of $H_2O/M(OR)_n = \gamma$ within the $\gamma = 0.2$–70 range intensify till $\gamma \approx 1$ (mean value is 19.3 $kJ \cdot mol^{-1}$). The heat effect corresponds to the substitution reaction of one alkoxyl group [63].

Critical factors are the optimization of the process conditions, use of catalyzers, including nucleophylic catalysis of NH_4F trifluoroacetic acid [64] and polystyrenesulfonic acid of a polymer origin [43]; nature of the metal and that of alkoxy group. The hydrolysis rate of $Ti(OBu)_4$ is almost 150 times less than that of $Ti(OEt)_4$ [65]. A perceptible effect is exerted by other ligands too in case of alkoxides of mixed type, especially of the chelate component: β-diketone, α and β-hydroxyacidic, polyolic, etc. in compounds like $Zr(OSiMe_3)_2(acac)_2$) applicable in metal–organic chemical vapor deposition (MOCVD) [66]. Besides, the process depends upon the degree of alkoxide association, e.g. $[Ti(OEt)_4]_n$ $n = 2$; 3. The rate of hydrolysis of oxide and alkoxy-cluster structures of the $[Ti_{18}O_{22}(OBu)_{26}(acac)_2]$ kind is by far slower than that of the initial $Ti(OR)_4$ hydrolysis. Many procedures for the described products are known today, e.g. production of TiO_2, highly important in industry, (Table 7.1) [67]; nevertheless the sol–gel process is believed to be the most promising.

The reactivity of alkoxides of four-valent metals $M(OR)_4$ augments in the series [48] $Si(OR)_4 \ll Sn(OR)_4$ (R_3Si groups are known to be nonhydrolysable) and $Ti(OR)_4 < Zr(OR)_4 < Ce(OR)_4$. The ion radius of the central atom increases in the same series 0.04, 0.06, 0.64, 0.087 and 0.102, respectively, as does its coordination number CN (4, 6, 6, 7, 8) as well as its nonsaturation degree (CN to valence difference 0, 2, 2, 3 and 4) [51]. Nevertheless, the γ ratio is the most important. On particular, for $VO(OPr)_3$ a homogeneous transparent gel with an alkoxide polymer lattice in n-propanol is derived at $n = 3$. At $n > 100$ the forming gel has a different structure incapable of generating the inclusion compounds [68]. Proceeding from the above, the sol–gel process consists of the gel-precursor polymerization (chemically controlled condensation) [69], tetramethoxysilane (TMOS) or tetraethoxysilane (TEOS) able to form a silica gel structure (host) round a dopant (guest) as if encapsulated into a specific cell-trap. In case the jellification (gel formation) runs quickly, the method is called a fast sol–gel method.

Sols are thermodynamically unstable systems possessing a high free surface energy and can exist only in the presence of the stabilizing colloids. One of the efficient means of their stabilization is an electrostatic method governed by the system pH value. Stabilization of highly concentrated

Table 7.1. The main production methods of TiO_2

Preparation Methods	Oxidation Process (CVD)	Classic Sulfate Process (WS)	Flame Synthesis (CVD)	Pyrolysis/ Hydrolysis of Organotitanium Compound (CVD)	Inert Gas Condensation, IGC, (PVD)	Sol-Gel Method (WC)
Reactant materials	$TiCl_4+O_2$	$TiOSO_4+H_2O$	$TiCl_4+O_2+H_2O$	$Ti(OC_4H_9)_4$	Ti, postoxidation to obtain TiO_{2-x}	$Ti(OC_4H_9)_4$
Procedure	Continuous	Batch	Continuous	Continuous	Batch	Batch
Preparative temperature	1670–1770	Hydrolysis below 370, calcination above 970	above 1270	1170		Hydrolysis below 370, calcination at high temperature
Particle size (nm)	100–400	100–400	30–60	<100	5–15	<15
Agglomeration	Weak	Strong	Weak or strong	Weak	Weak	Strong
Yield	Commercialized Cl_2, can be recycled	Commercialized Large quantity	Commercialized	Lab scale	Lab scale	Lab scale
Byproducts	recycled	H_2SO_4, acid water, SO_2, SO_3	Cl_2+ HCl	C_4H_8 or C_4H_9OH	–	C_4H_9OH & organic solvent
Main challenges	Severe $TiCl_4$ corrosion at high temperature	High energy consumption: waste processing and recycling	Relatively wide size distribution	Expensive precursor; possible residual carbon in particles	Very small yield	Expensive precursor and organic solvents

sols presents a complex problem. Stabilization of nanoparticles in the sol–gel is observed through binding of the sol surface by special molecules – monomers [53, 54, 70, 71].

Carboxylic acids, including polymer ones [72], have strong bonding with the surface of SiO_2, ZrO_2, TiO_2 or Al_2O_3 particles. Amines are capable of peculiar adsorption on metal particles of Pd and Au. More specific reagents for Au are thiols. It is possible to use bifunctional molecules incorporating a double bond along with hydrolysable silane or Zr–OR groups. Then the corresponding precursors can be obtained upon controlled hydrolysis with ZrO_2 particles of ~2 nm size prone to copolymerize via the double bond of methacrylic acid (MAA) functioning simultaneously as the surface modifier (see Scheme 7.1).

Moreover, zirconium alkoxide $Zr(OR)_4$ displays high reactivity towards hydrolysis, so during the direct process a sediment $ZrO_2 \cdot aq$ precipitates which is fit for production of a homogeneous composite material.

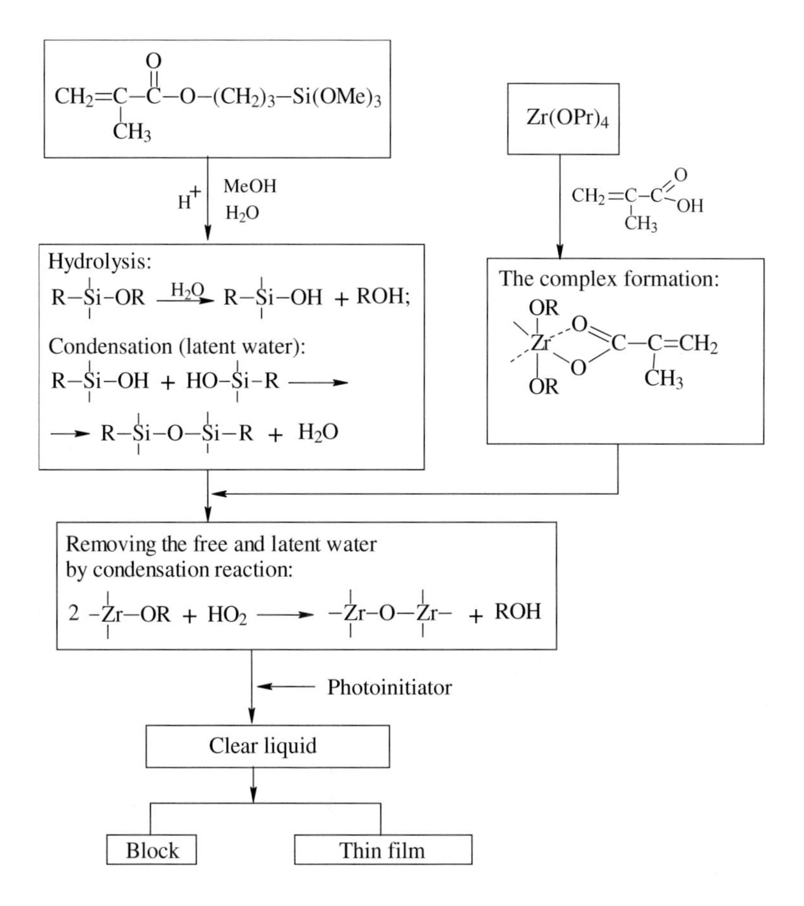

Scheme 7.1

Once the MAA is bound, the ability of $Zr(OR)_4$ to hydrolysis drops abruptly leading to formation of well dispersed $ZrO_2\cdot$ nanoparticles in the presence of latent water in the silane being hydrolyzed and condensed.

It should be noted that to deposit silica layers very often a well-known sequence of reactions is used [73]:

$$SiCl_4 + 4H_2O \rightarrow Si(OH)_4 + 4HCl$$
$$Si(OH)_4 \rightarrow Si_nO_{4n-x}H_{4n-2x} + xH_2O$$

Similarly, the anatase and rutile modifications of TiO_2 are obtained [74]. Their stability in aqua dispersions as well as in acidic and alkali solutions was studied in [75].

Oxopolymers synthesized by sol–gel synthesis acquire an ultrafine porous network similar to zeolites with pore size 1–10 nm. They are sometimes called nanoperiodical or mesostructural materials whose specific surface (130–1260 $m^2\cdot g^{-1}$) depends on the synthesis conditions. Of importance, especially for catalysis are titanium-containing silica mesoporous molecular sieves of the type of hexagonal MCM-41 and cubic MCM-48 materials (Mobil codes), Ti-MCM-41 [76–78]. The dimensions of their pores (above 2–3 nm) present a base for modification of their inner surface aimed at regulation of hydrophobic–hydrophilic and acidic properties and design of catalytic active centers [79,80]. The TiO_2-gel-monolite pore diameter is about 4 nm, and the material porosity reaches 50% depending on production conditions [81,82]. Their drying regime, during which the volatile products are removed, governs the product texture. For example, low-dispersed xerogels, upon long drying in air, may be formed due to coalescence of gel particles. The preparation temperature of TiO_2 by the sol–gel method plays a critical role too and its specific surface area (SSA, $m^2\cdot g^{-1}$), mean pore diameter (nm) and porosity (%) are given in Table 7.2 [83].

When the action of capillary forces is excluded by drying conditions, then high-dispersed aerogels are formed. CO_2 in a supercritical state is commonly used today for this purpose. The formation of the structure and texture of the

Table 7.2. SSA, porosity and mean pore diameter of titanium samples

Temperature, K	SSA (m^2g^{-1})	Porosity (%)	Mean pore Diameter (nm)
As prepared	332.8	57.6	4.4
570	123.7	54.4	1.4
670	118.4	53.6	10.5
770	90.5	45.6	10
870	25	18.6	9.8

target product are terminated at the stage of thermal treatment [8, 84–89].[2] The pyrolysis atmosphere affects their inert, vacuum or oxidative properties as well [100].

The sol–gel synthesis is used in production of vanadium oxide nanotubes [101–103], which implies that the sol–gel synthesis is run in the presence $VO(O\text{-}i\text{-}Pr)_3$ or α, ω-diamines, and is followed by a hydro-thermal reaction in an autoclave at 453 K for 2–7 days. After drying and washing the tube-shaped vanadium oxide with mixed valence represents nanoparticles with diameter 15–150 nm (5–50 nm inner diameter) and 15 µm long. The tubes consist of 2–30 layers of a crystalline vanadium oxide and amine molecules intercalated in between. The distances between the layers of about 1.7–3.8 nm are proportional to the amine length and are functioning as a structurally guiding template.

It may be concluded that the classification of materials produced by the sol–gel synthesis was formulated on the base of bonding type between the organic and inorganic components. According to [70, 71], the organic groups introduced into $R'_nSi(OR)_{4-n}$ may fulfill two functions: as modifying the network and forming the network in case of silica-based materials – ORMOSIL (Organically Modified Silicates) and ORMOCER (Organically Modified Ceramics). Low-percentage polycarbosilanes contain 2–30 vol.% of ceramics while those with high-percentage acquire 45–75 vol.%. For example, the material named "rubber ormosil" based on TEOS and polydimethylsiloxane with $M_w = 1700$–2000 contains more than 70 vol.% inorganic components [57].

Classification of the sol–gel technique [104] based on the structure, the forming network sources and bonding type is given in Table 7.3.

7.2 A Combination of Polymerization Reactions and In Situ Sol–Gel Synthesis of Nanocomposites

Convenient method for production of hybrid composites is based on polymerization-induced transformations of hybrid monomers. The compounds with an inorganic component may serve as precursor and a polymer-generating group.

[2] The sol–gel process with application of a highly porous silica gel in a polar solution can also be used for microencapsulation of various large molecules like dyes (e.g. [88, 90]), photochromic fluorescence substances, scintillators [15], porphyrins encapsulated into a sol–gel matrix generated in combined hydrolysis of $Si(OEt)_4$–$Zr(OBu)_4$, enzymes, proteins and so on [88, 91, 92]. The methods of their production and structure are akin to metal–polymer nanocomposites. In particular, a xerogel is impregnated into MMA [93, 94] in which an inorganic dye can also be dissolved (of perylene type) and upon UV or thermal polymerization a co-impregnated composition of SiO_2-polymethylmethacrylate-dye is received. Another variant is to topochemically incorporate dyes inside the silica with nanometer precision of the radial position [95–97]. These materials are termed organically doped/organically modified sol–gel materials [98, 99].

Table 7.3. Classification of the chief sol–gel methods [104]

Supporting Structure	Bonding Fashion	Source
(1) Colloid	Particles interconnected by Van der Waalse or hydrogen bonding	Mostly from oxide or hydroxide sols
(2) Metaloxane polymer	Inorganic polymers interconnected by Van der Waalse or hydrogen bonding	Hydrolysis and condensation of metal alkoxides
(3) Carbon-backbone polymer	Organic polymers interconnected by Van der Waalse or hydrogen bonding	Covalent polymers largely used in organic chemistry
(4) Metal–complexes	Associates weakly interconnected by Van der Waalse or hydrogen bonding	Concentrated metal complex solution; e.g. amorphous citrate method
(5) Polymer–complex		
in situ polymerizable complex (IPC)	Organic polymers interconnected by coordinate, Van der Waalse or hydrogen bonding	Polymerization between α-hydroxy carboxylic acid and polyhydroxy alcohol in the presence of metal complexes; e.g. Pechini method
polymer-complex solution (PCS)		Coordinating polymer, e.g. polyvinyl alcohol, and metallic salt solutions

The compounds of this kind are called metal-containing polymers whose synthesis and polymerization transformations was summarized in [105]. The pioneering work was devoted to the synthesis of titanium trialkoxide methacrylate monomers $CH_2{=}C(CH_3)COOTi(OR)_3$ in which R are Bu, i–Pr, t-Bu, t–amil or 2-ethylhexyl [106, 107]. Titanium alkoxides are inclined to associations with nucleophilic addition of a negatively charged OR group and positively charged metal atom M. The degree of association depends on the reaction conditions (especially the nature of the solvent and temperature) and the nature of alkyl groups. NMR ^{1}H, ^{13}C and IR spectroscopy proved the existence of different structures in which the methacrylate group formed a bridge bond (Scheme 7.2).

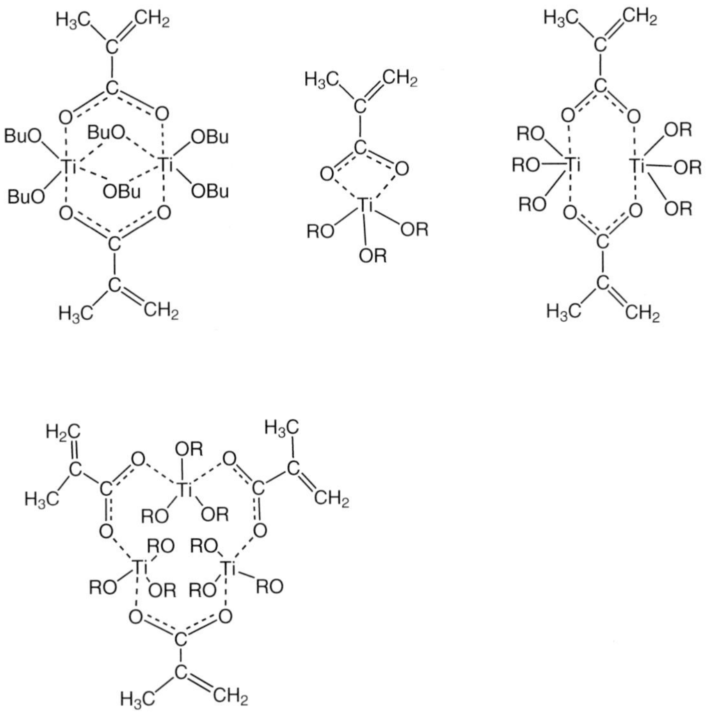

Scheme 7.2

It is relevant to use vinyl derivatives of the $CH_2{=}CHSi(OEt)_3$ type [90, 108, 109], including the single-stage production of organically functionalized mesoporous silica MCM-41 to produce ormosils (organically modified silica-based materials) [110]. Metal clusters like $[(RO)_nM]_xY$ where Y is a polymer-forming organic group $x \geq 2$ are the most applicable for these purposes [15, 111].

Furthermore, methacrylate-substituted tetranuclear titanium, zirconium and tantalum oxide clusters $Ti_4O_2(OPr^i)_6)OMc)_6$, $Zr_4O_2(OMc)_{12}$, $Ta_4O_4(OEt)_8(OMc)_4$ were synthesized. Their formation mechanism is rather intricate. They are brought to substitution of one or more alkoxide ligands for the methacrylate groups after which the liberated alcohol reacts with the excess of the acid yielding an ether and latent water. The latter hydrolyzes the unreacted alkoxide groups and forms oxide and hydroxyl groups in a cluster. Since these processes are comparatively slow, they allow one to control the growth of carboxylate-substituted oxometallic clusters. The nuclearity and shape of the cluster depend on the relationship of its initial components and the nature od the OR group in the alkoxide. Cluster monomers of other types have been generated, e.g. $Hf_4O_2(OMc)_{12}$, $Nb_4O_4(OPr^i)_8(OMc)_4$ [112] of higher nuclearity (Ti_6, Zr_6, Ti_9) and many mixed-metal oxide clusters, including titanium/zirconium by this procedure [113–115]. The structure of

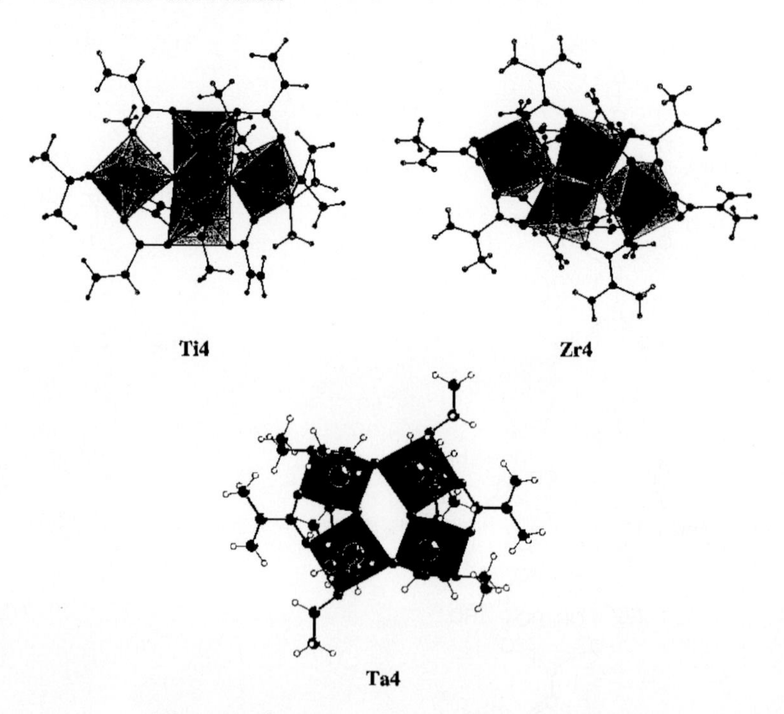

Fig. 7.1. Structure of polynuclear complexes

these formations has been comprehensively examined. The Ti_4 cluster consists of octahedral sections that are condensed to a larger extent than Ta_4 because two central octahedra use facets more readily than angles for construction (Fig. 7.1). The other two octahedra are linked via this facet through μ_3-oxygen [115, 116]. The lot of six methacrylate groups is to balance the charges and coordination places of metal atoms. The structure of Zr_4 is similar to titanium with the exception that its central atoms are seven-coordination and the outer ones are eight-coordination [117]. So far, the substitution degree by bidentant carboxylate groups is higher than in Ti_4. The molecular structure of Ta_4 is a centrosymmetrical cycle of four octahedra linked by angles [115, 116]. The methacrylate groups form a square planar surrounding perpendicular to the cluster backbone.

To obtain sol–gel products by the polymerization method along with unsaturated carbonic acids, unsaturated alcohols have come into use recently. One of the first investigations in this sphere was the production of alkoxy-derivatives $Ti(OR)_3(OR')$ [118, 119], $VO(OR)_{3-n}(OR')_n$ [120, 121], where OR′ is the residue of the unsaturated alcohol ($R{=}Pr^i$, $R' = \cdots CH_2C{\equiv}CH$, $CH_2CH{=}CH_2$, $(CH_3)_2CC{\equiv}CCH{=}CH_2$ and $(CH_2)_2OC(O)C(CH_3){=}CH_2$). Following the reaction $M(OR)_n + mR'OH \rightarrow M(OR)_{n-m}(OR')_m + R'OH$ in conditions of the isolating alcohol removal as, e.g. azeotropic solution, many polymerizing alkoxides were derived Ti^{4+} and V^{5+}.

Based on *i*-eugenol, 2-methoxy-4-propenylphenol the corresponding titanium alkoxides was synthesized [122, 123]. Using alkoxyzirconia derivatives even more complex products of the kind $\{(Zr(OPr^i)_2[(OC_6H_3)(OMe)_2(CH_2CH{=}CH_2)_4](\mu\text{-}OPr^i)\}_2$ have been obtained and their crystallographic structure examined [124].

A hybrid material can be produced by the reaction of metal amides as well. For example, it is possible to use $Ti(NMe_2)_4$ with hydroquinone [125], or $Zr(NMe_2)_4$ reacting with 2,6-dimethylphenol or 2-methoxy-4-propenylphenol [124,126], etc. [127]. Alcoholysis of $Ti(O\text{-}i\text{-}Pr)_4$ hydroquinone has a covalent three-dimensional Ti^{4+}–quinone network [128]. Finally, the hybrid material has unique electronic properties produced by yttrium isopropoxide condensation with organic diols [129] (Scheme 7.3). A specific interest for functionalization of alkoxy-derivatives of different metals or their nanoparticles presents 2-hydroxyethyl methacrylate (HEMA). Thanks to its OH side group, HEMA is a good co-solvent of TEOS and water and its viscosity remains low enough to achieve a good mixing level. To receive hybrid nanocomposites by polymerization, it is possible to use hybrid macromers [130–132]. Polyhedral oligosilsesquioxane-based (POSS) macromers containing a roughly spherical inorganic core (Si_8O_{12}) with approximately 1.4 nm diameter are mostly used (Fig. 7.2). Methacryloyl-functionalized POSS (MA-POSS) monomers are a

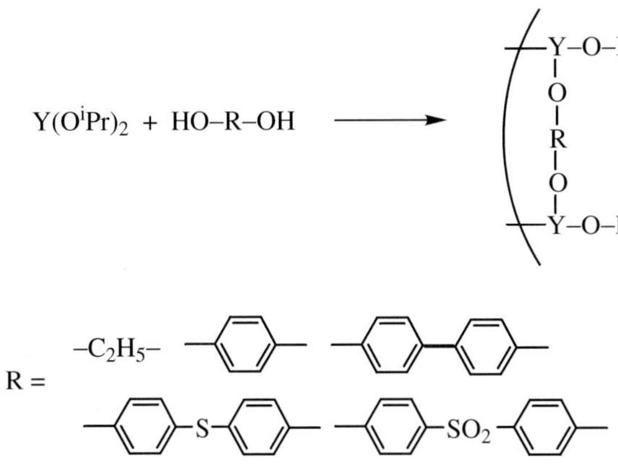

$Y(O^iPr)_2$ + HO–R–OH → $[$–Y–O–R–O–Y– (with –O–R–O– bridge) $]_n$

R = $-C_2H_5-$, –C₆H₄–, –C₆H₄–C₆H₄–, –C₆H₄–S–C₆H₄–, $-C_6H_4-SO_2-C_6H_4-$

Scheme 7.3

class of compounds which recently attracted interest for hybrid materials. MA-POSS monomer can be easily copolymerized with other methacrylate monomers (Fig. 7.3) (1,6-hexamethylene dimethacrylate, triethyleneglycol dimethacrylate, 2-hydroxyethyl methacrylate (HEMA)). It could be used to

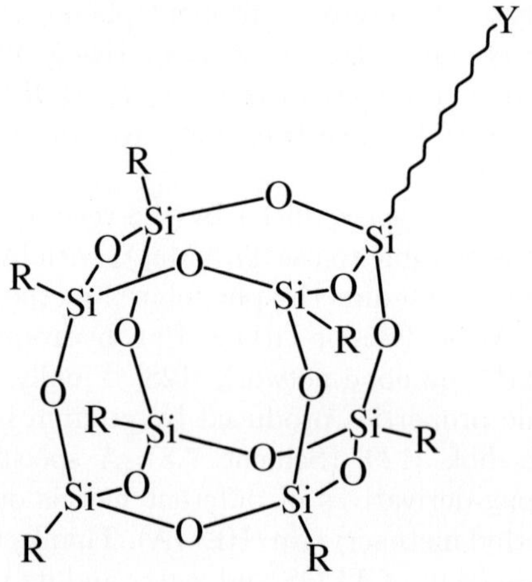

Fig. 7.2. Structure of POSS-based macromers

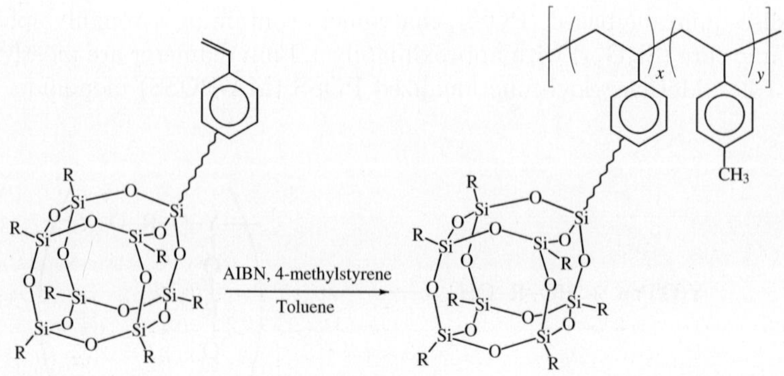

Fig. 7.3. Schematic diagram of producing a new type of hybrid POSS-based materials

decrease the shrinkage of dental composites on multi-methacrylates[3] [132]. Incorporating only 5 mass% MA-POSS reduced shrinkage of the prepared neat resins substantially. The formation of hybrid nanocomposites was reviewed in [133] and Fig. 7.4 shows the correlation of the radical polymerization reaction of N,N-dimethacrylamide (DMAAm) and acidic catalysis of TMOS hydrolysis-condensation. A polymer derived from DMAAm/TMOS = 1:2 blend and 1% of initiator has the number-average molecular mass 76 000. The composite displays high homogenization degree and forms aggregates

[3] Substantial reductions in volume (shrinkage) during polymerization of methacrylate monomers, which induces strain in the matrix or dental resorptive, and the low double bond conversion, are two factors which limit the clinical application of methacrylate-based dental composites.

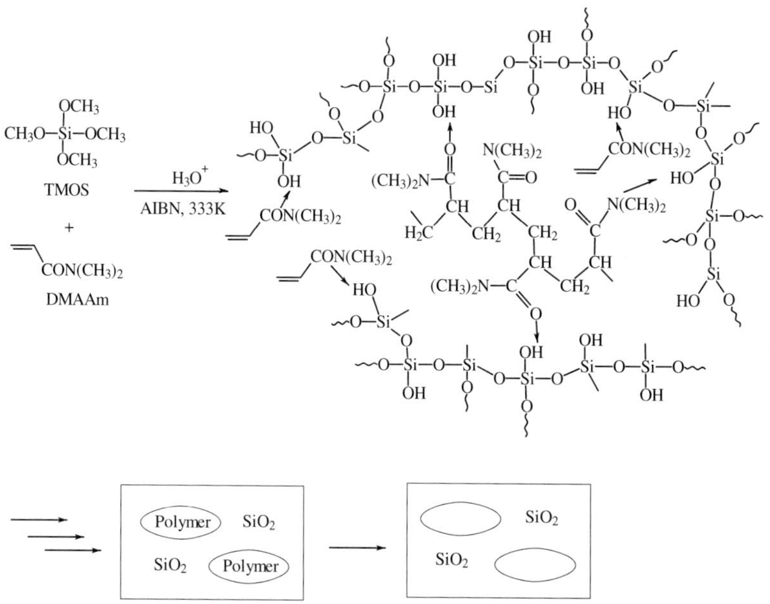

Fig. 7.4. Typical procedure for obtaining hybrid nanocomposites

with domains more than 30 nm in size. Its specific surface area and pore volume obtained by charring the polymer hybrids at 870 K depend on the production conditions of the composite, and the amount of acid, and reached 365 $m^2 \cdot g^{-1}$ and 83.9 $mL \cdot g^{-1}$, correspondingly. Two types of interpenetrating polymer networks (IPN) are formed during this process. One is generated through spontaneous polymerization of N,N′-dimethylacrylamide and N,N′-methylenebisacrylamide in methanol solution of TMOS. The other is derived by copolymerization of styrene with divinylbenzene in the presence of TMOS [134]. It is believed that the siloxane lattice is so rigid that it cannot be affected by temperature [28]. Therefore, the pore size was expected to be comparable to that of organic polymer domains in the original hybrids. This was confirmed for dendrimer polymer hybrids [135].

Polymer silanes can be easily synthesized by condensation of the carboxyl-terminated polymer with 3-(trietoxysilyl)propylamine and by the radical polymerization in the 3-(trimethoxysilyl)propylthiol as a chain transfer reagent [136, 137] (Scheme 7.4). The other, less widespread, group of monomers used for sol–gel synthesis in polymerization transformations represents such compounds as $(RO)_3M$–X–A in which the A group is capable of polymerization (polycondensation) or cross-linking reactions [138]. Most often a network-forming epoxy group is used for these purposes, e.g. 3-(glycidoxypropyl)trimethoxysilane [139–142] as a compatibilizing binder for TEOS with polyacids.

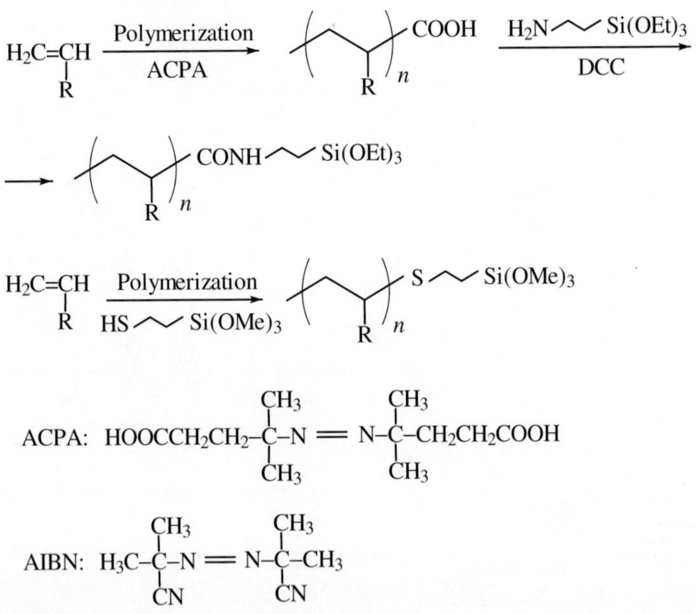

$$H_2C{=}CH(R) \xrightarrow[\text{ACPA}]{\text{Polymerization}} \left(CH_2{-}CH(R)\right)_n{-}COOH \xrightarrow[\text{DCC}]{H_2N{\sim\sim}Si(OEt)_3}$$

$$\longrightarrow \left(CH_2{-}CH(R)\right)_n{-}CONH{\sim\sim}Si(OEt)_3$$

$$H_2C{=}CH(R) \xrightarrow[\text{HS}{\sim\sim}Si(OMe)_3]{\text{Polymerization}} \left(CH_2{-}CH(R)\right)_n{-}S{\sim\sim}Si(OMe)_3$$

ACPA: $HOOCCH_2CH_2{-}C(CH_3)_2{-}N = N{-}C(CH_3)_2{-}CH_2CH_2COOH$

AIBN: $H_3C{-}C(CH_3)(CN){-}N = N{-}C(CH_3)(CN){-}CH_3$

DCC: N,N-dicyclohexylcarbodiimide

Scheme 7.4

PEO[4] is used rarely in network formation. Aluminum alkoxide-derivatives (Fig. 7.5) convert epoxysilanes into network polymers assisted by the catalytic effect of aluminum γ-oxide particles.

Isocyanate derivatives of alkoxides, e.g. 3-(triethoxysilyl)propyl isocyanate, are introduced and are often used to modify sol–gel products [143]. In [144] a method is described for substitution of an alkoxy-group of 3-isocyanatopropyl triethoxysilane for hexa(methoxymethyl)melamine to produce hybrid optical materials on its base.

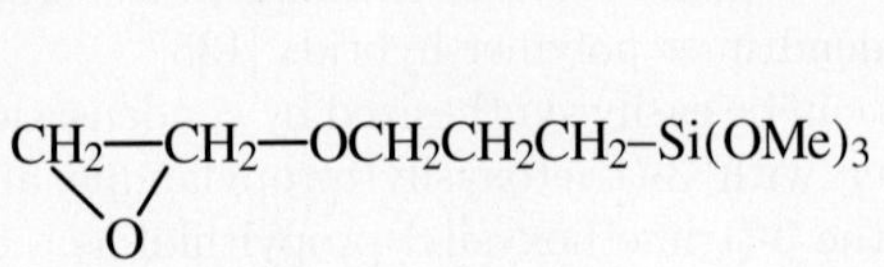

$$\underbrace{CH_2{-}CH}_{O}{-}CH_2{-}OCH_2CH_2CH_2{-}Si(OMe)_3$$

A free-radical copolymerization of methacrylate titanium trialkoxides with MMA was carried out at 313–343 K temperatures in toluene under the

[4] PEO is used to solvate minor cations like lithium. In combination with oxopolymers it is used in sol–gel synthesis to obtain polyelectrolytes with ionic conductivity.

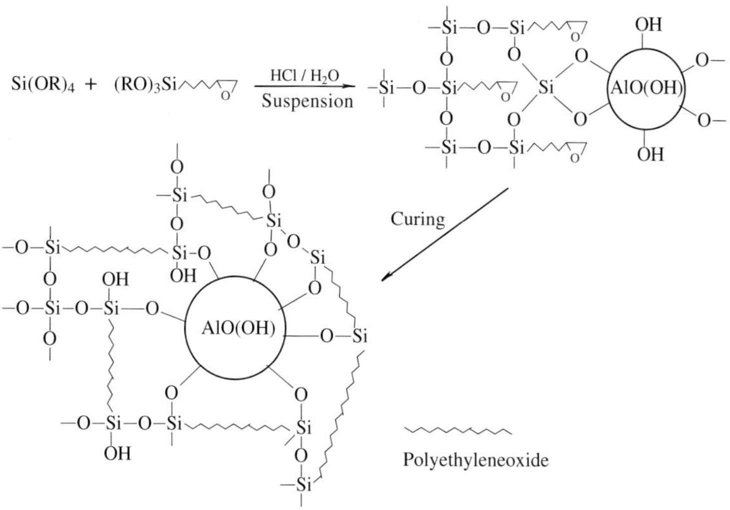

Fig. 7.5. Reactions and structural models of network systems based on Al^{3+}

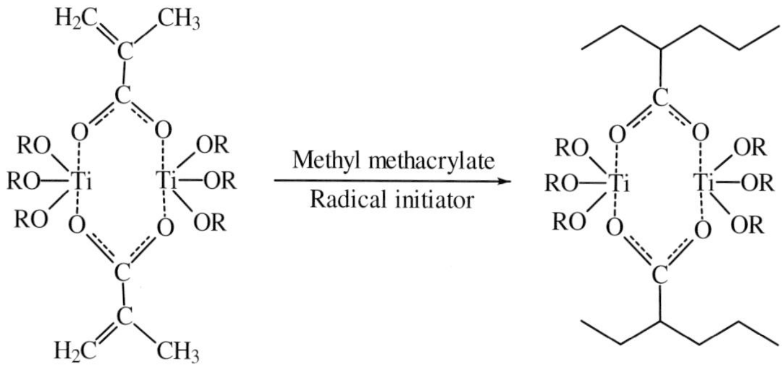

Methyl methacrylate
Radical initiator

Scheme 7.5

AIBN effect [44] when the total concentration of monomers is 0.5–2 mol·l^{-1}. 3D structures are derived when polynuclear clusters are used (Scheme 7.5).

The active inorganic surface furnishes a wide range of control over the radical polymerization process. Notice that structurally well-defined polymer–nanoparticle hybrids can be obtained through integrated nanoparticle synthesis with controlled radical polymerization [145–148]. In particular, a microinitiator is immobilized on the surface of SiO_2 particles and in the presence of CuCl/4,4′-di(5-nonyl)-2,2′-bipyridyl the atom transfer radical polymerization (ATRP) of styrene is performed [149]. Properties of the thus synthesized nanocomposite are significantly better than of the one obtained by traditional grafting polymerization of corresponding monomers. A similar method consisting of binding (CuX pyridylmethanimine) on SiO_2 particles was used for ATRP methyl methacrylate [150].

An uncommon variant of obtaining analogous materials is the thermal ring-opening polymerization route to high molecular weight poly(ferrocenylsilanes) – a novel class of ferrocenophane–silaferrocenophane materials [151]. A material with unusual magnetic properties (see Chap. 11) can be obtained by the anionic ring-opening polymerization with BuLi. This includes a polysiloxane block (Pt^0 catalyst, toluene) yielding a block-copolymer [152] (Scheme 7.6).

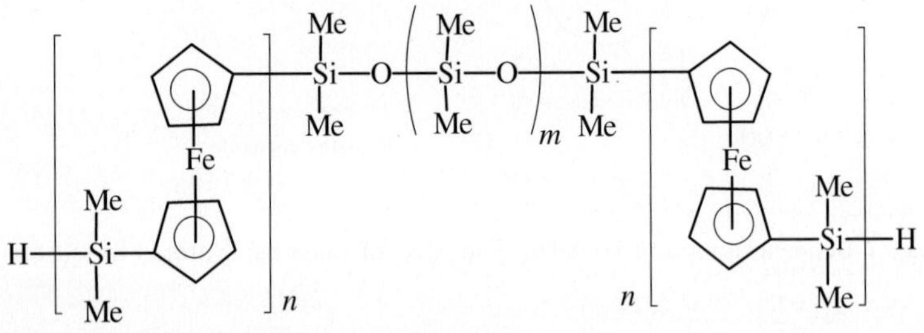

Scheme 7.6

When this poly(ferrocenylsilane)-block-poly(dimethylsilane) (PFS-b-PDMS) is dissolved in hexane it self-assembles into cylindrical, wormlike micelles with a PFS core and a corona of PDMS [153] (Fig. 7.6). Mesoporous materials are unique in their homogeneity degree (2D and 3D hexagonal symmetries, pore diameters 2.7 and 3.1 nm, surface areas 750 and 1170 $m^2 \cdot g^{-1}$). They are formed in case the polymer-forming, but inorganic sections are found within one molecule (organic fragments and inorganic oxide within the framework), which is exemplified in 1,2-bis(trimethoxysilyl)ethane $(CH_3O)_3Si\text{-}CH_2\text{-}CH_2\text{-}Si(OCH_3)_3$ [154].

A large group of monomers such as (trimethoxysilyl)ferrocene derivatives used in sol–gel synthesis is reviewed in [155]. The precursors of a more complex composition, like silylated chalcogen of the type $E(R)SiMe_3$ (E=S, Se, Te) [156] can be used too.

Although copolymerization of monomers, one of which contains trialkoxysilyl groups, was studied [157, 158], there is still a lack of products derived, hydrolytic instability of trialkoxysilyl groups introduced into the polymers. For instance, (methacryloxypropyl)trimethoxysilane gets hydrolyzed and condensed as a 0.5 N HCl is added. The product of interactions of equimolar amounts of $Zr(OR)_4$ and MAA is mixed with a controlled amount of condensed silane in necessary proportions and a rated content of H_2O is added. An optimum quantity of photoinitiator is introduced in the thus formed mixture and an amount of alcohol is added to regulate viscosity. A

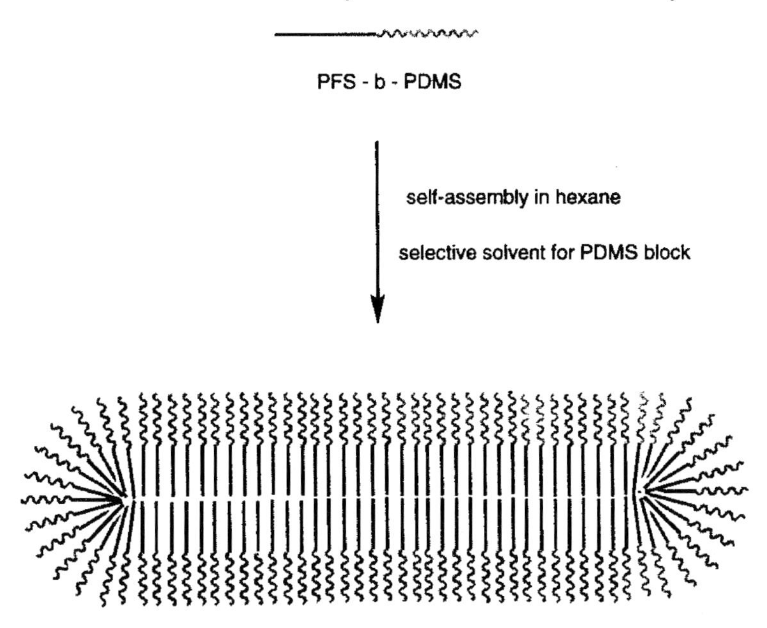

Fig. 7.6. Self-assembly of PFS-b-PDMS into cylindrical micelles in hexane

thin film of a photosensitive material is applied by dipping. The structure of the composite is illustrated in Fig. 7.7.

In a single stage the polytitanosiloxanes was produced via a combined controlled hydrolysis of $Si(OEt)_4$ and $Ti(O\text{-}Pr^i)_2(acac)_2$ [159]. A ladder-like polymer is lined up and contains the chain-forming sections of Si–O–Si and Si–O–Ti whose relationship depends upon synthesis conditions and can reach 10. It governs the jellification time and gives the chance of fiber spinning. The ceramic fibers are produced by the material annealing at 770–1170 K.

The dispersion polymerization of respective prepolymers is made by a more convenient method of synthesizing the described nanocomposites [49]. For example, silicone-organic monomers are synthesized and copolymerized (Fig. 7.8). Beads 100–500 nm in size are formed by statistic copolymerization, which function, on heating to 1270 K, as dimensional templates for removing organics. The final particles are about 100 nm in size. Hence, the general production scheme of nanoceramics with polymerizing of a respective monomer can be brought to the following [145] (Fig. 7.9).

7.3 Sol–Gel Syntheses in the Presence of Polymers

Polymer modification of colloidal particles, including adsorption, coating, or grafting, is a set of convenient and practical procedures intended to modify surface characteristics of particles. Many combining sol–gel processes with

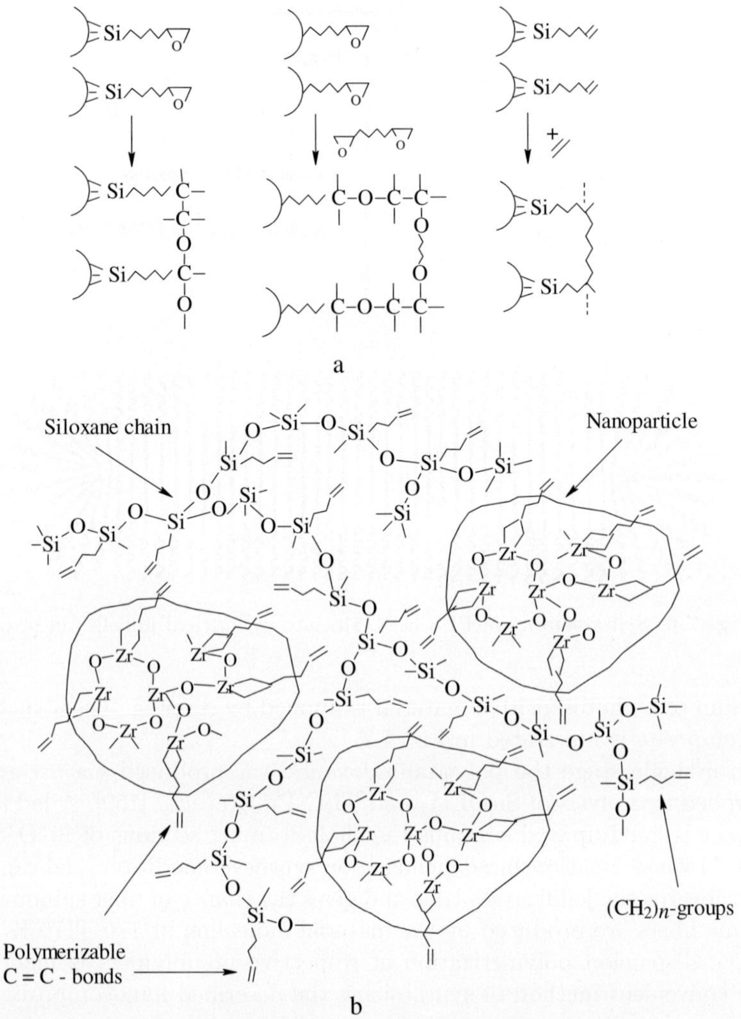

Fig. 7.7. General directions of reactions in case of functionalized silanes which are used for organo–inorganic cross-links (**a**), and the model of the ZrO_2–methacrylic acid–methacryloxysilane system (**b**)

organic polymers in their own solutions were developed. To this type of materials belongs class 5 of the classification set forth elsewhere [104] and presented in Table 7.3. It includes inorganic metal–oxide polymers and stiff organic polymers.

The organic polymer can be formed in situ (in situ polymerization complex, IPC) or by using a polymer capable of binding metal ions, polymer complex solution (PCS).

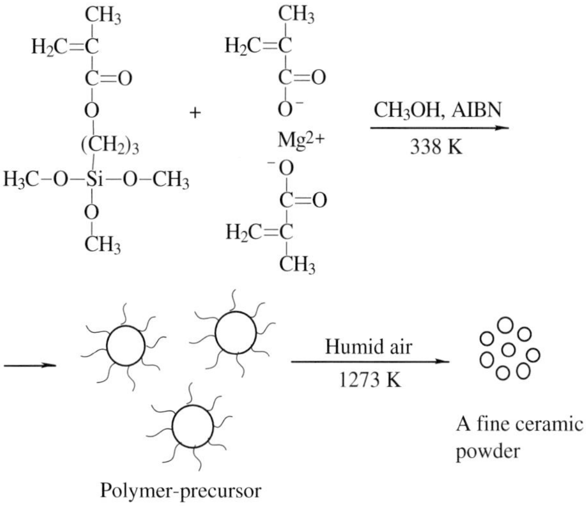

$CH_2=C(CH_3)-C(=O)-O-(CH_2)_3-Si(OCH_3)_3$ + $[CH_2=C(CH_3)-C(=O)-O^-]_2 Mg^{2+}$ $\xrightarrow[338 K]{CH_3OH, AIBN}$ Polymer-precursor $\xrightarrow[1273 K]{Humid air}$ A fine ceramic powder

Fig. 7.8. Nanocomposites produced by copolymerization of prepolymers

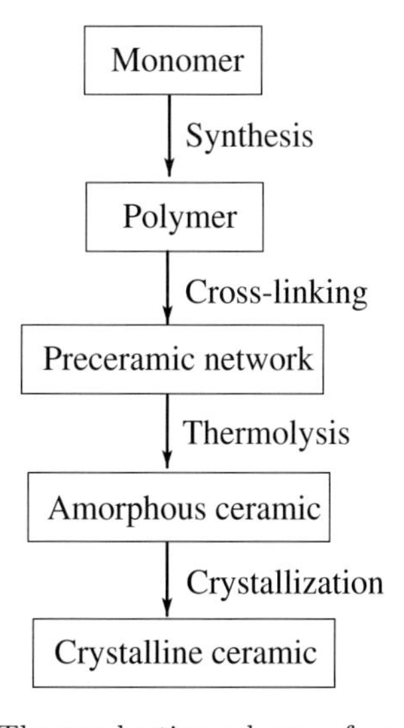

Fig. 7.9. The production scheme of nanoceramics

Systems such as polymer/TEOS can be used for sol–gel synthesis. They are based on various macromolecules: PMMA, poly(n-butylacrylate) [160, 161], PVA (polyvinyl alcohol) [162], PEO [163], PVA with (poly(vinyl acetate) [164–166], PVP [160, 167], poly(N,N-dimethylacrylate) [160], silicone rubber [168], poly(amide-ethylene oxide) [169], poly(ethylene oxide-co-epichlorhydrin) [170,171], Nafion [172–175], poly(ethyloxazoline), polyphosphazine [176] and many other.

The microstructure of these composites is regulated usually through formation of weak van der Waalse and hydrogen bondings or hydrophilic and hydrophobic interactions. An alternative to this is polymer embedding (or its precursor) into the oxogel during mixing of metal alkoxides, or its impregnation inside the lattice pores of the oxide xerogel (see Chap. 8). Many modifications of coating nanoparticles are offered today based on sol–gel synthesis with polymer latexes. A reverse process is described in [177]. It consists of covering latex particles with the forming nanoparticles, e.g. as a result of $Ti(OBu)_4$ hydrolysis. The coating thickness is regulated by the components. Hollow TiO_2 particles are derived upon calcination (520–1170 K) that changes their morphology with temperature rise within the series amorphous→ rutile → anastases.

The filler particles in conventional polymer–inorganic systems are usually subjected to some surface treatment with the coupling agents prior to mixing with the resins. This happens because of the thermodynamic incompatibility between the organic polymer matrices and inorganic fillers.

The hybrid materials produced during sol–gel synthesis are formed via interactions between surface hydroxyl groups of inorganic particles and active polymer groups. A schematic diagram of the interaction where polymers are a mediation agent is presented in Fig. 7.10. Interactions between phases of inorganic and polymer components play a leading part in the process [178, 179]. This path is a homogeneous formation of nanosize particles of the filler in the polymer medium. Strong interactions are observed in case with poly(vinylbutyral) – PVB – which is sturdy and flexible, with high impact strength at low temperature [180]. PVB has excellent adhesive properties with many materials such as glass, metal, plastics, and wood. PVB/titanium composites prepared by means of the sol–gel process showed good compatibility, strong interaction, and excellent mechanical properties.

Let us examine the hybrid network nanocomposites whose organic and inorganic components are strongly linked through covalent and ion chemical bonds. We shall consider two approaches used to synthesize networks of this type. The first of them is a sequential formation of secondary networks

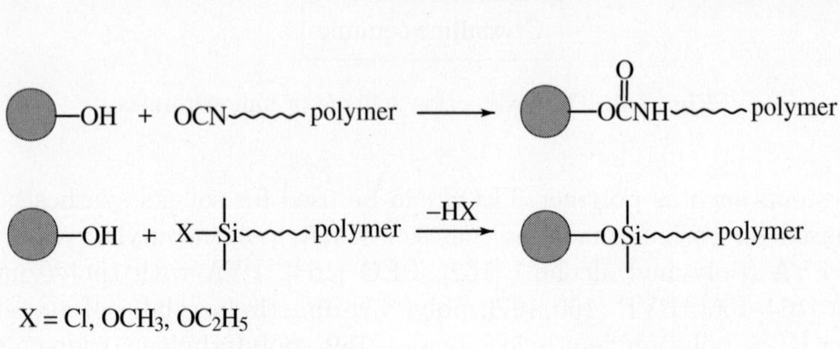

—OH + OCN~~~polymer ⟶ —OC(=O)NH~~~polymer

—OH + X—Si~~~polymer $\xrightarrow{-HX}$ —OSi~~~polymer

X = Cl, OCH_3, OC_2H_5

Fig. 7.10. A schematic diagram of interactions between functional groups of ingredients

in preliminary formed and properly functionalized initial ones. Rarely functionalized inorganic macromers or oxopolymers are used. The second procedure consists of the simultaneous formation of two different networks including interpenetrating ones from molecular precursors showing at once the organic and inorganic functionality. They react by different mechanisms: polyaddition, polycondensation, and hydrolysis–condensation. This was realized in synthesis of polyisobutylene/poly(dimethylsiloxane) bicomponent networks [181].

This problem is complex, and investigations in this field have just begun. Many aspects are far from being completely clear.

The primary approaches were based on the use of polymer solutions, in which case gelation was brought about uncontrolled phase separation and resulted in formation of an inhomogeneous material. The analogous path is taken for preparation of a thin interfacial layer between surface-oxidized aluminum, iron, porous silicon via alkoxyzirconium complex interface [182] as well as between these surfaces and poly(ethylene-*co*-acrylic acid) via tetra(*tret*-butoxy)zirconium [183]. The sol–gel materials are derived by dissolving HO–PES–OH (PES – polyestersulfone) and TEOS or TMOS in DMFA. The hydrolysis-condensation of tetraalkoxysilanes catalyzed by acids leads to formation of the products by the scheme presented below:

$$\equiv \text{Si–OR} + \text{HO–PES–OH} \rightarrow \equiv \text{Si–O–PES–OH} + \text{ROH}.$$

The fact of intensive interactions between inorganic and polymer chains complies with the results [184]. It has been proved, particularly, that poly(ester-sulfone) sections are cross-linked by alkoxysilane bridges wherefrom nanoparticles are formed during alkoxysilane condensation together with the bridge groups (Fig. 7.11).

When the polymer exhibits low surface energy and is non-porous, e.g. PTFE films, it is convenient to use [185] surface activation by silica deposition produced in situ from $SiCl_4$ and water without any chemical degradation or modification of the polymer bulk. This approach improves essentially silica adhesion to the fluorine polymer thanks to $SiCl_4$ and water diffusion into the polymer surface. Chemical interactions between the polymer organic and inorganic components yielding covalent bonding has been observed in the sol–gel process of the styrene copolymer with maleic anhydride TEOS in the presence of the binding agent (3-aminopropyl)triethoxysilane [187, 188]. The resultant particles were less than 20 nm in size (see Scheme 7.7). The organic copolymers act usually as compatibilizers for organic–inorganic composites. $Ti(OBu)_4$ or allyl acetylacetone (3-allyl-2,4-pentanedione) [189] may be used as a coupling agent. Novel organic-inorganic chemical hybrid fillers for dental composite materials prepared by the sol–gel reactions of poly[methyl methacrylate-*co*-3-(trimethoxysilyl)propyl methacrylate] [190] have been developed lately.

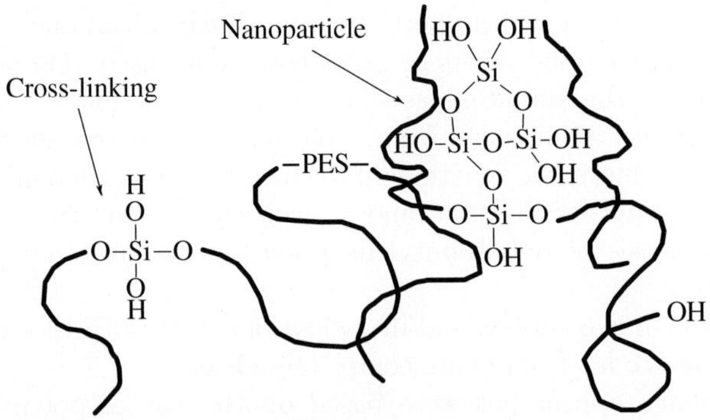

Fig. 7.11. Cross-linking of poly(estersulfone) sections via alkoxysilane bridges

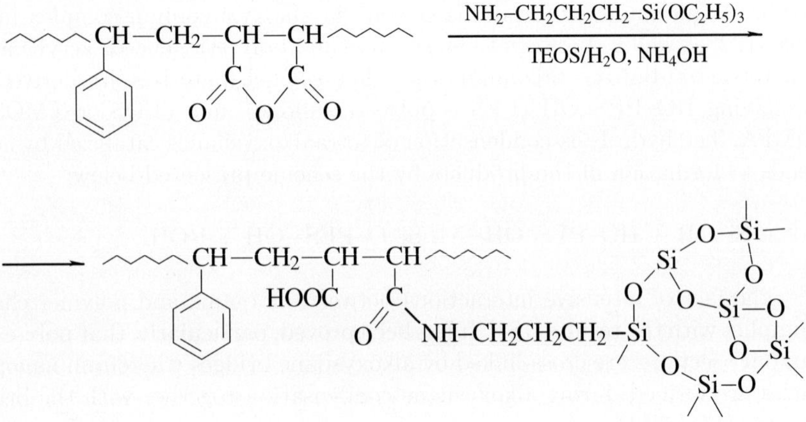

Scheme 7.7

The polymethacrylate chains are uniformly distributed in these hybrid fillers and covalently bonded to the silica networks at the molecular level without macrosopic organic-inorganic phase separation. In contrast to the standard dental resin based on 2,2-bis(p-2-hydroxy-3-methacryloxypropoxyphenyl)propane/triethylene glycol dimethacrylate filled by derived in the sol–gel procedure fine-dispersed there is a stronger interfacial bonding with the polymer matrix than with the pure silica fillers. This happens because the surface of the hybrid fillers inherently contains the polymer components that are structurally similar to the dental resins and polymer matrix. The requisite condition of homogeneity of the forming polystyrene-product hybrid in the sol–gel reaction of phenyltrinetoxysilane (PhTMOS) is, probably, the π–π interaction between phenyl groups of PS and those of the silica gel [191]. Evidently, composites based on poly(isoprene-b-ethylene-oxide) phase-separated block copolymer mixed with the alkoxides (3-glycidyloxypropyl)trimethoxysilane and aluminium butoxide in a sol–gel process [192–194] acquire still more structural organization. Polyethoxyethylene molecules are known to fix [75, 195] on metal oxide surfaces owing to H-bonds of oxygen atoms with OH surface groups even through the layers of coordinately bound aqua groups. A strong hydrogen-bonding interaction is observed usually between the inorganic and organic phases [196, 197]. Sometimes, both components are subjected to modification, the mineral precursor and the polymer to produce hybrid gels of silica/PEO type. To obtain efficient luminescent materials, often [198] 3-isocyanotriethoxysilane with O,O′-bis(2-aminopropyl)-poly(ethylene oxide) are used. When it is swollen in the ethanol solution of terbium nitrate and 2,2-bipyridine, it binds both lanthanide ions and the ligand. Similarly, hybrid materials are produced incorporating not only the luminescent but also redox-active or catalytic-active molecules [91, 199, 200], oxygen-selective inorganic–organic hybrid membranes containing salcomine as an oxygen carrier [201], etc. Formed in situ sol–gel process poly(amide-imide) (PI)/TiO_2 [202] (sizes domains of TiO_2 were from 5 to 50 nm, content increased from 3.7 to 17.9 wt.%) exhibits the formation of hydrogen bonds between the amide groups in PI and hydroxyl groups on the inorganic oxide.

Polyimide composite materials with SiO_2 or TiO_2 nanoparticles display high mechanical strength when the 3D inorganic networks are formed [203–206]. A specific variant of obtaining analogous composites is the use of precursors for yielding polyimide-polysilsesquioxane composites (PI-POSS) [207, 208]. In particular, 1,1-bis(4-aminophenyl)-1-phenyl-2,2,2-trifluoroethane is condensed with a pyrromelyte anhydride (PA) derivative and aminophenyltrimetoxysilane (simultaneously with imidization at 520–620 K) [209, 210] resulted in films of organic and inorganic hybrids (Scheme 7.8) containing 32–70% of homogeneously dispersed SiO_2 with particle size 0.5–7 nm.

Similar composites [15] were obtained based on analogues of polyimide polyoxazolines, along with those functionalized by triethoxysilane groups [211]. An optimum variant of SiO_2–polymer nanocomposite synthesis imposing minimum shrinkage is the metathesis polymerization with cycle opening and a free-radical addition of cyclic alcohols [45, 212, 213]. The interpenetrating networks are formed synchronously as a result of the competing polymerization and hydrolysis with silicone alkoxide condensation under the effect of a nucleophilic catalyst NaF (Fig. 7.12). The last component in PVP–TiO_2 composites [214] quickly catalyzes at the stage of calcination due to intensive interactions with the polymer. The same result is attained when organogelators are added at the sol–gel reaction stage [215]. These monomers get copolymerized in case of MAA-modified silanes. The silicone-containing polymer is used as a matrix material for the nanosize zirconium particles [216].

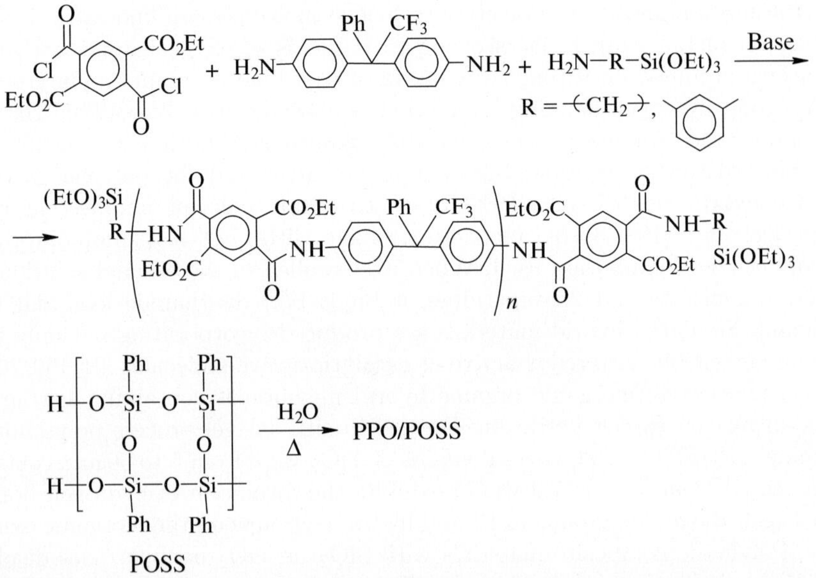

Base

$R = -(CH_2)_3-$,

PPO/POSS

POSS

Scheme 7.8

Furthermore, SiO_2/TiO_2 nanocomposites based on perfluorosulfonate ionomers are produced by sol–gel synthesis with particle size $\approx$ 10–50 nm [217]. Nanocomposites based on $Ti(OR)_4$, $Zr(OR)_4$ are derived from the styrene copolymer and 4-vinylphenol matrices [218].

Metal alkoxides (mostly of titanium and silicone) serve as cross-linking reagents for many natural polymers, such as polysaccharides, cellulose materials, and vegetable oil derivatives [219, 220]. They contain highly active hydroxyl groups capable of forming oxopolymers. Cocondensation of the macromers, containing trialkoxosilyl groups, with polymers like functionalized PS, polyoxozolin, PI, PEO, polyesteroketones, or PMMA containing end functional groups, is a convenient method of producing telechelate polymer networks [209, 221–226]. This also relates to polytetramethylenoxide whose end groups are modified by triethoxysilyl groups [227].

$$(EtO)_3Si–(CH_2)_3–O–[–(CH_2)_4–O–]_n–(CH_2)_3–Si(OEt)_3.$$

Esterification reactions of PAA and $Ti(OR)_4$ (Scheme 7.9) are proposed as an alternative method, consisting of polymer-analogous transformations, for titanium alkoxyl derivatives copolymerization with acrylic acid.

When the molar ratio of Ti/COOH < 1, i.e. there is a molar default $Ti(OR)_4$, then three-dimensional structures of cluster-reinforced polymers are formed (Scheme 7.10).

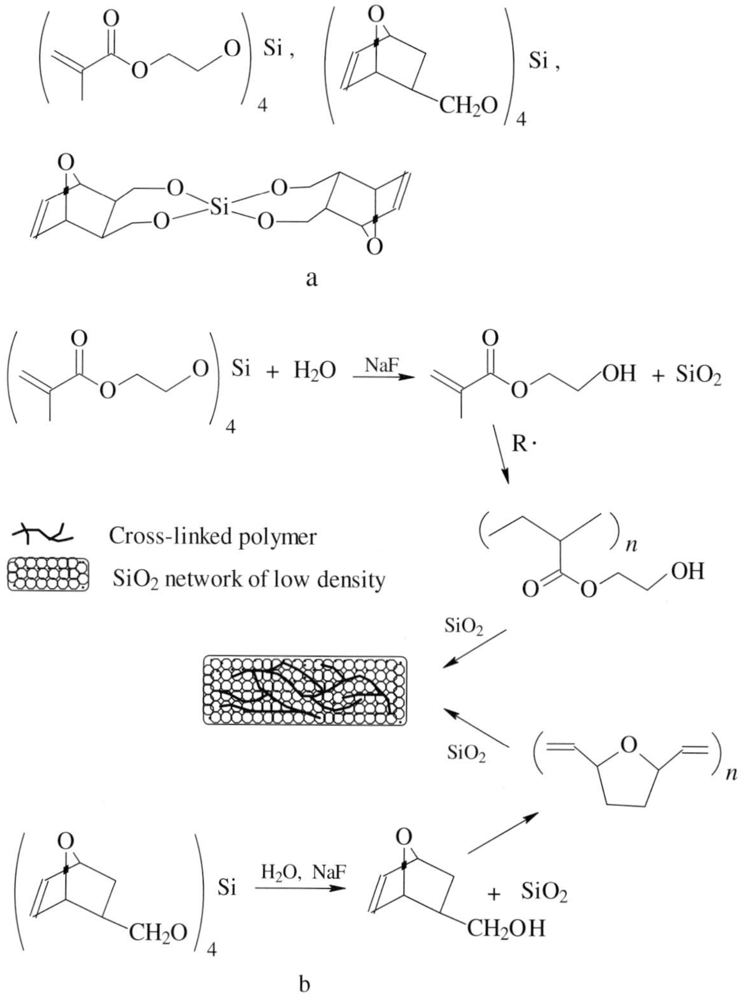

a

Cross-linked polymer

SiO_2 network of low density

b

Fig. 7.12. Modified Si alkoxide precursors (**a**) and the reaction scheme of the forming organo-inorganic interpenetrating networks (**b**)

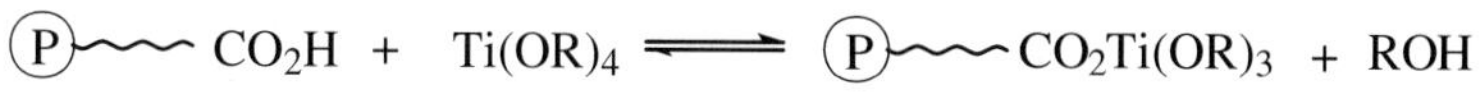

$$\text{(P)}\sim\sim CO_2H + Ti(OR)_4 \rightleftharpoons \text{(P)}\sim\sim CO_2Ti(OR)_3 + ROH$$

Scheme 7.9

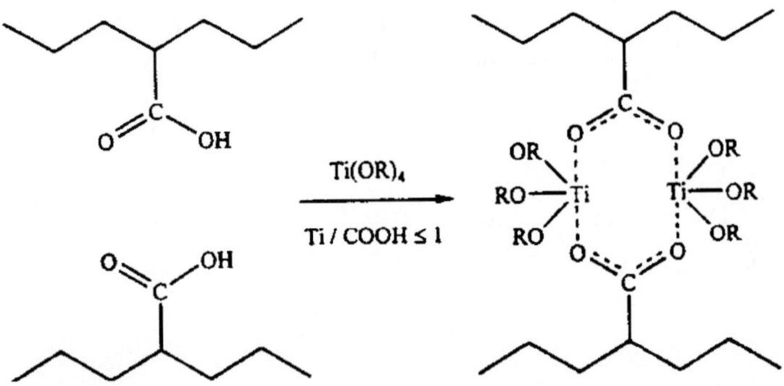

Scheme 7.10

7.4 Morphology and Fractal Model of Hybrid Nanocomposites

It follows from the morphological and fractal structure analysis of hybrid materials (see Chap. 1) that the structure of fractal aggregates is characterized by a mass fractal dimensionality D_m ($1 \leq D_m \leq 3$) which can be defined by the relation

$$N \approx R^D,$$

where N is the aggregate mass or the number of original particles in it, and R is a characteristic size equal to, e.g. the object inertia radius.

Self-similarity or the scale invariance of the fractal aggregate means that any of its parts resembles the aggregate as a whole.

The fractal dimension, D_s, of the surface expresses the relation between the object surface and its radius

$$S = r^{D_s},$$

where $2 \leq D_s \leq 3$. The value of $D_s = 2$ corresponds to a smooth surface, whereas $D_s = 3$ reflects maximum roughness.

The products of $Ti(OBu)_4$ hydrolysis represent a multilevel system whose structural element is a complex oligomer molecule of titanium oxobutyrate [228]. The system of polydispersed particles upon drying transforms into a system of homogeneous particles with fractal dimensionality $D_s = 3$. This complies with the case of a surface fractal with maximum developed surface [229]. Compounds that include a labile vinyl, methacrylate, epoxide or other groups can serve as a cross-linking agent as well.

Fractal dimensionalities of the polymer and colloidal samples of TiO_2 aerogel with high S_{sp} prepared by the sol–gel process and dried using CO_2 in a supercritical state [230] are in fact similar (2.6–2.8). The fractal surface of particles turned more irregular than that of the polymer aerogels. This size similarity suggests that the nanolevel morphology is shaped at the early stages of synthesizing and contribution of the synthesis process is insignificant. The mezolevel morphology, on the contrary, depends upon synthesis conditions. So, the extent of poly-ε-caprolactone incorporation into silica networks prepared by the sol–gel process depends on the HCl:TEOS molar ratio and the H_2O:TEOS molar ratio [231]. The structure of the material interface is mass fractal and it is in agreement with a co-continuous two phase morphology. The structure is more open when the acid content is lower or the water content higher, respectively (D_m varies from 1.4 to 2.0). This fact makes provision for regulating the surface parametrization.

The morphology of **A** group nanocomposites (30 wt.% SiO_2) obtained by block polymerization of HEMA in the presence of HEMA-functionalized preformed silica nanoparticles (~13 nm) was compared to the morphology of **B** group based on the in situ synthesis of the silica phase during free radical polymerization of HEMA (in this case formed from hydrolysis-condensation of TEOS) [232]. In **A** case the nanocomposite displays a common particle-matrix morphology where silica particles tend to form aggregates in the continuous PHEMA phases (Fig. 7.13, **A**). Material **B**, for which the mineral part has been synthesized *in situ,* exhibits a much finer morphology consisting of a very open mass-fractal silicate structure that is believed to be bicontinuous with the PHEMA phase at the molecular level. This very fine morphology is frozen by vitrification of the organic phase. Composite **B** can be named a molecular composite due to the molecular scale of its morphology [233]. Two types of nanoparticles are identified by the SAXS profile in material **A**: 1. the compact original nanoparticles forming a smooth distinct interface with the polymer matrix (mean diameter ~14 nm), and 2. nanoparticles which are the result of aggregation of the original ones. These aggregates exhibit a mass-fractal geometry whose dimension is below 2 ($D_s = 1.6$). Their equivalent spherical diameter approaches 30 nm, proving they include only a few original nanoparticles.

Investigations of TEOS condensation in the polysiloxane network has shown [12] that the typical fractal models observed in solutions aren't identified at polymerization in the matrix. The high molecular mass of the matrix and the forming polymer lowers the entropy of the blend and leads to phase

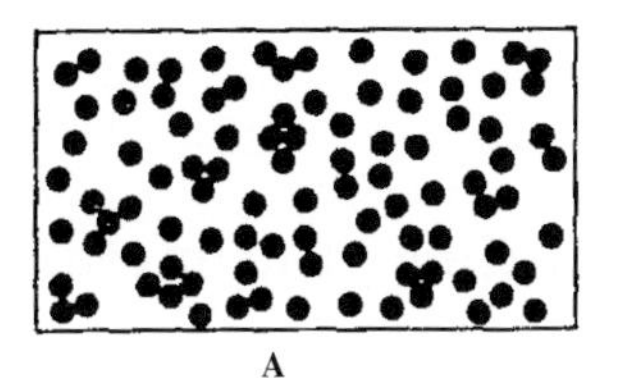

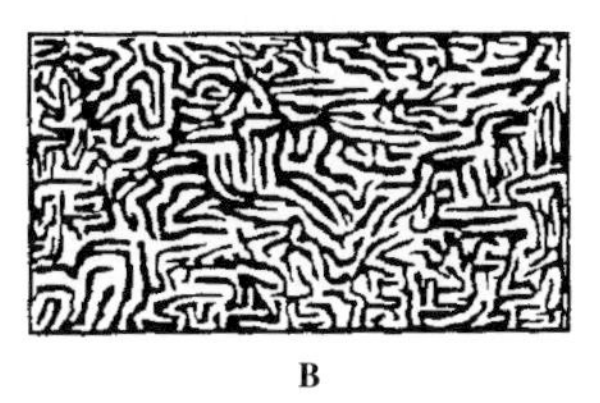

Fig. 7.13. Morphology of hybrid nanocomposites formed by different procedures (see text)

separation. Systems indivisible into phases are obtained only under specific conditions till the SiO_2 content starts to exceed 5%, when phase separation is induced and the composite mechanical strength improves. Maximum coarse domains are formed (to 5% of filling) when TiO_2 is deposited in situ and strength characteristics of such nanocomposites are inferior to the systems with SiO_2 [234]. Great attention has been paid lately to the mechanism of phase separation in hybrid materials which were prepared by the sol–gel process [161, 235, 236].

Two mechanisms are normally considered at phase separation: nucleation and growth (NG), and spinodal decomposition (SD). The NG mechanism is preferable for quenching into the metastable phase region, between the binodal and the spinodal curves. Nucleation initiates with local density or concentration fluctuation, forming nuclei with well-defined interfaces. It requires an activation energy, which depends on the interfacial energy for creating the nuclei. The process evolution with subsequent diffusion of macromolecules into the nucleated domains is spontaneous. The composition of the nucleus is constant over all the phase separation. The morphology observed during the NG phase separation is from the early to the later stages of the doped/matrix type. SD is the phase separation mechanism that predominates in quenching into the unstable region enclosed by the spinodal. Delocalized concentration fluctuations initiate spontaneously with a predominant wavelength, which is constant at the very early stages and depends on the quench depth. The fluctuation wavelength increases in the intermediate stages and phase coalescence is observed in the late stages. If coalescence can be avoided, a three-dimensional continuous morphology is obtained. As the solvent is removed from the hybrid poly(ethylene oxide-*co*-epichlorhydrin)-TEOS material (with composition equal to 71/29, 67/33, 60/40, 50/50,33/67 and 29/71), simultaneous growth of the inorganic network is observed by the mechanism of phase separation – spinodal decomposition [236]. The hybrid films based on poly(ethylene oxide–amide-6) containing sol–gel silicon or titanium oxide as inorganic fillers exhibit a similar morphology. The films containing titanium oxide were much more rigid [237]. The PEO–TEOS hybrid composite morphology and structure depend on the nature of the sol–gel catalyst as the polymer chains are linked to SiO_2 grains forming an ideal composite [238] when these materials are obtained under acidic conditions (as opposed to NH_4F). A simplified scheme of a hybrid composite, particularly of the composite incorporating cluster fragments, is presented in Fig. 7.14.

7.5 Nanocomposites Incorporating Multimetallic Ceramics

Films or powders incorporating binary oxides, e.g. ceramic mixtures of the oxides of Si–Ti [239], Ti, Ta and Nb [240–242], can be produced by sol–gel synthesis. They are of specific interest for catalytic applications [243–245],

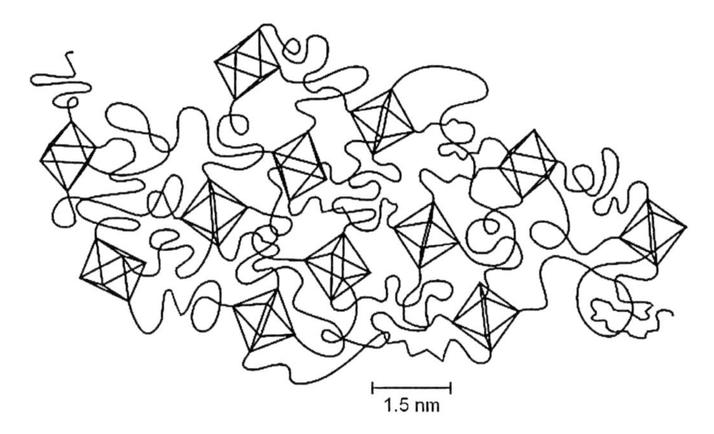

Fig. 7.14. Schematic diagram of a hybrid composite with polynuclear complexes

including Ti–Zr–Sn [246], Ti–Zr–Al [247] types as well as complex oxides like $Ba_{1-x}Sr_xTiO_3$, Ga–Si–O perovskites possessing acidic centers. They can be good precursors for novel Brönsted acidic catalysts [248]. Most important of them are zircon $ZrSiO_4$ (or $SiO_2{\cdot}ZrO_2$) produced from $ZrOCl_2{\cdot}$ $8H_20$ or $Zr(OPr^i)_4$ and $Si(OEt)_4$ [249] as well as V-zircon and Ni-zircon [250]. Using different methods $SrTiO_3$ [251] is produced (Scheme 7.11,a) and zircon with additions of 12 mol% of CeO_2 (Scheme 7.11,b).

The sol–gel synthesis [252] was used to receive the free-standing nanoparticle films of $Pb(Zr,Ti)O_3$. The properties of sol–gel products depend upon additionally impregnated modifying additives and high-dispersed fillers in the form of metal nanoparticles or their precursors.

Let us mention just a few of this type. These are xerogels $SiO_2 \cdot xM$ (M = Ni, Cu, Co, Cu/Ni, Co/Ni, Co/Cu) obtained via hydrolysis of the systems like metal acetate/Si((OEt)$_4$ [253], M = Pt [254–259], M = Gd/Pt [260], M = Au [261], M = Pd [262, 263], M = Au [264–266] M = Cr_2O_3 [267, 268], M = Co, Zn, Cu [269, 270].

The template-containing materials Ti-MCM-48 (S_{sp} ~1000 m^2/g, D_{por} ~ 2.5 nm), Cr-MCM-48 and V-MCM-48 [271] were produced based on TEOS, $Ti(OPr^i)_4$, surfactants and NaOH with addition of chrome or vanadyl salts ($CrCl_3{\cdot}6H_2O$, $VOSO_4{\cdot}3H_2O$). Their distinctive feature is the condensation degree of the silica skeleton increased by ~25% if ions of metal are present. Unfortunately, this phenomenon hasn't yet been explained. The sol–gel procedure was used also for growing ultra-fine barium ferrite powders, LiZn-ferrite thin films [272, 273] and cobalt ferrite $CoFe_2O_4$ [274, 275].

The polymerization method is most used in synthesis of heterometallic (polymetallic) ceramics, i.e with perovskite and ABO_3 structures. $CaTiO_3$ mineral perovskite found in nature has a pseudo-cubic crystalline lattice (Fig. 7.15) where large cations (A) are located in cellular angles, small ones (B) are in the center and oxygen ions are in the facet center.

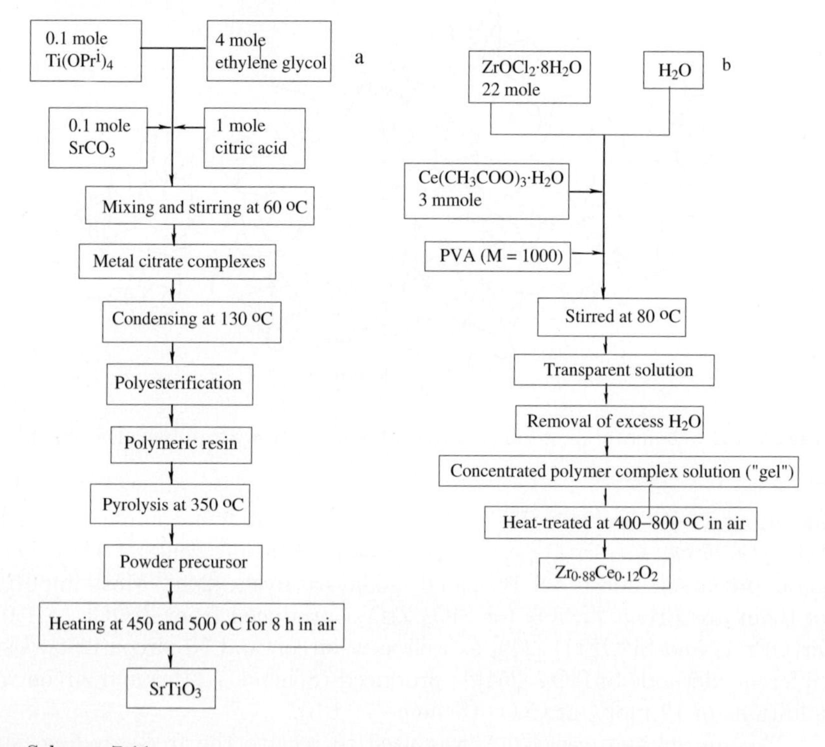

Scheme 7.11

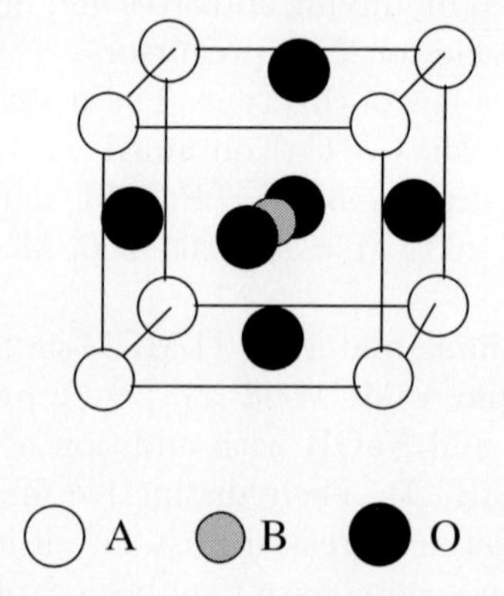

Fig. 7.15. Crystalline structure of $CaTiO_3$ perovskite (A = Ca, B = Ti)

These materials are extensively used in electronics thanks to their specific ferric, piezo- and pyroelectric properties [47]. A traditional method of production, e.g. $PbTiO_3(A^{2+}B^{4+}O_3$-ceramics), is the solid-phase mixing of PbO and TiO_2 in mills with further calcination at above 870 K, the so-called ex situ method. This process is accompanied by the formation of undesired toxic PbO. The sol–gel synthesis, during which the complex-formation is employed for components with citric acid and polyesterification by ethylene glycol followed by polymerization at 403 K and pyrolysis [276], is free of such drawbacks. As a result of the process the thin-film matter has properties of a block material.

It is possible to obtain barium-stannate via polymerization which acquires a perovskite $BaSnO_3$ structure [277] by pyrolysis of a preformed citrate complex $BaSn(C_6H_5O_7)_2$ (Scheme 7.12, a) with particle size 80–100 nm. The sol–gel method is used to synthesize some other multicomponent ceramics like perovskite, $BaTiO_3$ [278] $(Ba,Sr)TiO_3$, $Pb(Zr,Ti)O_3$, $NdAlO_3$ ($A^{3+}B^{3+}O_3$-ceramics), ZrO_2/X ($X = CeO_2$, Y_2O_3, Y_6WO_{12} [279,280]), $SrBi_2Ta_2O_9$ and even high-temperature superconducting ceramics $YBa_2Cu_3O_{7-\delta}$. Chemical methods of synthesizing analogous materials, use decomposition of metal-organic precursors are reviewed in [104, 281]. Polymer synthesis followed by decomposition of the organic phase at 570–670 K and calcination at 670–1170 K brings homogeneity at the molecular level and high purity oxides. The single-phase thin films of $KTiOPO_4$ can also be obtained by this method based on $Ti(OEt)_4$ and KOEt as precursors and various phosphorus sources (Scheme 7.12, b), the optimum of which is $(n\text{-}BuO)_2P(O)(OH)$ [282]. The resultant films exhibit high thermal stability and remarkable optical properties used in linear optics.

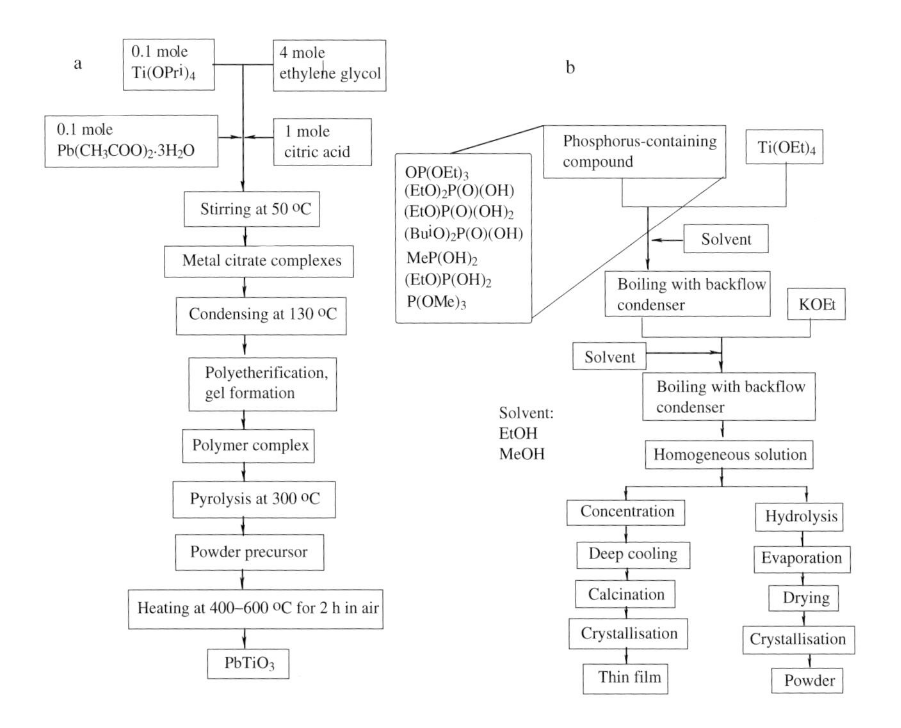

Scheme 7.12

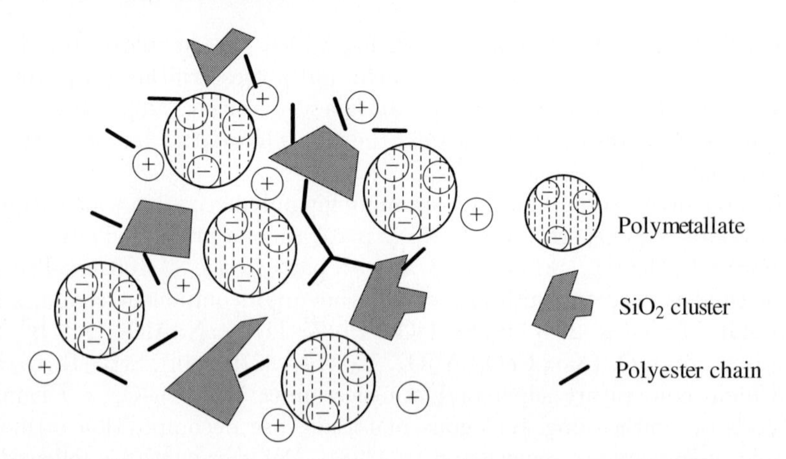

Fig. 7.16. Schematic diagram of organo-inorganic nanocomposites based on polymetallates

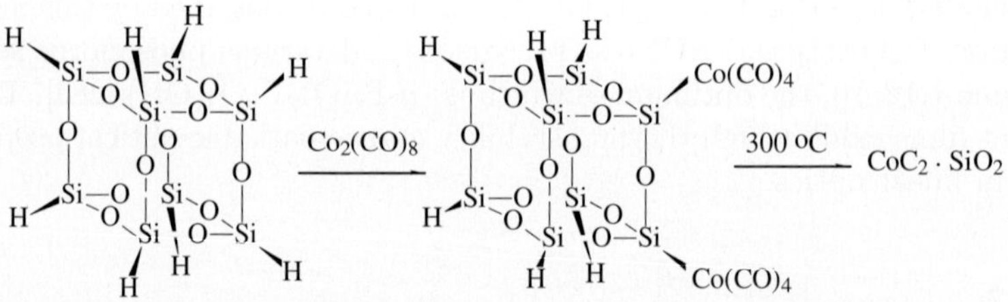

Fig. 7.17. Production route of $Co_2C{\cdot}SiO_2$

Heteropolymetallates of the Kegguin acid type $H_3PW_{12}O_{40}$, $H_4SiW_{12}O_{40}$ incorporated in organic–inorganic structures [53, 54] can be also referred to this group of nanocomposites. Their perfect electrochemical properties fit applications in holography. The polymetallates are spheres 1 nm in diameter. Their nanocomposites can be obtained in two ways: (i) by mixing these particles with TEOS ($W/Si_{alkoxide}$ of 0.2–0.6 ratio) and tetraethylene glycol or (ii) by impregnation with organosilanes. The structure of these materials in an idealized form is presented in Fig. 7.16.

The production of 3D oxovanadium boron phosphates with an open carcass [283–288] and precursors of polyborazinyl-amines is used to receive metal matrix composites incorporating metal borides and nitrides [289]. Cobalt carbide CoC_2 nanoparticles immobilized in the silicone matrix (Fig. 7.17) have been obtained via encapsulation of preliminary derived complex $[Si_8O_{12}H_6][Co(CO)_4]_2$ into amorphous silicone matrices followed by their heating [290].

One further way is polymerization transformations of specific clusters named oxometallates, whose W–O–Si chains are incorporated into the polymer

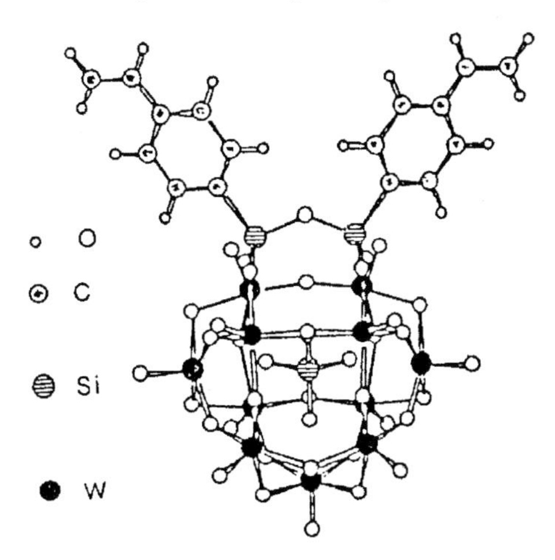

Fig. 7.18. Molecular structure of $SiW_{11}O_{40}Si(C_6H_5CH{=}CH_2)_2]^{4-}$ polyanion

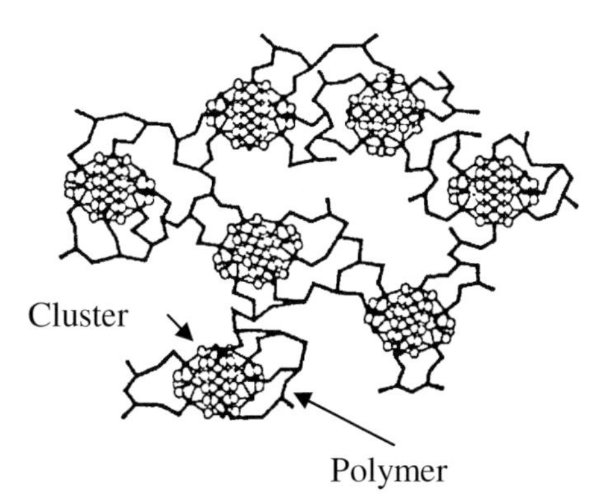

Fig. 7.19. Schematic diagram of a hybrid polymer-inorganic material

side chain. They are synthesized during interaction of substituted trichlorosilane $RSiCl_3$ (R – vinyl, allyl, methacryl, styryl and other) and polyanion $K_4SiW_{11}O_{40}$. The resultant cluster containing monomers consist of the metal oxide nuclei with two polymerizable ligands on the periphery (Fig. 7.18) [283–286].

The metal oxide nuclei are used to form cluster networks and nanohybrids representing ensembles of well-defined building blocks joined together. The polymer yield under the action of radical initiators and the chain length depends upon the reactivity of unsaturated groups and is diminished in the following sequence: $SiW_{11}O_{40}Si(vinyl) \ll SiW_{11}O_{40}Si(methacryl)_2 <$ $SiW_{11}O_{40}Si(styryl)_2$. The principal scheme of polyoxometallate polymers is shown in Fig. 7.19.

Nanoparticles formed by completion of a cluster of already known structure via addition of a hydroxyl group, e.g. $(\mu^3\text{-HC})Co_3(CO)_9]$, and its involvement into heterocondensation can be introduced into nanocomposites at the stage of sol–gel polymerization [291] (Scheme 7.13).

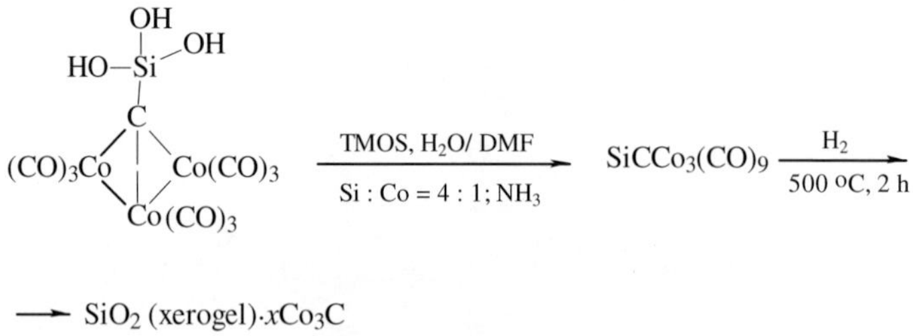

$$(HO)_3Si{-}C[Co(CO)_3]_3 \xrightarrow[\text{Si : Co = 4 : 1; } NH_3]{\text{TMOS, } H_2O\text{/ DMF}} SiCCo_3(CO)_9 \xrightarrow[500\ ^{o}C,\ 2\ h]{H_2} SiO_2\ (\text{xerogel})\cdot xCo_3C$$

Scheme 7.13

Particles 10–46 nm in size (the mean diameter of 25 nm) are distributed in a unimodal manner. Clusters such as Os_3, PtSn, Fe_2P, Co_2P, Ni_2P or Ge [292] are introduced into the composite by the same mechanism of binding into a complex with a bifunctional ligand containing either an alkoxy or hydroxyl group. Pd^{2+} [293], Cu^{2+} or other ions are introduced in poly(ethyltrimethylsilane).

One of the most interesting application fields of heterometallic hybrid nanocomposites is heterogenization of metal complex catalyzers. With this aim, metal compounds are combined with an inorganic matrix via a bidentate phosphorus-containing ligand of the type $[Ph_2PCH_2CH_2COO]EO_x(OR)_{3-2x}$ (E = Ti, Zr) [294]. During interaction of these ligands with tungsten carbonyl the products like $W(CO)_5[PPh_2\text{–}X\text{–}EO_x(OR)_{3-2x}]$ are isolated and characterized. These products can serve as potential precursors for the sol–gel syntheses. The hydrolytic sol–gel polymerization of benzyltricarbonylchrome derivative $[\eta^6\text{-}C_6H_5Si(OMe)_3]Cr(CO)_3$ with NH_4F as a catalyst leads to formation of a monophase organo-inorganic hybrid inside which a tricarbonylchromium ligand is preserved [295]. Three carbonyl groups split off only above 670 K and chromium of low oxidation degree is formed inside the gel. Tricarbonylmolybdenum groups may be impregnated into a ready polysilane by the method of polymer-analogous transformations – by treatment with the poly(methyl phenylsilane)pyridine complex $Mo(CO)_3(NC_5H_5)_3$ [296].

The composite materials based on inorganic structures and the polymers produced by the sol–gel process display higher thermal and mechanical characteristics as compared to the original ingredients. They can also be used in manufacturing optical waveguides [205, 297, 298]. These composites are formed in application of the composite polymerTiO_2 on a glass plate, which is then subjected to thermal treatment for 30 min at 573 K in an N_2 atmosphere. The advantage of manufacturing nanocomposites in polymer solutions is the possibility of obtaining transparent films thanks to elimination of titanium complexes that add a yellowish color. The thermal stability of these materials containing 4% of TiO_2 is almost the same as that of the initial polymer: 1% of mass loss in N_2 atmosphere at 734 and 736 K. This proves that impregnation of TiO_2 does not change the thermal characteristics of the host polymer. Thus, to form a polymer-inorganic composite, it is in some cases sufficient to mix the ingredients (polyimide solution and the sol–gel precursor) to have the phase of separation.

Of specific interest is the application of the sol–gel technique in template synthesis in which various processes are aimed at achieving certain products through the assembly of simple components in conditions of strictly observed stereochemical orientation of reagents. These conditions include the limited reaction volume due to which the reacting molecules may approach closer as opposed to the processes in solutions or solid phase and promote so-called soft template synthesis. As a result, the assembly runs under normal conditions and sometimes even at room temperature. The mesoporous molecular sieves like MCM-41 are high-organized substances whose formation includes self-organization of surfactant molecules in the solution and sol–gel processes inducing carcass formation round micellar structures functioning as templates (Fig. 7.20).

One more variant of manufacturing similar hybrid materials is synthesis with attraction of inverse micelles. This approach can be exemplified in formation of nanocomposites based on Au [261] and TiO_2 [205]. A surface-active substance (in the given case didodecyldimethylammonium bromide) is introduced into toluene thus forming the inverse micelles (see Chap. 3). In the latter $AuCl_3$ and tetraethoxysilane gel precursor are introduced. The reducing agent can be $LiBH_4$/TGF, the condensed substance – H_2O. TEOS hydrolysis and condensation lead to formation of Au nanoparticles ingrained into the moist gel. The mean nanoparticle size is 5–7 nm which depends on the H_2O:Si ratio and surfactant concentration H_2O:Si = 1:1–4:1.

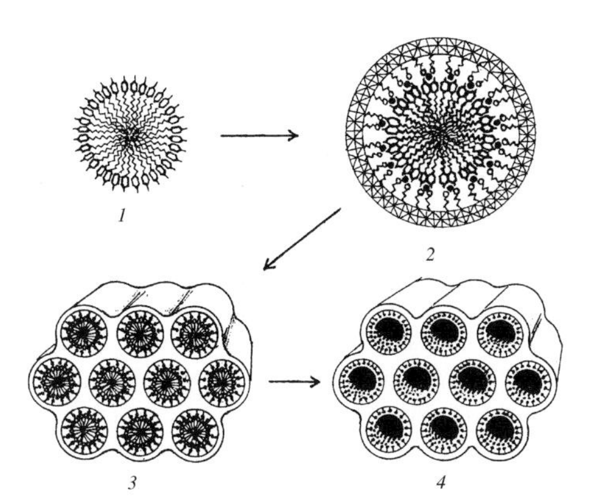

Fig. 7.20. Nanoperiodic structures and mesoporous molecular sieves formed via surfactant templated assembly. *1* – surfactant micelle, *2* – micelle coated with functionalized siliceous shell, *3* – functionalyzed nanoperiodic structure containing surfactant, *4* – surfactant free mesoporous molecular sieve with functionalized pores

To attain densely packed and organized two-dimensional Au layers (~10 nm) its colloidal particles are stabilized by a dodecanthiol surfactant [299] after immobilizing on the surface of high-dispersed SiO_2 through bifunctional aminosilane $(CH_3O)_3Si(CH_2)_3NHCH_2–CH_2NH_2$. Further treatment of particles with RSH makes them mobile because of broken links with SiO_2. Repetition of the immobilization–stabilization stages (Fig. 7.20,a) results in growth of 2D highly organized particles whose grains are joined together (Fig. 7.20,b).

The best surfactant used for inverse micelles is the sodium salt bis(2-ethylhexyl)sulfosiccine acid (AOT). TiO_2 nanoparticles are synthesized by hydrolysis of $Ti(O\text{-}i\text{–}Pr)_4$ dissolved in inverse micelles formed by isooctane and AOT containing a rated amount of water. $Ti(O\text{-}i\text{–}Pr)_4$ diffuses into micelles slowly, hydrolyzes and condenses in them like in microreactors and forms nanoparticles. The latter are extracted and dispersed in the solution of fluorinated polyimide (Fig. 7.21).

It is convenient to incorporate metal nanoparticles into such inorganic polymers where the reactions of formation of nanoclusters via reduction of corresponding salts in inverse micelles go together with the sol–gel synthesis. This is illustrated on the example of the formation of Pt/SiO_2 catalysts (Fig. 7.22) [300].

Metal nanoparticles of Pd, Rh, Ru, Ir get incorporated into the xerogel structure during sintering without any agglomeration [301, 302]. The spectrum of analogous transformations is extremely wide.

* * *

As the analysis showed, the combination of a polymer with an inorganic substance in the course of sol–gel synthesis takes several paths. Firstly, alkoxides modified by two different functional groups can be used. One of these groups reacts with a monomer macromolecule section, and the other enters into the reaction with the sol–gel precursors and results in a chemical bond between the polymer and the sol–gel network. Secondly, a special binding agent can be used for forming a chemical bond between those components. The surface of high-dispersed SiO_2 can be modified by, e.g. aminobutyric acid, after which the dried product undergoes dispersion in ε-caproamide and polymerized at 363 K [303]. In case the SiO_2 content is below 5 mass%, its particles are distributed evenly within the composite and the distance between them is about 50–110 nm. When the concentration reaches >10 mass%, the particles start to aggregate. This method of obtaining hybrid nanocomposites is unique as different modifying additives may be impregnated at the stage of formation of the sol–gel synthesis products. These additives of organic type can be 1,10-phenanthroline hydrochloride and n-butylamine as catalysts (<350 nm) translucent to visible light [304] and some other metal compounds. The methods of heterophase sol–gel synthesis and the deposition–precipitation based on the sol–gel synthesis products (Fig. 7.21) are used to produce highly active heterogeneous catalysts [305, 306].

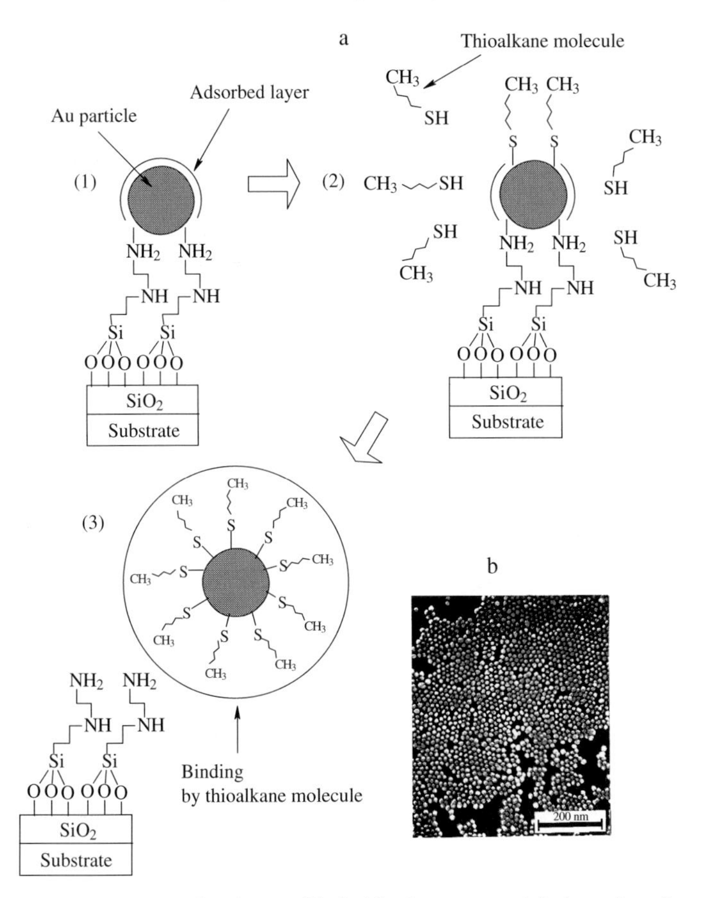

Fig. 7.21. Formation mechanisms of hybrid-phase materials based on inverse micelles (**a**) and electronic micrograph of highly organized 2D Au particle latex obtained by scheme (**b**): *1* – immobilization of colloidal Au particles; *2* – exchange reactions of alkanthiol molecules and breakage of Au–NH_2 bonds; *3* – fast restoration of the layer at complete binding of Au particles

Sol–gel processes often involve alumoxanes, i.e. carboxylates of general formula $[Al(O)_x(OH)_y(OOCR)_z]_n$. At this stage ceramics can be doped with additional introduction of acetylacetonate complexes like $M(acac)_n$ ($M = Ca^{2+}, Mn^{2+}, Y^{3+}$ and other [307, 308].

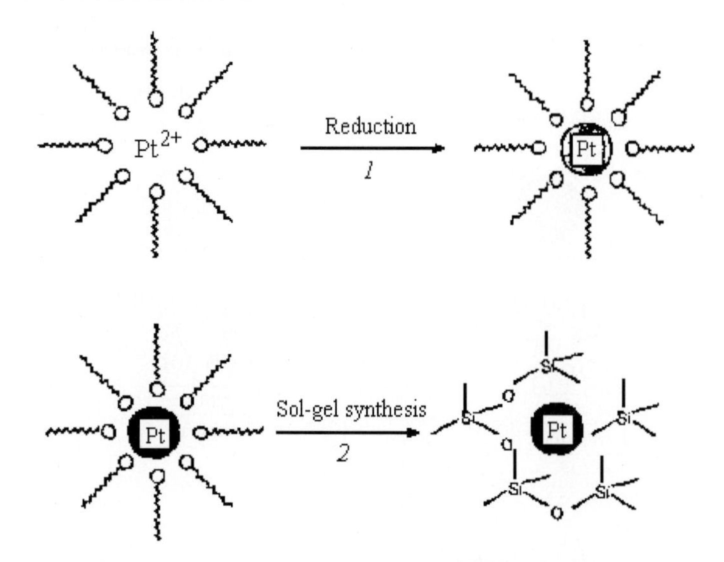

Fig. 7.22. Reduction of a metal salt combined with sol–gel processing. *1* – formation of nanoclusters by the inverse micelle technique, *2* – formation of the gel around nanoclusters

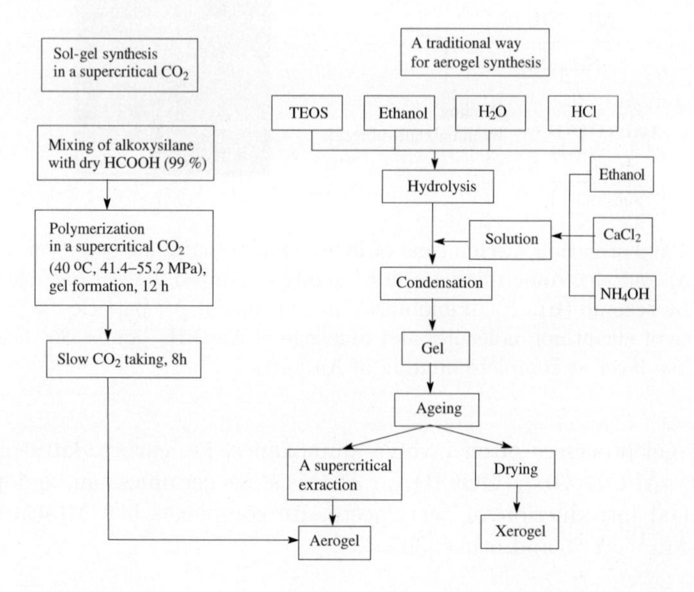

Scheme 7.14

Novel trends in this sphere are in the use of the carbon dioxide product in a supercritical state just as at the stage of sol–gel polymerization, so at extraction. Highly porous solid products exhibiting the meso- and macro-architecture of the pores are generated by this method [309]. It is possible to impregnate the components able to raise abruptly the adsorptivity of the products with respect to water (up to 30% of waterless $CaCl_2$) during the stage of SiO_2 formation [310]. The production of aerogels in conditions of supercritical CO_2 is illustrated in Scheme 7.14.

The application of alkogolates and $MO_n(OR)_m$ oxoalkogolates as well as those of other metals, e.g. Ta^{5+} or Nb^{5+} [311], rare-earth and trans-uranium elements is promising for sol–gel processes. Various types of copolymers are to be used as compatibilizers to raise compatibility of organic–inorganic composites. Styrene and 4-vinylphenol (50:50) block copolymers appear very useful for these purposes.

Some methods used to produce hybrid nanocomposites based on sol–gel synthesis will be analyzed further in Chaps. 8 and 9.

In conclusion it should be emphasized that the above inorganic-organic hybrid materials are biodegradable and biocompatible, particularly, the systems based on tetraethoxysilane-poly(ε-caprolactone) [197, 231, 312, 313].

References

1. N.J. Di Nardo: *Nanoscale characterization of surfaces and interfaces* (VCH, Weinheim 1994)
2. *Metal-Ceramic Interfaces* ed. by M. Ruhre, A.G. Evans, M.F. Ashby, J.P. Hizth (Pergamon, Oxford 1990)
3. *Inorganic and Organometallic Polymers with Special Properties, NATO ASI Ser.* ed. by R.M. Laine (Kluwer, New York 1992) Vol. 206
4. J. Lemmon, M. Lerner: Chem. Mater. **6**, 207 (1994)
5. T. Lan, P.D. Kaviratna, T.J. Pinnavaia: Chem. Mater. **6**, 1395 (1994)
6. J.-F. Nicoud: Science **263,** 636 (1994)
7. N. Guisaro, P. Lacan: New. J. Chem. **18**, 1097 (1994)
8. C.J. Brinker, G.W. Scherer: *Sol-Gel Science. The Physics and Chemistry of Sol-Gel Processing* (Academic Press, San-Diego, CA 1990)
9. *Sol-Gel Science and Technology* ed. by E.J.A. Pope, S. Sakka, L.C. Klein (Am. Ceram. Soc. Westerville, OH 1995)
10. *Better Ceramics Through Chemistry* ed. by C.J. Brinker, D.E. Clark, D.R. Ulrich (Materials Research Society, Pittsburg, Pa 1986)
11. *Ultrastructure Processing of Ceramics, Glasses and Composites* ed. by L.L. Hench, D.R. Ulrich (John Wiley & Sons, New York 1984)
12. D.W. McCarthy, J.E. Mark, D.W. Schaefer: J. Polym. Sci. B Polym. Phys. **36** 1167 (1998)
13. H. Smidt: In: *Chemistry, spectroscopy and applications of sol-gel glasses* (Springer-Verlag, Berlin 1992)
14. *Proceedings of the First European Workshop on Hybrid Organic-Inorganic Materials* ed. by C. Sanchez, F. Ribot (Paris, 1993)
15. C. Sanchez, F. Ribot: New J. Chem. **18**, 1007 (1994)
16. *Inorganic and Organometallic Polymers, ACS Symp. Ser. 572* ed. by P.W. Neilson, H.A. Allcock, K.J. Wynne (Am. Chem. Soc., Washington, DC 1994)

17. *Supramolecular Architecture, ACS Symp. Ser. 499* ed. by Th. Bein (Am. Chem. Soc., Washington, DC 1992)
18. A.J. Jacobson, M.S. Whittingham: *Intercalation Chemistry* (Academic Press, New York 1982)
19. *Sol-Gel Technology for Thin Films, Performs, Electronics and Especially Shapes* ed. by L.C. Klein (Noyes, Park Ridge, NJ 1988)
20. A.D. Pomogailo: Russ. Chem. Rev. **69**, 53 (2000)
21. *Science and Technology of Nanostructured Magnetic Materials* ed. by G.C. Hadjipanayis, G.A. Prinz (Plenum Press, New York 1991)
22. *Fine Particles Science and Technology from Micro to Nanoparticles* ed. by E. Pelizetti (Kluwer Academic Publishers, Netherlands 1996)
23. *Comprehensive Supramolecular Chemistry, Vol.9.* ed. by J.-P. Sauvage (Elsevier, New York 1996)
24. *Crystal Structures of Clay Minerals and their X-ray Diffraction, Vol.5.* ed. by G.W. Brindley, G. Brown (Mineral Society, London 1980)
25. J. Livage, M. Henry, C. Sanchez: Progress in Solid State Chemistry **18**, 259 (1988)
26. E. Ruiz-Hitzky: Adv. Mater. **5**, 334 (1993)
27. *Nanophase and Nanocomposite Materials. II. Mat. Res. Soc. Sympos. Proceed. Vol. 457* ed. by S. Komarneni, J.C. Parker, H.J. Wollenberger (Pen. Mat. Res. Soc., Pittsburgh 1997)
28. D.A. Loy, K.J. Shea: Chem. Rev. **95**, 1431 (1995)
29. *Proceeding of the First European Workshop on Hybrid Organic-Inorganic Materials, Château de Bierville, France, November 8–10, 1993* In: New J. Chem. **18**, 10 (1994)
30. *Nanostructured Materials* In: Chem. Mater. **8**, 8 (1996)
31. *Sol-Gel Derived Materials* In: Chem. Mater. **9**, 11 (1997)
32. *Prospects in Nanotechnology* In: Mendeleev Chem. J. **46,** N2 (2002)
33. C.R. Martin: Acc. Chem. Res. **28**, 61 (1995)
34. J.E. Mark, P.D. Calvert: Mater. Sci. Eng. C Biomim. Mater. Sens. Syst. **1**, 159 (1994)
35. X. Chen, K.E. Gonsalves, G.-M. Chow, T.D. Xiao: Adv. Mater. **6**, 481 (1994)
36. P. Judenstein, J. Titman, M. Stamm, H. Smidt: Chem. Mater. **6**, 127 (1994)
37. B. Brennan, T.M. Miller: Chem. Mater. **6**, 262 (1994)
38. B.R. Heywood, S. Mann: Chem. Mater. **6**, 311 (1994)
39. *Hybrid Organic-Inorganic Composites* ed. by J.E. Mark, C.Y.-E. Lee, P.A. Bianconi (Am. Chem. Soc., Washington, DC 1995)
40. *Photocatalytic Purification and Treatment of Water and Air* ed. by D.E. Ollis, H. Al-Ekabi (Elsevier, Amsterdam 1993)
41. J. Wen, G.L. Wilkes: *The Polymeric Materials Encyclopedia: Synthesis, Properties and Aplications* (CRC Press, Boca Raton, Fl 1995)
42. O. Lev, Z. Wu, S. Bharathi, V. Glezer, A. Modestov, J. Gun, L. Rabinovich, S. Sampath: Chem. Mater. **9**, 2354 (1997)
43. A.B. Bennan, H.H. Wang, G.L. Wikes: Polym. Prepr. **30**, 105 (1989)
44. M. Camail, M. Hubert, A. Margaillan, J.L. Vernet: Polymer **25**, 6533 (1998)
45. B.M. Novak: Adv. Mater. **5**, 422 (1993)
46. K.A. Mauritz: J. Appl. Polym. Sci. **40**, 1401 (1990)
47. C.D. Chandler, M.J. Hampden-Smith: Chem. Mater. **4**, 1137 (1992)
48. F. Ribot, P. Toledano, C. Sanchez: Chem. Mater. **3**, 759 (1991)

49. L.L. Beecroft, C.K. Ober: Adv. Mater. **7**, 1009 (1995)
50. Y. Morija, M. Sonoyama, F. Nishikawa, R. Hino: J. Ceram. Soc. Jpn. **101**, (1993)
51. D.C. Bradley, R.C. Mehrotra, D.P. Gaur: *Metal Alkoxides* (Academic Press, London 1978)
52. J. Wen, J.E. Mark: J. Appl. Polym. Sci. **58**, 1135 (1995)
53. P. Judenstein: Chem. Mater. **4**, 4 (1992)
54. P. Judenstein, P.W. Oliveria, H. Krug, H. Smidt: Chem. Phys. Lett. **220**, 35 (1994)
55. E. Bourgeat-Lami, P. Espiard, A. Guyot, C. Gauthier, L. David, G. Vigier: Angew. Macromol. Chem. **242**, 105 (1996)
56. N. Tsubokawa, K. Saitoh, Y. Shirai: Polym. Bull. **35**, 399 (1995)
57. J.D. Mackenzie, Y.J. Chung, Y. Hu: Non-Cryst. Solids **147/148**, 271 (1992)
58. L.C. Klein, B. Abramoff: Polym. Prepr. **32**, 519 (1991)
59. V.W. Day, T.A. Eberspacher, Y. Chen: Inorg. Chim. Acta, **229**, 391 (1995)
60. A.C. Pierre: Ceram. Bull. **70**, 1281 (1991)
61. T.A. Ring: Mater. Res. Soc. Bull. **12**, 34 (1987)
62. M.Z.-C. Hu, J.T. Zielke, C.H. Byers: J. Mater. Sci. **35**, 1957 (2000)
63. N.V. Golubko, M.I. Yanovskaya, I.P. Romm: Russ. J. Phys. Chem. **71**, 1747 (1997), **72**, 1023 (1998)
64. N.N. Khimich, B.I. Venzel, I.A. Drozdova, L.Ya. Suslova: Dokl. Phys. Chem. **366**, 361 (1999)
65. N.V. Golubko, M.I. Yanovskaya, S.G. Prutchenko, E.S. Obolonkova: Inorg. Mater. **34**, 1115. (1998)
66. M. Morstein: Inorg. Chem. **38**, 125 (1999)
67. B. Xia, W. Li, B. Zhang, Y. Xie: J. Mater. Sci. **34**, 3505 (1999)
68. M. Nabai, C. Sanchez, J. Livage: Eur. J. Solid. State Inorg. Chem. **28**, 1173 (1988)
69. G.W. Scherrer: Non-Cryst. Solids **87**, 199 (1986)
70. H. Smidt, B. Seiferling: Mat. Res. Soc. Symp. Proc. **73**, 739 (1986)
71. H.K. Smidt: ACS Symp. Ser. **360**, 333 (1988)
72. M. Nandi, J.A. Concin, L. Salvati, Jr., A. Sen: Chem. Mater. **2**, 772 (1990)
73. R.K. Iler: *The Chemistry of Silika* (John Wiley, New York 1979)
74. M. Anpo, T. Shima, S. Kodama, Y. Kubokawa: J. Phys. Chem. **91**, 4305 (1987)
75. B.V. Eremenko, T.N. Bezuglaya, A.N. Savitskaya, M.L. Malysheva, I.S. Kozlov, L.G. Bogodist: Colloid J. **63**, 194 (2001)
76. A.S. Kovalenko, V.G. Ilyin, A.P. Fillipov: Theoret. Exp. Chem. **33**, 53, 322 (1997), **35**, 177 (1999) **36**, 135 (2000)
77. A.S. Kovalenko, A.S. Korchev, J.V. Chernenko, V.G. Ill'in, A.P. Fillipov: Adsorption, Science and Technol. **17**, 245 (1999)
78. A. Corma: Chem. Rev. **97**, 2373 (1997)
79. J.M. Thomas: Faraday Discuss. **100**, C9 (1995)
80. A. Corma, M. Domine, J.A. Gaona: Chem. Commun. 2211 (1998)
81. J. Slunesco, M. Kosec, J. Holk, G. Drazic: J. Am. Ceram. Soc. **81**, 1121 (1998)
82. M. Gopai, W.J. Moberiy, Chan, L.C. de Jonghe: J. Mater. Sci. **32**, 6001 (1997)
83. B. Yao, L. Zhang: J. Mater. Sci. **34**, 5983 (1999)
84. L.L. Hench, J.K. West: Chem. Rev. **90**, 33 (1990)
85. P.G. Goswami: Chem. Rev. **89**, 765 (1989)

86. G.M. Pajonk: Appl. Catal. **72**, 217 (1991)
87. B. O/Regan, J. Moser, M. Anderson, M. Grätzel: J. Phys. Chem. **94**, 8720 (1990); Nature **335**, 737 (1991)
88. Y. Haruvy, S.E. Webber: Chem. Mater. **3**, 501 (1991) **4**, 89 (1992)
89. R. Zusman, C. Rottman, M. Ottolengi: Non-Cryst. Solids **122**, 107 (1990)
90. C. Rottman, G. Grader, D. Avnir: Chem. Mater. **13**, 3631 (2001)
91. B. Dunn, J.I. Zink: J. Mater. Chem. **1**, 903 (1991)
92. L.M. Ellerby, C.R. Nishida, F. Nishida, S.A. Yamanaka, B. Dunn, V.J. Selverstone, J.I. Zink: Science **255**, 113 (1992)
93. Reisfeld, D. Brusilovsky, M. Eyal, E. Miron, Z. Burshtein, J. Ivri: Chem. Phys. Lett. **160**, 43 (1989)
94. E.J.A. Pope, A. Asami, J.D. Mackenzie: J. Mater. Res. **4**, 1018 (1989)
95. H.K. Smidt: ACS Symp. Ser. **360**, 333 (1988)
96. Y. Hu, J.D. Mackenzie: Mat. Res. Soc. Proc. **271**, 681 (1992)
97. A. Blaaderen, A. Vrij: J. Colloid Interface Sci. **156**, 1 (1993)
98. E.M. Moreno, D. Levy: Chem. Mater. **12**, 128 (2000)
99. M.T. Reetz, R. Wenkel, D. Avnir: Synthesis 781 (2000)
100. N.S.M. Stevens, M.E. Rezac: Polymer **40**, 4289 (1999)
101. M.E. Spahr, P. Biterli, R. Nesper, M. Müller, F. Krumeich, H.-U. Nissen: Angew. Chem. Int. Ed. Engl. **37**, 1263 (1998)
102. F. Krumeich, H.-J. Muhr, M. Niederberger, F. Bieri, B. Schnyder, R. Nesper: J. Am. Chem. Soc. **121**, 8324 (1999)
103. R. Nesper, H.-J. Muhr: Chimia, **52**, 571 (1998)
104. M. Kakihana, M. Yoshimura: Bull. Chem. Soc. Jpn. **72**, 1427 (1999)
105. A.D. Pomogailo, V.S. Savost'yanov: *Synthesis and Polymerization of Metal-Containing Monomers* (CRC Press, Boca Raton, FL 1994)
106. A. Riondel, M. Camail, A. Margaillan, J.L. Vernet, M. Humbert: Fr. Pat. 96 10371; 96 10372 (1996)
107. M. Camail, M. Humbert, A. Margaillan, A. Riondel, J.L. Vernet: Polymer **39**, 6525 (1998)
108. D. Brunel: Microporous Mesoporous Mater. 27 (1999)
109. C. Rottman, G. Grader, Y. De Hazan. D. Avnir: Langmuir **12**, 5505 (1996)
110. N. Igarashi, Y. Tanaka, S. Nakata, T. Tatsumi: Chem. Lett. 1 (1999)
111. S. Gross, V. Di Noto, G. Kickelbick, U. Schubert: Mat. Res. Soc. Symp. Proc. **726**, Q4.1, 1 (2002)
112. L.G. Hubert-Pfalzgraf, V. Abada, S. Halut, J. Roziere, Polyhedron **16**, 581 (1997)
113. E. Ruckenstein, L. Long: J. Appl.Polym. Sci. **55**, 1081 (1995); Chem. Mater. **8**, 546 (1996)
114. L. Long, E. Ruckenstein: J. Appl. Polym. Sci. **67**, 1891 (1998)
115. B. Moraru, N. Hüssing, G. Kickelbick, U. Schubert, P. Fratzl, H. Peterlik: Chem. Mater. **14**, 2732 (2002)
116. B. Moraru, G. Kickelbick, U. Schubert: Eur. J. Inorg. Chem. 1295 (2001)
117. G. Trimmel, S. Gross, G. Kickelbick, U. Schubert: Appl. Organomet. Chem. **15**, 401 (2001)
118. G.I. Dzhardimalieva, A.D. Pomogailo, A.N. Shupik: Bull. Acad. Sci. USSR, Div. Chem. Sci. 451 (1985)
119. G.I. Dzhardimalieva, A.O. Tonoyan, A.D. Pomogailo, S.P. Davtyan: Bull. Acad. Sci. USSR, Div. Chem. Sci. 1744 (1987)

120. A.D. Pomogailo, N.D. Golubeva, A.N. Kitaigorodsky: Bull. Acad. Sci. USSR, Div. Chem. Sci. 482 (1991)
121. A.D. Pomogailo, N.D. Golubeva: Russ. Chem. Bull. **43**, 2020 (1994)
122. C. Barglik-Chory, U. Schubert: J. Sol-Gel. Sci. Technol. **5**, 135 (1995)
123. U. Schubert, F. Schwertfeger, C. Görsmann: In: *Nanotechnology. Molecularly Designed Materials* ed. by G.-M. Chow, K.E. Gonsalves (Am. Chem. Soc., Washington, DC 1996) p. 366
124. W.J. Evans, M.A. Ansari, J.W. Ziller: Inorg. Chem. **38**, 1160 (1999)
125. R. Burch: Chem. Mater. **2**, 633 (1990)
126. M.H. Chisholm, C.E. Hammond, J.C. Huffman: Polyhedron **7**, 2515 (1988)
127. K. Dahmouche, L.D. Carlos, C.V. Santilli, V.Z. Bermudes, A.F. Craievich: J. Phys. Chem. B **106**, 4377 (2002)
128. P. Valid, M. Tanski, M. Pette, B. Lobkovsky, T. Wolczanski: Inorg. Chem. **38**, 3394 (1999)
129. M. Yoshihara, H. Oie, A. Okada, H. Matsui, S. Ohshiro: Macromolecules **35**, 2435. (2002)
130. R.A. Mantz, P.F. Jones, K.P. Chaffee, J.D. Lichtenhan, M.K. Ismail, M. Burmeiste: Chem. Mat. **8**, 1250 (1996)
131. J.D. Lichenhan: In: *Polymeric Materials Encyclopedia* ed. by J.C. Salamone (CRC Pess, Boca Raton 1996) p. 7768
132. F. Gao, Y. Tong, S.R. Schricker, B.M. Culbertson: Polym. Adv. Technol. **12**, 355 (2001)
133. R. Tamaki, K. Naka, Y. Chujo: Polym. J. **30**, 60 (1998)
134. R. Tamaki, T. Horiguchi, Y. Chujo: Bull. Chem. Soc. Jpn. **71**, 2749 (1998)
135. Y. Chujo, H. Matsuki, S. Kure, T. Saegusa, T. Yasawa: Chem. Commun. 635 (1994)
136. K. Yoshinaga, R. Norie, F. Saigoh, T. Kito, N. Enomoto, H. Nishida, M. Kamatsu: Polym. Adv. Technol. **3**, 91 (1992)
137. K. Yoshinaga, K. Nakanishi: Compos. Interfaces **2**, 95 (1994)
138. U. Schubert, N. Hüsing, A. Lorenz: Chem. Mater. **7**, 2010 (1995)
139. L. Mascia: Trends Polym. Sci. **3**, 61 (1995)
140. M. Popall, H. Durand: Electrochim. Acta **37**, 1593 (1992)
141. R.L. Callender, C.J. Harlan, N.M. Shapiro, C.D. Jones, D.L. Callahan, M.R. Wiesner, D.B. McQueen, R. Cook, A.R. Barron: Chem. Mater. **9**, 2418 (1997)
142. A. Lee, D. Lichtenhan: J. Appl. Polym. Sci. **73**, 1993 (1999)
143. G.M. Kloster, C.M. Taylor, S.P. Watton: Inorg. Chem. **38**, 3954 (1999)
144. G.-H. Hsiue, R.-H. Lee, R.-J. Jeng: Polymer, **40**, 6417 (1999)
145. F. Aldinger, S. Prinz, N. Janakiraman, R. Kumar, M. Christ, M. Weinmann, A. Zimmermann: Int. J. Self-Propargating High-Temp. Synth. **10**, 249 (2001)
146. K. Matyjaszewski, T.P. Davis: *Handbook of Radical Polymerization* (Wiley-Interscience, Hoboken 2002)
147. T.E. Patten, K. Matyjaszewski: Adv. Mater. **10**, 1 (1998)
148. S.C. Hong, J.-F. Lutz, Y. Inoue, C. Strissel, O. Nuyken, K. Matyjaszewski: Macromolecules **36**, 1075 (2003)
149. T. Werne, T.E. Patten: J. Am. Chem. Soc. **121**, 7409 (1999)
150. D.M. Haddleton, D. Kukulj, A.P. Radigue: Chem. Commun. 99 (1999)
151. D.A. Foucher, B.Z. Tang, I. Manners: J. Am. Chem. Soc. **114**, 6246 (1992)
152. I. Manners: Chem. Commun. 857 (1999)

153. J. Massey, K.N. Power, I. Manners, M.A. Winnik: J. Am. Chem. Soc. **120**, 9533 (1998)
154. S. Inagaki, S. Guan, Y. Fukushima, T. Ohsuna, O. Terasaki: J. Am. Chem. Soc. **121**, 9611 (1999)
155. P. Audbert, P. Calals, G. Cerveau, R.J.P. Corriu, N. Costa: J. Electroanal. Chem. **372**, 275 (1994), **413**, 89 (1996)
156. A.I. Wallbank, J.F. Corrigan: Can. J. Chem. **80**, 1592 (2002)
157. Y. Wei, D.C. Yang, L.C. Tang, M.K. Hutchins: J. Mater. Res. **8**, 1143 (1993); Makromol. Chem. Rapid Commun. **14**, 273 (1993)
158. Z.H. Huang, K.Y. Qiu: Polym. Bull. **35**, 607 (1995); Acta Polym. Sin. 434 (1997); Polymer **38**, 521 (1997)
159. T. Gunji, I. Sopyan, Y. Abe: J. Polym. Sci. A Polym. Chem. **32**, 3133 (1994)
160. C.J.T. Landry, B.K. Coltrain, J.A. Wenson, N. Zambulyadis: Polymer **33**, 1486, 1496 (1992)
161. K.F. Silveira, I.V.P. Yoshida, S.P. Nunes: Polymer **36**, 1425 (1995)
162. F. Suzuki, K. Nakane, J.-S. Piao: J. Mater. Sci. **31**, 1335 (1996)
163. K. Nakane, F. Suzuki: J. Appl. Polym. Sci. **64**, 763 (1997)
164. J.J. Fitzgerald, C.J.T. Landry, R.V. Schillace, J.M. Pochan: Polym. Prepr. **32**, 532 (1991)
165. J.J. Fitzgerald, C.J.T. Landry, J.M. Pochan: Macromolecules **25**, 3715 (1992)
166. C.L. Beaudry, L.C. Klein: In: *Nanotechnology. Molecularly Designed Materials* ed. by G.-M. Chow, K.E. Gonsalves (ACS, Washington, DC 1996) p. 382
167. M. Toki, T.Y. Chow, T. Ohnaka, H. Samura, T. Saegusa: Polym. Bull. **29**, 653 (1992)
168. S.P. Nunes, J. Schultz, K.V. Peinemann: J. Mater. Sci. Lett. **15**, 1139 (1996)
169. R.A. Zoppi, C.R. Castro, I.V.P. Yoshida, S.P. Nunes: Polymer **38**, 5705 (1997)
170. R.A. Zoppy, C.M. Fonseca, M.-A. De Paoli, S.P. Nunes: Acta Polym. **48**, (1997)
171. S. Sakhora, L.D. Tickanen, M.A. Anderson: J. Phys. Chem. **96**, 11087 (1992)
172. Q. Deng, R.B. More, K.A. Mauritz: Chem. Mater. **7**, 2259 (1995)
173. Q. Deng, C.A. Wilkie, R.B. Moore, K.A. Mauritz: Polymer **39**, 5961 (1998)
174. R.A. Zoppi, I.V.P. Yoshida, S.P. Nunes: Polymer **39**, 1309 (1997)
175. R.A. Zoppi, S.P. Nunes: J. Electroanal. Chem. **445**, 39 (1998)
176. I.A. David, G.W. Scherer: Polym. Prepr. **32**, 530 (1991)
177. H. Shiho, N. Kawahashi: Colloid Polym. Sci. **278**, 270 (2000)
178. T. Sugimoto: *Fine Particles. Synthesis, Characterization and Mechanism of Grown* (Marcel Dekker, New York 2000)
179. K. Yoshinaga: Bull. Chem. Soc. Jpn. **75**, 2349 (2002)
180. K. Nakane, J. Ohashi, F. Suzuki: J. Appl. Polym. Sci. **71**, 185 (1999)
181. M.A. Sherman, J.P. Kennedy: J.Polym. Sci. A Polym. Chem. **36**, 1891 (1998)
182. Y.G. Aronoff, B. Chen, G. Lu, C. Seto, J. Schwartz, S.L. Bernasek: J. Am. Chem. Soc. **119**, 259 (1997)
183. S.K. VanderKam, A.B. Bocarsly, J. Schwartz: Chem. Mater. **10**, 685 (1998)
184. N. Juangvanich, K.A. Mauritz: J. Appl. Polym. Sci. **67**, 1799 (1998)
185. K.A. Mauritz, R. Ju: Chem. Mater. **6**, 2269 (1994)
186. C. Rehwinkel, V. Rossbach, P. Fischer, J. Loos: Polymer, **39**, 4449 (1998)
187. Z.D. Zhao, Y.C. Ou, Z.M. Gao, Z.N. Qi, F.S. Wang: Acta Polym. Sin. 228 (1996)
188. W. Zhou, J.H. Dong, K.Y. Qiu, Y. Wei: J. Polym. Sci. A Polym. Chem. **36**, 1607 (1998)

189. J. Zhang, S.C. Luo, L.L. Gui: J. Mater. Sci. **32**, 1469 (1997)
190. Y. Wei, D. Jin, G. Wei, D. Yang, J. Xu: J. Appl. Polym. Sci. **70**, 1689 (1998)
191. R. Tamaki, Y. Chujo: Chem. Commun. 1131 (1998)
192. M. Templin, U. Wiesner, H.W. Spiess: Adv. Mater. **9**, 814 (1997)
193. M. Templin, A. Franck, A. Du Chesne, H. Leist, Y. Zhang, R. Ulrich, V. Schädler, U. Wiesner: Science, **278**, 1795 (1997)
194. S.M. De Paul, J.W. Zwanziger, R. Ulrich, U. Wiesner, H.W. Spiess: J. Am. Chem. Soc. **121**, 5727 (1999)
195. B.A. Platonov, A.G. Bratunets, F.D. Ovcharenko, T.A. Polischuk: Dokl. Phys. Chem. USSR **259**, 1403 (1981)
196. A.D. Pomogailo, A.S. Rozenberg, U.E. Uflyand: *Nanoscale metal particles in polymers* (Khimia, Moscow 2000)
197. D. Tian, Ph. Dubois, R. Jerome: J. Polym. Sci. Polym. Chem. **35**, 2295 (1997)
198. V. Bekiari, P. Lianos, P. Judeinstein: Chem. Phys. Lett. **307**, 310 (1999)
199. D. Avnir: Acc. Chem. Res. **28**, 328 (1995)
200. M. Sykora, K.A. Maxwell, T.J. Meyer: Inorg. Chem. **38**, 3596 (1999)
201. K. Kuraoka, Y. Chujo, T. Yamazawa: Chem. Commun. 2477. (2000)
202. Q. Hu, E. Marand: Polymer **40**, 4833 (1999)
203. Y. Imai: J. Macromol. Sci. A **28**, 1115 (1991)
204. A. Morikawa, Y. Iyoku, M. Kakimoto, Y. Imai: Polym. J. **24**, 107 (1992)
205. M. Yoshida, M. Lal, N.D. Kumar, P.N. Prasad: J. Mater. Sci. **32**, 4047 (1997)
206. Z.-K. Zhu, Y. Yang, J. Yin, Z.-N. Qi: J. Appl. Polym. Sci. **73**, 2977 (1999)
207. W. Volksen, D.Y. Yoon, J.L. Hedrick, D.C. Hofer: Mater. Res. Soc. Symp. Proc. **277**, 3 (1992)
208. L. Hedrick, H.-J. Cha, R.D. Miller, D.Y. Yoon, H.R. Brown, S. Srinivasan, R. Di Pietro, R.F. Cook, J.P. Hummel, D.P. Klaus, E.G. Liniger, E.E. Simonyi: Macromolecules **30**, 8512 (1997)
209. M. Nandi, J.A. Gonkin, L. Salvati, Jr., A. Sen: Chem. Mater. **3**, 201 (1991)
210. A. Morikawa, Y. Iyoku, M. Kakimoto, Y. Imai: J. Mater. Chem. **2**, 679 (1992)
211. Y. Chujo, T. Saegusa: Adv. Polym. Sci. **100**, 11 (1992); Polym. Bull. **19**, 435 (1988); Macromol. Symp. **22**, 2040 (1989)
212. B.M. Novak, C. Davies: Macromolecules **24**, 5481 (1991)
213. M.W. Ellsworth, B.M. Novak: J. Am. Chem. Soc. **113**, 2756 (1991); Chem. Mater. **5**, 839 (1993)
214. M.-P. Zheng, Y.-P. Jin, G.-L. Jin, M.-Y. Gu: J. Mater. Sci. Lett. **19**, 433 (2000)
215. S. Kobayashi, K. Hanabusa, M. Suzuki, M. Kimura, H. Shirai: Bull. Chem. Soc. Jpn. **73**, 1913 (2000)
216. H. Krug, H. Smidt: New. J. Chem. **18**, 1125 (1994)
217. P.L. Sano, K.A. Mauritz, R.B. Moore: J. Polym. Sci. B Polym. Phys. **34**, 873 (1996)
218. Ch.J.T. Landry, B.K. Coltrain, D.M. Teegarden, T.E. Long, V.K. Long: Macromolecules **29**, 4712 (1996)
219. J. Kramer, R.K. Prud'homme: J. Colloid and Interface Sci. **118**, 294 (1987)
220. J. Kramer, R.K. Prudhomme, P. Wiltzius, P. Mirau, S. Knoll: Colloid. Polym. Sci. **266,** 145 (1988)
221. T.H. Mourey, S.M. Miller, J.A. Wesson, T.E. Long, L.M. Kelts: Macromolecules **25**, 45 (1992)
222. Y. Chujo, E. Ihara, S. Kure, K. Suzuki, T. Saegusa: Makromol. Chem. Macromol. Symp. **42/43**, 303 (1991)

223. G. Broze, R. Jerome, P. Tessie, C. Marko: Macromolecules **18**, 1376 (1985)
224. J.L.W. Noel, G.L. Wilkes, D.K. Mohanty, J. McGrath: J. Appl. Polym. Sci. **40**, 1177 (1990)
225. R.H. Glaser, G.L. Wilkes: Polym. Bull. **19**, 51 (1988); **22**, 527 (1989)
226. B.K. Coltrain, C.J.T. Landry, J.M. O'Reilly, A.M. Chamberlain, G.A. Rakes, J.S.S. Sedita, L.W. Kelts, M.R. Landry, V.K. Long: Chem. Mater. **5**, 1445 (1993)
227. H.H. Huang, G.L. Wilkes, J.G. Carlson: Polymer **30**, 2001 (1989)
228. A.N. Ozerin, E.Yu. Sharipov, L.A. Ozerina, N.V. Golubko, M.I. Yanovskaya: J. Phys. Chem. **73**, 277 (1999)
229. J.E. Martin, A.J. Hurd: J. Appl. Cryst. **20**, 61 (1987)
230. F. Meng, J.R. Schlup, L.T. Fan: Chem. Mater. **9**, 2459 (1997)
231. D. Tian, S. Blacher, R. Jerome: Polymer **40**, 951 (1999)
232. P. Hajji, L. David, J.F. Gerard, J.P. Pascault, G. Vigier: J. Polym. Sci. B Polym. Phys. **37**, 3172 (1999)
233. H. Kaddami, J.F. Gerard, P. Hajji, J.P. Pascault: J. Appl. Polym.Sci. **73**, 2701 (1999)
234. T.L. Porter, M.E. Hagerman, B.P. Reynolds, M.P. Eastman, R.A. Parnell: J. Polym. Sci. B Polym. Phys. **36**, 673 (1998)
235. K. Nakanishi, N. Soga: J. Non-Crystalline Solids **139**, 1, 14 (1992)
236. R.A. Zoppi, S.P. Nunes: Polymer, **39**, 6195 (1998)
237. R.A. Zoppi, S. Das Neves, S.P. Nunes: Polymer **41**, 5461 (2000)
238. M. Laridjani, E. Lafontaine, J.P. Bayle, P. Judeinstein: J. Mater. Sci. **34**, 5945 (1999)
239. J.-B. Tang, K.-Y. Qiu, Y. Wei: Chin. J. Chem. **19**, 198 (2001)
240. R.F. Cava, W.F. Peck, J.J. Krajewskii: Nature **377**, 215 (1995)
241. R.F. Cava, W.F. Peck, J.J. Krajewskii, G.L. Roberts: Mat. Res. Bull. **31**, 295 (1996)
242. M. Fang, C.H. Kim, B.R. Martin, T.E. Mallouk: J. Nanopart. Res. **1**, 43 (1999)
243. M.E. Raimondi, L. Marchese, E. Gianotti, T. Maschmeyer, J.M. Seddon, S. Coluccia: Chem. Commun. 87 (1999)
244. S.G. Zhang, Y. Fujii, H. Yamashita, K. Koyano, T. Tatsumi, M. Anpo: Chem. Lett. 659 (1997)
245. K. Kosuge, P.S. Singh: Chem. Lett. 9 (1999)
246. R.B. Van Dover, L.F. Schneemeyer, R.M. Fleming: Nature **392**, 162 (1998)
247. I. Ichinose, T. Kawakami, T. Kunitake: Adv. Mater. **10**, 535 (1998)
248. K. Waada, K. Yamada, T. Kondo, T. Mitsudo: Chem. Lett. 12 (2001)
249. M. Shoyama, N. Matsumoto, T. Hashimoto, H. Nasu, K. Kamiya: J. Mater. Sci. **33**, 4821 (1998)
250. G. Monros, J. Carda, M.A. Tena, P. Escribano, J. Alarcon: Mater. Res. Bull. **27**, 753 (1992)
251. M. Kakihana, T. Okubo, Y. Nakamura, M. Yashima, M. Yoshimura: J. Sol-Gel Sci. Technol. **12**, 95 (1998)
252. C. Liu, B. Zou, A.J. Rondinone, Z.J. Zhang: J. Am. Chem. Soc. **123**, 4344 (2001)
253. G. Trimmel, U. Schubert: J. Non-Crystal. Solids, 2001, **296**, 188.
254. M.A. Aramendia, V. Borau, C. Jimenez, J.M. Marinas, F.J. Romero: Chem. Commun. 873 (1999)
255. C. Lembacher, U. Schubert: New J. Chem. **22**, 721 (1998)

256. A. Corma, A. Martinez, V. Martinez-Soria: J. Catal. **169**, 480 (1997)
257. Z. Liu, Y. Sakamoto, T. Ohsuma, K. Higara, O. Terasaki, C.H. Ko, H.J. Shin, R. Ryoo: Angew. Chem. Int. Ed. **39**, 3107 (2000)
258. Y.-J. Han, J.M. Kim, G.D. Stucky: Chem. Mater. **12**, 2068 (2000)
259. H.J. Shin, R. Ryoo, Z. Liu, O. Terasaki: J. Am. Chem. Soc. **123**, 1246 (2001)
260. J.M. Watson, U.S. Ozkan: J. Catal. **210**, 295 (2002)
261. A. Martino, S.A. Yamanaka, J.S. Kawola, D.A. Loy: Chem. Mater. **9**, 423 (1997)
262. C.A. Koh, R. Nooney, S. Tahir: Catal. Lett. **47**, 199 (1997)
263. K. Okitsu, S. Nagaoka, S. Tanabe, H. Matsumoto, Y. Mizukoshi, Y. Nagata: Chem. Lett. 271 (1999)
264. U. Kreibig: J. Phys. F **4**, 999 (1974)
265. R. Vacassy, L. Lemaire, J.-C. Valmalette, J. Dutta, H. Hoffman: J. Mater. Sci. Lett. **17**, 16665 (1998)
266. C. Graf, A. Blaaderen: Langmuir **18**, 524 (2002)
267. O.A. Shilova, S.V. Hashkovsky, E.V. Tarasoyuk, V.V. Shilov, V.V. Shevchenko, Yu.P. Gomza, N.S. Klimenko: J. Sol-Gel Sci. Technol. **26**, 1 (2003)
268. O.A. Shilova, S.V. Hashkovsky: Mater. Techn. Tools, **6**, 64 (2001)
269. O.A. Shilova, S.V. Hashkovsky, L.A. Kuznetsova: J. Sol-Gel Sci. Techn. **26**, 687 (2003)
270. E.A. Trusova, M.V. Tsodikov, E.V. Slivinskii, G.G. Hermandez, O.V. Bukhtenko, T.N. Zhdanova, D.I. Kochubey, J.A. Navio: Mendeleev Commun. 102 (1998)
271. W. Zhang, T.J. Pinnavaia: Catal. Lett. **38**, 261 (1996)
272. V.K. Sankaranarayana, Q.A. Pankhurst, D.P.E. Dickson, C.E. Johnson: J. Magn. Magn. Mater. **125**, 199 (1993)
273. K.G. Broors, V.R.W. Amarakoon: J. Am. Ceram. Soc. **74**, 851 (1991)
274. V. Blaskov, V. Petkov, V. Rusanov, L.M. Marinez, B. Martinez, J.S. Munoz, M. Mikhov: J. Magn. Magn. Mater. **162**, 331 (1996)
275. J.-G. Lee, J.Y. Park, C.S. Kim: J. Mater. Sci. **33**, 3965 (1988)
276. M. Kakihana, T. Okubo, M. Arima, O.F. Uchiyama, M. Yashima, M. Yoshimura, Y. Nakamura: Chem. Mater. **9**, 451 (1997)
277. C.P. Udawatte, M. Kakihana, M. Yoshimura: Solid State Ionics **108**, 23 (1998)
278. S. O'Brien, L. Brus, C.B. Murray: J. Am. Chem. Soc. **123**, 12085 (2001)
279. O. Yokota, M. Yashima, M. Kakihana, A. Shimofuku, M. Yoshimura: J. Am. Ceram. Soc. **82**, 1333 (1999)
280. J. Ma, M. Yoshimura, M. Kakihana, M. Yashima: J. Mater. Res. **13**, 939 (1998)
281. R.W. Schwartz: Chem. Mater. **9**, 2325 (1997)
282. K. Noda, W. Sakamoto, K. Kikuta, T. Yogo, S. Hirano: Chem. Mater. **9**, 2174 (1997)
283. P. Bodo, J.E. Sundgren: Thin Solid Films **136**, 147 (1986)
284. N. Bowder, S. Brittain, A.G. Evans, J.W. Hutchinson, G.M. Whitesides: Nature **393**, 146 (1998)
285. D. Hagrman, C. Zubieta, D.J. Rose, J. Zubieta, R.C. Haushalter: Angew. Chem. Int. Ed. Engl. **36**, 795 (1997)
286. M.I. Khan, L.M. Meyer, R.C. Haushalter, A.L. Zubieta: Chem. Mater. **8**, 43 (1996)
287. C.J. Warren, R.C. Haushalter, D.J. Rose, J. Zubieta: Chem. Mater. **9**, 2694 (1997)

288. C.K. Narula, P. Czubarow, D. Seyferth: J. Mater. Sci. **33**, 1389 (1998)
289. G. Harrison, R. Kannengiesser: Chem. Commun. 2065 (1995)
290. J.P. Carpenter, C.M. Lukehart, S.R. Stock, J.E. Witting: Chem. Mater. **7**, 2011 (1995)
291. J.P. Carpenter, C.M. Lukehart, S.B. Milne, S.R. Strock, J.E. Witting, B.D. Jones, R. Glosser, J.G. Zhu: J. Organomet. Chem. **557**, 121 (1998)
292. M.V. Russo, A. Furlani, M. Cuccu, G. Polzonetti: Polymer **37**, 1715 (1996)
293. S. Bandyopadhyay, S. Roy, D. Chakravorty: Solid State Commun. **99**, 835 (1996)
294. A. Lorenz, G. Kickelbick, U. Schubert: Chem. Mater. **9**, 2551 (1997)
295. G. Cerveau, R.J.P. Corriu, C. Lepeyte: Chem. Mater. **9**, 2561 (1997)
296. A.J. Wiseman, R.G. Jones, A.C. Swain, M.J. Went: Polymer **37**, 5727 (1996)
297. C.J. Wung, Y. Pang, P.N. Prasad, F.E. Karasz: Polymer **32**, 605 (1991)
298. M. Yoshida, P.N. Prasad: Chem. Mater. **8**, 235 (1996)
299. T. Sato, D. Brown, B.F.G. Johnson: Chem. Commun. 1007 (1997)
300. A. Martino, A.G. Sault, J.S. Kawola, E. Boespflug, M.L.F. Phillips: J. Catal. **187**, 30 (1999)
301. R. Gomez, T. Lopez, S. Castillo, R.D. Gonzalez: J. Sol-Gel Sci. Technol. **1,** 205 (1994)
302. W. Zou, R.D. Gonzalez: Appl. Catal. **102**, 181 (1993)
303. Y. Ou, F. Yang, Z.-Z. Yu: J. Polym. Sci. B Polym. Phys. **36,** 789 (1998)
304. T. Matsumoto, Y. Murakami, Y. Takasu: Chem. Lett. 177 (1999)
305. M.A. Ermakova, D.Yu. Ermakov, G.G. Kuvshinov, L.M. Plyatova: J. Catal. **187**, 77 (1999)
306. M.A. Ermakova, D.Yu. Ermakov, L.M. Plyasova, G.G. Kuvshinov: J. Phys. Chem. **76**, 757 (2002)
307. A. Kareiva, C.J. Harlan, D.B. McQueen, R. Cook, A.R. Barron: Chem. Mater. **8**, 2331 (1996)
308. C.J. Harlan, A. Kareiva, D.B. McQueen, R. Cook, A.R. Barron: Adv. Mater. **9**, 68 (1997)
309. D.A. Loy, E.M. Russick, S.A. Yamanaka, B.M. Baugher, K.J. Shea: Chem. Mater. **9**, 2264 (1997)
310. J. Mrowieec-Bialon, A.B. Jarzebski, A.I. Lachowski, J.J. Malinowski, Y.I. Aristov: Chem. Mater. **9**, 2486 (1997)
311. E.P. Turevskaya, N.Ya. Turova, A.I. Belokon, D.E. Chebukov: J. Inorg. Chem. **43**, 1065 (1998)
312. D. Tian, Ph. Dubois, R. Jerome: Polymer **37**, 3983 (1996)
313. D. Tian, S. Blacher, Ph. Dubois, R. Jerome: Polymer **39**, 855 (1998)

8 Polymer Intercalation into Porous and Layered Nanostructures

The intercalation of polymers into inorganic layered materials (such as clay minerals) is used to construct novel inorganic-polymer nanoassemblies with unusual structures and interesting perspectives [1–5]. An original molecular architecture of these supramolecular formations has special interest for industry [6]. The intercalation of the various *guests* into the periodic mesoporous *hosts* is connected with serious problems [7]: sorption and phase transition, ion exchange and complexation, metal and semiconductor cluster and wires formation, oxide and sulfide cluster generation, inclusion of metal complexes, covalent grafting of ligands and functional groups, condensation hybrid materials obtained by in situ, a polymerization in mesoporous channels, etc.

8.1 A General Description of the Intercalation Process

A smectite group of clay minerals, such as montmorillonite, hectorite [8–10] and mica type layered silicates [11] was used in these systems for their good intercalation abilities. The layered alumosilicates (such as montmorillonite type clays, MMT) are very convenient for the polymer nanocomposite synthesis. The exfoliated lamellae of MMT may act as nanoscale *hosts* for polymer chains and may have an impact on the intercalated macromolecule structure and orientation. Other types of layered silicates are not so important for their low reactivity. This is true for the crystalline alkali metal polysilicates family (kanemite, makatite, octosilicate, magadiite and kenyaite) because their structure consists of one or more SiO_4 tetrahedra and their layers contain terminal oxygen ions, neutralized by sodium and/or a proton. In its natural form smectite clay, such as Na-MMT, has a layered structure. The interlayer dimension is determined by the crystal structure of the alumosilicate. This dimension is approximately 1.0 nm for dehydrated Na-MMT. The lateral dimensions of the clay particles are determined by the method of preparation. Clays prepared by milling usually have lateral platelet dimensions 0.1–1.0 mm. Delaminated clay possesses an extremely large surface area because of its colloidal size and large aspect ratio. It was supposed that large surface area, high aspect ratio, and good interfacial interaction were essential to produce enhanced solid-state properties in composite materials [12].

The pillared layered structures on the base of montmorillonites clays are widely used in various fields, including a sorption technique and catalytic processes [13,14]. It is known that the kaolinite type clay mineral is a unique layered material because its interlayer space is sandwiched by OH groups of octahedral aluminum hydroxide sheets and oxygen atoms of tetrahedral silicate sheets. For the presence of inherent hydrogen bonds between the layers, only a limited number of small polar guest molecules (such as dimethyl sulfoxide (DMSO) and N-methylformamide) can directly be intercalated [2].

The same type of materials can be produced if various organic molecules (monomers, solvents, haloimidamines, crown-ethers, cryptands, etc.) or polymers are introduced as *hosts* into the *guests*' structures. Usually these substances have a natural origin or are being produced by different synthesis methods (including sol–gel synthesis) and have intercalation properties. Sometimes it is even difficult to separate the products of the sol–gel processes (Chap. 7) and intercalation ones without considering their prehistory. The sol–gel processes include a gel-precursor polymerization (a chemically controlled condensation [15]), forming a *host* metalloxide structure around a dopant (*guest*) being concluded in a specific cage-trap.

The natural clays or silicate (so called smectite) structure and their active surface physical and chemical properties have been studied in details. These very widespread minerals crystals (among these are hectorite and montmorillonite with a mica-type structure) contain alternating layers of cations and negative charged silicates (Fig. 8.1). Cations in these minerals as opposed to mica can be interchanged (to the transition metal ions too). The exchange proceeds mainly into layers, containing solvated sodium cations. The interlayer cavities can swell under filling by organic solvents, monomers or polymers. The degree of swelling depends on the nature of the interlayer cations as well as the negative charge density on the silicate layers. Moreover many inorganic oxides have surface hydroxyl groups active in the metal ion bonding. The main properties of such materials are defined by the characteristics of their inner open pore systems (particularly by the pore size). The *guest* molecules in those intercalation compounds are arranged in the *host*-molecule crystallographic cavities. Such compounds have mostly a layered, sandwich structure or the one-dimensional channel substances (tubulates). The metal ions in these structures can be reduced under the action of various chemical reducers (as well as under thermal or photochemical action).

The models and properties of the intercalated systems were studied in many works [16–22] devoted to organic, inorganic and organometallic compound intercalation or their products. The main investigations began in the last few years. The *hosts*' rigid crystalline matrices with the controlled nanosize percolation pore system can be filled by the *guest* atomic or molecular structures: clusters, nanoparticles, inorganic coordination polymers (such as CdS), big molecules (such as C_{60} fullerene), etc. For example, a buckminsterfullerene (functionalized by ethylenediamine) intercalation to mica type

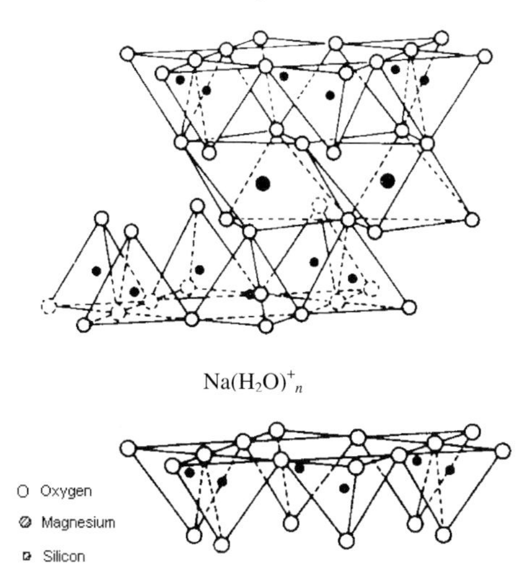

Fig. 8.1. A hectorite structure. The *upper* and *lower* layers of the tetrahedric cavities are filled by silicon; the *central* layers, by magnesium and lithium

fluorohectorite silicates was studied [23] using monomers and polymer units synthesized through an in situ process [24–26]. Such topotactic processes are characterized from the kinetic point of view by a low activation energy of diffusion. The metastable products cannot be synthesized by other methods, such as thermal synthesis (in particular because of the stratification processes, i.e. the polymer separation from an inorganic component).

The intercalation systems can be divided into two types by their architecture and pore properties [19].

The former is characterized by rigid pores with a constant volume, a parallel isolation of the lattice channels and the interconnected network channels. The *guest* concentration and spatial distribution are defined by the topology, chemical nature and reactivity of the *host* inner surface that can be additionally functionalized. The *guest* types are limited by a minimal cross-section of the linked channels system and the selective intercalation behavior in the matrices (similar to "molecular sieves").

The latter type presents a low-dimension *host* lattice, i.e. a layered or chain structure that provides "flexibility" with pore dimensions adapted to the *guest* sizes. The layer thickness in these systems (e.g. in perovskite) varies from 0.5 to 2.2 nm and the basal (interlayer, interplanar) spacing is ~5 nm. The *host* matrix lattice can have little or no effect on the intercalation–deintercalation (for insulating lattices, as in zeolites, layered alumosilicates, metal phosphates). Their intercalation behavior is characterized by the acid–base and exchange properties.

There are three main types of intercalated nanocomposites for clay particle dispersion in a polymer matrix [27] (Fig. 8.2).

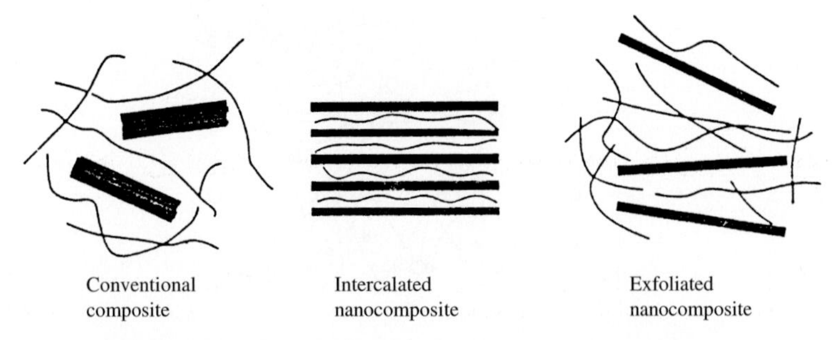

Fig. 8.2. The main types of intercalated nanocomposites

Conventional composites contain clay tactoids with layers aggregated in the intercalated face, where clay is dispersed as a segregated phase. The intercalation compounds of definite structure are formed by an insertion of one or more molecular layers of polymer into the clay *host* galleries. Their properties usually resemble the properties of the ceramic *host*. The exfoliated polymer-clay nanocomposites have a monolithic structure with a low clay content and the interplanar spacing between layers depends on the polymer content of the composite. The general material characteristics reflect those of the nanoconfined polymer.

The *host* lattices with an electronic conductivity (metals, semiconductors) have special role. Some redox reactions take place under intercalation with an electron or ion transfer, that can considerably change the *host* matrix physical properties. For example, the great intercalation capacity of xerogel based on V^{5+} is explained by its physical nature. Negative charge of the vanadium-oxygen layer in this polyvanadium acid is distributed along the V_2O_5 fibers [28, 29]. The fibers represent flat tapes (with width and length up to 10 and 100 nm correspondingly, as shown at Fig. 8.3, a) and include the

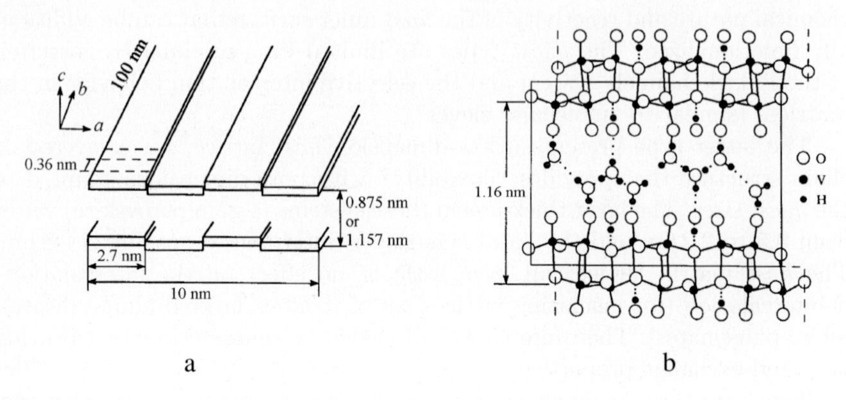

Fig. 8.3. A model of the layered xerogel $V_2O_5 \cdot nH_2O$ composition (**a**) and structure (**b**)

water molecules chemically connected with the vanadium atoms by various type bonds (Fig. 8.3, b) [30].

An intercalation can proceed through a dipole adsorption, ion exchange or a redox process. Vanadylphosphate $VOPO_4$ intercalation is similar to the described intercalation processes in $V_2O_5 \cdot nH_2O$ gels [31], because the sufficiently big cations can be introduced into a layered structure of $VOPO_4{\cdot}H_2O{\cdot}EtOH$ and during the process the interlayer spacing can rise from 0.75 to 1.03 nm [30]. It is shown that the aniline derivatives can be intercalated into the vanadyl phosphate interlayer space due to aminogroup protonation [31].

Other phosphates (such as α-$Zr(HPO_4)_2{\cdot}H_2O$ [32, 33] and $HUO_2PO_4{\cdot}4H_2O$) have intercalation properties too. For example the latter (hydrogen uranyl phosphate) is a bright yellow layered material with an interlayer distance of 0.869 nm, where the layers consist of dumbell-shaped UO_2^+-ion system with a uranium atom, coordinated by four equatorial oxygen atoms of four PO_4^{3-} tetrahedra forming two-dimensional sheets (Fig. 8.4).

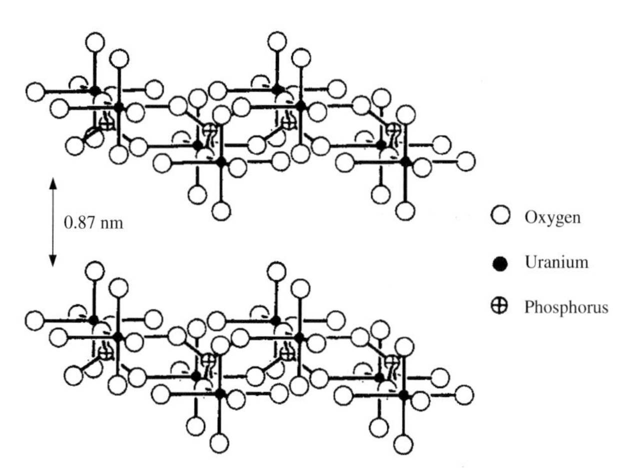

Fig. 8.4. Structure of $HUO_2PO_4{\cdot}4H_2O$

Layered nanohybrid materials can be produced [34] in $Zn(OH)_2$, ZnO or Zn/Al reactions with carbonic acids or their oxychlorides, and the *host/guest* ratio in these compositions is 1.5:0.5–1.8:0.2. The particle morphology changes depending on the nature of the reagents from fibrous to plate-like and the interlayer spacing increases from 1.61 to 2.01 nm. Silanization by the inorganic material (such as TiO_2, Al_2O_3, Fe_2O_3) is widely used. It includes the introduction of Fe_2O_3 into the silicate ($CaCO_3$, $BaSO_4$, etc.) mesopores [35]. A typical method is described in [36–39], when a thin coating is formed from silanol derivatives, dimethylsiloxane oil and hydroxyl groups on the inorganic surfaces. The solid coating is produced under a subsequent heating to 520–550 K. The coating formed is stable to hydrolysis and extraction by various solvents. The coating consists of the particles, formed through siloxane bond cleavage and chain terminal group reaction with the inorganic surface hydroxyl groups, as well as through chain cross-linking by methylene siloxane bridges. These materials have modified adsorption properties at the sacrifice of the hydrophobic layer and are stable in a wet atmosphere. The structures in the form of thin films on glass [40] can be formed by the functional layers (Ti and Au layers with thickness 5 nm and 50 nm, correspondingly) deposition on an activated PDMS surface [41, 42]. For this purpose some polyhedral oligosilsesquoixanes (POSS)-based-hybrid polymers can be used too [43–45]. The individual inorganic particle coatings (including metal and metal oxide ones) are often used in powder technology for production of *core–shell* structures, having additional useful characteristics. The polymeric coatings produced by a sol–gel method with subsequent burning-out have special properties and application. For example, water-swollen chelated polymers (containing In^{3+}–SiO_2, In^{3+}–Sn^{2+}–SiO_2 and Ag^+) were used for SnO_2 or In_2O_3 particle coatings that impart to them some new physical properties (surface conductivity, thermal resistance, antistatic property) [46, 47]. The semiconductor tubular nanostructures on the TiO_2 base were obtained using titanium tetraisopropoxide and the standard template Al_2O_3 membranes with the pore diameters 22 and 200 nm [48,49]. There is only little evidence about the target-oriented synthesis of hybrid nanocomposites with non-spherical particles. For example, the rod-like gold particles producing in the cylindrical pores of the membranes made from organic polymer or Al_2O_3 [50].

The galleries contain two water layers connected to each other by hydrogen bonds. The H^+ atom in the galleries can by exchanged with cations or be neutralized. In particular, it can be changed on the aniline molecules with the subsequent polymerization. Polyoxometalates are ideal models for molecular battery fabrication. The hybrid organic–inorganic electroactive material, composed of a conducting polyaniline matrix and the phosphomolybdate anion $(PMo_{12}O_{40})^{3-}$, is used as the active inorganic component [51].

The interesting field in sol–gel synthesis is vanadium oxide nanotube (i.e. nanotubular intercalation compounds) production [52–54]. Such sol–gel synthesis is carried out with $VO(OPr^i)_3$ in the presence of primary or α, ω-diamines and with the subsequent hydrothermal reaction in an autoclave at 453 K during 2–7 days. After washing and drying the tube-shaped vanadium oxide (having a mixed valence) presents a nanowire with $V_2O_5 \cdot 0.3\ H_2O$ composition with diameter from 15 to 150 nm (inner diameter from 5 to 50 nm) and length 15 μm [55–57]. The tubes consisted of 2–30 crystalline vanadium oxide layers with amine molecules intercalated between them. The interlayer spacing (1.7–3.8 nm) is proportional to the amine length and acts as a structure-directed template. Such nanocomposites are characterized by a turbulent morphology and well-structured walls. The vanadium-oxygen layer "spirals" are very flexible and can be stretched into a line and take part in the different exchange reactions. The structure flexibility provides potentials for a far-reaching tailoring of the nanotube structures by the modifications of their mechanical, electrical and chemical properties [58]. The most frequently studied compounds in this field are $LiMn_2O_4$, $LiCoO_2$ and $LiNiO_2$ [59]. Some metal oxides especially prepared by wet-chemical methods (such as hydrothermal or solvothermal synthesis) are structurally anisotropic. They can have different characteristics along the different crystallographic directions [60,61] (e.g. one-dimensional α-MoO_3 nanorods [62]). In this connection some other inorganic layered materials will be discussed below.

8.2 Polymerization into the Basal (Interlayer) Space

The most interesting type of intercrystalline chemical reaction is monomer molecule inclusion into the *host* pores followed by an inner controlled transformation to the polymeric, oligomeric or hybrid-sandwich products (the post-intercalation transformations, in situ process). Sometimes this method is called the "ship-in-the-bottle" polymerization approach or in situ polymerization of preintercalated monomers. Extensive studies in the field had long been conducted as can be seen from [63–65] or more recent works ([66–68], etc.). The polymer/alumosilicate nanocomposite frequently exhibit improved mechanical strength, high thermal stability, interesting barrier properties, and high chemical resistance [69–71]. The lamellar aluminosilicates act as host for polymerization of acrylic monomers [72–74]. The partial or total exfoliation of MMT aggregates depends on the nature of the monomer, medium and method of polymer clay composite formation [75,76]. The *guest* molecules are intercalated by forming new hydrogen bonds and the displacement method allowed producing various kinds of kaolin intercalation compounds.

For example, the intercalation compounds with methanol were mostly used as intermediates for *guest* displacement reactions. Thus a kaolinite-methanol intercalation compound works as an excellent intermediate that can be used for the intercalation of relatively large molecules, such as $p-$nitroaniline, ε-caprolactam, alkylamines, and poly(vinylpyrrolidone) [77, 78]. The simplest method of polymer intercalation into inorganic structure gives hybrid nanocomposites produced, for example, by emulsion polymerization of traditional monomers (usually styrene, methylacrylate, acrylonitrile, etc.) in the presence of various organophilic minerals [79]. It was discovered that the processes significantly improved many characteristics (heat and flame resistance, mechanical properties, thermal expansion coefficient, electronic and optical features) [80, 81].

Thus, even a simple one-step emulsion polymerization could offer a new approach to the synthesis of nanocomposites [82]. The physical mechanism of intercalation during emulsion polymerization is that monomer-containing micelles (2–10 nm in size) can penetrate into the swollen interlayer of MMT due to the clay swelling characteristics in an aqueous system. At the same time the monomer droplets are too big (10^2–10^4 nm) for a solution polymerization and they are adsorbed or bonded on the outer surfaces of the MMT particles. The emulsion polymerized latex particles are usually used in a 2D-colloidal array assembly [83]. There are many examples of PMMA–MMT, epoxy–MMT and other system usage. Nonextractable styrene–acrylonitrile copolymer–MMT nanocomposite material has been prepared by two different types of intercalation techniques: the usual one-step (in water) emulsion copolymerization and solution copolymerization (in cyclohexanone) with modified organophilic MMT [84]. The interlayer distance into the emulsion product is about 0.76 nm, while in the solution product (containing organic modified MMT) it is only about 0.39 nm (Table 8.1).

Table 8.1. XRD data of MMT and purified composites

Sample Code[a]	Interlayer Distance, nm	Δd, nm[b]
Na–MMT	0.96	–
Org-MMT	1.25	–
PESAN 5	1.72	0.76
PSSAN 5	1.64	0.39

[a]Org-MMT – organophilic MMT, PESAN 5 – composite styrene-acrylonitrile copolymer–MMT, prepared by an emulsion technique, PSSAN 5 – those of composites obtained from solution polymerization by employing organophilic clay
[b]Variation of distance of the montmorillonite induced by organic loading

For comparison, the expansion obtained for the PEO-MMT systems is 0.81 nm [85], which exceeds the value for the nylon-6 clay composite (0.6 nm) [86]. Moreover, the emulsion polymerization offers few advantages over solution polymerization, because the organic solvents are not used. The next step in this direction was some Ziegler–Natta catalyst intercalations in the silicate layers. Early work on olefin coordination polymerization was done in the presence of mineral fillers [87]. The metallocene catalysis investigations followed, when the internal surfaces were protected by methylaluminoxane and propylene was introduced to the system and intercalated oligomeric polypropylene was formed [88, 89]. Another and even more convenient method is based on usage of olefin polymerization one-component catalysts. For example, a specially synthesized chelated complex Pd^{2+} was intercalated into organically modified (by 1-tetradecylammonium cations) synthetic fluorohectorite [90]. The ethylene gas-phase polymerization (using this catalyst at 295 K) allows producing a high molecular PE (analysis of the toluene-extracted PE: $M_n = 159\,000$, $M_w = 262\,000$). The monomer consumption and a dramatic increase of the silicate–catalyst composite sizes was observed over a two hour period [91].

The absence of diffraction peaks on the roentgenogram of the silicate sheet structure in the PE polymer matrix composite (formed after 24 h) suggests the formation of an exfoliated polymer nanocomposite (Fig. 8.5). Thus, at the initial stage of in situ hybrid material formation a polymer-intercalated reaction takes place, and a reaction of silicate exfoliation happens at the later stages. Other nanocomposites of this type are materials from N-vinylcarbazole and ferric chloride impregnated MMT polymerization system [92] and an organosoluble polyimide–MMT hybrid material [93].

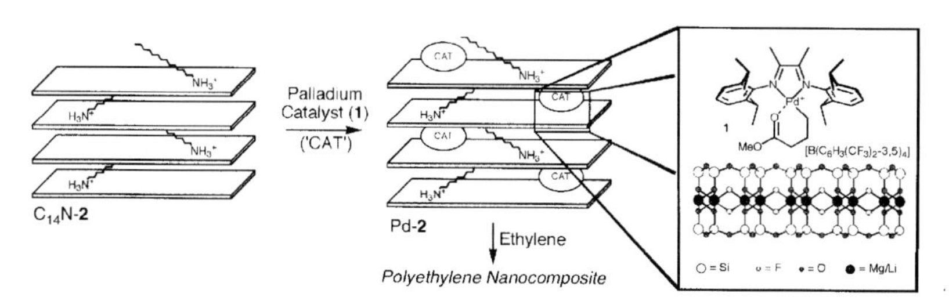

Fig. 8.5. Monomer introduction into an intercalated or exfoliated polymer nanocomposite

Apart from vinyl monomers in the mineral matrix some acetylene type monomers can be polymerized. Usually the substituted acetylenic monomers, in particalar, 2-ethynylpyridine (2-Epy), are polymerized in situ, being inserted into the galleries of MMT [94–96]. The polymer extracted from the nanocomposite P2-Epy/MMT complex is in an uncharged state. Some condensation type polymers are often fabricated by this method, e.g. the polymers on the base of ω-aminoacid with shorter carbon chains such as β-alanine [H_2N-$(CH_2)_2COOH$], and 6-aminohexanoic acid [H_2N-$(CH_2)_5COOH$]. Polymer synthesis into the interlayer space of clay was known from the early studies in this field [97–100] and later it was shown that nanocomposites of nylon-6 with kaolinite might have better properties than nylon-smectite hybrids [96,101,102]. The 6-aminohexanoic acid intercalation compound formation is eliminated for the large dimensions of the aminoacid and its low intercalation ability. Therefore, it is usually intercalated into vermiculite and montmorillonite [103] or by a *guest* displacement reaction, using a kaolinite–methanol intercalation compound as an intermediate [10]. The polymerization of the kaolin-intercalated 6-aminohexanoic acid is carried out under heating (1 h at 520 K) in a nitrogen flow. As shown in Fig. 8.6, the product of 6-aminohexanoic acid intercalation has the basal spacing 1.23 nm (the intensities of the peak at 10.6° (0.84 nm) due to crystalline 6-aminohexanoic acid adsorbed on the surface) which is larger than those kaolinite (0.72 nm) and the kaolinite-methanol intercalation compound (1.11 nm).

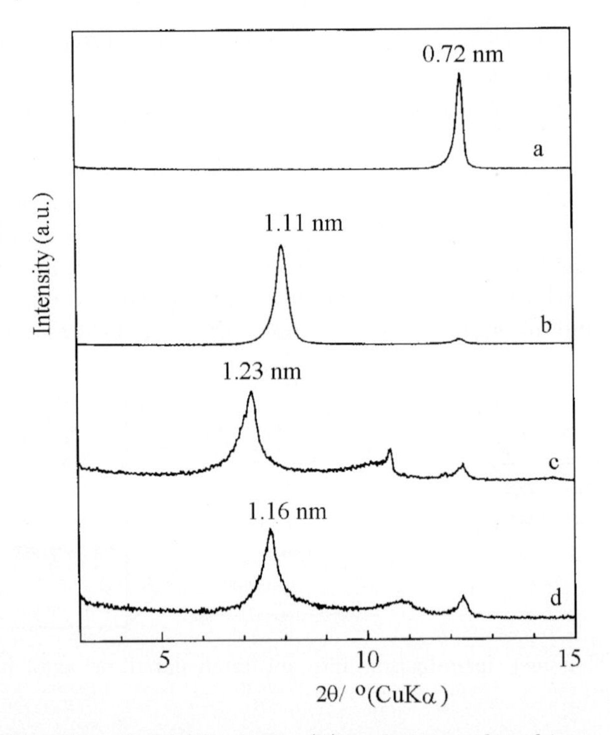

Fig. 8.6. XRD patterns of (**a**) kaolinite, (**b**) kaolinite-methanol intercalation compound under wet conditions, (**c**) kaolinite 6-aminohexanoic acid intercalation compound, (**d**) kaolinite-6-aminohexanoic intercalation compound heated to 520 K

The basal spacing of the thermally treated product (1.16 nm) is smaller than the spacing of regular product (1.23 nm). Probably, this decrease is caused by a polycondensation reaction and a simultaneous rearrangement of the hydrogen bonds of the OH groups and the *guest* species.

Polybenzoxazine-MMT hybrid nanocomposites were prepared using an organically modified MMT (treated by alkyl ammonium chloride) and a ring opening polymerization precursor (450–463 K). It is suggested that the modified MMT surface has some catalytic effect on the process of the ring opening polymerization [104]. The basal spacing of nanocomposites with up to 10 wt.% of MMT suggests the distortion and loss of structure registry of the MMT layers. Nanocomposites with a high content of organically modified MMT may have a mixed morphology (e.g. an exfoliation is combined with an intercalation).

The hybrid nanocomposites based on polyconjugated electrically conductive polymers, such as polyaniline, poly(2-ethylaniline), poly-*p*-phenylene, polythiophene, polypyrrole and polyacrylonitrile (with a subsequent pyrolysis), can be fabricated with the various mineral matrices [105, 106]. Such intrachannel reactions can even include polymerization of preadsorbed acrylonitrile that gives the confined filaments of common polymers. Pyrolysis of the included polyacrylonitrile led to the formation of both carbonized materials into the *host* material channels and a conducting material such as carbon [107].

These problems can be solved by various methods: polymer inclusion in the gels [108], usage of the polymer salt solutions, such as polyaniline-hydrochloride in an acid–methanol solution, poly-n-phenylene sulfonate (or other salts, soluble in many solvents), and introduction of monomers with their subsequent polymerization. For example, an aniline oxidative polymerization proceeds with the loss of two electrons and protons by the subsequent scheme:

$$n\mathrm{C_6H_5NH_2} \rightarrow (\text{–}\mathrm{C_6H_4NH}\text{–})_n + 2n\mathrm{H}^+ + 2ne.$$

An oxidative polymerization is described for pyrrole and dithiophene [22, 109], tetrahydrofuran [110] and aniline [111] in a FeOCl lattice. The intercalated aniline formed hydrogen bonds with the lattice chlorine atoms and the polymerization was performed along the lattice (101) diagonal (Fig. 8.7). The electrical conductivity of the forming polymer decreased in the process. This lattice is suitable for oxidative polymerization of aniline if the latter is introduced from an aprotic solvent [109, 112, 113]. The general composition of the product is $(\mathrm{An})_{0.28}\mathrm{FeOCl}$ and it can be isolated in a monocrystalline form. The zigzag polymeric chains[1] with $M_w = 6100$ (i.e. with a length comparable to the FeOCl lattice parameters) are arranged along the *host* crystal direction and the NH-group hydrogen bonds with chlorine atoms of the lattice layer. In this structure the ratio $\mathrm{Fe}^{2+}/\mathrm{Fe}^{3+}$ is ~1:9 and the polymeric intercalate is similar to a p-type semiconductor (the monocrystal specific conductivity is $1.5 \cdot 10^{-2}$ S cm^{-1}), forming a mixture of PAn and β-FeOOH under long oxidation by air [117]. The hybrid PAn-based nanocomposites can be used as electronic materials by virtue of their structures, special doping mechanisms, excellent environmental stability and good solution processability [118]. The most well-known and used variants are the PAn intercalates into V_2O_5, TiO_2, MoO_3, SnO_2, SiO_2, $BaSO_4$, HUO_2PO_4, Fe_2O_3, etc. [119–123].

A post-intercalation polymerization of aniline was carried out in air at 403 K (all the products were insulators) into phosphate layers (zirconium phosphate $\mathrm{Zr(HPO_4)_2}$ [124], vanadium phosphate $VOPO_4$ [125], uranyl hydrophosphate HUO_2PO_4 [115], layered double hydroxides [126, 127], $\mathrm{HMWO_6}\cdot$

[1] The method of quantitative polymer removal from an inorganic *host* with repeated dissolution and investigation may be an important tool for the analysis of polymer in hybrid materials. There are references on polymer extraction from the layered nanocomposites [113–115] but the extraction degree data are absent. In addition it is not correctly established that this polymer is presented in nanocomposite (i.e. that the process is not accompanied with polymer destruction). A fast and quantitative method of PEO extraction from $\mathrm{K}_x(\mathrm{C_2H_4O})_4\mathrm{M}_{l-x/2}\mathrm{PS_2}$ (M = Mn, Cd) is based on polymer extraction by tetraethylammonium aqueous salt under normal conditions [116].

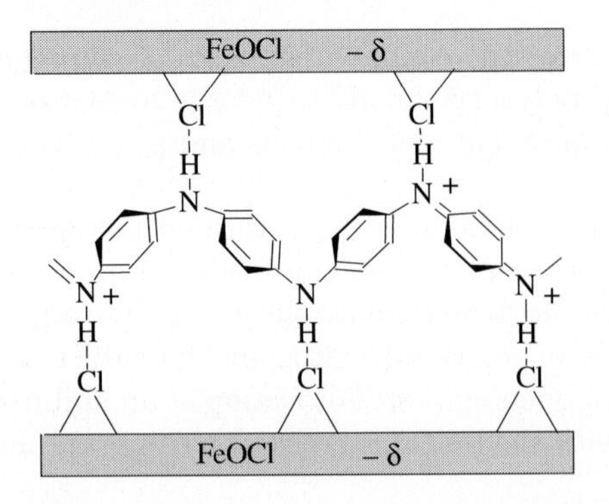

Fig. 8.7. Scheme of polyaniline bonding in a FeOCl lattice

H_2O (M = Nb, Ta), modernite type zeolites [128], and layered acidic zirconium-copper phosphates [129].

An aniline intercalation in layered Bronsted acids $HMMoO_6$ proceeds in air at 403 K and is accompanied by its polymerization with the scheme:

$$(C_6H_5NH_3)_xH_{1-x}MMo_6{\cdot}H_2O \xrightarrow{O_2} (PAn)_xH_{1-x}MMo_6 + (1+x)H_2O$$

The lattice parameters of the monomer intercalates ($x = 0.31$) show an expansion of ~0.69 nm as compared to anhydrous $HMMoO_6$, whereas the value for the polymer is only 0.5 nm (Fig. 8.8).

It is likely that the polymer chains are oriented in such a way that the C_2 axis is parallel to the inorganic stab (as in V_2O_5 gel) as well as probably in PAn/$HMWO_6$ type nanocomposites [130]. The in situ emulsion polymerization of a TiO_2 with encapsulated PAn nanocomposite allows us to utilize the functionalized protonic acid – dodecylbensene sulfonic acid both as a dopant and a surfactant to form the aqueous emulsion system [131].

Photochemical polymerization of diacetylene-3,5-octadiine into metal phosphate layers (M = Mg, Mn or Zn) was realized too [21]. It can be supposed that monomers under some conditions filled practically all the pore volume or the basal space. The following oxidative polymerization (so-called "redox-intercalation polymerization" [132]) was carried out by using molecular oxygen as an electron acceptor in the presence of an redox-active *host* (e.g. Fe^{3+}, Cu^{2+}), catalyzing the electron transfer. The layered silicates with metal ions capable of catalyzing intercalated monomer polymerization are interesting. For example, in hectorite (where sodium ions are changed to Cu^{2+} atoms) styrene polymerization was carried out both in the pores and on the surface [133]. The polymer has a "brush" type structure with an orientation effect of the inorganic surface, which is decreasing with a chain length growth and its separation from the surface. Surrounding shows itself in a higher degree of polymer ordering and its optical and mechanical properties. There are two forms of intercalated polystyrene [133]. The first one is similar

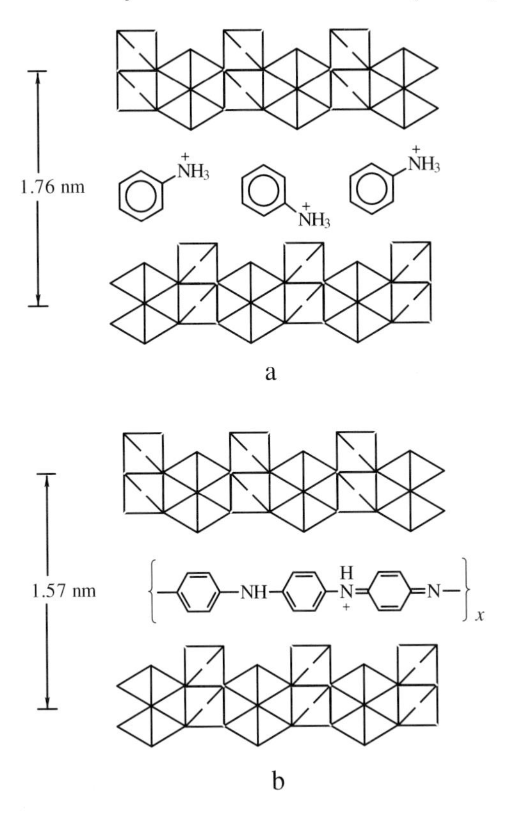

Fig. 8.8. Possible configuration of the intercalated aniline (**a**) and polyaniline (**b**) in $HMMoO_6$

to block polystyrene on the surface; the second one is more rigid (probably owing to a higher degree of polymer ordering).

The layered materials are formed with an electroconductive polymeric layer from the polyconjugated anisotropic laminar microstructures under the redox-intercalation polymerization of some polymers (as aniline, pyrrole, thiophene, 2,2′-thiophene) in V_2O_5 xerogel [134–138]. These structures arise in the oxidative-reductive reaction when the oxidizing monomer is polymerizing and V^{5+} ion is partially reducing to a tetravalent state.[2] In such systems intercalation is attended by polymerization. As a result in the aniline case, a deep-blue complex with composition $C_6H_4NH)_{0.44}{\cdot}V_2O_5{\cdot}0.5H_2O$ and d = 1.4 nm with a metallic shine is formed. The growth proceeds primarily into xerogel and it is critical for the intercalated polymer formation. Xerogels are connected with molecular oxygen transport, during which the V_2O_5 xerogel is

[2] Recently the binding energies for pyrrole were determined by both experimental and quantum chemical methods for a number of main groups and transition metals [139].

acting as a catalyst too. An electroconductive polymer with a changing ratio PAn/V_2O_5 is formed by the *host* mixed valence (V^{4+}/V^{5+}) lamellas ordered in one direction. The material is composed of vanadium oxide and polymer alternating layers (Fig. 8.9) and its conductivity is four orders of magnitude bigger than of the initial $V_2O_5 \cdot nH_2O$ xerogel and is equal to ~0.5 $ohm^{-1} \cdot cm^{-1}$ at room temperature. The distinctive characteristics of the process are that there are two different phases having composition $(C_6H_4NH)_{0.6} \cdot V_2O_5 \cdot nH_2O$ and $(C_6H_4NH)_{1.2} \cdot V_2O_5 \cdot nH_2O$ [140] with the interlayer distances 1.4 and 2.0 nm, corresponding to one and two monolayer intercalation into xerogel. Such structures allow one to dope intercalated PAn with a little increase of electroconductivity of the formed materials, being redox-active in both water acid media and aprotic organic electrolyte [141]. A salt form of PAn arises too. The material ageing in air undergoes partial oxidation of the inorganic skeleton and PAn oxidative bonding into the interlamellar space, which is attended with the further oxidative polymerization of aniline oligomers into the xerogel layers. The polymer is formed as it is "frozen" in the xerogel and it is aided by an interaction between organic and inorganic components due to hydrogen bond(–N–H $\cdots$ O–V–) formation.

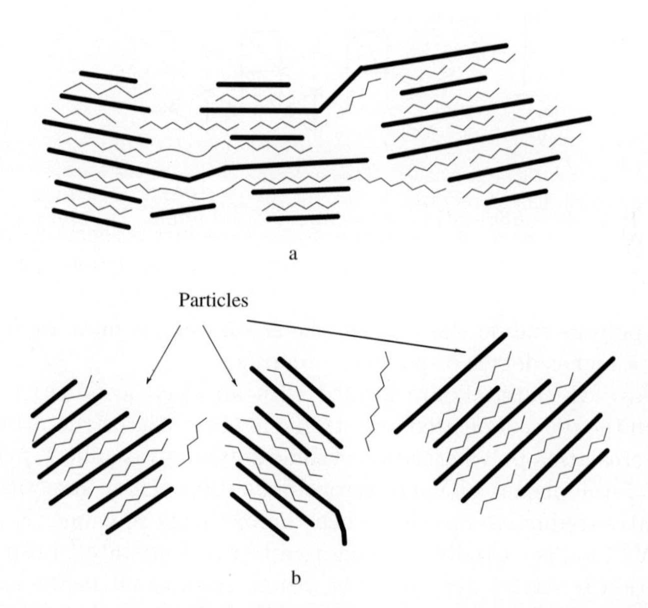

Fig. 8.9. Schematic presentation of the structural features on the nanometer scale for the samples with different morphology, being produced by intercalation polymerization in situ in the form of film and powder. **a** Film with electroconductive grains connected by the random polymeric bonds and V_2O_5 layers; **b** pressed powder with isolated electroconductive particles having distinct boundaries

The microstructure and properties of an aerogel composite V_2O_5–polypyrrole (PPy) depend on the synthesis methods. It can be produced under joint polymerization of pyrrole and vanadyl–alkoxide precursor in the mixture $VO(OPr)_3$ + pyrrole + water + acetone, in V_2O_5 + PPy gel. But a tablated nanocomposite, produced by "post-gelation polymerization", including the subsequent polymerization of the inorganic and organic phases [142–144] has the best conductivity.

A co-intercalation of various monomers in the hybrid material (including the two polymer components, as in polymeric blends) has not been studied. Only one method can be mentioned when a methanol solution of 2,5-dimercapto-1,3,4-thiadisole and aniline is mixed with $V_2O_5 \cdot nH_2O$ gel in molar ratio 1:2:2. The special studies proved that in this system a polymer is formed in the V_2O_5 matrix [145, 146].

An aniline intercalation process, its competitive polymerization in MoO_3 and the formation of $(PAn)_{0.24}{\cdot}MoO_3$-type structure composite were analyzed in [123, 132] using low-temperature intercalation methods. It was shown that polyaniline chains could widen the layers and change the potential surface by decreasing the lattice polarizability.[3] The following oxidation of the polyaniline chains intercalated into MoO_3 can be carried out using $(NH_4)_2S_2O_8$ [113]. The chemical synthesis of the hybrid material was carried out also by direct addition of pure aniline to solid phosphomolybdic acid $H_3PMo_{12}O_{40}$ [51,148]. The use of hybrid electrodes gives many advantages. It demonstrates the possibility of synthesizing hybrid materials based on a PAn matrix with the phosphomolybdate anion as the only doping species. It allows the establishment of the behavior of the hybrids upon cycling in terms of anion/cation insertion. Third, it allows exploring some of the possible applications of the bulk solid materials, incorporating as molecular species such phosphomolybdate anions [149].

It is important also that the oxidative polymerization of aniline, pyrrole and thiophen monomers, intercalated into the layered alumosilicates, gives the high-ordered *host–guest* layers.

The particles of TiO_2 (from $TiCl_3$ solution at pH 2.5) ~20 nm in size are formed on the photochemical electrodes and can be used in solid photosensitive solar cells (see Chap. 11). At the same time electrodeposition of PPy (formed during pyrrole electrochemical polymerization) proceeds after its adsorption into the electrode pores coated by a photosensitive ruthenium complex [150, 151]. Such structurally controlled "templates" are analogs of self-assembling supramolecular aggregates. The fabrication of these molecular recognition supramolecular systems (including chromophores, semiconductors, clusters and other structures with specific optical and electronic functions) is now a main approach to ferment modeling [21, 152, 153]. The

[3] The geometric structure and vibrational properties of polyene complex with aluminum atom (as a conductive model) were calculated by the quantum chemical Hartree–Fock ab initio method [147].

interconnected arrays of Au/polypyrolle colloids with 1D morphology were synthesized, using Al_2O_3 membranes as the template [154]. Some fibrous structures based on ZnO [155] and WO_3 [156] were produced. The diameters of these fibrillar and tubular materials correspond to Al_2O_3, but their length was significantly higher. The metastable monocrystals of anatase phase were observed for TiO_2 (such materials can be excellent photocatalysts). A general survey of such template sol–gel synthesis in fibers or nanotubes (as well as in micro- and nanoporous membranes) is given in [157]. In this way fibrous electrode materials were produced on the base of V_2O_5 [158], as well as oxide semiconductor materials such as MnO_2 [159], Co_3O_4 [160], ZnO, WO_3.

The *semiconductor/conductor* materials in the form of nanosize tubes or nanowire particles were produced [48, 161] by a template TiO_2 synthesis in Al_2O_3 pores with a subsequent polymerization of pyrrole [162]. The same type tubular composites were fabricated using a chemical vapor deposition (CVD) method [163]. The produced micro- and nanostructural materials and their synthesis methods were analyzed [164].

Structures having wide pores and big channels (with diameter ~3 nm) allow one to capsulate parallel polyaniline chains and form them in situ (Fig. 8.10) with the result that they gain a fibrous form and a microwave conductivity [165].

Thus intercalation polymerization has giant potential (now it is very far from realization) for polymer-inorganic nanocomposite fabrication and it is the reason for such intensive investigations in the field.

8.3 The Macromolecules Introducing into the Layered Host Lattices

This approach is very promising, first and foremost because it gives new possibilities for inorganic/organic polylayered composite fabrication and secondly because the physical and chemical features of intercalation are very interesting themselves. They allow one to fabricate systems with electronic conductivity (e.g. for using in the reversible electrodes [166]) or with improved physical and mechanical properties (as in layered nylon-silicate composites [167]), hybrid epoxy-clay nanomaterials [168], nanomaterials on the base of hectorite and polyaniline, polythiophene or polypyrrole [169]. In future such polymer/clay hybrids will find many new applications in industry and science [170] because of their excellent thermal, gas barrier and mechanical characteristics.

Furthermore, studies of such systems can give important information about polymer adsorption on the nanomaterials. For example, the PVP adsorption isotherm in suspensions of nanocrystalline CeO_2 (particle size about 9 nm, synthesis under hydrothermal conditions) were measured at different pH values [171, 172].

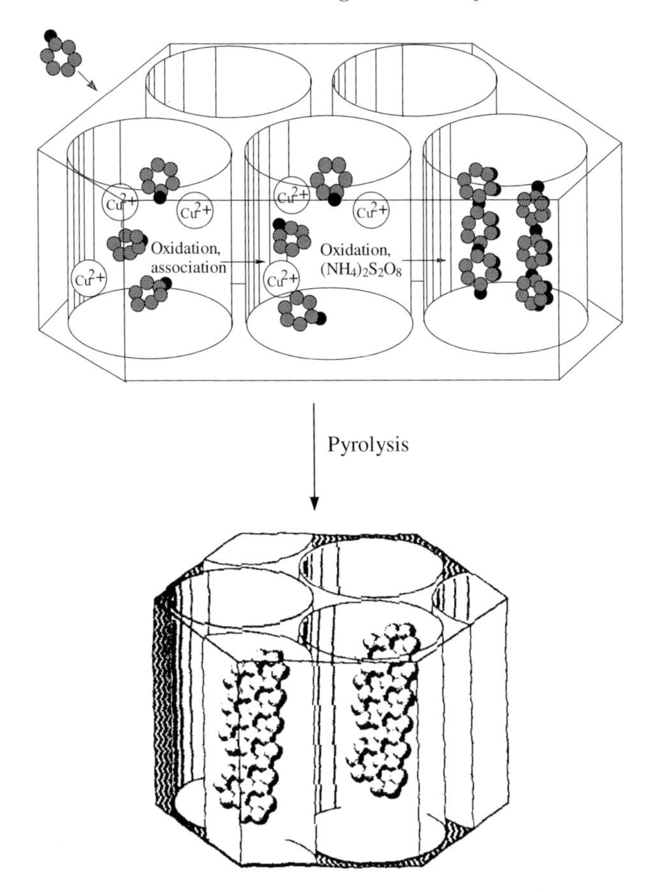

Fig. 8.10. Polyaniline incapsulation in nanochannels (details in text)

In general we can obtain two types of hybrid structures: intercalated (where one or more polymer is incorporated between the silicate layers) and delaminated (where the silicate layers are dispersed individually in the organic matrix).

These materials are usually made by using Na-montmorillonite clay (MMT), that was rendered organophilic by cation exchange with a long chain amine, such as stearylamine [173]. The layered anionic phyllosilicates were used as inorganic components due to their high aspect ratio and ease of ionic modifications, but the polymer must separate these silicate sheets, having either an intercalated morphology (silicate registry is retained) or preferably a delaminated (or exfoliated) structure that is microscopically isotropic.

Hybrid materials such as polypropylene, polystyrene, poly(ethylene oxide) and nylon/clay were made by melt mixing [174–178] and in the melt mixed with PP and a maleic anhydride functionalized PP additive – maleic anhydride as the compatibilizer. Such composites (they are marked as PP–

MA/clay) are nanocomposite sensitive of melt-state rheological measurements to interparticle structure and chemistry of the hybrid materials [179].

By TEM data, the silicate layer length is about 150 nm and thickness ~4 nm. They are finely dispersed in the PP-MA matrix [180]. It is supposed [181] that the correlation length of the dispersed clay particles in an intercalated PP-MA is about 30–50 nm and can be compared with a polymer radius of gyration. One or more of the extended PP-MA chains were inserted between the silicate galleries. The hierarchical structure of the intercalated system was viewed on scales from the structure of confined PP-MA chains in the space of the silicate galleries (intercalating sites) of 2–3 nm width to crystalline lamellae of 7–15 nm thickness and spherulitic texture 10 μm in diameter. The clay particles act as a nucleating agent for the PP-MA matrix. Polymer crystallization is one of the effective processes to control the extent of polymer chain intercalation into silicate galleries [182]. Since the interlayer spacing is limited (2–3 nm), the crystallization process must be hindered. One can expect that the extra heat will be diffused for that type of orientation inside the silicate galleries, when they are subjected to heating from room temperature to T_m.

A typical example of a polymer direct intercalation is introducing a PEO into the planar silicate (as mica) layers by an interaction of the melt polymer with the *host* Na^+- or NH_4^+-exchange lattice [183]. The general formula for the PEO intercalate in the layered polysilicate magadiite (both from a solution in a bipolar aprotonic solvent or from the polymer melt at 428 K) can be presented as $H_2Si_{14}O_{29}(-OCH_2CH_2-)_3$ [184]. This is in agreement with the structure where one unit cell of magadiite corresponds to three oxyethylene units, in other words the interlamellar region is not saturated with oxyethylene units completely.

The useful inorganic matrix is V_2O_5 [185] to which PEO [186], PVP, PPE [187, 188] are introduced from a water solution. It increased the interplanar distance to 1.32 nm. PEO can be introduced into the lamellar lattices of $V_2O_5 \cdot nH_2O$ and $CdPS_3$ too [188, 189]. Thus, PEO (mol. mass 10^5) water solution with an aqueous gel of $V_2O_5 \cdot nH_2O$ (interplanar distance – 1.155 nm) forms, after water removal, an xerogel composite with a total formula $(PEO)_xV_2O_5 \cdot nH_2O$ [190]. The interplanar distance increased from 1.32 to 1.68 nm (at $x = 0.5$–1.0) and from 1.76 to 1.83 nm (at $1 < x < 3$). It is of interest that at $x < 1$ the composites contain fibrous monolayers, but at $x \geq 1$ the composites contain double layers of PEO molecules, being closely adjoined to each other so that the xerogel interplanar space is completely filled by these fibrous mono- and double layers. The monophase composite (at $x < 0.8$) can easily form a flexible thin film and can take part in an oxidative-reductive intercalation, resulting in the formation of solid electrolytes and electrochromic materials [115, 188]:

$$y\text{LiI} + (\text{PEO})_x\text{V}_2\text{O}_5 \cdot n\text{H}_2\text{O} \rightarrow \text{Li}_y(\text{PEO})_x\text{V}_2\text{O}_5 \cdot n\text{H}_2\text{O} + (y/2)\text{I}_2.$$

Alkali metal ions with PEO mainly form inclusion compounds, which can be introduced into the ionic silicate, such as montmorillonite layers (Fig. 8.11). The intercalated salt complexes (PEO /Li^+–MMT) are separated by 0.8 nm and the PEO chain has a little stressed helicoidal conformation.

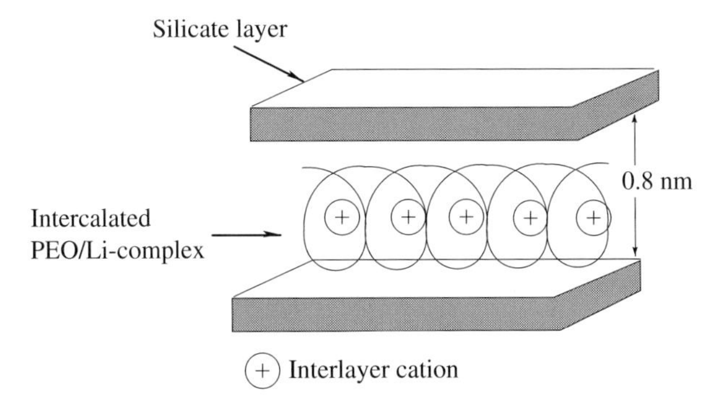

Fig. 8.11. Structure of PEO, intercalated into a monoionic silicate layer

The PEO–LiX systems (X – halogen, alkoxy-, aryloxy- or other group) are widely used as flexible solid electrolytes [191–194] and PEO have been modified to produce new ion-conductive materials [195–199]. The principal scheme of ionic conductivity in this type of self-organized systems is shown in Fig. 8.12.

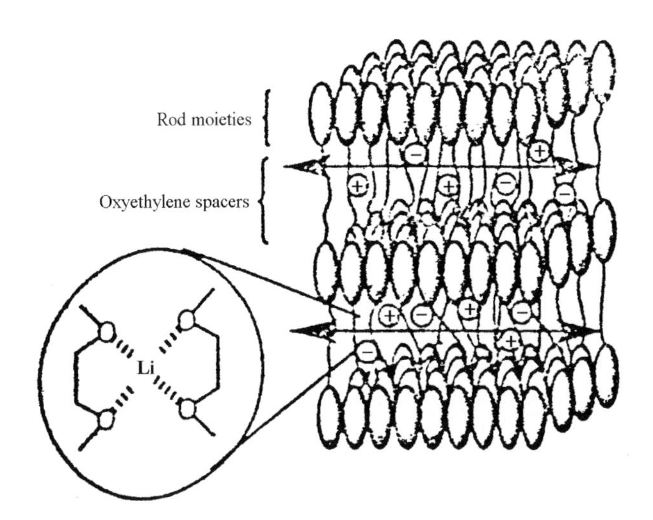

Fig. 8.12. Schematic illustration of anisotropic ion conduction in self-organized ion conductive materials

The size and charge number of the metal salts affect the mesomorphic behavior and ion-conductive properties of such systems [200]. The various lithium salts appear to be the most suitable for applications. Some other cations and particularly the PEO/Zn salt system were investigated as well. The compounds including the various number of PEO units, such as $(\text{–CH}_2\text{-CH}_2\text{–O–})_n\text{ZnBr}_2$ ($n = 6, 8, 15$), were described [201]. The local surroundings of complexes significantly affect the polymeric electrolytes properties, as was shown for the systems $(\text{–CH}_2\text{-CH}_2\text{-O-})_n[(\text{ZnBr}_2)_{1-x}(\text{LiBr})_x]$ ($n = 20$–80, $x = 0$–0.5) [202, 203].

Sometimes such systems are used with ceramic fillers (of $LiAlO_2$, Al_2O_3, TiO_2 type), as for example a polymer-inorganic composite was fabricated with conductivity 10^{-4} S/cm at 323 and 10^{-5} S/cm at 303 K (see Chap. 11) by introduction of PEO into a system where 10% of TiO_2 nanoparticles (13 nm) or Al_2O_3 (with mean size 5.8 nm) were dispersed in acetonitrile with $LiClO_4$ (molar ratio $LiClO_4$/ PEO = 1:8) [204].

The methods of sol–gel synthesis present an alternative approach for producing materials, where organic molecules or polymers are introduced as *guests* into structures having intercalation properties (see Chap. 7). For example in these processes poly(isoprene-block-ethyleneoxide), (3-glycidyloxypropyl)trimetoxysilane and aluminum *sec*-butoxide can be used [205], but the morphology of such self-organized systems differs from the traditional intercalated composites [206]. It is interesting to study the sol–gel polymerization of so-called "silanized" monomers the pyrrole alkoxide derivatives, e.g. N-[3-(trimethoxysilyl)propyl]pyrrole [207, 208] or 2,5-bis(trimethoxysilyl)thiophen [209]:

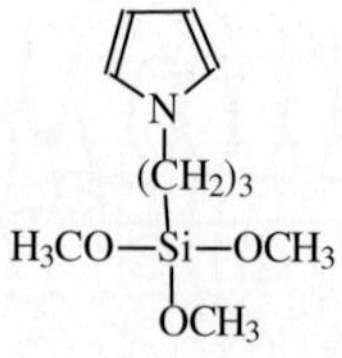

N
$(CH_2)_3$
H_3CO—Si—OCH_3
OCH_3

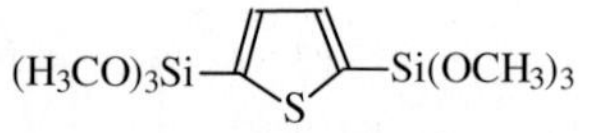

$(H_3CO)_3Si$—S—$Si(OCH_3)_3$

There is interest in nanocomposite material production on the base of water soluble polymers, such as copper tetrakis(N-methyl-4-pyridyl)porphin [210], PPO, PVP, methylcellulose [188], and poly(4-vinylphenol) [211]. The cellulose holds the greatest promise because it presents a high porous system of conjugated fibers. Microcrystalline cellulose is used as a matrix for metal nanoparticle intercalation, e.g. silver nanoparticles [212].

The intercalation of layered transition metal halides, contrary to the above described layered metal oxides, chalcogenides and thiophosphates, are not very stable and are subjected to the various chemical transformations under processing (hydrolysis, dissolution, decomposition). α-$RuCl_3$ with lamellar

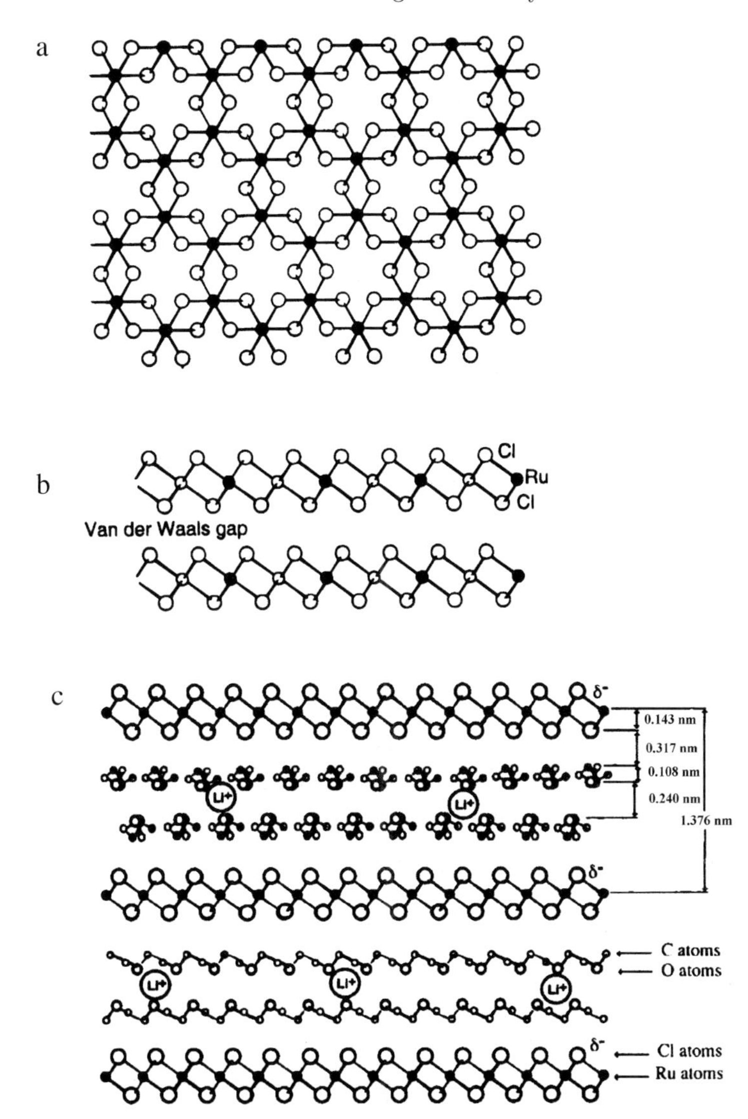

Fig. 8.13. The layered structure of α-$RuCl_3$. The coordination environment of the Ru atoms is octahedral. (**a**) View down the c-axis. (**b**) View parallel to the layers. (**c**) Structural model for $Li_x(PEO)_yRuCl_3$ derived from the one-dimensional electron density maps of $Li_x(PEO)_yRuCl_3$

structure (Fig. 8.13) and remaining stable under used test conditions is an expection [213].

The cations can be intercalated by a reduction or an ion exchange, while neutral polar molecules can be incorporated only through solvent exchange. For example, α-$RuCl_3$/PAn nanocomposites were synthesized with in situ redox intercalative polymerization [214]. They behave as an excellent polymer-intercalation *host* in addition the α-$RuCl_3$ group [215]. The method of encapsulative precipitation of exfoliated lamellar solid from a solution was used for water-soluble polymer (PEO, PVP, PEI) intercalation. The group $Li_xRuCl_3(x \approx 0.2)$ was produced under α-$RuCl_3$ interaction with $LiBH_4$ in an inert atmosphere. It can form intercalation compounds with these polymers of $Li_x(PEO)_yRuCl_3$ type (Fig. 8.13, a–c) and has an ion conductivity comparable with the best polymer electrolytes [215].

The above analyzed methods (Chap. 2) of nanoparticle production using ultrasonic treatment can be useful for nanoparticle intercalation into polymers too. An idea of the approach can be illustrated by the example of Ni nanoparticle (10–40 nm) deposition on a submicroscopic silica gel [216]. The initial nanoparticles are formed in decaline by US-irradiation of $Ni(CO)_4$, when cavitation combines the Ni–C bonds splitting coincidentally with SiO_2 surface activation (adsorbed water removal, Si–O–Si bond breaking and free silanol bond formation). An alternative approach includes a metallic nickel reaction with activated surface silanol groups and the Si–O–$Ni^{\delta+}$ type bond formation. Such centers become the nuclei for further Ni nanoparticle growth. A subsequent pyrolysis at 670 K converted the amorphous superparamagnetic Ni clusters into a crystalline ferromagnetic material. A similar mechanism is realized for other dehydrated surfaces, too, e.g. in the case of Al_2O_3 and metal carbonyls the Al–O–$M^{\delta+}$ structures are formed [217]. In a similar way the Au nanoparticles are stabilized in silica gel monolite [218] and MnO_2 nanocrystals (of KMn_8O_{16} type with the structure of hollandite [219] or mica [220]). Rhodium carbonyl thermolysis on a Al_2O_3 surface gives small clusters of rhodium [220]. A MMT modification by intercalation of mixed ferrochromic oligomeric polyhydrocomplexes allows increasing of its adsorptive and structural characteristics, as well as thermostability [221].

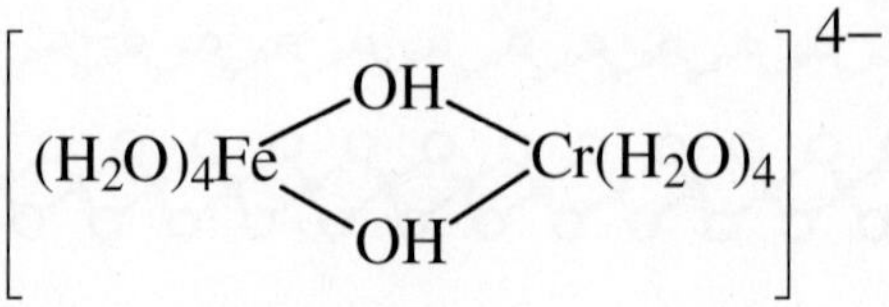

$$\left[(H_2O)_4Fe \begin{matrix} OH \\ \\ OH \end{matrix} Cr(H_2O)_4 \right]^{4-}$$

Self-assembling nanocomposites with a multilayered structure $(P/M)_n$, where M and P are nanoscale layers of inorganic component and polymer with opposite charges attract attention [222–224]. Such structures have a complex of valuable characteristics and now a diversity of approaches is realized for such "layered" assembling, for example from polyelectrolytes and clay [225, 226], exfoliated zirconium phosphates [227], and colloidal metal particles [228, 229]. The mechanisms of such material formation and defect generation are studied, as an example, in anionic montmorillonite

$[AlMg(OH)_2(O)_4]^{5-}[Si_2O_3]^{4+}[Na, nH_2O]^+$ and positively charged poly(diallylmethyl-ammonium chloride) on the plates of glass, quartz, silver, gold and even teflon [230]. Sequentially dipping these plates into the P solution and M suspension, one can increase the number of layers, n. Every such dipping extends the plate thickness to 1.6 nm (for P) and 2.5 nm (for M).

The multilayer formation is a multi-step process. At first the P-layer is adsorbed on the substrate surface by the electrostatic and Van der Waalse forces, during which a structural hierarchy of M on this template gives the numerous possibilities for the various molecule and cluster generation [231–235]. They can be introduced between the swelled layers and into the M plates and on these plate surfaces, as well as on the surface of isolated or coagulated M plates. The M layer adsorption on the opposite charged polyelectrolytes is strong and irreversible and also has a very dense planar orientation. The non-regular M-layers do not provide a complete coating of the intercalated P-layers and in this way the overlapping piles are formed. The interphase roughness is bigger than the P/M-layer thickness and does not depend on the nature of the substrate.

The layered nanostructures can be produced [236] using liquid-crystalline crystals with ionic groups and some inorganic components such as montmorillonite or hydrotalzite $[AlMg_2(OH)_2(OH)_4]^+[0.5CO_3, OH, Cl]^-$. By different estimations the mean thickness of a M/P pair is about 4.9 nm. Such electrostatic assembling provides a tight contact between components and allows producing new types of liquid-crystalline structures with unusual properties.

There is special interest in layered interstitial compounds in graphite (LCG) because the latter can be considered as an aromatic type macromolecule with about 1000 aromatic rings and with the identity period 0.335 nm. The parallel carbon planes in graphite are not connected by chemical bonds (the energy of the interplane interaction equals 16.8 J/mol), which makes possible the inclusion of monomolecular layers of various substances (including metal ions) and the formation of some layered (laminated) graphite compounds [237]. The methods of LCG compound production are based on the interaction of graphite with metal vapors or solutions into strong ionizing solvents, low-boiling chlorides or cationic metal complexes, followed by their introduction into the interplane space of the graphite lattice. These compounds are divided into the first, second and third steps of introducing products, according to the layer number, separating the two nearest layers of the introduced metal.

The bond nature in the layered graphite-metal compounds depends on the metal atom type. Thus, for Fe, Co, Ni, Mn and Cu the bond is carried out by Van der Waalse interactions. Sometimes the graphite π-electron density transfer occurs to the introduced atomic layer, so the graphite carbon grid in such situations presents an off-beat polymeric ligand. This bond for alkali metals is formed by an electron transfer from the metal atoms to the conduction band of a neighboring graphite layer, i.e. by an electrostatic interaction between the positive metal ions and free electrons of the graphite conduction band. The processes of the introduced metal ion reduction can be accompanied by their partial separation from the layered stacks and a reduction on the graphite outer surface with the formation of initial nanoparticles (for example, Ti particles), included in the imperfect graphite lattice [238]. The various processes of inclusion of many metals into graphite can proceed at high pressures in combination with a shearing deformation [239].

The intercalation chemistry gives practically boundless possibilities for hybrid nanocomposite fabrication. Now the main features of such materials formation are defined, as well as the structural organization and properties of the materials. Intensive research in the field of hybrid nanocomposites is initiated by growing demands and in the near future one could expect new types of such materials with the hierarchical structure of the intercalated system. For example, a new interesting class of hybrid materials including azomacrocycles/gallium-phosphates was synthesized and characterized [240]. The inorganic layers in such intercalation nanocomposites keep the structural characteristics of the parent polysilicate or other layered material. The organic polymeric monolayers are constrained in a flattened arrangement, resulting from their strong interaction with the interlamellar silicate surface. Moreover, such composites can be used for metal nanoparticle formation. Some electrically active polymers (polyaniline, polypyrrole) – SiO_2 were used for metal (gold and palladium) uptake [241]. Both polymers and hydrazine reduce these metals with the formation of immobilized nanoparticles that can be used as catalysts (Chap. 12). The other type of composition materials (without a polymeric ingredient) can be produced by these methods. For example, in [242] the intercalation compounds of layered double aluminum and lithium hydroxides were used. The compounds contain Ni, Co and Cu complexes with an organic ligand, namely ethylendiamine-tetraacetic acid with composition $[LiAl_2(OH)_6]_2[M(Edta)]\cdot nH_2O$ in the interplanar space. The materials under calcination in vacuum (at 670–720 K) include mono- and bimetallic nanoparticles (from 3–4 to 40–50 nm), that are stable to oxidation because the carburized matrix can protect them not only from air oxidation but even under endurance in nitric acid.

The monoclinic structure (m–WO_3), produced by the sol–gel method, can even be chemically bonded poly(propylene glycol) to triethoxysilane end-capping groups [243]. Such films gain gasochromic properties after impregnation by H_2PtCl_6/i-propanol solution and following heat treatment (at 650 K). Finally, this type of hybrid composite is seldom produced at the polymer processing stage, but a polyamide-6 hybrid composite with potassium titanate whiskers ($K_2Ti_6O_{13}$) was formed in twin-screw extrusion followed by injection moulding [244].

8.4 Intercalation Nanocomposites of Polymer/Metal Chalcogenide Type

The synthesis and characterization of special hybrid nanocomposites – periodic nanostructured semiconductors – are very interesting. Their unusual optical, electrical and other properties have great potential applications in microelectronics. These compounds are usually produced by wet chemistry methods including the sol–gel method [245, 246]. They are based on various systems: CdS nanocrystals entrapped in thin SiO_2 films with the $Co_6S_8(PPh_3)_x$ intercalated into MoS_2 [247], self-assembling CdS–Ag hybrid particles [248, 249], and semiconductors on Cd_3P_2. The color tunability of semiconductor nanocrystals as a function of size could lead to many applications that rely on color multiplexing [250].

The most attention was given to CdS encapsulation and stability, processability and important application of such systems. The first reports [251, 252] were about semiconductor/polymer systems with physically dispersed CdS grain arrangements, then many reports were published about various combinations: Nafion/CdS [253], PS/CdS [254], PVK/CdS [255], cellulose/CdS [256], CdTe/ PS–polyallylamine hydrochloride [257, 258], CdS/ PMMA–PMAA [259] and others [260, 261]. The different statistical and block-polymers, particularly the variant of CdS/PS-b-P2VPy [262, 263] and CdS/PS-b-PAA [264], were widely used for these purposes. As described in Chap. 5, a metal ion is absorbed by the polar part of the block-polymer into a THF solution with subsequent treatment of a formed macrocomplex by gaseous H_2S. The generating CdS nanoparticle size is regulated by the ratio 2-VPy/Cd^{2+} and is 4–7.5 nm. The micelles of amphiphilic triple block copolymer ABC allow the use of these copolymers as nanoreactors for inorganic nanocrystal production [265–267]. The CdS semiconductor nanocrystallites were synthesized in such reversed micelles with subsequent in situ enzymatic copolymerization of *p*-ethylphenol and 4-hydroxythiophenol (M_w below 2000). The polymer/CdS core was then dispersed into polycarbonate and the formed nanocomposites showed higher optical absorbance in the UV and visual range [268].

A simultaneous in situ polymerization-hydrolysis technique, used for polyacrylamide (PAm)/MS semiconductor preparation (M = Cd, Zn, Pd) [269], consists in a parallel regime of acrylamide radical polymerization in the presence of M^{2+} ions and hydrolysis of NH_2CSNH_2 (as sulfide ions precursor). The size distribution of CdS particles lies in the range 2–10 nm. In another approach hydrothermal emulsion polymerization and simultaneous sulfidation are carried out using a cadmium water solution, vinyl acetate and thioacetamide CH_3CSNH_2 [270].

A multicomponent semiconductor nanocomposite Cu_2S/CdS/ZnS was also successfully prepared in PS by ion exchange [271]. A standard way to produce hybrid nanocomposites with a two-component mixture (*p*- and *n*-type) of univalent copper sulfides and Cd^{2+}–CdS/Cu_2S [256] is to fabricate at the initial stage a CdS nanophase formation with a subsequent generation of Cu_2S (in a topochemical reaction of partly Cd^{2+} ions substitution to Cu^+ ions). The different methods of chalcopyrite ($CuInSe_2$) introduction into polymers are described [272] and the best is to combine its synthesis with acrylamide radical polymerization [273], when the forming nanoparticle mean size is 12 nm. The PAAm ionomers with a certain ratio of ionic and aliphatic chains can be attached to CdSe/ZnS nanocrystals by a phase transfer reaction [274], and the forming structures can be called core–shell structures (where the monodisperse submicrometric latex sphere presents a core and the nanocrystal presents a shell). The resulting polymer-coated nanocrystals can be dispersed in organic solvents and are characterized in terms of structure and photostability. The same systems have the CdSe particles coated by ZnS with a subsequent poly(N-acryloyloxysuccinimide) stabilization on the corresponding monomer polymerization [275].

The low-dimensional arrays of semiconductor nanoparticles as for example the self-organized 1D chains are observed in research [276]. The above mentioned hydrothermal polymerization and simultaneous sulfidation [270] allow us to fabricate many one-dimensional (1D) nanocomposites, where the CdS nanoparticles self-assembled in polymer nanorods are well dispersed in bulk.

Such materials have many applications in plastic semiconductors or superconductors: films with special properties or superconductive wires on a polymer matrix base, etc. [19]. The results are reviewed in [277, 278].

There are many methods of film composite production [279] but most frequently a mother solution (including the corresponding precursors) is used with a subsequent in situ synthesis, for example by mixing a zirconium tetrapropoxide solution with propanol and hydrolysation by an aqueous solution of acetic acid [280]. After addition of the cadmium acetate and ammonium thiocyanate solutions the composition is applied to a substrate (e.g. glass) and treated by the method described in Chap. 7 for the sol–gel synthesis.

The characteristic property of the layered materials (such as metal dichalcogenides MoS_2 or TaS_2 with a low density of the layer charge) is that they can decompose into nanoscale "blocks" under certain conditions and form colloidal solutions.[4] NbS_2 and TaS_2 (as well as MoS_2 and WS_2), produced by heating of amorphous MS_3 in a hydrogen stream, form fullerene-like polyhedra and nanotubes of the layered metal disulfides [282, 283]. For

[4] The MX_2 compounds have a layered structure if the anion is easily polarizable and the cation has strong polarizing power. Such structures have molybdenum dichalcogenides, where the molybdenum atom layer is placed between two layers of chalcogen (X). As a result three-layered packets are formed, where the inner bonds are significantly stronger than between two such packets (the latter bonds are Van der Waalse forces only). The molybdenum disulfide MoS_2 monolayers have a distorted octahedral configuration with a (2×1) superlattice and the coordination unsaturated Mo-centers in the MoS_2 prismatic centers [281].

example, NbS_2 gives individual nanotubes with hollow cores with diameters in the range 4–15 nm [284] and it is likely that TiS_2, ZrS_2 and HfS_2, produced by similar methods, have the same structure. The "exfoliation procedure" is well studied [285,286]. $NbSe_2$ is one of the best superconductors amongst the layered chalcogenides (its superconducting transition temperature T_c equals 7.2 K). A general scheme of PVP, PEO or PPO intercalation from aqueous solutions into the suspended $NbSe_2$ monolayers has the following form [287]:

$$NbSe_2 + x\text{LiBH}_4 \rightarrow \text{Li}_x\text{NbSe}_2 + x\text{BH}_3 + x/2\text{H}_2;$$

$$\text{Li}_x\text{NbSe}_2 + x\text{H}_2\text{O} \rightarrow \text{NbSe}_2 \text{ (monolayer)} + x/2\text{H}_2 + \text{LiOH};$$

$$\text{NbSe}_2 \text{ (monolayer)} + \text{polymer} \rightarrow \text{polymer–NbSe}_2.$$

The principle of such nanocomposites is given in Fig. 8.14 and their composition and some properties are listed in Table 8.2 (the indices relate to the mean number of the monomer units per NbS_2 molecule).

One of the promising applications of such nanocomposites may be the production of plastic superconductive electromagnetic materials [288]. The same mechanism corresponds to polymer introduction into the WS_2 phase [289].

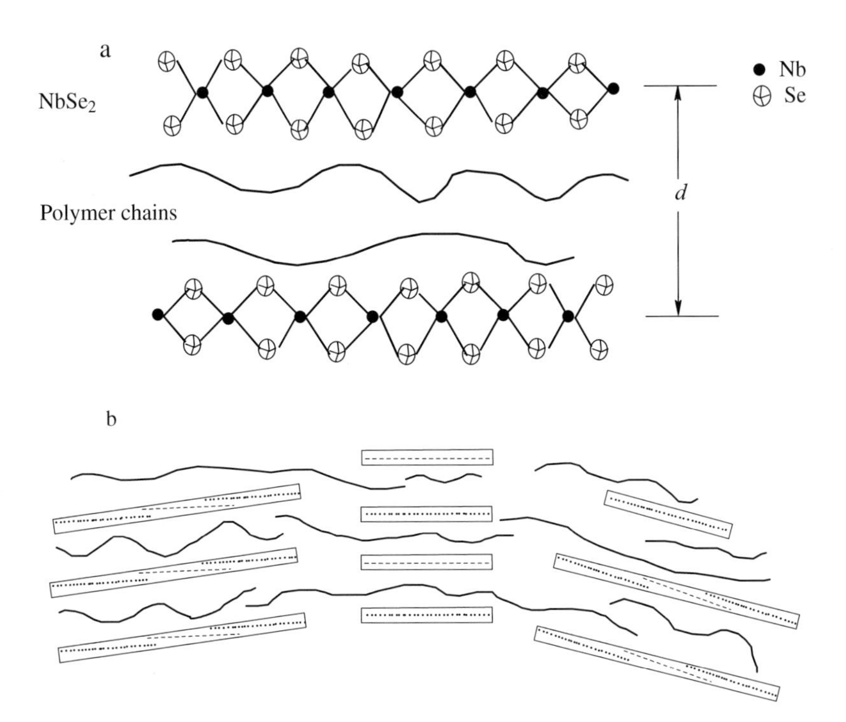

Fig. 8.14. Principle of polymer–$NbSe_2$ nanocomposite formation (**a**) and polymer-intercalated particle structure in the film (**b**)

Table 8.2. Some characteristics of intercalated polymer–$NbSe_2$ nanocomposites

Composite	d, nm	Thermal Stability in N_2, K	Specific Conductivity, S/cm	Superconducting Transition Temperature, T_c, K
$(PVP)_{0.14}NbS_2$	2.4	583	140	7.1
$(PEO)_{0.94}NbS_2$	1.96	497	250	6.5
$(PEO)_{0.80}NbS_2$	1.88	506	240	7.0

An interesting aspect of the delamination is the possibility of introducing the polymers by a *host–guest* mechanism in a moderate reaction time with a subsequent deposition of the intercalated systems by solvent removal or by an increase in the electrolyte concentration. It reminds us of the direct intercalation of polyaniline into a MoS_2 interlayer space (d = 1.037 nm) through colloidal suspensions [290] (Fig. 8.15).

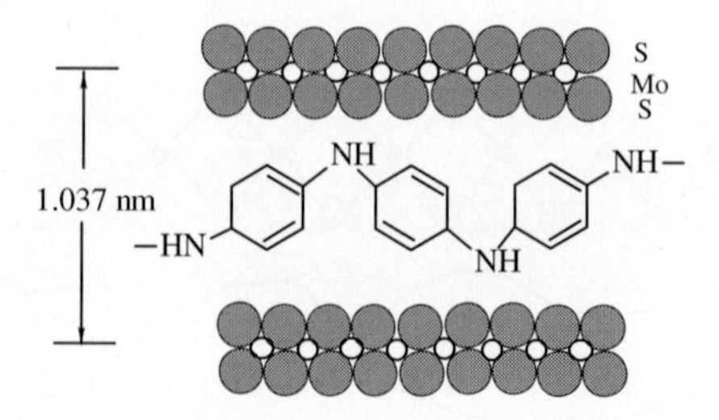

Fig. 8.15. The structure of polyaniline–MoS_2 nanocomposite

The polypyrrole(Ppy)/MoS_2 composite is a product of an in situ oxidative polymerization in kinetically controlled conditions. The composite is a p-type semiconductor with electronic conductivity by three orders of magnitude higher than in initial MoS_2 [132]. In the same way a PEO intercalation in delaminated TiS_2, TaS_2 and MoS_2 suspensions was carried out [132,287,291]. Interesting results were obtained in studies of alkali metal–PEO electrolytes intercalation in layered $MnPS_3$ and $CdPS_3$, proceeding by a two-step (ion-exchange solvation + shell-exchange) reaction [30, 189]:

$$CdPS_3 + 2xA^+_{aq} \rightarrow Cd_{1-x}PS_3A_{2x}(H_2O)_y + xCd^{2+}_{aq} \; (A = Li, K, Na, Cs)$$
$$Cd_{1-x}\ PS_3A_{2x}(H_2O)_y{+}PEO \rightarrow Cd_{1-x}\ PS_3A_{2x}(PEO) + yH_2O.$$

The lattice expansion on intercalation Δ is equal 0.8 nm; the PEO in both *trans* (T) and *gauche* (G) conformations : a helical conformation for the intercalated PEO may be ruled out. The experimental data suggest that the intercalated PEO adopts a planar zigzag structure (with at least two strands, complexing the alkali cation), i.e. its structure is similar to the PEO conformation into the complex with $HgCl_2$ [293, 294]. Under the same conditions, oligoethylene oxide oleyl ether intercalation (ZnS in poly(oxyethylene, $n = 23$) dodecyl ether [295]) in solid CdS, CdSe and ZnS were directly templated, while Ag_2S, CuS and PbS were produced only as non-featured solids [296].

Nanocomposites having semiconductive characteristics were synthesized from an aqueous solution of linear polyethylenimine (PEI) with the layered compounds MoS_2, $MoSe_2$, TiS_2, MPS_3 (M = Mn, Cd) [297] and with the TiO_2/PbS layers, using both PEI and poly(styrene-4-sulfonate) [298]. The formation in situ of the CuS and CdS dispersed phase in the film nanocomposites (produced in the swelled polymeric matrix – the mixtures of PAAc–PVA, PEI–PVA) proceeds through successive treatment by the proper metal salts and Na_2S. The process included the stages when the metal S-complexes, associates, amorphous clusters and finally crystalline nanoparticles are formed [299]. The systems on the base of CdS particles (~5 nm in size) are formed by a condensation mechanism (through a stage of the generation of critical nuclei), whereas the most probable mechanism for the CuS (with mean size 15 nm) nanophase formation is connected with an association of the sulfur-containing complexes. In general, the dispersion degree of the in situ produced nanoparticles increases with growth of the polymer matrix complexation ability in relation to the transition metal ions.

The semiconductor nanocrystals on a metal chalcogenide base when immobilized into a polymer matrix can form luminescent composites as for example CdS nanocrystals in polymers [300, 301], including poly(*p*-phenylene vinylene), PEI, poly(allylamine hydrochloride) and poly(styrenesulfonic acid) [302–305]. The same results were obtained for ZnS and ZnS/CuS [306]. The ZnS/CdS nanocrystal (~2 nm in size) in a polymeric matrix are interesting photo- and electroluminescent materials and can be produced by copolymerization of the metal acrylates with styrene and a subsequent treatment of the metal-copolymer solution with H_2S in chloroform [307]. It is interesting that a deposition did not precipitated from this practically colorless organosol even within a year. An original method of simultaneous synthesis and polymerization of metal-containing monomers was proposed in [308], whereby CdS, PbS or ZnS particles (~3 nm in size for ZnS) are formed from the proper halogenides and $Na_2S_2O_3{\cdot}5H_2O$ in aqueous solutions and then an emulsion polymerization of styrene (initiated by a ^{60}Co γ-ray source) was carried out with acrylic or methacrylic acids. An alternative approach is based on ultraviolet radiation polymerization (photolysis technique) [309]. For such nanocomposites an in situ copolymerization approach can be used or the decomposition technique as for the preparation of poly(acrylamide-*co*–styrene) semiconductor CdE (E = S,Se) [273]. Synthesized CdS quantum dots of low polydispersity formed within the spherical ionic cores of PS-block–poly(cadmium acrylate) reverse micelles [310]. Dendrimer-encapsulated CdS semiconductor quantum dots [311,312] are prepared by metal ion extraction into the dendrimer interior, followed by chemical reduction. The composition of hyperbranched conjugated polymers with nanosize cadmium sulfide particles was obtained by the same method [313].

These nanocomposites are a novel class of inorganic–organic hybrid photoconductive materials [314]. CdS/PVK nanocomposites [255] showed photorefractivity in the visible spectral range [315]. The composites were photosensitized at 1.31 μm through the inclusion of nanocrystals composed of PbS (~50 nm) or HgS [316].

One of the most important semiconductors is PbS, which due to the narrow band gap presents special interest for various nanocomposites preparation. The PbS electronic properties are very sensitive to crystal sizes from 1 to 30 nm) [317]. Therefore a variety of nanostructured systems containing PbS were prepared and investigated. For the fabrication different methods were used, e.g. the layer-by-layer self-assembly of lead sulfide nanoparticles and MMT platelets [318], m and PbS nanocrystals dispersing in various polymeric matrices. Sometimes CdS-based materials were used, but PbS nanocrystalline or PbS/polymer nanocomposites have displayed high characteristics [319–321]. The same can be said about PbS nanocrystals within a microphase-separated block copolymer matrix [265], including block copolymers of polystyrene and poly(vinylpyridine) with spherical morphology [262]. PbS/PVAc nanocomposites (with different morphology of the inorganic particles embedded in a polymer matrix – nanowires and nanotubes, respectively) were first synthesized upon γ-irradiation at room temperature and under ambient pressure in a simple system [322]. In a homogeneous ethanol system the inorganic PbS nanoparticle were homogeneously dispersed in a PVAc matrix, but in a heterogeneous water system they formed "nanocables" with a 30 nm diameter PbS core in a 80 nm PVAc sheath up to 10 μm in length. Another interesting method was used for poly(p-xylylene)/PbS nanocomposite film fabrication, which can be synthesized by simultaneous low-temperature cocondensation of p–xylylene and PbS vapors (see Chap. 5) [323]. The distribution curve for the PbS particles has a maximum at 4 nm and half-width about 2 nm.

There are some recent reports related to layered semiconductor materials such as PbI_2, BiI_3, HgI_2, Bi_2S_3 and Sb_2S_3. A composite containing PbI_2 appears to be an important material for digital X-ray imaging techniques. PbI_2 semiconductor colloidal particles stabilized by PVA (M_w = 86 000–100 000) are described in [324,325].

Nanoparticles with other ligand surroundings are synthesized and used rarely. The CdS/CdSe and CdSe/CdS/ZnS particles are known in the citrate-stabilized colloid variants [326], wurtzite-type CdS nanorods and nanowires were synthesized by organo-solution methods [327–329]. Stabilization of CdSe/CdS core/shell nanocrystals (the generation-3 dendrons) with following total cross-linking through a ring-closing metathesis is described in [330].

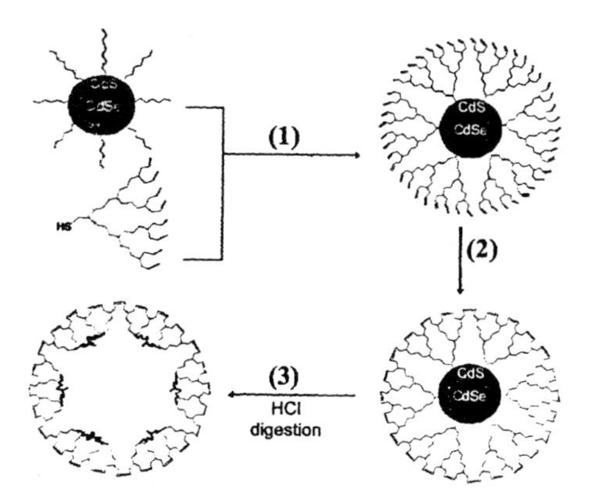

Fig. 8.16. Scheme of template cavity generation through shell formation from dendron on CdSe/CdS (1), cross-linking (2) and nanocrystal removal (3)

After the removal of inorganic nanocrystals, a nanosize cavity with a very thin shell is formed in the center of an empty dendron box (Fig. 8.16).

These nanosize capsule ions present a new class of mesoporous structures that are soluble in certain solvents, have a very narrow size distribution and can be used in guest–host chemistry. CdSe nanowires (mostly in zinc blended phase along CdTe ones in zinc blended form) were prepared through the redox reaction of selenite and tellurite with cadmium salt in the presence and with the assistance of PVA ($M_n = 1750$) at 433–453 K [331]. The structure of the CdSe and CdTe nanowires depends on the process temperature. PVA also plays an important role in nanowire formation by promotion of the oriented attachment growth in solvothermal conditions.

The metal chalcogenides with more complex layered structure (i.e., $PbNb_2S_5$ or $SmNb_2S_4$) can be split and intercalated too [332]. The one-dimensional *host* phases with different chalcogens of type MMo_3Se_3 (M = Li, Na) in polar solvents can form colloidal systems with the monodispersed and negative charged condensed cluster chains Mo_3Se_3 [333].

This short review shows that polymer intercalation into the chalcogenide interlayer spaces is a promising research field, though many problems (especially related to the introducing and host–guest interaction mechanisms) remain unsolved.

8.5 Langmuir–Blodgett Metallopolymer Films as Self-Organized Hybrid Nanocomposites

The synthesis, structure, properties and applications of many of the hybrid nanocomposites are well studied. Let us consider one more type of such materials for a molecular design, the nanosize metal particles in so-called Langmuir–Blodgett (LB) films. One-dimensional nanoscale building blocks, such as nanotubes, nanowires and nanorods, are used in fundamental research and potential applications [61, 328, 334].

These structures have long been described and it was known that the amphiphilic organic molecule monolayers on a water subphase surface can be transferred on a solid substrate under controlled surface tension [335–337]. The importance of this discovery appeared only within the past 25–30 years, when the usage of successive deposition of the different monolayers allows one to fabricate on a molecular level the new structures with the required physical and chemical properties. The recent development of self-organized inorganic/PVA composite synthesis opened new field in material science. The particle sizes in such materials are usually 2–10 nm. These materials can be used in microelectronics, nonlinear optics, and sensors [338–342]. Polymolecular LB films are used in the design of highly oriented ultrathin films with special characteristics arising due to their supramolecular structure. In such self-organized layers the various sensor groups (or their precursors) with nonlinear optical characteristics or even metallocomplexes and nanoparticles can be introduced. In addition, LB films can be used for modeling surface and biological processes [343]. Supramolecular assemblies with mesodimensional periodicity are the measure of a scale between the atomic and macroscopic dimensions.

The methodology and the Langmuir–Blodgett technique have been modified for new problems and aims. The heterogeneous polar nanolayer fabrication and transfer are provided in the subphase surface layers (deionized water) at fixed surface tension π (measured by a Wilhelmi balance in $nN \cdot m^{-1}$). The microprocessor and computer automatically monitored the technological process [344]. The liquid-crystalline layers are applied step-by-step on a solid substrate and the complex planar molecular structures were formed with various features. The application process is carried out by vertical (Langmuir–Blodgett) and horizontal (Langmuir–Schoffer) elevators. If a concentration of the metal particles on the LB surface is small (i.e. they do not interact with one another), the system presents a "two-dimensional gas". As this "gas" is compressed, three-dimensional states are created: "gaseous" (the distances between molecules are much greater than their dimensions), "liquid" (the distances between molecules are comparable with their dimensions), "liquid-crystalline" (the molecules conserve the mobility into the monolayer plane) and "solid".

Two ways of nanoparticle formation in such films can be identified. The first way is based on a combination of colloidal chemistry, self-organization and layer growth principles [345–347] when the nanoparticles are formed (for example, by a chemical or photochemical reduction of aqueous solutions of metal salts) in the presence of the stabilizers and LB-forming components. In these systems the arising layers play the role of the peculiar templates [347–349] and the approach may be considered as a method of biomineralization process modeling (see Chap. 9). It is connected with LB deposition on the stabilized nanoparticle surface with subsequent introducing into the polylayers (involving the functional groups), lamellar film formation, etc. The LB films can be formed on the surface of both the colloidal particles and water, with a subsequent transfer on a nanostructured substrate by the Langmuir–Blodgett method [350–352]. As an interesting example it can be noted that a gold hydrosol (stabilized by 4-carboxithiophen) was immobilized on an octadecylamine monolayer by an electrostatic mechanism [353] with the film charge controlled by means of the pH value. In this way, multilayered films with different density of Au-octadecylamine clusters can be produced with 2–20 layers (for every monolayer $\pi = 25$ nN/m, surface area $A = 0.37\,\mathrm{nm}^2 \cdot \mathrm{molecule}^{-1}$, thickness 10 ± 3 nm).

A reversed micelle templating method was used to synthesize uniform $BaWO_4$ particles (diameter $\sim$9.5 nm, length $\sim 1500 \pm 200$ nm) by the interaction of barium bis(2-ethylhexyl)sulfosuccinate [$Ba(AOT)_2$] with sodium tungstate Na_2WO_4 [354, 355] (Fig. 8.17).

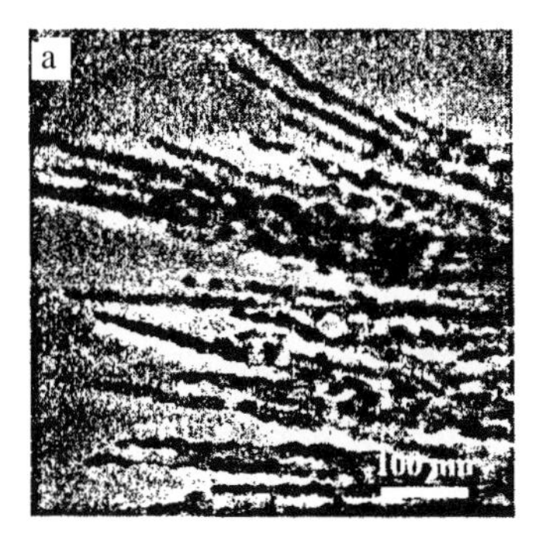

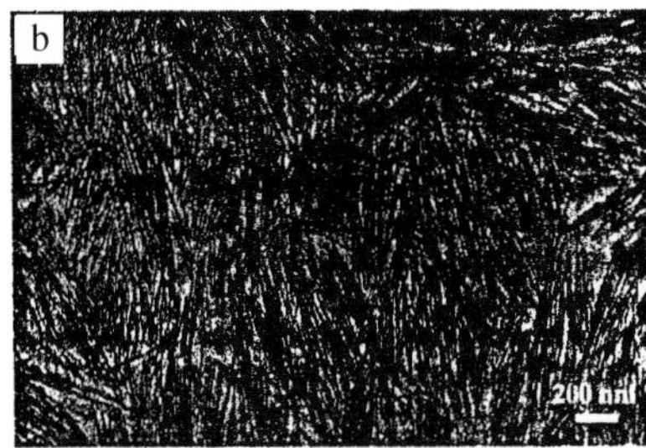

Fig. 8.17. TEM images of as-made $BaWO_4$ nanorods assemblies. (**a**) High magnification; (**b**) at the air–water interface after compression at low surface pressure

The $BaWO_4$ compounds with a scheelite structure are potential materials in the electrooptical industry (due to its blue luminescence) for designing all-solid-state lasers emitting in a specific spectral region and it may have application for medical treatment, up-conversion lasers and spectroscopy [356]. The penny-form $BaWO_4$ nanostructures were prepared in the cationic reversed micelles [357]. PEO-PMMA block-copolymer was combined with aqueous micelle solutions, after which the unusual inorganic hollow spheres were formed [358]. Recently a new polymer-directed synthesis of the $BaWO_4$ nanostructures in reverse micelles (by using a block-polymer as a directing agent) was reported [359]. The method allows one to carry out a process of hierarchical nanostructures (based on inorganic nanowires) by direct growth in the solutions.

It seems likely that a strategy of nanoparticle fabrication and introduction on LB films has many advantages over the usual chemical introduction of metal ions in LB films with the subsequent assembling of clusters. In particular, the films are not deformed, a lamellar phase order is conserved and the nanoparticle assortment is widened (due to introduction of the bi- and polymetallic particles by usage of the convenient hydrosol mixtures). It is important that in such systems a high level method recognition is used rather than a simple adsorption or chemical interaction. As an example let us consider the self-organization of alkylsiloxanes, fatty acids, dialkylsulfides and thiols on Al, Au, SiO_2 [360, 361], shown in Fig. 8.18. The two processes proceed on a mosaic Au–Al_2O_3 surface: the first is recognition of its "own" substrate and selective adsorption; the second is the self-organization of biphylic thiol (at the Au site) and acid (at Al_2O_3) on this substrate. A charged amphiphilic monolayer can by used as a template for organizing inorganic materials by ionic species adsorption over the interface. This strategy has been successfully employed to prepare LB films containing inorganic compounds like Keggin polyoxometallates [362], oxalato- [363] or cyano [364] complexes.

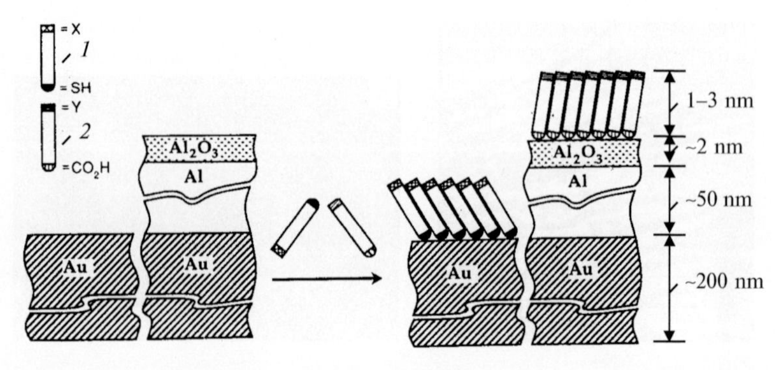

Fig. 8.18. Self-organized monolayer formation for diphilic molecules of alkane-thiol (1) and carbon acid (2) from the solution over a gold surface (textured by aluminum oxide)

The layer-by-layer self-assembly of opposite charged polyelectrolytes [365] and alumosilicates [230], zirconium phosphate [227], metal nanoparticles [228, 229], semiconductor quantum dots, graphite oxide [226] and other species can by considered as an alternative to LB deposition, chemical vapor deposition or spin-coating methods of advanced thin films preparation. For example, a promising gas-permeable material is a self-assembled multilayer system on the base of montmorillonite and poly(diallyl-dimethylammonium)-chloride [366]. Organic–inorganic films were found to be very flexible and crack-resistant even under considerable mechanical stress. The films are prepared either by attaching an inorganic layer onto a self-assembled organic anchor or alternatively by immersing a solid substrate into a polyelectrolyte solution and a suspension of inorganic layer compound [225, 367]. An amphiphylic complex $[Ru(phen)_2(4,4'\text{-dioctadecyl-}2,2'\text{-bipyridil})]^{2+}$ was intercalated from a chloroform solution by the Langmuir–Blodgett technique into a clay monolayer (lithium hectorite) [368, 369] and a floating film was transferred onto a hydrophilic glass plate [370]. Ru^{2+} complexes are widely used in the metal–polymer LB film fabrication because the tris(2,2′-bipyridine)ruthenium(II) complex has been intensively studied as a photosensitizer for various problems (solar energy transformation and storage, photocatalytic reduction of carbon dioxide, and other catalytic processes). Very often the polymer LB films are created by radical polymerization of the mixtures, containing $Ru(bpy)_3^{2-}$ complex with N-*tert* -pentylarylamide or N-dodecylacrylamidoine, in benzene as shown in Fig. 8.19 [371].

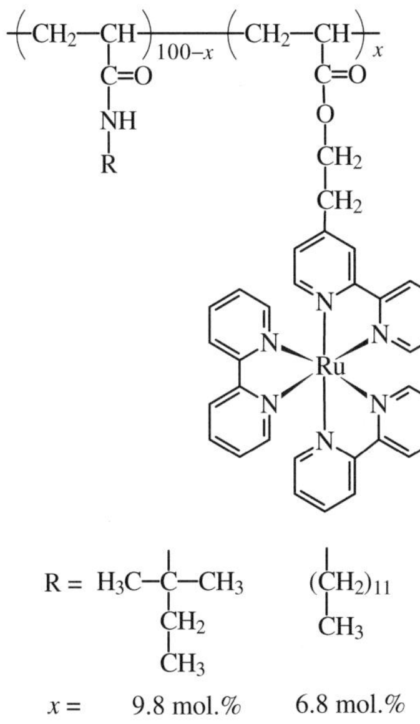

Fig. 8.19. Metal–polymer LB film composition

Such copolymers formed stable condensed monolayers on the air–water interface and can be successfully transferred onto solid supports and yield so-called Y-type LB films. By another method the LB film monolayers are prepared using the polymer-analogous method, i.e. by radical copolymerization of 4-vinyl-4′-methyl-2,2′-bypipyridine with N-dodecylacrylamide with the subsequent $Ru(bpy)_2Cl_2$ copolymer treatment in *n*-butanol within 72 hours [372]. The aim of combining PAn with the ruthenium complexes [$RuCl_3$(dppb)(py)] (the notation dppb relates to 1,4-bis(diphenylphosphine) is that the LB films are much less sensitive to the environment (as compared with the pure PAn films) and can have important applications in the various sensors of gas or taste [373]. Moreover the LB films formed from conducting polymers and ruthenium complexes (as well as self-assembled films of azobenzene-containing polymers) can be used as an artificial taste sensor [374]. Using the various types of LB films an electronic "tongue" can be created, being able to register well four basic tastes (sweet, sour, salt and bitter). Often the LB films are fabricated using the coordination complexes of metal ion and ligand monolayers, e.g. the K^+, Mg^{2+} and Fe^{3+} complexes with amphiphilic polymer poly(N-2-(4-imidazolil)ethyl)maleimide-*alt*-1-octadecene) [375].

The majority of investigations still relate to self-organized hybrid nanocomposites on the base of mononuclear complexes (specifically CdS) and to the methods of nanoparticle assembling, quantum scale effects (characteristic for semiconductor particles) and their applications in various fields [376]. For example, it was shown that CdS particle (2.65–3.4 nm in size, stabilized by dodecyl-benzenesulfonic acid) dispersion in chloroform spreads out on a water surface, forming stable monolayers of nanoparticles [377, 378]. It can be shown, using a compression isotherm π–A, that the rise in surface tension π leads to a transition from a "gaseous" state to a close-packed particle monolayer a finally to formation of a polylayer state. The CdS particles form LB films with limiting value of a specific surface area per particle $A = 0.65$–$1.1\,nm^2$ and this value is near to the meaning for hard sphere close packing (0.608–$0.887\,nm^2$). When the nanoparticle monolayer is transferred to a solid substrate by the Langmuir–Blodgett method at a transfer rate of about $3\,mm{\cdot}min^{-1}$ and surface pressure 15–30 $mN{\cdot}m^{-1}$, there the polylayers are formed from the dimensionally quantized CdS clusters and their optical density increases linearly with the number of transferred monolayers.

There is interest not only in LB transfer on the stabilized sol particles but also in the process of low-dimensional semiconductor particle formation into LB films in situ, that proceeds by the reactions of metal ions with H_2S or Na_2S [379–382]. By sulfidation of the cadmium, zinc or plumbum layers $(C_{21}H_{43}COO)_2M$, these metal sulfide films were fabricated with thickness 100 nm (34 layers). The films are anisotropic and their anisotropy increases under sulfidation, which allows one to suggest that the nanoparticles are formed layer-by-layer, generating so-called "cluster-layers" [383,384]. The system can be described as follows: the length of the acid molecule is 2.68 nm, the thickness of the cluster layer is 1.12 nm, cavities (pores) arise in the cluster layer during nanoparticle formation, and the nanoparticles are non-spherical with diameter 5–10 nm and thickness 1.1–1.3 nm. CdSe nanoparticle formation under cadmium arachidate $(C_{19}H_{31}COO)_2Cd(CdA)$ film treatment by vapors of H_2Se proceeds into the interlamellar space of the solid phase films and is accompanied by their considerable deformation or even by decomposition of their lamellar structure [385].

Most often the multilayer LB films are produced from metal stearates, such as cadmium [386], magnesium [387] or α-Fe_2O_3-stearate [388]. Self-organized structure formation was detected in hydrophobic layers of stearic acid from a silver stearate film (8–14 layers), being transferred to electrodes

($\pi = 25$ mN·m^{-1}) and electrochemically reduced in a neutral or acid solution, forming two-dimensional Ag clusters 20–30 nm in diameter [389]. The sandwich-type Ag clusters were found in such films too. In other research [390] barium stearate LB monolayers were deposited on plasma oxidized silicon substrates. In general, self-organized metal-containing assemblies of LB are often used for modification of the electrochemical behavior of the electrode surface on a molecular level [391, 392]. One of the examples is a self-organizing LB film (deposited on Au) on the base of C_8H_{17}–C_6H_4–N = N–$C_6H_4O(CH_2)_3COOH$ and X–$(CH_2)_2$–SH (X = OH, COOH) connected by a hydrogen bond. Such systems demonstrate stable and repeatedly reproducible electrochemical behavior.

Another promising direction of electrochemical synthesis is the formation of two-dimensional (Langmuir) monolayers of nanoparticles by kinetically controlled electroreduction. The process is carried out under monolayers of SAS matrices arranged on an electrolytic solution surface into a circular electrochemical cell [393, 394]. The silver two-dimensional aggregates are formed only when the SAS monolayer carries a negative charge. It is possible to use LB techniques to produce a new type of nanomaterial, the controlled deposition and hydrolysis of ferrum salts in the surfactant layer [395]. The layer thickness is defined by the different solubilities of Fe^{2+} and Fe^{3+} salts, as well as the solubility of their oxides in aqueous solutions and by the chemical processes, allowing us to control their transformation (using H_2O_2), i.e. a redox equilibrium $Fe^{2+} \leftrightarrow Fe^{3+}$. Such self-organizing iron/surfactant nanocomposites with chains of different length and one, two, three or six layers of iron oxide are shown at Fig. 8.20 (n is the number of carbon atoms). They have dimension-dependent properties (superparamagnetic characteristics) and are very important as a first step in the hierarchical organization of nanocomposite magnetic materials.

LB films of classical type can be fabricated using not only low molecular systems but polymeric systems as well. In this case the hydrophobic chains forming the monolayers can be functionalized either by polymer analogous transformation or by a grafting of side groups able to react on the external effects. As in the above-considered nanocomposite synthesis, the polymerization of functional monomers is used, their copolymerization with monomers being only the spacing units. In principle, the diphylic polymer-based LB films are favored over polymolecular films on the base of low molecular compounds (if for no other reason than the higher stability) to preserve the necessary monolayer density during the process. The accurate criterion of the monolayer thermodynamical stability impacts on their equilibrium spreading. The degree of molecular ordering in LB films is expected to help achieve better control over the size distribution, shape and distribution of clusters [396]. This approach can be extended to develop semiconductor cluster-doped polymers that is currently a novel composite material for nonlinear optical [397] and electroluminescent devices [398]. CdS-PAn composite films have been

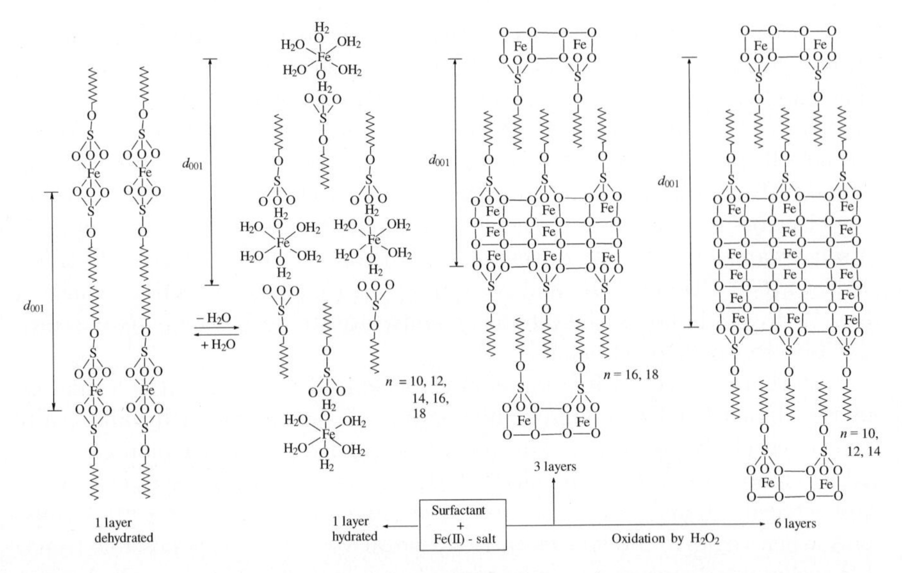

Fig. 8.20. Self-assembling scheme for multilayer iron/surfactant nanocomposites

prepared by co-depositing cadmium particles during electrochemical deposition of PAn and most often used for these purposes [399]. The standard procedure includes precursor usage (PAn-CdA LB films with 15–25 layers) with the subsequent treatment of H_2S flow, using a step-up [400, 401]. Arachidic acid presented in the H_2S-exposed films was selectively removed by organic solvent leaching, resulting in the formation of size quantized semiconducting CdS-doped PAn films.

Sometimes it is necessary to introduce into the LB film monolayers functional groups of at least two different types, for example the redox-pair components. In case of low-molecular reagents there are serious restrictions connected with the nonmiscibility of many diphylic substances on the molecular level, but this problem can be solved readily by polymer usage. The roles of polymer and copolymer structure or composition, as well as the external conditions, influence the formation and properties of the organized polymeric monolayers, and LB films on the solid and liquid surfaces are analyzed in the survey [402]. It is likely that the first attempts to introduce polymeric complexes into LB films were connected with complexes of porphyrin and phthalocyanide [403, 404]. The diphylic properties can be imparted to polymers by different methods, e.g. by copolymerization of long-chain α-olefins with maleic acid, maleic anhydride, etc. [405]. The metallocomplexes can be bonded by carboxyl groups. Donor–acceptor bonding of metallocomplexes (having coordination vacancies) in polymer groups can be created, using some reagents (such as 4-aminomethylpyridine) for anhydride cycle splitting. This method was used in [406] to bind tris(phenanthrolino)iron (II) sulfate

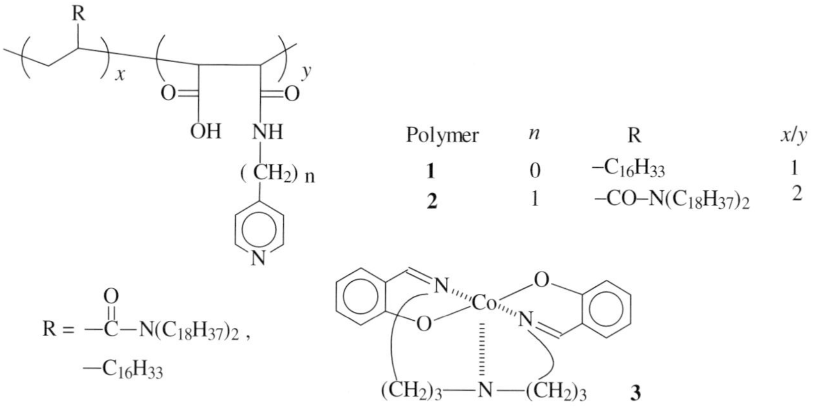

Polymer	n	R	x/y
1	0	$-C_{16}H_{33}$	1
2	1	$-CO-N(C_{18}H_{37})_2$	2

$R = -C(=O)-N(C_{18}H_{37})_2$, $-C_{16}H_{33}$

Fig. 8.21. Diphylic polymers and initial metallocomplex for LB films. **1** – Copolymer of pyridinemonoamidomaleic acid with octadiene; **2** – copolymer of picolinemonoamidomaleic acid with N,N-dioctadecylacrilamide; **3** – cobalt(II) (bis (salycyldenaminopropyl)amine)

or bis(salycylidenoamino- propyl)aminocobalt (II) by polymers as shown in Fig. 8.21.

As indicated at Fig. 8.22, the π–A film isotherms for the air–water phase border are defined by metal complex concentration. This can be explained by the larger volume of the metal complex as compared with the monomer units. It is supposed that the complex molecule is localized nearly parallel to the water surface. When the complex concentrations are above 10^{-7} mol·L^{-1}, they begins to play an important role in the organization of the monolayer. A stable and uniform monolayer is formed under compression (π = mN·m^{-1}), but the stability disappears at concentrations of about 10^{-4} mol·L^{-1} . It is important that two different metal complexes can be introduced into a single monolayer (being part of the multilayer system) and their bonding by two alternative functional groups prevents phase separation processes.

Films of monolayer protected Au_{140} clusters with mixed alkanethiolate and ω-carboxylate bridges were synthesized by $AuCl_4^-$ reduction with $NaBH_4$ in the presence of hexanethiolate and 11-mercaptoundecanoic acid linked by metal ion-carboxylate coordinative coupling (Fig. 8.23) [407–409].

Gold-containing nanoparticle of $Au_{140}(C_n)_{53}$ composition (n is the number of carbon atoms in the nonlinker ligand–alkanethiolate chain, average diameter 1.6 ± 0.8 nm) was employed as monolayer protected clusters and stabilizers of thiolate shells.

Mono- and multilayer LB films on the base of copolyimide (with carbazole groups included into the chain as an electron donor) with copper phthalocyanines were obtained [410]. Their pile structures and small interatomic distances give high charge mobility and high photoconductivity [411].

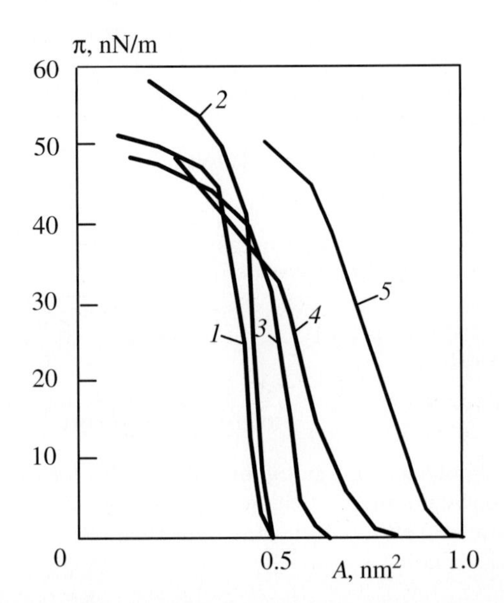

Fig. 8.22. The compression isotherm π–A for polymer **1** at various subphases: *1* – the aqueous solution of alkali (pH = 10); *2*–*5* – the aqueous solutions of cobalt **3** with concentrations 10^{-7} (*2*), 10^{-6} (*3*), 10^{-5} (*4*) and 10^{-4} (*5*) mol·L^{-1}

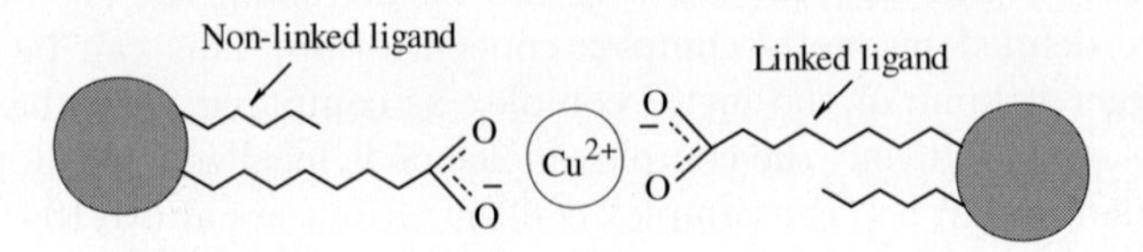

Fig. 8.23. The $Au_{140}(C_n)_{53}$ composition structure

Regular assemblies by organized multilayered LB composites can be produced using the electrostatic interaction between the charged sol nanoparticles (dispersed into the subphase) and the charged surface monolayers, e.g. between cationic electrolytes and anionic particles [412]. Using this method, a cross-linked polycationic polyvinylcarbazole (produced by an electrochemical polymerization of N-vinylcarbazole in the presence of sodium perchlorate) and thalium oxide (n-type semiconductor) sol there were synthesized some regular nanowire composites Tl_2O_3/PVC–arachidonic acid [413]. Such a polymeric layer (2.7 nm) is a template to generate regularly organized Tl_2O_3 particles (3.2 nm) and the formed multilayers (with thickness 5.5 nm) can be transferred layer-by-layer on a hydrophobic surface (with $\pi = 25$ mN·m^{-1}). It is likely that the method would have wide application, because electrochemical polymerization can be used for fabrication of various positively charged conductive polymers, e.g. polyaniline, polypyrrole, polythiophen, etc. [414].

Probably, such mechanisms are realized under the self-organizing formation of TiO_2 layers on a SO_3-functionalized surface [415] or ordered TiO_2 deposition on sodium poly(styrenesulfonate) PSSNa. The cation type TiO_2 particles with mean size ~3 nm produced by acidic hydrolysis of $TiCl_4$ organize the layered structures on cationic type polymers such as ultrathin (~1 nm) PSSNa or poly(allylamine-hydrochloride) [416] The optically transparent LB films organized on the molecular level with a thickness of 120 layers (i.e. ~60 bilayers) are deposited on the substrate surface. The bilayer thickness is about 3.6 nm (see Fig. 8.24). Metal, silicon and polymer can be used as a substrate. They are cleaned by a 5% solution of N-2-(2-aminoethyl)-3-aminopropyltrimethoxisilane. It is suggested that such a strategy allows one to produce very interesting combinations of semiconductive materials with metal/dielectric structure (i.e. with nanoscale sites of *p-n*, *p-n-p*, *n-p-n* types, etc.).

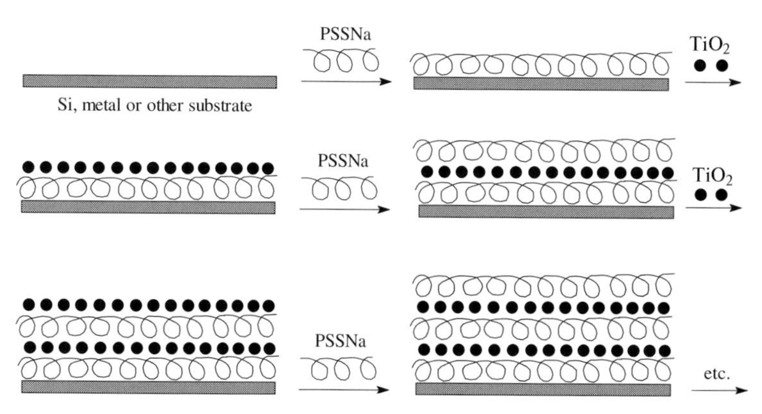

Fig. 8.24. Scheme of ultrathin multilayer TiO_2/polymer films fabrication

There is one more class of nanocomposites – clusters in LB monolayers considered promising for new metal film electronics and fixed catalyst modeling. For example, $Os_3(CO)_{11}(NCCH_3)$ interaction with the self-organized (3-mercaptopropyl)trimethoxisilane layers on a Au surface leads to the destruction of the well-ordered thiol surface and formation of other cluster aggregates with diameter 1–2.2 nm [417]. The fixation of metal clusters (in LB films) on a surface of highly oriented pyrolytitic graphite (HOPG) is connected with the problems of fabrication of new catalysts and production of stable and reproducible tunnel nanostructures. The $Rh_4(CO)_{12}$, $[NEt_4]_2[Pt_{12}(CO)^2_{24}]$ and other clusters sorbed on HOPG from organic solvents [418] under laser irradiation undergo cluster decarbonylation and transformation into highly dispersed Pt crystallites ($1.8 \times 0.5\,nm^2$ in size) connected to the graphite surface [419]. The techniques of single "naked" silver cluster deposition on the HOPG surface (with 3–5 nm in plane and 2–3 nm in height) and their ligand shell (from PPh_3) "assembling" is proposed in [420]. The current–voltage characteristics of the single clusters were measured on a fresh-spalled HOPG surface and the cluster-containing LB were formed on the solid HOPG surface by introducing the cluster into the stearic acid layers with transfer to the surface [421]. There were cluster molecules in the transferred layer, and an ordered two-dimensional lattice too.

These methods allow one to solve the problem of cluster fixation on a surface and to realize the one-electron tunneling mode at room temperature. The "build-in" of the cluster molecules (on the Pd_3, Pt_5, Pd_{10} base) into LBF and their current–voltage characteristics are analyzed in [422]. On the HOPG surface a multilayered LB film (composed from a stearic acid monolayer on an aqueous solution of $MgCl_2$) is formed in a "broken" state: the surface area per molecule is $0.15\,nm^2$, and $\pi = 36$ mN·m^{-1} [387]. LB films can be formed by fullerenes too [423–425] and, moreover, epitaxial layers of fullerenes (C_{60} and $C_{60/70}$) on Au^{3+} surfaces were prepared by simple transfer on LB films at the air–water interface [426].

These examples demonstrate that metal-containing LB films including nanoparticles and clusters are very interesting and promising materials for the nanohybrid organized composites.

* * *

Thus, the different aspects of solid state chemistry are developing. The main problems are connected with a justified selection of new types of lattices and *host–guest* combinations, optimization of the intracrystalline *host–guest* interactions and studies of the influence of the matrix on the electronic properties of complex systems. Other fundamental question of the physics and chemistry of polymers and composites is the orientation action of the pores and interlayer formation on the crystallinity and stereoregularity of the forming polymers. The above-considered approaches of inorganic–organic nanocomposites can be useful also for various monophase crystalline nanomaterials (metal–ceramics) with complex composition, as for example $MgAl_2O_4$ or $MgCr_2O_4$ in α-Al_2O_3 lattices with dimensions $<10\,nm$, including the use of Grignard reagents in their synthesis [427]. The nature of bonds between metal (Fe, Ni, Cu, Ag) and surface anionic oxygen has been theoretically studied for different cluster models. For practical purposes the best forthcoming achievements should be in the application of LB films (included clusters and nanoparticles as structure elements) in molecular electronics.

The establishment of the structure–property relationship on different hierarchial levels (from the molecular level up to supramolecules) is of the first importance in the design of new material with desirable properties.

The hybrid materials, considered in this chapter, are widespread in biological systems and the environment.

References

1. *Polymer-Clay Nanocomposites* ed. by T.J. Pinnavaia, G. Beall (J. Wiley & Sons, New York 2000)
2. B.K.G. Theng: *Formation and Properties of Clay-Polymer Complexes* (Elsevier, New York 1979)
3. D.W. Bruce, D. O'Hare: *Inorganic Materials*, 2nd edn. (J. Wiley & Sons, New York 1996)
4. *The Handbook of nanostructured materials and technology* ed. by H.S. Nalwa (Academic Press, San Diego 1998)
5. *Nanotechnology. Molecularly Designed Materials* ed. by G.-M. Chow, K.E. Gonsalves (Am. Chem. Soc., Washington, DC. 1996)
6. *Supramolecular Architecture: Synthetic Control in Thin Films and Solids* ed. by T. Bein (ACS Symp. Series, 1992) No 449
7. K. Moller, T. Bein: Chem. Mater. **10**, 2950 (1998)
8. G. Lagaly: Appl. Clay Sci. **15**, 1 (1999)
9. P.C. LeBaron, Z. Wang, T.J. Pinnavaia: Appl. Clay Sci. **15**, 11 (1999)
10. A. Matsumura, Y. Komori, T. Itagaki, Y. Sugahara, K. Kuroda: Bull. Chem. Soc. Jpn. **74**, 1153 (2001)
11. V. Mehrotra, E.P. Gianellis: Solid State Ionics **51**, 115 (1992)
12. H. Shi, T. Lan, T. Pinnavaia: J. Chem. Mater. **8**, 1584 (1996)
13. M.-Y. He, Z. Lin: Catal. Today 321 (1988)
14. M. Ogawa, K. Kuroda: Chem. Rev. **95**, 399 (1995)
15. A. Morikawa, Y. Iyoku, M. Kakimoto, Y. Imai: Polym. J. **24** (1992)
16. M. Vyunov, G.V. Isagulyants: Russ. Chem. Rev. **57**, 204 (1988)
17. *Progress in Intercalation Research* ed. by W. Müller-Warmuth, R. Schöllhorn (Kluwer, Dodrecht 1994)
18. *Intercalation Chemistry* ed. by M.S. Whittingham, A.J. Jacobson (Academic Press, New York) 1982
19. R. Schöllhorn: Chem. Mater. **8**, 1747 (1996)
20. R. Tamaki, Y. Chujo: Chem. Mater. **7**, 1719 (1999)
21. G.A. Ozin: Adv. Mater. **4**, 612 (1992)
22. D. O/Hare: New. J. Chem. **18**, 989 (1994)
23. V. Mehrotra, E.P. Giannelis, R.F. Ziolo, P. Rogalskyj: Chem. Mater. **4**, 20 (1992)
24. M.W. Anderson, J. Shi, D.A. Leigh, A.E. Moody, F.A. Wade, B. Hamilton, S.W. Carr: Chem. Commun. 533 (1993)
25. A. Gügel, K. Müllen, W. Schmidt, G. Schön, F. Schüth, J. Spickermann, J. Titman, K. Unger: Angew. Chem. **105**, 618 (1993); Angew. Chem., Int. Ed. Engl. **28**, 359 (1989)
26. T. Ogoshi, H. Itoh, K.-M. Kim, Y. Chujo: Macromolecules **35**, 334 (2002)
27. T.J. Pinnavaia, T. Lan, Z. Wang, H. Shi, P.D. Kaviratna: In *Nanotechnology. Molecularly Designed Materials* ed. by G.-M. Chow, K.E. Gonsalves (Am. Chem. Soc., Washington, DC. 1996)
28. J. Livage: Coord. Chem. Rev. **180**, 999 (1998)
29. K. Brandt: Solid State Ionics **69**, 173 (1994)
30. G.S. Zakharova, V.L. Volkov: Russ. Chem. Rev. **72**, 346 (2003)
31. H. Nakajima, G. Matsubayashi: J. Mater. Chem. **4**, 1325 (1994); **5**, 105 (1995)
32. K.J. Chao, T.C. Chang, S.Y. Ho: J. Mater. Chem. **3**, 427 (1993)

33. Y.-J. Liu, M.G. Kanatzidis: Inorg. Chem. **32**, 2989 (1993)
34. S. Ogata, Y. Tasaka, H. Tagaya, J. Kadokawa, K. Chiba: Chem. Lett. 237 (1998); J. Mater. Chem. **6**, 1235 (1996); J. Solid. State Chem. **117**, 337 (1995)
35. T. Abe, Y. Tachibana, T. Uematsu, M. Iwamoto: Chem. Commun. 1617 (1995)
36. R.F. Soares, C.A.P. Leite, W. Bottler, Jr., F. Galembeck: J. Appl. Polym. Sci. **60**, 2001 (1996)
37. K.W. Allen: J. Adhesion Sci. Technol. **6**, 23 (1992)
38. F.D. Osterholtz, E.R. Pohl: J. Adhesion Sci. Technol. **6**, 127 (1992)
39. W. Botter, Jr., R.F. Soares, F. Galembeck: J. Adhesion Sci. Technol. **6**, 781 (1992)
40. Y.K. Hwang, K.-C. Lee, Y.-U. Kwon: Chem. Commun. 1738 (2001)
41. P. Bodo, J.-E. Sundgren: Thin Solid Films **136**, 147 (1986)
42. N. Bowder, S. Brittain, A.G. Evans, J.W. Hutchinson, G.M. Whitesides: Nature **393**, 146 (1998)
43. R.K. Bharadwaj, R.J. Berry, B.L. Farmer: Polymer **41**, 7209 (2000)
44. L. Zheng, R.J. Farris, E.B. Coughlin: Macromolecules **34**, 8034 (2001)
45. H. Xu, S.-W. Kuo, J.-S. Lee, F.-C. Chang: Macromolecules **35**, 8788 (2002)
46. E. Ruckenstein, L. Hong: J. Appl. Polym. Sci. **55**, 1081 (1995); Chem. Mater. **8**, 546 (1996)
47. L. Hong, E. Ruckenstein: J. Appl. Polym. Sci. **67**, 1891 (1998)
48. B.B. Lakshmi, P.K. Dorhout, C.R. Martin: Chem. Mater. **9**, 857 (1997)
49. Y. Hamasaki, S. Ohkubo, K. Murakami, H. Sei, G. Nogami: J. Electrochem. Soc. **141**, 660 (1994)
50. C.R. Martin: Science **26**, 1961 (1994)
51. M. Lira-Cantu, P. Gomez-Romero: Chem. Mater. **10**, 698 (1998)
52. J.P. Carpenter, C.M. Lukehart, S.R. Stock, J.E. Witting: Chem. Mater. **7**, 2011 (1995)
53. J.P. Carpenter, C.M. Lukehart, S.B. Milne, S.R. Strock, J.E. Witting, B.D. Jones, R. Glosser, J.G. Zhu: J. Organomet. Chem. **557**, 121 (1998)
54. M.V. Russo, A. Furlani, M. Cuccu, G. Polzonetti: Polymer **37**, 1715 (1996)
55. H.-J. Muhr, F. Krumeich, U.P. Schonholzer, F. Bieri, M. Niederberger, L.J. Gauckler: Adv. Mater. **12**, 231 (2000)
56. F. Krumeich, H.-J. Muhr, M. Niederberger, F. Bieri, R. Nesper: Z. Anorg. Allg. Chem. **626**, 2208 (2000)
57. X. Chen, X.M. Sun, Y.D. Li: Inorg. Chem. **41**, 4524 (2002)
58. F. Krumeich, H.-J. Muhr, M. Niederberger, F. Bieri, B. Schnyder, R. Nesper: J. Am. Chem. Soc. **121**, 8324 (1999)
59. T. Ohsuku, A. Ueda: Solid State Ionics **69**, 201 (1988)
60. C. Pcholski, A. Kornowski, H. Weller: Angew. Chem. Int. Ed. **41**, 1188 (2002)
61. G.R. Patzke, F. Krumeich, R. Nesper: Angew. Chem. Int. Ed. **41**, 2446 (2002)
62. X.W. Lou, H.C. Zeng: Chem. Mater. **14**, 4781 (2002); J. Am. Chem. Soc. **125**, 2697 (2003)
63. C.R. Smith: J. Am. Chem. Soc. **56**, 1561 (1934)
64. A. Blumstein: Bull. Soc. Chim. 899 (1961); J. Polym. Sci. A **3**, 2653 (1965)
65. A. Blumstein, S.L. Malhotra, A.C. Waterson: J. Polym. Sci. A-2 **8**, 1599 (1970)
66. A. Akelah, N. Salahuddin, A. Hiltner, E. Baer, A. Moet: NanoStruct. Mater. **4**, 965 (1994)
67. A. Usuki, M. Kawasumi, Y. Kojima, A. Okada, T. Kurauchi, O. Kamigaito: J. Mater. Res. **8**, 1174, 1179, 1185 (1993)

68. A. Usuki, M. Kawasumi, Y. Kojima, A. Okada, T. Kurauchi, O. Kamigato: J. Polym. Sci. A Polym. Chem. **31,** 983 (1993)
69. M. Eckle, G. Decher: Nano Lett. **1**, 45 (2001)
70. Gomez-Romero: Adv. Mater. **13**, 163 (2001)
71. J.M. Garces, D.J. Moll, J. Bicerano, R. Fibiger, D.J. McLeod: Adv. Mater. **12**, 1835 (2000)
72. M. Alexandre, P. Dubois: Mater. Sci. Eng. R **28**, 1 (2000)
73. X. Huang, S. Lewis, W.J. Brittain, R.A. Vaia: Macromolecules **33**, 2000 (2000)
74. D.W. Kim, A. Blumstein, J. Kumar, S.K. Tripathy: Chem. Mater. **13**, 243, 1916 (2001)
75. H. Ishida, S. Campbell, J. Blackwell: Chem. Mater. **12**, 1260 (2000)
76. M. Okamoto, S. Morita, H. Taguchi, Y.H. Kim, T. Kotaka, H. Takeyama: Polymer **10**, 3887 (2000)
77. Y. Komori, Y. Sugahara, K. Kuroda: Appl. Clay Sci. **15**, 241 (1999)
78. Y. Komori, A. Matsumura, T. Itagaki, Y. Sugahara, K. Kuroda: Appl. Clay Sci. **11**, 47 (1999)
79. M. Laus, M. Camerani, M. Lelli, K. Sparnacci, F. Sandrolini, O. Francescangelli: J. Mater. Sci. **33**, 2883 (1998)
80. E.P. Giannelis: Adv. Mater. **8**, 29 (1996)
81. Z. Wang, T.J. Pinnavaia: Chem. Mater. **10**, 3769 (1998)
82. D. Lee, L.W. Jang: J. Appl. Polym. Sci. **61**, 1117 (1996); **68**, 1997 (1998)
83. S. Tandon, R. Kesavamoorthy, S.A. Asher: J. Chem. Phys. **109**, 6490 (1998)
84. M.H. Noh, D.C. Lee: J. Appl. Polym. Sci. **74**, 2811 (1999)
85. J. Wu, M.M. Lerner: Chem. Mater. **5**, 835 (1993)
86. Y. Kojima, A. Usuki, M. Kawasumi, A. Okada: J. Polym. Sci. A Chem. Ed. **31**, 983 (1993)
87. F.S. Dyachkovskii, L.A. Novokshonova: Russ. Chem. Rev. **53**, 200 (1984)
88. J. Tudor, L. Willington, D. O'Hare, B. Royan: Chem. Commun. 2031 (1996)
89. J. Tudor, D. O'Hare: Chem. Commun. 603 (1997)
90. L.K. Johnson, S.M. Mecking, M. Brookhart: J. Am. Chem. Soc. **118**, 267 (1996)
91. J.S. Bergman, H. Chen, E.P. Giannelis, M.G. Thomas, G.W. Coates: Chem. Commun. 2179 (1999)
92. S.S. Ray, M. Biswas: J. Appl. Polym. Sci. **73**, 2971 (1999)
93. Z.-K. Zhu, Y. Yang, J. Yin, X.-Y. Wang, Y.-C. Ke, Z.-N. Qi: J. Appl. Polym. Sci. **73**, 2063 (1999)
94. A. Blumstein, S. Subramanyam: Macromolecules **25**, 4058 (1992)
95. D.W. Kim, A. Blumstein, H. Liu, M.J. Downey, J. Kumar, S.K. Tripathy: J. Macromol. Sci. Pure Appl. Chem. A **38**, 1405 (2001)
96. S.K. Sahoo, D.W. Kim, J. Kumar, A. Blumstein, A.L. Cholli: Macromolecules **36**, 2777 (2003)
97. O.L. Glaveti, L.S. Polak: Neftechim. **3**, 905 (1963)
98. H.Z. Friendlander, C.R. Frick: J. Polym. Sci. B **2**, 475 (1964)
99. D.H. Solomon: J. Appl. Polym. Sci. **12** 1253 (1968)
100. Y. Ou, Z. Yu, Y. Feng: Chin. J. Polym. Sci. **11**, 31 (1993)
101. T. Watari, T. Yamane, S. Moriyama, T. Torikai, Y. Imaoka, K. Suehiro, H. Tateyama: Mater. Res. Bull. **32**, 719 (1997)
102. Y. Kojima, A. Usuki, M. Kawasumi, A. Okada, T. Kurauchi, O. Kamigaito: J. Polym. Sci. A Polym. Chem. **31**, 983 (1993)

103. C. Kato, K. Kuroda, M. Misawa: Clays. Clay Miner. **27**, 129 (1979)
104. T. Agag, T. Takeichi: Polymer **41**, 7083 (2000)
105. G.A. Ozin, A. Kupermann, A. Stein: Angew. Chem. **101**, 373 (1989); Angew.Chem. Int. Ed. Engl. **32**, 556 (1993)
106. Th. Bein: Chem. Mater. **4**, 819 (1992)
107. C.G. Wu, T. Bein: Science **266**, 1013 (1994)
108. B.R. Mattes, E.T. Knobbe, P.D. Fuqua, F. Nishida, E.W. Chang, B.M. Pierce, B. Dunn, R.B. Kaner: Synth. Met. **41**, 3183 (1991)
109. M.G. Kanatzidis, L.M. Tonge, T.J. Marks, H.O. Kannewurf: J. Amer. Chem. Soc. **109,** 3797 (1987)
110. C.G. Wu, M.O. Marcy, D.C. De Groot, J.L. Schindler, C.R. Kannewurf, W.Y. Leung, M. Benz, E. Le Goff, M.G. Kanatzidis: Synth. Met. **41**, 797 (1991)
111. M.G. Kanatzidis, C.G. Wu, M.O. Marcy, D.C. De Groot, C.R. Kannewurf, A. Kostikas, V. Papaehthymiou: Adv. Mater. **2**, 364 (1990)
112. G. Wu, D.C. De Groot, H.O. Marcy, J.L. Schindler, C.R. Kannewurf, T. Bacas, V. Papaehthymiou, W. Nirpo, J.P. Yesniowski, Y.L. Liu, M.G. Kanatzidis: J. Am. Chem. Soc. **117**, 9229 (1995)
113. R. Bissessur, D.C. De Groot, J.L. Schindler, C.R. Kannewurf, M.G. Kanatzidis: Chem. Commun. 687 (1993)
114. J. Lemmon, M. Lerner: Chem. Mater. **6**, 207 (1994)
115. Y. Liu, D. De Groot, J. Schindler, C. Kannewurf, M. Kanatzidis: Chem. Mater. **3**, 992 (1991)
116. C.O. Oriakhi, M. Lerner: Chem. Mater. **8**, 2016 (1996)
117. E. Ruiz-Hitzky, P. Aranda, B. Casal, J.C. Galvan: Adv. Mater. **7**, 180 (1995)
118. Y.N. Xia: Plast. Eng. **45**, 359 (1998)
119. C. Dearmitt, S.P. Armes: Langmuir, **9**, 652 (1993)
120. P.R. Somani, R. Marimuthu, U.P. Mulik, S.R. Sinkar, D.P. Amalnerkar: Synth. Met. **106**, 45 (1999)
121. R.E. Partch, S.G. Gangolli, E. Matijevic, W. Cai, S. Arajs: J. Colloid. Interface. Sci. **27**, 144 (1991)
122. Y.-J. Liu, M.G. Kanatzidis: Chem. Mater. **7**, 1525 (1995)
123. T.A. Kerr, H. Wu, L.F. Nazar: Chem. Mater. **8**, 2005 (1996)
124. D.J. Jones, R. El Mejjad, J. Roziere: In: *Supramolecular Architecture: Synthetic control in thin films and solids. ACS Symp.Ser. N 449* ed. by T. Bein (Am. Chem. Soc., Washington, DC 1992) p. 220
125. H. Nakajima, G. Matsubayashi: Chem. Lett. 423 (1993)
126. T. Challier, C.T. Slade: J. Mater. Chem. **4**, 367 (1994)
127. N.S.P. Bhuvanesh, J. Gopalakrishan: Mater. Sci. Eng. B **53**, 2676 (1998)
128. C.-G. Wu, Th. Bein: Science **264,** 1757 (1993)
129. *Supramolecular Architecture. ACS Symp. Ser. N 499* ed. by Th. Bein (Am. Chem. Soc., Washington, DC 1992)
130. B.E. Koene, L.F. Nazar: Solid State Ionics **89**, 147 (1996)
131. W. Feng, E. Sun, A. Fujii, K. Niihara, K. Yoshino: Bull. Chem. Soc. Jpn. **73**, 2627 (2000)
132. M.G. Kanatzidis, R. Bissessur, D.C. De Groot, J.L. Schindler, C.R. Kannewurf: Chem. Mater. **5**, 595 (1993)
133. T.L. Porter, M.E. Hagerman, B.P. Reynolds, M.P. Eastman, R.A. Parnell: J. Polym. Sci. B Polym. Phys. **36**, 673 (1998)

134. M.G. Kanatzidis, C.-G. Wu, H.O. Marcy, C.R. Kannewurf: J. Am. Chem. Soc. **111**, 4139 (1989); Chem. Mater. **2**, 221 (1990)
135. G. Wu, D.C. DeGroot, H.O. Marcy, J.L. Schindler, C.R. Kannewurf, Y.-J. Liu, W. Hirpo, M.G. Kanatzidis: Chem. Mater. **8**, 1992 (1996)
136. Y.-J. Liu, D.C. DeGroot, J.L. Schindler, C.R. Kannewurf, M.G. Kanatzidis: Chem. Commun. 593 (1993)
137. G.R. Goward, F. Leroux, L.F. Nazar: Electrochim. Acta. **43**, 1307 (1998)
138. D.R. Rolison, B. Dunn: J. Mater. Chem. **11**, 963 (2001)
139. A. Gapeeva, C.N. Yang, S.J. Klippenstein, R.C. Dunbar: J. Phys. Chem. A **104**, 3246 (2000)
140. M. Lira-Cantu, P. Gomez-Romero: J. Solid State Chem. **147**, 601 (1999)
141. V.D. Pokhodenko, V.A. Krylov, Ya.I. Kurys: Theor. Exp. Chem. **31**, 361 (1995)
142. J. Harreld, H.P. Wong, B.C. Dave, B. Dunn, L.F. Nazar: J. Non-Cryst. Solids **225**, 319 (1998)
143. J. Harreld, B. Dunn, L.F. Nazar: Int. J. Inorg. Mater. **1**, 135 (1999)
144. G.J.-F. Demets, F.J. Anaissi, H.E. Toma, M.B.A. Fontes: Mater. Res. Bull. **37**, 683 (2002)
145. E. Shouji, D.A. Buttry: Langmuir **15**, 669 (1999)
146. N.-G. Park, K.S. Ryu, Y.J. Park, M.G. Kang, D.-K. Kim, S.-G. Kang, K.M. Kim, S.-H. Chang: J. Power Sources **103**, 273 (2002)
147. V. Parente, C. Fredriksson, A. Selmani, R. Lazzaroni, J.L. Bredas: J. Phys. Chem. B **101**, 4193 (1997)
148. P. Gomes-Romero, M. Lira-Cantu: Adv. Mater. **9**, (1997)
149. T. Ohtake, K. Ito, N. Nishina, H. Kihara, H. Ohno, T. Kato: Polym. J. **31**, 1155 (1999)
150. K. Murakashi, G. Kano, Y. Wada, S. Yamagida, H. Miyazaki, M. Matsumoto, S. Murasawa: J. Electroanal. Chem. **396**, 27 (1995)
151. K. Murakashi, R. Kogure, Y. Wada, S. Yamagida: Chem. Lett. 471 (1997)
152. G.D. Stucky: In: *Progress in Inorganic Chemistry* ed. by J. Lippard (J. Willey, New York 1992) vol. 40, p. 99
153. V. Ramanmurthy: J. Am. Chem. Soc. **116**, 1345 (1994)
154. S.M. Marinakos, L.C. Brousseau, A. Jones, D.L. Feldheim: Chem. Mater. **10**, 1214 (1998)
155. S. Sakhora, L.D. Tickanen, M.A. Anderson: J. Phys. Chem. **96**, 11087 (1992)
156. T. Nishide, H. Yamaguchi, F. Mizukami: J. Mater. Sci. **30**, 4946 (1995)
157. B.B. Lakshmi, C.J. Patrissi, C.R. Martin: Chem. Mater. **9**, 2544 (1997)
158. J.B. Bates, G.R. Gruzalski, N.J. Dudney, C.F. Luck, X. Yu: Solid State Ionics. **70/71**, 619 (1994)
159. S. Bach, M. Henri, N. Baffier, J. Livage: J. Solid State Chem. **88**, 325 (1990)
160. E. Zhecheva, R. Stoyanova, M. Gorova, R. Alcantara, J. Morales, J.L. Tirado: Chem. Mater. **8**, 1429 (1996)
161. G. Kiss: Polym. Eng. Sci. **27**, 410 (1987)
162. R.V. Parthasarathy, C.R. Martin: Nature **369**, 298 (1994)
163. *Chemical Vapor Deposition* ed. by M.L. Hitchman, K.F. Jensen (Academic Press, San Diego, CA 1993)
164. V.M. Cepak, J.C. Hulteen, G. Che, K.B. Jirage, B.B. Lakshmi, E.R. Fisher, C.R. Martin: Chem. Mater. **9**, 1065 (1997)
165. C.-G. Wu, Th. Bein: Chem. Mater. **6**, 1109 (1994)

166. T. Fujinami, K. Sugie, K. Mori, M.A. Mehta: Chem. Lett. 619 (1998)
167. A. Usuki, M. Kawasumi, Y. Kojima, A. Okada, T. Kurauchi, O. Kamigato: J. Polym. Sci. A Polym. Chem. **31,** 983 (1993)
168. T. Lan, P.D. Kaviratna, T.J. Pinnavaia: Chem. Mater. **6**, 573 (1994)
169. M.P. Eastman, J.A. Attuso, T.L. Porter: Clays Clay Miner. **44**, 769 (1996)
170. R. Dagani: Chem. Eng. News **77**, 25 (1999)
171. Y.C. Zhou, M.N. Rahaman: J. Mater. Res. **8**, 1680 (1993)
172. S. Lakhwani, M.N. Rahaman: J. Mater. Sci. **34**, 3909 (1999)
173. M. Kawasumi, N. Hasegawa, M. Kato, A. Usuki, A. Okada: Macromolecules **30**, 6333 (1997)
174. R. Krishnamoorti, R.A. Vaia, E.P. Giannelis: Chem. Mater. **8**, 1728 (1996)
175. R.A. Vaia, B.B. Sauer, O.K. Tse, E.P. Giannelis: J. Polym. Sci. B Polym. Phys. **35**, 59 (1997)
176. P. Reichert, H. Nitz, S. Klinke, R. Brandsch, R. Thomann, R. Mülhaupt: Macromol. Mater. Eng. **375**, 8 (2000)
177. *Interfaces in Polymer, Ceramic, and Metal Matrix Composites* ed. by H. Ishida (Elsevier, New York 1988)
178. H.G. Jeon, H.T. Jung, S.W. Lee, S.D. Hudson: Polym. Bull. **41**, 107 (1998)
179. M.J. Solomon, A.S. Almusallam, K.F. Seefeldt, A. Somwangthanaroj, P. Vauadan: Macromolecules **34**, 1864 (2001)
180. N. Haasegawa, H. Okamoto, M. Kato, A. Usuki: J. Appl. Polym. Sci. **78**, 1981 (2000)
181. P.H. Nam, P. Maiti, M. Okamoto, T. Kotaka, N. Hasegawa, A. Usuki: Polymer **42**, 9633 (2001)
182. P. Maiti, P.H. Nam, M. Okamoto, N. Hasegawa, A. Usuki: Macromolecules **35**, 2042 (2002)
183. R.A. Vaia, S. Vasudevan, W. Krawiec, L.G. Scanlon, E.P. Giannelis: Adv. Mater. **7**, 154 (1995)
184. M.-J. Binette, C. Detellier: Can. J. Chem. **80**, 1708 (2002)
185. M. Nabai, C. Sanchez, J. Livage: Eur. J. Solid. State Inorg. Chem. **28**, 1173 (1988)
186. E. Ruiz-Hitzky, P. Aranda, B. Casal: J. Mater. Chem. **2**, 581 (1992)
187. E. Ruiz-Hitzky: Adv. Mater. **5**, 334 (1993)
188. Y. Liu, D. De Groot, J. Schindler, C. Kannewurf, M. Kanatzidis: Adv. Mater. **5**, 369 (1993)
189. I. Lagadic, A. Leaustic, R. Clement: Chem. Commun. 1396 (1992)
190. Y.-J. Liu, J.L. Schindler, D.C. DeGroot, C.R. Kannewurf, W. Hiro, M.G. Kanatzidis: Chem. Mater. **8**, 525 (1996)
191. M. Yoshizawa, H. Ohno: Chem. Lett., 889 (1999)
192. P.V. Wright: Brit. Polym. J. **7**, 319 (1975)
193. *Solid State Electrochemistry* ed. by P.G. Bruce (Cambridge University Press, Cambridge 1997)
194. F.M. Gray: *Solid Polymer Electrolytes – Fundamentals and Technical Applications* (VCH, Weinheim 1991)
195. F.M. Gray: *Polymer Eldectrolytes* (Springer-Verlag, New York 1997)
196. F.B. Dias, S.V. Batty, A. Gupta, G. Ungar, J.P. Woss, P.V. Wright: Electrochim. Acta **43**, 1217 (1998)
197. C.T. Imrie, M.D. Ingram, G.S. McHattie: Adv. Mater. **11**, 832 (1999)
198. T. Ohtake, K. Ito, N. Nishida, H. Kihara, H. Oho, T. Kato: Polym. J. **31**, 1155 (1999)

199. T. Ohtake, M. Ogasawara, K. Ito-Akita, N. Nishina, S. Ujiie, H. Ohno, T. Kato: Chem. Mater. **12**, 782 (2000)
200. T. Ohtake, Y. Takamitsu, K. Ito-Akita, K. Kanie, M. Yoshizawa, T. Mukai, H. Ohno, T. Kato: Macromolecules **33**, 8109 (2000)
201. J. McBreen, I. Lin: J. Electrochem. Soc. **139**, 960 (1992)
202. S. Chintipalli, R.E. Frech, B.P. Grady: Polymer **25**, 6189 (1997)
203. B.P. Grady, C.P. Rhodes, S. York, R.E. Frech:. Macromolecules **34**, 8523 (2001)
204. F. Croce, G.B. Appetecchi, L. Persi, B. Scrosati: Nature **394**, 456 (1998)
205. M. Templin, A. Franck, A. Du Chesne, H. Leist, Y. Zhang, R. Ulrich, V. Schädler, U. Wiesner: Science **278**, 1795 (1997)
206. S.M. De Paul, J.W. Zwanziger, R. Ulrich, U. Wiesner, H.W. Spiess: J. Am. Chem. Soc. **121**, 5727 (1999)
207. C. Sanchez, B. Alonso, F. Chpusot, F. Ribot, P. Audbert: J. Sol-Gel Sci. Technol. **2**, 161 (1994)
208. H. Krug, H. Smidt: New J. Chem. **18**, 1125 (1994)
209. R.J.P. Corriu, J.J.E. Moreau, P. Thepot, M.W. Chi Man, C. Chorro, J.-P. Lere-Porte, J.-L. Sauvajol: Chem. Mater. **6**, 640 (1994)
210. H.P. Oliveira, C.F.O. Graeff, J.M. Rosolen: Mater. Res. Bull. **34**, 1891 (1999)
211. J.Z. Wang, K.E. Gonsalves: J. Comb. Chem. **1**, 216 (1999)
212. N.E. Kotelnikova, G. Vegner, T.A. Pakkari, P. Serima, V.N. Demidov, A.S. Serebryakov, A.V. Shchukarev, A.B. Gribanov: Russ. J. Gen. Chem. **73**, 447 (2003)
213. W. Nonte, M. Lobert, W. Müller-Warmuth, R. Schöllhorn: Synth. Met. **34**, 665 (1989)
214. L. Wang, P. Brazis, M. Rocci, C.R. Kannewurf, M. Kanatzidis: Chem. Mater. **10**, 3298 (1998)
215. L. Wang, M. Rocci-Lane, P. Brazis, C.R. Kannewurf, Y. Kim, W. Lee, J.-H. Choy, M.G. Kanatzidis: J. Am. Chem. Soc. **122**, 6629 (2000)
216. S. Ramesh, Y. Koltypin, R. Prosorov, A. Gedanken: Chem. Mater. **9**, 546 (1997)
217. D.A. Hucul, A. Brenner: J. Phys. Chem. **85**, 496 (1981)
218. L.M. Liz-Marzan, M. Giersig, P. Mulvaney: Langmuir **12**, 4329 (1996)
219. S. Yamamoto, O. Matsuoka, I. Fukada, Y. Ashida, T. Honda, N. Yamamoto: J. Catal. **159**, 401 (1996)
220. S. Yamamoto, O. Matsuoka, S. Sugiyama: Chem. Lett. 809 (1998)
221. A.S. Panasyugin, A.I. Ratko, N.P. Masherov: Russ. J. Inorg. Chem. **43**, 1437 (1998)
222. F. Capasso: Thin Solid Films **216**, 59 (1992)
223. J.H. Fendler: *Advances in Polymer Science, V. 113* (Springer-Verlag, Berlin 1994)
224. J.H. Fendler, F.C. Meldrum: Adv. Mater. **7**, 607 (1995)
225. E.R. Kleinfeld, G.S. Ferguson: Chem. Mater. **7**, 2327 (1995); Mat. Res. Soc. Symp. Proc. **369**, 697 (1995)
226. N.A. Kotov, L. Dekany, J.H. Fendler: J. Phys. Chem. **99**, 13065 (1995); Adv. Mater. **8**, 637 (1996); Langmuir **10**, 3797 (1994)
227. S.W. Keller, H.N. Kim, T.E. Mallouk: J. Am. Chem. Soc. **116**, 8817 (1994)
228. J. Schmitt, G. Decher, W.J. Dressik, S.L. Branduo, R.E. Geer, R. Shashidhal, J.M. Calvert: Adv. Mater. **9**, 61 (1997)

229. R.G. Freeman, K.C. Grabar, K.J. Allison, R.M. Bright, J.A. Davis, A.P. Guthrie, M.B. Hommer, M.A. Jackson, P.C. Smith, D.G. Walter, M.J. Natan: Science **267**, 1629 (1995)
230. N.A. Kotov, T. Haraszti, L. Turi, G. Zavala, R.E. Greer, I. Dekany, J.H. Fendler: J. Am. Chem. Soc. **119**, 6821 (1997)
231. *Fine Particles Science and Technology from Micro to Nanoparticles* ed. by E. Pelizetti (Kluwer Academic Publishers, Netherlands 1996)
232. R. Lakes: Nature **361**, 511 (1993)
233. A. Laschewsky, E. Wischerhoff, P. Bertrand, A. Delcorte, S. Denzinger, H. Ringsdorf: Eur. Chem. J. **3**, 28 (1997)
234. A.C. Fou, M.F. Rubner: Macromolecules **28**, 7115 (1995)
235. Y. Lvov, K. Agira, I. Ichinose, T. Kunitake: Langmuir **12**, 3038 (1996); Chem. Lett. 257 (1996)
236. D. Cochin, M. Passmann, G. Wilber, R. Zentel, E. Wischerhoff, A. Laschewsky: Macromolecules **30**, 4775 (1997)
237. M.E. Vol/pin, Yu.N. Novikov, N.D. Lapkina, V.I. Kasatochkin, Yu.T. Struchkov, M.E. Kazakov, R.A. Stukan, V.A. Povitskij, Yu.S. Karimov, A.V. Zvarikina: J. Am. Chem. Soc. **97**, 3366 (1975)
238. V.L. Solozhenko, I.V. Arkhangelskii, A.M. Gaskov: Russ. J. Phys. Chem. **57**, 2265 (1983)
239. V.A. Zhorin, N.I. Alekseev, I.N. Groznov, V.D. Kuznetsov, A.S. Bakman, V.G. Nagornyi, V.I. Goldanskii: Dokl. AN USSR **266**, 391 (1982)
240. D.S. Wragg, G.B. Hix, R.E. Morris: J. Am. Chem. Soc. **120**, 6822 (1998)
241. K.G. Neoh, K.K. Tan, P.L. Goh, S.W. Huang, E.T. Kang, K.L. Tan: Polymer **40**, 887 (1999)
242. K.A. Tarasov, V.P. Isupov, B.V. Bokhonov, Yu.A. Gaponov, R.P. Mitrofanova, L.E. Chupakhina, B.P. Tolochko, S.S. Shatskaya: Chem. Steady State **8**, 291 (2000)
243. U.Opara-Krasovec, R. Jese, B. Orel, G. Drazic: Monat. Chem. **133**, 1115 (2002)
244. S.C. Tjong, Y.Z. Meng: Polymer **40**, 1109 (1999)
245. K. Hu, M. Brust, A.J. Bard: Chem. Mater. **10**, 1160 (1998)
246. J.M. Stipkala, F.N. Castellano, T.A. Heimer, C.A. Kelly, K.J.T. Livi, G.J. Meyer: Chem. Mater. **9**, 2341 (1997)
247. J. Brenbner, C.L. Marshall, L. Ellis, N. Tomczyk.: Chem. Mater. **10**, 1244 (1998)
248. I. Honma, T. Sano, H. Komiyama: J. Phys. Chem. **97** 6692 (1993)
249. S. Chen, K. Kimura: Chem. Lett. 233 (1999)
250. A.D. Yoffe: Adv. Phys. **50**, 208 (2001)
251. D. Meissner, R. Memming, B. Kastening: Chem. Phys. Lett. **96**, 34 (1983)
252. A.K. Atta, P.K. Biswas, D. Ganguli: 'CdS-Nanoparticles in Gel Film Network: Synthesis, Stability and Properties'. In: *Polymer and Other Advanced Materials: Emerging Technologies and Business Opportunities* ed. by P.N. Prasad, J.E. Mark, T.J. Fai (Plenum Press, New York 1995) p. 645
253. M. Krishnan, J.R. White, M.A. Fox, A. Bard: J. Am. Chem. Soc. **105**, 7002 (1983)
254. J.M. Huang, Y. Yang, B. Yang, S.Y. Liu, J.C. Shen: Polym. Bull. **36**, 337 (1996)
255. Y. Wang, N. Nerron: Chem. Phys. Lett. **200**, 71 (1992)

256. A.V. Volkov, M.A. Moskvina, S.B. Zezin, A.L. Volynskii, N.F. Bakeev: Polym. Sci. Ser. A **45**, 283 (2003)
257. A.S. Susha, F. Caruso, A.L. Rogach, G.B. Sukhorukov, A. Kornowski, H. Möhwald, M. Giersig, A. Eychmüller, H. Weller: Colloid Surf. A **163**, 39 (2000)
258. I.L. Radtchenko, G.B. Sukhorukov, N. Gaponik, A. Kornowski, A.L. Rogach, H. Möhwald: Adv. Mater. **13**, 1684 (2001)
259. J. Zhang, N. Coombs, E. Kumacheva: J. Am. Chem. Soc. **124**, 14512 (2002)
260. J. Rockenberger, L. Troger, A. Kornowski, T. Vossmeyer, A. Eychmuller, J. Feldhaus, H. Weller: J. Phys. Chem. **101,** 2691 (1997)
261. C.B. Murray, C.R. Kagan, M.G. Bawendy: Science **270**, 1335 (1995)
262. M. Möller: Synth. Met. **41–43**, 1159 (1991)
263. H. Zhao, E.P. Douglas, B.S. Harrison, K.S. Schanze: Langmuir **17**, 8428 (2001)
264. M. Moffitt, H. Vali, A. Eisenberg: Chem. Mater. **10**, 1021 (1998)
265. R.S. Kane, R.E. Cohen, R. Silbey: Chem. Mater. **8**, 1919 (1996)
266. M. Antonietti, S. Forster, J. Hartmann, S. Oestreich: Macromolecules **29**, 3800 (1996)
267. R.E. Cohen, J. Yue: Supramol. Sci. **1**, 117 (1994)
268. J. Wang, D. Montville, K.E. Gonsalves: J. Appl. Polym. Sci. **72**, 1851 (1999)
269. Y. Zhou, S. Yu, C. Wang, X. Li, Y. Zhu, Z. Chen: Chem. Commun. 1229 (1999)
270. J.-H. Zeng, J. Yang, Y. Zhu, Y.-F. Liu, Y.-T. Qian, H.-G. Zheng: Chem. Commun. 1332 (2001)
271. J. Huang, Y. Yang, B. Yang, S. Liu, J.C. Shen: Polym. Bull. **37**, 679 (1996)
272. J. Vedel: Adv. Mater. **6**, 379 (1994)
273. Y. Zhou, L. Hao, Y. Hu, Y. Zhu, Z. Chen: Chem. Lett. 136 (2001)
274. N. Gaponic, I.L. Radtchenko, G.B. Sukhorukov, H. Weller, A.L. Rogach: Adv. Mater. **14**, 879 (2002)
275. I. Potapova, R. Mruk, S. Prehl, R. Zentel, T. Basche, A. Mews: J. Am. Chem. Soc. **125**, 320 (2003)
276. A. Chemseddine, H. Jungblut, S. Boulmaaz: J. Phys. **100**, 12546 (1998)
277. *Intercalated Layered Materials* ed. by F. Levy (Reidel Publishing, Dodrecht 1979)
278. *Nanoparticles in Amorphous Solids and Their Nonlinear Properties, Advances and Applications. NATO ASI Series. B. Physics, Vol. 399* ed. by B. Di Bartolo (Plenum Press, New York 1994)
279. L. Spanhel, E. Arpac, H. Smidt: Non-Cryst. Solids **147/148**, 657 (1992)
280. M. Zelner, H. Minti, R. Reisfeld, H. Cohen, R. Tenne: Chem. Mater. **9**, 2541 (1997)
281. P. Mottner, T. Butz, A. Lerf, G. Ledezma, H. Knözinger: J. Phys. Chem. **99**, 8260 (1995)
282. R. Tenne: Adv. Mater. **7** 965 (1995)
283. A. Rothschild, J. Sloan, R. Tenne: J. Am. Chem. Soc. **122**, 5169 (2000)
284. M. Nath, C.N.R. Rao: J. Am. Chem. Soc. **123**, 4841 (2001)
285. P. Joensen, R.F. Frindt, S.R. Morrison: Mater. Res. Bull. **21**, 457 (1986)
286. M.A. Gee, R.F. Frindt, P. Joensen, S.R. Morrison: Mater. Res. Bull. **21**, 543 (1986)
287. H.-L. Tsai, J.L. Schindler, C.R. Kannewurf, M.G. Kanatzidis: Chem. Mater. **9**, 875 (1997)

288. S.G. Haup, D.R. Riley, C.T. Jones, J. Zhao, J.T. McDevit: J. Amer. Chem. Soc. **115**, 1196 (1993); **116**, 9979 (1994)
289. H.-L. Tsai, J. Heising, J.L. Schindler, C.R. Kannewurf, M.G. Kanatzidis: Chem. Mater. **9**, 879 (1997)
290. M.T. Pham, D. Möller, W. Matz, A. Mücklich: J. Phys. Chem. B **102**, 4081 (1998)
291. E. Ruiz-Hitzky, R. Jimenez, B. Casal, V. Manriquez, A.A. Santa, G. Gonzalez: Adv. Mater. **5**, 738 (1993)
292. P. Jeevanandam, S. Vasudevan: Chem. Mater. **10**, 1276 (1998); Solid State Ionics **104**, 45 (1997)
293. A. Blumberg, S.S. Pollack: Polym. Sci. Part A **2**, 2499 (1964)
294. H. Todokoro, T. Yoshihara, Y. Chetani, S. Takora, S. Murahashi: Macromol. Chem. **73**, 109 (1964)
295. T. Hanaoka, T. Tago, K. Wakabayashi: Bull. Chem. Soc. Jpn. **74**, 1349 (2001)
296. P.V. Braun, P. Osenar, V. Tohver, S.B. Kennedy, S.I. Stupp: J. Am. Chem. Soc. **121**, 7302 (1999)
297. Ch.O. Oriakhi, R.L. Nafshum, M.M. Lerner: Mater. Research Bull. **31**, 1513 (1996)
298. Y. Sun, E. Hao, X. Zhang, B. Yang, M. Gao, J. Shen: Chem. Commun. 2381 (1996)
299. A.V. Volkov, M.A. Moskvina, A.L. Volynskii, N.F. Bakeev: Polym. Sci. Ser. A **40**, 1441 (1998); **41**, 963 (1999)
300. V.L. Colvin, M.C. Schlamp, A.P. Alivistato: Nature **370**, 354 (1994)
301. B.O. Dabbousi, M.G. Bawendi, O. Onitsuka, M.F. Rubner: Appl. Phys. Lett. **66**, 1316 (1995)
302. M. Gao, B. Richter, S. Kirstein, H. Möhwald: J. Phys. Chem. B **102**, 4096 (1998)
303. T. Cassagneau, T.E. Mallouk, J.H. Fendler: J. Am. Chem. Soc. **120**, 7848 (1998)
304. F. Hide, B.J. Schwartz, M. Diaz-Garcia, A. Heeger: J. Chem. Phys. Lett. **256**, 424 (1996)
305. M.P.T. Christiaans, M.M. Wienk, P.A. van Hal, J.M. Kroon, R.A.J. Janssen: Synth. Met. **101**, 265 (1999)
306. M. Yang, S.Y. Huang, S.Y. Liu, J.C. Shen: J. Mater. Chem. **7**, 131 (1997); Appl. Phys. Lett. **69**, 377 (1996)
307. Y. Yang, J. Huang, B. Yang, S. Liu, J. Shen: Synt. Met. **91**, 347 (1997)
308. B. Cheng, W.Q. Jiang, Y.R. Zhu, Z.Y. Chen: Chem. Lett. 935 (1999)
309. Y. Zhou, L.Y. Hao, Y. Hu, Y.R. Zhu, Z.Y. Chen: Chem. Lett. 1308 (2000)
310. M. Moffitt, L. McMahon, V. Pessel, A. Eisenberg: Chem. Mater. **7**, 1185 (1995)
311. B.I. Lemon, R.M. Crooks: J. Am. Chem. Soc. **122**, 12886 (2000)
312. J.R. Lakowicz, I. Gryczynski, C.J. Murphy: J. Phys. Chem. B **103**, 7613 (1999)
313. J. Yang, H. Lin, Q. He, L. Ling, C. Zhu, F. Bai: Langmuir **17**, 5978 (2001)
314. C.G. Granqvist: *Handbook of Inorganic Electrochromic Materials* (Elsevier, Amsterdam 1995)
315. J.G. Winiarz, L. Zhang, M. Lal, C.S. Friend, P.N. Prasad: J. Am. Chem. Soc. **121**, 5287 (1999); Chem. Phys. **245**, 417 (1999)
316. J.G. Winiarz, L. Zhang, J. Park, P.N. Prasad: J. Phys. Chem. **106**, 967 (2002)
317. Y. Wang, A. Suna, W. Mahler, R. Kasovski: J. Chem. Phys. **87**, 7315 (1987)
318. J.H. Fendler, N.A. Kotov, I. Dekany: NATO ASI Ser. **12**, 557 (1996)

319. V. Sankaran, C.C. Cummins, R.R. Schrock, R.J. Sibey: J. Am. Chem. Soc. **112**, 6858 (1990)
320. M. Mukherjee, A. Datta, D. Chakravorty: Appl. Phys. Lett. **64**, 1159 (1994)
321. Z. Zeng, S. Wang, S. Yang: Chem. Mater. **11**, 3365 (1999)
322. Z. Qiao, Y. Xie, M. Chen, J. Xu, Y. Zhu, Y. Qian: Chem. Phys. Lett. **321**, 504 (2000)
323. E.V. Nikolaeva, S.A. Ozerin, A.E. Grigoriev, E.I. Grigoriev, S.N. Chvalun, G.N. Gerasimov, L.I. Trakhtenberg: Mater.Sci.& Engin. C Biomimetic and Supramol. Syst. **8-9**, 217 (1999)
324. M.V. Artemyev, Yu.P. Rakovich, G.P. Yablonski: J. Cryst. Growth **171**, 447 (1997)
325. A. Sengupta, B. Jiang, K.C. Mandal, J.Z. Zhang: J. Phys. Chem. B **103**, 3128 (1999)
326. L. Manna, E.C. Scher, L.S. Li, A.P. Alivisatos: J. Am. Chem. Soc. **124**, 7136 (2002)
327. L. Manna, E.C. Scher, A.P. Alivisatos: J. Am. Chem. Soc. **122**, 12700 (2000)
328. X.G. Peng, L. Manna, W.D. Yang, J. Wickham, E. Scher, A. Kadavanich, A.P. Alivisatos: Nature **404**, 59 (2000)
329. Z.A. Peng, X. Peng: J. Am. Chem. Soc. **123**, 183 (2001); **124**, 3343 (2002)
330. W. Guo, J.J. Li, A. Wang, X. Peng: J. Am. Chem. Soc. **125**, 3901 (2003)
331. Q. Yang, K. Tang, C. Wang, Y. Qian, S. Zhang: J. Phys. Chem. B **106**, 9227 (2002)
332. P. Bonneau, J.L. Mansot, J. Rouxel: Mater. Res. Bull. **28**, 757 (1993)
333. J.H. Golden, F.J. Di Salvo, J.M. Frechet: J. Chem. Mater. **6**, 844 (1994)
334. J. Hu, T.W. Odom, C.M. Lieber: Acc. Chem. Res. **32**, 435 (1999)
335. I. Langmuir: J. Am. Chem. Soc. **39**, 1848 (1917)
336. K.B. Blodgett: J. Am. Chem. Soc. **57**, 1007 (1935)
337. K.B. Blodgett, I. Langmuir: J. Phys. Rev. **51**, 964 (1937)
338. A.J. Bard: *Integrated Chemical Systems: A Chemical Approach to Nanotechnology* (J. Wiley & Sons, New York 1994)
339. A. Ulman: In: *An Itroduction in Ultrathin Organic Films from Langmuir-Blodgett to Self-Assembly* (Academic Press Inc., New York 1991)
340. *Langmuir–Blodgett Films* ed. by G. Roberts (Plenum Press, New York 1990)
341. C.T. Kresge, M.E. Leonowitz, W.J. Roth, J.C. Vartuli, J.S. Beck: Nature **359**, 710 (1992)
342. R.H. Tredgold: Rep. Progr. Phys. **50**, 1609 (1987)
343. R.H. Tredgold, R.A. Allen, P. Hodge, E. Khoshdel: J. Phys. D. Appl. Phys. **20**, 1385 (1987)
344. S.G. Yudin, S.P. Palto, V.A. Kravrichev: Thin Solid Films, **210/211**, 46 (1992)
345. V.L. Colvin, A.N. Goldstein, A.P. Alivisatos: J. Am. Chem. Soc. **114**, 5221 (1992)
346. G. Chumanov, K. Sokolov, B.M. Gregory, T.M. Cotton: J. Phys. Chem. **99**, 9466 (1995)
347. J. Yang, F.C. Meldrum, J.H. Fendler: J. Phys. Chem. **99**, 5500 (1995)
348. K.C. Yi, Z. Horvolgi, J.H. Fendler: J. Phys. Chem. **98**, 3872 (1994)
349. B.R. Rajam, B.R. Heywoood, J.B.A. Walker, S. Mann, R.J. Davey, J.D. Birchall: J. Chem. Soc. Faraday Trans. **87**, 727 (1991)
350. H.S. Mansur, F. Grieser, R.S. Urquhart, D.N. Furlog: J. Chem. Soc. Faraday Trans. **91**, 3399 (1995)

351. J. Leloup, A. Ruadel-Teixier, A. Barraud: Thin Solid Films **210/211**, 407 (1992)
352. F.C. Meldrum, N.A. Kotov, J.H. Fendler: J. Chem. Soc. Faraday Trans. **90**, 673 (1994); Langmuir **10**, 2035 (1994)
353. K.S. Maya, V. Patil, M. Sastry: J. Chem. Soc. Faraday Trans. **93**, 3377 (1977)
354. S. Kwan, F. Kim, J. Akana, P. Yang: Chem. Commun. 447 (2001)
355. F. Kim, S. Kwan, J. Akana, P. Yang: J. Am. Chem. Soc. **123**, 4360 (2001)
356. P. Cerny, P.G. Zverev, H. Jelinkova, T.T. Basiev: Opt. Commun. **177**, 397 (2000)
357. H. Shi, L. Qi, J. Ma, H. Cheng: Chem. Commun. 1704 (2002)
358. D. Zhang, L. Qi, J. Ma, H. Cheng: Adv. Mater. **14**, 1499 (2002)
359. H. Shi, L. Qi, J. Ma, H. Cheng: J. Am. Chem. Soc. **125** 3450 (2003)
360. P.E. Laibinis, J.J. Hickman, M.S. Wrighton, G.M. Whitesides: Science **245**, 845 (1989)
361. S. Schacht, Q. Huo, I.G. Voigt-Martin, G.D. Stucky, F. Schüth: Science **273**, 768 (1996)
362. M. Clemente-Leon, C. Mingotaud, B. Agricole, C.J. Gomez-Garcia, E. Coronado, P. Delhaes: Angew. Chem. Int. Ed. Engl. **36**, 1114 (1997)
363. M. Aiai, J. Ramos, C. Mingotaud, J. Amiell, P. Delhaes: Chem. Mater. **10**, 728 (1998)
364. C. Lafuente, C. Mingotaud, P. Delhaes: Chem. Phys. Lett. **302**, 523 (1999)
365. G. Decher, Y. Lvov, J. Smith: Thin Solid Films **244**, 772 (1994)
366. N.A. Kotov, S. Magonov, E. Tropsha: Chem. Mater. **10**, 886 (1998)
367. D. Rong, Y. Kim, T.E. Mallouk: Inorg. Chem. **29**, 1531 (1990)
368. Y. Hotta, K. Inukai, M. Taniguchi, A. Yamagishi: Langmuir **12**, 5195 (1996)
369. A. Yamagishi, Y. Goto, M. Taniguchi: J. Phys.Chem. **100**, 1827 (1996)
370. K. Tamura, H. Setsuda, M. Taniguchi, T. Nakamura, A. Yamagishi: Chem. Lett. 121 (1999)
371. T. Taniguchi, Y. Fukasawa, T. Miyashita: J. Phys. Chem. B **103**, 1920 (1999)
372. N. Fukuda, M. Mitsuishi, A. Aoki, T. Miyashita: J. Phys. Chem. B **106**, 7048 (2002)
373. M. Ferreira, K. Wohnrath, R.M. Torresi, C.J.L. Constantino, R.F. Aroca, O.N. Oliveira, Jr., J.A. Giacometti: Langmuir **18**, 540 (2002)
374. A. Riul, Jr., D.S. dos Santos, Jr., K. Wohnrath, R. Di Tommazo, A.C. Carvalho, F.S. Fonseka, O.N. Oliveira, Jr., D.M. Taylor, L.H.C. Mattoso: Langmuir **18**, 239 (2002)
375. H. Jeong, B.-J. Lee, W.J. Cho, C.-S. Ha: Polymer **41**, 5525 (2000)
376. X.K. Zao, S. Xu, J.H. Fendler: J. Phys. Chem. **98**, 4913 (1994); Langmuir **7**, 250 (1991)
377. N.A. Kotov, F.S. Meldrum, C. Wu, J.H. Fendler: J. Phys. Chem. **98**, 2735 (1994);
378. F.C. Meldrum, N.A. Kotov, J.H. Fendler: J. Phys. Chem. **98**, 4506 (1994)
379. X.G. Peng, S.G. Guan, X.D. Chai, Y. Jiang, T. Li: J. Phys. Chem. **96**, 3170 (1992)
380. X. Ji, C.Y. Fan, F.Y. Ma: Thin Solid Films **242**, 16 (1994)
381. J.K. Pike, H. Byrd, A.A. Morrone, D.K. Talham: J. Am. Chem. Soc. **115,** 8497 (1993)
382. L.L. Sveshnikova, S.M. Repinskii, A.K. Gutakovskii, A.G. Milekhin, L.D. Pokrovskii: Chem. Steady State **8**, 265 (2000)

383. F.N. Dultsev, L.L. Svechnikova: Thin Solid Films **288**, 103 (1996)
384. F.N. Dultsev, L.L. Svechnikova: Russ. J. Struct. Chem. **38**, 803 (1997)
385. R.S. Urquhart, D.N. Furlong, T. Gegenbach, N.J. Geddes, F. Grieser: Langmuir **11**, 1127 (1995)
386. Y.H. Park, B.I. Kim, Y.J. Kim: J. Appl. Polym. Sci. **63**, 619, 779 (1997)
387. Z. Gu, X. Yang, Z. Lu, Y. Wei: J. Southeast Univ. **24**, 103 (1994)
388. X. Peng, Y. Zhang, J. Yang, B. Zou, L. Xiao, T. Li: J. Phys. Chem. **96**, 3412 (1992)
389. Y. Zhang, Z. Xie, B. Hua, B. Mao, Y. Chen, Q. Li, Z. Tian: Science in China Ser. B **40**, (1997)
390. B. Pignataro, C. Consalvo, G. Compagnini, A. Licciardello: Chem. Phys. Lett. **299**, 430 (1999)
391. Z.F. Liu, C-.X. Zhao, M. Tang, S.M. Cai: J. Phys. Chem. **100**, 17337 (1996)
392. C.-X. Zhao, J. Zhang, Z.-F. Liu: Chem. Lett. 473 (1997)
393. X.K. Zao, J.H. Fendler: J. Phys. Chem. **94**, 3384 (1990)
394. N.A. Kotov, M.E.D. Zaniquelli, F.C. Meldrum, J.H. Fendler: Langmuir **9**, 3710 (1993)
395. S.H. Tolbert, P. Sieger, G.D. Stucky, S.M.J. Aubin, C.-C. Wu, D.N. Hendrickson: J. Am. Chem. Soc. **119**, 8652 (1997)
396. S. Vitta, T.H. Metzger, S. Magor, A. Dhanabalan, S.S. Talwar: Langmuir **14**, 1799 (1998)
397. Y. Wang, N. Herron, W. Mahler, S. Suna: J. Opt. Soc. Am. **B6**, 808 (1989)
398. K. Kamitani, M. Uo, H. Inoue, A. Makishima, T. Suzuki, K. Horie: J. Lumin. **64**, 291 (1995)
399. H. Yoneyama, M. Tokuda, S. Kuwabata: Electrochem. Acta **39**, 1315 (1994)
400. A. Dhanabalan, R.B. Dabke, S.N. Dutta, N.P. Kumar, S.S. Major, S.S. Talwar, A.Q. Contractor: Thin Solid Films **295**, 255 (1997)
401. A. Dhanabalan, S.S. Talwar, A.Q. Contractor, N.P. Kumar, S.N. Narang, S.S. Major: J. Mater. Sci. Lett. **18**, 603 (1999)
402. K. Yase, S. Schwiegk, G. Lieser, G. Wegner: Thin Solid Films **210/211**, 22 (1992)
403. M. Rikukawa, M.F. Rubner: Langmuir **10**, 519 (1994)
404. V.V. Arislanov: Russ. Chem. Rev. **63**, 3 (1994)
405. C.S. Winter, R.H. Tredgold, A.J. Vickers, E. Koshdel, P. Hodge: Thin Solid Films **134**, 49 (1985)
406. J. Nagel, U. Oertel: Polymer **36**, 381 (1995)
407. J.F. Hicks, F.P. Zamborini, A.J. Osisek, R.W. Murray: J. Am. Chem. Soc. **123**, 7048 (2001)
408. A.C. Templeton, F.P. Zamborini, F.P. Wuelfing, R.W. Murray: Langmuir **16**, 6682 (2000)
409. F.P. Zamborini, M.C. Leopold, J.F. Hicks, P.J. Kulesza, M.A. Malik, R.W. Murray: J. Am. Chem. Soc. **124**, 8958 (2002)
410. Y.S. Chen, Z.K. Xu, B.K. Xu, B.K. Zhu, Y.Y. Xu: Chem. J. Chin. Univ. **18**, 973 (1997)
411. Z.-K. Xu, Y.-Y. Xu, M. Wang: J. Appl. Polym. Sci. **69**, 1403 (1998)
412. E.R. Kleifield, G.S. Ferguson: Science **265**, 370 (1994)
413. J.F. Liu, K.Z. Yang, Z.H. Lu: J. Am. Chem. Soc. **119**, 11061 (1997)
414. A.F. Diaz, J.F. Rubinson, H.B. Mark, Jr. Adv. Polym. Sci. **84**, 11 (1988)
415. H. Shin, R.J. Collins, M.R. De Jquire, A.H. Heuer, C.N. Sukenik: J. Mater. Res. **10**, 692 (1995)

416. Y. Liu, A. Wang, R. Claus: J. Phys. Chem. B **101**, 1385 (1997)
417. A. Morneau, A. Manivannan, C.R. Cabrena: Langmuir **10**, 3940 (1994)
418. T. Fujimoto, A. Fukuoka, J. Nakamura, M. Ichikawa: Chem. Commun. 845 (1989)
419. T. Fujimoto, A. Fukuoka, M. Ichikawa: Chem. Mater. **4**, 104 (1992)
420. S.P. Gubin, E.S. Soldatov, A.S. Trifonov, S.G. Yudin: Inorg. Mater. **32**, 1265 (1996)
421. S.A. Yakovenko, S.P. Gubin, E.S. Soldatov, A.S. Trifonov, V.V. Khanin, G.B. Khomutov: Inorg. Mater. **32**, 1272 (1996)
422. S.P. Gubin, V.V. Kolesov, E.S. Soldatov, A.S. Trifonov, S.G. Yudin: Inorg. Mater. **33**, 1216 (1997)
423. R. Castillo, S. Ramos, J. Ruiz-Garcia: J. Phys. Chem. **100**, 15235 (1996)
424. Y. Tomioka, M. Ishibashi, H. Kajiyama, Y. Taniguchi: Langmuir **9**, 32 (1993)
425. M. Yanagida, T. Kuri, T. Kajiyama: Chem. Lett. 911 (1997)
426. S. Uemura, A. Ohira, T. Ishizaki, M. Sakata, M. Kunitake, I. Taniguchi, C. Hirayama: Chem. Lett. 536 (1999)
427. K.H. Johnson, S.V. Pepper: J. Appl. Phys. **53**, 6634 (1982)

9 Nanobiocomposites

The particles and primarily the supramolecular bifunctional systems connected with the different biological processes and products (ferments, liposomes, cells) have a special meaning for nanochemistry [1–6]. The research field is located at the crossroad of materials investigation, nanoscience and molecular biotechnology.

It is known that the metal-containing particle interaction with biopolymers (proteins, nucleic acids, polysacharides, etc.) plays an important role in the processes of fermentative catalysis, biosorption, geobiotechnology, biohydrometallurgy, and biomineralization. Natural polymers bond not only polynuclear complexes but create an optimal environment for the effective functioning of the active centers. Some noncovalent or weak interactions also play vital roles in molecular recognition and structural organization of many biological systems. Entering into a living organism, silver ions are bonded to argentaffine biopolymers and rapidly reduced by biological substrates to the metal state [7]. The same processes take place during ion silver accumulation by some bacteria. The amount of absorbed metal can reach 25% of dry weight of microbe cells [8]. Water-soluble metal-polymer nanocomposites containing silver and other metal nanoparticles are potentially useful as antiviral and antimicrobic agents.

The newly emerging area of inorganic particles entrapping biomolecules has already exhibited its diversity and potential for applications in many frontiers of modern material science, including biosensors, biocatalysts, immunochemistry, and materials for environmental sciences. Transient metal mixed valence clusters function in many ferments or redox-active metalloproteins and support electron transport over long distances.

Life-scientists have always been delighted with the perfection, organization and self-regulation of biological processes. For example, ribosomes (particles 20 nm in size) are effective supramolecular machines that can spontaneously be self-assembled from more than 50 individual building blocks (proteins and nucleic acids) and thereby can impressively demonstrate the power of biologically programmed molecular recognition [9]. The investigators tried modeling these properties in laboratory conditions and getting synthetic analogs of different natural substances. It became a base for a new intensively developing field in chemical science (called *biomimetics*), connected with the modelling or imitation of biological functions, and in this science metalloclusters and nanoparticles are important model objects for many conceptions [10–15]. The first notions of biomimetic inorganic chemistry were established in 1872 by P. Harting, who thought that the structures and activities of living organisms (ideas, methods, reaction mechanisms) had to be studied in the corresponding chemical structures, systems and processes [16].

Almost all the above-described approaches and methods (including sol–gel synthesis, intercalation, LB films, etc.) can be used for nanoparticle formation in biopolymers and their analogs. Chemistry is the central science for the development of applied disciplines such as materials research and biotechnology [17]. The most important problems are connected with biomineralization, polynuclear structures and clusters with mixed valency in biology (especially, oxoferric, oxomolybdenum or oxomanganese) and the activation or biosorption of small molecules. There are a number of surveys and special journals issues devoted to these problems [18–20].

In this chapter we analyze polynuclear metalloproteins and make an attempt to find interconnections between biosorption, biomineralization and nanoscale particle chemistry. Particular attention will be given to problems mainly connected with material science aspects.

9.1 Basic Notions of Metal-Containing Protein Systems

The efficiency of metalloenzymes owe much to weak interactions involving the side chains of aminoacid residues, when a central metal ion serves as a site of intermolecular interactions. It allows the coordination of molecules in various geometries and brings them close together for mutual interaction [21–23]. Enzyme nano-encapsulation by inorganic materials is used for sustained drug release in vivo or for target drug delivery to the given body organs and specific sites and is also expected to have potentiality in enzyme therapeutics. Such systems play an important role in biocatalysis and metabolism. They can be divided [24] into four main groups: multinuclear metalloferments (including Mn, Fe, Co, Cu, Mo), redox-proteins, transport and spare or stocking proteins (regulating a metal concentration in a given site of the organism). Sometimes kidney proteins (that carry out bonding and removal of heavy metals from the organism) are related to them too. The biological functions and mutual interactions of these compounds are diverse. For example, as a result of the interaction between metals and stocking proteins (the most characteristic of which is ferritin) there are some supramolecular inclusion compounds [25]. The iron-stocking protein particle contains 24 proteinic subunits and every one of them has an inorganic nucleus (with diameter 7.3 nm) of 4500 iron atoms in the form of basic hydrophosphate $(FeOOH)_8(FeOOPO_3H_2)$ [26].

The special feature of many oxidizing–reducing catalysts is the presence of several interacting transient metal atoms with mixed valence on one protein molecule. At the present time more then a hundred known ferments sometimes contain several atoms of iron, copper, molybdenum and manganese. For many of them, especially for metalloproteins, containing polynuclear non-heme iron complexes, connected by bridges of oxo-, hydroxo- and alkoxy-groups, arrangements into a multinuclear assemble are established [27–29]. For example, binuclear iron centers (the $Fe_2O(OCOR)_2$ nucleus) are presented in hemerythrin (Hr), ribonucleotide reductase, methane monooxygenase (MMO) hemerythrine, ribonucleotidrenuclease and acid phosphatase. The polynuclear complexes Fe^{2+}–Fe^{3+} may be present (in the Fe_3O nucleus) in many ferments, Fe_2S_2-groups – in ferredoxine and oxo-centers of high nuclearity. They may be included in iron-storage proteins (such as ferritin and hemosiderin), as well as in some magnitotactic organisms analysed below. Significant success was achieved in studies of the structure and action of MMO ferments. The ferment carries out mild and selective methane oxidation into methanol by the scheme:

$$CH_4 + O_2 + NADH + H^+ \quad \rightarrow \quad CH_3OH + H_2O + NAD^+$$

where NADH is reduced nicotinamide-dinucleotide. The reduced protein contains a $[Fe_2(\mu\text{-}OH)(\mu\text{-}O_2CR)_2(His)_5]^+$ core and binds O_2 in the form of a terminal hydroperoxide ligand that forms a hydrogen bond to the oxo bridge of the resulting (μ-oxo)diiron (III) unit [30] (Scheme 9.1).

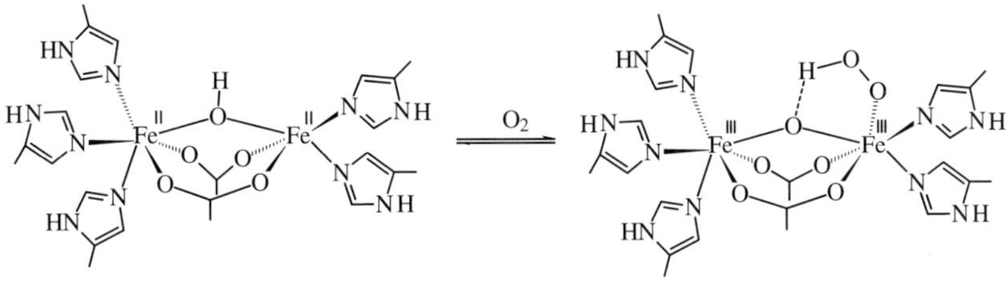

Scheme 9.1

On the basis of numerous investigations of carboxylate-bridged non-geme binuclear centers it is supposed that the structure of the MMO active centers can be presented as iron μ-carboxylate complexes, similar to the active centers of the oxygen-binding protein hemerythrite (Hr) (a non-geme analog of hemoglobin) being a redox-ferment ribonucleotide-reductase and taking part in ribonucleic acid (RNA) synthesis (Scheme 9.2).

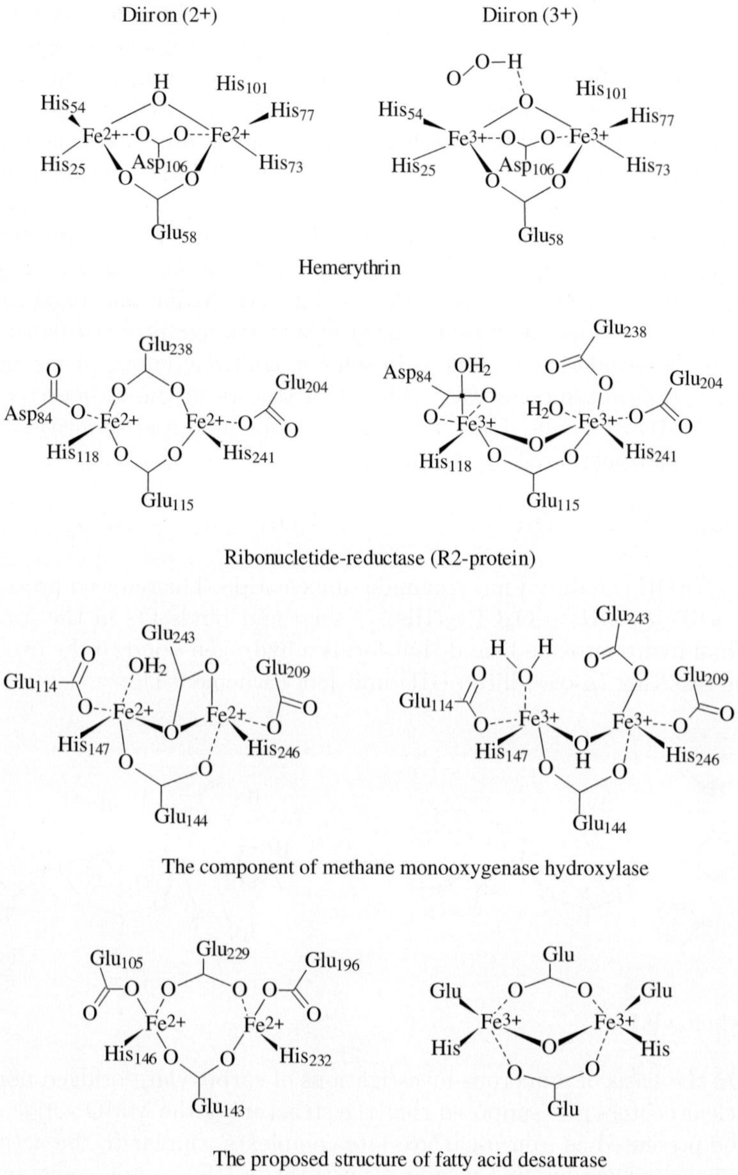

Diiron (2+) Diiron (3+)

Hemerythrin

Ribonucletide-reductase (R2-protein)

The component of methane monooxygenase hydroxylase

The proposed structure of fatty acid desaturase

Scheme 9.2

The achievements in ferment structure studies stimulated new approaches to chemical modeling and synthesis of such complexes [31]. The structures of MMO [32] and NiFe-hydrogenase [33] were deciphered, and this became an important step in synthetic analog modeling. For example, the synthesis and research of polyiron oxo-complexes with mixed valence $Fe_4^{3+}Fe_8^{2+}(O)_2(OCH_3)_{18}$

$(O_2CCH_3)_6(CH_3OH)_{4.67}$ present a useful model of the iron-containing proteins that play an important role in biomineralization processes [34].

This is true for the design of the ferment-type oxygen emission photosystems (systems II of natural photosynthesis) including polynuclear metallocomplexes as both electron-transport chain elements and catalytic centers for direct control of the substrate redox transformations.

The binding of the tetranuclear manganese complex with the oxygen-emitting center (OEC) peptides of the green plant photosystem II (that provides water oxidation to molecular oxygen) includes many stages [34,35,37]. The cubane-like manganese nucleus is immobilized into the protein complex by carboxylic groups of the reactive centers D_1 and D_2 polypeptides. These ligand groups are more suitable for binding because they can form stable complexes with manganese in (3+) and (4+) oxidation states into aqueous media distinct from the polypeptide side chains with amide coordination centers. These groups can be hydrolized into carboxylates by coordination with Lewis acids and in this case the hydrolysis is assisted by the weak acidic medium of the inner side of the thilacoid membrane where the manganese complex OEC is placed. It was supposed [33] that the most probable sites of manganese cluster binding were located at the sections Asp_{308}–Val–Val–Ser–Gln_{304} and Ser_{304}–Val–Phc–Asp_{300} (in D_1 and D_2 polypeptide chains, correspondingly), each of which gave two carboxylate ligands – asparagate (Asp) and glutamate (Glu, hydrolyzed glutaminate Gln), as shown in Fig. 9.1.

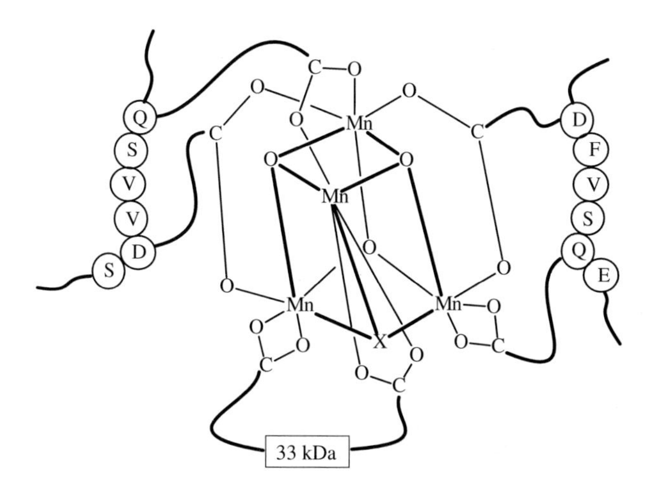

Fig. 9.1. Scheme of manganese cluster binding by polypeptides (details in text)

The build-up of the coordination sphere of manganese ions is finished when the two terminal carboxyl groups of protein with molecular mass 33 kDa are attached.

In every catalytic cycle of water oxidation the manganese complex accumulates four one-electron oxidation equivalents (transferring the electrons to a light-generated primary acceptor of the photosystem II), which are used to oxidize two coordinated oxoligands with the formation of the O–O bond of molecular O_2. When it is removed from the Mn coordination sphere, the carboxylate side chains connected with the complex guard against the destruction of the complex, and this was verified by model manganese polynuclear complexes [37, 38]. The oxygen yield increases when a H_2O oxidation manganese catalyst is fixed on the lipid vesicle surface. There are many mixed valence manganese compounds of various types, which function in photosystems II. Some of these oxidized states are isolated and characterized (for example, Mn^{2+}, Mn^{3+}, Mn^{4+}, binuclear Mn^{2+}/Mn^{3+}, tetranuclear $3Mn^{3+}/Mn^{4+}$).

A ferritin–cubane cluster skeleton $[Fe_4S_4(SEt)_4]^{2+}$ is a synthetic analog of a sulfite–reductase, i.e. a ferment catalyzing a six-electron sulfite–sulfate reduction [39]. Protein-connected heterometallic cluster subsystems $[MFe_3S_4]^+$ (M = Zn, Ni, Co, Mn) that take part in electron transfer during photosynthesis probably have similar cubane-type structures [40, 41]. That can be supposed from the comparison of isoelectronic synthetic clusters (as $NiFe_3O_4$) with clusters separated from *Pirococcus furiosus*, where the nickel cluster includes a delocalized Fe^{3+}–Fe^{2+} pair and a localized Fe^{3+}-center and its electronic density is shifted to the nickel ion.

An active form of the biological nitrogen fixation ferment is connected with the combined action of two components, namely the ferrous- and ferrous-molybdenum proteins (the latter includes 32–34 atoms of Fe and two atoms of Mo) under natural electron-donor agents (ferrodoxines, flavodoxine) in the presence of ATP. For the ferrous-molybdenum nitrogenase cofactor a possible structure $MoFe_6(SR)_6L_3$ is supposed and MoFe- and VFe-proteins are separated from *Azotobacter vinelandii* and *Azotobacter chroocaccum* bacteria [6]. There were many attempts (including complex quantum chemical calculations) to establish with various degrees of approximation the crystallographic structure and electronic composition of both the model clusters (the nitrogenase analogs) and MoFe-protein ones (e.g. see [42, 43]).

From other bimetallic systems the Cu^{2+}–Zn^{2+}-containing superoxidedismutases from erythrocites, Cu_2Co_2-superoxidedismutases, etc. can be mentioned and some advantages of such structures with generalized nuclear skeletons are surveyed in [27, 28]. The general route to protein-metal colloid conjugation involves a simple solution-mixing process of protein with metal colloidal sols formed prior to the mixing [44]. In the first stage a bioconjugate macromolecular complex with thiols or amine functional groups (in the protein), where usually there arises an inhomogeneous dispersion between the protein and the metal colloid, always results in the formation of large aggregated conjugate structures. The formation of spherical core–shell Au-silk fibroins and biopolymers (with ~100 nm size metal cluster-containing nanoparticles) was reported. For example, a Ru_3 cluster [45] with $[Ru_3(\mu_3\text{-}O)$

$(\mu_2\text{-OAc})_6(H_2O)_3]^+$ structure can effectively bond the proteins and is very convenient for controlling their aggregation, due to the stability of the trinuclear structure with labile water molecules in the axial positions. Under the interaction with cytochrome a cluster is formed with the proteinic oxopolypyridylruthenium complex as a ligand, being a selective splitting reagent of calf thymus DNA [46]. Ru^{2+}-based emission, while almost entirely quenched in a Ru^{2+}/Os^{2+} heterodimetallated DNA hairpin, is dramatically restored upon hybridization to a complementary olygonucleotide [47]. The ribosomes and transport RNA can be modificated by the clusters (containing at least 10 atoms of gold) and the functionality of the formed products is studied [48]. Novel core–shell nanostructured gold-containing colloid silk fibroins are bioconjugated by the in situ protein redox technique [49] and it was shown that the process mechanism includes Au(III) macrocomplex formation with a silkworm *Bombyx mori* tyrosine residue component, contained in its silk fibroin, having strong electron donating properties and in situ reducing $HAuCl_4$ to Au colloids. Simultaneously the oxidized silk fibroin macromolecules, attached to the surface of the Au nanoparticles, form the core–shell nanostructured bioconjugate, which stabilized the gold nanoparticles and prevented the close approach of the particles via both steric and electrostatic stabilization mechanisms [50]. This approach allows us to fabricate stable and highly monodispersed gold nanoparticle cores (with average size 15 nm) and spherical core–shell Au-silk fibroin nanostructures (with average size 45 nm) as shown in Fig. 9.2.

Fig. 9.2. Typical transmission electron microscopy image of the core–shell nanostructured Au colloid–silk fibroin bioconjugate produced by the present protein in situ redox technique at room temperature

There are other variants of this method and the first of them [51] is based on colloid metal formation by using aminodextranes (40–150 kDa) as the reducing agents. The formed gold nanoparticles are coated from the outset by an aminodextranes stabilizing layer and therefore at the next stage the gold particles can be bound with biospecific macromolecules of immunoglobulines by a bifunctional reagent (glutaric aldehyde). The principle of the biological approach that was proposed for Au nanowire production by using the sequenced histidine-rich peptide nanowires as templates is shown in Fig. 9.3 [52]. This nanowire incorporates the binding sites, having a high affinity to biological molecules such as DNAs and proteins [53].

The metal-containing protein systems and their biological roles are summed up in Table 9.1.

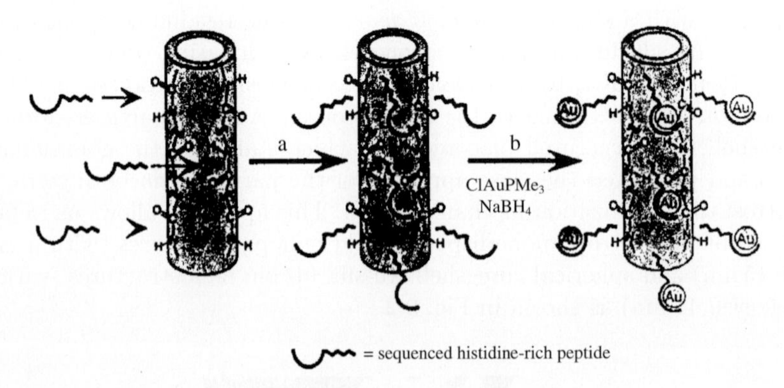

Fig. 9.3. Scheme of Au nanowire fabrication. (**a**) Immobilization of sequenced histidine-rich peptide at the amide binding sites of the heptane dicarboxylate nanowires. (**b**) Au coating nucleated at the histidine sites of the nanowires

9.2 Metal Nanoparticles in Immunochemistry, Cytochemistry and Medicine

In recent years nanoparticles have been widely used in the immunochemical methods of experimental biology and medicine. For example, gold nanoparticles (being effective optical transformers of some biospecific interactions) are successfully employed to make biochips and biosensors in biology (determination of nucleic acids, proteins and metabolites) and in medicine (drug screening, antibodies and antigens recognition, diagnostics of infections). In particular, the color changes induced by association of nanometer-sized gold particles provide a basis of a simple, highly selective, method for detecting specific biological reactions between anchored ligand molecules and receptor molecules in the milieu [54]. The formed conjugate is used as a marker able

Table 9.1. The distribution of the main metals and their biological role in different tissues

Metal	Total Content, mg/g	Blood, mg/l	Liver, mg/g for Dehydrated Tissue	Muscles, mg/g for Dehydrated Tissue	Metal-Ligand Interaction Type	Metal Functions
Na	1.5	1960	300	2600–8000	Very weak	Osmotic balance. Charge neutralization
K					Very weak	Osmotic balance Charge neutralization Enzyme structures stabilization and activation
Mg	0.3	24	590	900	Mean	Enzyme structures stabilization and activation
Ca	15	61	100–360	140–700	Mean	Enzyme structures stabilization and activation
Cr	0.03	2–5	0.02–3.3	0.024–0.084		Oxidation-reduction processes catalysts
Fe	60	450	250–1400	180	Very strong	Oxygen carriers
Cu					Very strong	Oxygen carriers
Zn	20–40	0.29	0.23–2.3	0.33–2.4	Strong	Lewis acids Structures stabilization
Ni					Strong	Lewis acids
Other transient metals					Very strong	Oxidation-reduction processes catalysts

"to recognize" some particular chemical structures. The use of colloidal gold particles as a label for biospecific markers became a revolution in cytochemistry [55]. Of prime importance and interest is nucleic acid detection based on the three-dimensional ordered structures forming under hybridization in solvents with complementary oligonucleotides covalently linked with the gold nanoparticles [56,57]. For instance, DNA-linked Au trimers have been synthesized [58]. The main methods are based on the attachment of a contrasting label to the biospecific probe molecule (antibody, lectin, enzyme) [55, 59]. The new approach is called a "programmed assembly" strategy for organizing nanoparticles into periodic functional materials. Its essence can be illustrated by an original method of pathogenic and diagnostics of genetic diseases [60, 61]. Au nanoparticles (with diameter 13 ± 2 nm), modified with (3′- or 5′-) alkanethiol-capped series of oligonucleotide (single-stranded DNA) and linkers ranging from 24 to 72 base pairs (∼8–24 nm) in length, are used as the labels. They are connected to the nanoparticles "head-to-tail" (Fig. 9.4,a) or "head-to-head" rearrangement (Fig. 9.4,b). On a defect site of DNA-target (disease-disposing gene) complementary label binding proceeds, resulting in nanoparticle aggregation (Fig. 9.4,c), and manifests itself in the colloid color change from red to blue.

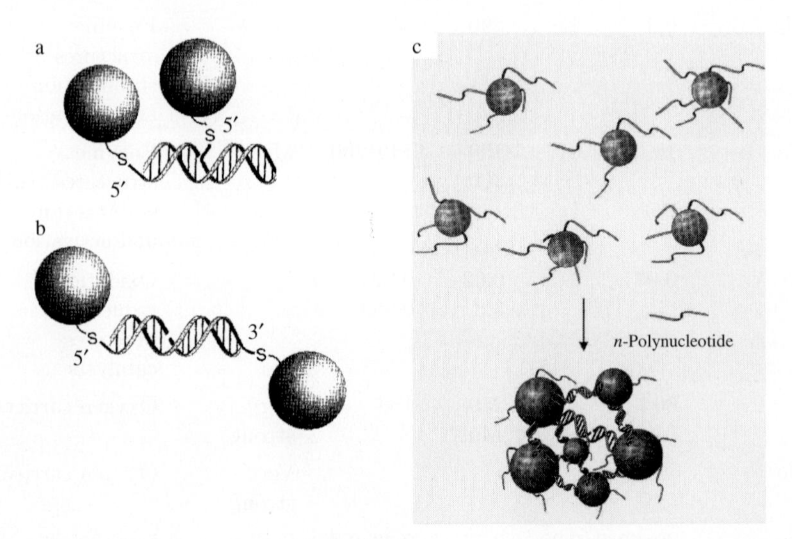

Fig. 9.4. The scheme of Au nanoparticles, modified by (3′- or 5′-) alkanethiol (24- or 28-) oligonucleotides, connected with the particle by "head-to-tail" (**a**) or "head-to-head" arrangement (**b**). Particle aggregations (**c**) in the presence of complementary oligonucleotides

The change of the colloid color results from the formation of big three-dimensional DNA-bonded aggregates of Au nanoparticles that leads to a red shift of the plasmon resonance band from $\lambda_{max} = 524$ nm to $\lambda_{max} = 576$ nm

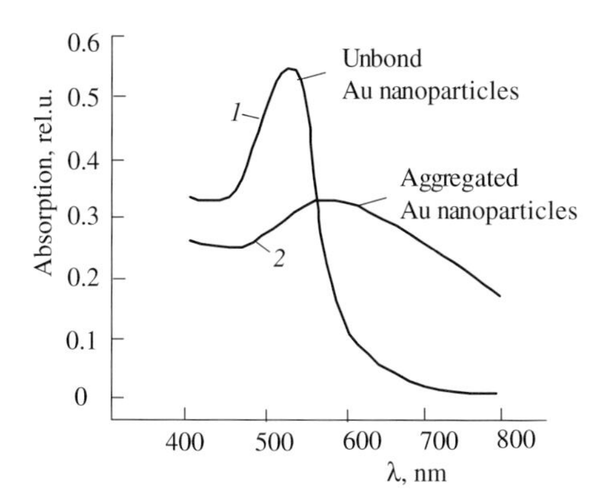

Fig. 9.5. UV-spectra of Au nanoparticles, modified by 5′-hexanethiol-12-oligonucleotide before (*1*) and after (*2*) treatment by complementary 12-oligonucleotide

(Fig. 9.5). The orange sol (corresponding to particles 5 nm in size) adsorbs at $\lambda_{max} = 513$ nm; the characteristic adsorption zones are $\lambda_{max} = 320–340$ nm (for aurum halogenides) and $\lambda_{max} = 500–550$ nm (for metallic colloid gold, depending on the particles size).

The red color of gold colloidal systems arises due to the fact that the distance between the particles is considerably greater than the particle diameter, but when in aggregations these distances decrease to a minimum (i.e. to the particle size) the system color becomes blue. The techniques of colorimetric diagnostics are very selective, particularly in combination with other methods, such as the improved scattered light differential spectroscopy [62], optical studies [57, 63], angle-dependent light scattering and fractal dimension analysis [64], eletrophoretic and structural [65], luminescence and CD [66] analysis. Therefore the method of mercaptoalkyl-oligonucleotides modification by Au nanoparticles has superiority over other methods mostly for applications connected with gene mapping and microarray analysis [67]. The same approach was realized by a metathesis polymerization of 1-mercapto-10-(exo-5-norbornen-2-oxy)decane with the subsequent immobilization of the formed polymer on Au particles 3 nm in size [68].

Thus the general scheme of inorganic nanoparticle and biomolecule coupling methods (FG – functional coupling group) can be presented as follows (Fig. 9.6): the sets of nanoparticles are functionalized with an individual recognition FG that are directly complementary to each other. In a recent review [17] problems of biomolecule-directed nanoparticle organization (including the protein-based recognition systems and DNA-based nanoparticle aggregates) were analyzed.

The usage of silver nanoparticles for the described immunochemical purposes is very tempting, firstly, because they (as distinct from Au nanoparticles, adsorbing in the range 520–580 nm) exhibit a surface plasmon band

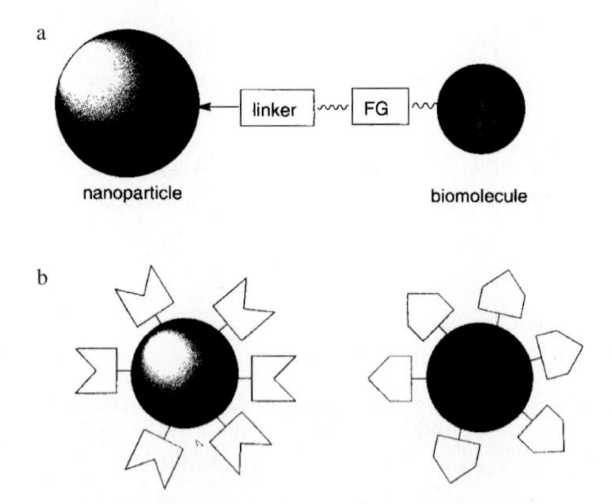

Fig. 9.6. General schematic representation of methods to couple inorganic nanoparticles and biomolecules (**a**) and two sets of nanoparticles are functionalized with individual recognition groups that are complementary to each other (**b**) (FG = functional coupling groups)

in a more convenient region between 390 and 420 nm, depending on the particle size (see Chap. 5). Secondly, the extinction coefficient of the surface plasmon band for an Ag particle is about four times larger than for an Au particle of the same size [69]. But Ag nanoparticles cannot be effectively passivated by alkylthiol-modified-oligonucleotides using the established protocols for Au particle modification [70]. Therefore core–shell approach was used [71], when an Au shell has been grown upon an Ag nanoparticle, forming a particle with an Au outer surface, that can be easily modified with alkylthiol-oligonucleotides. Such a complex particle consists of an 11.8 nm Ag core and a monolayer Au shell. It is supposed that this approach can be generalized for preparing other types of particles (such as Cu and Pt) and to create a series of such core–shell particles with tailorable physical properties. Their surface chemical properties and stability are defined by native pure gold particles that can be modified by oligonucleotides. An effective solution of many problems is based on the use of polymers, such as aminodextranes [40], poly-L-lysine [72], PEI, PEO and PVP [73], polyvinyladenine and vinyladenine-*alt*-maleic acid copolymer [74]. Another interesting approach is the reversible lectin-induced association of gold nanoparticles modified with α-lactosyl-ω-mercapto-PEO [75] that includes the synthesis of various types of heterobifunctional PEOs using the ring-opening polymerization of ethylene oxide with a protected functional group [76]. As a model ligand, lactose was introduced in the distal end of the PEO chain and acetale-PEO-SH was in reversible association of gold nanoparticles under physiologically relevant conditions. The degree of particle aggregation was proportional to the added lectin concentration. For an immunoanalysis of estradiol [77] it was marked by a $Co_3(CO)_9$ fragment with high cross-reactivity relative to the specific antibodies.

The anticancer activity of platinum complexes is well-known and recently it was reported in addition that platinum (II) amine complexes, containing a dicarba-*closo*-dodecaborane(12) in vitro DNA-binding can be used for tagging of plasmid DNA [78]. There is evidence of existence of polynuclear metalloproteins, including the clusters Rh_6, Pt_{15}, etc. The combination of DNA and $PdAcAc_2$ with the subsequent reduction of the DNA-bound Pd ions led to the formation of DNA-associated one- and two-dimensional arrays from the well-separated Pd particles 3–5 nm in size [80]. Sometimes CdS and CdSe quantum dots are used for these purposes.

Biocompatible magnetic nanospheres are used in medicine and biology [81, 82].

Metalloprotein preparates please check 'preparates' are often used in medicine for control of the fermentative activity and for new, narrow-directed drugs production [25]. For example, calcium and magnesium colloidal compounds with vegetable proteins are utilized for regulation of gastric juice acidity; the microdispersed Fe_3O_4 complex is injected with the serum albumin transfusion as a contrast substance for roentgenography. Colloidal gold and its biopreparates present ancient medical agents[1] and had the various applications (many of them have a strong antiarthritis effect). Such colloidal systems are well-known contrast substances in immunochemical studies, when they are used as an electron-dense cytochemical marker for detecting the surface-active antigen localizations in electron microscopy [84]. The specific markers with many macromolecules (antibodies, lectins, ferments, various proteins, glycoproteins and polysaccharides) can be produced using hydrophobic particles and the electrostatic charge of their surface [85].

At present synthetic oligopeptide models of the complexes are being studied intensively. It will give additional information on the action mechanisms and the electron transfer in native proteins. Particularly, some histidine-containing peptides have been studied for their high affinities to metal ions. The problem is connected with damage to the central nervous system by altering protein conformations into abnormal forms via histidine-metal complexation and this protein deformation, which may cause Parkinson's and Alzheimer's diseases [86]. The same mechanism of immobilization in polypyrrole microcapsules is used for different ferments (glucosooxidase, catalase, trypsin, subtiline, alcoholdehydrogenase [87]), chemical entities (Pd nanoparticles [88], ruthenium complexes [89, 90]), and catalytic systems, incorporated into a protein scaffold or L-tyrosine esters, and antibody-metalloporphyrin [91].

[1] Although the first references on medical applications of colloidal gold can be found in medieval treatises, until now it is one of a few elements for which the involvement in biochemical processes of the living cell is not established but it was reported [83] that some microbial metabolites transform it into a solution in the presence of oxidizers (see below).

9.3 Biosorption, Selective Heterocoagulation and Bacterial Concentration of Metal Nanoparticles

The interaction of biopolymers and microorganisms with metal sols and metal oxide surfaces is highly specific and different for various metals. In this connection alternative templates (such as bacteria) have been examined to increase the diversity of compositions and morphologies [92]. This interaction is studied for gold colloids [93, 94], and other noble metals, as well for iron, copper and manganese (Table 9.2).

Table 9.2. Metal sorption by microorganisms [95]

Solutions	Microorganisms	Metal Content per 1 g of Cell Dry Mass
Radioactive elements containing solutions	Denitrifying bacteria	Up to 140 mg of uranium and thorium
	Rhizopus arrhizus	Up to 180 mg of uranium and thorium
	Pseudomonas aeruginosa	Up to 560 mg of uranium
	Saccharomyces cerevisiae	From 100 to 150 mg of uranium
	Biosorbent M (*Penicillum chrysogenum)*	Up to 80–120 mg of uranium (from water with elements content below 0.05 mg/L)
Silver contaning solutions	Consortium of *Ps. maltophila, Staphylococcus aureus* bacteria and non-identified forms	Up to 300 mg of silver

The bacterial concentration of metals begins from metal atom adsorption on the cell surface with the subsequent assimilation. Coarsening of the sol particles can proceed without any considerable adsorption on the cell surface (flocculation), as well as the adsorption and deposition of large metal–bacterial formation (heterocoagulation). The sol with particles ~10 nm is the most stable one. The most important feature of the colloid–bacterial interaction is the diversity of forces and mechanisms (electrostatic, molecular and chemical). Significant differences were revealed between various taxonomic groups, species and strains of microorganisms in relation to the rate of gold sol accumulation. Living cells (for example, *Bacillus subtilis*) can attract colloidal particles, attach them and form their aggregates (as opposed to an inactivated mass of microorganisms, having a low sorption activity). The cells have a high power to adsorb metal particles but they are very sensitive to many other factors [96]. For example, preferrential particle adhesion is observed in their contact points and the single cells are deposited by a metal layer. For example the coccus bacteria *M. leuteus* (with mean size 2000 nm), separated from the gold bearing stratum, produce packets of 4–8 cells, which can validly carry out adsorption of gold (up to 10 mg per 1 g of the wet biomass) [93]. The living cells of blue-green algae *Chlorella vulgaris* are even more effective (up to 65–70 Au mg per 1 g of biomass; every cell binds about 10^4 colloidal gold particles) but they don't cause flocculation.

It can be supposed that the electrostatic and chemosorption attachment of the colloid metal particles and their compounds to a cell surface is influenced by the cell metabolism products that includes many functional groups, such as –S–, –COO–, –PO(OH)O–, $-NH_3^+$. The chemical binding of metals by the M–N, M–C, M–O, M–S bonds is well-established experimentally [97]. The negatively charged proteins (human serum albumin, soya inhibitor of trypsin) did not cause colloidal gold flocculation under experimental conditions (pH 7.6), contrary to the positive charged proteins (trypsin, papain, cytochrome *c*, RNAase). The lysozime systems have special characteristics, sol flocculation proceeds only into a narrow concentration range of the ferment (1–10 μg/mL) and the concentration increase assists sol stabilization and the aggregative stability of the system. The successive flocculation and stabilization stages of the metal colloids are characteristic of some synthetic polymer systems too (e.g. cationic polyelectrolite PEI and nonionogenic PEO).

A significant difference exists between proteins and synthetic polymers: the proteins are selective to the various dispersed phases at flocculation. It is connected not only with electrostatic factors but also with the specific binding by the functional groups of aminoacid residues, especially by sulfhydrylic and amino groups [98]. For example, the addition of *n*-chloromercury-benzoate (which is specifically blocking proteins SH-groups) decreases their flocculation activity. The native structure of biopolymers is very important for sorption processes. Thus partial destruction of the polysacharide frame of the bacterial envelope (by lysozyme ferment) can considerably decrease their sorption power.

There are specific differences between living and dead cells relative to the metal biosorption, sedimentation stability, electrolytic coagulation thresholds, and heterocoagulation effectiveness with nanoparticles [99]. There are correltions between the cell accumulating ability and the flocculating activity of some preparates, separated from these cells. The metal particles are bonding to the cell surface by some protein, hydrocarbon or glycoprotein structures from various cells (including *Bacillus sb.*). A glycoproteid with molecular mass about 50 kDa was extracted from various cells by the butanol extraction method that was able to induce colloid particle coagulation and the binding specificity of proteins changes in the series Au $\gg$ Pt > MnO_2 > Al_2O_3. Cultures of *Bacillus sb.* can be arranged by their accumulating ability to the different metal ions in the following order: Au > Mn > Cu > Fe > Ni > Co. An elevated gold accumulation (Table 9.3) is attributable to a cell's previous adaptation to gold. Moreover, the liposomes as the nearest model for biomembranes adsorb the metal particles after modification by glycoproteides.

Table 9.3. Metal accumulation by biomass of cultures *Bacillus sb*

Solution (5 mmol/L)	Metal Content (μg/g of Dry Residue) at the Contact Time			
	30 min	2.5 h	1 day	7 days
$MnCl_2$	10.7	11.7	23.6	73.7
$FeCl_3$	17.2	17.4	17.7	24.4
$CoSO_4$	7.5	7.7	11.7	11.7
$NiCL_2$	8.6	8.8	10.4	12.4
$CuCl_2$	16.8	22.9	22.1	28.8
$AuCl_3$	81.4	149.2	177.6	–

It is possible that glycoproteide is connected with lipids in the bacterial envelope and the survival ability of *M. luteus* under high concentration of Au is explained by increased glycophospholipids content and branching of fatty acids residues [100]. The sorptive capacity of *Chlorella vulgaris st. 18* cells increases in the metal ion series ($Ag^+ < Cu^{2+} \ll Au^{3+}$) [97]. The sorbed metals are in a colloidal state with particles fixation both into the cells and on their surface. The cell bactericide action rises in reverse order.

The high variability of living cells is defined not only by biosorbent type, but also by the nature of the sorbing metals, their concentration and even by the prehistory of the sorption system. For example, in the presence of concurrent ions (Co^{2+}, Mn^{2+}) the value of ionic gold adsorption by living cells of blue-green algae *Spirulina platensis* is considerably changed during the process [99] because at high concentrations the concurrent ions can inactivate the initially living cells (as a result they become chemically similar to the dead ones). Such regularities were observed also for Cu^{2+}, Ag^+, Ni^{2+} ions, depressing the activity of many bacteria and aquatic plants that correlates with the absolute value of the electrokinetic potential of the one-cell organisms. It is supposed that the electrosurfacial or bioenergetic characteristics of the microbial cells can be used as a sensitive indicator for the evaluation of heavy metal toxicity [99].

The interaction at the initial stages proceeds due to electrostatic binding. It explains the reversibility of heterocoagulation and the fact that the dispersed metal particles escape from the cell surface if the formed product is being introduced into a protonophor incubation medium. There are two processes: further particle assimilation proceeds into the cell associates (in the points of contact) with their coarsening, but on the single cells the particle growth occurs without previous aggregation because the formed particles are gradually rejected from the cell surface. Thus, branched fibrillar associates (with no less than 2–3 particles) were observed in gold sol in the presence of proteins, with the polymeric bridges between the sol particles.

By analogy with the reductive synthesis of polymer-immobilized nanoparticles (see Sect. 4.4) the colloid adsorption and heterocoagulation into biopolymers have its continuation in the recrystallization and coarsening of the metal particles placed on the living cells surface. These processes result from the intensive reduction-oxidation (specifically, for gold particles) with participation of carboxyl-, aminogroups and dialkylphosphoric esters. At the first stage the colloidal gold particle oxidation proceeds [93]; at the second stage the metal ions are reduced by endogenic reducers and in this process the cytoplasmatic membrane is involved. The process is feasible only under a considerably intensive ionic flux. It is supposed that gold dissolution (oxidation) by *Bacillus sb.* cells is determined by cationic groups $-NH_3^+$, and gold complex (with the metabolite products) reduction is by anionic groups, such as carboxylic and phosphoanhydrylic acids [101]. It is known that moulding fungus can reduce gold to a metallic state and similar processes run in many region with participation of humic and fulvic acids, that creates a bioelement "shop" and presents an important and main stage in soil formation. Iron enrichments are carried out by unicellular algae (for example, from acidic waste waters of uranium mines).

There are many independent sites of a metal particle [M] binding on bacteria surfaces and therefore the binding process is described by an equilibrium relation:

$$\mathrm{M} + \mathrm{L} \leftrightarrow \mathrm{ML},$$

where L is a binding site on the bacteria surface and ML is a "bacteria–sol particle" complex. The total number of binding sites (n), the number of the bound metal particles (ν) as well the association constant $K = [\mathrm{ML}]/([\mathrm{M}][\mathrm{L}])$ are connected by the Sketchard equation:

$$\nu/[\mathrm{L}] = nK - \nu K. \tag{9.1}$$

It was found that there are two independent mechanisms of gold binding in *M. luteus* cells by hydrophilic sections and electrostatic interaction ($K_1 = 3.85{\cdot}10^9$ L/mol, $\nu_1 = 4.7{\cdot}10^3$) and by oxidative complexation of the surface layer with particles ($K_2 = 4.6{\cdot}10^8$L/mol, $\nu_2 = 4.3{\cdot}10^3$) [93]. Metal particle coarsening proceeds after oxidative complexation and reduction, and it is connected with accumulation (association) and recrystallization. Thus, particle redissolution occurs in biopolymer–colloid particle systems and results in a bigger particle formation (in the case of gold simultaneously the gold assay value was increasing).

The model of bacterial concentration and crystallization is rather complex. The concentration can proceed through the intergrowth (assimilation) of the little needle-shaped polyhedra 10–20 nm in size, connecting together and forming the rather big complexes (visible to the naked eye) that can be separated by a gravimetric method. In addition the colloidal particles present the nucleus of a new phase and can serve as the crystallization centers under permanent cooperative oxidation and reduction of the formed ions. The cell surface is heterogeneous. The mean sizes of cell and nanoparticles are $(1–5)\cdot10^3$ nm and 15–20 nm, correspondingly. Electron microscopy studies had fixed the essential qualitative and quantitative distinctions between the Au particle distributions over the various bacteria stain cells [99]. As shown in Fig. 9.7, there are several types of particles in the system. There are the randomly distributed gold nanoparticles unbonded to cells and forming agglomerates along with the free cell surfaces (Fig. 9.7, a), the agglomerated particles, localized between the cells (Fig. 9.7, b) or particles more or less uniformly distributed on the cell surface or near it (Fig. 9.7, c). Some particles are not uniformly distributed over the cell surface. They are localized near one of the cell ends (Fig. 9.7, d) and finally there are nanoparticles, uniformly coating all the cell surface (Fig. 9.7, e). The bacterial associates from the different cell types can participate in the aggregation processes and concentrate the gold particles on their surface (Fig. 9.7, f).

The reactivity of bacteria relative to the nanoparticles varies from the total inertness to an effective adsorption and concentration in accordance with a chemical composition of the cell surface.

Probably by this mechanism endogenic gold (may be, not only gold) transformation proceeds under conditions of the formation of gold-fields.[2]

The problems of the selective metallophility of microorganisms are now widely discussed. It is clear that the property is connected with many factors, the selective directions of the colloidal particle motion relative to the microorganisms, the character and strength of the chemical binding on the cell surface, the coordination bond formation with considerable covalent bonding, etc.

[2] The gold dissolution, deposition and new formations on its initial grain surface are characteristic features of the whole hypergenic history of nugget gold [102]. The microscopic processes of microorganism interaction with metal nanoparticles in nature lead to macroscopic changes, their bacterial concentration and coarsening in a geological timescale.

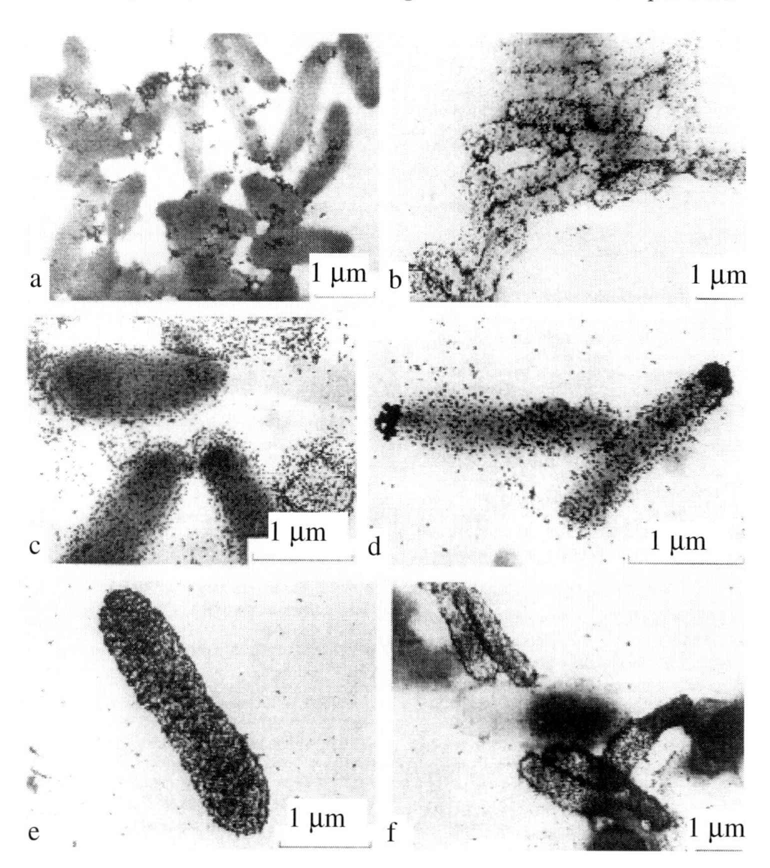

Fig. 9.7. Electron microscope photographs of bacterial cells (of individual or associated cultures) after incubation with colloidal gold (details in text)

The most important cell components are low molecular (5–10 kDa) metal-binding proteins – thioneines, containing many –SH groups and accumulating both the essential (Cu, Mn, Fe, Co, Cr, Mo) and nonessential metals for an organism. It relates to such important microelements as iron and manganese, because their deposits are forming in a biogenic way on the continents and in oceans. The macromolecules of such proteins consist of two separate ferments. Each of them is able to bind metal ions to a polynuclear metal-thiol cluster surrounded by cysteine aminoacid residues. For example, Cd- and Zn-thioneines contain seven metal atoms, four of which are connected with 11 cysteine residues (M_4Cis_{11}) in a carboxyl-terminal A-cluster and three with nine cysteine residues (M_3Cis_9) in an amino-terminal B-cluster [103]. At the same time Ag and Cu atoms form M_{12}-metal-thioneines with six metal atoms in every domain.

The biological functions of thioneines appear to be accumulation, storage and detoxication of metals as well as support of resistance of the organism to heavy metals. Probably, the metal accumulation is attributed to the intensive reduction, localization and adhesion of the forming metal particles on the external side of the membrane. During biological adaptation cells can gain a resistance plasmid to the metal and this mechanism is realized through an intensive metal ion reduction and natural selection itself had led to the forming of metal-tolerable microorganism stains.

There are two important features in the problem of metal nanoparticle immobilization by biopolymers. First, most heavy metals can be connected with several protein macroligands simultaneously in a given three-dimensional space surrounding and conformation state with forming some triple complexes, for example protein–M–nucleic acid (M = Au, Pt, Pd) [104]. It is possible that such formations give an anticancer effect in an organism. Second, the mixed-metal thionenines can play an important role in many processes, in particular an Au(I) thiomalate can interact with native Zn-, Cd-, Co-thionenines and selectively substitute the metals (firstly Zn, but at excess in equilibrium Cd atoms too) selectively. The protein conformations under the process are minimal.

A consequence of the selective metallophility is the genetic specialization of a microorganism relative to the interaction with some metals. In natural conditions strongly determined systems, namely the given cell – M (or a group of similar metals), are formed. Cell flocculation by colloidal metal particles leads to the formation of considerably greater bio-bone aggregates (50–60 μm in size) deposited in an aquatic medium.

Thus, the living cells are able to interact with colloidal particles and attract to the cell surface (the primary act is carried out by some components of the cell wall or plasmatic membrane), to attach with aggregate formation, dissolution, recrystallization and new, bigger particle formation. The processes are complicated by the nature of the biopolymer complex and the high molecular mass of proteins. The selective heterocoagulation of living cells with mineral particles is based on colloid chemistry interactions having a selective adhesion. Moreover a mosaic of the cell surface ("functional heterogeneity" results from a nonhomogenic distribution of the ion exchange ferments) is an important factor for the whole process of the heterocoagulation of highly dispersed particles [99]. The mentioned primary act of nanoparticles attraction to a cell surface or to the protein functional groups is important for many biological problems and for an understanding of the sorption processes on solids. For example, the long-chain alkane-thiols are selectively adsorbed on the Au particle surface and can form a stable monomolecular layer that is spontaneously self-organized.

It is often used for protein adsorption modeling. For example, a protein monomolecular layer from the immobilized glutathione (GSH) and L-cysteine (L-cys) was organized (Fig. 9.8) on the surface of Au nanoparticles

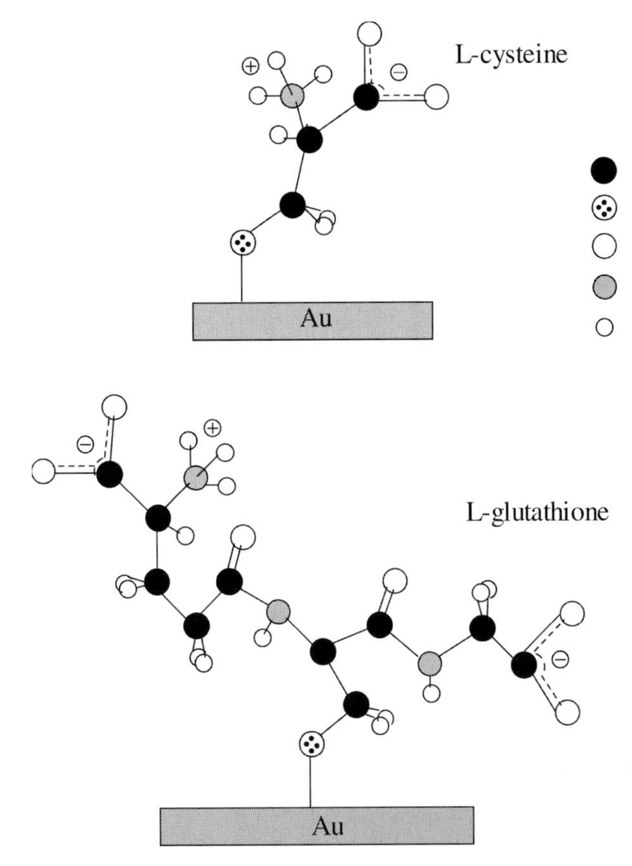

Fig. 9.8. Structural scheme of glutathione and L-cysteine, being immobilized on the Au surface

[107–111] in the form of a layer with thickness 100–200 nm, obtained by vacuum evaporation with rate ~0.5 nm/sec.

Such modified nanoparticles are convenient for different protein adsorption (human serum albumin, plasmoprotein). An approach based on antigen immobilization is used for the fabrication of immunosensors. The modified surface of nanoscale Au films with thickness (~50 nm) produced by Au deposition on the glass plates is formed with the Au-mercaptide bonds. Such layers are submerged into a 2-aminoethanethiol solution. The self-organized monolayers are strongly bound with oxidized dextrane-hydrosulfate (mol. mass ~6100) by the formed in situ Schiff bases (the surface layer and the polymer formyl groups). It can be used for fabrication of human serum albumin (HSA) (Fig. 9.9).

The repeated processes of anti-HSA association and dissociation with bound HSA were studied by plasmon resonance [111] and many other useful and effective experimental methods in the field.

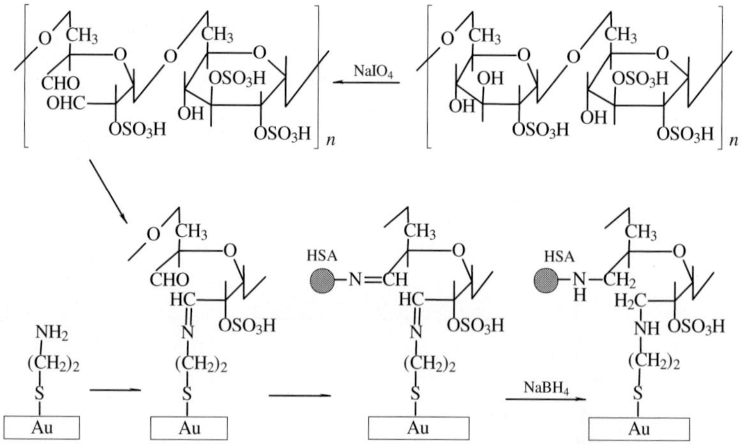

Fig. 9.9. Antigen immobilization scheme

Let us consider several variants of biosorption use in engineering biotechnology. For example, there are two complex problems in using the catheters (especially when they contact with blood): protein adsorption on the surface (it leads to a serious danger of thrombosis) and possible infection [112]. Both problems can be solved if the catheter material (a composition of polyurethane, silicone, PEO, PVA) is imparted with antibacterial characteristics due to Ag nanoparticles (from 10–15 to 50–70 nm) deposited by thermal vaporization or introduced by a reduction mechanism.

The lipid structures are fragile [113] and template technologies are used to produce some strong structures of envelope type for engineering applications [114, 115]. One of them is based on polymer binding with phospholipid and the formation of modified and nonmodified phospholipid mixture (with several percent content of modificated ones), that allows one to form tubular templates on the polymer surface [116]. Such surfaces are sensitive to the adsorption of colloidal metal suspensions (e.g. to Pd colloids) and after the subsequent electrochemical deposition there arise the lipids, having a metal coating in the form of tubes (hollow microcylinders with inner diameter ~0.5 μm and wall thickness ~50 nm). In another variant a sorbent – sacharose – was coated by magnetite particles 10 nm in size [117]. The coating was carried out through either the porous sorbent by a ferromagnetic liquid suspension or by formation of a mixed sol (Fe_3O_4 + agarose) with subsequent generation of the sorbent particles. Such a sorbent coated by the magnetic particles allows one to separate very small proteins (12 kDa). On the other hand the DNA-modified superparamagnetic 3-nm iron oxide particles can be used as magnetic nanosensors, after self-assembling in the presence of complementary single-stranded DNA [118].

Many unicellular algae in an initial state are able to take part in nanoparticle binding and accumulation. There is a wide choice of bioobjects useful for practical applications in the field.

Heavy metal (such as Pb^{2+}, Hg^{2+} and Cd^{2+}) removal and retreatment are important as they are due to the intrinstically persistent nature and are the major contributors to biosphere pollution and have a large environmental, public health and economic impact [119, 120]. For their treatment many approaches are used. One of the main ones is metal complexation by polymeric ligands. Now polymer filtration is an emerging technology, which employs water-soluble polymers for the chelation of heavy metals and ultrafiltration membranes to concentrate the polymer–metal complex and produce the target metals [121]. The protein-based biopolymers use is an alternative to synthetic polymers, especially the elastin-based biopolymers consisting of repeating peptides [122, 123]. Removal of the heavy metals (such as Cd^{2+}) from dilute waste streams, was proposed with a protein–protein interaction being tailored specifically for the target metals. It offers the advantages of being easily regenerated and reused in many repeating industrial cycles [124].

9.4 Sol–Gel Process as a Way of Template-Synthesized Nanobioceramics

The chemical methods of sol–gel synthesis analyzed in Sect. 7.1 present a general methodology of nanostructured material fabrication. They also embrace some mechanisms of the inclusion of biologically active macromolecules during the formation stage of ceramics, glasses and other inorganic composites [125, 126]. The first glass bioceramic, Bioglass, was used in the early 1970s [127]. The possible systems include the important biotechnical ferments, designed biosensors, fermentative electrodes, bioactive optical sensor components, lipid bilayer vesicles, and encapsulating agents for drug delivery [128–137]. Now such approaches are used for biological fabrication of ceramic–metal composites – cermets [138]. The encapsulating ceramic particles can by used as improved biosensor devices and for vaccine formulation [139].

Bioactive material entrapment by ceramic gels is mostly performed by sol–gel methods. The first encapsulation of an active ferment into a sol–gel matrix was carried out in 1990 by mixing biomolecules with sol–gel precursors [140] and after a few years more than 35 various types of hybrid bioceramic materials were elaborated [126]. The inorganic matrices include silicon, titanium and zirconium oxides, and TiO_2–cellulose composites [141]. The glassy mass obtained by this method is dried to a xerogel and is ground to powder before application [142]. Bioceramics are considerably active and can be directly bound to living bones and teeth in the human body. This is an ideal material since the articles are very hard and suitable for human bones in terms of mechanical strength and fracture toughness. Most organic polymers (which are the mainstream materials for the fabrication of soft-tissue substitutes) are only biotolerant, but not bioactive, and are usually surrounded by a fibrous tissue after embedding in the body [143–146]. The state of the art in the field of bioceramic sensors, solid electrolytes, and electrochemical biosensors is reviewed in [147]. Hydrolysis and condensation-polymerization of monomeric metal or metalloid alkoxides are characterized by moderate temperatures and mild conditions (see Chap. 7) that allows them to trap the proteins without their denaturation at the stage of matrix formation. The high stability of such "caged" biomolecules together with the matrix inertness, the large value of S_{sp}, porosity and optical transparency makes the homogenization procedure easier and does not require protein covalent bonding. These advantages make the sol–gel immobilization methods very convenient and attractive in general and for proteins (including even cellular structures and cells) especially. Below we give an analysis of the main approaches and some materials.

The proteins (such as copper-zinc-superoxide-dismutase, cytochrome, myoglobin, hemoglobin and bacterio-rhodopsin) in the considered systems are encapsulated into a porous silicagel matrix that is formed by sol–gel synthesis. The biomolecules are strongly retaining in the matrix and do not lose their fermentative activities or other characteristics [148]. Such matrices ensure transport of the small molecule to a reactive center, product removal, as well as protein molecule fixation into pores. In this way glucosooxidase and peroxidase were heterogenized and used as an active solid-phase element of glucososensor. The spectral changes were observed in gels, containing oxalatoxidase and peroxidase, during holding in aqueous solutions with oxalic acid.

Antibodies are bound by a similar method for potential use in medicine, immunochromatography, and immunosensorics. For example, the antibody immunoglobulines trapped by the sol–gel method retain the ability to bind the external antigens (2,4-dinitrophenylhydrazine) from solution [149]. The atarazine-binding property is studied for a sol–gel matrix, doped by 10% PEO, and including monoclonal antiatarazine antibodies [150]. Such a matrix can "recognize" the widely used atarazine herbicides in solution and bind them. It is important that under the process neither antibody leaching nor nonspecific physical adsorption of atrazine (by the ceramic matrix) were observed. An activity decrease was not observed either, at least during two months (whereas the activity in solution in the same conditions had decreased to 40%). An additional advantage of the sol–gel methodology is the need for immunoglobuline purification lacks. Aantibody encapsulation by the described particles can be used in the sensors of specific antigen detection [151]. The first attempts to trap catalytic antibodies by the sol–gel matrices and use them in practice were successful [152], the 14D9 antibodies were encapsulated into such matrices, and they catalyzed many reactions, including hydrolysis of cyclic acetals, ketals, and epoxides. Peroxidase entrapped into silica nanoparticles shows a higher stability under temperature and pH changes as compared with a free enzyme molecule [153]. The enzyme encapsulation in silica nanoparticles and their introduction into the animal system can replenish the enzyme deficiency in the organism (as well as the use of enzymes as medicine). Moreover, it will not create any risk of allergic or proteolytic reactions of these enzymes (due to their practically zero leachability). Some other advantages of such materials are their high thermal and pH-stability, the prevention of the trapped protein leaching, the ease and convenience of fermentative reaction control (the reactions can be controlled by spectral methods both in the pores and in the matrix volume), easy storage, and possible repeated use. In addition such systems present a wide choice (based on the molecular design of components) for effective control of aggregate morphology, particle size, composition and various physical properties.

Although the investigations are mostly devoted to enzymes, trapped by ceramic materials made of silica gel, another method has also been used based on reverse micelle use of nanosize porous silica particle preparation [153–155]. The lipid bilayer vesicles with an inner aqueous compartment were used as biomembrane models in the field of supramolecular chemistry. They are often used as nanocapsules for drug delivery or gene transfection systems, artificial cell membranes, and enzymes [156–158]. Such lipid bilayer vesicles seem to be a nanomaterial candidate for designing functional supramolecular devices [159]. There is an interesting approach to the production of synthetic peptide lipids (bearing an amino acid residue interposed between a polar head moiety and a hydrophobic double-chain segment through the peptide bond) based on using cationic and anionic Cerasomes (the organoalkoxysilane proamphilites), as shown in Scheme 9.3 [160, 161].

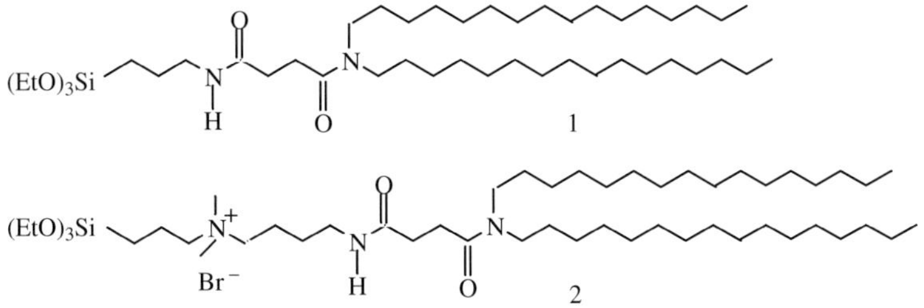

$(EtO)_3Si$ N H O O N 1

$(EtO)_3Si$ N^+ Br^- N H O O N 2

Scheme 9.3

In a sol–gel reaction by a layer-by-layer adsorption method a novel bioinspired organic–inorganic hybrid can be produced composed from a liposome membrane and a ceramic surface. The particles have mean diameters 70–300 and 20–100 nm (prepared from Cerasomes 1 and 2, respectively) and form a three-dimensional packed vesicular assembly.

The ferments can also be immobilized by the sol–gel method and work as a "bioreactor" [87] if the chemically active terminal groups and the ceramic dopant active bonds such as Sn–Cl are used. The immobilization and synthesis mechanisms can be described by Scheme 9.4 [162].

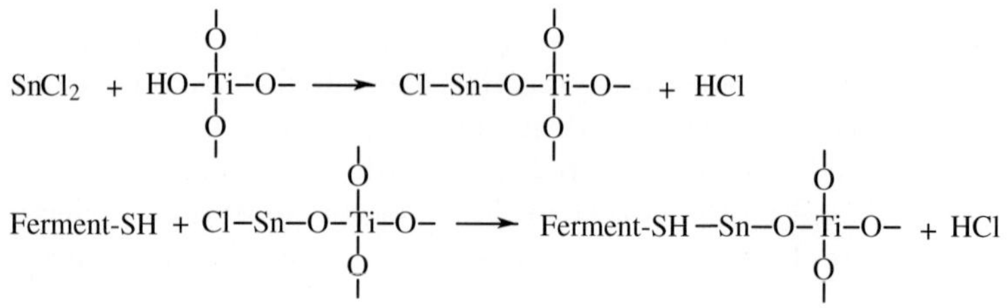

$$SnCl_2 + HO{-}\underset{\substack{| \\ O \\ |}}{\overset{\substack{| \\ O \\ |}}{Ti}}{-}O{-} \longrightarrow Cl{-}Sn{-}O{-}\underset{\substack{| \\ O \\ |}}{\overset{\substack{| \\ O \\ |}}{Ti}}{-}O{-} + HCl$$

$$\text{Ferment-SH} + Cl{-}Sn{-}O{-}\underset{\substack{| \\ O \\ |}}{\overset{\substack{| \\ O \\ |}}{Ti}}{-}O{-} \longrightarrow \text{Ferment-SH}{-}Sn{-}O{-}\underset{\substack{| \\ O \\ |}}{\overset{\substack{| \\ O \\ |}}{Ti}}{-}O{-} + HCl$$

Scheme 9.4

For example, into a template-synthesized TiO_2 nanotube alcohol dehydrogenase was immobilized, which (cofactor NAD^+, phosphate buffer, pH 8) remains active in methanol oxidation for more than 4 days [163]. The nanotubes are open at both ends and this configuration allows the use of the structure as a flow-type reactor. There are many other examples of such systems, including antibody covalent binding for sol–gel film functionalization [149].

At the same time such processes lead to uncontrolled particle formation with size distribution far from the uniform distribution of entrapped molecules in ceramic matrices. The entrapped enzymes in both dry and wet gel monoliths don't obey Michaelis–Menten kinetics and quantitative assay sometimes becomes unreliable [153, 164–166]. The organized matter formation problems are solved in sol–gel synthesis by different approaches [167]. It is possible to use the self-organizing organic template formation (transcription synthesis), the cooperative assembling of templates and other building blocks (synergetic synthesis), and morphosynthesis with generation of some specially organized nonlinear chemical surroundings [16]. Finally, there is the combination of the listed methods (so-called integrated synthesis) too. This strategy (reaction assembly → replication → methamorphism) is similar to the general scheme of mineralization and can be illustrated with the examples of the template-directed synthesis of ordered mesoforms, organoclays and macroskeleton structures, including the use of bacterial templates. The scheme is very obvious for the reproduction of hierarchical macrostructurated organized silica gels, as can be illustrated with an example of multicellular filaments of *Bacillus subtilis* as large-scale organic templates [168, 169] (Fig. 9.10).

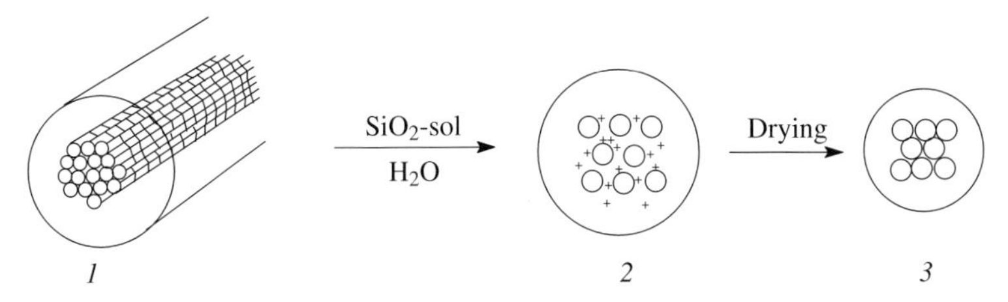

Fig. 9.10. A scheme of an organized macroporous SiO_2 structure, formed by the bacterial templates: bacterial filaments with multicellular fiber structure) (*1*); a mineralization of the inter lament space (*2*) and a macroporous replica formation by drying at 873 K (*3*)

9.5 Biomineralization and Bioinorganic Nanocomposites

The interactions of metal ions and nanoparticles with biological macromolecules and cells play a most important role not only in fermentative catalysis, but in geomicrobiology in general, because they represent the greatest research opportunities in the chemistry of natural metal and metal oxide surfaces [1, 170, 171]. The inorganic crystal formation is controlled in biomineralization processes by organic biomolecules such as proteins. Although the content of the organic component in a typical composite is very low (of the order of 1 wt. %), it exerts tremendous control of the mineralization process which provides the uniform size of particles, crystal morphology and specific crystallographic orientations [172]. In addition, from the viewpoint of material chemistry, the biomineralization methods with use of abundant and relatively low-cost sources may lead to the fabrication of a new type of organic–inorganic composite material with high performance and environmental benignancy.

Natural biominerals such as calcite ($CaCO_3$ in shells), apatite [$Ca_5(OH,F)(PO_4)_3$ in bones] or even magnetite (Fe_3O_4 in bacteria), greigite (Fe_3S_4), maghemite (γ-Fe_2O_3) and various ferrites ($MO{\cdot}Fe_2O_3$, where M = Ni, Co, Mg, Zn, Mn) are crystallized by living organisms under ambient conditions. Biomineralization in living organisms leads to inorganic–organic nanocomposite formation. The most abundant biogenic mineral is silica (opal) in a stable amorphous phase, but generally the ratio of the crystalline and amorphous phases in biogenic minerals is approximately 80 and 20%. Biomineralization terms are used for the description of a variety of minerals, produced by biological organisms, or their composition, structure, morphology, mechanical properties and functions. The growth of these minerals in an organism is controlled in such a way that they obtain optimal characteristics as required constructive materials [1, 2]. The process is controlled by the interaction of nuclei and growing crystals with biopolymers, acting as surfactants (usually peptides and glicopeptides). Poly-L-aspartate was considered as an analogue of acidic macromolecules, which are bond to an insoluble protein matrix with gelling properties [173]. The foregoing relates to many other bioinorganic processes, especially those connected with geobiotechnology, and biohydrometallurgy.

Microorganisms (both natural and created by gene engineering methods) living in extreme conditions of high temperatures, pressures, pH and salinity values and toxic metal concentrations are of special interest for practical problems, connected with the geomicrobiological transformations of heavy metals. By analogy with bacterial species adapted to unusual conditions (such as thermophiles, halophiles, acidophiles), the considered bacteria are related to the representative group called metallophilic microorganisms [93,174,175]. Metal binding is provided not only by bacteria, but by many yeasts, fungi, algae and even protozoa, that considerably widens the possibilities of the biochemical transformation and transport of metals both in nature and in technology.

The important appearance of chemical-colloidal binding is metal sol coagulation. The biopolymer-immobilized particles form mixed (biobone) aggregates of the living microorganisms and metal nanoparticles, many of which have special features. The characteristics can be determined by the diffusion of cell layers or by the electrical inhomogeneity of the cell surface. The sol particle binding proceeds together with decrease of the coagulation threshold and such high-dispersed biogenic formations are widespread in nature and environment: industrial metal stocks, and sedimentary facies [176]. The silver-accumulating bacterial strain *Pseudomonas stutzeri*, which was originally isolated from a silver mine, was used to produce a silver-carbon composite material promising for the fabrication of optical thin film coatings.

Metalloclusters play an important role in the bacterial mineralization processes [1, 177–180]. A new science – bioinorganic solid-phase chemistry – studies the formation mechanisms and structures of nanoscale inorganic materials in biological surroundings [1].

An organic matrix (template) under *biomineralization–bioaccumulation* regulates the nucleation, growth and formation of the inorganic materials with perfect morphology. The complex hierarchical composite structures are produced with unusual physical or chemical properties and are imitated in artificial conditions. There were many attempts to imitate such systems in laboratory conditions [181–183, 185].

Two issues of biomineralization should be understood – how such strictly organized inorganic materials are produced in nature (morphogenesis) and how the processes can be realized in a biomimetic system (morphosynthesis and mineralization in situ). Therefore one major approach to the problem of nucleation and crystal growth control is to mimic natural processes. The modeling of biomineralization is a promising approach to controlled crystal growth for seeking materials analogous to those produced by nature. The majority of these efforts have focused on exploring the promoting effect of templates on crystal nucleation and growth. The anionic poly(amidoamine)

Table 9.4. Main types and functions of biominerals

Mineral	Composition	Organism	Functions
Calcite	$CaCO_3$	Mollusk *Crustacea Aves*	Exoskeleton Mechanical strength Protection
Magnesium calcite	$(Mg,Ca)CO_3$	*Octocorallia Echinoderms*	Mechanical substrate Strength and protection
Aragonite	$CaCO_3$	*Scleractinian corals* Mollusks Fish	Exoskeleton Exoskeleton Attraction tool
Vaterite	$CaCO_3$	*Gastropoda*	Exoskeleton
Hydroxy-apatite	$Ca_{10}(PO_4)_6(OH)_2$	Vertebrates	Exoskeleton/ions stock
Calcium-phosphate	$Ca_8H_2(PO_4)_6$	Vertebrates	Bone precursors
Amorphous silica gel Amorphous silica gel	$SiO_2 \cdot nH_2O$	*Diatoms*	Exoskeleton
Magnetite	Fe_3O_4	Bacteria *Chitons* *Type/salmon*	Magnitotacticity Mechanical strength Magnetic orientation

dendrimers modified the metastable vaterite surface and inhibited the further growth of vaterite particles by changing the addition time during the crystallization of calcium carbonate [184].

Practically all metals take part in the biomineralization, but the most studied are processes with Mg, Ca, Sr, Ba, Mn, Fe and Si ions (Table 9.4). The problems of mineralization control and the various metal nanocrystal synthesis into biological systems or organisms are considered in [186].

The traditional viewpoint is that inorganic materials are permanent and inert substances with limited geometrical forms or structures generally characterized or described by the space groups and elementary cells, connected with the crystal symmetry. Now it must be considerably revised. Such notions come into contradiction with well-known plasticity and the form-function interconnection of biological processes. Inorganic components, forming and incorporating in some components of living organisms, are involved in the extraordinary morphological structures (spiral, curved, screwed, network) from the helix form of mollusk shells to the functional morphology of human skeleton and extracellular bone formations.

The symbiosis and complex structures are preserving not only on the macroscopic scale, but are characteristic for biomineralized system organization on the microscopic, mesoscopic and nanoscopic levels too. The different control forces are used at different scales and levels to support a hierarchy order of the structure [187, 188]. For example, the development of the above mentioned extracellular structures of skeleton bones on a macroscopic level is defined by global physical fields (gravitation, mechanical load). But the microstructure depends on the place and local activity of bone cells under hormonal regulation on mesoscopic and nanoscopic levels. The interphase processes, the supramolecular growth and the collagen fibers or calcium phosphate crystal organization are chemically and biochemically controlled by the tissue matrices. The control includes manipulation of the local concentrations of the depositing substances in the presence of the nucleation surfaces and the inhibitors in the solution (they bind to the specific faces of the growing mineral). The regulation also relates to the particle size, form and orientation (i.e. there is a need for the epitaxial matching of protein or polysacharide matrix and mineral). For example, the equation for apatite precipitation demonstrates the effect of the dissolved Ca^{2+} ions increasing the degree of supersaturation in solution.

$$5Ca^{2+} + 3PO_4^{3-} + OH^- \rightarrow Ca_5(PO_4)_3OH$$

The mineralization picture for unicellular organisms appears in the intracellular structures such as the *coccolith scales* and *diatom frustules* radiolarian microskeletons, being commensurate with the individual cells or the vesicular ensembles. In 1995 the first artificial microskeletons of radiolaria and diatoms were synthesized. As described in the survey of their structural organization and biomimetic form morphogenesis [16], all the chemical and cell processes (the formation of collagen fibrils, nucleation templates, cavities) in these structures are controlled during the lifetime of the organism. These dynamics are retained through the prolonged time of evolutionary adaptation of the biomineralized structures to the permanent changing of a given biological niche.

A general scheme of mineralization includes modeling the building up of the organic supramolecular structures functionalized by a proteins (or other macromolecules), nucleation and growth [177]. In the modeling space the mineralization produces an inorganic "replica" of the ensemble organized beforehand (called "replica on stone" or "chemical Medusa principle"). The process can proceed through a one-stage mechanism ("form at template") and by several synergistic stages (see the dotted line on Scheme 9.5), depending upon the system morphogenesis.

Molecular recognition and molecular tectonics are the most important aspects of biomineralization. But the genetic foundation of biomineralogical evolution remains unknown (just as we haven't an answer to the most

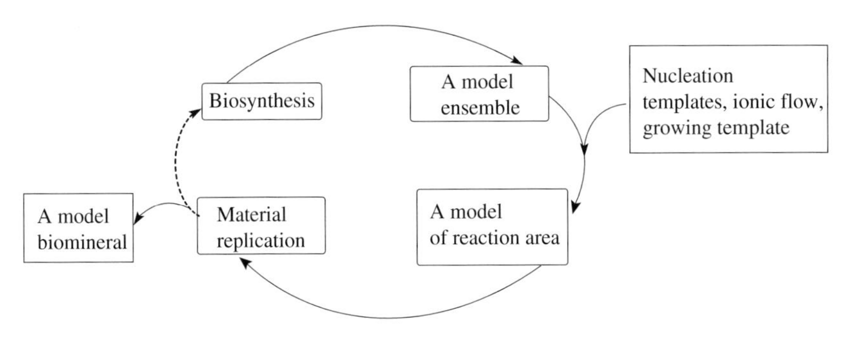

Scheme 9.5

principal question – how morphogenic compatibility is achieved at the *biology–inorganic chemistry* frontier). Most often the form of biominerals is prederminated by the given cell conformation or cell (Fig. 9.11).

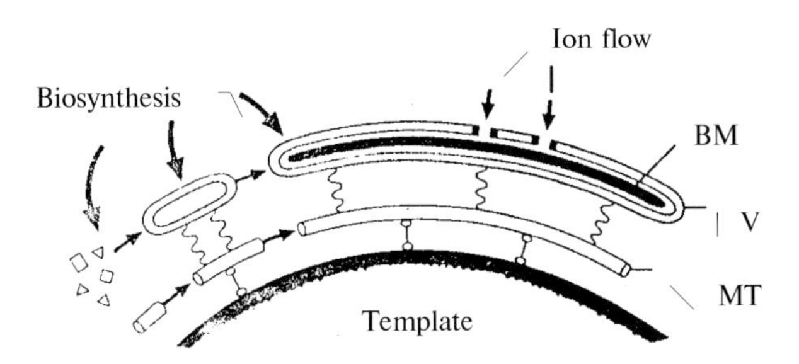

Fig. 9.11. An illustrative scheme of programmed self-assembling in biomineralization. The cell walls, organelles and cellular assemblies present the templates for the microtube (MT) formations, being directing agents for the model vesicles (V) involved in the biomineralization (BM) process

The interactions and relations between the equilibrium crystallization processes and the forming biological structures with complicated composition allow them to provide the very complex but reproducible morphological patterns in geological timescales. Such process modeling (morphosynthesis of the biomimetic forms) includes three main stages [188]. The first stage is modeling of a liquid–solid interface (some unusual morphoforms are produced in the mineralization due to chemical fluctuations, arising in local perturbations). In the second stage it is necessary to carry out template-directed assembling, when the bioinorganic forms are localized along, around and into the outer surfaces of the organic formations and assemblies (acting as the templates) or can take the forms of intratemplate voids. Finally controlled replication must take place (Table 9.5).

Table 9.5. Modeling of biomineralization into membrane-bound vesicles

Template	Directing Agents	Form	Architecture	Examples
Inner cellular wall	Microtubes	Bent SiO_2–rod	Tissue network	*Hoano-flagellates*
	?	Formed Fe_3O_4-crystals	Linear structure	*Magnetotactic bacteria*
	Microtubes	Bent SiO_2-frame	Hollow cyst	*Chrysophites*
	Microtubes/ areolar vesicles	Perforated SiO_2	Hollow frame	*Diathomias/ radiolarie*
Endoplasmatic network	Microtubes Microtubes	Bent SiO_2-leaf	Connected-scale (chain-mail) structure	*Chrysophites*
Nuclear envelope	Microtubes	Formed $CaCO_3$-crystals	Enfolliated coccospheric	*Coccolytho-phorides*
Cytoplasmatic envelope	Microtubes	Network SiO_2/ $SrSO_4$	Microskeleton	*Radiolarie/ acantarie*
Cellular assemblies	Syncytium/ microtubes	Bent/ figured $CaCO_3$	Spycules/hollow frame	*Holoturia/ Sea Urchin*
Cellular organization (without vesicle)	Biopolymers/ force fields	Hybrid composites	Macroskeleton	Vertebrates/ mollusks

Many approaches (sol–gel synthesis, layered inclusion compounds, Langmuir–Blodgett layers and films, nanoparticle deposition and polymer surface decoration) can be considered as approximate variants of biomimetic mineralization. One of the most realistic models is a metastable solution into which a polymeric matrix is dipped. It can induce a deposition in the polymer or in its swollen layer (but not in the solution itself). Such system can be used for the following mineralization. The polymer reagents are often used as polyelectrolytes [189–191], especially those that can reduce the crystal growth rate or change the crystal habits [192]. The films of some hydroxyl-containing polyacrylates (in particular, 4-hydroxybutylacrylate and 3-hydroxypropylacrylate copolymers with tetrahydroxyfurfurylmethacrylate, etc.) in calcium oxalate solution are mineralized slowly [181, 193]. Some double-block copolymers are of special interest [194] (see Chap. 5). For example, crystalline ZnO (zincite) was prepared by zinc nitrate hydrolysis in the presence of PMMA, PEO or PEO-*br*-PMMA [191] and both the particle habits and their size distribution can be controlled in the presence of PEO-*br*-PMMA.

Material produced from ormosil (used for rubber processing, Chap. 7) with dispersed Ca^{2+} ions after a sol–gel synthesis may serve as a bioactive soft tissue substitute [195].

The material relates to the composites SiO_2-poly(methyl methacrylate), which involved Ca^{2+} ions, exhibited bioactivity [196]. Silanol group formation on the ormosils can be a predominant factor in bioactivity control, while the effect of dissolved Ca^{2+} ions to increase the degree of supersaturation in the simulated body fluid is secondary [197].

The forming product morphology changes with the copolymer composition. The observed effect is specific, because other tested copolymers on the base of methacrylic acid, butylmethacrylate were inefficient. It seems that the nucleation of CaC_2O_4 crystals begins at the interface, and the growth lasts into the polymer film. Amorphous calcium carbonate is significantly less stable than the other five forms of $CaCO_3$ (aragonite, vaterite, monohydrocalcite and calcium carbonate hexahydrate) [198] and in this connection a biogenic "amorphous" calcium carbonate phase (formed by the ascidian *Pyura pachydermatina*) was analyzed [199]. It was shown that in principle the amorphous calcium carbonate should be considered as a group of different mineral phases.

Proteinic derivatives and synthetic solid substrates are often used to grow $CaCO_3$ in various combinations. For example, sulfated polystyrene with sorbed polyasparaginic acid can induce oriented calcite formation with a higher density than in the cases when the components are used separately [200]. Such consistency of actions can be connected, on one hand, with a strong electrostatic interaction between calcium ions and sulfogroups and, on the other hand, with the weak orientating influence of the β-layered structures of polyasperaginic acid (the so-called "ionotropic effect").

A specially prepared matrix from the *Nautilius pompilius* envelope produces in vitro spherolitic $CaCO_3$ (assumed to be vaterite[3]). The oriented growth of same forms of minerals was observed for polystyrene and calcite on

[3] There are several polymorphic mineral types of calcium carbonate: vaterite (originated in the formation of some shells, gastropods, etc.), which is transformed at ageing to aragonite (and later to calcite) and can be produced artificially; calcite (calcareous spar), being the main component of rocks (limestones), and aragonite, which presents the main component of pearls and pearly layers of some sea mollusks.

glass, but the nuclei originated both in the solution and on the substrate by physical adsorption. It seems that for the orientation a specific structural interconnection between crystal nucleation and substrate is not necessary and the electrostatic and van der waalse forces favorable for the colloid sorption on polymers play the main role. It is not inconceivable that the substrate composition (including its swollen layer) provides the phase transitions of the forming mineral. In support of this interpretation it was shown that some acidic macromolecules extracted from aragonitic or calcitic mollusks shell layers were indeed responsible for the selective precipitation of aragonite and calcite *in vitro* [173, 201, 202].

The precipitation process demands some interface between two solid electrolytes and a soluble glycoprotein. At the same time the macromolecules, extracted from the aragonite and calcite parts of mollusk shells, must be able to induce growth in vitro of the aragonite and calcite mineral forms, immobilized on β-chitine or silk fibroine correspondingly. The structure of the polyelectrolyte-gelatin assembly can be easily changed through the mechanical deformation modifying the microenvironmental impact on the nucleation and growth sites [203]. The molecular interactions between functional groups of chitin fibers and acid-rich macromolecules [(PAA, poly(L-aspartate) and poly(L-glutamate)] play a key role in inorganic crystallization control of the thin film of $CaCO_3$ crystals coating formation from a $CaCO_3$ solution [204]. The same can be related to the polymer/$CaCO_3$ composite films prepared in a combined system of chitosan and PAA [205].

It was supposed [181] that surface synthetic mineralization modeling had to be based on the biological macromolecule role determination in the process. Thus a bone-bonding mechanism in such materials indicated that the spontaneous deposition of the apatite layers (with composition and crystallinity similar to bone structure) proceeds only when there is contact with the blood plasma [144]. Vertebrate mineralization is of special interest, connected with the biomineralization process of the dental enamel matrix with a carbonated hydroxyapatite mineral [206], including visible light-curable dental composites [207] and other dental biomaterials.

The biomimetic methods of semiconducting material nanoparticles (ZnS, PbS, CdS, CuS, Cu_2S, In_2S_3) are studied widely, including various membrane usage (polyurethane films [208], panathion membranes [209–211]). The main methods are based on fabrication of the bacterial crystalline cell surface layers (S-layers) with their subsequent applications ranging from bio- and molecular nanotechnology to diagnostics and vaccine techniques [212]. The regular pore system within the two-dimensional protein crystals with S-layers have been used as templates to generate nanoparticle arrays. For example, by this method the synthesis of CdS superlattices was carried out using the recrystallized S-layers obtained by *Bacillus stearothermophilus* with a subsequent mineralization through the incubation in $CdCl_2$ solution and subsequent treatment with H_2S [213]. The monodisperse 5-nm CdS nanocrystals form a regular array located within the nanopores of the protein matrix. The specific peptide sequences were used for mineralizing some metals and semiconductors and producing the high-crystalline nanocrystals [214, 215]. The method is based on the effect that a negatively charged phosphate backbone of the DNA double helix can accumulate Cd^{2+} ions [216, 217] and usually for these purposes the DNA-templated metal and semiconductor nanoparticle arrays are used.

Another interesting approach is based on a fermentative synthesis of thiol-containing polymers (M_n = 1750–4390) with the subsequent Cd^{2+} ions binding. The case in point is an oxidative binding of 4-ethylphenol with 4-hydroxythiophenol in the presence of H_2O_2, catalyzed by peroxidase ferment [218] with the subsequent covalent binding of CdS. The latter is synthesized in the microstructured environment of the reverse micelles. The formed particles have sizes 5–12 nm. Such nanocomposites were produced by biocatalytic methods in the form of film, coating or microspheres. They have high luminescent properties (so it is promising for sensor and display techniques) and, what is of special interest, is biodegradable. Semiconductor nanoparticles are also powerful fluorescent probes and can be used for the labeling of bioogical components [17, 219, 220]. For this purpose the biopolymer-immobilized nanophase Mn^{3+}-oxides (with particle diameter 8 nm) were obtained using the transformations Mn^{2+} – Mn^{3+}. This method is effective also for nanoscale material production on the base of Ag_2O and especially the calcium-phosphates of $Ca_{10}(PO_4)_6(OH)_2$, $Ca_8H_2(PO_4)_6$ and other types, forming part of the bone tissues.

One of the most characteristic biomineralization examples is the formation of biogenic magnetic nanoparticles. Ferritin-based materials include the Fe_3S_4 nanoparticles (Fig. 9.12) deposing into an inner void of the protein envelope [221]. The biological self-assembly of superparamagnetic iron-oxo coresin ferritin presents a unique example of matrix-assisted formation of magnetic nanomaterials and, therefore, fabrication of an artificial ferritin analog still remains an elusive goal for many specialists [222]. Magnetotactic bacteria contain one or more linear protein chains and nanosized magnetite Fe_3O_4(40–100 nm) or greigite Fe_3S_4 (40–100 nm), which are usually arranged in protein chains [223, 224].

Many model systems were used for iron-involving mineralization studies including swelling alginate fibers [225] and Fe_2O_3 particles with sizes from 4 to 15 nm, produced by the reduction of in situ cross-linked alginate gel [226]. A critical requirement in modeling of the fundamental aspects of ferritin core assembly is direct chemical involvement of a template (analogous to ferritin) using the coordinating residues from the protein shell for metal ion binding and directing their growth [227]. For to produce a model, able to mimic all the essential aspects of natural ferritin, bis-[3-(trimethoxysilil)-propyl]ethylenediamine with composition $(CH_3O)_3Si(CH_2)_3NH(CH_2)_2NH$ $(CH_2)_3Si(OCH_3)_3$ was used after an organic modification [228, 229]. The

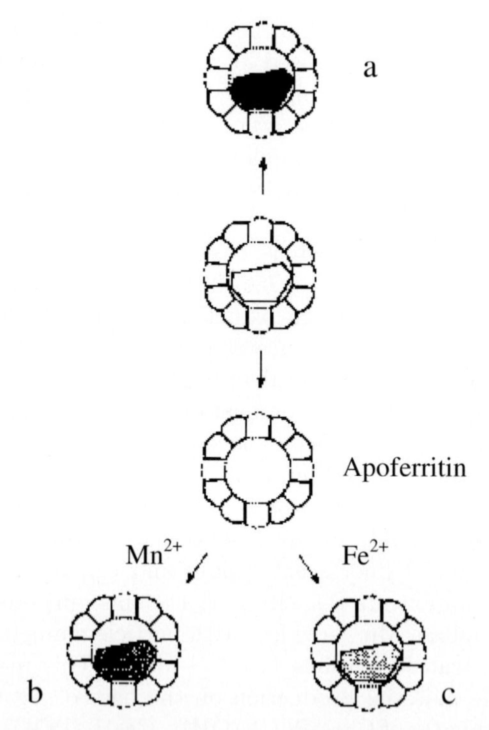

Fig. 9.12. Nanophase material synthesis on the base of ferritin: (**a**) – Fe_3S_4; (**b**) – Mn_2O_3; (**c**) – Fe_3O_4

network amino groups provide considerably high internal pH for Fe–O–Fe link formation from iron-aqua complexes.

This approach allows getting the biomimetic assemblies of iron-oxo clusters (as the synthetic analog of ferritin) with core diameters 4–7 nm and the number of iron atoms from 700 to 4000 (Fig. 9.13). The magnetic moments per cluster range from 100 to 2700 μB (the values for natural ferritin are 300–400 μB). For comparison, magnetic nanoparticles (activated with 3-aminopropyltriethoxysilane) have been used for the immobilization of various enzymes, antibodies and proteins after glutaraldehyde treatment [230]. It is suggested that these substances would find many useful technological applications as transparent magnetic materials, contrast agents for magnetic imaging and for magnetically targeted drug delivery [229].

The process can be even more simplified if a "metal-less" apoferritin is used. That is especially convenient for formation into the ferritin the in situ depositions of Fe_3O_4 with given crystallochemical characteristics [231]. It is a way to produce the ferromagnetic proteins important for clinical applications, such as magnetic imaging or the immobilization of various enzymes (e.g. glucose oxidase and uricase), antibodies, and oligonucleotides [232, 233]. There

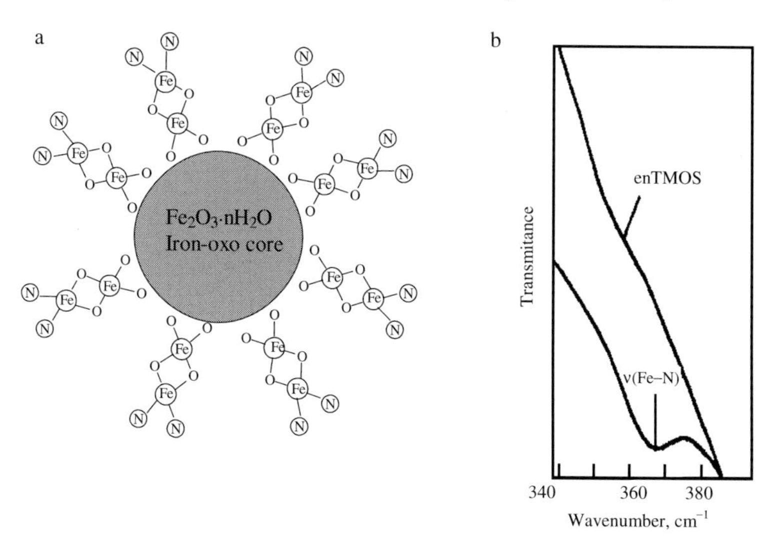

Fig. 9.13. (**a**) Schematic two-dimensional representation of matrix-assisted assembly of iron-oxo clusters inside bis-[3-(trimethoxysilyl)-propyl]ethylenediamine (enTMOS) sol–gel pores; (**b**) FTIR spectra of enTMOS gels with and without the iron-oxo clusters showing the appearance of ν(Fe–N) vibrational mode at 367 cm^{-1} in the sample containing the nanoclusters due to binding of metal ions to the amino groups of the sol–gel network

are other magnetic materials such as the dextran-based biocompatible magnetic nanoparticles (50 nm in diameter), PEO-enzyme conjugated to magnetic nanoparticles, magnetically modified lipase, *L*-asparaginase, urokinase, and magnetic labeled antibodies [234]. The bioinorganic reactions are very complicated and precise, because the requirements for the products are connected not only with the particle arrangements and sizes, but with their morphology too. The problems can be illustrated by the example of a bacterial magnetite, an intracellular structure of Fe_3O_4 crystals, forming a chain parallel to the longitudinal axis of *Coccus* or *Spirillum* type bacteria (Fig. 9.14) [235]. The magnetic moment of bacteria (imparted to them by the magnetic crystals) enables them to align and subsequently migrate along geomagnetic field lines. Such magnetoordered or magnetotactic bacteria can be oriented and move even in the very weak magnetic field of the Earth (0.25–0.50 G), providing a navigation function in the living organisms. The imitation and mimicking of these structures and functions are interesting problems. There were many attempts to synthesize such inorganic materials and inorganic–organic composites in vitro [236]. In particular, good results were obtained [237] in studies of a magnetic force driven orientation of the submicronic ferromagnetic Co–B arrays $Co_{23}B_{10}$ (in soluble starch), inspired by magnetotactic bacteria. Many theoretical and practical problems of magnetic nanoparticle applications in biosciences (the immobilization and modification of biologically

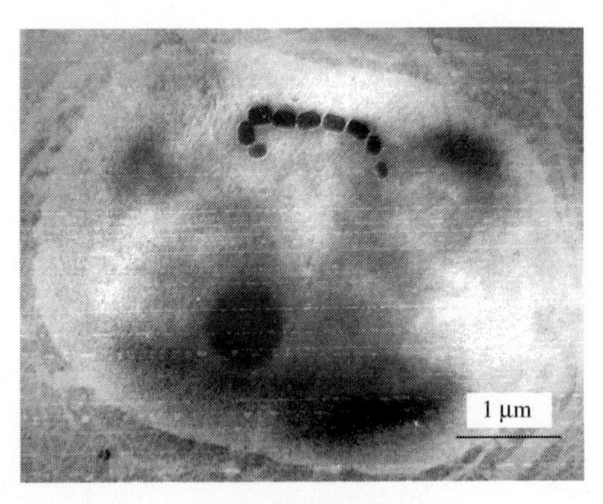

Fig. 9.14. Electron microscope photography of *Coccus* bacteria with a chain, consisting of nine intracellular magnetic crystals (cell length ∼3 μm, every crystal ∼50 × 50 nm)

active compounds, drug and radionuclide targeting, magnetic fluid hyperthermia, contrast-increasing materials, cancer cell detection) are analyzed in the overview [238].

Microbiological geotechnology, metal separation from ores, concentrates, rocks and solutions by using microorganisms or their metabolism products [95], are very interesting. Many such processes (e.g. the bacterial leaching of copper and uranium) were used in industry. The process includes many stages: ore milling and pulping, bacterial leaching of a target metal, filtration, solid residue neutralization and cyanidation.

The above presented analysis demonstrates that metal nanoparticles are widespread, play a paramount role in living nature and often are present in biological objects as important components. The nanoparticles take part in the binding and decomposition of various organic substances and they give rise and cause matter and energy turnovers in nature, including formation of many mineral resources. In the field of "mimic" chemistry only the first (but very encouraging) steps are made in the understanding of the functionality of living matter. In near future one can expect considerable successes in the design and fabrication of the self-assembling highly organized molecular systems including nanomaterials. For example, recently a new biomaterial class had been discovered on the basis of nanotubes, including some self-assembling cylindrical peptide subunits [239–242] that generates novel paradigms for material science and biology [243]. The environment of a protein can assist the generation of complex products. For example, the $BaTiO_3$, $SrTiO_3$, $NaNbO_3$ particles can be involved in bioaggregation processes [244], which present perovskites with ABO_3-type structure (their synthetic analogs were considered in Sect. 7.1).

Bacterial methods can even lead to the formation of the fine monodispersed precursors for high-temperature superconducting ceramics [177].

The survey [245] is devoted to metal oxide surfaces and their interaction with microbial organisms and considered surface chemistry in search of new insights (environmental mineral formation and dissolution), cell attachment and biofilm formation, cooperative interactions, mechanisms of metal dissolution, reduction and oxidation, and structural and reactive properties of some biogenic oxide surfaces. The solutions of these problems are interesting in connection with new technological advances in surface chemistry, geochemistry, molecular biology and biophysics. Thus, the interface chemistry and problems of nanoparticles, proteins and nucleic acids meet each other in biotechnology and materials science. Moreover, it is possible that further research will lead to the fantastic problems of the genetically coded biological growth of artificial inorganic materials.

References

1. *The Biomineralisation of Nano- and Micro-Structures* ed. by E. Bauerlein (Wiley-VCH, Weinheim, 2000)
2. *Colloidal Drug Delivery Systems* ed. by J. Kreuter (Marcel Dekker, New York 1994)
3. C.M. Niemeyer: Angew. Chem. Int. Ed. **36**, 585 (1997)
4. R.F. Service: Science **277**, 1036 (1977)
5. C.A. Mirkin, T.A. Taton: Nature **405**, 626 (2000)
6. M. Yokoyoma, T. Okano: Adv. Drug Delivery Rev. **21**, 77 (1996)
7. F. Gallays: Hystochem. **64**, 87 (1979)
8. T. Klaus, R. Joerger, E. Olsson. Proc. Nat. Acad. Sci. USA **96**, 13611 (1999)
9. E. Westhof, N. Leontis: Angew. Chem. Int. Ed. **39**, 1591 (2000)
10. *Biomineralization: Chemical and Biochemical Perspectives* ed. by S. Mann, J. Webb, R.J.P. Williams (VCH Publ., Weinheim 1989)
11. *Biomimetic Materials Chemistry* ed. by S. Mann (VCH Publ., Weinheim 1996)
12. M. Sarikaya, I.A. Aksay: *Biomimetics: Design and Processing of Materials* (AIP Press, New York 1995)
13. A.E. Shilov: *Metal Complexes in Biomimetic Chemical Reactions. N_2 Fixation in Solution, Activation and Oxidation of Alkanes Chemical Models of Photosynthesis* (CRC Press Inc., Boca Raton 1997)
14. C.D. Garner: J. Chem. Soc. Dalton Trans. 3903 (1997)
15. *Bioinorganic Chemistry: Transition Metals in Biology and Coordination Chemistry* ed. by A. Trautwein (Wiley-VCH, Weinheim, 1997)
16. G.A. Ozin: Acc. Chem. Res. **30**, 17 (1997)
17. C.M. Niemeyer: Angew. Chem. Int. Ed. **40**, 4128 (2001)
18. J. Chem. Soc. Dalton Trans. (1997)
19. Chem. Rev. *Bioinorganic Enzymology* **96**, 7 (1996)
20. *Problems of the molecular engineering and biomimetics* Mendeleev Chem. J. **39**, 1 (1995)
21. *Molecular Recognition* ed. by S.H. Gellman In: Chem. Rev., **97**, N5 (1997)
22. J.P. Gallivan, D.A. Dougherty: Proc. Nat. Acad. Sci. USA **96**, 9459 (1999)
23. O. Yamauchi, A. Odani, S. Hirota: Bull. Chem. Soc. Jpn. **74**, 1525 (2001)

24. A.L. Leninger: *The bases of biochemistry. V. 1* (Mir, Moscow 1985)
25. M.G. Bezrukov, A.M. Belousova, V.A. Sergeev: Russ. Chem. Rev. **51**, 696 (1982)
26. R.R. Crichton: *Inorganic Biochemistry of Iron Metabolism* (Horwood, New York 1991)
27. G.I. Likhtenstein: *Multinuclearic metal enzymes* (Nauka, Moscow 1979); In: *Oxidation-reduction metal enzymes and their models. The theoretical and methodology aspects V. 1* (Institute of Chemical Physics AN USSR, Chernogolovka 1982) p. 7
28. *Metal Cluster in Proteins* ed. by L. Que, Jr. (ACS, Washington, DC 1988)
29. L. Que, Jr.: J. Chem. Soc. Dalton Trans. 3933 (1997)
30. D.D. LeCloux, A.M. Barrios, T.J. Mizoguchi, S.J. Lippard: J. Am. Chem. Soc. **120**, 9001 (1998)
31. A.A. Steiman: Russ. Chem. Bull. 1011 (1995)
32. A.C. Rosenzweig, S.J. Lippard: Acc. Chem. Res. **27**, 229 (1994)
33. Y. Montet, E. Garcin, A. Volbeda, C. Hatchikian, M. Frey, J.C. Fontecilla-Camps: Pure Appl. Chem. **70**, 25 (1988)
34. K.L. Taft, G.C. Papaefthymiou, S.L. Lippard: Science **259**, 1302 (1993)
35. G.C. Dismukes: Chem. Scr. A **28**, 99 (1988); Chem. Rev. **96**, 2909 (1996)
36. G.W. Brudvig, R.H. Crabtree: Proc. Nat. Acad. Sci. USA **83**, 4586 (1986)
37. J.E. Sheats, R.S. Czernuszewicz, G.C. Dismukes, A.L. Rheingold, V. Petrouleas, J.A. Stubbe, W.H. Armstrong, R.H. Beer, S.J. Lippard: J. Am. Chem. Soc. **109**, 1435 (1987)
38. G.C. Dismukes, J.E. Sheats, J.A. Smegal: J. Am. Chem. Soc. **109**, 7202 (1987)
39. E. Bradshaw, S. Moghaldas, L.J. Wilson: Gazz. Chem. Ital. **124**, 159 (1994)
40. K.K.P. Srivastava, K.K. Surerus, R.C. Conover, M.K. Johnson, J.-B. Park, M.W.W. Adams, E. Munck: Inorg. Chem. **32**, 927 (1993)
41. S. Ciurli, S. Yu, R.H. Holm, K.K.P. Srivastava, E. Munck: J. Am. Chem. Soc. **112**, 8169 (1990)
42. D.C. Rees, Y. Hu, C. Kisker, H. Schindlin: J. Chem. Soc. Dalton Trans. 3909 (1997)
43. A.-M. Liu, T.-J. Zhou, H.-L. Wan, K.-R. Tsai: Chem. J. Chin. Univ. **14**, 996 (1993)
44. A. Gole, C. Dash, V. Ramakrishnan, S.R. Sainkar, A.B. Mandale, M. Rao, M. Sastry: Langmuir **17**, 1674 (2001)
45. G. Joonen, A. Vessieres, I.S. Butler: Acc. Chem. Res. **26**, 361 (1993)
46. T.W. Welch, G.A. Neyhart, J.G. Goll, S.A. Ciffan, H.H. Thorp: J. Am. Chem. Soc. **115**, 9311 (1993)
47. H.S. Joshi, Y. Tor: Chem. Commun. 549 (2001)
48. S. Weinstein, W. Jahn, M. Lascherer, T. Arad, W. Tichelaar, M. Haider, C. Glotz, T. Boeckh, Z. Berkovitch-Yellin, F. Franceschi, A. Yonath: J. Cryst. Growth **122**, 286 (1992)
49. Y. Zhou, W. Chen, H. Itoh, K. Naka, Q. Ni, H. Yamane, Y. Chujo: Chem. Commun. 2518 (2001)
50. A. Dalmia, C.L. Lineken, R.F. Savinell: J. Colloid Interface Sci. **205**, 535 (1998)
51. O. Siiman, A. Burshteyn: US Pat. WO 93/15117
52. R. Djalali, Y. Chen, H. Matsui: J. Am. Chem. Soc. **124**, 13660 (2002)
53. H. Matsui, R. McCuspie;. Nano Lett. **1**, 671 (2001)

54. S. Mann, W. Shenton, M. Li, S. Connolly, D. Fitzmaurice: Adv. Mater. **12**, 147 (2000)
55. W. Faulk, G. Taylor: Immunochemistry **8**, 1081 (1971)
56. C.A. Mirkin, R.L. Letsinger, R.C. Mucic, J.J. Storhoff: Nature **382**, 607 (1996)
57. J.J. Storhoff, A.A. Lazarides, R.C. Mucic, C.A. Mirkin, R.L. Letsinger, G.C. Schatz: J. Am. Chem. Soc. **122**, 4640 (2000)
58. A.P. Alivisatos, K.P. Johnsson, X. Peng, T.E. Wilson, C.J. Loweth, M.P. Bruchez, P.G. Schultz: Nature **382**, 609 (1996)
59. D. Polak, S. Van Norden: *Introduction in immunocytochemistry: the modern methods and problems* (Mir, Moscow 1987)
60. R. Elghanian, J.J. Storhoff, C.R. Mucic, R.L. Letsinger, C.A. Mirkin: Science **277**, 1078 (1997)
61. J.J. Storhoff, R. Elghanian, R.C. Mucic, C.A. Mirkin, R.L. Letsinger: J. Am. Chem. Soc. **120**, 1959 (1998)
62. V.A. Bogatyrev, L.A. Dykman, Ya.M. Krasnov, V.K. Plotnikov, N.G. Khlebtsov: Colloid J. **64**, 745 (2002)
63. L.M. Demers, S.-J. Park, T.A. Taton, Z. Li, C.A. Mirkin: Angew. Chem. Int. Ed. **40**, 3071 (2001)
64. G.R. Souza, J.H. Miller: J. Am. Chem. Soc. **123**, 6734 (2001)
65. D. Zanchet, C.M. Micheel, W.J. Parak, D. Gerion, S.C. Williams, A.P. Alivisatos: J. Phys. Chem. B **106**, 11758 (2002)
66. M. Taki, H. Murakami, M. Sisido: Chem. Commun. 1199 (2000)
67. D. Gerion, W.J. Parak, S.C. Williams, D. Zanchet, C.M. Micheel, A.P. Alivisatos: J. Am. Chem. Soc. **124**, 7070 (2002)
68. K.J. Watson, J. Zhu, S.T. Nguyen, C.A. Mirkin: J. Am. Chem. Soc. **121**, 462 (1999)
69. S. Link, Z.L. Wang, M.A. El-Sayed: J. Phys. Chem. B **103**, 3529 (1999)
70. T.A. Taton, G. Lu, C.A. Mirkin: J. Am. Chem. Soc. **123**, 5164 (2001)
71. Y.W. Cao, R. Jin, C.A. Mirkin: J. Am. Chem. Soc. **123**, 7961 (2001)
72. E. Skutelsky, J. Roth: J. Histochem. Cytochem. **34**, 693 (1986)
73. L.A. Dykman, A.A. Lyakhov, V.A. Bogatyrev, S.Yu. Shchegolov: Colloid J. **60**, 757 (1998)
74. M. Akashi, H. Iwasaki, N. Miyauchi, T. Sato, J. Susamoto, K. Takemoto: J. Bioactive and Compatible Polymers **4**, 124 (1989)
75. H. Otsuka, Y. Akiyama, Y. Nagasaki, K. Kataoka: J. Am. Chem. Soc. **123**, 8226 (2001)
76. Y. Akiyama, H. Otsuka, Y. Nagasaki, K. Kataoka: Bioconjugate Chem. **11**, 947 (2000)
77. D. Osella, O. Gambino, C. Nervi, E. Stein, G. Jaouen, V.A. Gerard: Organometallics **13**, 3110 (1994)
78. S.L. Woodhouse, L.M. Rendina: Chem. Commun. 2464 (2001)
79. *Bioinorganic Catalysis* ed. by J. Reedijk (Marcel Dekker, New York 1993)
80. J. Richter, R. Seidel, R. Kirsch, M. Mertig, W. Pompe, J. Plaschke, H.K. Schackert: Adv. Mater. **12**, 507 (2000)
81. C. Grüttner: 'Preparation and Characterization of Magnetic Nanospheres for in vivo applications'. In: *Scientific and Clinical Applications of Magnetic Carriers* (Plenum Press, New York, London 1997)
82. C. Grüttner, M. Connah: Intern. Labor. News **28**, 1A, 26 (1998)
83. R. Paddefet: *Chemistry of gold* (Mir, Moscow 1982)

84. *Immuno-gold labelling in cell biology* ed. by A.J. Verkley, J.L.M. Leunissen (CRC Press, Boca Raton 1989)
85. M. Horisberger: Cellulaire **36**, 253 (1979)
86. G. Pappalardo, G. Impellizzeri, R.P. Bomono, T. Campagna, G. Grasso, M.G. Saita: New J. Chem. **26**, 593 (2002)
87. *Immobilized Cells and Enzymes – A Practical Approach* ed. by J. Woodward (IRL Press, Washington, DC 1985)
88. R.V. Parthasarathy, C.R. Martin: J. Appl. Polym. Sci. **62**, 875 (1996)
89. Y.-Z. Hu, S. Thinkai, I. Hamachi: Chem. Lett. 517 (1999)
90. M. Suzuki, M. Sano, M. Kimura, K. Hanabusa, H. Shirai: J. Polym. Sci. A Polym. Chem. **37**, 4360 (1999)
91. S. Nimri, E. Keinan: J. Am. Chem. Soc. **121**, 8978 (1999)
92. S.A. Davis, H.M. Patel, E.L. Mayes, N.H. Mendelson, G. Franko, S. Mann: Chem. Mater. **10**, 2516 (1998)
93. S.A. Marakushev: *Geomicrobiology and biochemistry of gold* (Nauka, Moscow 1991)
94. D.I. Nikitin, M.S. Oranskaya, A.S. Savichev, P.V. Mikheev: Bull. Russ. Acad. Sci. Ser. Biol. 302 (1986)
95. G.I. Karavaiko: Vestn. AN USSR N1, 72 (1985)
96. V.I. Karamushka, T.G. Gruzina, Z.R. Ulberg: *Microbiology* **64**, 192 (1995)
97. Z.R. Ulberg, L.G. Marochko, A.G. Savkin, N.V. Pertsev: Colloid J. **60**, 836 (1998)
98. E.M. Kinsch, S.W. Douglas: Inorg. Chim. Acta **91**, 263 (1984)
99. I. Karamushka, T.G. Gruzina, Z.R. Ulberg: Colloid J. **60**, 327, 331, 775 (1998)
100. A.M. Melin, M.A. Carboneau, N. Rebeyrotte: Biochim. **68**, 1201 (1986)
101. T.J. Beveridge, R.G.E. Murray: J. Bact. **141**, 876 (1980)
102. E.D. Korobushkina, I.M. Korobushkin: Dokl. AN USSR **287**, 978 (1986)
103. K.B. Nielson, D.R. Winge: J. Biol. Chem. **260**, 8698 (1985)
104. S. Kasselouri, A. Garoufis, N. Hadjiliadis: Inorg. Chim. Acta **135**, 123 (1987)
105. A. Jordan, R. Scholz, P. Wust, H. Fähling: J. Magn. Magn. Mater. **201**, 413 (1999)
106. G. Wu, R.H. Datar, K.M. Hansen, T. Thundat, R.J. Cote, A. Majumdar: Nat. Biotechnol. **19**, 856 (2001)
107. M.D. Porter, T.B. Bright, D.L. Allara, C.E.D. Chidsey: J. Am. Chem. Soc. **109**, 3559 (1987)
108. R.G. Nusso, B.R. Zegarski, L.H. Duboin: J. Am. Chem. Soc. **109**, 733 (1987); **112**, 570 (1990)
109. K.L. Prime, G.M. Whitesides: Sciences **252**, 1164 (1991)
110. P. Tengvall, M. Lestelius, B. Liedberg, I. Lundström: Langmuir **8**, 1236 (1992)
111. M. Lestelius, B. Liedberg, I. Lundström, P. Tengvall: J. Coll. Interface Sci. **171**, 533 (1995)
112. M.C. Coen, J. Kressler, R. Mülhaupt: Macromol. Symp. **103**, 109 (1996)
113. J.M. Schnur: Science **262**, 1669 (1993)
114. T.A. Kavassalis, J. Noolandi: Macromolecules **22**, 2709 (1989)
115. J.P. Schneider, J.W. Kelly: Chem. Rev. **95**, 2169 (1995)
116. J. Noolandi, A.-C. Shi: Macromol. Rapid Commun. **17**, 471 (1996)
117. L. Nixon, C.A. Koval, R.D. Noble, G.S. Slaft: Chem. Mat. **4**, 117 (1992)
118. L. Josephson, J.M. Perez, R. Weissleder: Angew. Chem. Int. Ed. **40**, 3071 (2001)

119. G.M. Gadd, C. White: Trends Biotechnol. **11**, 353 (1993)
120. Ji Gy, S. Silver: J. Ind. Microbiol. **14**, 61 (1995)
121. J.A. Thompson, G. Jarvinen: Filtr. Sep. **36**, 28 (1999)
122. D.W. Urry: Progr. Biophys. Mol. Biol. **57**, 23 (1992)
123. D.W. Urry, D.C. Gowda, T.M. Parker, C.H. Luan: Biopolymers, **32**, 1243 (1992)
124. J. Kostal, A. Mulchandani, W. Chen: Macromolecules **34**, 2257 (2001)
125. *Biochemical Aspects of Sol-Gel Science and Technology* ed. by D. Avnir, S. Braun (Kluwer Academic Publishers, Boston 1996)
126. D. Avnir, S. Braun, O.F. Lev, M. Ottolenghi: Chem. Mater. **6**, 1605 (1994)
127. L.L. Hench, R.J. Splinter, W.C. Allen, T.K. Greenlee: J. Biomed. Mater. Res. Symp. **2**, 117 (1972)
128. E.J.A. Pope, K. Braun, M. VanHirtum, C.M. Peterson: J. Sol-Gel Sci. Technol. **8**, 635 (1997)
129. K. Kawakami: Biotech. Technol. **10**, 491 (1996)
130. P. Andebert, C. Demaille, C. Sanchez: Chem. Mater. **5**, 911 (1993)
131. A. Wiseman: *Handbook of Enzyme Biotechnology* (Horwood, Chichester 1985)
132. F. Akbarian, A. Lin, B.S. Dunn, J.S. Valentine, J.I. Zink: J. Sol-Gel Sci. Technol. **8**, 1067 (1997)
133. C.J. Brinker, G. Schrer: *Sol-Gel Science: The Physics and Chemistry of Sol-Gel Processing* (Academic Press, San Diego 1990)
134. D.D. Lasic, D. Papahadjopoulos: Curr. Opin. Solid State Mater. Sci. **1**, 392 (1996)
135. I.M. Hafez, P.R. Cullis: Adv. Drug. Deliv. Rev. **47**, 139 (2001)
136. *Protein Architecture: Interfacing Molecular Assemblies and Immobilization Biotechnology* ed. by Y. Lvov, H. Möhwald (Marcel Dekker, New York 2000)
137. R. Jelinek, S. Kolusheva: Biotechnol. Adv. **19**, 109 (2001)
138. R. Joerger, T. Klaus, C.G. Granqvist: Adv. Mater. **12**, 407 (2000)
139. J. Kreuter: 'Nanoparticles as Adjuvant for Vaccines' In: *Vaccine Design: The Subunit and Adjuvant Approach* ed. by M.F. Powell, N.J. Newman (Plenum Press, New York 1995)
140. S. Braun, S. Rappoport, R. Zusman, D. Avnir, M. Ottolenghi: Mater. Lett. **10**, 1 (1990)
141. C. Dave, B. Dunn, J.S. Valentine, J.L. Zink: Anal. Chem. A **66**, 1120 (1994)
142. *Better Ceramics Through Chemistry* ed. by C. Sanchez, M.L. Mecartney, C.J. Brinker, A. Cheetham (VI. Mater. Res. Soc. Sympos. Proc. 1994)
143. L.L. Hench:. J. Am. Ceram. Soc. **74**, 1487 (1991)
144. T. Kokubo: J. Ceram. Soc. Jpn. **99**, 965 (1991)
145. *Introduction to bioceramics* ed. by L.L. Hench, J. Wilson (World Scientific, Singapore 1993)
146. D.M. Dabbs, I.A. Aksay: Annu. Rev. Phys. Chem. **51**, 601 (2000)
147. O. Lev, Z. Wu, S. Bharathi, V. Glezer, A. Modestov, J. Gun, L. Rabinovich, S. Sampath: Chem. Mater. **9**, 2354 (1997)
148. J.I. Zink, J.S. Valentine, B. Dunn: New. J. Chem. **18**, 1109 (1994)
149. R. Collino, J. Jherasse, P. Binder, F. Chaput, B.-P. Boilot, Y. Levy: J. Sol-Gel Sci. Technol. **2**, 823 (1994)
150. A. Bronshtein, N. Aharonson, D. Avnir, A. Turniansky, M. Altstein: Chem. Mater. **9**, 2632 (1997)
151. C. Roux, J. Livage, K. Farhati, L. Monjour: J. Sol-Gel Sci. Technol. **8**, 663 (1997)

152. D. Shabat, F. Grynszpan, S. Saphier, A. Turniansky, D. Avnir, E. Keinan: Chem. Mater. **9**, 2258 (1997)
153. T.K. Jain, I. Roy, T.K. De, A. Maitra: J. Am. Chem. Soc. **120**, 11092 (1998)
154. F.J. Arriagada, K. Osseo-Asare: J. Colloid Interface Sci. **170**, 8 (1995)
155. K. Kusunoki, K. Kawakami: Bull. Fac. Eng. Kyushu Sangyo Univ. **33**, 48 (1996)
156. *Artificial Self-Assembling System for Gene Delivery* ed. by P.L. Felgner, M.J. Heller, P. Lehn, J.P. Behr, F.C. Szoka (ACS, Washington, DC 1996)
157. Y. Murakami, J. Kikuchi, Y. Hisaeda, O. Hayashida: Chem. Rev. **96**, 721 (1996)
158. S.K. Davidson, S.L. Regen: Chem. Rev. **97**, 1269 (1997)
159. *Precision Polymers and Nano-Organized Systems* ed. by T. Kunitake, S. Nakahama, S. Takahashi, N. Toshima (Kodansha, Tokyo 2000)
160. Y. Murakami, J. Kikuchi, T. Takaki: Bull. Chem. Soc. Jpn. **59**, 3145 (1986)
161. K. Katagiri, R. Hamasaki, K. Agira, J. Kikuchi: J. Am. Chem. Soc. **124**, 7892 (2002)
162. J.F. Kennedy, J.M.S. Cabral: *Solid Phase Biochemistry. Vol. 66* (J. Wiley & Sons, New York 1983)
163. B.B. Lakshmi, C.J. Patrissi, C.R. Martin: Chem. Mater. **9**, 2544 (1997)
164. S. Shtelzer, S. Braun: Biotech. Appl. Biochem. **19**, 293 (1994)
165. S.A. Yamanaka, F. Nishida, L.M. Ellerby, C.R. Nishida, B. Dunn, J.S. Valentine, J.I. Zink: Chem. Mater. **4**, 495 (1992)
166. I. Gill, A. Ballestens: J. Am. Chem. Soc. **120**, 8587 (1998)
167. S. Mann, S.L. Burkett, S.A. Davis, C.E. Fowler, N.H. Mendelson, S.D. Sims, D. Walsh, N.T. Whilton: Chem. Mater. **9**, 2300 (1997)
168. S.A. Davis, S.L. Burkett, N.H. Mendelson, S. Mann: Nature **385**, 420 (1997)
169. *Geomicrobiology: Interactions between Microbes and Minerals. Vol. 35* ed. by J.F. Banfield, K.H. Nealson (Mineralogical Society of America, Washington, DC 1997)
170. S.K. Lower, M.F. Hochella, Jr., T.J. Beveridge: Science **292**, 1360 (2001)
171. A.P. Alivisatos: Science **289**, 736 (2000)
172. A.M. Belcher, P.K. Hansma, G.D. Stucky, D.E. Morse: Acta Mater. **46**, 733 (1998)
173. G. Falini, S. Albeck, S. Weiner, L. Addadi: Science **271**, 87 (1996)
174. M.A. Glazovskaya, N.G. Dobrovolskaya: *Geochemical function of microoganisms* (Moscow State University, Moscow 1984)
175. A.N. Ilyaletdinov: *Microbiological transformations of metals* (Nauka, Alma-Ata 1984)
176. A.S. Monin, A.P. Lisitsyn: *Ocean biogeochemistry* (Nauka, Moscow 1983)
177. S. Mann: J. Chem. Soc. Dalton Trans. 3953 (1993)
178. T. Douglas, D.P.E. Dickson, S. Betteridge, J. Charnock, C.D. Garner, S. Mann: Science, **269**, 54 (1995)
179. L. Addadi, S. Weiner: Angew. Chem. Int. Ed. **31**, 153 (1992)
180. S. Mann: Angew. Chem. Int. Ed. **39** (2000)
181. P. Calvert, P. Rieke: Chem. Mater. **8**, 1715 (1996)
182. *Biomimetics* ed. by M. Starikaya, I.A. Aksay (AIP Press, Woodburg, New York 1995)
183. *Calcification in Biological Systems* ed. by E. Bonucci (CRC Press, Boca Raton 1992)

184. J.F. Banfield, S.A. Welch, H. Zhang, T. Thomson-Ebert, R.L. Penn: Science **289**, 751 (2000)
185. Dong-Ki Keum, K. Naka, Y. Chuji: Bull. Chem. Soc. Jpn. **76**, 1687 (2003)
186. S. Brown, M. Sarikaya, E. Johnson: J. Mol. Biol. **299**, 725 (2000)
187. S. Mann: Nature **332**, 119 (1988); **365**, 499 (1993)
188. S. Mann, G.A. Ozin: Nature **382**, 313 (1996)
189. M. Öner, P. Calvert: Mater. Sci. Eng. C **2**, 93 (1994)
190. F. Grases, J.J. Gill, A. Conte: Colloid Surf. **36**, 29 (1989)
191. M. Öner, J. Norwig, W.H. Meyer, G. Wegner: Chem. Mater. **10**, 460 (1998)
192. Z. Amjad: Langmuir **9**, 597 (1993)
193. P. Calvert: Mater. Res. Soc. Symp. Proc. **330**, 79 (1994)
194. J.M. Marentette, J. Norwig, E. Stoeckelmann, W.H. Meyer, G. Wegner: Adv. Mater. **9**, 647 (1997)
195. Y. Hu, J.D. Mackenzie: J. Mater. Sci. **27**, 4415 (1992)
196. S.M. Jones, S.E. Friberg, J. Sjoblom: J. Mater. Sci. **29**, 4075 (1994)
197. K. Tsuru, C. Ohtsuki, A. Osaka, T. Iwamoto, J.D. Mackensie: J. Mater. Sci.: Mater. in Medicine **8**, 157 (1997)
198. L. Brecevic, A.E. Nielsen: J. Cryst. Growth **98**, 504 (1989)
199. Y. Levi-Kalisman, S. Raz, S. Weiner, L. Addadi, I. Sagi: J. Chem. Soc. Dalton Trans. 3977 (2000)
200. L. Addadi, J. Moradian, E. Shay, N.G. Maroudas, S. Weiner: Proc. Nat. Acad. Sci. USA **84**, 2732 (1987)
201. G. Falini, S. Albeck, S. Weiner, L. Addadi: Science **271**, 87 (1996)
202. X.Y. Shen, A.M. Belcher, P.K. Hansma, G.D. Stucky, D.E. Morse: J. Biol. Chem. **272**, 32472 (1997)
203. G. Falini, S. Fermani, M. Gazzano, A. Ripamonti: J. Chem. Soc. Dalton Trans. 3983 (2000)
204. T. Kato, T. Amamiya: Chem. Lett. 199 (1999)
205. T. Kato, T. Suzuki, T. Amamiya, T. Irie, M. Komiyama, H. Yui: Supramol. Sci. **5**, 411 (1998)
206. A.G. Fincham, J. Moradian-Oldak, J.P. Simmer: J. Struct. Biol. **126**, 270 (1999)
207. F. Gao, Y. Tong, S.R. Schricker, B.M. Culbertson: Polym. Adv. Technol. **12**, 355 (2001)
208. D. Meissner, R. Memming, B. Kastening: Chem. Phys. Lett. **96**, 34 (1983)
209. M. Krishnan, J.R. White, M.A. Fox, A.J. Bard: J. Am. Chem. Soc. **105**, 7002 (1983)
210. B.H. Kuczynski, B.H. Milosavjevic, J.K. Thomas: J. Phys. Chem. **88**, 980 (1984)
211. J.H. Fendler: Chem. Rev. **87**, 877 (1987)
212. C.M. Niemeyer, U.B. Sleytr, P. Messner, D. Pum., M. Sara: Angew. Chem. Int. Ed. **38**, 1034 (1999)
213. W. Shenton, D. Pum, U.B. Sleytr, S. Mann: Nature **389**, 585 (1997)
214. J. Ziegler, R.T. Chang, D.W. Wright:. J. Am. Chem. Soc. **121**, 2395 (1999)
215. S.R. Whaley, D.S. English, E.L. Hu, R.F. Barbara, A.M. Belcher: Nature **405**, 665 (2000)
216. S.R. Bigham, J.L. Coffer: Colloids Surf. A **95**, 211 (1995)
217. J.L. Coffer: J. Cluster Sci. **8**, 159 (1997)

218. R. Premachandran, S. Banerjee, V.T. John, G.L. McPherson, J.A. Akkara, D.L. Kaplan: Chem. Mater. **9**, 1342 (1997)
219. H. Mattoussi, J.M. Mauro, E.R. Goldman, G.P. Anderson, V.C. Sundar, F.V. Mikulec, M.G. Bawendi: J. Am. Chem. Soc. **122**, 12142 (2000)
220. A.P. Alivisatos: Pure Appl. Chem. **72**, 3 (2000)
221. R.B. Frankel, R.P. Blakemore: *Iron Biominerals* (Plenum Press, New York, London 1991)
222. P.M. Harrison, P. Arosio: Biochim. Biophis. Acta **1275**, 161 (1996)
223. H.A. Lowenstam: Bull. Geol. Soc. Am. **73**, 435 (1962)
224. R.P. Blakemore: Science **190**, 377 (1975)
225. J.R. Revol, D.H. Ryan, R.H. Marchessault: Chem. Mater. **6**, 249 (1994)
226. E. Kroll, F.M. Winnik, R.F. Ziolo: Chem. Mater. **8**, 1594 (1996)
227. N.D. Chasteen, P.M. Harrison: J. Struct. Biol. **126**, 182 (1999)
228. M.S. Rao, B.C. Dave: J. Am. Chem. Soc. **120**, 13270 (1998)
229. M.S. Rao, I.S. Dubenko, S. Roy, N. Ali, B.C. Dave: J. Am. Chem. Soc. **123**, 1511 (2001)
230. H. Shinkai, H. Honda, T. Kabayashi: Biocatalysis **5**, 61 (1991)
231. F.C. Meldrum, B.R. Heywood, S. Mann: Science **257**, 522 (1992)
232. T. Matsunaga, M. Kawasaki, X. Yu, N. Tsujimura, N. Nakamura: Anal. Chem. **68**, 3551 (1996)
233. T. Matsunaga, H. Nakayama, M. Okochi, H. Takeyama: Biotechnol. Bioeng. **73**, 400 (2001)
234. *Scientific and Clinical Applications of Magnetic Carriers* ed. by U. Häfeli, W. Schütt, J. Teller, M. Zborowski (Plenum, New York, London 1997)
235. S. Mann: Chem. Brit. **23**, 137 (1987)
236. K.M. McGrath: Adv. Mater. **12–13**, 989 (2001)
237. X. Cao, Y. Xie, F. Yu, Z. Yao, L. Li: J. Mater. Chem. **13**, 893 (2003)
238. I. Šafaik, M. Šafaikova: Monatsh. Chem. **133**, 737 (2002)
239. M.R. Ghardiri, J.R. Granja, R.A. Milligan, D.E. McRee, N. Khazanovich: Nature **366**, 324 (1993)
240. H. Nishijima, S. Kamo, Y. Nakayama, K.I. Hohmura, S.H. Yoshimura, K. Takeyasu: Appl. Phys. Lett. **74**, 4061 (1999)
241. S. Akita, H. Nishijima, Y. Nakayama, F. Tokumasu, K. Takeyasu: J. Phys. D Appl. Phys. **32**, 1044 (1999)
242. K.I. Hohmura, Y. Itokazu, S.H. Yoshimura, G. Mizuguchi, Y. Masamura, K. Takeyasu, Y. Shiomi, T. Tsurimoto, H. Nishijima, S. Akita, Y. Nakayama: J. Electron Microscopy **49**, 415 (2000)
243. M.R. Ghadiri: Adv. Mater. **7**, 675 (1995)
244. H.A. Pohl: In: *Coherent Excitation in Biological Systems* ed. by H. Frolich, F. Kremer (Springer, Heidelberg 1983) p. 199
245. G.E. Brown, Jr., V.E. Henrich, W.H. Casey, D.L. Clark, C. Eggleston, A. Felmy, D.W. Goodman, M. Grätzel, G. Maciel, M.I. McCarthy, K.H. Nealson, D.A. Sverjensky, M.F. Toney, J.M. Zachara: Chem. Rev. **99**, 77 (1999)

Part IV

The Main Applications of Metal–Polymeric Nanocomposites

Polymeric hybrid materials have attracted a great deal of interest recently due to their high efficiency and performance. Typical hybrid materials may contain a cross-linked inorganic phase bonded mostly covalently with an organic phase. Nanocomposites are a class of composites in which the reinforcing phase dimensions are on the order of nanometers. The preparation of organic/inorganic hybrid materials composed of organic (co)polymers and inorganic colloids is a route to combine the advantageous properties of both classes of macromolecules into one material. Polymer coated inorganic nanoparticles are an important class of materials that have good coupling between the organic and inorganic phases. The mechanical properties, such as elastic modulus, hardness, and compressive yield strength, depend on the nature, volume fraction, and particle size and distribution of the fillers employed. The filler particles are usually subjected to the same surface treatment with the coupling agents prior to mixing with the resins because of the thermodynamic incompatibility between the organic polymer matrices and the inorganic fillers. The most interesting properties of such materials are their good dispersibility in organic media and their suitability for compression-molding operations.

The synthesis of nanocomposite materials from (co)polymers and nanoparticles was widely investigated. In situ polymerization is a good method where at first nanometer scale inorganic fillers or reinforcements are dispersed in the monomer, then this mixture is polymerized using a technique similar to bulk polymerization. The most important factors that affect the properties of composites are the dispersion and the adhesion at the polymer and filler interfaces. Inorganic particles may disperse homogeneously in the polymer matrices when they are premodified by a coupling agent. Furthermore, the resulting materials obtained by this method also can be easily processed since they have good flowing properties. Colloids of uniform size and precise morphology have been synthesized from a variety of emulsion, precipitation and sol–gel approaches.

Controlling surface properties of colloidal materials is a major technological subject in the wide area of applications in pharmaceutical and cosmetic drug improvement, pigmented paint films, semiconductors, catalysis, printing, food, biological, and medical fields. Achievements in nanoparticle or the cluster synthesis and structure for immobilization into a polymer matrix allowed workers to find many promising practical applications (sometimes very unexpected ones) of the formed nanocomposite materials in physics, chemistry, and biology. The small sizes of the particles give them many properties, unusual for bigger particles (structural, magnetic, catalytic and biological). The new materials are very important for modern information technologies. But the first significant commercial applications for these nanocomposites are to be a basis of new emerging construction materials, including tougher yet lighter automotive parts, having improved gas-barrier properties in packaging, tribological, enhanced flame-retardant, etc.

Another interesting category of polymer encapsulated inorganic particles are pigments used in the manufacture of cosmetics, inks and paints (of special interest in water-based paints). Thus, organic–inorganic hybrid materials display enhanced properties by bridging the characteristics of two dissimilar types of materials.

10 Nanoparticle Modifying Action on Polymers

At present a standard way to get the necessary properties or to modify polymer compositions is to introduce some substances (pigments, inhibitors, antioxidants, plasticizers, fillers), but the required concentrations are often so great that they considerably change (the whole complex of the material characteristics). The mechanical properties tend to be improved with a general increase of the filler volume. The filler particles are traditionally classified into three broad categories based on their size: macrofillers (~5–50 µm), microfillers or nanofillers (~0.01–0.1 µm) and physical hybrids (physical mixture of macro- and microfillers) [1]. The decrease of the filler sizes usually reduces the stress concentration, maintains a smooth sample surface, and extends the durability of the composites. However, the smaller the particles are, the more difficult the polymer volume loading process is because of the increase of the surface area limiting the wetting of small particles by resins and rising to the stress concentrations around the unwetted interfacial defects and filler particles aggregates. An usual result is the failure of composite materials.

Therefore one of the most actual problems of high-molecular compound science is to create systems able to recognize the outer influence or action at the molecular level and to give a corresponding response. Nanocomposites are a class of materials with a filler-induced increase of the characteristics [2–5]. A change of the phase dimensions on the nanometer level leads to reinforcing of the material properties. Our studies are devoted to the influence of nanoscale (single or cluster) particles on conventional polymer properties, primarily an increase of the physical, mechanical and operational characteristics of the polymer. The influence is based on the particles potential ability to form an ionic or coordination cross-linking, which can limit molecular chains or mobility of their segments and also create cohesion and adhesion cross-linkings. That allows the production of materials with a new architecture [6], to improve considerably their properties relative to conventional nanocomposites due to maximizing the interfacial adhesion. Such improved properties can be attained in nanocomposites in which the various building blocks (nanoscale metal particles, silica, ceramic sheets like layered silicates) are dispersed in a polymer matrix.

The introduction of nanoparticles (mostly the high-dispersed SiO_2 particles) into a polymerizing system is widely used (see Chap. 7) for the production of hybrid materials. For example, it takes place in the (polyaniline + silica gel) or (polypyrrole + silica gel) combinations [7, 8]. Due to their high specific surface the spherical SiO_2 particles act as special dispersion agents in depositing of aqueous media and "glueing" the polymer that allows it to form raspberry-like composite particles 100–300 nm in size. In an oxidative–reductive system [$FeCl_3{\cdot}6H_2O$ – $(NH_4)_2S_2O_8$] initiated polymerization. One can introduce not only spherical SiO_2 particles into the forming polypyrrole, but fibrous SiO_2 formations too [9]. The particle size changes from 80 to 200 nm, depending on the SiO_2:polypyrrole relation. A similar effect was observed also in SnO_2-based composites used for the production of thin lithographic films.

The introduction of metal-containing nanoscale particles into a polymer or polymeric envelope can lead to two principal results [10–13]: a) changing the properties of the polymer matrix itself, and b) adding new characteristics to the immobilized nanoparticles. A detailed analysis of such changes is very important and demands serious separate research, therefore here we describe the most essential features.

10.1 The Control of Physico-Mechanical Properties of Nanocomposites

The incorporation of an inorganic phase into a polymer matrix may increase mechanical strength and provide improvements in other specific properties, during which the preparation methods of these hybrid systems have a pronounced influence on their physical and mechanical properties. The composite systems for the various applications (that usually require the significant improvement in the optical clarity or strength, as compared to the base polymer) must be homogeneous. The phase separation is undesirable because it can result in a degradation of both optical and mechanical properties. As a rule, in such nanocomposites a strong covalent interaction is observed between the organic and inorganic phases. However it was shown that homogeneous and optically transparent hybrid materials could be obtained even in the absence of a primary chemical bonding (e.g. between aliphatic polyesters and silica) due to the strong hydrogen-bonding interaction between the organic and inorganic phases [14].

Composite sheets of polymer coated by carbon powders are used as excellent thermal radiation protection for many electronic devices (transistors, diodes, integrated circuits) or as electroconductive additives to plastics. Polymer-coated inorganic nanoparticles have several applications, for example, for the improvement of the tensile properties of rubbers, industrial paints, and the manufacture of diaphragm materials for loudspeakers [15].

Polymers with colloidal particles exhibit nonlinear rheological behavior. A variety of suspensions exhibit nonlinear viscoelastic effects, such as shear thinning, thixotropy, rheopexy and dilatancy (e.g., see [16]), but early rheological studies of nanocomposites have focused on oscillatory and steady-shear studies of nylon and polystyrene (PS)/polyisoprene (PI) block copolymer materials [17, 18].

Polymer nanocomposites (especially polymer-layered silicate, PLS) represent a radical alternative to conventionally filled polymers. New nanoscaled materials exhibit markedly improved properties (as compared to their macrocomposite counterparts) and these enhancements were achieved for low clay particle loadings, typically in the range 1–10 wt.%. For example, a doubling of the tensile modulus and strength is achieved for nylon-layered silicate nanocomposites containing 2 vol.% of inorganic material. The tensile modulus of epoxy/clay hybrids at 10 wt.% concentration is a factor of 6 greater than for the neat epoxide [19]. It can be added that the heat distortion temperature of the nanocomposites (with 5 wt.% clay loading) increases by up to 373 K extending the use of the composite to higher temperature environments, such as automotive under-the-hood parts [20].

The first survey of nanoparticle usage as polymer fillers was devoted to the intercalations of clay minerals with polymers [21] and contained more than 70 references, though a general conception of the future of such materials in applied organo-clay chemistry was formulated even earlier [22]. Materials of these types have various applications in agriculture, foundation engineering and industry. The high dispersed clay minerals are used as an active filler if the polymer is able to form strong hydrogen bonds with surface hydroxyl groups of the clay particles in the nascent state or after modification of the particle surface. The filler particle sizes in the process have considerably more influence on the elastomer mechanical properties than their chemical composition or surface chemical pretreatment. In this connection it is interesting to consider the thermomechanical curve of PMMA, filled by octadecyl-ammonia bentonite (a mineral presenting nearly pure montmorillonite in calcic form), being previously dispersed to the primary particles in a vibrating mill [23]. At the filler content 20% the material is not yielding even at 530 K, i.e. the yield temperature rise is about 70° (Fig. 10.1). The material hardness also increases considerably and achieved 28.2 10^7 Pa (28.2 kg/mm^2), as compared with 19.2 10^7 Pa (19.2 kg/mm^2) for initial, nontreated PMMA. In principle in the smectite clays there are two diametrically opposite morphologies for the design and production of such nanomaterials [24–32]. The first morphology presents well-ordered and stacked multilayers, formed by the intercalated polymeric chains between the host-silicate clay layers. Another structure corresponds to delaminated exfoliation materials, where the host layers had lost their ordering and are well dispersed into a continuous polymer matrix (see Chap. 8). The latter type of nanocomposites achieve as a rule high degrees of hardness (and other mechanical properties) at significantly low values of the degree of filler volume than in conventional filled systems. The proper monomer intercalation usually promotes delamination and dispersion of the host layers that

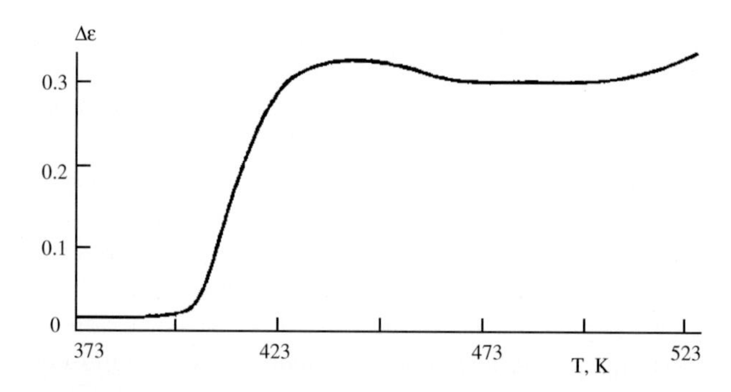

Fig. 10.1. A thermomechanical curve for polymethylmethacrylate containing 20% octadecylammonium bentonite

leads to the formation of a linear or cross-linked polymer matrix. As mentioned above intercalation often is preceded by preliminary silicate treatment by a long-chain amine. For example, by this method, from an organo-modified MMT (with trade mark I 30P or Nanocor) many nanocomposites are produced on the base of PA-6 (fosta nylon), PA-66, polyesters, epoxides, PVP, PEO, PI, polyamide-imide, polysulfone, polyacrylate, etc. (e.g. see [33–46]). The delaminated nylon–clay nanocomposites have better physical and mechanical properties (as compared with conventional nylon) without loss of impact strength. Such materials also have improved characteristics of thermal degradation resistance that considerably widens their application fields. It must noted, too, that the nanocomposites synthesized at room temperature by copolymerization of N-isopropyl acrylamide and methylene bisacrylamide (cross-linking monomer) in an aqueous suspension of Na-montmorillonitrile (containing 3.5 wt.% of montmorillonite) exhibited a low critical solution temperature (LCST), near to unmodified polymer hydrogel (305 K), but did not exhibit LCST if the aqueous suspension contained 10 wt.% of montmorillonite [47].

The analysis of the mechanical properties of a complex system (clay-based pigmented coating films bonded with carboxylated styrene-butadiene latex) has shown that the macromolecule segmental mobility is significantly limited in the presence of clay particles, though the NMR data demonstrate only a weak interaction between these components [48]. At the same time a strong interaction exists between $CaCO_3$ and the polymer. The difference is connected first of all with the form of the filler nanoparticles, because the clay particles are similar to sheets, but the calcium carbonate particles have a spherical form, and as a result the polymer can cover (envelope) the $CaCO_3$ particles more effectively than the clay particles. On one hand the interaction between carbonate and copolymers is stronger than with clay, but on the other hand the clay is a hydrophilic component and carbonate is a hydrophobic one. Thus the mechanical properties (like tensile strength of rubbers) will improve when an interfacial bond between the rubber and $CaCO_3$ filler exists [49].

An alternative to the above-considered variants is a direct intercalation of a polymer melt and/or the compatibilizing additives into an organophilic host-silicate lattice [50]. Increased intercalation and exfoliation are observed if the organophilic clays possess a sufficient affinity to the host and additive. In this way nanocomposites on the base of PP, PS, PEO, PE, PP, PA-6, PA-12, polymethylsiloxane, etc. can be produced [51–57]. From other methods the solution-casting, melt-extrusion and solid-state drawing technique is used. For example, hybrid nanocomposites were prepared from PA-6, reinforced with potassium titanate ($K_2Ti_6O_{13}$) whiskers, and liquid crystalline polymer in a twin-screw extruder followed by injection molding [58]. Static tensile measurements showed that the tensile strength and modulus of the hybrid composites tend to increase with the whisker content, and the Young's modulus was considerably higher than those predicted from the rules for hybrid mixtures. At the same time the Izold impact test indicated that the hybrid composites show a slight decrease in the impact strength initially with whisker content. It is interesting that in the drawn nanocomposites an optical anisotropy was observed, connected with the uniaxial orientation [59].

It is instructive to study the rheology of polymer–clay hybrids for two reasons. First, the rheological properties are indicative of melt-processing behavior in the separate operations, such as injection molding, etc. Second, since the rheological properties of particular suspensions are sensitive to the many parameters (structure, particle size, shape and surface characteristics of the dispersed phase) the rheology data potentially give the possibility of studying nanocomposite dispersions directly in the melt state. Rheology can be envisaged as a tool that is complementary to traditional methods of material characterization, such as electron microscopy, X-ray scattering, mechanical testing, etc.

10.2 The Peculiarity of Nanocomposites, Synthesized by Sol–Gel Methods

The main methods of fabrication of such nanocomposites were described above (Chap. 7) and here we will only briefly sum up the characteristics of the forming products. Morphology and phase separation control are critical in the generation of hybrid organic–inorganic composite materials via in situ polymerization of metal alkoxides in organic polymers. These materials generally exhibit higher stiffness and hardness, although their toughness is reduced. High degrees of homogeneity have been achieved using organic polymers, functionalized with trialkoxysilane or titanium alkoxides groups, that can react with added inorganic monomers, such as tetraethoxysilane, tetraetoxytitanium, thereby preventing a phase separation. This approach is very effective for nanocomposites, including polymers with appropriate backbone structures, which are able to interact with the growing inorganic oxide network (some examples are given in references [60–77]). PMMA, PVAc, PVP, PEO, PVA, and some others, can be used as such polymers, and sometimes a third component or compatibilizer can be added into the compositions to improve the blend properties [78, 79]. The compatibilizer can diffuse to the interface between the two immiscible components and reduces the interfacial tension thereby reducing the dispersed-phase particle size, i.e. in other words it can increase the adhesion between the two phases. This effect was demonstrated by the example of polystyrene–titanium composite systems, where surprisingly poly(4-vinylphenol) homopolymer turns out to be an effective compatibilizer [62]. A very promising compatibilizer (especially for its high impact strength at low temperatures) is poly(vinylbutyral) or (PVB), which is sturdy and flexible and shows excellent adhesive properties with different materials (glass, metal, plastics, wood) and has special value as a film sandwiched material in safety glass for automobiles [80]. The nanocomposite PVB (P_n 700–1700)-TiO_2 (synthesized by the sol–gel method) has improved values of Young's modulus, yield stress, elongation at break, and ultimate strength obtained from the stress–strain curve for samples with different composition [81]. In Fig. 10.2 (a,b) the results of the dynamic mechanical property measurements are shown. The $tg\delta$ decreased with increasing titanium content and the peak, corresponding to the T_g of PVB, became broader and shifted to a high temperature gradually. Moreover, with rise of the titanium content, the decrease of the storage modulus near T_g became gentle, and high values were maintained. The results confirm the supposition that titania was dispersed homogeneously in PVA and the thermal motion of the molecular chain was prevented by the interaction between the particles.

The mechanical properties of the hybrid organic–inorganic films poly(ethylene oxide-*b*-amide-6) (PEOA) and silicon oxide or titanium oxide prepared by the sol–gel process were evaluated by stress–strain tests [82]. In Fig. 10.3 the force is shown as a function of deformation for pure PEOA and hybrid systems. Independently of the nature of the precursors, in all cases the hybrids (with higher than 20 wt.% content of inorganic precursor) were much more rigid that pure polymer, and the films with an inorganic precursor content higher than 50 wt.% were extremely fragile as well as other nanocomposites, as, for example, poly(amide-imide) (PAI)/TiO_2, prepared by an in situ sol–gel process [83]. The high T_g values were obtained for composites as a whole always when the nanosize metal oxide rich domains (their size increased from 5 to 50 nm when the TiO_2 content was increased from 3.7 to 17.9 wt.%) were incorporated into the matrix with high T_g. The systems with nanosized TiO_2 domains was attributed to having hydrogen bonding interactions between the amide group polymer and the hydroxyl groups on the inorganic oxide. These interactions also seem to be responsible for disrupting the polymer crystallinity during the film formation stage. Moreover, the TiO_2 domains

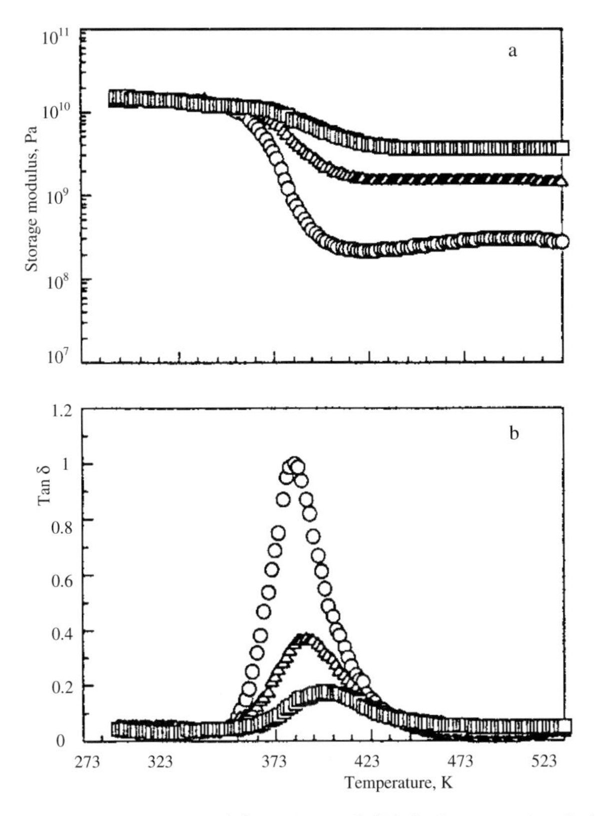

Fig. 10.2. The storage modulus (**a**) and tan δ (**b**) behavior of poly(vinyl)butyral ($\overline{P} = 1200$)/titania composites: titania 5 wt.% (○), 10 wt.% (Δ), 15 wt.% (□)

also serve as physical cross-links, decreasing the segmental mobility of the PAI chains, which led to changes in the apparent thermal and mechanical properties of the system (Fig. 10.4).

The elastic modulus of silicone rubbers, PDMS (poly(dimethyl siloxane), prepared by different methods, are compared in Table 10.1 [84].

It is of interest to consider also the methods based on direct hydrothermal crystallization of a silicate from the polymer-containing gels [24] that can be demonstrated on the example of the systems clay-polyaniline (PANI), polyacrylonitrile (PAN), and the cationic polymer polydimethyldiallilammonium chloride (PDMACl), as well as polymer reinforcing by ceramics, formed under the sol–gel processes (see Chap. 7).

The methods, connected with the polymer binding at the surface of oxide nanoparticles, are now widely used (such systems are usually called inorganic core–polymer shell nanocomposites). Polymer encapsulated oxide nanoparticles not only improve the product properties, but can also give many advantages in processing, because they can both be well dispersed in organic

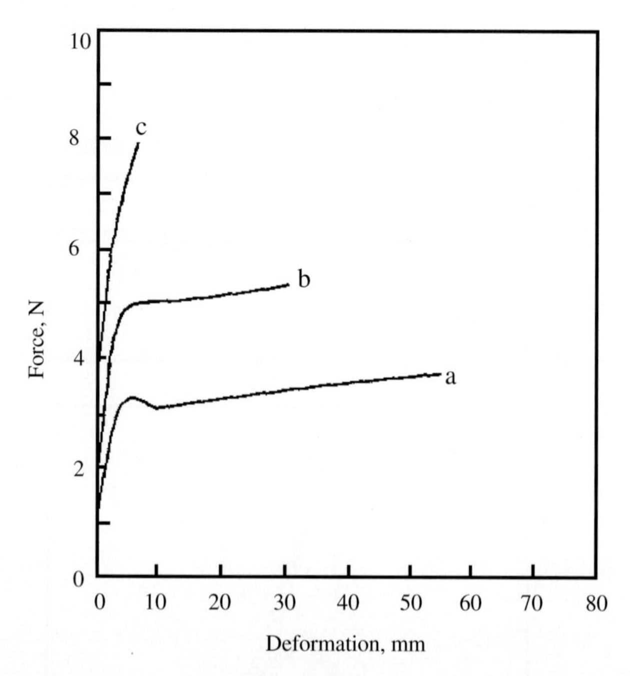

Fig. 10.3. Force as a function of deformation of (**a**) pure PEOA, (**b**) 80/20 and (**c**) PEOA/TEOS films

media and be directly molded. Many such products are used in catalysis, as toners in electro- and photographic applications, etc. The processing methods of polymer-encapsulated oxide nanoparticles can be divided into two main groups [85]: in the former case already existing polymer chains are coupled to the inorganic surface (usually modified or activated); in the latter, polymerization is carried out at the particle surface. For example, polymer-coated TiO_2 particles were obtained by polymerization of monomers adsorbed on the oxide surface and used as substrates for paint production. Some similar problems and possible solutions were considered in previous chapters.

10.3 Polyolefin-Based Nanocomposites

Nanocomposites and filled materials based on a polyolefin (especially on polypropylene [86]) matrix now have wide commercial applications. These materials usually consist of micron size fibers, flakes or powders (chalk, mica, silica), but it is important that the functional groups are absent from their polyolefinic matrices, being able to connect the polymer and filler particles. The new possibilities of changing polyolefin properties arise due to the replacement of microsize filler particles with nanosize particles. Most popular fillers of this type are nanopowders of abundant minerals, such as $CaCO_3$ [87],

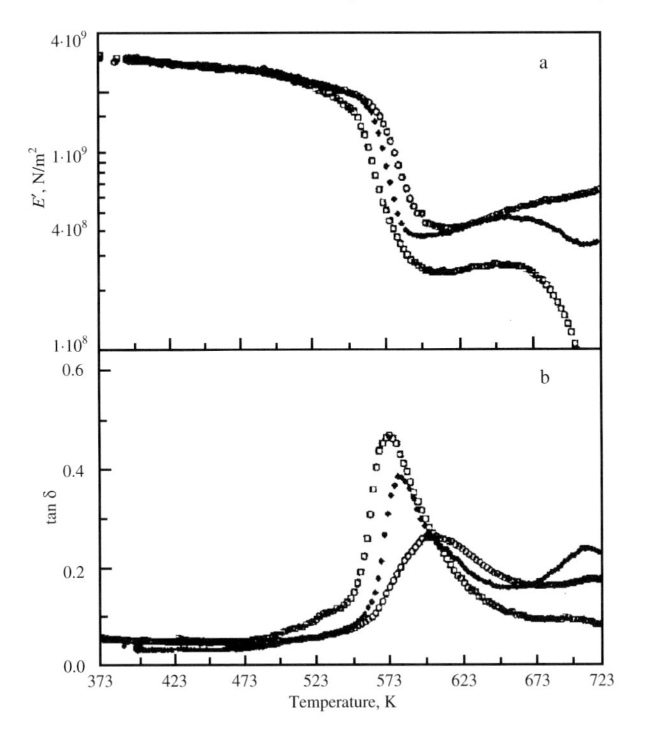

Fig. 10.4. Dynamic mechanical thermal analysis (plateau modulus (**a**) and tan δ (**b**)) of pure poly(amide-imide) (PAI) and PAI/TiO_2 composites (□ – unfilled PAI; ◇ – PAI/TiO_2 (5.3%); ○ – PAI/TiO_2 (10.1%)

$CaPO_3$ [88], SiO_2 [89] and MMT type clays [57, 90–92]. It is known that inorganic micron size fillers may promote crystallization of PP, change the size and number of spherulites and as a result lead to different new properties [93]. The nucleating ability of these fillers after decreasing their size to nanometers scale has not yet been fully studied.

It is very important that the forming nanocomposite structure is changing if the used fillers are produced in situ. Most often polyolefin composites are formed by melt mixing with Brabender industrial mixer conditions (for polypropylene, $T = 453\,K$, t = 8 min, 60 rpm), and the nanofiller concentrations, as a rule, fall within the range 0.01–5 wt.%. In the process a two-step procedure is applied: first a MMT is mixed with maleic anhydride (1 wt.%) grafted PP (PP-gr-MA) or with acrylic acid (6 wt.%) grafted PP (PP-gr-AA). Al_2O_3 or ZnO in compositions with the nanoparticles occasionally form agglomerates roughly 100 nm in diameter and the tendency to agglomeration increases with the increase of powder contents.

Thus, introducing 10% nanosized $CaSO_4$ in PEO/PP (from melt) leads to δ-form isotactic PP fabrication during the melt isothermal crystallization, whereas introducing coarse (commercial) $CaSO_4$ in the same conditions gives

Table 10.1. Comparison of mechanical properties of samples prepared by different methods

Sample	Weight % Silica	Elastic Modulus (N/mm^2)
Silicone rubber	0	0.481
Silica filled silicone	10	0.62
Silica filled silicone	20	0.85
Silicone filled via TEOS	14.6	3.99
Precipitation		
$T_p = 473\,K$, $T_o = 533\,K$	14	11 ± 1
$T_p = 523\,K$, $T_o = 533\,K$		14.5 ± 0.1
$T_p = 623\,K$, $T_o = 533\,K$	21–26	20 ± 1
TEOS Modified with PDMS	72	192

in general the α-form of PP.[1] The PEO enveloping of $CaSO_4$ has an impact on the PP crystallization rate. Fine powders (~800 nm) of PEO (the rare earth oxides, such as Y_2O_3, Nd_2O_3, Ho_2O_3) act as the nucleators and influence the isothermal crystallization and melt behavior, spherulite growth rate and mechanical and thermal characteristics of PP [94, 95]. Addition of such oxides (with fraction PP/PEO = 100:1) tends to increase the α-form crystallite sizes and PP crystallinity. It is worth noting that Y_2O_3 and Nd_2O_3 are also the nucleating agents for the β-form of isotactic PP under isothermal conditions [96].

The rheology of polypropylene/MMT hybrid materials has been studied in sufficient detail (e.g., see [97]). It was then melt mixed with polypropylene and maleic anhydride (0.43 wt.%) functionalized by a polypropylene additive. The polymer–clay hybrid materials were prepared by melt mixing of organophilic clay, polypropylene and compatibilizer. The organophilic clay was prepared by cation exchange of natural counterions with amine surfactant by a very convenient method [57]. The interlayer spacing of natural clay MMT (initial value 1.1 nm) is changed to 2.1 nm after amine (stearylamine) treatment and reached 2.9 nm in a nanocomposite with 4.8 wt.% of inorganic content. MMT was mixed with PP by melting in the presence of a compatibilizer. It resulted in an intercalation or exfoliation, depending on the compatibilizer type. The effectiveness of the MMT layer exfoliation is influenced by the viscosity of the compatibilizer and the interfacial tension modified by the existence of

[1] Three crystal forms (α, β and γ) of isotactic PP are known. The α-form is the most stable, while the β- and γ-forms of PP are usually produced under special conditions or by using special nucleators (for β-form crystallization).

functional groups. The exfoliation was not so effective for the higher MMT concentration (6 and 10 wt.%). Some mechanical properties of PP with the MMT clay (PP/PP-gr-MA/MMT 87:10:3) are presented in Table 10.2.

Table 10.2. Mechanical properties of PP/MMT composites with 3 wt.% of MMT

Composition	Young's Modulus, MPa	Max. Stress, MPa	Elongation to Break, %
PP	0.96	30.7	1100
PP/MMT	1.08	33.0	16
PP/PP-gr-MA(PB3200)/MMT	1.20	35.1	12
PP/PP-gr-MA(PB3150)/MMT	1.04	31.5	11
PP/PP-gr-AA/MMT	1.18	33.8	14

The composites with PP break immediately after necking, with elongation at rupture less than 20%. At present this field of polymer nanocomposites is progressing rapidly and there is a lot of interesting information in patents and journals.

10.4 Polymer Matrix Structurization in Nanocomposites

Studies of interfaces in composites are very important because the internal contact surfaces of the various components play a special or determining role in the fabrication of materials with the tailored characteristics. This transitional region can be considered as a particular nanocomposite component with special properties different from the properties of the polymer matrix and nanoparticles. Even small particles can considerably change the polymer structure [98]. The depth of the structural changes depends on the nanoparticle concentration. The structural model of nanocomposites includes four main parameters: (a) an initial polymer matrix (PM); (b) the metal-containing particles; (c) a polymer connected with the particles into the surface (border) space and (d) the border space itself. The (c) component has a complicated structure, including both the absorbed near-surface layer and a series of transitional layers with specific characteristics (their structures and properties change with the distance from the nanoparticle and are distinct from the polymer matrix properties). Thus, the two-site model was proposed for the compositions of TiO_2 or $CaCO_3$ with poly(dimethylsiloxane) (PDMS). The process goes on within the particle pores in such a slow-relaxing polymer, while the faster relaxation is due to polymer immobilized in between the particles [99].

The nanoparticles introduced into the polymer matrix from a solution or crystalline melt polymers can change the relations between the amorphous and crystal phases. In the nanocomposite transient region of developed crystalline structure is formed and the nanoparticle structurizing power improves thermodynamic conditions for the crystallization of molecular chains into the transition layer, and as a result many of the important performance characteristics are increased (impact strength, tensile resistance, cleavage resistance). At the same time in PDMS–$BaTiO_3$ systems the rheological and dielectric behavior can be explained in terms of the break-up of colloidal particle aggregates under a shear field [100], when the aggregates size and structure is changed depending on the shear rate (Fig. 10.5). A relationship is traced between the rheological properties and the structure: the chain-like aggregates of $BaTiO_3$ particles can be formed by a dipole–dipole interaction. For comparison it can be noted that the dipole–dipole interaction in hybrid polymers influences their thermal characteristics and T_g values [101–104].

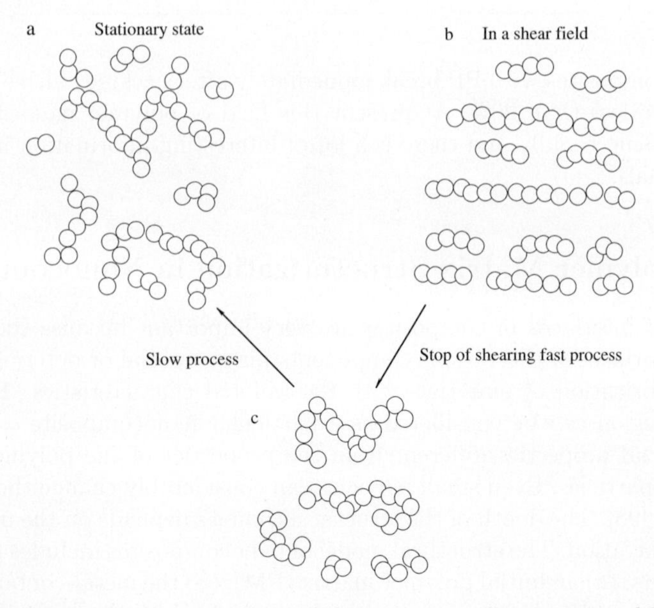

Fig. 10.5. Schematic representation of the structure of aggregates of $BaTiO_3$ in suspensions: (**a**) the stationary state; (**b**) in a shear field; (**c**) after stopping the shearing

The structurization processes in the polymer phase can be considered as the following: (1)nanoparticle coagulation by their mutual interaction especially at high concentrations; (2) morphological changes in a polymeric matrix itself at low nanoparticle concentrations without any coagulation formations are absent). Such transformations usually lead to increasing mechanical and thermomechanical parameters (i.e. temperature of thermal destruction, glass transition) that can be connected with the suppression of the molecular mobility, interaction with the particle surface, and the forming of secondary structure. It is important that when the nanofiller fraction increases considerably, the whole polymer matrix volume gradually transforms into a border layer "state" and this volume should be considered under special conditions. This transition is similar to a phase transition and is accompanied by a change in the nanocomposite properties for both the particle concentration increase and polymer fraction increase into the border layer. High concentrations (about 80–95 vol.% of nanofiller content) of nanocomposites have special relations of nanocomposite-interface volume. If a polymer layer is very thin, then polymer breakdown proceeds in the system and direct contacts between the nanoparticles are formed.

The mechanical behavior of the composite is determined by the polymer matrix. The influence of metal nanoparticles appears mainly in the quantitative changes of characteristics (dynamometrical properties are shifted to the low-stress region, the initial modulus value is changed, rupture characteristics are decreasing). The main composite characteristics are connected with the nature and peculiarities of the polymer matrix, as well as with the nanofiller size and dispersity, and rupture characteristics. The interactions of the elementary units with the matrix and its surface on a molecular level for small particles are very substantial and lead to new effects connected with the macroscopic properties of the polymer matrix and the filler parameters. Thus, a strong interaction between nanoparticles and the polymer matrix, and an increase in the particle dispersity (the number of contacts increases) gives a dramatic increase of the mechanical characteristics and modulus of the composite [105, 106]. As a rule, the mechanical characteristics of crystalline polymers are changed over a wider range as compared with amorphous polymers. This can be explained by the change of the ratio between the ordered and disordered polymer regions.

There are many examples of such behavior, but we will give only the most representative ones. In so-called cluspols (a good abbreviation for clusters in polymers), obtained by salt thermal decomposition into polymer melts (PE, PTFE), as shown in Chap. 5, the mobility of the molecular chains is decreased and as a result the thermostability of polymer compositions increases[2] [108].

Nanocomposites based on a porous PP matrix previously elongated in adsorptive active media and spherical Ni particles (~30–40 nm) can be considered as a complicated heterogeneous system [109], for which the nonuniformity of the nanoparticle distribution must be taken into account, as well as the existence of the fibrillar structures able to various transformations. Fine palladium hydroxide dispersion in PVC under decomposition [110] was used to form network structures with the high-dispersed metal in the nodes

[2] Colloidal Pt and Pd, stabilized by polymers, are always stable with respect to heating [107].

which increases the temperature of the complete dehydrochlorination of the polymer. The nanoparticles formed in copper oxalate thermodestruction increase the polyamide crystallite sizes but decrease the general degree of structure crystallinity [111]. The crystallization water evolved in cadmium trihydrate oxalate decomposition can act as a specific plastisizer. It can accelerate the structural elements, decrease the cohesion energy and lead to loosening of the polymer matrix. The ferrum carbonyl π-complexes with polybutadiene–styrene copolymers in the region of softening temperatures can influence the segmental mobility of the polybutadiene microphase and can have an organizing effect even at concentrations <2 mass % of Fe, which results in an increase of the rigidity of the macromolecules [112]. At higher metal concentrations the effect becomes stronger due to the development of a physical network. For example, even at 1% content of a highly dispersed mixture (Fe + FeO) formed under decomposition of ferrous oxalate in PELD one can change the size of the structural elements, the density of the fluctuating network of the molecular "catching" and the ratio of the oriented polymer chains [113]. The formation of dispersed metal in situ (sometimes the method is called "rolling nanoparticles up into polymers") is now a promising modification of polymer processing methods [114].

Some equimolar quantities of bisphenol A and epoxy prepolymer in an emulsion medium were polymerized in the presence of Na^+-MMT [115]. The strong fixation of the polymer to the inorganic surfaces leads to the cooperative formation of the ion–dipole force acting between the polar functional groups of the polymer chain and the interlayer ions. The char yield (Fig. 10.6 (a, b)) is 1070 K for a (polybenzoxazine-MMT) nanocomposite and it increased by approximately 20% for 5% MMT content in the nanocomposite [31]. The thermal stability of the nanocomposites was improved by the presence of dispersed MMT nanolayers in comparison with the presence of pristine polybenzoxazine. This is probably connected with the MMT nanolayers acting as barriers to minimize the permeability of volatile degradation products out from the material. The isothermal TGA curves (Fig. 10.6, b) gave more evidence about the improvement of the thermal stability of polybenzoxazine in the presence of MMT (the same regularities were observed for other nanocomposites too [116–121]). Thus, the PLS nanocomposites exhibit many such advantages including:

(a) they have lower weight and density;
(b) they exhibit outstanding diffusional barrier properties without requiring a multipolymer layered design allowing recycling;
(c) their mechanical properties are potentially superior to unidirectional fiber reinforced polymers because reinforcement from the inorganic layers will occur in two rather than in one dimension;
(d) increased solvent resistance and thermal stability;
(e) they have fireproof properties.

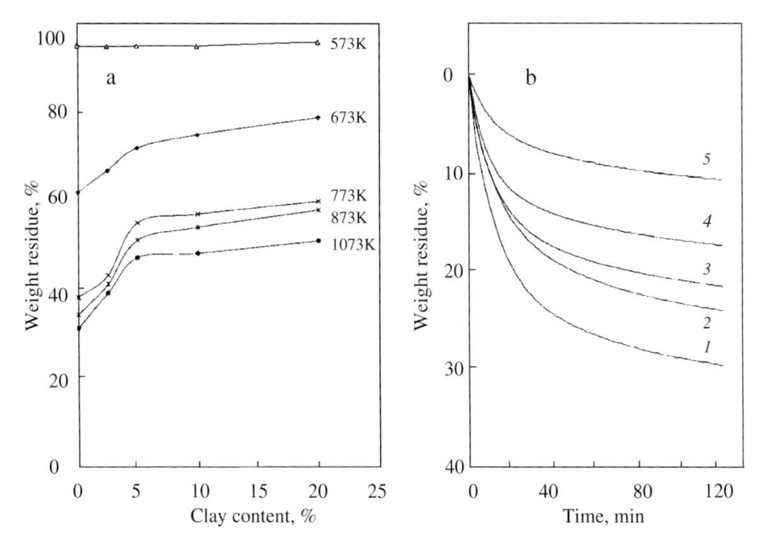

Fig. 10.6. Effect of stearyl-MMT content on the weight residue at different temperatures (**a**) and isothermal TGA of polybenzoxazine–stearyl-MMT nanocomposite with various stearyl-MMt content at 573 K: (*1*) 0%; (*2*) 2.5%; (*3*) 5%; (*4*) 10%; and (*5*) 20% (**b**)

Such polymer/clay hybrids will be used as new materials in automotive, packaging and aerospace applications [122].

The more rigid dispersed fillers – the nanosize particles introducing into elastomers or crystalline polymers – the more the initial nanocomposite modulus rise [105, 106]. For example [123], the TiO_2 surface can be modified against the aggregation by special apprets, i.e. by the Ti(O-*iso*–Pr)$_2$(OCO–CH_2–$C_{16}H_{33}$)(OCO–C(CH_3)=CH_2) type. These agents react with the hydroxyl groups on the TiO_2 surface, transforming them to Ti–OCO–CH_2–$C_{16}H_{33}$ and Ti–OCO-methacrylate hydrophobic groups. Such appreted nanoparticles disperse into SAS (sodiumdodecylsulfate) aqueous solutions and create micellar structures with an inorganic core particle. MMA polymerization allows one to obtain a substance with *core–shell* morphology and PMMA connected with TiO_2 by covalent bonds. This strategy is used for preventing nanoparticle aggregation (amorphous or crystalline, including high valence metal particles, such as oxides CeO_2 and ZrO_2) [124, 125].

Conventional methods of introducing fillers into elastomer matrices are based on mechanical mixing. These methods allow one to get particles of $<1\,\mu m$ in size [126], but an increase is achieved by larger amounts of filler (40–50 mass parts of technical carbon are used per 100 mass parts of elastomer). At the same time, in situ methods allow one, with different ultradispersed mineral fillers ($CaCO_3$, $CaSO_4$, $BaSO_4$, 1–40 nm in size), to achieve considerable changes of the elastomer structure with a substantial improvement of the material characteristics even at content 0.3–0.8 mass % in composition.

Thus, modification of the elastomer matrix by internal synthesis of nanoparticles ($CaSO_4$, particle size 1–4 nm) increases the mechanical properties of the polymer, decreases the relaxation rate and increases cohesion strength not only for rubber, but for its derivatives too. The Mooney viscosity increases from 65 to 77 (at 373 K), the flow index decreases at the same temperature from 0.45 to 0.29, the SRRT rigidity rises from 21.4 to 45.7, while the SRRT relaxation rate decreases from 0.86 to 0.59 and the cohesion strength increases by three times (from 0.14 to 0.42). The influence of the nanoparticles on the structure, viscoelastic and relaxation properties of the rubbers and resins on their base are connected with polymer structure regularity and its molecular-mass distribution [127, 128]. The influence of nanofiller particles on the elastomer structure and properties is greater for a low regular microstructure and a low molecular mass. An even more considerable effect is achieved in isoprene and butadiene synthetic rubbers if colloid metals are used as the physical modificators.

It is of interest to form nanoparticles with the participation of both the polymer matrix and a previously introduced polymer, being able to control the particle growth, size and morphology (so-called polymer-controlled nanoparticles growth). For example, $CaSO_4$ nanoparticles can be formed through a complex of $CaCl_2$ and PEO (mol. mass $3{\cdot}10^5$), dissolving $CaCl_2$ and PEO and methanol and adding stoichiometric amounts of K_2SO_4 [129]. In the formed structure the monoclinic orthorhombic structure of a dehydrated type is characteristic (as distinct from orthorhombic structure in the anhydride type in commercial samples). The crystal morphology and the nanoparticle size distribution (the sizes decrease with the polymer concentration) depend on the ratio of precursor concentrations. Such regularities can be explained by the strong interactions in the considered systems.

The polymer matrix can undergo significant chemical transformations leading to partial chain destruction and cross-linking, to changes corresponding to thermodestruction. Usually the interaction of metal atoms proceeds with the products of matrix decomposition or oxidation. Such regularities must be taken into account at the design and fabrication of metallopolymer nanocomposites.

10.5 The Physical and Mechanical Properties of Metallopolymer Nanocomposites

Obviously the quantitative relations between nanoparticle concentration (usually the volume ratio φ is used as the main parameter) and the physical and mechanical characteristics of nanocomposites are of great importance. The widely used evaluation of the effective elastic properties of traditional composites are based on the linear equations of elasticity theory with various methods of averaging or special boundary conditions. This description of the effective elastic properties proceeds from the given parameters of the composite components (matrix and filler) and their relations, but the approach does not take into account their interaction on a molecular level. The Kerner equation [130] is widely used for the analysis of filled systems, describing the elastic modulus under simple tension over the whole range of the φ parameter as follows:

$$\frac{E'_{\mathrm{c}}}{E'_{\mathrm{o}}} = \frac{G'_F \varphi_F / [(7-5\nu)\,G'_{\mathrm{o}} + (8-10\nu)\,G'_F] + (1-\varphi_F)\,/[15\,(1-\nu)]}{G'_{\mathrm{o}} \varphi_F / [(7-5\nu)\,G'_{\mathrm{o}} + (8-10\nu)\,G'_F] + (1-\varphi_F)\,/[15\,(1-\nu)]}\,, \quad (10.1)$$

where E'_{c} is the elasticity modulus of the composite; E'_{o} is the matrix elasticity modulus; ν is the matrix Poisson coefficient; G'_F is the real part of the filler shear modulus; G'_{o} is the real part of the nonfilled matrix shear modulus; and φ_F is the volume ratio of the filler. This equation cannot describe systems with a strong interaction between the nanoparticles and the matrix, connected with changes in the parameter φ_F. If the polymer is closely coupled with the nanoparticles on an interface (a phase boundary, formed under the specific physico-chemical interactions between nanofiller and polymer), the volume ratio of “nanoparticle + interphase material” (parameter φ_e in Fig. 10.7) becomes greater than φ_F for some factor. The real volume ratio can be defined from the equation

$$\varphi_e = \varphi_F(1 + \Delta R/R)^3\,, \quad (10.2)$$

where the ratio $\Delta R/R$ corresponds to the relative increase of the spherical particles radius.

This modification allows one to match the better correspondence of the experimental results with the Kerner equation [131], and in particular, for

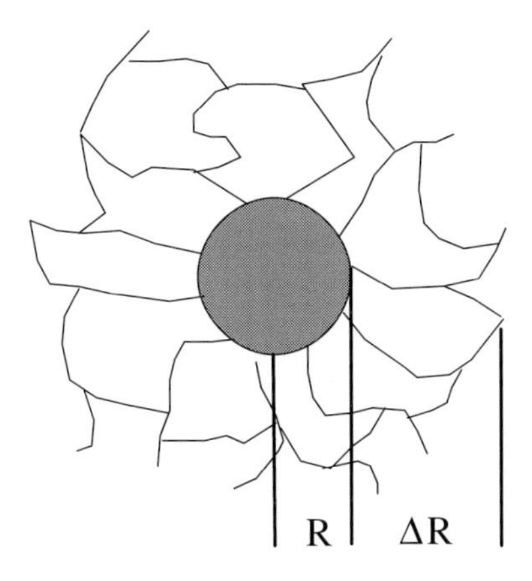

Fig. 10.7. The structure of cross-linked PMMA, surrounding a Pt particle

elastomer matrices, with $(1 + \Delta R/R)^3 = B$, $\nu = 0.5$ and $G'_F \gg G'_o$, the equation takes the form

$$E'_c/E'_o = (1 + 1.5\varphi_F B)/(1 - \varphi_F B)\,. \tag{10.3}$$

For example, the relative increase of the elasticity modulus of nanocomposite (on the base of PMMA and Pt particles with interphase boundary width $\Delta R = 22.5\,\text{nm}$) as a function of the logarithm of the real volume ratio (φ_e) is described [132] by this equation (Fig. 10.8).

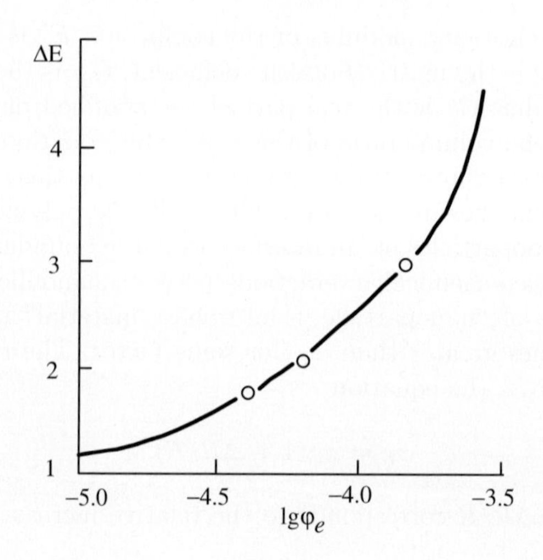

Fig. 10.8. The dependence of the relative increase of the modulus of the nanocomposite (on the base of PMMA and Pt particles with interphase boundary width $\Delta R = 22.5\,\text{nm}$) as a function of the logarithm of the real volume ratio (φ_e)

Tsai–Halpin equation is commonly used to predict the elasticity modulus of a discontinuous short fiber reinforced composite from the modulus. of the individual components [133]. If the fiber cross-section is circular, the fibers are arranged in a square array and are uniformly distributed throughout the matrix and moreover the matrix is free of voids, and the equation

$$E_c/E_m = (1 + \xi\eta\varphi_f)/(1 - \eta\varphi_f) \tag{10.4}$$

is correct, where E_c and E_m are the elastic moduli of the composite and matrix, respectively; φ_f is the volume fraction of short fibers, and the constants ξ and η are given by

$$\xi = 2(L/D), \eta = (E_f - E_m)/(E_f + \xi E_m)\,, \tag{10.5}$$

where L/D is the aspect ratio (length/diameter) of the reinforcing fibers, and E_f is the modulus of the fibers. Many nanocomposites satisfied the Tsai–Halpin equation, although their length is relatively short.

Such modification of the equations seems to be a formal one because it does not take into account any real mechanisms of the microscopic interactions between the matrix elements and the filler particle surface.

Sometimes the nanocomposite properties are described by an even more simplified equations, e.g. the Smallwood equation

$$E_c = E_p(1 + 2.5\varphi + 14\varphi^2) \tag{10.6}$$

or the Takayanagi equation

$$E_c = E_p E_f / (E_p \varphi_f + E_f \varphi_p) \,, \tag{10.7}$$

where E_c, E_p, E_f are the mechanical moduli of the composite, polymer and filler, correspondingly.

From contact-angle analysis, some thermodynamic parameters (such as surface tension) can be obtained as a function of the organic–inorganic composition of the hybrid materials. According to the theory of fractional polarity [134] the various molecular forces are linearly additive and therefore the surface tension (γ, $mJ{\cdot}m^{-2}$) can be separated into two components: the dispersion (γ^d) and polar (γ^p) components:

$$\gamma = \gamma^d + \gamma^p \,.$$

The cohesive energy of polymers is much lower than that of the nanofiller. The contact angle and surface tension parameters indicate that hybrid fillers have better wetting properties with the polymer (Fig. 10.9) and stronger interfacial bonding with the polymer matrix, than pure silica fillers.

In recent years there have been many attempts to calculate specific interactions between macromolecule segments and active centers of nanoparticles to describe the viscoelastic properties of the composite. For example, on the basis of scaling theory, the mechanical properties of colloidal particles and the macromolecule complex were described [135] in equilibrium, when the segments are partially adsorbed on the active centers, which are the complex (polyconjugated) nodes of the nanocomposite [136, 137]. Under a deformation or temperature change some of the segments leave the particle surface and as a result the total number of segments connecting the surface of the two particles increases. This process leads to two nontrivial conclusions: 1) the nanocomposite elastic properties, as well as its volume filling ratio, are inversely proportional to the particle sizes; 2) in extrapolation of the temperature to zero the nanocomposite shear modulus does not tend to zero (which is opposite to the situation for nonfilled polymeric networks). These conclusions correspond well with the experimental results of the relaxation of mechanical properties and with the dependence of the composite viscosity on the structural and molecular parameters, such as the energy of the interaction of the macromolecule segment with nanoparticle active centers, the number and sizes of segments, the volume filling ratio and particle size ($D \leq 30$ nm),

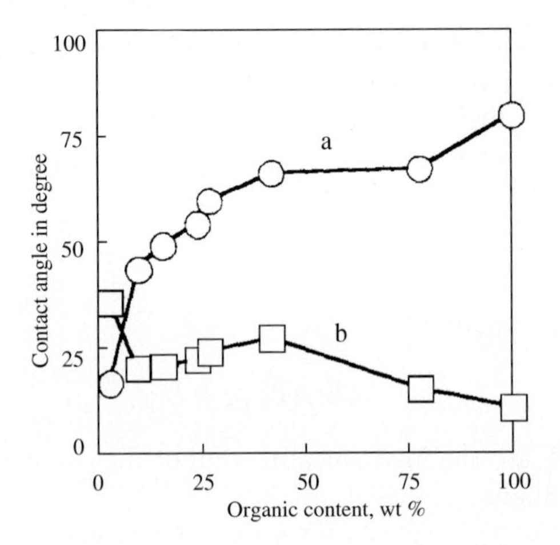

Fig. 10.9. Contact angles of (**a**) water and (**b**) the dental resin (2,2-bis(*p*-2-hydroxy-3-methacryloxypropoxyphenyl)propane/triethylene glycol dimethacrylate) on the surface of polymethacrylate–silica hybrid materials at various organic contents

and the temperature [136]. All this supports the general idea that nanohybrid materials are macroscopically homogeneous, though they are obviously inhomogeneous (microheterogeneous) on a molecular level. The physical and chemical characteristics of microheterogeneous materials are quite distinct from the properties of mixed composites of the same composition. This is connected with the specific inhomogeneity dimensions (comparable with macromolecule sizes) that leads to a high developed surface of the phase interface, where these hybrid microheterogeneous material properties are formed [138]. These results can be used for the optimization of the mechanical properties of nanocomposites, especially rubbers [139–141].

10.6 Nanocomposites in Adhesion Compounds (Contacts) and Tribopolymers

Practically all methods of nanocomposite fabrication are connected with the deep and strong interactions between the metal atoms and the matrix, especially on the interface boundaries formed in situ. The formation of chemical bonds between the metal atoms and the separate functional polymer groups creates a high adhesion strength because the physical interaction (due to the van der waalse or electrostatic forces) gives only weak adhesion [142–144]. Studies of the metallization of the polymer surface layer have shown that there is a strong correlation between the metal reactivity and its diffusion behavior [145]. The metallization is accompanied by a strong and reliable adhesion of the metal coating, but due to inertness of the polymers in question, conventional methods often fail to provide coating with sufficient adhesion. Some metals such as Cu and Au were chosen because their adhesion to untreated polymer surfaces is expected to be very poor as a result of the low reactivity.

One of the methods for strengthening the interphase interactions in the PEHP–steel system is based on introducing polyfunctional modificators (additives), in particular primary aromatic amines [146]. Under the coating formation process the amine interacts with PE by both an addition mechanism of the multiple end bonds and by the formation of hydrogen bonds with the PE oxygen groups, which arise during its thermooxidation (in this way the thermooxidative processes are simultaneously inhibiting). The amine reacts with steel by acid–base interactions due to a donor–acceptor mechanism that allows the raising of the adhesion ability with sacrifice of the chemical bonds between the adhesive and adherent by the amine. In this case a free surface energy (γ) increase is observed for the modified polymer (being in a contact with steel). The copolymers of the poly(ethylene-*co*-acrylic acid) type have been used for more than 30 years (usually they contain about 15 wt.% of acrylic acid) for enhancing adhesion of PE to a variety of inorganic surfaces, such as Al, Cu, Pb, Fe, and Si [147–151]. The adhesion strength of photosensitive epoxy with a copper or ceramic substrate was (depending on the conditions and the copper content in the composition) about 6.0–10.5 MPa for copper and 15.0–22.7 MPa for ceramics [152]. Strong adhesion was observed in the blend Al–foil–rubber (epichlorohydrin or carboxylated nitrile rubber), and besides, the changes in hysteresis loss due to the Al foil can be correlated with the peel strength of Al–rubber–Al joints [153]. The time of adhesion connection between polymer and aluminum (or stainless steel) with a deposited coating (formed by plasmopolymerization, see Chap. 6) increases [154] by several times (Table 10.3).

Table 10.3. The time (hrs) of destruction of an adhesion bond (the thickness of the plasmopolymerization coating is 200 nm) in water at 334 K [154]

Substrate Material		Plasma Polymerization Coating							
A	B	Without Coating	Methane			Ethylene			Acetylene
		1*	1.0*	2.0*	10.0*	1.0*	2.0*	10.0*	2.0*
PE	Al	20	180.5	178	38	40	53	19	23
PE	HC	20	169.5	141	19	79.8	65	19	23
PTFE	Al	20	103.7	127	28.5	40	32	19	23
PTFE	HC	20	129	103	19	79.8	91	19	23

* – the monomer feed rate, mL·min^{-1}.

One of the main directions of metallopolymer nanocomposite fabrication in situ is connected with tribology, i.e. with so-called friction polymer fabrication of polymeric antifriction coatings or lubricating material components. Such studies are important for increasing the antifriction and antiwearing characteristics of the friction nodes and units in numerous machines and mechanisms [155].

Polymers are used as low-friction materials, coatings or lubricating fluids in a growing number of mechanical devices where low adhesion, friction and wear are important factors for determining their performance and lifetime. There are two main variants: the polymer surface may slide against another polymer surface (a symmetric system [156]) or against another material (an asymmetric trybological system). In the latter case one must distinguish between the sliding surface and the nominally stationary substrate surface.

The friction of polymer materials is a very complicated process with the conditions changing with time in the widest ranges of the temperature, rates, deformations, chemical compositions and physical properties of the surface layers, including complex tribochemical reactions [157], triboelectrization [158–160] and changes of the real contact surface area. In practical conditions there always exist some interlayers (lubricants) or other adsorbed films (they are called servovitic films) [161, 162]. Wearing results in material destruction throughout the friction surface. The surface layers, by a cohesion–adhesion mechanism, impose severe limitations for the serviceability of most machines and devices. Polymer antifriction coatings (especially on PETF fibers, such as bearings, bushings and inserts from polyamides) are more wearproof than metallic ones and are now widely used in many friction nodes, especially in low-speed devices with small sliding rates. Some of these workpieces have very high tribotechnical characteristics: a wide temperature service range (170–570 K), low friction coefficient (0.04–0.05 without lubricants and 0.02–0.03 with lubricants), chemical inertness, and high processability [163,164]. However, the poor mechanical strength, excessive viscoelastic deformation under load and a high wear rate greatly limit the applications of PTFE in practice. An improvement of such coatings is connected first of all with widening the velocity range by introducing metal-containing fillers into the compositions. Usually for this purpose Cu, CuO, PbO and Pb_3O_4 are used in concentrations up to 30 vol.% (Table 10.4) [165]; or graphite, MoS_2, Al_2O_3 may be used.

Cu_2O and Pb_3O_4 additives reduce the coefficient of friction of the composites. Some other metals and their oxides can improve the thermal modes of the coating service and sometimes can even provide autocompensation of wearing. The possible control of the tribotechnical characteristics is connected with a specific self-organization of the friction system as a whole. For example, the industrial sliding bearings of nylon and α-iron have high service characteristics. The main characteristic is the coefficient of friction of the interacting surface system and now special standards (so-called

Table 10.4. The friction and wear test (in dry friction conditions) results of PTFE composites filled with metal oxides

Material	Friction Coefficient	Wear (mg)
PTFE	0.257	385.4
PTFE + 30% (v) Pb_3O_4	0.232	1.1
PTFE + 30% (v) Cu_2O	0.245	5.9
PTFE + 30% (v) PbO	0.261	125.6

Sliding speed: 1.5 m/s; sliding against GCr-15 bearing steel; Load: 100 N; Time: 30 min.

"start–stop" tests) are carried out for the value of the sliding friction of a magnetic head with magnetic discs at a given load (for digital high-density magnetic recording). The friction forces exhibit complex time, temperature, loading and velocity dependencies and cannot be properly described in terms of conventional parameters, such as the friction coefficient or shear stress [156]. Some metals (Cu, Au) were chosen because their adhesion to untreated polymer surfaces is expected to be very poor, as a result of the low reactivity. However, if Cu^{2+} macrocomplexes are added to PETF (as antiwear and antiscorching agents), tribochemical reactions proceed and some copper complexes are formed with the perfluorinated macroligands as the products of PETF destruction and provide metal transport between the friction surfaces [166]. Sometimes such reactions are called tribocoordination, and in some friction modes stationary regimes can be established in systems due to such auto-oscillating chemical reactions under tribocoordination. Copper film, formed from single atoms under continuous deformation at the friction nodes, is supersaturated by vacancy-type defects (Fig. 10.10).

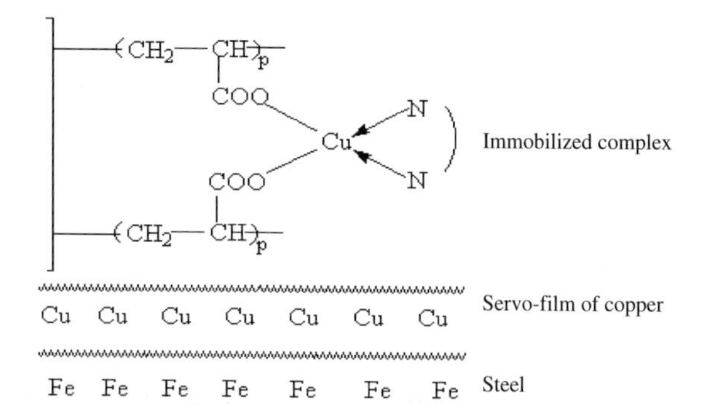

Fig. 10.10. The friction contact structure

Such behavior can change the film form without the accumulation of defects that is characteristic of fatigue processes. Such servovital film under wearless mode conditions can be considered as a quasi-liquid, providing wear protection of the friction surface (for example, steel) by localization of the defects in the copper film. Of course, the role of other lubricant components (especially, organic components) must be taken into account for the formation of such metal-polymeric films in such complex chemical conditions.

For electrochemical formation of tribological polymeric Ni-PTFE coatings (from an electrolyte-suspension mixture) a homocoagulation interaction is characteristic between the polymeric (about 10 nm in size) and metallic particles at the instant of deposition on an electrode [167]. Heterocoagulation is absent for low values (ξ = 9–20 mV) of the electrokinetic potential, but if the potential is higher the polymer deposition rate exceeds the metal electrochemical deposition rate. Therefore there is only a narrow range of conditions for the metal–polymer nanocomposite fabrication. In recent years an ion beam technique (see Chap. 6) was intensively developed and now it is very efficient in achieving good adhesion between materials and various types of polymers [168, 169]. In these metallization methods, many metal can be implanted at a depth of about 100 nm (the implantation depth in Cu and Fe samples is ~50 nm for poly(ether sulfone)/metal interfaces) with the metallization deposition rate kept at ~0.1 $nm \cdot c^{-1}$ [170]. Such interfaces of metallized, ion implanted and as-deposited high-temperature resistant polymer films are increasingly important for many modern applications. Ni nanoparticles are of interest as tribological additives in paraffin oils [171].

Special analysis is needed for so-called metal-clad lubricators, and for liquid or plastic lubricants those tribotechnical characteristics are connected with the metal particles (usually nanoscale) or their compounds, included into the lubricant composition and providing wear autocompensation. Use of such lubricators in friction nodes and metal processing allow one to form servovital films on the working surfaces, to decrease the friction coefficients and to realize a diffusion-vacancy exchange mechanism of the deformation, but this problem is far outside the scope of our considerations.

10.7 New Trends in Material Science Connected with Metallopolymeric Nanocomposites

In recent years new approaches have arisen in nanomaterial design, e.g. connected with bicomponent networks. One of the first such examples was the synthesis of three-arm star telechelic polyisobutylene [172] with the subsequent stoichometric addition of SiY-ditelechelic PDMS HSi-PDMS-SiH groups with the end allyl groups (~$CH_2CH{=}CH_2$) by a hydrosilation method (H_2PtCl_6 was used as catalyst), that allows us to get an average molecular chain weight between cross-links of ~20 000. In such nanocomposites a very efficient mixing of the PIB and PDMS domains takes place and the composites must have in no question the important properties. Another important research direction is based on the methodology of the fabrication of organic/inorganic hybrid materials in living or controlled polymerization, e.g. for control of the precise molecular mass, composition, functionality to inorganic substrate [173]. This approach seems to be very promising in the special controlled/living radical polymerization process [174,175], called atom transfer radical polymerization (ATRP), with simultaneous synthesis of the hybrid materials. In the method various colloidal particles are used, allowing for subsequent application as initiators for the ATRP of styrene, MMA, etc. The latest research survey is given in [176]. As a rule, colloidal initiators are prepared by inorganic surface functionalization, e.g. by the scheme of Fig. 10.11, where the nanoparticle serves as a unit, providing compatibility of the incompatible segments of various polymers.

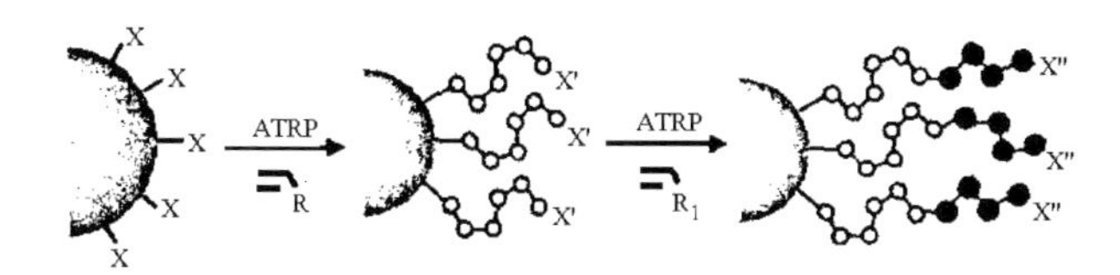

Fig. 10.11. Synthetic methodology to prepare hybrid nanoparticles (X–2-bromoisobutyrate groups)

Very interesting materials are membranes constituted from new hybrid organic–inorganic materials for gas separation, of thus poly(ethylene oxide-*bψ*-amide-6) or PEOA, containing silicon oxide as an inorganic filler and having gas (N_2, CH_4, O_2, H_2, CO_2) permeability coefficients decreased as a function of the inorganic component content [60]. When TiO_2 was used as a filler, the gas permeability is considerably lower than that measured for films containing SiO_2 (80:20 PEOA/MO_2), and titanium dioxide based nanocomposites are more rigid than with SiO_2. The selectivity of PEOA/TiO_2 to CO_2/N_2 was higher than that of pure PEOA. These permeability and selectivity properties are a function of the morphological and mechanical characteristics of the films [177–179].

There are some very promising uses of hybrid nanocomposites. An effective way to improve the control polymer flammability and fire resistance has relied on the introduction of highly aromatic rings into the polymer structure. The applications are limited because of processing difficulties that might inadvertently affect the mechanical properties. The problem is to use the nanocomposites as a new, environmentally benign approach to improve fire resistance of polymers. Thus it was shown [180] that polyetherimide-layered silicate nanocomposites with increased chat yield can simultaneously raise the material fire retardancy. Improvements in the barrier properties of poly(ε-caprolactone) nanocomposites and the fire resistance of nylon-6/clay hybrids have been reported [181, 182].

Multilayered polymeric composites typically fall into one of two categories [183]. First, those that have only a few layers and function as a barrier or provide impact resistance [184]. As a specific example of adhesion control in this case, some carbonated beverage bottles have five polymeric layers that must remain adhered during use, yet be separable for recycling. Second, they may consist of hundreds of submicron layers, possessing unique optical properties [185]. For polymeric composites with specialized optical behavior to retain the advantage over traditional materials, the layers must remain adhered when processed into unique shapes. Difficulties in providing these types of controlled adhesion have required alternative designs and have led to significant delays in manufacturing.

Many of these substances can be used as novel biodegradable and biocompatible hybrid materials, especially those prepared by the sol–gel process, poly ε-caprolactone (PCL)-SiO_2 [186–188]. Such systems and their volume-interface ratio are the mass fractal (they have a fractal dimension depended on the synthesis conditions, e.g. $D_m = 1.6 - 2.0$) and the particle sizes (2.9–12.5 nm) are in agreement with a co-continuous two-phase morphology. The very interesting approach of "nonchain" inhibition of the thermodestruction of thermostable polymers [189] is based on oxygen bonding by the very effective acceptors, which are metal nanoparticles formed in a polymeric matrix (the oxygen acceptors are generated directly in the matrix).

Dental composites are superior to amalgams in several areas, such as esthetics, elimination of galvanic action, thermal conductivity, tensile strength, and toughness [190]. Introduced to dental clinical practices a few decades ago, dental composites are usually made using an organic polymer matrix, pure inorganic fillers and interfacial coupling agents. Silicon dioxide is one of the most common fillers in dental composites and it is prepared by sol–gel processing [191, 192]. The fillers improve the mechanical properties of the composites and reduce the volume shrinkage (often for this purpose spiro monomers are used). In particular, methacryloyl-functionalized polyhedral oligomeric silsesquioxanes (POSS) are a class of compounds which have received a considerable amount of interest for hybrid materials – incorporation within the resin matrix. Nanostructural hybrid materials of this type have decreased density and increased usage temperature [193]. It is supposed [194] that modification of methacrylate POSS-MA monomer and the oligomer can be used for fully formulated resins and highly filled systems. Work along this line is now in progress, directed towards preparing hybrid of nanocomposite dental restoraties. For the same purposes, a series of new polymethacrylate-silica chemical hybrid dental fillers is promising. They were prepared by the sol–gel reactions of poly[methyl methacrylate-co-3-(trimetoxysilil)propyl methacrylate] with $Si(OEt)_4$ [195]. Such TiO_2-based nanocomposites have wide applications in the production of ceramic pieces, electronic ceramics, and pigments (for lacquers and paints). Titanium oxides are widely used for pigment and opacifiers in prints, make-up foundations in cosmetics, and coating of various substrates as a UV-absorbed, photocatalytic, antibacterial or light-induced amphiphilic material. Organo-titanium acrylic polymers [196, 197] like titanium alkoxides (as distinct of stannum-organic ones) can form self-polishing coatings free of toxicity (release of biocide) and are completely nontoxic for the marine environment. Thus the combinations of polymers and nanoparticles, considered in this chapter, enables the preparation of a hybrid material composed of well-defined colloidal systems and copolymers capable of organization into distinct domains [198, 199].

References

1. M. Gross, W.H. Douglas, R.P. Fields: J. Dent. Res. **62**, 850 (1983)
2. L. Mascia: Trends Polym. Sci. **3**, 61 (1995)
3. *The handbook of nanostructured materials and technology* ed. by H.S. Nalwa (Academic Press, San Diego 1998)
4. F.S. Bates, G.H. Fredrickson: Annul. Rev. Phys. Chem. **41**, 32 (1990)
5. M. Lee, B.K. Cho, W.C. Zin: Chem. Rev. **101**, 3869 (2001)
6. D. Awnir, S. Braun, M. Ottolenghi: *Supramolecular Architecture. Synthetic Control in Thin Films and Solids. ACS Symposium Ser. N499* ed. by T. Bein (ASC, Washington 1992)
7. M. Gill, J. Mykytiuk, S.P. Armes, J.L. Edwards, T. Yeates, P.J. Moreland, C. Mollett: Chem. Commun. 108 (1992); Langmuir **8**, 2178 (1992)
8. S. Maeda, S.P. Armes: J. Colloid Interface Sci. **159**, 257 (1993); J. Mater. Chem. **4**, 935 (1994); Chem. Mater. **7**, 171 (1995)
9. R. Fitton, J. Johal, S. Maeda, S.P. Armes: J. Colloid Interface Sci. **173**, 135 (1995)
10. H. Ishida: Polym. Compos. **5**, 101 (1984)
11. *Interfaces in Polymer, Ceramic and Metal Matrix Composites* ed. by H. Ishida (Elsevier, New York 1988)
12. G. Kiss: Polym. Eng. Sci. **27**, 410 (1987)
13. F. Yang, R. Pitchumani: Polym. Eng. Sci., **42**, 424 (2002)
14. D. Tian, Ph. Dubois, R. Jerome: J. Polym. Sci. Polym. Chem. Ed. A **35**, 2295 (1997)
15. T. Nakatsuka, H. Kawasaki, K. Yamashita: J. Colloid Interface Sci., **82**, 298 (1981)
16. M. Tirado-Mirande, A. Schmitt, J. Callejas-Fernandez, A. Fernandez-Barbero: Progr. Colloid Polym. Sci. **104**, 138 (1997)
17. R. Krishnamoorti, E.P. Giannelis: Macromolecules **30**, 4097 (1997)
18. J. Ren, A.S. Silva, R. Krishnamoorti: Macromolecules **33**, 3739 (2000)
19. T. Lan, T. Pinnavaia: J. Chem. Mater. **6**, 2216 (1994)
20. Y. Kojima, A. Usuki, M. Kawasumi, A. Okada, Y. Fukushima, T. Kurauchi, O. Kamigaito: J. Mater. Res. **8**, 1185 (1993)
21. B.K.G. Theng: Clays. Clay Minerals **18**, 357 (1970)
22. P.G. Nahin: Clays. Clay Minerals **10**, 257 (1961)
23. I.A. Uskov, E.A. Kusnitsyna: Polym. Sci. USSR. **2**, 728 (1960)
24. K.A. Carrado, L.Q. Xu: Chem. Mater. **10**, 1440 (1998)
25. K.A. Carrado: Appl. Clay Sci. **17**, 1 (2000)

26. P.C. LeBaron, Z. Wang, T.J. Pinnavaia: Appl. Clay Sci. **15**, 11 (1999)
27. M. Zanetti, S. Lomakin, G. Camino: Macromol. Mater. Eng. **279**, 1 (2000)
28. M.A. Nour: Polimery **47**, 326 (2002)
29. M. Alexandre, P. Dubois: Mater. Sci. Eng. R **28**, 1 (2000)
30. M.H. Choi, I.J. Chung, J.D. Lee: Chem. Mater. **12**, 2977 (2000)
31. T. Agag, T. Takeichi: Polymer **41**, 7083 (2000)
32. M. Kato, A. Usuki, A. Okada: J. Appl. Polym. Sci. **66**, 1781 (1997)
33. G.S. Sur, H.L. Sun, S.G. Lyu: Polymer **42**, 9783 (2001)
34. P.C. LeBaron, T.J. Pinnavaia: Chem. Mater. **13**, 3760 (2001)
35. F. Ranade, N.A. D'Souza, B. Gnade: Polymer **43**, 3759 (2002)
36. X.H. Liu, Q.J. Wu, L.A. Berglund: Polymer **43**, 4967 (2002)
37. J.C. Huang, Z.K. Zhu, J. Yin: Polymer **42**, 873 (2001)
38. E.P. Giannelis, R. Krishnamoorti, E. Manias: Adv. Polym. Sci. **138**, 107 (1999)
39. G. Lagaly: Appl. Clay Sci. **15**, 1 (1999)
40. Z. Wang, T.J. Pinnavaia: Chem. Mater. **10**, 1820 (1998)
41. M.J. Binette, C. Detellier: Can. J. Chem. **80**, 1708 (2002)
42. Z.H. Chen, C.Y. Huang, S.Y. Liu: J. Appl. Polym. Sci. **75**, 796 (2000)
43. H. Ishida, S. Campbell, J. Blackwell: Chem. Mater. **12**, 1260 (2000)
44. S. Ray, A.K. Bhowmick: Rubber Chem. Technol. **74**, 835 (2001)
45. T.B. Tolle, D.P. Anderson: Compos. Sci. Technol. **62**, 1033 (2002)
46. C.S. Triantafillidis, P.C. LeBaron, T.J. Pinnavaia: Chem. Mater. **14**, 4088 (2002)
47. P.B. Messersmith, F. Znidarsich: Mat. Res. Soc. Symp. Proc. **457**, 507 (1997)
48. M. Parpaillon, G. Engström, I. Petterson, I. Fineman, S.E. Svanson, B. Dellenfalk, M. Rigdahl: J. Appl. Polym. Sci. **30**, 581 (1985)
49. D. Avnir, D. Levy, R. Reisfeld: J. Phys. Chem. **88**, 5956 (1984)
50. R.A. Vaia, H. Ishii, E.P. Giannelis: Chem. Mater. **5**, 1694 (1993)
51. J.W. Cho, D.R. Paul: Polymer **42**, 1083 (2001)
52. T. McNally, W.R. Murphy, C.Y. Lew: Polymer **44**, 2761 (2003)
53. M. Biswas, S.S. Ray: Adv. Polym. Sci. **155**, 167 (2001)
54. G.M. Chen, Y.M. Ma, Z.N. Qi: Scripta Mater. **44**, 125 (2001)
55. C.O. Oriakhi: J. Chem. Educ. **77**, 1138 (2000)
56. P. Reichert, H. Nitz, S. Klinke, R. Brandsch, R. Thomann, R. Mülhaupt: Macromol. Mater. Eng. **375**, 8 (2000)
57. M. Kawasumi, N. Hasegawa, M. Kato, A. Usuki, A. Okada: Macromolecules **30**, 6333 (1997)
58. S.C. Tjong, Y.Z. Meng: Polymer **40**, 1109 (1999)
59. Y. Dirix, C. Bastiaansen, W. Caseri, P. Smith: J. Mater Sci. **34**, 3859 (1999)
60. J.M. Mark, C.Y. Jiang, M.Y. Tang: Macromolecules **17**, 2613 (1984)
61. C.J.T. Landry, B.K. Coltrain, M.R. Landry, J.J. Fizgerald, V.K. Long: Macromolecules **26**, 3702 (1993)
62. C.J.T. Landry, B.K. Coltrain, D.M. Teegarden, T.E. Long, V.K. Long: Macromolecules **29**, 4712 (1996)
63. I.A. David, G.W. Scherer: Polym. Prepr. **32**, 530 (1991)
64. M. Toki, T.Y. Chow, T. Ohnaka, H. Samura, T. Saegusa: Polym. Bull. **29**, 653 (1992)
65. F. Suzuki, K. Nakane, J.-S. Piao: J. Mater. Sci. **31**, 1335 (1996)
66. K. Nakane, F. Suzuki: J. Appl. Polym. Sci. **64**, 763 (1997)

67. A.B. Brennan, B. Wang, D.E. Rodrigues, G.L. Wilkes: Inorg. Organomet. Polym. **1**, 167 (1991)
68. K. Nakanishi, N. Soga: J. Non-Crystalline Solids **139**, 1, 14 (1992)
69. L. Carrido, J.E. Mark, C.C. Sun, J.L. Ackerman, C. Chang: Macromolecules **24**, 4067 (1991)
70. H. Hatayama, T. Swabe, Y. Kurokawa: J. Sol-Gel Sci. Tech. **7**, 13 (1996)
71. K.A. Mauritz, C.K. Jones: J. Appl. Polym. Sci. **40**, 1401 (1990)
72. K.E. Gonsalves, G. Carlson, X. Chen, S.K. Gayen, R. Perez, M. Jose-Yacaman: Nanostruct. Mater. **7**, 293 (1996)
73. K.F. Silveria, I.V.P. Yoshida, S.P. Nunes: Polymer **36**, 1425 (1995)
74. J. Hyeon-Lee, L. Guo, G. Beaucage, M.A. Macip-Boulis, A.J.M. Yang: J. Polym. Sci. Polym. Phys. B **34**, 3073 (1996)
75. J. Mackenzie, Q. Huang, T.J. Iwamoto: Sol-Gel Sci. Tech. **7**, 151 (1996)
76. Y. Xia, B. Gates, Y. Yin, Y. Lu: Adv. Mater. **12**, 693 (2000)
77. F. Caruso: Adv. Mater. **13**, 11 (2001)
78. L.A. Utracki: *Polymer Alloys and Blends* (Oxford University, New York 1990)
79. N.G. Gaylord: Macromol. Sci. Chem. A **26**, 1211 (1989)
80. T.A. Blomstrom, P. Concise: In: *Encyclopedia of Polymer Science and Engineering* ed. by J.I. Kroschwitz (Wiley, New York 1990)
81. K. Nakane, J. Ohashi, F. Suzuki: J. Appl. Polym. Sci. **71**, 185 (1999)
82. R.A. Zoppi, S. Das Neves, S.P. Nunes: Polymer **41**, 5461 (2000)
83. Q. Hu, E. Marand: Polymer **40**, 4833 (1999)
84. N.S.M. Stevens, M.E. Rezac: Polymer **40**, 4289 (1999)
85. C.H.M. Hofman-Caris: New J. Chem. **18**, 1087 (1994)
86. *Polypropylene. Structure, blends and composites* ed. by J. Karger-Kocsis (Chapman& Hall, London 1995)
87. C.-M. Chan, J. Wu, J.-X. Li, Y.-K. Cheung: Polymer **43**, 2981 (2002)
88. C. Saujanya, S. Radhakrishnan: Polymer **42**, 6723 (2001)
89. M.Z. Rong, M.Q. Zhang, Y.X. Zheng, H.M. Zeng, R. Walter, K. Friedrich: Polymer **42**, 167 (2001)
90. Y. Kurokawa, H. Yasuda, M. Kashiwagi, A. Oyo: J. Mater. Sci. Letters **16**, 1670 (1997)
91. H. Wang, C. Zeng, M. Elkovitcn, L.J. Lee, K.W. Koelling: Polym. Eng. Sci. **41**, 2036 (2001)
92. J.M. Gloaguen, J.M. Lefebvre: Polymer **42**, 5841 (2002)
93. A. Pawlak, J. Morawiecz, E. Piorkowska, A. Galeski:Solid State Phenomena **94**, 335 (2003)
94. G. Qu, J. Liu, G. Tang: Polym. Mater. Sci. Eng. **7**, (6) 102 (1991)
95. J. Liu, G. Tang, G. Qu, H. Zhou, Q. Guo: J. Appl. Polym. Sci. **47**, 2111 (1993)
96. C. Ye, J. Liu, Z. Mo, G. Tang, X. Jing: J. Appl. Polym. Sci. **60**, 1877 (1996)
97. M.J. Solomon, A.S. Almusallam, K.F. Seefeld, A. Somwangthararoj, P. Varadan: Macromolecules **34**, 1864 (2001)
98. T. Wan, K.A. Taylor, D.L. Chambers, G.T. Susi;. In: *Metallized Plastics* ed. by K.L. Mittal (Plenum Press, New York 1991) V. 2, p. 81
99. F.M. Vichi, F. Galembeck, T.K. Halstead, M.A.K. Williams: J. Appl. Polym. Sci. **74**, 2660 (1999)
100. D. Khastgir, K. Adachi: Polymer **41**, 6403 (2000)
101. H. Xu, S.-W. Kuo, J.-S. Lee, F.-C. Chang: Macromolecules **35**, 8788 (2002)
102. R.K. Bharadwaj, R.J. Berry, B.L. Farmer: Polymer **41**, 7209 (2000)

103. H. Xu, S.W. Kuo, J.L. Lee, F.C. Chang: Polymer **43**, 5117 (2002)
104. L. Zheng, R.J. Farris, E.B. Coughlin: Macromolecules **34**, 8034 (2001)
105. Yu.S. Lipatov: *Physical chemistry of the filled systems* (Khimiya, Moscow 1977)
106. D. Mensen, L. Sperling: *Polymer blends and composites* (Mir, Moscow 1979)
107. C. Paal, C. Amberger: Ber. Dtsch. Chem. Ges. **37**, 124 (1904)
108. I.D. Kosobudskii, L.V. Kashkina, S.P. Gubin, G.A. Petrakovskii, B.N. Piskorskii, N.M. Svirskaya: Polym. Sci. USSR A **27**, 689 (1985)
109. V.V. Stakhanova, N.I. Nikonorova, G.M. Lukovkin, A.V. Volynskii, N.F. Bakeev: Polym. Sci. Ser. B. **34** (7) 28 (1992)
110. Yu. Khimchenko, L.S. Radkevich: Plast. Massy 53 (1975)
111. A.I. Savitskii, Sh.Ya. Korovskii, V.I. Prosvirin: Colloid. J. USSR. **41**, 88 (1979); **42**, 998 (1980)
112. V.S. Eremin, L.M. Bronshtein, A.F. Klinkykh, V.P. Dyachkova, V.S. Voishchev, P.M. Valetskii, S.V. Vinogradova, V.V. Korshak: Dokl. AN USSR **301**, 896 (1988)
113. N.I. Mashukov, L.G. Kazaryan, A.E. Azriel, V.A. Vasilev, A.A. Zezina: Plast. Massy 18 (1991)
114. V.G. Elson, Yu.D. Semchikov, D.N. Emelyanov, N.L. Khvatov: Polym. Sci. USSR B **21**, 609 (1979)
115. D.C. Lee, L.W. Jang: J. Appl. Polym. Sci. **68**, 1997 (1998)
116. H.L. Tyan, C.M. Leu, K.H. Wei: Chem. Mater. **13**, 222 (2001)
117. N. Hasegawa, H. Okamoto, M. Kato et al.: Polymer **44**, 2933 (2003)
118. J.H. Park, S.C. Jana: Macromolecules **36**, 2758 (2003)
119. G.S. Zhang, Y.J. Li, D.Y. Yan et al.: Polym. Eng. Sci. **43**, 204 (2003)
120. D.M. Delozier, R.A. Orwoll, J.F. Cahoon et al.: Polymer **43**, 813 (2002)
121. Q.H. Zeng, D.Z. Wang, A.B. Yu et al.: Nanotechnology **13**, 549 (2002)
122. R. Dagani: Chem. Eng. News **77**, 25 (1999)
123. C.H.M. Caris, L.P.M. van Elven, A.M. van Herk, A.L. German: Brit. Polym. J. **21**, 133 (1989)
124. M. Chatry, M. In, M. Henry, C. Sanchez, J. Livage: J. Sol-Gel Sci. Technol. **1**, 233 (1994)
125. H. Vesteghem, A. Lecomte, A. Dauger: J. Non-Cryst. Solids **147/148**, 503 (1992)
126. R.S. Saifullin: *Physico-Chemistry of polymer and composite materials* (Khimiya, Moscow 1990)
127. B.S. Grishin, T.I. Pisarenko, G.I. Esenkina, V.P. Tarasov, A.K. Khitrin, V.L. Erofeev, I.R. Markov: Polym. Sci. Ser. A. **34**, 91 (1992)
128. L.N. Erofeev, A.V. Raevskii, O.I. Kolesova, B.S. Grishin, T.I. Pisarenko: Chem. Phys. Reports **15**, 427 (1996)
129. C. Saujanya, S. Radhakrishnan: J. Mater. Sci. **33**, 1063 (1998)
130. E. Kerner: Proc. Roy. Soc. B **69**, 808 (1956)
131. K.D. Ziegel, A. Romanov: J. Appl. Polym. Sci. **17**, 1119 (1973)
132. M. Kryszewski: Polimery **43**, 65 (1998)
133. J.C. Halpin, J.L. Kardos: Polymer Eng. Sci. **16**, 344 (1976)
134. S. Wu: J. Polym. Sci. C **34**, 19 (1971)
135. V.E. Zgaevskii: Dokl. Phys. Chem. **341**, 758 (1995); Colloid J. **57**, 679 (1995); Dokl. Phys. Chem. **363**, 42 (1998)
136. V.E. Zgaevskii., Yu.G. Yanovskii: Mechanics of compositional materials and constructions **4**, 106 (1998)

137. V.E. Zgaevskii: Dokl. Phys. Chem. **385**, 625 (2002)
138. A.D. Pomogailo, V.E. Zgaevskii, Yu.N. Karnet: Mechanics of compositional materials and constructions **8**, 351 (2002)
139. P. Dreyfuss, Y. Eckstein: Ind. Eng. Chem. Prod. Rev. Dev. **22**, 71 (1983)
140. D.C. Edwards: J. Mater. Sci. **25**, 4175 (1990)
141. V.E. Zgaevskii, A.D. Pomogailo, L.N. Paspopov: In: *First Russian Conference on Rubber and Resins* (Moscow 2002) p. 52
142. *Fundamtntals of Adhesion* ed. by H.-Y. Lee (Plenum Press, New York 1991)
143. N.J. Chou, C.H. Tang: J. Vac. Sci. Technol. A **2**, 751 (1984)
144. *Metallized Plastics. Fundamental and Applied Aspects* ed. by K.L. Mittal, J.R. Susko (Plenum Press, New York 1989, 1991, 1992) Vols. 1,2,3
145. F. Faupel, T. Strunskus, M. Kiene, A. Thran, C. Bechtolsheim, V. Zaporojtchenko: Mat. Res. Soc. Symp. Proc. **511**, 15 (1998)
146. O.V. Stoyanov, I.A. Starostina, N.A. Mukmeneva, R.Ya. Deberdeev: In: *The fundamental problems of polymer science, Intern. Conference* (Moscow State University, Moscow 1997) pp 2-79
147. C.K. Cho, K. Cho, C.E. Park:. J. Adhes. Sci. Technol. **11**, 433 (1997)
148. A. Stralin, T. Hjertberg: Surf. Interface Anal. **20**, 337 (1993)
149. P.J. Ludwig: Mod. Plast. **60**, 78 (1983)
150. Y.G. Aronoff, B. Chen, G. Lu, C. Seto, J. Swartz, S.L. Bernasek: J. Am. Chem. Soc. **119**, 259 (1997)
151. S.K. VanderKam, A.B. Bocarsly, J. Schwartz: Chem. Mater. **10**, 685 (1998)
152. S.-M. Lian, T.-H. Wang, A. Hung: J. Appl. Polym. Sci. **67**, 1639 (1998)
153. T. Bhattacharya, S.K. De: J. Polym. Sci. Polym. Phys. B **33**, 2183 (1995)
154. N. Inagaki, H. Yasuda: J. Appl. Polym. Sci. **26**, 3333 (1981)
155. *Friction and Wear of Polymer Composites* ed. by K. Friedrich (Elsevier, Amsterdam 1986)
156. A. Heuberger, G. Luengo, J.N. Israelachvili: J. Phys. Chem. B **103**, 10127 (1999)
157. G. Heinicke: *Tribochemistry* (Akademie-Verlag, Berlin 1984)
158. A.F. Klimovich, A.I. Sviridenok: *Electrical Phenomena at the Friction of Polymers* (Nauka and Tekhnika, Minsk 1994)
159. J. Szczerba, T. Prot, K. Paciorek-Gontek: Polimery **42**, 100 (1997)
160. Yu.F. Deinega, Z.R. Ulberg: *Electrophoretic Compositional Covers* (Khimiya, Moscow 1989)
161. Y. Ikada, Y. Uyma: *Lubricating Polymer Surfaces* (Technomic Publ. Inc., Lankaster, PA. 1993)
162. V. Stepina, V. Vesely: *Lubricants and Special Fluids* (Elsevier, Amsterdam 1992)
163. K. Fallmann: Techn. Rep. **9**, 16 (1982)
164. Y. Yamaguchi: *Tribology of plastic Materials* (Elsevier, Amsterdam 1990)
165. Z.-Z. Zhang, W.-M. Liu, Q.-J. Xue, W.-C. Shen: J. Appl. Polym. Sci. **66**, 85 (1997)
166. I.E. Uflyand, A.S. Kuzharov, M.O. Gorbunova, V.N. Sheinker, A.D. Pomogailo: React. Polym. **13**, 145 (1990)
167. N.M. Teterina, G.V. Khaldeev: Russ. J. Appl. Chem. **65**, 778 (1992); **67**, 1528 (1994)
168. N. Tegen, J. Wartusch, K.-H. Merkel: Nucl. Instrum. Methods Phys. Res. B **80/81**, 1055 (1993)

169. A.A. Galuska: J. Vac. Sci. Technol. A **10**, 381 (1992)
170. N. Tegen, S.A. Morton, J.F. Watts: J. Vac. Sci. Technol. A **15**, 544 (1997)
171. S. Qiu, Z. Zhou, J. Dong, G. Chen: J. Tribology **123**, 441 (2001)
172. M.A. Sherman, J.P. Kennedy: J. Polym. Sci. Polym. Chem. A **36**, 1891 (1998)
173. O.W. Webster: Science **251**, 887 (1991)
174. K. Matyjaszewski, J. Xia: Chem. Rev. **101**, 2921 (2001)
175. V. Coessens, T. Pintauer, K. Matyjaszewski: Prog. Polym. Sci. **26**, 337 (2001)
176. J. Pyun, S. Jia, T. Kowalewski, G.D. Patterson, K. Matyjaszewski: Macromolecules **36**, 5094 (2003)
177. D. Khastgir, H.S. Maiti, P.C. Bandyopadhyay: Mater Sci. Eng. **100**, 245 (1988)
178. M. Smaihi, T. Jermoumi, J. Marigan, R.D. Noble: J. Membr. Sci. **116**, 211 (1996)
179. J.C. Shrotter, S. Goizet, M. Smaihi, C. Guizard: Proc. Euromembrane (in Bath) **6**, 1313 (1995)
180. J. Lee, T. Takekoshi, E.P. Giannelis:. Mat. Res. Soc. Symp. Proc. **457**, 513 (1997)
181. P.B. Messersmith, E.P. Giannelis: J. Polym. Sci. Polym. Chem. A **33**, 1047 (1995)
182. J.W. Gilman, T. Kashiwagi, J.D. Lichtenhan: SAMPE J. **33**, 40 (1997)
183. P.J. Cole, R.F. Cook, C.W. Macosco: Macromolecules **36**, 2808 (2003)
184. A.J. Hsieh, J.W. Song: J. Reinf. Plast. Compos. **20**, 239 (2001)
185. M.F. Weber, C.A. Stover, L.R. Gilbert, T.J. Nevitt, A.J. Ouderkirk: Science **287**, 2451 (2000)
186. D. Tian, Ph. Dubois, R. Jerome: Polymer **37**, 3983 (1996)
187. D. Tian, S. Blacher, Ph. Dubois, R. Jerome: Polymer **39**, 855 (1998)
188. D. Tian, S. Blacher, R. Jerome: Polymer **40**, 951 (1999)
189. G.P. Gladyshev, N.I. Mashukov, A.K. Mikitaev, S.A. Eltsin: Polym. Sci. USSR B **28**, 62 (1986)
190. *Symposium on Dental Polymers, Composite and Adhesives. Polym. Preprint* ed. by J.M. Antonucci, A.B. Brennan, B.M. Culbertson **38**, 82 (1997)
191. M. Taira, H. Susuki, K. Wakasa, M. Yamaki, A. Matsuri: Brit. Ceram. Trans. J. **89**, 203 (1990)
192. M. Taira, H. Toyooka, H. Miyawaki, M. Yamaki: Dent. Mater. **9**, 167 (1993)
193. T.S. Haddad, J.D. Lichtenhan: Macromolecules **29,** 7302 (1996)
194. F. Gao, Y. Tong, S.R. Schricker, B.M. Culbertson: Polym. Adv. Techn. **12**, 355 (2001)
195. Y. Wei, D. Jin, G. Wei, D. Yang, J. Xu: J. Appl. Polym. Sci. **70**, 1689 (1998)
196. M. Camail, M. Hubert, A. Margaillan, J.L. Vernet: Polymer **25**, 6533 (1998)
197. L.E. Nielsen: *Mechanical Properties of Polymers and Composites*, 2nd edn. (Marcel Dekker, Inc., New York 1994)
198. R. Ulrich, A. Du Chesne, M. Templin, U. Wiesner: Adv. Mater. **11**, 141 (1999)
199. M.J. MacLachlan, I. Manners, G.A. Ozin: Adv. Mater. **12**, 675 (2000)

11 Magnetic, Electrical and Optical Properties of Metal–Polymer Nanocomposites

Metal–polymeric nanocomposites are now widely used as drug carriers, magnetooptical media, magnetic liquids, and information recording films. In particular, organic–inorganic hybrid composites [1–4], are very promising photonic materials (for optical data storage, optical waveguides, sensors, electrochromic smart windows, solid-state lasers and screen displays) and for other supramolecular and photonic systems and devices [5–7]. Engineering of colloidal surfaces is a current topic of applied chemistry in the field of developing new materials. Composite colloidal particles (core–shell structures) have various applications expected in the area of coatings, electronics, photonics and catalysis [8].

Various kinds of optical components are required for telecommunications networks in addition to light sources and detectors. The silica-polymer based technologies are convenient for these purposes. Improvement of information technology is connected with the use of innovations in material science. The new nanocomposites allow one to increase the recording density, the information exchange rate and fidelity (modern computers with changeable optical information storage have in fact infinite storage).

The electrochemical and electrochromic properties of nanocomposite films promise wide applications in electrochromic display panels, smart windows and rear-view mirrors for cars [9]. Metallopolymers based on complex ferroelectric oxides such as $BaTiO_3$, $Pb(Zr,Ti)O_3$, and $(Ba,Sr)TiO_3$ have applications in the electronics industry for transducers, actuators, and high-k dielectrics. The nature of the ferroelectricity effects in composites at nanosize level depends on the critical sizes of the particles and thin films [10]. Such nanoparticle assemblies open the door to "tunable" materials, in which optical and electronic properties depend on both initial cluster sizes and the manner in which clusters organize to form larger structures.

This chapter is a survey of various investigations in different fields of material science, united only by the idea of the study of nanoparticles and their use in polymer systems with special interest in their electromagnetical and optical properties.

11.1 Metallopolymeric Composites with Magnetic Properties

Magnetically responsive particles consist of a metal/metal oxide or metal alloy core, coated with polymer. Such encapsulated particles can be used in magnetic recording media in tape or disc form, oil spill clean-up and moldable magnetic powders and as toners in electrophotography [11,12]. The nanoparticles have unique physical characteristics, connected with surface or quantum-dimensional effects. The magnetic nanoparticles buried in a dielectric or nonmagnetic matrix have unusual properties such as a giant magnetoresistive effect [13, 14], and an abnormally great magnetocaloric effect.

There are three types of magnetic effects: diamagnetism, paramagnetism and cooperative magnetism (including ferromagnetism, antiferromagnetism and ferrimagnetism). Diamagnetic magnetization in the absence of an external field is negligible for the compensation of the orbital and spin moments of the substance. Diamagnetic atoms under an external magnetic field gain magnetic moments for the circular motion of all electrons and the resultant magnetization vector has a direction opposite to the external field and weakens it. In paramagnetics the spin and orbital magnetic moments are not partially compensated and therefore a nonzero resulting magnetic moment (for all atoms and molecules of the compounds) is observed. The mutual compensation of the moment in the absence of an external field takes place (for a distribution over the particle magnetic moment directions). In such conditions paramagnetics do not have a magnetic moment and are nonmagnetic, but under an external field the magnetic moments of the atoms and molecules are oriented and the substance as a whole becomes magnetizing in the external field direction. In this case an increase in the field intensity H ($\text{A}\cdot\text{m}^{-1}$) stimulates a proportional rise of magnetization $J = \chi\text{H}$, where χ is the magnetic susceptibility of the substance and $\chi = \lim_{\Delta H\to 0, H\to 0}(\Delta J/\Delta H)$. For diamagnetics $\chi < 0$, and for paramagnetics $\chi > 0$, its value is very low (about 10^{-3}–10^{-4}). Some substances are able to a strong magnetization even under weak magnetic fields and they are related to cooperative magnetics.

Ferromagnetics present a rather limited group of substances characterized by unfilled inner electronic shells and noncompensation of spins (in particular, for *d*-shells of transition metals). Among these are iron, cobalt,[1] chromium, gadolinium, dysprosium, some of their alloys (including manganese and chromium alloys) and some simple and complex metal oxides, for example ferrites, which are complex oxides of $Fe_2O_3\cdot MO$ type (M = Co, Ni, Mn, etc.). Ferromagnetic properties are shown only in certain crystal lattices, where a strong electrostatic exchange interaction appears, and is connected with the given types of lattice symmetry. This interaction in the *domains*

[1] Cobalt has the highest Curie temperature (1390 K) amongst all magnetic materials, its saturation magnetization at room temperatures is ≈1.79 T (≈1400 CGSM units cm^3 ≈160 emu·g) and its coercive force is ≈ 800 $\text{A}\cdot\text{m}^{-1}$ [15, 16].

(the small regions of space with order of size 10^{-3}–10^{-4} cm) is determined by a spontaneous magnetization for parallel orientation of the spins. The crystal space is divided by the boundaries (walls, having a limited width) of the regions with different directions of the magnetization vectors.[2]

Sometimes the interaction between the magnetic atoms in matter can give an antiparallel and mutually compensating arrangement of the neighboring atomic spins and results in the absence of a magnetic moment (it is called *antiferromagnetism*). If the resulting magnetic moment is not equal to zero, the noncompensated antiferromagnetism is called *ferrimagnetism* (the corresponding spin arrangements are shown at Fig. 11.1).

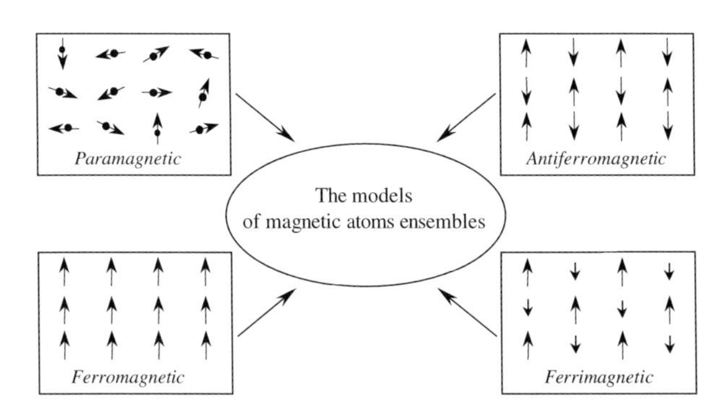

Fig. 11.1. Spin arrangement models

It is interesting to study the magnetic parameters of composites, such as the *magnetization curve* (i.e. the relationship between J and A or B and H, where B is the magnetic induction,[3] measured in Tesla, T). The relationship between magnetic induction (B), magnetic field strength (H) and magnetization (M) is given by $B = H + 4\pi M$, with B in gauss, Gs (1 Gs = 10^4 T), *relative permeability* ($\mu/\mu_0 = dB/dH = 1 + \chi$), *magnetic hysteresis*

[2] The magnetic properties of the domains depend strongly on temperature and its rise decreases the domain magnetization (as a result of the weakening of the exchange interaction). In the limit the domain structure breaks down and the corresponding temperature (when the energy of the atomic chaotic motion is equal to the exchange interaction energy) is called the Curie point. Its value does not depend on the structural defects of the material and is specific to every ferromagnetic.

[3] The magnetic induction can be defined as the magnetic flux density F, i.e. $B = F/S$ T. The B and J values are connected by the relation $B = \mu_o J + \mu_o H$, where μ_o is a magnetic constant or vacuum magnetic permeability and is equal to $12.8 \cdot 10^{-7}$ T·m/A. In the literature either SI or CGS units systems are used. In the latter case the magnetization M is measured in emu·cm^{-2} (electromagnetic units), the magnetic field strength H is measured in Oe (oersted), and the magnetic induction in teslas, T.

loop (a closed S-type curve on the Fig. 11.2 for a dependence of J or B on amplitude H for previously demagnetized material under a periodical and slow change of the magnetic field), hysteresis loop rectangularity coefficient ($\gamma = J_r/J_s$ is a ratio of the saturation residual magnetization J_r to a limiting magnetization J_s), *coercive force*[4] H_c (the magnetic field strength when the magnetization equals zero), etc. It allows one to get a full picture of the behavior of ferromagnetic materials.

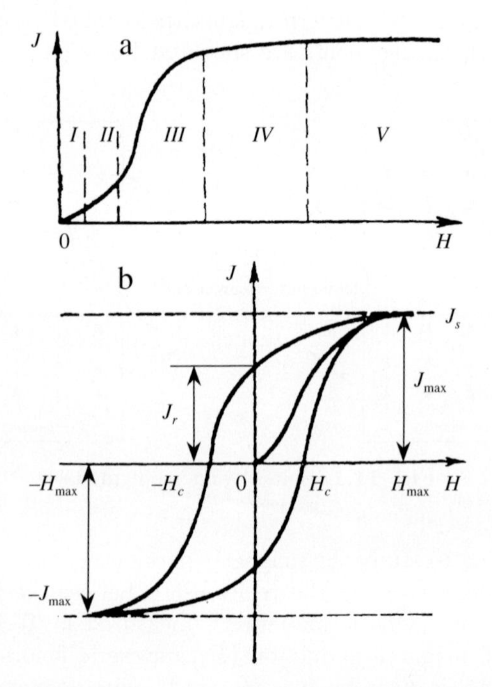

Fig. 11.2. Five regions of an initial magnetization curve: (**a**) *I* – elastic shift of the domain boundaries (a reversible magnetization process, weak fields); *II* – shift of domain boundaries (an irreversible magnetization process, strong fields); *III* – irreversible shift of domain boundaries (the region of maximum permeability); *IV* – the approach to saturation; *V* – saturation (dependence is vanishing); (**b**) a magnetization curve and a magnetic hysteresis loop for a ferromagnetic over one cycle + $J_{\max}$, + J_r, 0, – $J_{\max}$, – J_r, 0, + $J_{\max}$. The J_s value is the saturation magnetization (a maximal J value that does not change with an increase of field H); J_r is the residual saturation magnetization (at $H_c = 0$)

[4] Ferromagnetic materials, according to their nature, are divided into two groups. The first includes so-called magnetically soft ferromagnetics, e.g. crystalline or amorphous binary alloys Fe–Si, Fe–Ni, Fe–Co, etc. (with different dopants, such as Mo, Mn, V, Nb, Cu, Al). These materials have a low coercive strength (<4/m) and relatively low magnetic permeability.

The various characteristics of magnetic matrices depend on the diversity of interconnected factors (magnetic field direction, material prehistory, nature and temperature of the polymer, nanoparticle concentration and structure). These characteristics can be changed over wide limits [17–19].

Ferromagnetic metal particles can be formed in a polymer matrix (ferroplastics or ferroelastics) by practically all methods described in Chaps. 4–6. Many ferromagnetic polymeric nanocomposites are already synthesized with nanosize transition metal particles [20,21], ferrites [22] and nanosize magnetic particles of Fe [23], Co [24,25], Ni [26], γ-Fe_2O_3 [27–31], and Fe_3O_4 [28]. An ensemble of cobalt nanoparticles, isolated into a polymer matrix, has the same magnetic properties as in molecular beams. But the average magnetic moment per atom arises at 13–20% stipulated by the decrease in atomic coordination number, or an increase in the interatomic distances [32,33].

Magnetic nanocomposites were obtained also by introducing metal containing nanoparticles into dielectric zeolite [34], intercalated graphitic [35] and aluminosilicate [36] matrices, into the surface silica layer [37]. For example, using regulated thermodestruction of $Co_2(CO)_8$ into NaY zeolite voids (H_2, 470 K), nonoxidized cobalt particles (0.6–1 nm in diameter) can be formed [38] and into the molecular sieves MCM-41 (an impregnation by aqueous solution of $CoCl_2$, calcinations at 1020 K in an O_2 current) an antiferromagnetic material is formed in the amorphous silicate matrix (with periodic structure shown in Fig. 11.3), containing 6 mol.% of Co_3O_4 (with mean particle sizes 1.6 nm) [39]. Organosilane coatings can be formed, including α-Fe (70 at.% of Fe), magnetite Fe_3O_4 and Fe_2O_3 additives [40].

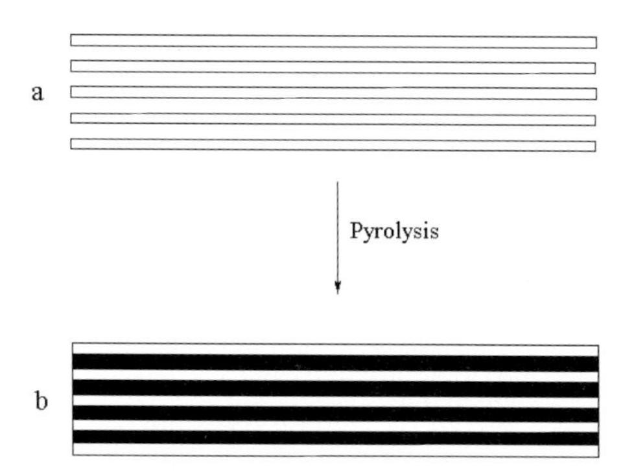

Fig. 11.3. Formation of nanostructured magnetic composites (**a**) via the pyrolysis of nanoscale cylinders of poly(ferrocenylsilane) within the pores of mesoporous silica (MCM-41) (**b**)

The various physical and mechanical parameters of metallopolymeric nanocomposites can be considerably improved with retention or sometimes even with increase of the conventional magnetic characteristics (saturation magnetization, coercive force, etc.) that show considerable promise for various

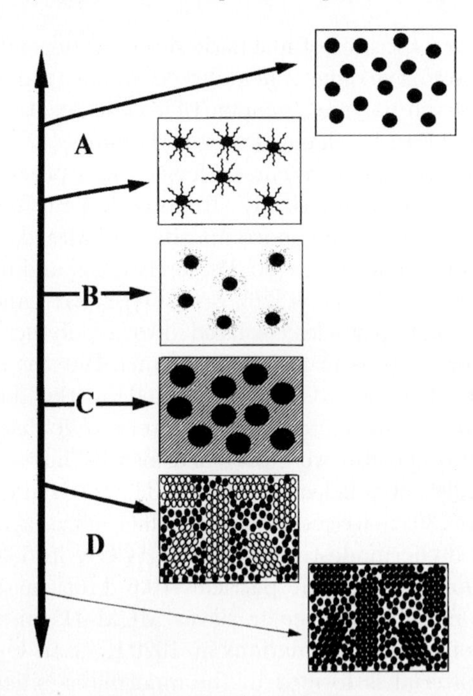

Fig. 11.4. Schematic representation of the different types of magnetic nanostructures (details in text)

applications. Such substances are very promising as high-density information media, materials for permanent magnets, magnetic cooling systems or fabrication of magnetic sensors. They can be used also for medical and biological applications, e.g. for magnetic resonance tomography, magnetic targeted drug delivery systems, magnetic resonance imaging (MRI) contrast enhancement, color imaging, and cell sorting [41,42]. There is a correlation between the size and form of nanoparticles and their magnetic characteristics. Small particles (<1 nm) are nonmagnetic, particles with diameter 1–10 nm exhibit superparamagnetic properties, and bigger ones are ferromagnetic (at least this is true in the case of Co particles, for which the change in magnetic properties shows itself as a change of the ferromagnetic resonance spectra parameters). There is a classification of the magnetic nanostructured materials (Fig. 11.4), based on the correlations[5] between nanostructure morphologies and magnetic properties [18,44].

Type A relates to systems of isolated particles with nanoscale diameters (noninteracting system) with usually unique magnetic properties; type B relates to materials with ultrafine particles and a core–shell morphology in the matrix; such a shell, in the real systems, is formed via oxidation and may itself be magnetic (e.g. Co-core/CoO-shell); type C relates to compositions

[5] There are another systems of magnetic materials classification described in [43].

of small magnetic particles embedded in a chemically dissimilar matrix; type D relates to materials consisting of small particles dispersed in noncrystalline (including polymeric ones) matrices. Type E nanostructure morphology represents a magnetic nanoparticle core with a polymer coating formed during polymerization (in situ polymerizable method). The typical situation relates to Fe, Fe_2O_3 or Fe_3O_4 nanoparticles, coated by a nanoscale layer of the polymer [45–48]. Thus, under MMA polymerization a core of mixed-phase iron oxides ~10 nm is formed (the total nanoparticle size is about 130 nm).

The magnetic properties of in situ polymerizable materials are better than those produced by the standard solid-state reaction method [49] (Fig. 11.5), but in the former case the nanoparticles may or may not include a magnetic core (the majority of polymer particles do not have a magnetic core). An original method was proposed for production of uniform material with mean particle size <15 nm [44]. It is based on the use of atom transfer radical polymerization (ATRP, see Chap. 10) of styrene and modified magnetic $MnFe_2O_4$ (≈9 nm) as chemically attached macroinitiators. For this purpose a reverse micelle microemulsion procedure can be used too [50]. Received core–shell nanoparticles show a core of 9.3 ± 1.5 nm with a 3.4 ± 0.8 nm polystyrene shell. Polymer particles without a magnetic core have not been observed. In such materials a decrease in coercivity was observed, which is consistent with the reduction of magnetic surface anisotropy upon polymer coating.

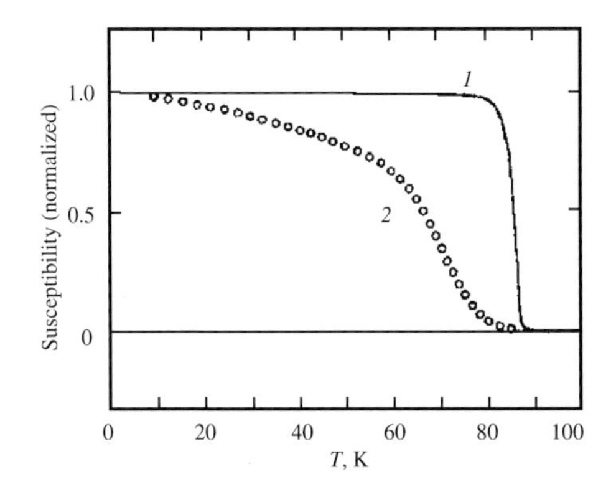

Fig. 11.5. Real parts of the complex magnetic susceptibility for $Bi_2Sr_2Ca_{0.8}Y_{0.2}Cu_2O_{8.2}$ prepared by the in situ polymerizable complex method (1) and by the conventional solid-state reaction method (2)

It is interesting to form polymeric matrices including Fe_3O_4 and γ-Fe_2O_3 nanoparticles (≤ 20 nm) by the exchange reactions of Fe-containing salts with some perfluorated ion-exchange membranes (such as Nafion® 117, Du Pont) with a subsequent alkaline hydrolysis of exchange product [51]. The sonochemical decomposition of $Fe(CO)_5$ was carried out in the presence of different surfactants, including polymeric surfactants, such as poly(vinylamine).

It allows us to form coated Fe_2O_3 nanoparticles 5–16 nm in diameter [52]. In another method [53] Co^{2+} were adsorbed on an ion-exchange resin and later reduced by an excess of $NaBH_4$ that allows one to form (Co–B) alloys nanocomposites, having a very wide distribution over the sizes (3–30 nm). These cobalt–boron alloys are intensively studied owing to their superior properties (such as high strength, high magnetic permeability combined with high resistivity and large resistance to corrosion). One other similar method is based on reduction of Co salts in a PVP (or THF) solution [32, 54].

A five times greater coercive force of (Co in PS)-nanomaterial can be achieved by low-temperature (below 720 K) annealing [27]. Cobalt carbonyl $Co_2(CO)_8$ was introduced into a styrene copolymer matrix from solutions (in DMF or in *iso*-propanol) and thermally treated at 470 K. The content of the spherical Co nanoparticles with mean diameter 2.6 nm increases to 8 wt.%, but this fact and the nature of the precursor had no considerable effect on the FMR (ferromagnetic resonance) spectrum [55]. The magnetic behavior of Ni–PELD and Co–PELD systems depends on the synthesis conditions [30]. In the former system at high temperatures (>370 K) the Ni clusters irreversibly interact with the matrix; in the latter (Co–PELD) the temperature rise to 470 K changes the thermomagnetic curve radically (as a result of coarsening of the Co particles).

The magnetic hysteresis of the samples obtained in a crazing mode of PP-composites filled by nickel shows that the coercive force does not depend on the magnetic field direction (the parallel and perpendicular directions to the stretching axis) and equals ~2,5 $kA{\cdot}m^{-1}$, though there is a considerable difference between the J_r and J_s values [56].

The magnetic behavior was studied for systems obtained by reduction of metal salts in various polymeric media, such as a swollen interpolymeric PPA complex with urea formaldehyde oligomer [57], P4VPy and styrene copolymer with 4-vinylpyridine [58]. In these cases the composites synthesized under a reduction of $FeCl_3$ molecules (coordinated with the pyridine groups) by hydrazine and their subsequent oxidation, contain spherical Fe_3O_4 nanoparticles (20–200 nm) with H_c values up to 31.5 $kA{\cdot}m^{-1}$. Nanocomposites with spherical γ-Fe_2O_3 nanoparticles (~5 nm in diameter), obtained by alkaline-hydrolytic "development" of $FeCl_3$ macrocomplexes in block-copolymers, also have similar characteristics [59].

Magnetic polymer composites were synthesized by thermodestruction of arenecyclopentadienyl-ferrum (or its arylene-carbonic oligomers) cationic complexes [61] or by a thermodestruction of Co and Fe carbonyls in PVF matrices [62]. Decarboxylated polymer composites were obtained by thermolysis of metal-containing carboxylate monomers (acrylates of Co^{2+}, Fe^{3+}, Ni^{2+} and 1:1 or 1:2 cocrystallizates of Co^{2+}-, Fe^{3+}-acrylates) (Fig. 11.6). The composites contain ferromagnetic particles (6–13 nm) uniformly distributed into a matrix volume with mean distance between them of ~8 nm [60, 63–65]. The magnetic characteristics of some such composites are given in Table 11.1.

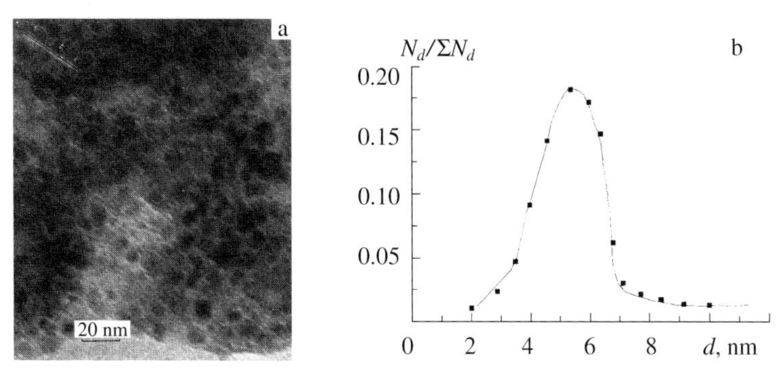

Fig. 11.6. (**a**) Electron microscope photograph of Co^{2+}–Fe^{3+}-acrylate 1:2-cocrystallizates and (**b**) its particle size distribution

Metal-containing polymer matrices with magnetic properties can be synthesized by hydrogen reduction of cobalt–organic precursors [66] and Ni carbonyl in epoxy films in various hardening modes [67], as well as by tetrahydroborate reduction of nickel compounds into stretched and crazed PP polymeric matrices (with Ni crystallite size <40 nm) [56]. In some other methods polymeric composites with monodispersed Co nanoparticles (with diameter 1 nm) were obtained by metal spraying [29] or by introducing magnetoactive particles into various gels, such as silicone [68], polyacrylamide [69], polyethyleneoxide [70], etc. [18].

Table 11.1. Magnetic properties of some metal acrylates ($MAcr_n$)

$MAcr_n$	T, K	$\chi_{sp}\cdot 10^5$, cm^3/g	$J_{sp}\cdot 10^3$, (kA/m) (cm^3/g)	H_c, kA/m	γ	Products
$FeAcr_3$	292	26.1	33.45	1.751	0.051	Fe_3O_4
	77	26.1	35.94	17.753	0.250	
$NiAcr_2$	292	16.7	14.36	0.712	0.020	Mixture of Ni, NiO, Ni_3C
	77	12.3	14.95	4.267	0.133	
CoFeAcr	292	41.2	26.16	49.756	0.310	$Fe_2O_3\cdot CoO$
	77	31.1	13.91	49.756	0.420	

Note: the product characteristics were determined by roentgenography and electron diffractometry.

It is known that only atoms in nanoparticles can have long-range order in magnetic nanomaterials, because it is very hard to order spatially the particles in themselves. On the other hand, there are approaches allowing the forming of some polymer-mediated self-organized magnetic nanoparticles with control of the assembly dimensions and thickness. For example, one of these methods relates to magnetic materials based on binary FePt [41] or CoPt [71] nanoparticles, formed by $Pt(AcAc)_2$, $FeCl_2$ reduction. Recently a method of $Fe_{50}Pt_{50}$ nanoparticle production using PVP or PEI was described [72]. Such multilayer self-assembled particles were localized on a PEI-modified silicon oxide surface. The formation rate of the annealed thin film depends on the assembly thickness and increases with the annealing time and temperature (the optimum is achieved during 30 min at 770–870 K). Such layer-by-layer assembled nanoparticles have great potential for practical applications.

Another very interesting method is the forming of polymer-immobilized particles *in statu nascendi* in magnetic fields when ferroplastics are first magnetically oriented in the superimposed fields. For example, neodium ferrite can be orientated in fields with magnetic induction of $3 \cdot 10^{-4} - 3 \cdot 10^{-3}$ and form the final ferroplastics after the subsequent mechanochemical treatment, such as thermal pressing [69,73–75], Brigman anvil treatment [18] or a combination of different actions [76,77]. Thus for Fe–Pt-based nanodevice fabrication ordered submicron ferromagnetic Co–B arrays (about 30 nm, aggregated up to 450–600 nm in size) were obtained [78] by $NaBH_4$ reduction, using the orientation in the magnetic field of the soluble macrocomplexes (Co^{2+} with starch). The driving of the magnetic field (1 T) and the mediation of soluble starch play a critical role in the formation of well-defined patterns. The prepared chains with composition $Co_{23}B_{10}$ exhibit strong ferromagnetic signals in the range 100–420 K. It is supposed that such a convenient strategy should be helpful in the production of ordered arrays of other magnetic materials. This process can be inspired by magnetotactic bacteria, which provide excellent templates to direct the synthesis of advanced materials. Such methods in principle allow the production of nanocomposites with maximal magnetic characteristics, especially if the anizometric form nanoparticles are treated, as for example magnetically hard barium and strontium ferrites, having a hexagonal crystalline lattice with monoaxial magnetic anisotropy (the single axis of easy magnetization in the monocrystal is directed perpendicular to the hexagonal basal plane). The orientation by linearly or circularly moving magnetic fields can be used practically for improvement of the electrophysical or magnetic characteristics of the composites and for the production of magnetic matrices with oriented chain-like structures that can be used in varnish and film technologies for information recording systems. The coating formation in the moving magnetic fields allows one to prevent the filler particle deposition and to concentrate them either nearly on the film surface or uniformly through the film volume.

A very promising method is connected with the electrochemical formation of modified thin (3–7 nm) magnetic coatings on the above mentioned self-assembling Langmuir–Blodgett films [79]. The charged amphiphilic monolayer can by used as a template to organize inorganic materials by adsorption of ionic species along the interface. A special technique for studying the unusual magnetic properties of so-formed objects (on the base of $Cu_3[Fe(CN)_6]_2{\cdot}12H_2O$) was demonstrated in [80]. The results show that those multilayers present a spin ordering below 25 K.

The coercive force of powder ferromagnetic materials depends on the particle sizes and forms, but the dimensional dependence of H_c is very peculiar. In systems with very big multidomain particles the coercive force increases with decrease of the particle size (Fig. 11.7). It achieves the maximum value $H_{c,\max}$ when the mean particle size reaches one domain dimension. For spherical particles the critical diameter (when the coercive force value equals $H_{c,\max}$) can be calculated from the following equation [81]:

$$d_{\mathrm{cr}} = (1.9/J_s)[10cI/(\mu_0 a N_p)]^{1/2} , \qquad (11.1)$$

where c is a constant depending on the lattice crystalline structure; I is the so-called exchange constant (for γ-Fe_2O_3, $I \approx 1.3 \cdot 10^{-20}$); a is the lattice constant; $N_p = -H_p/\mathrm{J}$ is a demagnetizaion coefficient (for a demagnetization field H_p, is directed opposite to the external field).

For monodomain Fe and Co particles the d_{cr} value is ~20 nm, and for Ni it is ~60 nm at $H_{c,\max} \approx 10^3$–10^4 A·m^{-1}.

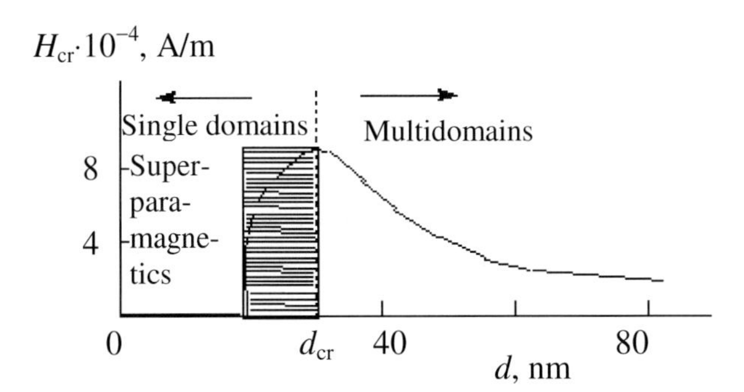

Fig. 11.7. The dependence of the coercive force H_c on the ferrous nanoparticle diameter d [76]

When the monodomain particle size becomes less than critical (every such particle can be considered as a ferromagnetic particle, but with variable magnetization) a drop of the H_c value is observed. The total magnetization (arising under an external magnetic field H sufficient for saturation) relaxes to zero after removing the external field according to the law:

$$J = J_s \exp\left(-t/\tau\right) , \tag{11.2}$$

where $\tau = \tau_0 \exp[E_M/(k_B T)]$ is the relaxation time; τ_0 is a constant, in sec; $E_M = KV$ is the potential barrier, J; K is the constant of uniaxial anisotropy, $J \cdot m^{-3}$; and V is the particle volume.

Simultaneously small particles are similar to a paramagnetic substance with a large paramagnetic moment and can transfer to the superparamagnetic state. This process is a magnetic phase transition of the second order, characterized by continuous change of the first derivatives of the thermodynamic potential. The properties of such supermagnetic systems can be characterized by the values of τ_o and K.

Manifestation of supermagnetism is often indirect evidence for the ferromagnetic phase low content compared with the filler total content. For example, in the Ni–crazed PP system [56] there is a large number of Ni superparamagnetic nanoparticles (<10 nm) in the samples, whose magnetization (at $H = 75\ kA \cdot m^{-1}$) is considerably higher than for block nickel, which leads to an apparent (illusory) decrease of nickel mass in the whole. The same situation appears to occur in the case of decarboxylated thermodestruction products of Fe^{3+} and Ni^{2+} acrylates [63–65].

Superparamagnetic particles were found in many metallopolymeric nanocomposites, for example in the systems: Nafion 117 – Fe_3O_4 [51]; poly(dimethylphenylenoxide) – α-Fe (spherical particles with diameter $d \leq 2.5$ nm, $K_{eff} = 1.83 \cdot 10^6\ J \cdot m^{-3}$ at 100 K) [82]; in block-copolymers – γ-Fe_2O_3($d \approx$ 5 nm, $K = 1.58 \cdot 10^5\ J \cdot m^{-3}$, $\tau_0 = 4.2 \cdot 10^{-12}$ sec) [83] and in many polymer-immobilized catalysts (see Chap. 12).

Possible ways to increase the relative magnetic permeability were studied by scanning a polymer composite with various dispersed hybrid fillers [84]. Such compositions were obtained on the base of composites PE – amorphous metallocomposite (69% Co, 12% B, 4% Si, 4% Ni, 2% Mo), PE – nickel-zinc ferrite, PE – metallocomposite ferromagnetic (80% Ni, 5% Mo, 0.5% Mg, 0.15% Si, 0.1% C) with maximum density and they are of great interest for the fabrication of permanent metallopolymeric magnets with optimal technical parameters.

Another group of magnetic nanocomposites must be especially noted. We are dealing with the problem of the electromagnetic wave–matter interactions in the microwave range that is created mainly by the need to improve the stealthiness of aircraft, missiles and ships, against detection by radar receivers. The best way to produce such materials is to form magnetic nanocomposites to cover the metallic surface with microwave radar absorption (0.1–18 GHz range, in general, in X-band – 8.2–12.4 GHz and P-band, 12.4–18 GHz) [85]. A ceramic microwave absorbent can be obtained by three main methods [36]: (a) mixing of micronic or even submicron metal powders within a liquid matrix precursor; (b) preparation of a porous host matrix impregnated by a concentrated solution of transition metal nitrates; (c) mixing of alkoxides with an aqueous solution of metal ions. The last method was used [86] for polymeric nanocomposite fabrication from complex ferrite $Ni_{0.5}Zn_{0.4}Cu_{0.1}Fe_2O_3$ or $CoFe_2O_3$ by adding conducting materials (5 wt.% of graphite powder). This complex ferrite was obtained as a coprecipitant at various pH values after drying and silanization by 3-aminopropyltrimetoxysilane or 3-methacryloxypropyltrimethoxysilane until a three-dimensional polysiloxane network was formed (see Chap. 7). Finally, the silanized $Ni_{0.5}Zn_{0.4}Cu_{0.1}Fe_2O_3$ or $CoFe_2O_3$ was dispersed in MMA and its radical suspension polymerization was carried out. The permittivity is affected by the sample thickness (4–10 nm) and nanofiller concentration in the polymeric matrix, as well as by the thermal treatment conditions. The optimal variant presents nanocomposites (with 20 wt.% of complex ferrite, thickness ~10 nm) that show strong microwave absorption at 15.2 GHz with magnitude of reflection loss of −25.4 dB. Polymer composites, containing conducting fillers or ferroelectric particles, are effective microwave adsorption materials important to suppress microwaves reflected from metal structures (in both civil and stealth defense system for military platforms). These multicomponent materials can be effective absorbers of electromagnetic waves due to the large imaginary part of the complex dielectric permittivity responsible for dissipation of electromagnetic energy and due to their low bulk conductivity.

Fibrous magnetic materials can also be produced using the technology of joint spraying in a gaseous current (the melt-blowing technique) of polymer (PELD) melts and a highly dispersed (~1000 nm) filler of strontium or barium ferrite [87]. Such composites can be used for gas and liquid filtration, oil product sorption, industrial and domestic waste, etc. In recent years there has been increased interest in so-called magnetic fluids, being colloidal solutions with very strong magnetic properties. Polymers such as oligoorganosiloxanes can be used as carriers capable of working in extreme conditions (in a wide temperature range, in vacuum, in aggressive or biologically active media). Recent investigations of composites with conducting and ferroelectric particles showed that they really possess some specific characteristic properties of interest. Such materials have also been proposed for use as sound absorbers and damping materials. In this case the elastic waves interact with ferroelectric particles and elastic energy is transformed into electrical energy that is dissipated in carbon black particle chains [88].

Other alternative ways in the problem of magnetic conductivity in the metal-containing polymeric systems is the creation of ferromagnetic element-organic compounds [89], and multispinous paramagnetic molecules on the base, for example, of polymetalorganosiloxanes [90].

The role of magnetic materials in biopolymers was discussed in Chap. 9. Their value has increased because many of them are now produced commercially, especially materials used as magnetic resonance contrast agents or magnetic labels (Table 11.2). The selective combinations of biomolecules

Table 11.2. Some commercially available biocompatible magnetic materials with nanoparticles [95]

Product Name	Composition	Particle Size, nm	Application	Manufacturer or Supplier
Combidex	Magnetic iron oxides-dextran	17–20	Magnetic resonanse contrast agent	Advanced Magnetics, USA
Endorem[a]/ Feridex[b]	Magnetic iron oxides-dextran	100–250	Magnetic resonanse contrast agent	Advanced Magnetics, USA
MicroBeads	Magnetic iron oxides-dextran	50	Separation and labelling of cells and molecules	Miltenyi Biotec, Germany
Nanomag	Magnetic iron oxides-dextran	100	Magnetic labelling	Micromod Partikeltech-nologie, Germany
Resovist	Magnetic iron oxides-dextran	57	Magnetic resonanse contrast agent	Schering AG, Germany

[a]Commercial product name in Europe; [b]commercial product name in the United States

with magnetic nanoparticles may enhance the ability for biomolecular recognition [91,92] and magnetic field-assisted drug delivery [93,94].

Thus both the science and practice of polymeric metal-containing magnetic nanocomposites are now intensively developing and this trend is kept going for the new possibilities of synthesis, new methods of composite forming and the great practical interest. The final scientific aim of the synthesis of such nanostructured materials and studies is modeling of their structures for a correct interpretation of the experimental facts. On the other hand, the modeling and deeper understanding of the formation and properties of such nanomaterials will open new horizons both for their design and practical application.

11.2 Electrical Characteristics of Metallopolymer Composites

The electrical and magnetic properties of all materials are closely connected because the elementary carriers for these fields are electrons with electrical charge and magnetic moment. Even the interatomic and intermolecular interactions are determined by the mutual orientation of orbital and spin moment. Theoretically the electroconductivity σ ($\mathrm{ohm}^{-1}\cdot\mathrm{cm}^{-1}$) of a two-component inhomogeneous system is connected with the value of the electroconductivity $C_{\mathrm{elc,cr}}$ by the following relation:

$$\sigma = \sigma_{\mathrm{eff}}[(C_n - C_{\mathrm{elc,cr}})/(1 - C_{\mathrm{elc,cr}})]^{\theta} , \tag{11.3}$$

where θ is the critical conductivity index and σ_{eff} is some effective electrical conductivity (see Sect. 2.6).

The σ_{eff} value for a metallopolymeric composite is defined by the particle concentration, nature, morphology, oxide and presence of other films on the surfaces, as well as by the electroconductive properties of the polymeric interlayers between the conducting particles, i.e. by the transient resistance in the contact zone (mainly for oxide film formation with thickness up to tens of nanometers). These layers with a transient conductivity provide the difference between nanocomposite electroconductivity and total electroconductivity of all elements of such complicated conductive object. The spatial distribution of the current-conducting particles in a polymer can be conventionally described [17] by the following structures (see Fig. 11.8):

– a matrix structure, for which a considerably rigid spatial distribution is characteristic, as in the zeolite voids (Fig. 11.8,a);

– a statistical structure, as under mechanical dispersion of nanoparticles into the polymer (Fig. 11.8,b);

– a chain structure, when the nanoparticles are spatially, as in ferromagnetic material formation in a magnetic fields or in nanolithographic processes (Fig. 11.8,c);

– a layered structure, including intercalation (see Sect. 7.2) into an interlayer space of layered substances (such as graphite, hexagonal boron nitride, MoO_2, transition metal dichalcogenides – TiS_2 and MoS_2, natural clays, etc.), electrodeposition of high-density, ordered arrays of nanowires $Bi_{0.85}Sb_{0.15}$ in alumina templates (thermoelectric materials) [96] or introducing alkaline (Li, Na, K, Rb), alkali-earth (Ca) and transition (Ti, V, Cu, etc.) metals into fullerene nanotubes [97–100] (Fig. 11.8,d);

– a globular structure with particles placed between polymer granules (Fig. 11.8,e).

The electrophysical properties of nanocomposite are strongly influenced by the nature of the nanoparticle–polymer interaction. Unlike massive bodies in noncontinuous metallic media (where an interparticle distances are considerably greater, than interatomic ones), the role of the interparticle space

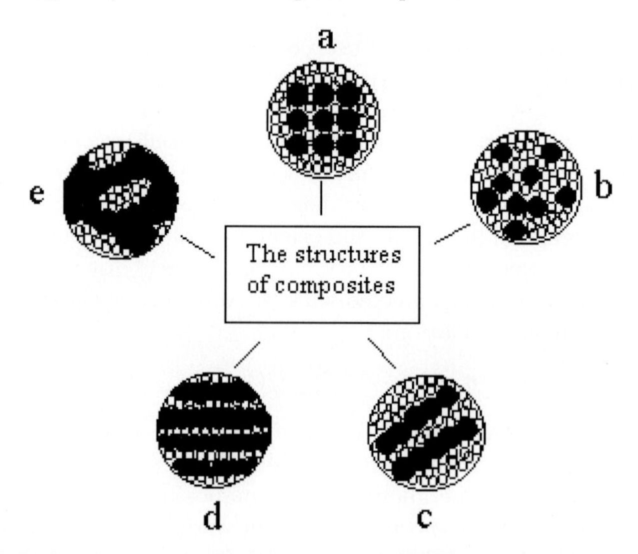

Fig. 11.8. Structural types of conductive particle distributions in nanocomposites (details in text)

nature and a borders structural disordering in nanocomposites becomes very important. At the same time conductivity can exist in every single (discrete) metal-containing nanoparticle. Some dimensional effects must be taken into account (the nanoparticle must be commensurable with the characteristic length of the electron processes, e.g. with the free path length). Metallopolymeric media present systems of metallic conductive regions, divided (in the general case) by dielectric interlayers (i.e. polymeric matrix). It is a problem to present a general electroconductivity analysis of the structures and systems (Fig. 11.8) in terms of electric contact theory (when applied to polymers with metallic fillers). Electro-physics of such systems is near to a behavior of so called island-type metallic films, for which the basic theory of electrophysical processes is well studied. In particular, for metallopolymeric composites there are two percolation thresholds [101, 102], the first of which relates to classic electron transfer over a barrier (if the electron energy exceeds the barrier height), and the second corresponds to a transition between a nonconductive state and a state with tunnelling conductivity through the mentioned polymeric interlayers. The process of electron "tunnelling" through the barrier is essentially a quantum-mechanical process caused by the electron wave nature. Electroconductive objects in such compositions have no direct contacts and the conductance is carried out by the second mechanism. Charge carrier transport proceeds by a combination of quantum-mechanical tunnelling and the so called "jumping mechanism" through the localized energetic states. It is not inconceivable that the localized breakdowns of interlayers (between particles) take place under the effect of weak external electrical influence, which can abruptly increase the composition conductivity ("electrode effect"). The fractal geometry of the percolation cluster (PC) does not change and the first current flow threshold corresponds to the formation of a primary PC. After studies of Cu–PS nanocomposites it was proposed [103] that the copper particles could form "an infinite cluster" in the PS volume, providing a very high conductivity.

The traditional method of nanocomposite formation is to introduce into a polymeric matrix some metal powders (including stabilized powders). Sometimes the nanoparticles "rolling in polymer" method is used with the subsequent mixing of such materials with the polymer matrix. Now, for the fabrication of electroconductive nanocomposites a mixing or combination of different substances, such as metals (Ag, Cu, Ni, Cr, Fe, Al, etc.), their compounds (e.g. oxides, such as V_2O_3 [25], In_2O_3 [104] or $FeCl_3$ [105]) or organic compounds with a metallic conductance, such as colloidal particles (50–450 nm) of polypyrrole and polyaniline [106], is widely used.

Electrical conductivity (σ, $\text{ohm}^{-1}\cdot\text{cm}^{-1}$) of polymeric composites filled by dispersed metals (Ag, Ni, Cu, Al) depends on the filler volume content φ and the dependence is usually nonlinear [17] because the small metal particles form monolayers on the large particles of the polymer. Electroconductive properties appear themselves only at a given component ratio in the *metal–polymer* system, when in the polymeric matrix the conductive channels (i.e. the continuous chains of metal-containing clusters) in polymer matrix there are the conductivity channels. The biggest conductivity is achieved if the composite becomes a network of such interconnected conductive chains, i.e. acquires a percolation structure. There is some critical value (φ_{cr}) of filler volume content (the value is called the percolation threshold), above which (at $\varphi > \varphi_{\text{cr}}$) an abrupt rise of the electrical conductivity is observed.

The percolation threshold value linearly depends on the maximum filling F (i.e. a theoretical limit of the system filling):

$$\varphi_{\text{cr}} = X_{\text{cr}} F \, , \tag{11.4}$$

where X_{cr} is a critical parameter defined by the number of conductive knots in the solid body lattice. F is 0.64 for monodispersed spherical particles of every size statistically packed into a composition [107]. The value of φ_{cr} is important for composition optimization and allows the prediction of a minimum ratio of electroconductive filler required for percolation. The electroconductivity of nanocomposites was studied [108] for systems based on various modified resins, filled by dispersed spherical nickel particles. The studies of epoxide (epoxydiane ED-20, epoxysilicone T-10, epoxyresorcine VII-637) and oligoester (nonsaturated oligoglycolmaleine-phthalates with styrene PH-1 and methylmethacrylate PHM) resins showed that the percolation thresholds of these systems were refistrated at $\varphi_{\text{cr}} = 0.32$ (ED-20); 0.25 (T-10); 0.36 (VII-637); 0.50 (PH-1) and 0.55 (PHM) at $X_{\text{cr}} = 0.42$; 0.30; 0.50; 0.56; 0.62 correspondingly. The percolation threshold for the oligoester matrix was higher than for epoxides. The conductive composites with filler content up to 60 vol.% and $\varphi_{cr} \approx 0.1$ were obtained by hot pressing (413–423 K) of dispersed Al powders with PMMA [109]. The introduced filler content can be brought up to 90 vol.%. The conductivity and dielectric constant studies [110] of the three-component composite (epoxy resin-barium titanate-carbon black) had shown that the $BaTiO_3$ concentration influenced the percolation threshold, critical index and the mechanism of conduction. At low concentrations the dominating transport mechanism is anomalous diffusion in clusters of carbon black particles (average size 30 nm); at high concentrations of barium titanate the conductivity and dielectric permittivity are mainly determined by interphase polarization.

In keeping with the ratio of conductive nanoparticles and the nature of the interparticle region, as well as on their relative sizes, the electrophysical properties of the composite change from dielectric (specific electrical resistance $\rho_v = 10^{11}–10^{14}$ ohm·m) to conductive, and it is very interesting that sometimes conductivity had been increased by changing the composite composition during thermal treatment. Thus in the course of successive thermolysis of *m*-carbodicarboxylic acid oligosalts (being initially dielectrics) with a composition $(^{-}OOC–CB_{10}H_{10}C–COO^{-}M^{2+})_n$ (where M = Cu, Mg, Ca, Zn, Cd, Pb) oxidation of the carborane nuclei proceeds attended by hydrogen evolution and M^{2+} oligosalt ion reduction to a free metal state [111]. As a result, in the system the electroconductive layers are formed, which can considerably decrease ρ_v value (Table 11.3). The electrical conductivity of the oligosalts increased by 3–9 orders (after thermolysis at 773 K) and even by 6–11 orders (after pyrolysis at 1173 K). The most important fact is that such electrophysical properties are retained over a long period of time (for years).

The metallopolymer conductive characteristics depend considerably on conductive component dispersity and its volume distribution. For example, layered polypyromellitimide films (imidazed by heating to 573 K) with silver particles of various dispersity have different conductance depending on the metal content [17]. The percolation threshold for a composite with a high dispersed silver powder is registered in the filler content >9 mass.%. Introducing uniformly dispersed silver nanoparticles with size 10–15 nm, obtained by prepolymer thermolysis of silver acetate solution in polypyromellitimide acid, does not create any conductance at the same filler content, and the films preserve high dielectrical characteristics ($\sigma = 10^{-15} - 10^{-12}\,\text{ohm}^{-1}\cdot\text{cm}^{-1}$), that can be explained by the presence of a considerable content of dielectric polymer interlayers between the filler particles.

A very important role is played by the methods of introducing conductive particles into a matrix, as well as internal changes to the composite due to ageing, annealing or external actions (for example, due to treatment in an external electric or magnetic fields or due to use of statical pressure, adsorbing agents) that change the conductivity parameters too [112].

Table 11.3. Thermal and electrophysical characteristics of oligosalts $[^{-}OOC–CB_{10}H_{10}C–COO^{-}M^{2+}]_n$

M (T_{dec}, K)*	Thermotreatment Conditions	ρ_v, ohm·cm	Products Composition (XRD Data)
Cu (523)	293 K	$9 \cdot 10^{11}$	Oligosalt
	473 K, 1 hr, argon	–	Oligosalt
	523 K, 1 hr, argon	$5 \cdot 10^{6}$	Cu, Cu_2O
	1173 K, 10 hrs, air	$9 \cdot 10^{5}$	Cu
Mg (673)	293 K	$4 \cdot 10^{12}$	Oligosalt
	293 K, 4 hrs, argon	–	Oligosalt
	473 K, 4 hrs, argon	–	Oligosalt
	523 K, 4 hrs, argon	–	–
	1173 K, 10 hrs, air	$6 \cdot 10^{6}$	Undefined structure
Ca (673)	293 K	$1 \cdot 10^{13}$	Oligosalt
	493 K, 1 hr, argon	–	Oligosalt
	573 K, 1 hr, argon	–	Oligosalt
	673 K, 1 hr, argon	$4 \cdot 10^{12}$	–
	673 K, 1 hr, argon	–	–
	1173 K, 10 hrs, air	$2 \cdot 10^{4}$	Undefined structure
Zn (733)	293 K	$1 \cdot 10^{14}$	Oligosalt
	293 K, 1 hr, argon	$2 \cdot 10^{15}$	Oligosalt
	573 K, 1 hr, argon	$1 \cdot 10^{13}$	–
	653 K, 4 hrs, argon	$6 \cdot 10^{10}$	Zn, ZnO
	1173 K, 10 hrs, air	$6 \cdot 10^{3}$	Zn, ZnO
Cd (583)	293 K	$1 \cdot 10^{11}$	Oligosalt
	293 K, 1 hr, argon	$1 \cdot 10^{13}$	Oligosalt
	553 K, 4 hrs, argon	$1 \cdot 10^{14}$	–
	673 K, 4 hrs, argon	$1 \cdot 10^{11}$	Cd, CdO
	1173 K, 10 hrs, air	$1 \cdot 10^{4}$	Amorphous structure
Pb (543)	293 K	$3 \cdot 10^{14}$	Oligosalt
	493 K, 1 hr, argon	$6 \cdot 10^{14}$	Undefined structure Pb
	523 K, 1 hr, argon	$6 \cdot 10^{13}$	Pb
	1173 K, 10 hrs, air	$1 \cdot 10^{4}$	

* DTA data (T_{dec} – decomposition temperature).

The specific conductivity ρ_v of metallopolymer composites decreases with increasing temperature by the standard formula:

$$\rho_v = C \exp\left[-E_a/(k_B T^\alpha)\right] , \tag{11.5}$$

where C is a temperature-dependent constant, E_a is the effective activation energy of the conductance process and $\alpha \leq 1$, but it is evident that it should be studied in detail and measured for every concrete system.

The electrical properties of Ag clusters in plasma polymer matrices were studied [113] and it was established that there was a strong cluster–matrix interaction, connected with electron transfer from the metal into the matrix and the formation of a dipole layer around the clusters as a consequence. The measured conductivity of the films can be explained by an electron hopping process between separated clusters. Probably the same mechanism is observed in conductive Ag-containing films obtained by $AgNO_3$ reduction in PVA [114].

The electrical conductivity of epoxy resins containing gold and silver particulate fillers was studied in [115] as a function of the interparticle distance (Fig. 11.9). The large jump in conductance (indicated by the arrow) is thought to correspond to the establishment of a physical contact between the conducting surfaces.

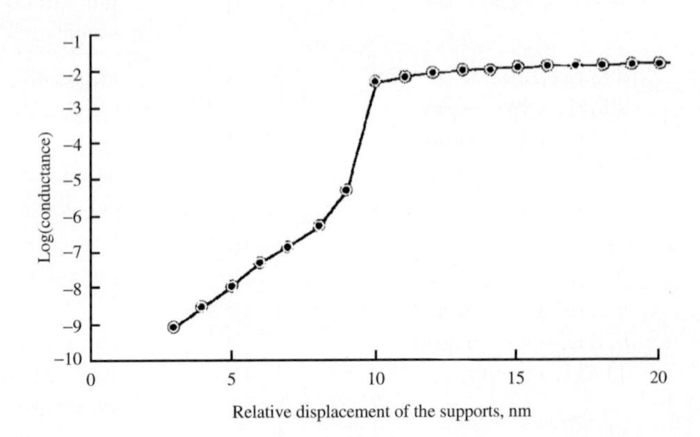

Fig. 11.9. Electrical conductance between two gold particles

The same effect was observed [116] also in thin-layer heterogeneous nanocoatings obtained by cocondensation of Pd and poly-*p*-xylylene (see Chap. 5), where self-organization of Pd nanoparticles was discovered, as well as a direct connection between the particle morphology and the electrophysical and gas-sensitive properties of the material (Table 11.4). An electric current in such systems runs due to the jumping conductivity between the metal particles in the samples, containing 1 and 2.5% of Pd.

At low content the temperature dependence of the electroconductivity is similar to that of the above mentioned island-type metallic films. But for

Table 11.4. Structural characteristics and properties of metal–polymer coatings with different content of palladium nanoparticles [116]

C_m (Pd content, vol.%)	ρ, ohm·cm	D ±0.01, nm
1	$5.2 \cdot 10^{11}$	1.99
2.5	$2.1 \cdot 10^{8}$	1.87
8	$1.8 \cdot 10^{3}$	1.93

Notations: C_m – Pd volume concentration; D – a fractal dimensionality.

samples with 8% Pd it is characteristic for massive metal blocks or continuous metal films. Such a sample structure is intermediate between a fractal and closed-packed structures (Table 11.4).

The conductivity and percolation parameters of the statistically distributed systems depend on a hardening mechanism and the connected sedimentation processes and the rate relations of the hardening and deposition [109].

It is interesting to consider the characteristics of two types of metallic silver reduced composites (10^{-5}–10^{-3}g·cm^{-3}) in an imidized polyimidic (PI) matrix [117]. For the "silver-on-surface" composites a tangential growth of the silver nuclei is observed, but for the "silver-in-volume" systems in PI volume a radial growth is preferable and the σ value is about 10–10^3 ohm^{-1}·cm^{-1}, which is only two order of magnitude less than for pure silver ($\sigma = 6.8 \cdot 10^5$ ohm^{-1}·cm^{-1}). Electroconductive polymeric films can be formed on electrodes enduring a reduction electrochemical polymerization [118]. For this purpose, for example, pyrrole was polymerized in the presence of Ni nanoparticles, stabilized by surface active substances (see Chap. 5), after which the polypyrrole-coated Ni particles were mixed with PEHD (in xylol solution at 383 K), which allows one to cast on a mercurium surface films 0.6–0.9 mm thick [119]. The obtained nanocomposite demonstrates conductivity at the filler contents above the percolation threshold without any considerable change of the thermal and mechanical characteristics.

If the nanocomposites are formed on the base of ultradispersed copper powders having mean sizes 20–30 nm (as under $CuSO_4 \cdot 5H_2O$ thermodestruction in glycerin at 423–458 K) with formaldehyde or epoxy resins as the binders, the systems are synthesized with a stable conductivity near to the metallic value ($\rho_v = 10^{-2}$–10^{-4} ohm·cm) [120]. There was only partial oxidation of the surface layer after years of exposure in air that caused some decrease of the composition conductivity (the conductivity can be easily restored by formic acid treatment). Ester-based nanocomposites filled by high dispersed nickel powders [121] have conductivity values $\rho_v = 10^{-1}$–10^{-3}

ohm·cm and can be used as electrodes for some electrochemical reactions in alkaline media.

Today metallization and surface finishing are important fields of applications in the macroscopic world as well as in nanotechnology [122]. Conventional methods for the deposition of metallic conducting films use the galvanic technique, but its application is limited by the conducting surfaces or surfaces treated with a conducting precursor film and, moreover, not every metal is suited for a proper deposition [123]. Besides, the spatially defined structuring of the metal surface is only possible via a subsequent process step and requires great expenditure. Alternatively, the use of template polymers offers possibilities for the defined positioning of metal clusters.

Many of these problems can be solved when the nanoparticles are formed in the presence of a polymer. The methodological problems of the development of such technologies are now intensively studied, especially the problems related to metal filler dispersion rise and its uniform distribution throughout a matrix [56, 124–126]. One such system organization method, called *cluspol* [124] and mentioned above in Chap. 5, is based on the thermolysis of a metal-forming compound (usually metal carbonyl) and polymer in a joint solution. Thus, by the thermolysis at 443 K of copper formate triethylenediaminic complex $[Cu(En)_3](HCOO)_2$ in a polystyrene matrix the metal–polymer compositions (less than 10 mass % of metal) with the Cu particles (the mean sizes up to 28 nm) distributed in the polymer natural voids [127] were obtained. Localization of such particles eliminates the possibility of their direct contact, which is verified by the conductivity value ($\rho_v = 3.2 \cdot 10^5$ ohm·cm). However the additional introduction of carbonyl nickel particles (mean sizes ~1 µm) by mechanical mixing in the copper-containing matrix leads to a rise in conductivity and allows getting the conductive compositions of Co/Ni/xNi. The values of ρ_v (in units 10^{-3} ohm·cm) are 14, 9.3 and 6.7 for the nickel contents 20, 50 and 80 mass %, correspondingly. Probably the systems are similar to metallized polymer films by plating out and a variant of their forming can be a micron size powder dispersion in PVA with the subsequent treatment by metal salt solutions having metal ions with higher reduction potential (positive net electromotive force value- E^0_{net} (e.g., Zn + $Cu^{2+} \rightarrow Cu + Zn^{2+}$; $E^0_{net} = 1.100V$) [114]. Such metallized PVA films exhibited a low surface resistivity around 10^0–10^1 ohm·cm^2.

The polymer composition conductivity can be raised by forming structures of special type – *nuclei* (conductor or dielectric) + a *sheath* (conductor). The method is of great practical interest for lacquer or glue composition, because it allows decreasing the nanocomposite cost (as a rule the sheath is made of plastic and oxidation-resistant substances, such as silver or gold) and specific density. For example, the conductivity of the system with dielectric SnO_2 particles, coated by a silver layer (8 vol.%) by thermal treatment of a precursor (Ag^+-containing polymer, a water-soluble system of $AgNO_3$ – ethylene glycol – EDTA), is very high and achieves $\sigma = 1 \cdot 10^{-3}$ ohm^{-1}·m^{-1}, whereas for a mechanical mixture of the SnO_2 and Ag powders (16 vol.%) this value is only $\sigma = 2 \cdot 10^{-7}$ ohm^{-1}·m^{-1} [104].

Another promising way of forming conductive polymer film composites is based on metal salt reduction (Ag, Cu, Ni, Co, Fe) with sodium tetrahydroborate, using counter-flow diffusion directly into the voids of the crazed polymer matrix (PP, PETF, PELP, etc.) [52, 124]. As a result, new metallopolymer systems were synthesized with the nanocrystal sizes 5–40 nm (in accordance with the natures of the metal and matrix) and a wide complex of electrophysical characteristics (the materials can become conductors even at low filler content) (Fig. 11.10). The percolation threshold position and composite dielectrical permeability can be varied by regulation of the metal layer sizes and compactness into the given polymer matrix. For example, using some water-swollen polymer matrices, such as PVA ($M_n \approx 6 \cdot 10^4$) or PVA–PAA ($M_n \approx 3 \cdot 10^5$) and PVA–PEI mixtures, many polymer composite films were synthesized in situ [128] with CuS nanoparticles (5–150 nm) having conductivity values up to σ_v = 1–10^{-2} ohm^{-1}·m^{-1} and percolation threshold $\varphi_{cr} < 40$.

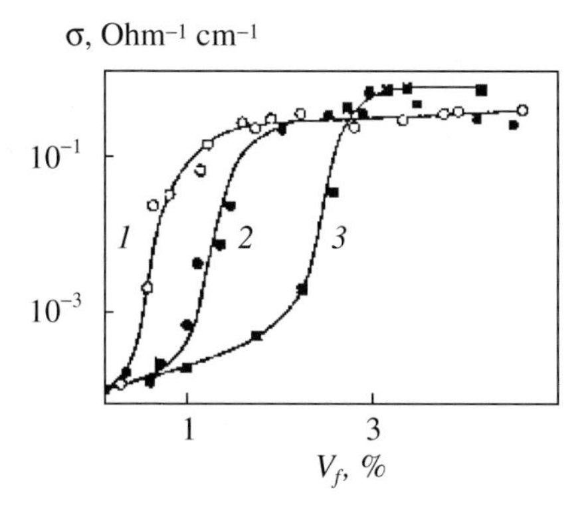

Fig. 11.10. The specific conductivity σ (of a nickel-containing PP-composite with a uniform porosity distribution through the matrix volume) dependence on the nanofiller volume content V_f: at 293 K (*1*); for alcohol–glycerin solutions at 293 K (*2*) and 276 K (*3*) [63]

The production of polymer nanocomposites with spatial orientation and improved electrophysical properties is a very important practical problem, especially for adhesive and film technologies. One of the solutions can be connected with the composites forming by hardening with the imposition of an external magnetic or electric field. Thus nonhardened epoxide mixtures, containing up to 50 vol.% of highly dispersed nickel, are usually nonconductive [17], but after hardening in a magnetic field their specific conductivity along the field decreases and the rate of ρ_v decrease arises with the field strength.

A new stage of development of film and membrane technologies is connected with composition formed by successive "layering" of the conductive and dielectric layers. For example, by cryochemical (77 K) synthesis from chromium vapors polymeric Cr films (4–5 mass %) with an anisotropy of electrophysical characteristics, the material is an insulator in the direction perpendicular to the film surface ($\rho_{v,\perp} = 10^{12}$–10^{13} ohm·m), but it is a conductor in the longitudinal direction ($\rho_{v,||} = 3 \cdot 10^3$–$10^4$ ohm·m) [129]. Very promising approaches are used for the forming of anisotropic Ag-clustered films with a layered morphology (Fig. 11.11), which were obtained by thermal reduction (393 K, 24 hrs) of Ag-block-copolymer layers and have, after thermolysis, size <10 nm [59]. Films with such layered structure are characterized by a very high anisotropy; the conductivity values differ by four orders of magnitude for the parallel and perpendicular directions to the film surface ($\sigma_\perp < 10^{-13}$ ohm^{-1}·cm^{-1} and $\sigma_{||} \approx 10^{-9}$ ohm^{-1}·cm^{-1}).

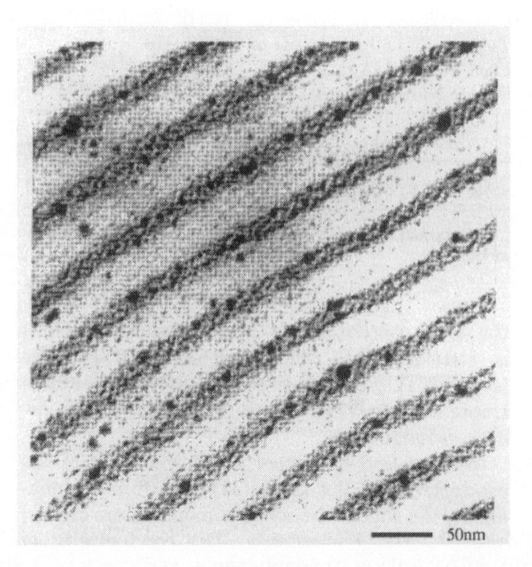

Fig. 11.11. Electron microscope photograph of Ag nanoparticles after thermoreduction of $[Ag]_{60}[MTD]_{300}$ (MTD- methyltetracyclododecene)

The external pressure influences primarily the resistance during the formation of *particle–matrix* and *particle–particle* contact. The electric conductivity of this system decreases with the pressure value P by the law:

$$1/\rho_v \sim KP^\varepsilon \; , \qquad (11.6)$$

where K is an effective constant and $0.5 < \varepsilon < 1$.

Usually the conductive properties are changed and reversible with pressure, but sometimes after pressure reduction a resistance hysteresis is observed. This can be explained by conductive particles merging in places where there are no isolwating films. If the pressure effects on nanocomposite conductivity are irreversible, one must solve the problem of optimization the composite pressing. A pressure rise for several systems with metal fillers can lead initially even to a decrease in conductivity (it is connected with introducing the polymer between the conductive particles). The value of ρ_v begins to rise only after some minimum. Therefore sometimes piezoelectric ceramic nanofillers are used for new materials, in acoustical transducers and medical diagnostics, items with advanced pyroelectric and piezoelectric properties [130]. For example, after the roll sheeting (413 K, 20–30 MPa) of metalloceramic mixtures of PVDF with $PbTiO_3$ (up to 65 vol.%) the coatings were formed with a dense uniform microstructure and high piezoelectric characteristics [131].

In recent years some multicomponent metal-containing polymer precursors and alloys have attracted special interest as promising electrode materials for the galvanic battery [132], and high-temperature superconductive ceramics. It was shown in Chap. 5 that the new materials with superconductive transition temperature about 90 K were produced on the base of polycondensation type Y-, Ba- and Cu-chelates [133]. This temperature is even higher and is 110 K for composites based on bismuthous ceramics, obtained by a prepolymer synthesis (including condensation products of polyol based on castor oil and diisocyanate) (Bi+Pb):Sr:Ca:Cu:O=2223 [134–136]. At last case the superconductor phases present 2201, 2212, 2223 and Ca_2PbO_4. As shown in Fig. 11.12, the composite conductivity [134] is nine orders of magnitude higher than the pure polyurethane conductivity (10^{-15} $ohm^{-1} \cdot m^{-1}$).

Temperature-resistant ceramic coatings formed by the sol–gel method from sols are extremely promising for applications such as electronic insulating, radiotolerent, light-diffusion-reflecting and light-absorbing coatings for electrical, space engineering and power industries. An important technological scheme for formation of such coatings is given in Fig. 11.13 [137]. They retain their stable optical and mechanical characteristics at temperatures up to 1300 K in intensive electromagnetic, solar and penetrating radiation.

Another promising field of the application of metallopolymer coatings is their use as ion-selective electrodes to define metal ion content in aqueous media. For example, it was reported [138] that highly conductive polymer materials, containing conformationally rigid W^{6+}-capped calixarenes, were obtained, and their conductivity can be influenced by added *p*-xylene. It is supposed that these materials can be effectively used as sensors for aromatic analytes, because generally calixarenes are highly versatile scaffolds for 3D cavity design [139].

But the main application field of metallopolymer nanocomposites remains their usage as various polymeric solid electrolytes (PSE), especially for lithium cells (many of which present solid component dispersions into high molecular compound solutions). In electrical fields there is structural polymerization (connected with polymer network deformation without its

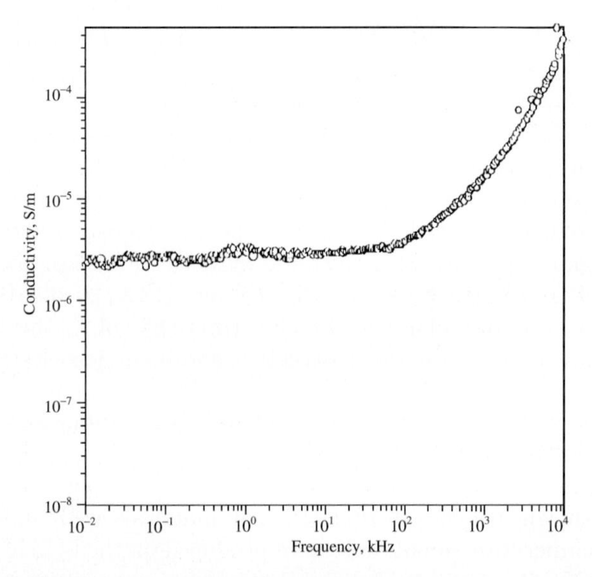

Fig. 11.12. Electrical conductivity of a composite with 10% ceramic content as a function of frequency at room temperature

completeness and entity violations). The same approach was used for ion-conductive PSE membrane production due to the potential in the solid electrochemical devices, especially in lithium batteries with Li^+-inserted compounds as cathodic materials [140–143]. PEO-salt systems were investigated over the past 10–20 years because of their possible use as polymeric electrolytes [144]. The light weight and processability of these polymers (PAn, PPy, polythiophene) represented at the time great promise for improving the technical characteristics of rechargeable lithium cells and "plastic" batteries [145–147]. The lithium ion transfers by jumps from one chain fragment (1.93 nm) to another in the electrolyte with composition $(\text{–O}-CH_2-CH_2-)_8$ $LiClO_4$. However at room temperatures the conductivity of this material is very low (about $10^{-8}\text{ohm}^{-1}\text{cm}^{-1}$) for high crystallinity of the polymer matrix, but the practically interesting values are distinguished only at 333–353 K ($\sigma \approx 10^{-4}$ $\text{ohm}^{-1}\text{cm}^{-1}$) [148]. The crystallization temperature and values σ $\sim 10^{-4}$ $\text{ohm}^{-1}\text{cm}^{-1}$ (323 K) and $\sim 10^{-5}$ $\text{ohm}^{-1}\text{cm}^{-1}$ (303 K) can be achieved by adding ceramic powder nanoparticles to the $LiClO_4$–PEO electrolytes. The ceramic nanoparticles (10 mass %, 5.8–13 nm in size) of TiO_2, Al_2O_3 or $LiAlO_2$ work as rigid plastisizers that can kinetically inhibit PEO crystallization due to complexation reactions [149–153]. The conductivity of materials with nanoparticles is one order of magnitude higher than for micron-size particles. The mechanical strength of polymer electrolytes is considerably higher too. Polymeric electrolytes on the base of comb-shaped branched polymers with inorganic SiO_2 [154], poly(dimethylsiloxane-ethyleneoxide) or polysilane skeleton allow one to reach very high ionic conductivities (up to $2 \cdot 10^{-3}$

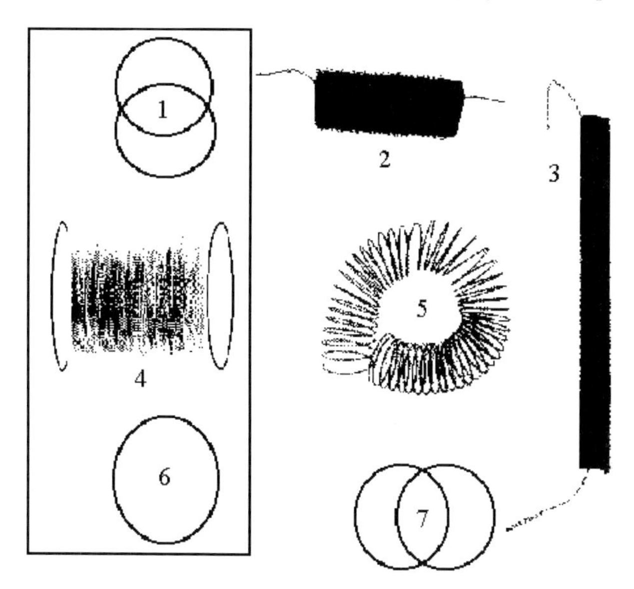

Fig. 11.13. Elements of electrical devices having temperature-resistant electric insulation prepared from TEOS- or H_3PO_4-derived suspensions: (1) elements of a chromel versus alumel thermocouple (silicate coating); (2) element of a heater with an open nichrom spiral (silicate coating); (3) element of a heater with a closed nichrom spiral (phosphate coating); (4) nichrom wire (silicate coating); (5) nichrom spiral (silicate coating); (6) wire of aluminum alloy (phosphate coating); and (7) elements of a tungsten versus rhenium thermocouple (silicate coating)

$ohm^{-1}cm^{-1}$), as well as in composites from lithium-ion mixtures of conductive ceramics $Li_{1.3}Al_{0.3}Ti_{1.7}(PO_4)_3$ with a polymer electrolyte on the base of polyesterurethane [155]. The ceramic and polymeric phases in such electrolytes are divided by thin interlayers. Their conductivity increases considerably under the action of solvent vapors and even more under treatment by agents able to increase the adhesion on the phase boundaries. Searches and studies of polymer electrolytes for lithium-ion batteries are intensively continuing, especially connected with the possible partial lithium substitution by zinc or with the use of $(PEO)[(ZnBr_2)_{1-x}(LiBr)_x]$ compositions [156]. At the same time the optimization of known compositions, such as the microparticles doped NaCl, $CaCl_2$ [157] or $LiN(SO_2CF_3)_2$ [158], are being continued.

Investigations of the electrical characteristics metallopolymer nanocomposites relate to the very interesting and intensively developing field of nanomaterial science. Experience that has been accumulated over a number of years gives a lot of information and at some future date we can await a very interesting generalization of the experimental facts and their practical applications connected with electroconductive polymer nanocomposites.

11.3 Optical and Semiconductive Properties of Metal–Polymer Nanosystems

The described nanocomposites have a unique complex of optical and semiconductive properties (as well as plasmon sensitivity to dimensional effects). It opens wide promising in many directions of physics and environmental sciences. For example, they can be used in such important fields as optomicroelectronics problems and practical diagnostics and control of nature). Such polymer composition materials are easily processable and have many film-forming properties, that allow one to produce various optical elements, band light filters and many other valuable products.

Nanoparticles are well known for their attractive optical properties: strong optical resonance and large and fast nonlinear polarizability associated with the plasmon frequency of the conduction electrons in the particle [159]. In addition, due to the plasmon resonance the large-scale aggregation of the nanoparticles may also be avoided with polymer-stabilized particles using appropriate materials and processing routes. Transparent materials can be obtained, due to the small size of the nanoparticles, which prevents excessive light scattering [160–163]. There is a lot of interest in low dimensional structures, especially 1D arrays of metal and semiconductor nanoparticles, their potential applications in optical and electronic sensors, nanoscale electronics and catalysis [164,165]. Polymer nanosystems with inorganic semiconductive components (chalcogenides, transition metal oxides) have novel properties (electrophysical, photochemical, magnetic). They have high potential in such applications as solar cells, electroluminescent devices, and magnetic memory units [166–169].

Optoelectronics has been an intensively developing science in recent years. One of its directions (integral optics) includes the fabrication of multifunctional integral circuits and devices for information treatment, storage and transfer in the optical range of frequencies. Such devices can be autonomous or conjugated with optical fiber connection lines. The elemental bases of these so-called optoelectronic circuits are planar optical waveguides and therefore intense investigations are carried out of various optical materials: semiconductive, dielectrical, crystalline and amorphous (including inorganic, polymeric and hybrid, inorganic–polymeric ones). The investigation range is very wide – from synthesis and technology methods of material fabrication to optical and physical studies in the fields of a nonlinear optics, photophysics, and the theory of systems.

The first investigations in the field of optical properties of nanocomposites were carried out by Faraday [170]. In a recent survey [171] the dependence of optical properties of metal nanoparticles on particle size, shape and dielectric environment was studied. The photoconductive properties of many organic systems have been studied and it was shown that the polymeric systems are especially attractive due to their ease of processability.

The nanoparticle sizes can be related to the Bohr radius of excitons in semiconductors. It defines many optical, luminescent and redox characteristics of nanocomposites. The distinctive features of optical composition materials are the functional interconnections between an optical carrier and medium and the dependence of optical parameters on external irradiation [172–174]. A general scheme of material "pumping" by exciting photons and its "reaction" to such action is shown in Fig. 11.14 [175].

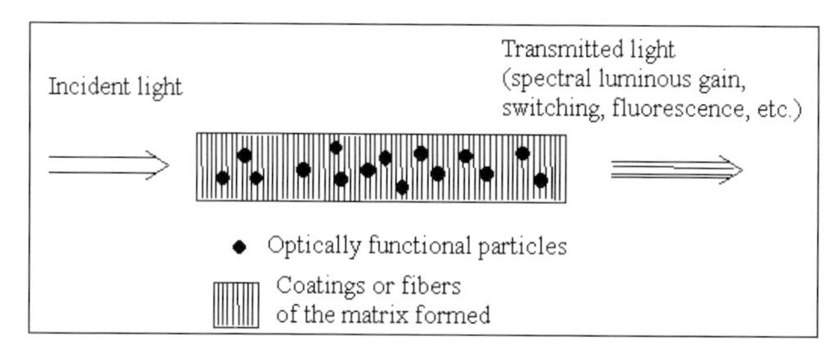

Fig. 11.14. An optical composite scheme

In many optically active systems (e.g. in colloidal suspensions of ultradispersed metal particles in liquid or solid inorganic and polymeric matrices) some resonance adsorption bands (so-called UV-Vis spectra) arise in the visible (400–800 nm) and near-infrared ($\lambda = 800$–1700 nm) spectral regions, as well as in the UV-region ($\lambda = 250$–400 nm), as shown in Fig. 11.15. The color tunability of semiconductor nanocrystals as a function of size could lead to many applications that rely on color multiplexing [176]. For example, $(PEO)_xV_2O_5 \cdot nH_2O$ films are photosensitive and their color becomes light-blue on irradiation by a mercury-discharge lamp. Such coloring changes are caused by a photoinduced reductive reaction, when PEO is oxidized by pentavalent vanadium, whereas an initial $V_2O_5 \cdot nH_2O$ xerogel film is not light sensitive [177]. It was shown that the process was accompanied by an increase of V^{4+} concentration (that increases the magnetic momentum up to 0.8 μB per formal unit at 300 K). After irradiation the conductivity increases by two orders of magnitude. $(PEO)_xV_2O_5 \cdot nH_2O$ films exhibit semiconductive properties and its conductivity at room temperature is 10^{-2} ohm^{-1}cm^{-1}.

Some organic dyes are used for photoactive nanocomposite material production. Thus it was found that reduction of methylene blue (MB^+) by V^{4+} particles was the driving force of its intercalation into the $V_2O_5 \cdot nH_2O$ gel [178]. The reduction was followed by protonation, reoxidation and dimerization of methylene blue. The metal-containing nanoparticles introduced into matrix media can produce some anomalous optical properties under the action of the electric field of an exciting light wave. These properties are connected first with the dimensional effects on the inorganic nanoparticle level and it can sometimes be used even for determining the particles size by the

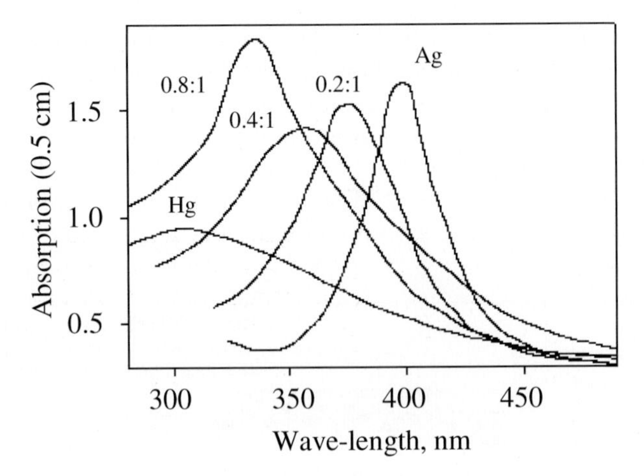

Fig. 11.15. Adsorption spectra for clusters of silver (∼5–10 nm), mercury (∼5–10 nm) and their bimetallic colloids (Hg–Ag) in polyethyleneimine. The relations Hg:Ag at constant Ag content are given on the curves [122]

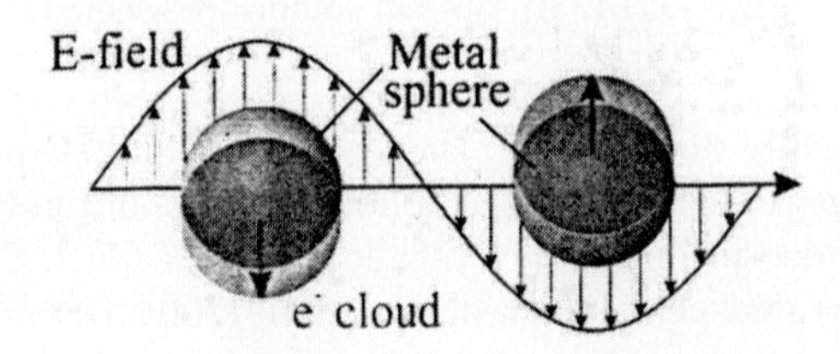

Fig. 11.16. Schematic presentation of plasmon oscillation for a sphere, showing the displacement of the conduction electron charge cloud relative to the nuclei

material color. For example, particles of zero valence silver Ag_3-Ag_{11} in a cellulose matrix are yellow-colored; $Ag_{11}-Ag_{22}$ particles are redish and brown; but $Ag_{>5000}$ particles are gray or green-colored [179]. Similar examples were observed for other mono- and bimetallic nanoparticles (Fig. 11.15).

The optical characteristics are studied for systems such as silver nanoparticles [180–183]); bimetals Ag–Hg [184]); semiconductors of II–VI (CdS, CdSe) [185–192] and the III–V (InP, InAs) [193–201] groups of the Periodical Table; and pigments, such as $PbCrO_4$ [202].

Optical adsorption (so-called "pumping") of materials is caused by excitation of the electron plasma oscillations (plasmons) in the nanoparticles. The process parameters (frequency, intensity, bandwidth and form) depend on the particle temperature, size and form, as well as on the nature and morphology of the matrix or composite [171, 202–205]. When a small spherical metallic nanoparticle is irradiated by light, the oscillating electric field causes the conducting electrons to oscillate coherently. The plasma frequency usually decreases with the particle size, because the number of oscillating free electrons is decreasing [171] (Fig. 11.16).

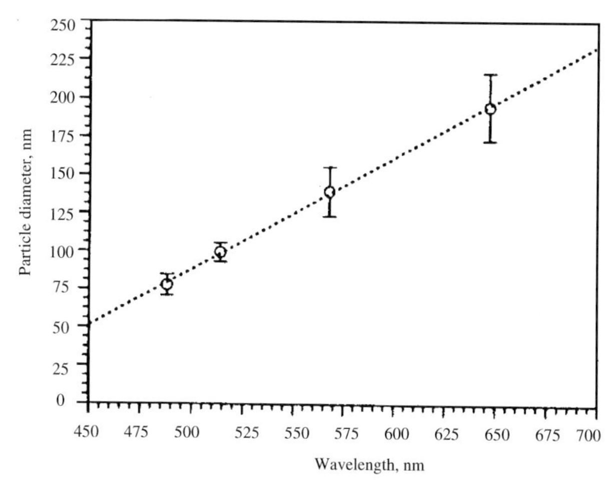

Fig. 11.17. Plot of Ag particle size as a function of excitation wavelength

Correlated optical and morphological studies reveal that the optically active nanoparticles identified at different wavelengths have different sizes. The relationship between particle size and excitation wavelength is found to be approximately linear for surface-enhanced Raman scattering Ag colloidal particles [206] (Fig. 11.17).

Thermokinetic studies of the evolution of nanocomposite adsorption spectra allow one to find and optimize the most stable states of the optical nanocarriers. In particular, the forming conditions of thin coatings with high reflecting ability were determined after adsorbing power investigations of some composites (from Ag-containing particles (30–60 nm) and polydiacetylene [183], copolymers of 3,3′,4,4′-benzophenone-tetracarbonic acid dianhydride with 4,4′-oxidianiline [180, 181]). Such materials can be used in *concentrator–reflector* systems for solar energy accumulation and transformation both in space [207] and on Earth [208], in reflecting devices of γ-telescopes [209], and in bactericide films [210]. Metallodielectric colloidal core–shell particles (small gold nanoclusters with a variable core radius and controlled gold shell with thickness of about 20–80 nm, attached to silica spheres) were prepared for photonic applications by the sol–gel method [211]. Such thin (5–20 nm) dielectric envelopes of metal (metal oxides) nanoparticles offer a chance to study their thickness by simple optical methods competitive with electron diffraction methods.

A novel impact for such systems design is connected with the use of powerful optical lasers for nanocomposite studies. The properties of the NLO effects (NLO – nonlinear optical) were discovered in an amplification of the local electrical field of the light waves [175, 212–215]. The NLO interactions are now widely used in spectroscopy (for example, for combination scattering spectroscopy), in local nonperturbing methods of fast process diagnostics, and for coherent short-waves generation by optical frequency transformation. Thus a combination of organic chromophore and nanoparticle gold was adapted for incorporation into PMMA-based composite films, whose electronic coefficient is larger than the sum of the individual contributions of dye and nanoparicles [216].

Induced system polarization P (dipole momentum per unit volume) arising under the action of the electrical field of the light wave with strength E ($V \cdot m^{-1}$) can be for simplicity presented as a Taylor expansion:

$$P = \varepsilon(\chi^{(1)}E + \chi^{(2)}E^2 + \chi^{(3)}E^3 + \ldots) , \tag{11.7}$$

where ε is the dielectric constant of the medium.

If the polarization depends linearly on the field strength under usual conditions (i.e. at values $E < 10^8$–10^9 $V \cdot m^{-1}$), the main contribution in P gives the first term of the expansion with coefficient $\chi^{(1)}$ (polarizability) and the following terms, $\chi^{(2)}$ (hyperpolarizability, second order susceptibility), $\chi^{(3)}$ (third order susceptibility), etc. are very small. All the usual effects (diffraction, reflection, interference) are defined by known linear optics laws. The nonlinear effects become important only when the field strength is 10^9 $V \cdot m^{-1}$, i.e. under high excitation intensity.

The effect in optical $\chi^{(2)}$-composites (so-called second order NLO materials) is observed near the doubled frequency of the excitation line (the line position in the composite spectrum is determined by the corresponding frequencies of nanoparticles). Such $\chi^{(2)}$ composites are now widely used in electrooptical transducers [175] and photonic devices (e.g. for frequency doubling and electrooptical modulation) because of their large optical nonlinearity, excellent processibility, low dielectric constants, and high damage thresholds [172, 217]. The $\chi^{(2)}$-characteristics are well studied and optimized over temperature for many composites, containing small particles in zeolites [218], as well as for polymer matrices, such as epoxides [219], siloxanes [220, 221], polyimides [222], melamines [223, 224], and P4Vpy [225]. Nonlinear optical effects have also been observed in layer-by-layer assembled films of magnetic and metallic nanoparticles [226,227]. Materials synthesized with inserted into transparent layered MPS_3 (M=Mn, Cd, Zn) are active for second harmonic generation [228].

Metal nanoparticles dispersed in a dielectric medium have a unique optical effect called surface plasmon resonance that serves as the basis for high third-order optical nonlinear susceptibility. The optical effects connected with $\chi^{(3)}$ NLO materials relate mainly to nanocomposites including metallic (Au, Ag [215, 229–232]) and semiconductive [185–188] nanosized particles. They are usually based on specific resonance conditions, when an energetic transition frequency (for example, a plasmon frequency in the composite optical inorganic component) coincides with the exciting laser frequency. As a result, an excitation is localized in the highly polarizable nanoparticles that drastically amplifies the local field generated by the induced irradiation.

In particular, such effects are used in coherent anti-Stokes Raman spectroscopy[6] when two laser beams (with frequencies ν_1 and ν_2 and the given conditions of phase consistency) are simultaneously interacting with the medium and, as a result, directed coherent laser-like radiation with frequency $\nu = 2\nu_1 - \nu_2$ is generated. Using a laser with tunable frequency ν_2 and changing the intensity of ν (as a function of $\nu_1 - \nu_2$), it is possible to get the resonance conditions at $\nu \approx \nu_1 - \nu_2$. Materials with NLO characteristics are very important for use in electrooptical devices and various starting processes.

The sizes of the local excitation region (resonance domain) depend on the morphology of the composite structure and the wavelength of the exciting light. The ratio of particle sizes (particles group or domain, effectively interacting with light) to the wavelength in composites is usually bigger than for conventional molecular media. This is true for sol–gel derived gold nanoclusters in silica glass (the mean diameters of Au nanoparticles in gels and glasses vary from 10 to 20 nm and from 27 to 38 nm, respectively) possessing large optical nonlinearities [234]. The third-order optical nonlinearities χ^3 exhibit a higher value of $2.2 \cdot 10^{-9}$ esu.

The Au particles in such nanocomposites are condensed with a large excess of diacetylene monomer (diphenylbutadiine) or its polymer [229], after which the metallopolymer is produced (containing 7–16% of metal particles with a mean diameter of ~2 nm). This composite demonstrates a 200-fold amplification of third-order optical coefficient. The Au- and Ag-containing plasma polymerized polyfluorocarbon thin films have good optical characteristics too [230, 231].

Giant nonlinear optical activity was observed in stabilized PVP fractal aggregates of silver nanoparticles (~10 nm) excited by light with $\lambda = 532$ nm and intensity 2 mV·cm^{-2} [215]. An amplification effect (~10^2 times) was registered in nanoparticle aggregation (with a number >1000) into a fractal complex ($D_m \approx 1.7$). Another interesting approach (that can be considered as an alternative to conventional galvanic technologies) is connected with the use of photoaddressable diazosulfonate terpolymer films, covalently attached on glass or silicon substrates as the template surface for selective complexation of silver salts and subsequent reduction of the salt to silver nanoparticles [235, 236].

The properties of NLO materials can be quantitatively characterized by a third order susceptibility ($\chi^{(3)}$, esu), a nonlinear refractive index (n_2, cm^2/W) or nonlinear adsorption coefficient (β, cm^2/W). The values of n_2 and β are connected with the real (Re) and imaginary (Im) parts of the complex-valued $\chi^{(3)}$ by the next equations:

$$\mathrm{Re}[\chi^3] = n_2 n_0/(12\pi); n_0 = (1 + 4\pi\mathrm{Re}[\chi^{(1)}])^{1/2} \tag{11.8}$$

$$\beta = 96\pi^2\omega\mathrm{Im}[\chi^{(3)}]/(n_0 c)^2 \, , \tag{11.9}$$

[6] Nanoparticles use in Raman spectroscopy have enormous enhancement factors of 10^{15} per aggregate of silver particles [233].

where n_0 is a linear refraction coefficient, ω is the angular frequency of light wave and c is the velocity of light.

The properties of the $\chi^{(3)}$ NLO make themselves evident not only in metal-containing polymer composites but also in these precursors. The third-order susceptibility coefficients [$\chi^{(3)}$] were studied for a large class of metal-containing compounds: halogenides, acetylides of Ti-, Zr- and Hf-metallocenes, alkinylzirconocenes [171], octapolar alkinylruthenium complexes, including $Ru(C{\equiv}CPh)(C{\equiv}C{-}C_6H_4{-}C{\equiv}CPh)(Ph_2P)_4$, $ClRu(C{\equiv}CPh)(Ph_2P)_4$ [187]. The values of $\chi^{(3)} \cdot 10^{36}$ were about 1–5 esu with the third harmonic excitation in laser irradiation with 1908 nm for metallocene halogenides and rhuthenium complexes, while for the majority of acetylide metallocenes these values were practically one order of magnitude higher (for titanocene biphenyl-acetonyl $\chi^{(3)} \cdot 10^{36} = 92$ esu and for $Cp_2Zr(Cl)CH{=}CH{-}1{,}4{-}C_6H_4{-}CH{=}CH(Cl)ZrCp_2$ – $\chi^{(3)} \cdot 10^{36} = 154$ esu).

In recent years there is special interest in nanosized semiconductor structures, such as quantum dots (QD), quantum wells and quantum wires. The interest is connected with their peculiarities relative to conventional crystalline structures. The so-called quantum-dot polymers are very interesting examples of NLO properties [214]. This effect arises when the sizes of metallic or semiconductor particles introduced into the rigid matrices (glasses, polymers, etc.) are considerably less than the exciting field wavelength ($\ll \lambda/20$). A localized free charge polarization in the nanoparticles under the effect of an electric field can essentially modify the electron quantum states which results in a change of the dielectric constant of the composite medium as a whole. There are some effects of the local field due to a difference in the constants between the QD and matrix volume, which leads to an amplification (up to several order of magnitude) of the optical response. There is a rigid connection between the exciting light wavelength and the nanoparticle diameter (an optical response arises with decrease in the particles size). In this connection QD-polymeric composites are very interesting objects for forming active layers in photodiodes [237,238]. An optical excitation in the case of fluorophores results in the appearance of a narrow luminescence band (electroluminescence). Polymers (on the organic side) and metallic complexes (on the inorganic side) were made to investigate electroluminescent properties [239–242]. It relates to organic conjugated polymers, such as poly(*p*-phenylene vinylene)(PPV) and its derivatives. They presented many advantages in electroluminescence applications because of their high fluorescence efficiency and their good processability [243–247]. The peculiarities of luminescence quenching may reflect some sophisticated details of the interaction process between photoexcited semiconductor nanoparticles and polymers [248].

The most widely studied semiconductor QD-materials are CdS and CdSe nanoparticles, stabilized by various polymers [101,102,185,189,190,214,249], glasses [250], and Langmuir–Blodgett films [251]. Uniform composition materials with high reflectivity and adsorption coefficients suitable for production of optical elements used polymer glasses (doped by transition metal sulfoselenides), photochromic glasses (with Ag and Cu halogenides), or holographic emulsions of halogen-silver nanoparticles in gelatin. As a rule, the semiconductive component content in the polymer matrix (P4VPy, PVC, etc.) does not exceed 0.1%, i.e. practically only "dilute solutions" are used.

It is of special interest to introduce (dope) metal ions into polymeric fibers to produce colored light guides that can have a very wide range of applications: from logical or memory storage elements of optical computers [252,253] to sensitive detectors for medical diagnostics [254]. Thus doping of Nd^{3+} (neodium octanoate) into PMMA allows one to get (by optimization of the fiber length and Nd^{3+} concentration) high-quality green-coloured optical fibers ($\lambda = 532\,\text{nm}$) [255].

A very important practical problem was to design a polymer with a specific refractive index precisely in the range 1.420–1.460. The aim was achieved by using copolymers from methacrylate monomers containing fluorine atoms [256]. The technology of forming such hybrid nanomaterials and their waveguide characteristics are described in [257]. A comparison of the refractive indices (measured by the prism coupling method) versus wavelength for silica and copolymers is given in Fig. 11.18.

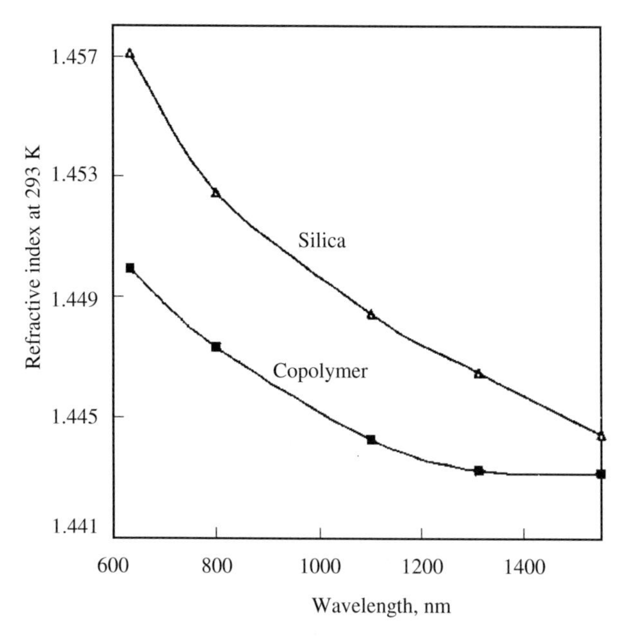

Fig. 11.18. Examples of chromatic dispersion, measured for copolymers and for pure silica

The majority of trivalent lanthanide ions have long-lived luminescent 4f state (f–f transition generated by direct metal excitation or more commonly via the "antenna effect" - energy transfer from ligand excited states to metal, which exhibit sharp emission bands at characteristic wavelengths ranging from green (Tb^{3+}) to near-infrared light (Nd^{3+}, Er^{3+}) [258]. Therefore an effective method of fabrication of luminescent nanocomposites is incorporation of lanthanide salts (usually Eu^{3+}, Tb^{3+}) in sol–gel derived silica/PEO hybrid gels [259–262]. Other types of polymers are used too, i.e. poly(2,5-methoxy-propyloxysulfonate)phenylene vinylene [263, 264], poly(p-phenylene vinylene)-silica composite (a novel sol–gel processed non-linear optical material for optical waveguides [265]) or polyester macroligands [266].

Such composites are now used as organic light emitting diode materials for new display technologies [267], in layer-by-layer assembly of thin film Zener diodes from conducting polymers and nanoparticles [268], and in diodes in biological contexts in time-resolved fluoroimmunoassays [269].

Hybrid polymer-inorganic type nanomaterials are widely used also as diffractive optical elements (in application such as optical signal processing), optical interconnects, optical data storage, optical computing, binary optics and integrated optics devices. Several techniques including holographic interference, electron-beam lithography, photolithography and variable acoustic-optics modulated laser beam devices, are used for fabrication of the diffractive gratings [270–272]. Here are a few examples: patternable TiO_2/Ormosils – hybrid films for photonics applications [273]; TiO_2/PVP(polyvinylpyrrolidone), that is one of the n-type semiconductors which could be applied to photochemical electrodes for the conversion of solar energy into other forms (such as hydrogen gas and electricity, see for example [274–276]), and CdS nanocrystals (7 nm) entrapped in thin (100 nm) SiO_2 films [277]. Such products are used for light-induced charge separation in sensitized sol–gel processed semiconductors [278–280] and are discussed in Chap. 12.

The numerous variants of Ru^{2+}-complexes use in film nanocomposites must be mentioned, especially because the compound is a well-known redox-active photosensitizer with a large molar absorption coefficient in the visible light region. They are often used in photoelectric conversion systems such as photoelectrochemical cells [281–285], Zn-metalloporphyrin doped alumosilicate sol–gel hosts [286], and metal–organic π-conjugated oligomers with (L)$Re(CO)_3Cl$ chromophore, having a relatively long-lived red photoluminescence [287].

One of the modern directions of material science is connected with forming and studies of molecular level organized systems, because it is a way to construct a new generation of molecular electronic photodevices. The above described Langmuir–Blodgett film technique is very attractive for its simplicity and efficiency. Molecularly thin films of polymer–metal complexes have attracted interest because of their potential application in sensors, catalytic systems and molecular electronic devices, and generation of effective photocurrent devices, based on a polymer LB monolayer on a silver electrode excited by surface plasmon resonance [288–291]. A very suitable practical method of production of such materials is exposure of fatty acid salt LB films to H_2S gas that results in the formation of semiconductor chalcogenide nanoclusters in the LB matrix[7] [294–296]. Polyaniline (PANI)-cadmium sulfide films have been prepared by co-depositing cadmium particles during the electrochemical deposition of PANI [297], and imidazole-containing polymers with metal ions at the air–water interface and in LB films were studied [298]. This self-supporting capability of LB films should be important for practical applications in sensors and selective permeation of gases or biomolecules, and in nonlinear optical and electroluminiscent devices [299]. LB multilayers of a copolyimide incorporating both copper-phthalocyanine and carbazole as electron donors were produced in a bilayer photoreceptor as the charge-generating layer [300]. This type of LB film deposited on metal electrodes can regulate the photocurrent behavior and photoinduced surface [301]. It is interesting to use indium tin oxide (ITO) electrodes coated with LB film carotenoids immersed in an electrolyte solution [302], as well as coated by a mixture of TiO_2 nanoparticles, carotenoid and PVK [303]. Polyaniline (PANI)-TiO_2 nanocomposite [304] has high photoconducting properties, as shown in Table 11.5.

Table 11.5. The characteristics of PANI-TiO_2 composites

PANI Content, wt.%	Density, $G \cdot cm^{-3}$	Conductivity, $ohm^{-1} \cdot cm^{-1}$	Colour	Dimension, nm
100	1.16	1.36	bottlegreen	–
88.5	1.20	0.96	emerald	50–60
72	1.31	0.63	turquoise	50
53	1.72	0.39	aqua	40–45
39	2.04	0.19	celadon	30–35

There is continuing interest in the photochemical properties of nanocomposites, including metal chalcogenides. Apart from high transparency, such nanocomposites also possess a variety of other potentially useful properties. Polymer-based nanocomposites containing well-dispersed inorganic particles can exhibit semiconducting properties [305, 306], quantum dot effects [307],

[7] In reference [292] thermodynamical analysis is given of Zn, Pb and Cd behenate multimolecular layers sulfidation and nanoclusters forming (with mean sizes 3–6 nm), corresponding to sulfides into LB-films matrix. The electron structure of the metal sulfide nanoclusters is discussed by many authors (e.g., see [293]).

non-linear optical properties [308, 309] and extremely low [310] or high [311] refractive index.

Nanocomposites with inorganic semiconductive compounds (sulfides, transition metal oxides, etc.) can be produced by the exchange reactions of metal-containing compounds with formation of a dispersed component directly into the volume of a swollen polymeric matrix (see Chap. 5). Such composites have unusual electrophysical, photochemical and other properties that can be used in photocatalysis, electrophotography, magnetic memory device fabrication, and laser techniques.

The kinetic approach to forming CdSe and InAs particles (by "focusing" their sizes and distribution) allowed one to produce monodispersed particles [192] (Fig. 11.19).

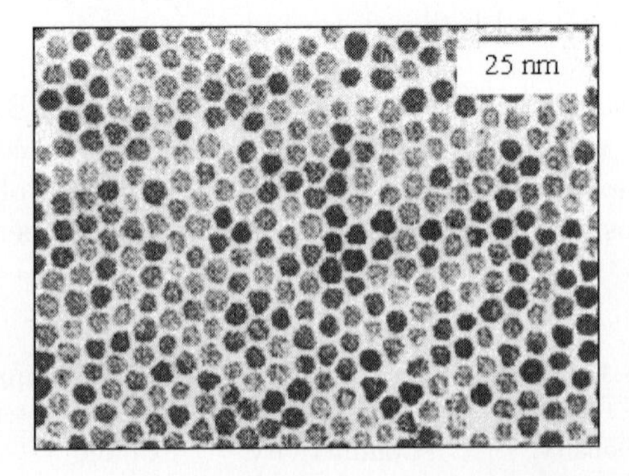

Fig. 11.19. Electron microscope photograph of CdSe nanocrystals (with diameter 8.5 nm), formed by the "distributed focusing" method [192]

Optical excitation of diblock-copolymer composites $[MTD]_{300}[NBE–CH_2O(CH_2)_5P(Oct)_2]_{20}$ (where MTD is methyltetracyclododecene and NBE is 5-norbornene-2-yl) with practically monodispersed filler (CdSe nanoparticles) showed (Fig. 11.20) that the radiation frequency in the luminescence spectrum was 436 nm (for CdSe nanoparticles with sizes ~3.7 nm) and 478 nm (for particles with sizes ~4.5 nm) [190]. The NLO characteristics of some semiconductor-based nanocomposites and comparison with conventional materials are given in Table 11.6.

Semiconductor clusters are a potentially useful class of polymer photosensitizers for many applications that require materials with photoconductive and photorefractive properties. While there are several polymers which show reasonable charge generation in the ultraviolet region, it is often desirable to enhance their optical response in the visible portion of the spectrum. The physical and chemical properties of small (sizes <10 nm) semiconducting particles are very peculiar and they can be considered as a big single molecule with high energy of intramolecular interactions. Electronic excitation of such nanocrystals leads to the formation of a weakly coupled electron–hole

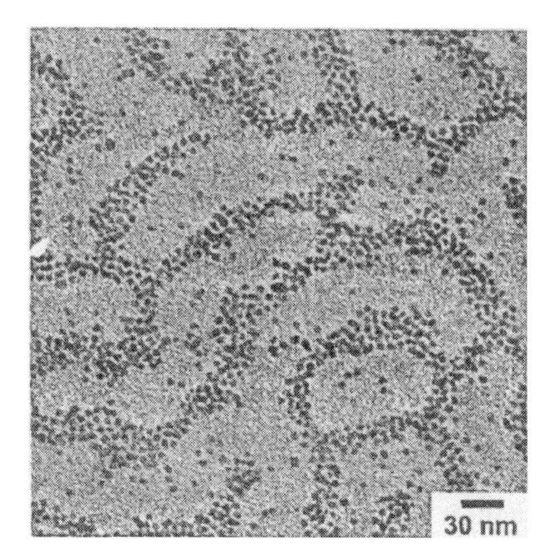

Fig. 11.20. Electron microscope photograph of CdSe nanoparticles (~4.5 nm), stabilized into a $[MTD]_{300}[NBE\text{–}CH_2O(CH_2)_5P(Oct)_2]_{20}$ matrix [190]

pair with a delocalization region considerably bigger than the crystal lattice constant, which has a strong influence on the nanocomposite electronic characteristics [312]. The physical properties of semiconductor nanocrystallites in such systems are dominated by quantum confinement, the widening HOMO–LUMO gap with decreasing particle size which directly affects the photophysics of the material. The proper control of particle size is critical in any investigation involving these materials, but simultaneously the surface chemistry of the nanocrystals is of major importance, because functional surface ligands can serve as linkers between the particles and an appropriate docking site. For example, increase in electrical conductivity of the photoconductive materials upon illumination is important in an array of commercial application such as transistors and detectors. The following outstanding features can therefore be expected when particles are used as an emitting layer in an organic–inorganic hybrid device:

(a) since the emission wavelength can be controlled by the particle size and their chemical properties, the process of device fabrication remains unaltered for the fabrication of different colors;

(b) various polymers may be used in combination with the particles to optimize the transport properties of electrons and holes, respectively;

(c) the high fluorescence efficiency of semiconductor nanoparticle materials will contribute to improving the external quantum efficiency of such organic–inorganic hybrid devices [313].

Thin films of semiconducting polymers are used in electronic and optoelectronic devices such as field-effect transistors, light-emitting diodes, solar

Table 11.6. Third-order nonlinear optical properties of some optical nanocomposite materials [175]

Nanocomposite	NLO Characteristics of Material, esu	NLO Parameter Expression	λ, nm
CdS–Nafion	$-6.1 \cdot 10^{-7}$*	$\mathrm{Im}[\chi^{(3)}]/\beta$	480
CdS–Nafion/NH_3	$-8.3 \cdot 10^{-7}$*	$\mathrm{Im}[\chi^{(3)}]/\beta$	450
CdSe–PMMA	$1.2 \cdot 10^{-5}$*	$\mathrm{Im}[\chi^{(3)}]$	544–560
Glasses CdS_xSe_{1-x}	$1.3 \cdot 10^{-8}$	$\{\mathrm{Re}[\chi^{(3)}]^2 + \mathrm{Im}[\chi^{(3)}]^2\}^{0.5}$	532
CdS in sol–gel glasses	$5 \cdot 10^{-12}$	$\{\mathrm{Re}[\chi^{(3)}]^2 + \mathrm{Im}[\chi^{(3)}]^2\}^{0.5}$	380
PFV in SiO_2	$3 \cdot 10^{-10}$	$\{\mathrm{Re}[\chi^{(3)}]^2 + \mathrm{Im}[\chi^(3)]^2\}^{0.5}$	602
PFV in V_2O_5	$6 \cdot 10^{-10}$	$\{\mathrm{Re}[\chi^{(3)}]^2 + \mathrm{Im}[\chi^{(3)}]^2\}^{0.5}$	602
GaAs in Vicor glasses	$-5.6 \cdot 10^{-12}$*	$\mathrm{Re}[\chi^{(3)}]$	1064
Standard NLO-materials			
Fused quartz	$8.5 \cdot 10^{-14}$	$\mathrm{Re}[\chi^{(3)}]$	1064
SF_6	$8 \cdot 10^{-13}$	$\mathrm{Re}[\chi^{(3)}]$	1064
CdS	$-5 \cdot 10^{-11}$	$\mathrm{Re}[\chi^{(3)}]$	610

*unit of measurement – cm^2/W

cells, and xerography photoreceptors. Various attempts were made to characterize the kinetics of photoexcited charge transfer by several methods in a wide range of timescales, down to femtoseconds. Laser flash photolysis, resonance Raman spectroscopy, diffused reflection and microwave absorption have been demonstrated to be useful in investigating the mechanism of charge injection from the excited state of the sensitizer to the conduction band of the inorganic semiconductor. Semiconductor investigations are mostly interdisciplinary and are related to metallopolymers based on well-studied semiconductor materials, such as CdS (for which a "blue shift" of exciton adsorption band is observed from the nanocrystal sizes 5–6 nm), CdSe, PbS, HgS, MoS_2, GaP, TiO_2, SiO_2, ZnO, BiJ_3, PbJ_2, and Fe_2O_3. The properties of CdS-based metallopolymer nanocomposites have significant technological applications as nonlinear optical materials [314–317]. Moreover, the absorption characteristics of CdS nanocomposites are affected by the crystalline size, and their luminescence is a strong function of the surface characteristics. For example, using polymers with aminogroups for CdSe or ZnS bonding increases the fluorescence quantum yield up to 70–80%, if the solution contains an excess of these ligands [318, 319]. In photoconductive characteristics of inorganic (CdS)/organic·PVK the latter component serves as a charge-transporting matrix and a QD composed of surface passivated CdS serves as a charge-generating sensitizer [168]. The hyperbranched (the third generation) conjugated polymers-CdS relative fluorescence quantum efficiencies (in dichloromethane solutions) were determined for nanocomposites [320] from the corrected spectra against anthracene as a standard (ϕ = 0.36, excited at 323 nm). The efficiency was within the limits $\phi = 0.22 - 0.46$ and a two-exponential decay profile fluorescence lifetime ($\tau_1 = 1.1 - 2.6$ and $\tau_2 = 6.8 - 8.9$ ns was observed depending on the polymer molecular mass. Macroligands with aminofunctionalities protect the CdSe/ZnS inorganic core and, as a result, a long-term quantum yield is observed [321]. The same is true for CdTe nanocrystals too [242]. Nanoscale size affects the photoconductivity of semiconducting polymer thin films, for example poly(2,5-pyridylenebenzobisthiazole) [322]. The direct measurements of ultrafast electronic relaxation dynamics in PbI_2 colloidal nanoparticles in a PVA matrix in various solvents were carried out using femtosecond transient absorption spectroscopy [323]. Oscillations appeared in such systems at early stages (in alcohol dominated by 75 ps decay) with the period changing with solvent but not with particle size (3–100 nm). Probably, this happens because the relaxation is dominated by surface characteristics that do not vary significantly with size and spatial confirment is not significant in affecting the relaxation process. The electronic properties with excess charges on the polymer nanoparticles are predicted analogously to those of semiconductor QDs and hence these structures are referred to as quantum drops [324]. Chalcopyrite $CuInSe_2$ is a promising candidate for the anode materials of photochemical devices due to its high performance and high output stability [325]. $CuInS_2$ is also a suitable absorbed material for polycrystalline thin film solar cells with energy conversion efficiency as high as 17% [326].

One important problem in the field is connected with photoconductivity and photorefractive polymeric nanocomposites. Photoconductivity in materials is based on two processes: photogeneration of charge carriers and their subsequent transport. Multifunctional properties are derived from the combination of photoconductivity and electrooptic activity. Inorganic–organic hybrids are a novel class of materials for potential application to real-time optical information processing [327]. Photoconductive materials have drawn given involvement in interesting fields such as xerography and photorefractivity [328–331]. Besides the thermoplastic behavior in holographic xerography applications or optical nonlinearity in photorefractivity and photoconductivity determines the optimum work of the material. Such materials are suitable for infrared wavelengths from 1.31 or 1.55 μm commonly used in optical communications.

PVK composites with their extraordinary properties are used as charge-transporting matrices (hole transporting). Thus the increase in photoconductivity in the CdS/PVK system is probably due to the ability to achieve relatively higher loading for encapsulated CdS (> 1%) [332]. The main factor is a higher dielectric constant associated with inorganic materials. This leads to a decrease in the elective space-charge field, experienced by the chromophores and to a decrease in photorefractive properties. More convenient for these applications are PVK composites with nanocrystal PbS (mean diameter ~50 nm) or HgS (mean diameter ~10 nm) [333]. The quantum efficiency of the photocharge generation process in these materials was characterized at this wavelength. The photosensitive properties of such nanocomposites are given in (Table 11.7).

Table 11.7. Photosensitivity of PVK/SiO_2 IPN photoconductive materials and for some polymeric compounds[a]

S^a_{ph} Cm/(Ω/W)	Type of Material	Reference
$1.4 \cdot 10^{-11}$	PVK/TNF/SiO_2IPN	[328]
10^{-12}	PVK/CdS-nanocrystal polymer composite	[168, 332]
10^{-11}	Full-functionalized polymer composite	[331]
10^{-14}	Carbazole discrete units within a sol-gel matrix	[334]
10^{-10}	PVK/TNF/chromophore polymer composite	[335]

[a]The S_{ph} value has been calculated units, σ_{ph} – photocurrent density per unit of applied electric field, normalised by the illumination irradiance. TNF is 2,4,7-trinitro-9-fluorenone, IPN – interpenetrated network

Interesting results were obtained [336] in investigations of the poly-*p*-xylylene-PbS (nanocrystal size ~4 nm) system by simultaneous low-temperature condensation of *p*-xylylene and PbS vapors (Chap. 6). The dark conductivity of this nanocomposite is characterized by the linear *current–voltage* dependence (Fig. 11.21, curve 1). The *photocurrent–voltage* relationship is nonlinear (Fig. 11.21, curve 2) and is well described by the relation $\lg I \sim V^{1/2}$. This seems to be due to the tunneling of the charge carriers between semiconducting nanoparticles of PbS, when, as a result of UV irradiation, the increase in conductivity of samples occurs (this effect is reversible).

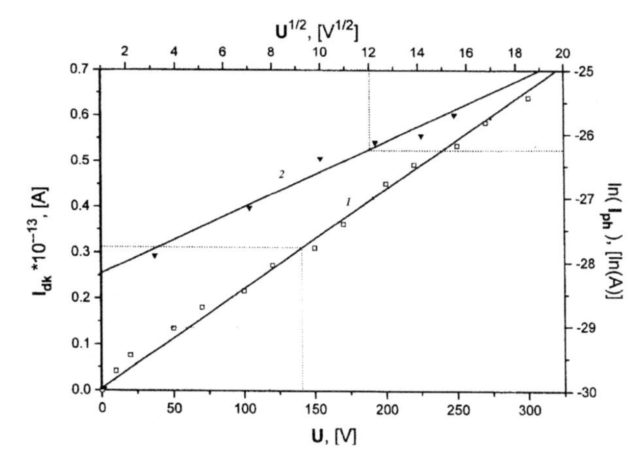

Fig. 11.21. Dark- and photoconductivity of poly-*p*-xylylene-PbS composite film (4.7 vol.% of PbS): the current–voltage dependence of staring films (*1*) and photocurrent–voltage relationship (*2*)

The hypothesis is proved also by low activation energy of photoconductivity (less then 0.1 eV), by a nonlinear volt–ampere characteristic under UV irradiation, and phenomena of photoconductivity at quantum energy less than 2 eV.

Metallopolymeric nanocomposites dispersed in polymer chalcogenide glasses were used as sensor units [337]. A simple and facile means of anchoring different ligand molecules onto particle surfaces allows their utilization as colloidal sensors [338, 339]. It is possible to produce analogues of human receptors of taste able to distinguish the four basic types of taste detected by the human tongue (sweet, salty, sour and bitter tastes) [340]. In particular, attempts to use as the artificial taste sensors some LB films of conducting polymer and Ru^{2+} complexes were successive [341]. Such materials can serve as a tool to improve quality control in the food and beverage industry. This is particularly important in cases where humans cannot be used in tests owing to the toxic nature of the contaminants. This is refereed to us the global selectivity concept.

In recent years there has been an expansive development in the field of the investigations of optical properties of metal-containing matrices, and in the near future some new fundamental effects and phenomena are likely to be discovered. Moreover, some novel classes of materials (such as dendrimers, for example) have also been developed [342–344].

References

1. C. Sanchez, F. Ribot, B. Lebeau: J. Mater. Chem. **9**, 35 (1999)
2. H. Law, T. Tou: Appl. Opt. **37**, 5694 (1998)

3. *Physics and Chemistry of Luminescent Materials. VI* ed. by C. Ronda, T. Welker (San Francisco, CA 1998)
4. C. Roth, R. Kobrich: J. Aerosol Sci. **19**, 939 (1988)
5. M.D. Greaves, V.M. Rotello: J. Am. Chem. Soc. **119**, 10569 (1997)
6. Q.R. Huang, H.-C. Kim, E. Huang, D. Mecerreyes, J.L. Hedrick, W. Volksen, C.W. Frank, R.D. Miller: Macromolecules **36**, 7661 (2003)
7. M.L. Brongersma, J.W. Hartman, H.A. Atwater: Phys. Rev. B **62**, 16356 (2000)
8. R. Davies, G.A. Schurr, P. Meenan, R.D. Nelson, H.E. Bergna, C.A.S. Brevett, R.H. Goldbaum: Adv. Mater. **10**, 1264 (1998)
9. C.G. Gtanqvist: *Handbook of Inorganic Electrochromic Materials* (Elsevier, Amsterdam 1995)
10. P. Ghosez, K.M. Rabe: Appl. Phys. Lett. **76**, 2767 (2000)
11. J.M. Oton, A. Serrano, C.J. Serna, D. Levy: Liq. Cryst. **10**, 733 (1991)
12. E.J.A. Pope, M. Asami, J.D. Mackenzie: J. Mater. Res. **4**, 1018 (1989)
13. A.E. Berkowitz, J.R. Mitchell, M.J. Carey, A.P. Young, S. Zhang, F.E. Spala, F.T. Parker, A. Hutten, G. Thomas: Phys. Rev. Lett. **68**, 3745 (1992)
14. S. Takahashi, S. Maekawa: Phys. Rev. Lett. **80**, 1758 (1998)
15. J.M.D. Coey: J. Magn. Magn. Mater. **226**, 2107 (2001)
16. *Funcctionalization and Surface Treatment of Nanoparticles* ed. by M.I. Baraton (Am. Sci. Publ., Los-Angeles 2002)
17. V.E. Gul, L.Z. Shenfel: *Electroconducting polymer compositions* (Khimiya, Moscow 1984)
18. D.L. Lesli-Pelecky, R.D. Rieke: Chem. Mater. **8**, 1771 (1996)
19. A.Yu. Vasilkov, P.V. Pribytkov, E.A. Fedorovskaya, A.A. Slinkin, A.S. Kogan, V.A. Sergeev: Dokl. Phys. Chem. **331**, 179 (1993)
20. El-Awansi, N. Kinawy, M. Emitwally: J. Mater. Sci. **24**, 2497 (1989)
21. H.S. Göktürk, T.J. Fiske, D.M. Kaylon: J. Appl. Polym. Sci. **50**, 1891 (1993)
22. J. Yacubowicz, M. Narkis: Polym. Eng. & Sci. **30**, 46 (1990)
23. D.E. Nikles, J.L. Cain, Ap.P. Chacko, R.I. Webb: IEEE Trans. Magn. **30**, 4068 (1995)
24. D.C. Edwards: J. Mater. Sci. **25**, 4175 (1990)
25. X-S. Yi, G. Wu, Y. Pan: Polym. Int. **44**, 117 (1997)
26. C. Laurent, D. Mauri, E. Kay, S.S.P. Parkin: J. Appl. Phys. **65**, 2017 (1989)
27. D.L. Leslie-Pelesky, X.Q. Zhang, R.D. Ricke: J. Appl. Phys. **79**, 5312 (1996)
28. R.H. Marchessault, S. Ricard, P. Rioux: Carbohyd. Res. **224**, 133 (1992)
29. J.K. Vassilion, V. Mehrotra, M.W. Russell, E.P. Gianneis: Mater. Res. Soc. Symp. Proc. **206**, 561 (1991)
30. R.F. Ziolo, E.P. Gianneis, B. Weinstein, M.P. O'Horo, B.N. Ganguly, V. Mehrotra, M.W. Russell, D.R. Huffman: Science **257**, 219 (1992)
31. R.F. Ziolo, E.P. Gianneis, R.D. Shull: NanoStruct. Mater. **3**, 85 (1993)
32. M. Respaud, J.M. Broto, H. Racoto, A.R. Fert, L. Thomas, B. Barbara, M. Verelest, E. Snoek, P. Lecante, A. Mosset, J. Osuna, T.O. Ely, C. Amiens, B. Chaudret: Phys. Rev. B **57**, 2925 (1998)
33. M. Respaud, J.M. Broto, H. Racoto, J.C. Ousset, J. Osuna, T.O. Ely, C. Amiens, B. Chaudret, S. Askenary: Physica B **247**, 532 (1998)
34. F.J. Lozaro, J.L. Garcia, V. Schünemann, A.X. Trautwein: IEEE Trans. Magn. **29**, 2652 (1993)
35. J.J. Host, V.P. Dravid: Mat. Res. Soc. Symp. Proc. **457**, 225 (1997)

36. Ph. Colomban, V. Vendange: Mat. Res. Soc. Symp. Proc. **457**, 451 (1997)
37. V.M. Smirnov, N.P. Bobrysheva, V.B. Aleskovskii: Dokl. Chem. Phys. **356**, 492 (1997)
38. Z. Zhang, Y.D. Zhang, W.A. Hines, J.I. Badnick, W.M.H. Sachtler: J. Am. Chem. Soc. **114**, 4843 (1992)
39. Y. Yayakawa, S. Kohiki, M. Sato, T. Babasaki, H. Deguchi, A. Hidaka, H. Shimooka, S. Takahashi: Physica E **9**, 250 (2001)
40. O.M. Mikhailik, V.I. Povstugar, S.S. Mikhailova, A.M. Lyakhovich, O.M. Tedorenko, G.T. Kurbatova, N.I. Shklouskoya, A.A. Chuiko: Colloids Surf. **52**, 315; 325; 331 (1981)
41. S. Sun, C.B. Murray, D. Weller, L. Folks, A. Moser: Science **287**, 1989 (2000)
42. *Applications of Magnetic Carriers* ed. by U. Höfeli, W. Schütt, J. Teller, M. Zborowski (Plenum, New York 1997)
43. R.W. Siegel: Nanostruct. Mater. **3**, 1 (1993)
44. C.R. Vestal, Z.J. Zhang: J. Am. Chem. Soc. **124**, 14312 (2002)
45. H. Srikanth, R. Hajndl, C. Chirinos, J. Sanders, A. Sampath, T.S. Sudarshan: Appl. Phys. Lett. **79**, 3503 (2001)
46. S. Lopez, I. Cendoya, F. Torres, J. Tejada, C. Mijiangos: Polym. Eng. Sci. **41**, 1845 (2001)
47. S.A. Gomez-Lopera, R.C. Plaza, A.W. Delgado: J. Colloid. Interface Sci. **240**, 40 (2001)
48. D. Portet, B. Denizot, E. Rump, J.J. Lejeune, P. Jallot: J. Colloid. Interface Sci. **238**, 37 (2001)
49. H. Mazaki, M. Kakihana, H. Yasuoka: Jpn. J. Appl. Phys. **30**, 38 (1991)
50. C. Liu, B. Zou, A.J. Rondinone, Z.J. Zhang: J. Phys. Chem. B **104**, 1141 (2000)
51. L. Raymond, J.-F. Revol, D.H. Ryan, R.H. Marchessault.: J. Appl. Polym. Sci. **59**, 1073 (1996)
52. R.V.P.M. Shafi, A. Ulman, X. Yan, N.-L. Yang, C. Estournes, H. White, M. Rafailovich: Langmuir **17**, 5093 (2001)
53. T. Ji, H. Shi, J. Zhao, Y. Zhao: J. Magn. Magn. Mater. **212**, 189 (2000)
54. J. Ramos, A. Millan, F. Palacio: Polymer **41**, 8461 (2000)
55. S.N. Sidorov, L.M. Bronstein, V.A. Davankov, M.P. Tsyurupa, S.P. Solodovnikov, P.M. Valetsky, E.A. Wilder, R.J. Spontak: Chem. Mater. **11**, 3210 (1999)
56. N.I. Nikonorova, S.V. Stakhanova, I.A. Chmutin, E.S. Trofimchuk, P.A. Chernavskii, A.L. Volynskii, A.T. Ponomarenko, N.F. Bakeev: Polym. Sci. B. **34**, 487 (1998)
57. I.M. Papisov, Yu.S. Yablokov, A.I. Prokofev, A.A. Litmanovich: Polym. Sci. A. **35**, 515 (1993); **36**, 352 (1994)
58. L. Chen, W.-J. Yang, C.-Z. Yang: J. Mater. Sci. **32**, 3571 (1997)
59. H.B. Sohn, R.E. Cohen: Chem. Mater. **9**, 264 (1997)
60. A.S. Rozenberg, G.I. Dzhardimalieva, A.D. Pomogailo: Polym. Adv. Technol. **9**, 527 (1998)
61. A.I. Aleksandrova, A.I. Prokofev, V.N. Lebedev, E.B. Balagurov, N.N. Bubnov, I.Yu. Metlenkova, S.P. Solodovnikov, A.N. Ozerin: Russ. Chem. Bull. 2355 (1995)
62. R. Tannenbaum, C.I. Flenniken, E.P. Goldberg: J. Polym. Sci. B. Polym. Phys. **28**, 2421 (1990)

63. M. Lawecka, M. Kopcewicz, A. Slawska-Waniewska, M. Leonowicz, J. Kozubowski, G.I. Dzhardimalieva, A.S. Rosenberg, A.D. Pomogailo: J. Nanopart. Res. **5**, 373 (2003)
64. A.D. Pomogailo, G.I. Dzhardimalieva, A.S. Rosenberg, D.N. Murav'ev: J. Nanopart. Res. **5**, 497 (2003)
65. A.D. Pomogailo, A.S. Rosenberg, G.I. Dzhardimalieva: Solid State Phenomena **94**, 313 (2003)
66. J. Osuna, D. Caro, C. Amiens, B. Chaudret, E. Snoeck, M. Respaund, J.-M. Broto, A. Fert: J. Phys. Chem. **100**, 14571 (1966)
67. M.V. Shamurina, V.I. Roldugin, T.D. Pryamova, T.D. Vysotskii: Colloid. J. **56**, 450 (1994); **57**, 580 (1995); **61**, 473 (1999)
68. T. Shiga, A. Okadda, T. Kurauchi: J. Appl. Polym. Sci. **58**, 787 (1995)
69. T. Ktapeinski, A. Galeski, M. Kruszewski: J. Appl. Polym. Sci. **58**, 1007 (1995)
70. J. Twomey, S.H. Chew, T.N. Blanton, A. Schmid, K.L. Marshall: J. Polym. Sci. B Polym. Phys. **32**, 1687 (1994)
71. J.H. Park, J. Cheon: J. Am. Chem. Soc. **123**, 5473 (2001)
72. S. Sun, S. Anders, H.F. Haman, J.-U. Thiele, J.E.E. Baglin, T. Thomson, E.E. Fullerton, C.B. Murray, B.D. Terris: J. Am. Chem. Soc. **124**, 2884 (2002)
73. M. Fujiwara, T. Matsushita, K. Yamagichi, T. Fueno: Synth. Met. **41–43**, 3267 (1991)
74. M. Fujiwara, W. Mori, K. Yamagichi: Mol. Cryst. Liq. Cryst. **274**, 175 (1995)
75. V.N. Kestelman, A.D. Stadnik: *Thermomagnetic treatment of polymer composite materials* (NIITEKHIM, Moscow 1989)
76. A.G. Golubkov, N.P. Evrukov: Plast. Massy (3) 22 (1998)
77. Y. Osada: Adv. Mater. **3**, 107 (1991)
78. X. Cao, Y. Xie, F. Yu, Z. Yao, L. Li: J. Mater. Chem. **13**, 893 (2003)
79. M.R. Bryce, M.C. Petty: Nature **374**, 771 (1995)
80. C. Lafuente, C. Mingotaud, P. Delhaes: Chem. Phys. Lett. **302**, 523 (1999)
81. D.D. Mishin: *Magnetic materials. Manual for colleges* (High school, Moscow 1991)
82. D. de Caro, T.O. Ely, A. Mari, B. Chaudret, E. Snoeck, M. Respaund, J.-M. Broto, A. Fert: Chem. Mater. **8**, 1987 (1996)
83. B. Sohn, R.E. Cohen, G.C. Papaefthymiou: J. Magn. Magn. Mater. **182**, 216 (1998)
84. T.J. Fiske, H. Gokturk, D.M. Kaylon: J. Appl. Polym. Sci. **65**, 1371 (1997)
85. S. Praveen, V.K. Babbar, A. Razdan, R.K. Puri, T.K. Goel: J. Appl. Phys. **89**, 4362 (2000)
86. D.-K. Kim, M.S. Toprak, M. Mikhaylova, Y.-S. Jo, S.J. Savage, T. Tsakalakos, M. Muhammed: Solid State Phenomena (in press) (2004)
87. V. Goldade, L. Pinchuk, A. Kravtsov: *In: Polymers Friendly for the Environment, 7th Eur. Polym. Federation Symp. Polym. Mater.* (1998)
88. K. Uchino: MRS Bull. **18**, 42 (1993)
89. A.L. Buchachenko: Russ. Chem. Rev. **59**, 529 (1990)
90. M.M. Levitskii, A.L. Buchachenko: Russ. Chem. Bull. **46**, 1367 (1997)
91. D.K. Kim, Y. Zhang, J. Kehr, T. Klason, B. Bjelke, M. Mahammed: J. Magn. Magn. Mater. **225**, 256 (2001)
92. L. He, M.D. Musick, S.R. Nicewarner, F.G. Salinas, S.J. Benkovic, M.J. Natan, C.D. Kreating: J. Am. Chem. Soc. **122**, 9071 (2000)
93. A. Jordan, R. Scholz, P. Wust, H. Fahling, R. Felix: J. Magn. Magn. Mater. **201**, 413 (1999)

94. E.E. Carpenter: J. Magn. Magn. Mater. **225**, 17 (2001)
95. I. Safarik, M. Safarikova: Monatsh. Chem. **133**, 737 (2002)
96. A.L. Prieto, M. Martin-Gonzalez, J. Keyani, R. Gronsky, T. Sands, A.M. Stace: J. Am. Chem. Soc. **125**, 2388 (2003)
97. S. Iijima, T. Ichihashi: Nature **363**, 603 (1993)
98. T. Yamabe: Synth. Metals **70**, 1511 (1995)
99. Y.K. Tsang, Y.K. Chen, P.J. Harris: Nature **372**, 152 (1994)
100. T.W. Ebbesen: Physics Today 49, 26 (1996)
101. T. Trindade, P. O'Brien, X. Zhand: Chem. Mater. **9**, 523 (1997)
102. H. Liu, A. Bard:. J. Chem. Phys. **93**, 3232 (1989)
103. A.A. Khoroshilov, K.N. Bulgakova, Yu.N. Sychev: Izv. Vuz. Khim. Khim. Tekhnol. **43**, 124 (2000)
104. L. Hong, E. Ruckenstein: J. Appl. Polym. Sci. **67**, 1891 (1998)
105. E. Punkka, J. Laakso, H. Stubb, P. Kuivalainen: Synth. Metals **41–43**, 983 (1991)
106. M.C. Aronson, M. Aldissi, S.P. Armes, J.D. Thompson: Synth. Metals **41–43**, 837 (1991)
107. *Fundamentals of Adhesion* ed. by H.Y. Lee (Plenum Press, New York 1991)
108. Yu.N. Anisimov, L.P. Dobrova, A.Yu. Anisimov: Russ. J. Appl. Chem. **71**, 790 (1998)
109. V. Simgh, A.N. Tiwar, A.R. Kulkarni: Mater. Sci. & Eng. B **41**, 310 (1996)
110. I.A. Tchmutin, A.T. Ponomarenko, V.G. Shevchenko, N.G. Ryvkina, C. Klason, D.H. McGueen: J. Polym. Sci. B Polym. Phys. **36**, 1847 (1998)
111. V.A. Sergeev, A.A. Askadskii, M.A. Surikov, V.V. Kazantseva, N.I. Bekasova, E.A. Baryshnikova: Polym. Sci. B **39**, 334 (1997)
112. V.V. Vysotskii, V.I. Roldugin: Colloid J. **58**, 312 (1996); **60**, 729 (1998); **61**, 190 (1999)
113. D. Salz, M. Wark, A. Baalmann, U. Simon, N. Jaeger: Phys. Chem. Chem. Phys. **4**, 2438 (2002)
114. C.-C. Yen: J. Appl. Polym. Sci. **60**, 605 (1996); **71**, 1361 (1999)
115. Y. Zweifel, C.J.G. Plummer, H.-H. Kausch: J. Mater. Sci. **33**, 1715 (1998)
116. M.Yu. Yablokov, S.A. Zavyalov, E.S. Oblonkova: Russ. J. Phys. Chem. **73**, 219 (1999)
117. K.L. Levin, V.I. Frolov, Yu.M. Boyarchuk, T.I. Borisova: Polym. Sci. USSR. B **41**, 363 (1991)
118. N.A. Surridge, F.R. Keene, B.A. White, J.S. Facci, M. Silver, R.M. Murray: Inorg. Chem. **29**, 4950 (1990)
119. W.B. Genetti, W.L. Huan, B.P. Grady, E.A. O'Rear, C.L. Lai, D.T. Glatzhofer: J. Mater. Sci. **33**, 3085 (1998)
120. I.I. Obraztsova, O.A. Efimov, N.K. Eremenko, G.Yu. Simenyuk: Inorg. Mater. **31**, 798 (1995); **35**, 937 (1999); Russ. J. Appl. Chem. **75**, 1772 (2002)
121. M.M. Dávila, M.P. Elizalde, R. Silva: J. Mater. Sci. **32**, 3705 (1997)
122. A.N. Shipway, E. Katz, I. Willner: Phys. Chem. Chem. Phys. **1**, 18 (2000)
123. D.M. Kolb: Angew. Chem. Int. Ed. **40**, 1162 (2001)
124. S.P. Gubin, I.D. Kosobudskii: Russ. Chem. Rev. **52**, 1350 (1983)
125. N.I. Nikonorova, E.V. Semenova, V.D. Zanegin, G.M. Lukovkin, A.L. Volynskii, N.F. Bakeev: Polym. Sci. A **34**, 123 (1992)
126. B.H. Sohn, R.E. Cohen: J. Appl. Polym. Sci. **65**, 723 (1997)
127. T.Yu. Ryabova, A.S. Chirkov, L.S. Radkevich, N.V. Evtushok: Ukr. Chem. J. **59**, 1329 (1993)

128. A.V. Volkov, M.A. Moskvina, I.V. Karachevtsev, O.V. Lebedeva, A.L. Volynskii, N.F. Bakeev: Polym. Sci. A **40**, 970 (1998)
129. V.A. Sergeev, L.I. Vdovana, Yu.V. Smetannikov, A.Yu. Vasilkov, E.M. Belavtseva, L.G. Radchenko, V.N. Guryshev: Organomet. Chem. USSR. **3**, 916 (1990)
130. M.H. Lee, R.E. Newham: Ferroelectrics **87**, 71 (1988)
131. X. Cai, C. Zhong, S. Zhang, H. Wand: J. Mater. Sci. Lett. **16**, 253 (1997)
132. A. Bukowski: Polimery **61**, 139 (1996)
133. S. Maeda, Y. Tsurusaki, Y. Tochiyama, R. Naka, A. Ochi, T. Ohgushi, T. Takeshita: J. Polym. Sci. Polym. Chem. A **32**, 1729 (1994)
134. W.K. Sakamoto, D.H.F. Kanda, C.L. Carvalho: J. Mater. Sci. Lett. **19**, 603 (2000)
135. A.D. Pomogailo, V.S. Savostyanov, G.I. Dzhardimalieva, A.V. Dubovitskii, A.N. Ponomarev: Russ. Chem. Bull. **44**, 1056 (1995)
136. A.D. Pomogailo, G.I. Dzhardimalieva. Polym. Sci. A **46**, 250 (2004)
137. O.A. Shilova, S.V. Hashkovsky, L.A. Kuznetsova: J. Sol-Gel Sci. Technol. **26**, 687 (2003)
138. A. Vigalok, Z. Zhu, T.M. Swager: J. Am. Chem. Soc. **123**, 7917 (2001)
139. C.D. Gutsche: *Calixarene* (The Royal Soc. Chem., Cambridge 1989)
140. *Fast Ion Conduction in Solid* ed. by P. Vashishta, N.J. Mundy, G.K. Shenoy (North-Holland,. Amsterdam 1979)
141. M.A. Rarner, D.F. Shriver: Chem. Rev. **88**, 109 (1988)
142. M. Watanable, J. Ikeda, I. Shinihara: Polym. J. **15**, 175 (1983)
143. S.B. Fang, Y.Y. Jiang: Polym. Bull. **19**, 81 (1988)
144. F.M. Gray: *Polymer Electrolytes* (Springer, New York 1997)
145. L.W. Shacklette, T.R. Jow, M. Maxfield, R. Hatami: Synth. Metal. **28**, C655 (1989)
146. G. Pistoia: *Lithium Batteries. New Materials, Developments and Perspectives* (Elsevier, Amsterdam 1994)
147. *Lithium Batteries with polymer electrodes* ed. by N. Furukawa, K. Nishio (Chapman & Hall, United Kingdom 1993)
148. F. Croce, G.B. Appetecchi, L. Persi, B. Scrosati: Nature **394**, 456 (1998)
149. S.H. Chung, Y. Wang, L.L. Persi, S.G. Greenbaum, B. Scrosati, E. Plichta: J. Power Sources **97–98**, 644 (2001)
150. F. Groce, L. Persi, F. Ronci, B. Scrosati: Solid State Ionics **135**, 47 (2000)
151. M.C. Borghini, M. Mastagostino, S. Passerini, B. Scrosati: J. Electrochem. Soc., **142**, 2118 (1995)
152. Q. Li, H.Y. Sun, Y. Takeda, N. Imanishi, J. Yang, O. Yamamoto: J. Power Sources **94**, 201 (2001)
153. G.B. Appetecchi, G. Dautzenberg, B. Scrosati: J. Electrochem. Soc. **143**, 6 (1996)
154. J. Hou, G.L. Baker: Chem. Mater. **10**, 3311 (1998)
155. K.M. Nairn, A.S. Best, P.J. Newman, D.R. MacFarlane, M. Forsyth: Solid State Ionics **121**, 115 (1999)
156. B.P. Grady, C.P. Rhodes, S. York, R.E. Frech: Macromolecules **34**, 8523 (2001)
157. J.V. Ford, B.G. Sumpter, D.W. Noid, M.D. Barnes: Polymer **41**, 8075 (2000)
158. K. Hayamizu, Y.S. Aihara, W.S. Price: J. Chem. Phys. **113**, 4785 (2000)
159. J.Y. Bigot, J.C. Merle, O. Cregut, A. Daunois: Phys. Rev. Lett. **75**, 4702 (1995)

160. D.Yu. Godovski: Adv. Polym. Sci. **119**, 79 (1995)
161. G. Maier: Progr. Polym. Sci. **26**, 3 (2001)
162. *Handbook of Multilevel Metallization of Integrated Circuits* ed. by S.R. Wilson, C.J. Tracy, J.L. Fteeman (Noyes Publ., Park Ridge, NJ 1993)
163. T. Kyprianidou-Leodidou, P. Margraf, W. Caseri, U.W. Suter, P. Walther: Polym. Adv. Technol. **8**, 505 (1997)
164. H. Feilchenfeld, G. Chumanov, T.M. Cotton: J. Phys. Chem. **100**, 4937 (1996)
165. *Science and Engineering of one- and Zero-Dimentional Semiconductors. v.214* ed. by S.P. Beaumont, C.N. Sotomayor-Torres (Plenum Press, New York 1990)
166. E. Hoo, T. Lian: Langmuir **16**, 7879 (2000)
167. N.C. Greenham, X. Peng, A.P. Alivisatos: Synth. Met. **84**, 545 (1997)
168. J.G. Winiarz, L. Zhang, M. Lal, C.S. Friend, P.N. Prasad: Chem. Phys. **245**, 417 (1999)
169. H.S. Mansur, W.L. Vasconcelos, F. Grieser, F. Caruso: J. Mater. Sci. **34**, 5285 (1999)
170. M. Faraday: Philos. Trans. R. Soc. London **147**, 145–153 (1857)
171. K.L. Kelly, E. Coronado, L.L. Zhao, G.C. Schatz: J. Phys. Chem. B **107**, 668 (2003)
172. P.N. Prasad, D.J. Williams: *Introduction to Nonlinear Optical Effect in Molecules and Polymers* (Wiley, New York 1991)
173. *Poled polymers and their application to SHG and EO devices* ed. by S. Miyata, H. Sasabe (Gordon& Breach Sci. Publ., Japan 1997)
174. D.W. Bruce, D. O′Hare: *Inorganic Materials; Metal-Containing materials for Nonlinear Optics Series* (Wiley, New York 1992)
175. L.L. Beecroft, C.K. Ober: Chem. Mater. **9**, 1302 (1997)
176. A.D. Yoffe: Adv. Phys. **50**, 208 (2001)
177. Y. Liu, D. De Groot, J. Schindler, C. Kannewurf, M. Kanatzidis: Chem. Mater. **3**, 992 (1991)
178. B. Ackermans, R. Schoonheydt, E. Ruiz-Hitzky: J. Chem. Soc. Faraday Trans. **92**, 4479 (1996)
179. N.E. Kotelnikova, G. Vegner, M. Stol et al.: Russ. J. Appl. Chem. **76**, 121 (2003)
180. A.V. Loginov, L.V. Mikhailova, V.V. Gorbunova, G.A. Shagisultanova: Russ. J. Appl. Chem. **63**, 1070 (1990); **67**, 803 (1994)
181. R.E. Southward, D.W. Thompson, A.K.St. Clair: Chem. Mater. **9**, 501 (1997)
182. R.E. Southward, D.S. Thompson, D.W. Thompson, A.K.St. Clair: Chem. Mater. **9**, 1691 (1997)
183. H.S. Zhou, T. Wada, H. Sasabe, H. Komiyama: Synth. Metals **81**, 129 (1996)
184. M. Ajayan, O. Stephan, C. Colliex, D. Trauth: Science **265**, 1212 (1994)
185. M.L. Steigerwald, L.E. Brus: Annu. Rev. Mater. Sci. **19**, 471 (1989)
186. C.B. Vurray, D.J. Norris, M.G. Bawendi: J. Am. Chem. Soc. **115**, 8706 (1993)
187. T. Vossmeyer, L. Katsikas, M. Giersig, I.G. Popovic, K. Diesner, A. Chemseddine, A. Eychmüller, H. Weller: J. Phys. Chem. **98**, 7665 (1994)
188. J.E.B. Katari, V.L. Colvin, A.P. Alivisatos: J. Phys. Chem. **98**, 4109 (1994)
189. D.E. Fogg, L.H. Radzilowski, B.O. Dabbousi, R.R. Schrock, E.L. Thomas, M.G. Bawendi: Macromolecules **30**, 8433 (1997)
190. D.E. Fogg, L.H. Radzilowski, R. Blanski, R.R. Schrock, E.L. Thomas: Macromolecules **30**, 417 (1997)
191. M. Zelner, H. Minti, R. Reisfeld, H. Cohen, R. Tenne: Chem. Mater. **9**, 2541 (1997)

192. X. Peng, J. Wickham, A.P. Alivisatos: J. Am. Chem. Soc. **120,** 5343 (1998)
193. A.A. Guzelian, J.E.B. Katari, A.V. Kadavanich, U. Banin, K. Hamad, E. Juban, A.P. Alivisatos, R.H. Wolters, C.C. Arnold, J.R. Heath: J. Phys. Chem. **100**, 7212 (1996)
194. A.A. Guzelian, U. Banin, A.V. Kadavanich, X. Peng, A.P. Alivisatos: Appl. Phys. Lett. **69**, 1432 (1996)
195. M.A. Olshavsky, A.N. Goldstein, A.P. Alivisatos: J. Am. Chem. Soc. **112**, 9438 (1990)
196. O.L. Micic, C.J. Curtis, K.M. Jones, J.R. Sprague, A.J. Nozik: J. Phys. Chem. **98**, 4966 (1994)
197. O.L. Micic, J.R. Sprague, C.J. Curtis, K.M. Jones, J.L. Machol, A.J. Nozik, H. Giessen, B. Fluegel, G. Mohs, N. Peyghambarian: J. Phys. Chem. **99**, 7754 (1995)
198. O.L. Micic, A.J. Nozik: J. Luminescence **70**, 95 (1996)
199. O.L. Micic, J.R. Sprague, Z.H. Lu, A.J. Nozik: Appl. Phys. Lett. **68**, 3150 (1996)
200. T. Douglas, K.H. Theopold: Inorg. Chem. **30**, 594 (1991)
201. S.S. Kerr, R.L. Wells: Nanostructured Mater. **7**, 591 (1996)
202. V.V. Sviridov, G.P. Shevchenko, E.M. Afanaseva, A.N. Ponyavina, N.V. Loginova: Colloid J. **58,** 390 (1996)
203. A. Henglein: Chem. Rev. **89**, 1861 (1989)
204. N. Arul Dhas, H. Cohen, A. Gedanken: J. Phys. Chem. B **101**, 6834 (1997)
205. T.B. Boitsova, V.V. Gorbunova, A.V. Loginov: Russ. J. Gen. Chem. **67**, 1741 (1997)
206. S.E. Emory, W.E. Haskins, S. Nie: J. Am. Chem. Soc. **120**, 8009 (1998)
207. R. Rafaeloff, Y.-M. Tricot, F. Nome, P. Tundo, J.H. Fendler: J. Phys. Chem. **89**, 1236 (1985)
208. G. Jorgensen, P. Schissel: In: *Metallized Plastics* ed. by K.L. Mittal, J.R. Susko (Plenum Press, New York 1989) Vol. 1, p. 79
209. V.B. Hueggle: Soc. Photo-Opt. Instrum. Eng. Proc. Refl. Opt. II, **79**, 1113 (1989)
210. H. Liedberg, T. Ludeberg: Urol. Res. **17**, 359 (1989)
211. C. Graf, A. Blaaderen: Langmuir **18**, 524 (2002)
212. V.M. Shalaev: Phys. Rep. **272**, 61 (1996)
213. R.J. Gehr, R.W. Boyd: Chem. Mater. 8, 1807 (1996)
214. P. Chakraborty: J. Mater. Sci. **33**, 2235 (1998)
215. V.P. Drachev, S.V. Perminov, S.G. Rautian, V.P. Safonov: JETP Lett. **68**, 618 (1998)
216. I. Vargas-Baca, A.P. Brown, M.P. Andrews, T. Galstian, Y. Li, H. Vali, M.G. Kuzyk: Can. J. Chem. **80**, 1625 (2002)
217. G.A. Lindsay, K.D. Singer: *Polymer for Second-Order Nonlinear Optics. Advances in Chemistry Series* (Am. Chem. Soc., Washington, DC 1995)
218. S.D. Cox, T.E. Gier, G.D. Stucky: Chem. Mater. **2**, 609 (1990)
219. R.J. Jeng, Y.M. Chen, J. Kumar, S.K. Tripathy: J. Macromol. Sci. – Pure Appl. Chem. A **29**, 1115 (1992)
220. R.J. Jeng, Y.M. Chen, A.K. Jain, J. Kumar, S.K. Tripathy: Chem. Mater. **4**, 972 (1992)
221. R.J. Jeng, Y.M. Chen, A.K. Jain, S.K. Tripathy, J. Kumar: Opt. Commun. **89**, 212 (1992)

222. R.J. Jeng, Y.M. Chen, A.K. Jain, J. Kumar, S.K. Tripathy: Chem. Mater. **4**, 1141 (1992)
223. R.J. Jeng, G.H. Hsiue, Y.M. Chen, S. Marturunkakul, L. Li, X.L. Jiang, R.A. Moody, C.E. Masse, J. Kumar, S.K. Tripathy: J. Appl. Polym. Sci. **55**, 209 (1995)
224. G.-H. Hsiue, R.-H. Lee, R.-J. Jeng: Polymer **40**, 6417 (1999)
225. U. Caruso, A. Maria, B. Panunzi, A. Roviello: J. Polym. Sci. Polym. Chem. A **40**, 2987 (2002)
226. W. Schrof, S. Rozouvan, E. Van Keuren, D. Horn, J. Schmitt, G. Decher: Adv. Mater. **10**, 338 (1998)
227. T.V. Murzina, A.A. Nikulin, O.A. Aktsipetrov, J.W. Ostrander, A.A. Mamedov, N.A. Kotov, M.A.C. Devillers, J. Roark: Appl. Phys. Lett. **79**, 1309 (2001)
228. T. Coradin, R. Clement, P.G. Lacroix, K. Nakatani: Chem. Mater. **8**, 2153 (1996)
229. A.W. Olsen, Z.H. Kafafi: J. Am. Chem. Soc. **113**, 7758 (1991)
230. G. Kampfrath, A. Heilmann, C. Hamann: Vacuum **38**, 1 (1988)
231. J. Perrin, B. Despax, E. Kay: Phys. Rev. B Condens. Mater. **32**, 719 (1985)
232. M. Nogami, Y. Abe: J. Am. Ceram. Soc. **78**, 1066 (1995)
233. S. Nie, S.R. Emory: Science **275**, 1102 (1997)
234. S.T. Selvan, T. Hayakawa, M. Nogami, Y. Kobayashi, L.M. Liz-Marzan, Y. Hamanaka, A. Nakamura: J. Phys. Chem. B **106**, 10157 (2002)
235. H. Aert, M.V. Damme, O. Nuyken, U. Schnöller, K.-J. Eichhorn, K. Grundke, B. Voit: Macromol. Mater. Eng. **286**, 488 (2001)
236. Ch. Loppacher, S. Trogisch, F. Braun, A. Zherebov, S. Grafström, L.M. Eng, B. Voit: Macromolecules **35**, 1936 (2002)
237. J.Y. Burroughes, D.D.C. Bradley, A.R. Brown, R.N. Marks, R. Mackay, R.H. Friend, P.L. Burn, A.B. Holmes: Nature **347**, 539 (1990)
238. M. Herold, J. Gmeiner, C. Drummer, M. Schwoerer: J. Mater. Sci. **32**, 5709 (1997)
239. A. Wu, J.-K. Lee, M.F. Rubner: J. Am. Chem. Soc. **121**, 4883 (1999)
240. E.S. Handy, A.J. Pal, M.F. Rubner: J. Am. Chem. Soc. **121**, 3525 (1999)
241. A. Eychmüller: J. Phys. Chem. B **104**, 6514 (2000)
242. I.L. Radtchenko, G.B. Sukhorukov, N. Gaponik, A. Kornowski, A.L. Rogach, H. Möhwald: Adv. Mater. **13**, 1684 (2001)
243. X.G. Peng, M.C. Schlamp, A.V. Kadavanich, A.P. Alivisatos: J. Am. Chem. Soc. **119**, 7019 (1997)
244. W.B. Stockton, M.F. Rubner: Macromolecules **30**, 2717 (1997)
245. U. Kreibig, M. Vollmer: *Optical Properties of Metal Clusters. Springer Series in Material Science, No. 25* (Springer-Verlag, Berlin 1995)
246. *Handbook of Optical Properties. Vol.2. Optics of Small Particless, Interfaces and Surfaces* ed. by R.E. Hummel, P. Wissmann (CRC Press, Boca Raton 1997)
247. P.W. Barber, P.K. Chang: *Optical Effects Associated with Small Particles* (World Scientific, Singapore 1988)
248. J. Xiang, C. Chen, B. Zhou, G. Xu: Chem. Phys. Lett. **315,** 371 (1999)
249. Y. Wang, W. Mahler: Opt. Commun. **61**, 233 (1987)
250. Y. Yang, J. Huang, B. Yang, S. Liu, J. Shen: Synthetic Metals **91**, 347 (1997)
251. A.I. Ekmov, A.L. Alfros, A.A. Onushenko: Sol. State Commun. **56**, 921 (1985)

252. D.J. Welker, M.G. Kuzyk: Appl. Phys. Lett. **66**, 2792 (1995)
253. H. Suzuki, T. Higurashi, A. Morinaka, T. Shimada, K. Sukegava, D. Haarer: J. Luminescence **56**, 125 (1993)
254. S. Muto: In: *Proceeding of Second International Conference on Plastic Optical Fiber and Application* **62**, 887 (1993)
255. Q. Zhang, H. Ming, Y. Zhai: J. Appl. Polym. Sci. **62**, 887 (1996)
256. D. Bosc, Y. Pietrasanta, R. Rigal, A. Rousseau: J. Fluor. Chem. **26**, 369 (1984)
257. D. Bosc, N. Devoldere, M. Bonnel, J.L. Favennec, D. Pavy: Mater. Sci. Engn. B **57**, 155 (1999)
258. G.F. Sa, O.L. Malta, C. Mello-Donega, A.M. Simas, R.L. Longo, P.A. Santa-Cruz, E.F. Silva: Coord. Chem. Rev. **196**, 165 (2000)
259. L.D. Carlos, Y. Messaddeq, H.F. Brito, R.A. Sa Ferreira, Z.V. Bermudez, S.J.L. Ribeiro: Adv. Mater. **12**, 594 (2000)
260. K. Dahmouche, L.D. Carlos, C.V. Santilli, Z.V. Bermudez, A.F. Craievich: J. Phys. Chem. B **106**, 4377 (2002)
261. V. Bekiari, P. Lianos, P. Judeinstein: Chem. Phys. Lett. **307**, 310 (1999)
262. V. Bekiari, P. Lianos: Adv. Mater. **10**, 1455 (1998)
263. R. Hernandedez, A.-C. Franville, P. Minoofar, B. Dunn, J.I. Zink: J. Am. Chem. Soc. **123**, 1248 (2001)
264. C.J. Wung, Y. Yang, P.N. Prasad, F.E. Karasz: Polymer **32**, 604 (1991)
265. H. Krug, H. Smidt: New J. Chem. **18**, 1125 (1994)
266. J.L. Bender, P.S. Corbin, C.L. Fraser, D.H. Metcalf, F.S. Richardson, E.L. Thomas, A.M. Urbas: J. Am. Chem. Soc. **124**, 8526 (2002)
267. N.A.H. Male, O.V. Salata, V. Christou: Synth. Metals **126**, 7 (2002)
268. T. Cassagneau, T.E. Mallouk, J.H. Fendler: J. Am. Chem. Soc. **120**, 7848 (1998)
269. M. Elbanowski, B. Makowska: J. Photochem. Photobiol. Chem. A **99**, 85 (1996)
270. *Photonics and Nonlinear Optics with Sol-Gell Processed Inorganic Glass: Organic Polymer Composite* ed. by L.C. Klein (Kluwer Avademic, Boston 1994)
271. H.J. Jiang, C.-C. Yuan, Y. Zhou, Y.C. Lam: Opt. Connun. **185**, 19 (2000)
272. J.T. Rantala, R. Levy, L. Kivimaki, M.R. Descour: Electron. Lett. **36**, 530 (2000)
273. W. Que, X. Hu, Q.Y. Zhang: Chem. Phys. Lett. **369**, 354 (2003)
274. M. Yashida, P.N. Prasad: Appl. Optics **35**, 1500 (1996)
275. A. Fujishima, K. Honda: Nature **238**, 37 (1972)
276. M.-P. Zheng, Y.-P. Jin, G.-L. Jin, M.-Y. Gu: J. Mater. Sci. Lett. **19**, 433 (2000)
277. M.T. Pham, D. Möller, W. Matz, A. Mücklich, S. Oswald: J. Phys. Chem. B **102**, 4081 (1998)
278. A. Slama-Schwok, D. Avnir, M. Ottolenghi: Photochem. Photobiol. **54**, 525 (1991)
279. F.N. Castellano, G.J. Meyer: Prog. Inorg. Chem. **44**, 167 (1997)
280. J.M. Stipkala, F.N. Castellano, T.A. Heimer, C.A. Kelly, K.J.T. Livi, G.J. Meyer: Chem. Mater. **9**, 2341 (1997)
281. N. Fukuda, M. Mitsuishi, A. Aoki, T. Miyashita: J. Phys. Chem. B **106**, 7048 (2002)
282. K. Yamada, N. Kobayashi, K. Ikeda, R. Hirohashi, M. Kaneko: Jpn. J. Appl. Phys. **33**, L544 (1994)

283. J. Serin, X. Schultze, A. Andronov, J.M.J. Frechet: Macromolecules **35**, 5396 (2002)
284. J.L. Bourdelande, J. Font, G. Marques, A.A. Abdel-Shafi, A.F. Wilkinson, D.R. Worrall: Photochem. Photobiol. Chem. A **138**, 65 (2001)
285. A.C. Franville, D.Z. Zambon, R.M. Mahiou, S. Chou, Y. Troin, J.C. Cousseins: J. Alloys Compd. **275–277**, 831 (1998)
286. K. Dou, X. Sun, X. Wang, R. Parkhill, Y. Guo, E.T. Knobbe: Solid State Commun. **107**, 101 (1998)
287. K.A. Walters, K.D. Ley, C.S.P. Cavalaheiro, S.E. Miller, D. Gosztola, M.R. Wasielewski, A.P. Bussandri, H. Willigen, K.S. Schanze: J. Am. Chem. Soc. **123**, 8329 (2001)
288. A. Ulman: *Organic thin films and surfaces: directions for the nineties* (Acad. Press, London 1995)
289. *An introdution to molecular electronics* ed. by M.C. Petty, M.R. Bryce, D. Bloor (Edward Amold., London 1995)
290. *Langmuir-Blodgett films* ed. by G. Roberts (Plenum Press, New York 1990)
291. H. Byrd, C.E. Holloway, J. Pogue, S. Kircus, R.C. Advincula, W. Knoll: Langmuir **16**, 10322 (2000)
292. L.L. Sveshnikova, S.M. Repinskii, F.K. Gutakovskii, A.G. Milekhin, L.D. Pokrovskii: Chem. Steady State **8**, 265 (2000)
293. R.S. Kane, R.E. Cohen, R. Silbey: J. Phys. Chem. **100**, 7928 (1996)
294. H. Dhanabalan, H. Kudrolli, S. Major, S.S. Talwar: Solid State Commun. **99**, 859 (1996)
295. S. Vitta, T.H. Metzger, S. Major, A. Dhanabalan, S.S. Talwar: Langmuir **14**, 1799 (1998)
296. H. Dhanabalan, S. Major, S.S. Talwar, A.Q. Contractor, N.P. Kumar, S.N. Narang, S.S. Major: J. Mater. Sci. Lett. **18**, 603 (1999)
297. H. Yoneyama, M. Tokuda, S. Kuwabata: Electrochem. Acta **39**, 1315 (1994)
298. H. Jeong, B.-J. Lee, W.J. Cho, C.-S. Ha: Polymer **41**, 5525 (2000)
299. B.O. Dabbousi, M.G. Bawendi, O. Onitsuka, M.F. Rubner: Appl. Phys. Lett. **66**, 1316 (1995)
300. Z.-K. Xu, Y.-Y. Xu, M. Wang: J. Appl. Polym. Sci. **69**, 1403 (1998)
301. Y. Nishikata, S. Fukui, M. Kakimoto, Y. Imai, K. Nishyama, M. Fujihira: Thin Solid Films **210/211**, 296 (1992)
302. L. Sereno, J.J. Silber, L. Otero, V. Bohorquez, A.L. Moore, T.A. Moore, D. Gust: J. Phys. Chem. **100**, 814 (1996)
303. G. Gao, Y. Deng, L.D. Kispert: J. Chem. Soc. Perkin Trans. **2**, 1225 (1999)
304. W. Feng, E. Sun, A. Fujii, H. Wu, K. Niihara, K. Yoshino: Bull. Chem. Soc. Jpn. **73**, 2627 (2000)
305. W. Mahler: Inorg. Chem. **27**, 435 (1988)
306. I.A. Akimov, I.Yu. Denisov, A.M. Meshkov: Opt. Spectrosc. **72**, 558 (1992)
307. Y. Wang, N. Nerron: J. Phys. Chem. **95**, 525 (1991)
308. S. Ogawa, Y. Hayashi, N. Kobayashi, T. Tokizaki, A. Nakamura: Jpn. J. Appl. Phys. **33**, L331 (1994)
309. G.Y. Yang, R.M. Briber, E. Huang, R.M. Rice, W. Volksen, R.D. Miller: Polym. Mater. Sci. Eng. **85**, 18 (2001)
310. L. Zimmermann, M. Weibel, W. Caseri, U.W. Suter, P. Walther: Polym. Adv. Technol. **4**, 1 (1992)
311. T. Kyprianidou-Leodidou, W. Caseri, U.W. Suter: J. Phys. Chem. **98**, 8992 (1994)

312. *Advances in Photochemistry. Vol. 19* ed. by D.C. Neckers, D.H. Volman, G. Von Bünau (Wiley, New York 1995)
313. M. Gao, B. Richter, S. Kirstein, H. Möhwald: J. Phys. Chem. B **102**, 4096 (1998)
314. P. Poussiganol, D. Ricard, J. Lucasik, C. Flytzanis: J. Opt. Soc. Am. B **4**, 5 (1987)
315. C. Flytzanis, F. Hache, M.C. Klein, D. Richard, P. Poussiganol: Progr. Opt. **29**, 323 (1991)
316. J. Wang, D. Montville, K.E. Gonsalves: J. Appl. Polym. Sci. **72**, 1852 (1999)
317. T. Hirai, T. Watanabe, I. Komasawa: J. Phys. Chem. B **103**, 10120 (1999)
318. D.V. Talapin, A.L. Rogach, A. Kornowski, M. Haase, H. Weller: Nano Lett. **1**, 207 (2001)
319. O. Schmelz, A. Mews, T. Basche, A. Herrmann, K. Müller: Langmuir **17**, 2861 (2001)
320. J. Yang, H. Lin, Q. He, L. Ling, C. Zhu, F. Bai: Langmuir **17**, 5978 (2001)
321. I. Potapova, R. Mruk, S. Prehl, R. Zentel, T. Basche, A. Mews: J. Am. Chem. Soc. **125**, 320 (2003)
322. X. Zhang, S.A. Jenekhe, J. Perlstein: Chem. Mater. **8**, 1571 (1996)
323. A. Sengupta, B. Jiang, K.C. Mandal, J.Z. Zhang: J. Phys. Chem. B **103**, 3128 (1999)
324. K. Runge, B.G. Sumpter, D.W. Noid, M.D. Barnes: Chem. Phys. Lett. **299**, 352 (1999)
325. G. Cahen, G. Dagan, Y. Mirovsky, G. Hodes, W. Giriat, M. Lubke: J. Electrochem. Soc. **132**, 1065 (1985)
326. L.C. Yang, H.Z. Xiao, W.N. Shafarmann, R.W. Birkmire: Sol. Energy Mater.Sol. Cells. **36**, 445 (1995)
327. P. Yeh: *Introduction to Photorefractive Nonlinear Optics* (Wiley, New York 1993)
328. G. Ramos, T. Belenguer, E. Bernabeu, F. Monte. D. Levy: J. Phys. Chem. B **107**, 110 (2003)
329. W.E. Moemer, S.M. Silence: Chem. Rev. **94**, 127 (1994)
330. K.S. West, D.P. West, M.D. Rahn, J.D. Shakos, F.A. Wade, K. Khand, T.A. King: J. Appl. Phys. **84**, 5893 (1998)
331. U. Gubler, M. He, D. Wright, Y. Roh, R. Twieg, W.E. Moemer: Adv. Mater. **14**, 313 (2002)
332. J.G. Winiarz, L. Zhang, M. Lal, C.S. Friend, P.N. Prasad: J. Am. Chem. Soc. **121**, 5287 (1999)
333. J.G. Winiarz, L. Zhang, J. Park, P.N. Prasad: J. Phys. Chem. B **106**, 967 (2002)
334. B. Darracq, F. Chaput, K. Lahlil, J.P. Boilot, Y. Levy, V. Alain, L. Ventelon, M. Blanchard-Desce: Opt. Mater. **9**, 265 (1998)
335. K.B. Sandalphon, N. Peyghambarian, S.R. Lyon, A.B. Padias, H.K. Hall: Opt. Lett. **19**, 68 (1994)
336. E.V. Nikolaeva, S.A. Ozerin, A.E. Grigoriev, E.I. Grigoriev, S.N. Chvalun, G.N. Gerasimov, L.I. Trakhtenberg: Mater. Sci. Eng. C **8–9**, 217 (1999)
337. Y. Vlasov, A. Legin, A. Rudnitskaya, A. D'Amico, C. Di Natale: Sens. Actuators B **65**, 235 (2000)
338. M. Sastry, N. Lala, V. Patil, S.P. Chavan, A.G. Chittiboyina: Langmuir **14**, 4138 (1998)

339. D. Fitzmaurice, S. Connolly: Adv. Mater. **11**, 1202 (1999)
340. L. Duran, E. Costell: Food Sci. Technol. Int. 299 (1999)
341. A. Ruil, D.S. Santos, K. Wohnrath, R. Tommazo, A.C. Carvalho, F.J. Fonseca, O.N. Olivera, D.M. Taylor, L.C.H. Mattoso: Langmuir **18**, 239 (2002)
342. P.-L. Kuo, W.-F. Chen: J. Phys. Chem. B **107**, 11267 (2003)
343. M. Zhao, L. Sun, R.M. Crooks: J. Am. Chem. Soc. **120**, 4877 (1998)
344. T. Konishi, A. Ikeda, M. Asai, T. Hatano, S. Shinkai, M. Fujitsuka, O. Ito, Y. Tsuchiya, J. Kikuchi: J. Phys. Chem. B **107**, 11261 (2003)

12 Metallopolymeric Nanocatalysis

In recent years the prefix nano has become common in catalytic studies. We define the term "nanocatalysis" (more accurately, it must be called metallo-nanocatalysis) as a definition of a catalytic process induced by nanosize metal particles, their clusters, oxides or other compounds, presented initially in the reaction volume or forming during the reaction or even formed as a result of catalyst deactivation products. The definition is correct also for catalysis by polyatomic clusters or mononuclear metallocomplexes with nanosize carriers including catalytic processes in nanopores (for example, in zeolite pores, such as ZSM-5, MCM-41). Catalysis serves as a nexus or bridge that brings together numerous disciplines: solid state physics, chemical engineering, theoretical physics, and chemistry. Now it becomes obvious that the main meaning in catalysis relates to surface chemical composition and the methods of its control. Not long ago it was considered that rather large nanoparticles including more then 20 atoms are similar to massive, consolidated pieces and their catalytic characteristics arise only due to an increase of the catalyst surface area [1]. But now it is clear that the catalytic abilities of nanoparticles are connected with their electronic properties (the electron work of exit, forbidden bandwidth, energetic spectra) due to dimensional quantum or electronic effects. Moreover, the electronic properties of nanoparticle can be changed after interaction with a polymer or inorganic matrix surface that results in cooperative effects [2]. Therefore modification of a catalyst(e.g. by creating a stabilizing envelope around the metal particle) can be used only for colloidal metal, which gives them great advantages over other catalysts [3].

The catalysis kinetics of the processes involving nanoparticles are very specific because their sizes are comparable with the catalyzed substance molecules. For the particles adsorbed on a surface the catalytic reaction can be considered in an infinite space with constant molecule concentrations. The chemical kinetics for such particles are not classical, and must be described by other methods, for example, using stochastic approaches, connected with the statistical fluctuations of the number of reacting molecules (Monte Carlo or cell automates methods) [4,5]. Usually the classical kinetics equations can be used only for processes with linear dimensions 2–7 times bigger then the characteristic scale of the reacting molecules or their effective interaction radius.

12.1 A General Survey of Metallopolymer Catalysis

Heterogeneous catalysis might be the oldest area of nanotechnology. As a rule, the metal catalysts were prepared as nanoparticles dispersed on the surfaces of high surface area materials, like SiO_2, MgO, or Al_2O_3 activated by carbon and other elements. This allows us to raise the ratio of metal involved into the catalytic actions [6–8]. It can be realized in two different ways. The first one is to apply the metal or its oxide nanoparticles on a dispersed carrier (film-type, fibrous) as encapsulated, intercalated, entrapped catalysts, catalysts anchored on functionalized solids, by crystalline mesoporous molecular sieves [9–11]. The second way is connected with metallocomplexes (MX_n) bonded with the carrier particles including the polymeric particles.

Group VIII metals in a colloidal state are very effective catalysts in hydrogenation of organic compounds, but for the active coagulation of nanoparticles their catalytic activity is not sufficiently reproducible. On the other hand, there is the problem of regeneration of the catalyst from the reactive medium. Therefore at first various polymers (PVA, PVAc, PMMA in water, water–alcohol mixtures and organic solvents) were used as protective colloids for Pd and Pt nanoparticles, which are widely used catalysts for hydrogenization of nitrobenzene, $m-$bromonitrobenzene, aldehydes, propiolic acid, quinone, as well as for the water–gas shift reaction ($CO + H_2O \rightarrow CO_2 + H_2$) [12,13]. In [12] the aim of the investigation of the formulation was stated up as follows: "The particle size of the protective colloid and the rate of hydrogenation have been related by studies of PVA (polyvinyl alcohol) in various degrees of polymerization as the protecting agent". Thereafter series of investigations followed, in those for the more complex compounds were used, such as silk fibroin, nylon, PETF, etc. An important achievement in the field was the creation of hydrogenization catalysts on the basis of highly dispersed Pd (its content achieves 88–94 mass %), stabilized by epoxidianic resin or by PVAc-electrolysis in a two-layer bath (see Chap. 5) [14]. The many aspects of the problem are considered and discussed in recent reviews and monographs [15–19].

At present the study of polymer-stabilized metal complexes or/and nanoparticles is an independent field of catalysis [20]. The main feature of such systems is that they have an open porosity and therefore practically all nanoparticles are accessible to the action of an external medium. Such three-dimensional systems[1] are considerably more stable than the surface island-type films and deposited catalysts. The most interesting cases are when the nanoparticle catalysis acquires new characteristics, as can be illustrated by the example of gold. Historically it was thought that gold is a bad or very weak catalyst. Now it is clear that gold in nanosized forms (particles size <3–5 nm, optimal size 3.2 nm) in combination with matched carriers

[1] The general methods of the design, fabrication and use of polymer resin beads are presented in [21,22].

is an even stronger catalyst than well-known Pd and Pt in many oxidative hydrogenization processes (CO oxidation at low and room temperature, propylene oxide synthesis, nitrogen oxide reduction by hydrogen at room temperature, NO reduction by propylene, and dienes selective hydration to alkenes) [23–27]. Propylene oxidation in a gas phase by an O_2 and H_2 mixture with a Au/TiO_2 catalyst and epoxidation of propylene proceeds with 100% selectivity, whereas with decrease of the gold particle size to <2 nm the reaction route was changed and its hydration product was formed (i.e. propane) with the same 100% selectivity. Such results cause growing interest in the study of nanocatalysts and nanoheterogeneous thin-film coatings.

Polymer-stabilized nanoparticles are interesting as a theoretical object. They are excellent models allowing one to demonstrate and study the features of the main theoretical catalytic schemes, especially connected with the influence of dimensional effects on the catalyst activity (such as multiple catalysis or active assembly theories, and the skeleton catalysts model). There were many attempts to study the influence of dimensional effects on the catalytic activity. The dimensional effects for nickel were studied from atoms to bulk metal [28], and for Ni_n/SiO_2 and $(Ni_n–Cu_m)/SiO_2$ assemblies a quantitative estimation of the catalytic activity in hydrogenization reactions was performed [29], and for polynuclear complexes (on the carrier surface) in olefin hydration and isomerization reactions the activity was studied in [30–32]. Heterogeneous contact catalyzed reactions can be divided into two main groups: structure-insensitive and structure-sensitive. This idea was proposed by Pauling, who made nanostructured solids using silica gels able to a molecular recognition and moleculary templating [33]. After that the nanostructured silica gels were prepared by imprinting the formation of the silica with homologues of methyl orange [34]. The hydrogenization reactions relate to the most structure-insensitive, whereas the oxidation reactions are often structure-sensitive (at temperatures below 600 K oxygen is adsorbed on the carrier defect centers, the substrate becomes partially oxidized and so the catalyst takes some specific selectivity).

12.2 Hydrogenization Processes

Most often nanocatalysts are used in hydrogenization processes. When the catalyst is formed and hydrogen is adsorbed, the dislocations are attached at interblock interfaces of the polymer-stabilized nanoparticles (those surface and inner structures are very active), whereas the polymer matrix prevents multiple defects "smoothing" under aging of the catalytic system.

Macromolecular complexes including nanoparticles of platinum or first transition groups in mild conditions can hydrogenate alkenes, aromatic or heteroaromatic substrates and provide partial or total hydrogenization of fats and unsaturated acids [35, 36], reduce various functional groups (nitro-, carbonyl-, nitril-), effectively produce polyol sugars, and transform alkynes to cis- and trans- alkenes [20].

The important role in such processes has not only a general content of the polymer-bounded metal particles, but their distribution characteristics into the polymer matrix (especially, their topography). As a rule, the maximal specific reaction rate (related to the total metal content in catalyst w_{sp}) corresponds only to the considerably rare cases, when all the nanoparticles are active and available to reactants. An idealized variant is when all the particles are removed to the polymer carrier surface. Such or near distributions can be designed and realized on the stage of catalyst formation and only seldom is formed during the catalyzed reaction running. For example, Rh-containing catalysts based on phosphinated CSDVB with different content of cross-linking agent (2–60 %) formed under conditions, excluding matrix swelling (i.e. metal particle penetration into the matrix), were 1.5–4 times more active than on the base of swollen matrixes [18].

Usually the w_{sp} value depends on the total or surfacial metal concentration in the polymer matrix. The value w_{sp} first increases with concentration, passing through a maximum and then decreases (Fig. 12.1).

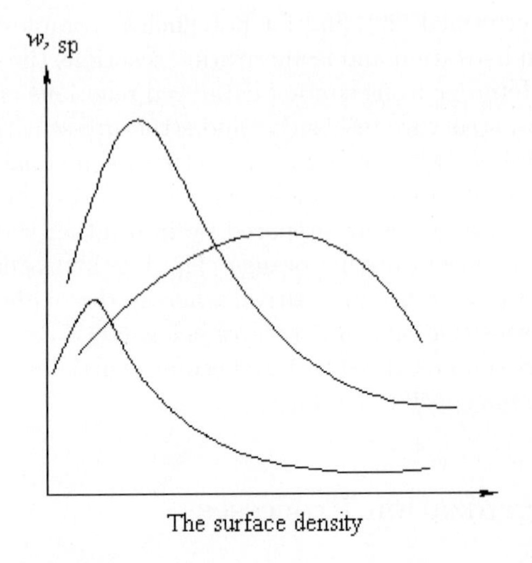

Fig. 12.1. Typical dependencies of the specific rate of hydrogenation on the surface density of the metal on a polymer

Similar behavior is observed in many other catalytic reactions (hydroformylation, oxidation, polymerization). The descent of the curve can be explained by nanoparticle association that decreases the catalyst activity. There is no good explanation for the ascending part of the curve. It is possible that the catalyst active centers are localized at the cluster boundaries and stabilized by their electronic systems (with a rise of the nanoparticle surface density, the number of such centers at first increases, but later decreases). This idea is confirmed by the fact that the w_{sp} value is greater for the hexene-1 hydrogenation by rhodium catalysts with low Rh content (0.5–1.2% of metal) with all other factors the same. In these contacts the main part of the metal is localized on the polymer surface and its cross-linked structure prevents nanoparticle aggregation at the stage of catalyst formation and during reaction.

At the same time the rates of 1,5-COD, butadiene or acetylene hydrogenation, the particle dispersity Rh_0, their ratio Rh_s on the carrier surface, and their part in the total Rh content in the catalyst, depend on the P/Rh ratio [37] (Table 12.1). A general tendency is that the maximal rate of butadiene-1,3 hydration is achieved at $P/Rh_s \sim 1$ and does not depend on this ratio for $P/Rh_s > 2$. The Rh_s nanoparticle size grows with phosphorus content in macroligand and the P/Rh_s ratio decreases and for low values of these parameters the size is near to the size of the Rh_s particles, formed on the polymer without any functional groups (i.e. nonmodified CSDVB).

Table 12.1. Some characteristics of rhodium complexes attached to phosphorylated CSDVB (2% DVB) and their catalytic activity in butadiene and acetylene hydrogenation

P, %	Rh, %	Rh_s, %	Rh Particles Size, nm	P/Rh_s	$N^a \times 10^2$ Butadiene	Acetylene
1.6	1.1	1.0	1.5	5.8	1.3	1.0
1.7	0.6	0.5	1.7	10.7	1.0	12
1.7	1.6	1.3	1.9	4.3	0.8	–
1.2	0.5	0.5	1.2	8.4	1.4	–
1.2	1.0	0.9	1.65	4.5	1.3	0.8
1.2	2.0	1.7	1.8	2.3	1.5	–
1.2	5.0	3.3	2.9	1.2	4.2	5.0
1.7	3.0	2.4	1.9	2.3	1.1	–
0.5	0.5	0.4	1.6	4.6	3.9	–
0.5	1.0	0.7	2.9	2.4	3.8	21
0.5	2.0	1.1	4.4	1.6	6.3	–
0.5	5.0	1.7	6.7	0.9	30	–
0 (CSDVB)	1.0	0.5	4.8	–	28	42

$^a N$ – substrate molecules number $\cdot$(at $[Rh_s]$ s)$^{-1}$; reaction conditions – 348 K; $P_{H_2} = 67$ kPa; $P_{C_4H_6} = 0.11$ kPa; $P_{C_2H_2} = 0.11$ kPa (368 K)

The formation of the catalyst active center is accompanied with a deep interaction of reduced atoms with the polymeric matrix. For example, in the reduction of Pd^{2+} on polyheteroarylenes (PBIA) the palladium cluster particles are formed with mean sizes 2–3.5 nm depending on the bound palladium content, including two forms of Pd_0 with bond energies at $Pd_{3d_{5/2}}$ – 338.0 and 336.0 eV [38]. Such small particles have a very strong interaction with the polymeric matrix and (as a result of electron transfer) a positive charge $Pd^{\delta+}$ arises on the palladium atoms that favor coordination of the substrate molecules. The process is assisted by the developed π-electron system of macroligand (PBIA) that pass to an ion-radical state. Thus, a strong metal–polymer interaction is accompanied by electron transfer and its reduction to a leuco-state that naturally affects the catalyst activity and has initiated numerous investigations of metal particle dispersion or condensation of their ions to nanoparticles, stabilized by these polymer matrices [39,40]. The role of the Pd particle size was studied for ethylene and propylene (390 K, P_{H_2}= 0.2 MPa) hydration rates and their mixtures (in the molar ratio H_2:C_2:C_3 = 4:1:1). The influence of Pd-1 catalysts with mean size ~3 nm and Pd-2 with mean size ~1.4 nm is that the hydration rate for Pd-1 is 0.012 mole·(mole Pd · c)$^{-1}$, and it is 1/8 times that for Pd-2. This can be explained by the corresponding changes of the specific surfaces of these catalyst, because the Pd-1 particle size was not changed after the catalytical reaction, whereas the Pd-2 particles grew from 1.4 to 1.9 nm.

Colloidal dispersions (1.8–5.6 nm) of Pd_0 connected with a polymeric matrix demonstrate very high activity, stability and selectivity in the processes of cyclopentadiene reduction to cyclopentane or 1,5-, 1,4-, 1,3- and 2,4-hexadienes, during which the conjugated diene hydrogenation runs in 1,2 and 1,4 positions. The palladium particles with mean size 1.8 nm in a PVP protective envelope are very active catalysts of many processes, such as linoleic acid methyl ether hydrogenization, partial hydrogenation of soy-bean oil under mild conditions, and selective hydrogenation of the dihydrolinalole (3,7-dimethyloctene-6-in-1-ol-3) triple bond to the double bond in the production of aromatic substances [41]. Such dispersions have narrow size distribution and are stable (the colloidal particles are not release from the gel, when the pH changes from 2 to 13). The catalyst is easily separated from the reaction medium by usual decantation and keeps its activity under repeated use. The nanoparticle sizes do not affect the rate of linear olefin hydration; the cyclohexene hydrogenation rate decreases with increase of the catalyst particle size [42] (Table 12.2).

In recent years novel approaches have been developed for the dispersion of noble metal nanoparticles into polymer matrices, forming so-called near-monodisperse nanoparticles (see Chaps. 4–6). The catalysts on the basis of the individual immobilized metal clusters (Sect. 5.2), usually the polymer derivatives of Os_3, Ir_4, Ru_4, Rh_4, Rh_6 clusters, have more definite and often tailored structures [43,44]. Early studies [45] demonstrated that the ethylene

Table 12.2. Colloidal particle sizes Rh^0 and their activity for olefin hydrogenation

System	Rh Particles Mean Size, nm	Activity, (Mole H_2/g-atom Rh·c)	
		Hexene-1	Cyclohexene
$RhCl_3/PVA/CH_3OH/H_2O$	4.0	15.2	3.1
$RhCl_3$/polymethyl-ester/CH_3OH/H_2O	4.3	22.5	9.6
$RhCl_3/PVPr/CH_3OH/H_2O$	3.4	15.8	5.5
$RhCl_3/PVPr/C_2H_5OH$	2.2	15.5	10.3
$RhCl_3$/ PVPr/iso-C_3H_7OH	2.4	–	9.3
$RhCl_3$/PVPr/-CH_3OH/ NaOH	0.9	16.9	19.2

Conditions: 303 K, pressure – H_2 0.1 MPa, [olefin] = 25 mmole; [Rh] = 0.01 mmole.

hydrogenation rate by tetrairidium (Table 12.3) and tetraruthenium clusters (connected with phosphinated polymers) decreased with an increase of the electrodonor phosphine ligand content into the ligand. That is a result of the bond strength between the substrate (ethylene) and catalyst.

The hydrogenation is totally inhibited even in the presence of trace amounts of CO (but with its removal the catalyst activity is resumed) which points to the absence of coordination vacancies in the immobilized carbonyl

Table 12.3. The influence of the P/Ir ratio in clusters on the ethylene hydrogenation rate

Catalyst	P/Ir Ratio	$w \times 10^3$ (Mole of C_2H_4/g-atom Ir·c)
Poly-$PPh_2Ir_4(CO)_{11}$	4.54	21.3
Poly-$PPh_2Ir_4(CO)_{11}$	14.3	25.3
Poly-$PPh_2)_2Ir_4(CO)_{10}$	3.4	3.76
Poly-$PPh_2)_2Ir_4(CO)_{10}$	6.9	3.60

Conditions: Pressure of ethylene – 20 kPa, pressure H_2 – 80 kPa, 313 K; the sign "Poly" relates to polymer (CSVDB).

cluster centers. At the same time physical and chemical studies demonstrate that the M–CO bonds do not split during catalysis and consequently the coordination vacancies are formed for the L→M bonds breaking (most probably the M–M bonds, such as Ru–Ru, Rh–Rh, Os–Os). It was proposed for unsaturated aminoacid hydrogenization (with $Rh_4(CO)_{12}$ or $Rh_6(CO)_{16}$ clusters) that a catalytic activity was revealed by the monomer $HRh(CO)_2L$ particles, formed as a result of the oxidative addition of H_2 and cluster segregation [46]. The mechanism for styrene hydrogenation by tetranuclear osmium clusters was postulated, including cluster fragmentation, when a low concentration of the highly active particles (with lower degree of nuclearity, even monomeric particles) is observed, and these particles are the true catalysts of the reaction [47]. The different variants of the immobilization were studied for various cluster types including hybrid, multiligand or heterometallic clusters. For example, heterometallic $RhOs_3$-clusters in a phosphinated matrix (an initial $H_2RhOs_3(AcAc)(CO)_{10}$ cluster) in the model butene-1 hydrogenation and isomerization reactions were segregated. They formed polymer-connected three-osmium clusters (responsible for the stable activity under isomerization) and metallic rhodium particles 1.0 nm in diameter, active in butene-1 hydrogenation [48].

Methods were developed for production of such catalysts by copolymerization of the proper cluster-containing monomers with traditional monomers (Sect. 5.2) [43,44]. The correct *metal:functional group ratio* in such systems is achieved at the stage of formation of the catalyst. Therefore these catalysts may have a high activity and selectivity, for example, the catalysts on the base of vinylpyridine or allyldiphenyl copolymers with Rh_6 or Os_3 complexes in the cyclohexene hydrogenation model reactions [44]. There are some other interesting examples of the use of nanocatalysts. The traditional catalysts of many processes (Ag_n/C, Pd_n/C, Pt_n/C) can be formed as polymer-bonded nanoparticles, using the combined cryochemical methods (Chap. 6) for Ag, Pd and Pt atoms with liquid polyesters or polydienes and "critical loading" (when the metal concentration is higher than a ligand coordination ability) and with a following application of carbon powder. The same catalysts are based on platinum or other metal nanoparticles incorporated into the various carbon-glass matrices [49].

There is interest in catalysts based on organic–inorganic hybrid materials with catalytically active metal particles, intercalated into a material oxide lattice. Such highly dispersed heterogeneous catalysts and their applications are reviewed in [50]. The three-stage mechanism of their formation into a SiO_2 lattice through substituted alkoxysilanes is given in Scheme 12.1.

It is beyond reasons to wait for a change of an action mechanism for the cluster-based or nanosize catalysts, introduced into polymer or inorganic matrix. Such systems show higher activity and selectivity in reactions of ionic-coordination polymerization, hydrogenization, and oxidative transformations [18]. Their activity depends essentially on the nature of the polymer

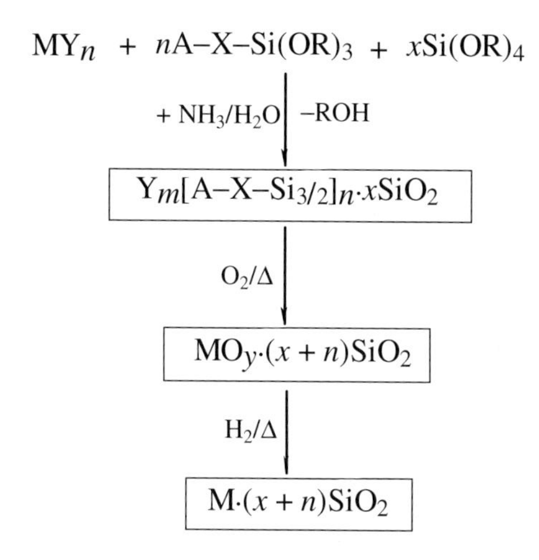

Scheme 12.1

and the nanoparticle characteristics. Such catalysts most often are used as powders, less common as solutions (suspensions), and sometimes as fibers or films.

As a rule, diffusional limitations are not significant in nanocatalyzed reactions, because many organic compounds (substrates) can easily cross the protective layer of the stabilizing polymer. These limitations are essential in rare cases and can control the transport of reagents and products to the catalytically active centers. At first glance this ability is a negative factor, but in fact it can be of interest for improving the catalyst selectivity, because one can provide a considerable difference in the various substrate diffusion to the catalyst (i.e. increase its so-called substrate selectivity) using a proper polymeric stabilizer. The protective layer thickness can be comparable or even greater than the nanoparticle sizes, as for example the thickness of a PVP layer (adsorbed on palladium particles with sizes 2.0–2.5 nm) is from 1.9 to 7.8 nm, depending on the polymer molecular mass (6000–574 000) [51].

The transport of reagents through a polymer matrix is the limiting stage of the hydration of olefins and dienes under the action of monodispersed Pd nanoparticles, formed from a reduced copolymer, obtained by a metathesis polymerization of metal-containing monomer of the type Pd[*endo*-2-(cyclopentadienylmethyl)norborn-5-en] -η^3- 1-phenylallyl (see Chap. 5) [52]. The coefficients of diffusion ($D\cdot 10^{11}$, $cm^2\cdot s^{-1}$), solubility ($S\cdot 10^3$, $g\cdot cm^{-3}\cdot at$) and permeability [$p = DS\cdot 10^{14}$ mole (mole Pd)$^{-1}\cdot g\cdot cm^{-1}\cdot s^{-1}\cdot$ at] for ethylene, propylene and 1,3-butadiene at room temperatures are equal to, respectively: 9.1, 0.4 and 0.6; 0.28, 0.08 and 0.015; 0.25, 0.03 and 0.1. These values explain the lower hydration rate of propylene as compared with ethylene. At the same time catalyst exposure to a mixture of C_4 gases increases the ethylene hydration rate by ~4 times. This means that the transport characteristics are changed in the polymer matrix. Probably, the inner voids are opened, the mean path for gaseous reagents and product penetration becomes shorter and, as a result, the catalytic activity increases. The effect is observed only for ethylene hydration and the propylene hydration rate remains nearly the same. This shows that, apart from the transport properties, many other characteristics play an important role (for example, the difference between the adsorption coefficients of ethylene and propylene or the steric effects of the Pd particle polymer environment). As a rule the ethylene hydration rate is 40% higher then for propylene. The conditions of the formation of near-monodisperse 4 ± 0.6 Rh [53] and 2 ± 0.3 $Ir(0)_{\sim 300}$ nanoclusters were analyzed [54], as well as their influence on the hydrogen mass-transfer limitations and the chemical-reaction rate-limiting region.

Hexene and cyclohexene hydrations are very useful models and were well studied. For example, the dependence of the cyclohexene conversion to cyclohexane on the nature of the reducing agent was studied for colloidal dispersed Pd particles, obtained in block-copolymer PS-P4VPy [55]. The activity increased as follows (the conversion degree is given in brackets, %): phenylhydrazine (0.2) < hydrazinehydrate (1.0) < without reducer (4.2) < $(C_2H_5)_3SiH$ (10.8) < $NaBH_4$ (28.3) < superhydride (32.7) < Pd (1%)/C (40.0). As in many other processes, a synergic action was found for the bimetallic nanoparticles. Thus the Pd/Au combinations in block-copolymer exhibit catalytic properties, depending on their composition:

Au/Pd (atoms)	1/6	1/5	1/4	1/3	1/2	Pd/C
Cyclohexene yield, %	55.5	71.3	65.7	68.7	39.4	40.0

The cyclohexene yields (%) for nanosize catalysts with Rh, Pd, Pt, as well as for Pd/Pt (4:1), Rh/Pt (4:1) were equal to 26.4, 32.7, 60.7 and 55.5, 49.1% correspondingly.

Hydrophobic latex dispersions can adsorb Pt nanoparticles just at the moment of their generation [56,57], forming a novel type of nanocatalyst, the metal nanoparticles attached to the carrier nanoparticles. Various systems for water dispersed latex were used: N-vinyl-2-pyrrolidone copolymer with styrene (40/60), PVDC, PVAc, and vinylchloride copolymer with methylmetacrylate. The large particles were formed under slow reduction, the considerably bigger ones – at the H_2PtCl_6 boiling in the ethanol : water = 1:1 (vol.) mixture, but under a KBH_4 action (i.e. at a fast reduction) only the small particles were formed. Their catalytic activity depended on the latex type and the (polymer:platinum) mass ratio (Table 12.4). As a rule, the values of the ratio rise in the series 10:1, 25:1 and 50:1 and lead to a drastic increase of the catalytic activity.

Table 12.4. Cyclohexene hydration on Pt and Pd nanocatalysts, immobilized in polymers

Polymer	Reduction 1 – in Boiling Ethanol 2 – KBH_4	Pt, nm	Cyclo-Hexane, %	Pd, nm	Cyclo-Hexane, %
PVP	1	2.7	73.0	1.5	98.6
(mol. mass 360 000)	2	1.5	70.8	5.5	80.2
Poly(2-ethyl-2-	1	2.6	55.8	2.4	90.8
oxazoline)	2	1.7	56.0	2.0	77.2
(mol. mass 200 000)					
Poly(1-	1	2.2	61.7	3.0	97.1
vinylpyrrolidone-co-	2	1.9	61.3	3.7	84.2
vinylacetate) (60/40)					
PVDC	1	1.8	22.4	5.0	2.3
	2	2.2	23.6	1.0	3.2
Poly(methacryl-	1	1.4	22.7	5.0	2.4
amidopropyl)trimethyl-	2	2.1	25.4	1.2	7.7
ammoniumchloride					

Notes: reaction conditions – 10 mL methanol; 0.05 mL cyclohexene; P_{H_2} = 0.07 MPa; room temperature; reaction time – 30 min; 0.09 % (mass) Pd or 0.16 % (mass) Pt to cyclohexene content.

Particle accumulation takes place (an adsorption from the solution) along with nanoparticle immobilization on latex (the immobilization is assisted by the lower hydrophobicity of the latex and a low reduction rate). This leads to an agglomeration (particularly in catalytic processes) and this factor can considerably decrease the activity. To prevent this effect, the *latex–metallic* particles can be stabilized by introduction of a protective polymer, using soluble nonionic homopolymer PVP, polystyrenesulfonic acid or cationic polyelectrolyte – (PVDC). In such variants the nanoparticles remain stable over several weeks or even months. The protective polymer creates a specific environment around the active center (electrostatic, hydrophobic or steric, see Chap. 3) that can affect the catalyst activity and its selectivity.

Latexes (PS, PVAc, PVDC) similar in their hydrophobicity influence the nanoparticle accumulation and immobilization processes differently [56]. The strong interactions between cationic polyelectrolytes and anions, which are $[PtCl_6]^{2-}$ or $[PdCl_4]^{2-}$ nanoparticle precursors [58], can also assist in the formation of very small particles with a narrow size distribution under reduction processes.

Pt nanoparticles attached to styrene-divinylbenzene copolymer, modified by iminodiacetic acid can be used for cycloolefin and cyclodiene hydrogenation [59]. The catalyst production includes additional mechanochemical immobilization of Al^{3+}, Mg^{2+} or Na^{+} ions in swollen matrices with a subsequent treatment by a H_2PtCl_6 aqueous solution and consequent reduction by $LiBH_4$. The specific surface area of the catalysts was 16.2, 13.7 and 1.5 $m^2 \cdot g^{-1}$, correspondingly, and the mean size of particles was ~2.6 nm. For an optimal catalyst combination (Al/Pt) the initial rates of cyclohexene, cycloheptene and cyclooctene hydrogenation (under correlated conditions) were 0.615, 0.982 and 0.260 mL $H_2 \cdot min^{-1}$, correspondingly.

The described nanocatalysts can very effectively hydrogenizate not only the double bonds but the triple bonds and the conjugated double bond systems as well. The selective partial hydration of alkynes and dienes is a potentially promising field of metallopolymer nanocomposites. The industrial processes of this type include selective synthesis of alkenes and dienes from alkynes in flow [60], partial hydrogenization of benzene [61], cyclopentadiene, isoprene and butadiene [62]. The selectivity has a special role in such reactions that proceed at low temperatures and pressures, as well as under limited mass transfer conditions.

Pt-bimetallic catalysts can selectively hydrogenizate cyclooctadiene (COD), whereas the reaction rate is proportional to the catalyst surface and increases in the series of additionally introduced metals as Na $<$ Mg $<$ Al (0.057, 0.404 and 0.543 mL $H_2 \cdot min^{-1}$, correspondingly). Pd/Pt nanoparticles stabilized by polymers (usually PVP) are used [63–66] with maximal activity (1.4 times greater than monometallic catalyst activity) at atomic ratio Pd/Pt = 4:1 (Fig. 12.2) and the optimal mean diameter of particles equals 1.4 nm. It was shown experimentally that an alloy with the same ratio as Pd/Pt is characterized by a large negative value of formation enthalpy (there arises a very unusual cluster with a nucleus of 13 Pt atoms and an envelope of 42 Pd atoms). The protective function of H_2 in such alloy formation is based on an ability to be adsorbed and decomposed at these noble metal atoms [67]. The selectivity degree of 1,3-*cis*-COD hydrogenation to cyclooctene achieves 99%.

A similar situation is observed for mono- and bimetallic Pd/Au colloids (stabilized by styrene block-copolymer with 4-vinylpyridine) for the selective hydrogenation of 1,3-cyclohexadiene and 1,3-COD [55]. For the substrate/catalyst ratio = 500:1, 100% conversion of 1,3-COD to cyclooctadiene is achieved, i.e. the selectivity is greater than for the conventional catalyst Pd/C (the conversion degree is 97.3%). The selective hydrogenation of 1,3-cyclohexadiene is a more complicated process, because the reaction has an additional route and the substrate is disproportioned to cyclohexene and benzene. The main factor of selectivity is the substrate/catalyst ratio. These reaction rates must depend on the steric surroundings of the active center, because the hydrogenation reaction is preceded by a coordination activation

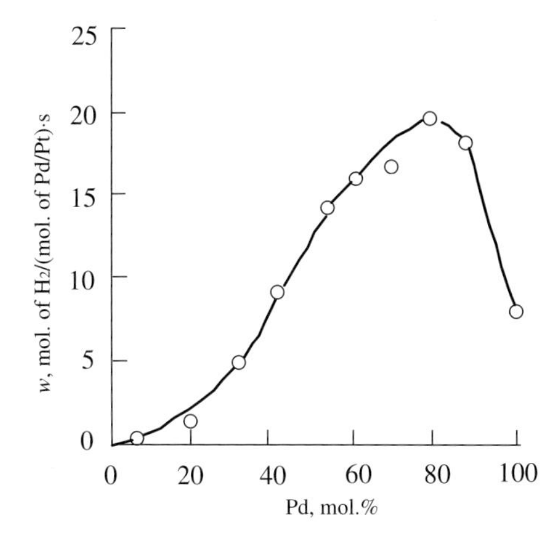

Fig. 12.2. Initial rate (w) of 1,3-cyclooctadiene (25 mmole) hydration at 303 K as a function of polymer-immobilized Pd/Pt-bimetallic cluster composition (0.01 mmole), prepared in an N_2 atmosphere

of the substrate and H_2 molecules. It is necessary for the disproportionation to have a few big coordination vacancies (near the active center) and to coordinate and activate two considerably bulky substrate molecules. The difference in selectivity was found to be small (from 28.6 to 41.5% of benzene) which can be interpreted as a similarity of the nature of the active centers in both the considered catalysts and Pd/C. The highest selectivity (only 11.5% of benzene) was obtained with the catalyst on the base of the Au/Pd (1:5) nanoparticles applied on Al_2O_3. It proves the need to create proper steric surroundings of the active centers for their selectivity rise. The considerable rise of the bimetallic Pd/Rh nanocluster activity, as compared with monometallic catalysts, was registered for *cis*-cyclopenta-1,3-diene hydrogenation to cyclopentene [68]. New opportunities for heterogeneous catalysis processes on the nanoscale bimetallic system Pt–M (where M is a 4d transition metal) arose due to X-ray synchrotron radiation techniques and theoretical studies [69].

The influence of the polymer matrix on the selectivity of acenaphthylene hydrogenization by polymer-stabilized palladium (obtained by Pd vapor deposition onto cooled polymeric matrices; particle size is 3.2 ± 1.3 nm) was demonstrated in [70]. The substrate contains an activated olefinic acid and has unsaturated aromatic double bonds that are hydrogenated successively (Scheme 12.2).

The process with this catalyst is completed during 15 minutes, while with a conventional catalyst (palladium black) it runs in no less than 5 hours (Fig. 12.3). One other advantage is that the protective polymer envelope prevented catalyst poisoning by the potential catalyst poisons (such as dibenzothiophen).

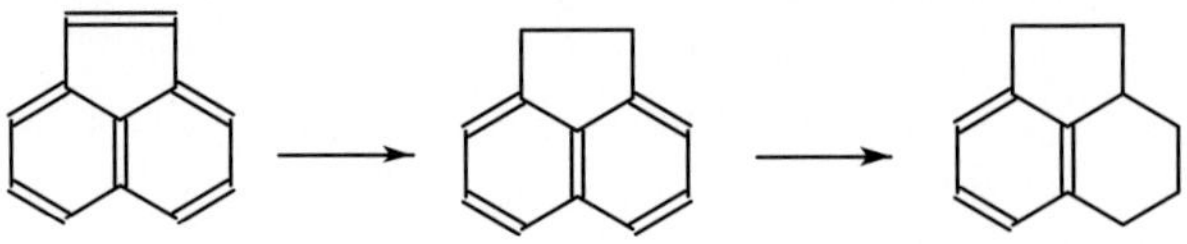

Scheme 12.2

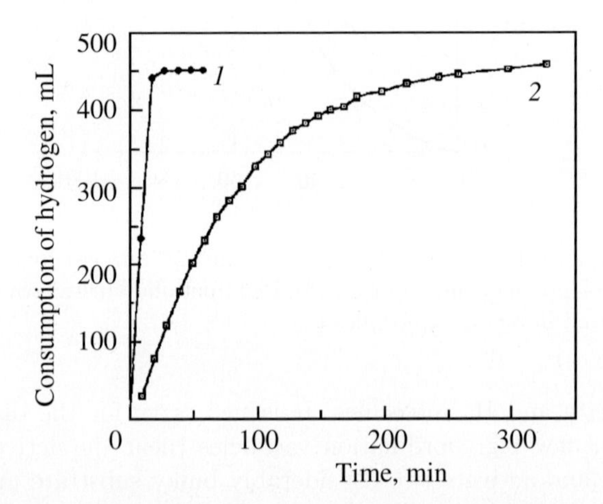

Fig. 12.3. Hydrogen consumption as a function of time for hydration of acenaphthylene (20 mmole) by 0.31 mmole polymer-stabilized Pd colloid (*1*) and Pd/C (*2*) (50 ml methylcyclohexane; 298 K; 0.1 MPa H_2)

The protective envelope of PS where the Pt nanoparticles (mean sizes 7 nm) are introduced, does not prevent an effective reduction of methylviologen (dimethyl-4,4′–bipyridine), which is a (MV^{2+})-acceptor of electrons in photocatalytic systems. The reduction is carried out by molecular hydrogen [71] by the Scheme 12.3:

$$2MV^{2+} + H_2 \quad \rightarrow \quad 2MV^{+} + 2H^{+}$$

Scheme 12.3

In recent years, so-called "intelligent" materials have attracted great attention. They can react to the various external influences or factors (heat, light, electrical field, pH, chemical action). The activity in nanocatalysis with polymer-immobilized particles can be controlled by so-called temperature-dependent ligands [72]. For example, Pt particles (1–5 nm, mean value 2.38 nm) in poly-N-isopropylacrylamide (P*i*PA) with average molecular mass ~6000) exhibit temperature-dependent properties in allyl alcohol hydrogenization to 1-propanol [73]. As long as the PVP-stabilized nanoparticles are not deposited from water at low temperatures without electrolyte [74], the temperature dependence of the hydrogenation rate is Arrhenius. The situation for P*i*PA is more complicated. When either a solution of initial P*i*PA or a solution of Pt nanoparticles stabilized by this polymer are heated to a temperature higher than the low critical solution temperature (LCST) (305 K), the homogeneous Pt dispersion is rejected from the solution (Fig. 12.4) and the catalytic activity reduces by 30%.

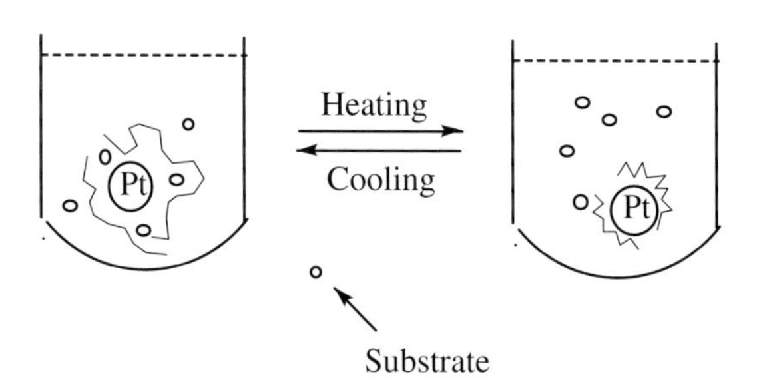

Fig. 12.4. The main scheme of action of temperature-dependent catalysts

The effect is reversible, i.e. on cooling to 293 K the catalyst is again dissolving, its activity is restoring and allylic alcohol is hydrated faster at temperature below LCST. Catalyst deactivation was not observed after these temperature cycles. This means that the P*i*PA-protected Pt nanoparticles also exhibit temperature-dependent solubility. The effects depend also on the composition of solvent used for Pt^{2+} reduction (ethanol:water in ratio 9:1–6:4), changing the conformation of the nanoparticle polymeric environment.

The catalytic activity of metallopolymeric nanocomposites can be very high, thus the activity of Pd_0/P4VPy in allylic alcohol hydration is one order of magnitude greater than the platinum black activity [75]. The same metallopolymers applied onto a previously calcinated (at 570 K) MgO are even more active in the selective hydration of allyl and cinnamic alcohols, crotonic and cinnamic aldehydes and isoprene (the hydration product is a mixture of isomeric butenes). The isoprene hydration catalysis demonstrates that addition of H_2 to a conjugated bond system can proceed in both the 1,4- and 1,2-positions.

The bimetallic Pd/Ni catalysts obtained by joint reduction of metal salts in glycol at 400 K and pH 9–11 (the alloyed particles have sizes 1.6–2.0 nm [76]) can hydrogenize nitrobenzene to aniline under mild conditions. The Pd/Ni clusters containing from 40 to 80% of Pd exhibit a higher activity than Ni- and Pd-containing monometallic clusters. The highest activity (4.2 times greater than the Pd nanoparticle activity) was observed with an alloy containing 60% of Pd (Fig. 12.5). The hydrogenation rates for the substituted nitrocompounds under the action of a palladium-polymer catalyst (a graft

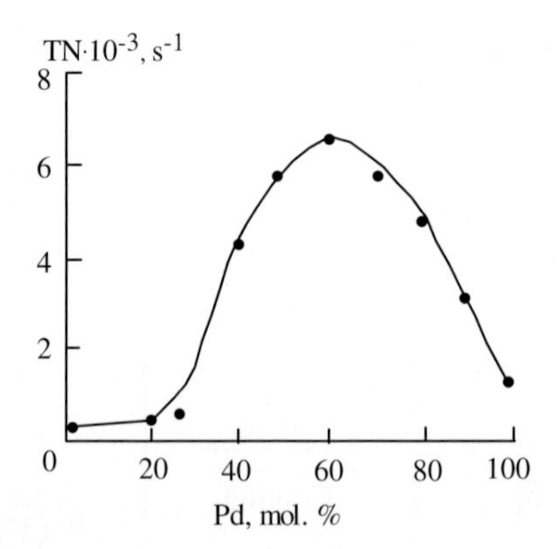

Fig. 12.5. Hydrogenization rate dependence (TN – number of catalyst turning) for nitrobenzene ($4.7 \cdot 10^2$ mole·L^{-1}, P_{H_2}= 0.1 MPa) on Pd/Ni cluster composition [$1 \cdot 10^{-2}$ (g-at)·L^{-1}) in ethanol at 303 K

copolymer of PE and P4VP) are correlating with Hammett constants of their substituents [77].

Let us analyze some specific characteristics of the described polymer-immobilized nanocatalysts. As a rule, nitrogroup reduction to an aminogroup runs with small rates, and benzene ring cannot be hydrated at all. Sometimes carbonyl group (benzaldehyde or acetophenone) reduction proceeds with a high rate and does not stop at the stage of alcohol formation but runs to the formation of aromatic hydrocarbons. At the same time, polymer-protected bimetallic Pt/Co colloids exhibit high (up to 99.8%) selectivity in cinnamic aldehyde to the proper alcohol. Some noble metal nanoparticles with sizes 0.5–10 nm are very active and selective catalysts of nitrile hydrogenization [78].

Dehydrolinalool hydrogenation (DHL, 3,7-dimethyloctene-6-in-1-ol-3) is an important industrial process, because the hydrogenation products are intermediates in the synthesis of vitamins A, E and K and fragrant substances (linalool) for some aromatic substances. The catalysts based on palladium nanoparticles obtained in block-copolymer micelles of PS-P4VPy allows one to reach the selectivity 99.5% in triple bond hydrogenation to double catalysts [41]. It is possible that in this case the selectivity change is connected with modification of the metal surface by pyridinic units of the P4VPy nucleus. If the second metal atoms are introduced into the nucleus and a bimetallic nanoparticle is formed (Pd/Au, Pd/Pt, Pd/Zn), the electronic structure and surface geometry of palladium, resulting in an additional increase in selectivity up to 99.8%, are modified. Pt, Pd, Rh, Ir and Ru nanoparticles, stabilized in polymers by a “coordination trapping” method [79, 80] (Chap. 5) can reduce citronellal (3,7-dimethyl-6-octenal) to citronellol (3,7-dimethyl-6-octen-1-ol) with selectivity up to 80–90%. The reaction rates and their selectivities are practically stable during the many hydrogenization cycles. The thin-layered metallopolymer composites on the base of Pd particles (mean diameter ∼7 nm), obtained by cocondensation of *p*-xylylene and Pd vapors (Chap. 5), can catalytically isomerize 3,4-dichlorobutene-1 into *trans*- and *cis*-1,4-dichlorobutene-2 [81]. Nanoparticle interaction can be changed by changing the volume metal content in the nanocomposite which leads to a change of the catalyzed reaction selectivity (the ratio of *trans*- and *cis*-isomers).

The regioselectivity effect was most clearly observed in comparison of 10-undecenoic or oleic acids by Pt and Pd nanoparticles, stabilized by nonionic surface active substances (e.g., by PEO–monolaurate) or polymers [82, 83]. The nanoparticle hydrophobic surface is protected by hydrophobic laurate chains and the polyethyleneglycol hydrophilic chains form an external layer. Such polymeric micelles can not only stabilize the particles, but at the same time they can give orientation to unsaturated fatty acids and control the substrate orientation as a whole (Fig. 12.6).

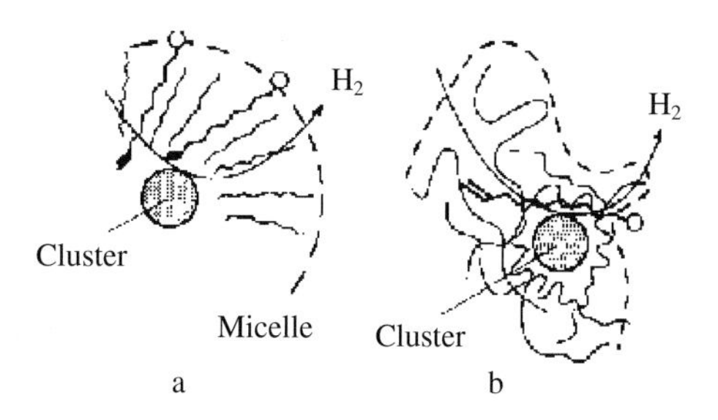

Fig. 12.6. The supposed mechanisms of undecenoic acid hydrogenation in micellar (**a**) and polymeric (**b**) systems

In the formed intermediates the substrate bond position defines the hydrogenation rate, because the molecule is oriented in the micelle walls in such a way that the polar carboxyl group is placed in the micelle hydrophilic part, while the hydrophobic part is amongst the hydrophobic laurate chains, forming a nucleus. The hydration rates for the described acids decrease in the series (undec-10-enoic > oleic > undec-2-enoic), showing that the terminal double bond is placed most favorably for contact with the catalyst centers, whereas the inner double bond has only limited access to the Pt nanoparticles. The access of a substrate double bond is not limited for linear polymer PVPy. Therefore the Pt colloids do not exhibit any regioselectivity.

Other interesting effects in nanocatalysis arise when metal ions are introduced into a system too. For example, introducing Nd^{3+} ions in Pd nanoparticles stabilized by a water-soluble polymer promotes the catalytical activity in hydrogenization of acrylic acid, methylacrylate or allylamine [84]. The mechanism of this effect is based on the substrate concentration due to the additional coordination interaction with neodium ions (this effect is similar to substrate enrichment in biocatalysis).

Another interesting direction in nanocatalysis is to perform reactions in supercritical CO_2. Thus, olefin hydrogenation catalyzed by palladium nanoparticles in a water-in-CO_2 microemulsion is a very effective process [85, 86] and has wide promising in the conversion of a nitro group to an amine group.

12.3 Nanocatalysis of Reactions, Traditionally Perfomed with Homogeneous Metallocomplexes

Pharmaceuticals, agrochemicals, flavors and fragrances are the principal areas that require the synthesis of optically pure chiral compounds [87]. Most of them are still synthesized starting from the chiral pool or in their racemic form, followed by resolution. Homogeneous asymmetric catalysis was used dominantly in this field. Nevertheless, only few examples have been developed to an industrial scale. The use of heterogenated polymer-supported catalysts is reviewed in [88–91].

It is very likely that such polymer-protected nanoparticles should attach novel features to enantioselective catalysis. The chiral complex was heterogenized in a PDMS film [92, 93], entrapment of metal complexes in silica matrices by sol–gel methods [94]. Platinum sol and Pt/Al_2O_3 (well-known enantioselective catalysts for hydrogenization processes [95]) are now studied by this approach. The colloidal platinum stabilized by PVP and modified by cinchonine is a very effective and selective catalyst of peruvic acid ethyl ether enantioselective hydrogenization to lactic acid ethyl ether [96]. The cluster $[Pt_{15}(CO)_{30}]^{2+}$ attached to Sefadex is a good catalyst for selective reduction by hydrogen of redox-active cofactors and proteins [97].

Impregnation of hyper-cross-linked PS solution of Pt(II) and its subsequent reduction by H_2 leads to the formation of Pt nanoparticles with mean diameter 1.3 nm. They are effective and selective catalysts of L-sorbose oxidation to 2-keto-L-gulonic acid which serves as an intermediate in vitamin C production (Scheme 12.4) [98].

Polymer-immobilized systems are very active in low-temperature CO oxidation processes [99, 100]. The substrates themselves can serve as reducing agents. For example, cations (Ag, Cu, Rh, Ru, Pt and Ir) connected with perfluoroethylensulfonic acid or with partially sulfated PS can oxidize CO, NO, NH_3, N_2H_4, C_2H_2 at 373–473 K, which is accompanied by the formation of colloidal particles with sizes 2.5–4.0 nm. Now the processes of CO catalytical oxidation are being studied intensively, mainly for ecological and environmental problems connected with their removal from waste and exhaust gases. Many results of these investigations are in patents. There is information about the use of copper polymeric chelates and platinum metals on the fluorinated polymer Nafion (the metal particles, formed after reducing

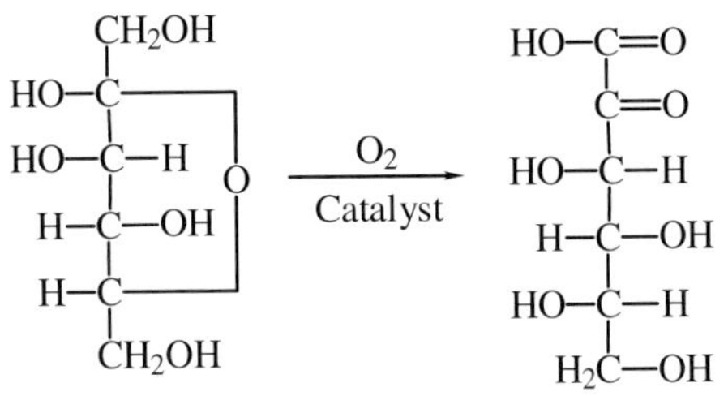

O_2, Catalyst

L-sorbose → 2-keto-L-gulonic acid

Scheme 12.4

by hydrogen, have sizes 2.5–4 nm) [101]. It is supposed that in such systems the reaction of CO oxidation to CO_2 is diffusion-limited.

The catalytic epoxidation of ene-compounds must be noted from the most characteristic reactions of organic oxidation [102]. Metals (in particular, Au and Pt) were used in an ultrafine colloidal state in the presence of protective reagents (synthetic polymers and artificial liposomes). Another example is liquid-phase cyclohexene oxidation by organic peroxides under the action of osmium carbonyl clusters, anchored into polymeric matrices. This process runs with the preferential formation of unsaturated alcohol and the initial structure and composition of Os_3-clusters are not changed during the numerous cycles [103].

Metal nanoparticles are widely used in several so-called "named" reactions. For example, palladium nanoparticles in the well-known and important Suzuki reaction [104] were obtained by metal salt reduction in the presence of polymeric stabilizers, such as dendrimers of poly(amido-amine) with terminal hydroxyl groups (G*n*-OH, where Gn is the *n*-th generation), sodium polystyrene-*block*-polyacrylate (PS-PANa) and PVP. Such nanostructurated systems were used in catalytic reaction of cross-combination between phenylboric (or 2-thiophenboric) acid and iodobenzene in aqueous medium at 373 K during 24 hrs. The Suzuki reaction is a good test for the stability of nanoparticles, which can be evaluated by palladium black formation during a catalytic cycle. The metallopolymeric nanocatalysts were effective under these conditions, though strong capsulation of the Pd nanoparticles into the dendrimers decreased their catalytic activity. Probably, these two properties are mutually opposite, i.e. higher stability means lower catalytic activity, which can be easily connected with strong nanoparticle capsulation into the substrate.

Interesting results were obtained for the Mannich reaction (aldol condensation of silyl enolates with imines) in a supercritical CO_2-PEO system [105] and many others. Let us describe a specific catalytic reaction, the Heck reaction (vinylation of aryl halides), shown in Scheme 12.5 and Table 12.5 [106, 107].

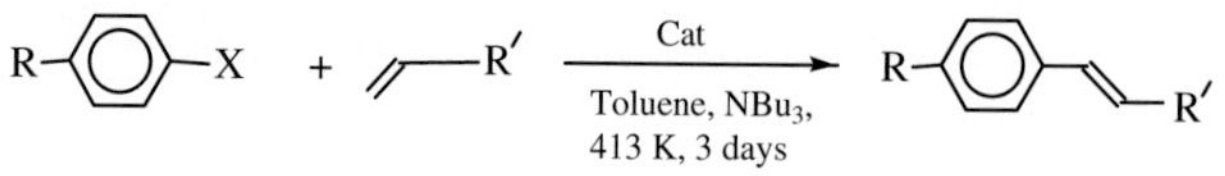

$$R-C_6H_4-X + CH_2=CH-R' \xrightarrow[\text{Toluene, NBu}_3\text{, 413 K, 3 days}]{\text{Cat}} R-C_6H_4-CH=CH-R'$$

Scheme 12.5

Table 12.5. Heck coupling reactions of aryl iodides, catalyzed by Pd-polymer catalyst

Aryl Iodide	Alkene	t, h	Yield, % *trans*	Yield, % *cis*
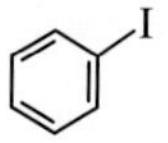 Iodobenzene (C_6H_5I)	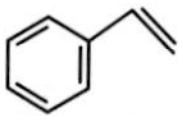 Styrene	6	93	7
"	 $CH_2=CH-CN$	6	51	48
"	 $CH_2=CH-COOCH_3$	2	90	10
4-Iodotoluene ($H_3C-C_6H_4-I$)	Styrene	4	89	11
"	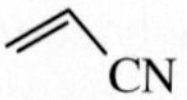 $CH_2=CH-CN$	6	61	29
"	 $CH_2=CH-COOCH_3$	4	90	10

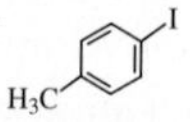

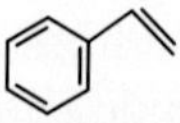

Reaction conditions: a molar ratio of 1:1.2:2.4 was used for the aryl iodide (5 mmol.):alkene:base in 6ml DMF; T = 363 K; molar ratio Pd/ArI = 1:1000

The coupling between aryl iodides and bromides with olefins are mostly catalysed by expensive palladium complexes in homogeneous solution in the presence of phosphorus ligands, which are toxic and unrecoverable. The problem can be substantially solved by the use of a heterogeneous system made up of supported palladium complexes, which are generally phosphine-free. Pd nanoparticles, stabilized by PS-*block*-P4VPy, did not exhibit a considerable rise of the catalytic activity in this reaction, but were more stable than homogeneous systems [108]. At the same time, the palladium metal colloids, stabilized by the presence of the polymer or mineral supports, exhibited a great stability and activity (even after 50 000 cycles) and in optimal variants the product yield exceeded 90% per cycle [109].

The results given in Table 12.5 relate to the reactions with Pd-polymer catalysts, obtained by copolymerization of Pd(2-(acetoacetoxy)ethylmethacrylate) with ethyl methacrylate and ethylene glycol dimethacrylate [110]. The Pd nanoparticles, coated by an amorphous carbon envelope and obtained after ultrasonic radiation of the proper precursors, have similar catalytic properties [112]. Recently dendrimer-encapsulated nanoparticles Pd were described and shown to be versatile catalysts for both the hydrogenation of styrene and Heck heterocoupling of iodobenzene and methacrylate in supercritical CO_2 [113]. Pd-modified zeolites are active and recyclable catalysts for the Heck reaction of aryl bromides with olefins [114]. Polyacrylamides with hydrophobic alkyl groups are effective handles for the recovery of a homogeneous catalyst and sequence of trace metals in the nonpolar phase of a biphasic solvent mixture [115].

We do not analyze many dehydrogenization and desulfuration reactions (including reactions in bimetallic catalytic systems), the Fischer–Tropsch synthesis reactions, the electrophilic oxidation reactions, etc. In general, it must be concluded that the most important organic synthesis processes, including olefin acetoxylation, alcohol oxidation in mild conditions (for example, the electro-oxidation of alcohols is described with modified electrodes, which have been fabricated from polypyrrole films with incorporated monophosphinic ruthenium complexes [116]), can be effectively catalyzed by polynuclear, cluster and nano-sized metal particles being incorporated into a polymeric matrix [18]).

Particles of Cu, Fe, Ni/Cr, Al and other metals, arising as a result of the erosion or wear of reactors, can play an important role in nanocatalysis. Such particles can influence the oxidation stability of nanocomposites, especially the active centers of catalytic oxidation [117].

12.4 Nanocatalysis in Polymerization Processes

In Chap. 10 the methods of fabrication of metallopolymeric nanocomposites were analyzed for the catalytic or radical polymerization of the proper monomers, when an initiating agent was chemically bonded or adsorbed on the nanoparticle. This approach was advantageously realized in structurally well-defined polymer-nanoparticle hybrids through integrating colloid and nanoparticle synthesis with controlled/living radical polymerization [118]. The main scheme of the process was given in Chap. 10 (see Fig. 10.11). All nanoparticles are coated by polymer, formed under atom transfer radical polymerization (ATRP), for which the special silica particles, with an average diameter of 70 nm, surface-modified nanoparticles were then used as macroinitiators for styrene ATRP [119, 120].

$TiCl_3$ (Chap. 4) and VCl_3 submicrocrystals with particle sizes 70–100 nm, encapsulated into polymeric matrices, are very effective catalysts of propylene stereospecific polymerization [121]. Such particles can be produced in different ways: coprecipitation on MCl_3 polymers (M = Ti or V) in situ, reduction on polymer-chemiadsorbed MCl_4, various methods of joint grinding and dispergation of MCl_3 crystals and polymers, including functionalized PS and CSDVB, and anion exchange (in the form of NH_4^+-) or cation exchange (with carboxylic or sulfoacid groups) resins. MCl_3 can be easily deposited on polymeric surfaces (in particular, if the latter contain excessive electrons) and the electron donors assist in the grinding of MCl_3. In titanium-magnesium metallopolymeric catalysts (heterosubmicrocrystals with sizes 40–75 nm) the part of isolated ions Ti^{3+} (stabilized by $TiCl_3$ or $TiCl_2$ crystallites) and likely responsible for the polymerization, achieves 50% from the heterogenized titanium content and the same samples exhibit maximal activity in ethylene polymerization (550 kg PE/g of Ti·MPa·h) [122, 123].

Catalysts based on zirconocene derivatives in combination with methylalumoxane (MAO) have been used for the last 20 years. The use of bridged zirconium bi(fluorenyl) complexes $[Zr(C_{13}H_4C_2H_4C_{13}H_8)Cl_2]$ in ethylene polymerization allowed the achievement of a PE yield of 300 PE tons/g of Zr·h [124]. Moreover use of this catalyst allows us to control the formation of stereospecific polymers (e.g. propylene) and polymeric branching (both short and long ones), as well as the generation of block-copolymers (for this purpose so-called oscillating catalysts are used). This makes it possible to obtain polyolefins having specific characteristics (for example, to produce materials, such as LPELD – linear polyethylene of low density).

The main aim of catalyst immobilization on polymer carriers is not to raise their activity (many of them are highly effective and stereospecific ones) but to preserve the technological properties of these unstable and nonconventional agents. Novel catalysts used in many cases for industrial applications allow one to change the working temperature, substitute aromatic solvents to aliphatic ones, simplify the technological processes of gaseous or suspension polymerization and regulate the polymer molecular mass, morphology or other characteristics.

The highly effective nanocatalysts (metallocene/MAO) of olefin polymerization were obtained using the supported-metallocene on zeolites [125] or molecular sieves of MCM-41 type [126]. The activities are correlated with the higher zeolite surface acidity [127]. There is a tendency to employ other nanosized hybrid materials for these purposes, including directly in nanoparticle form (e.g. see [39,40,128]), which can be fabricated using other catalysts or cocatalysts. They can be produced from functionalized nanoparticles, such as alumoxane *p*-hydroxybenzoate with macroligands strictly identical in dimensions [129]. On the other hand, inexpensive aluminum oxide nanoparticles can be prepared by the reaction of the mineral boehmite with carboxylic acids with size 10–100 nm [130] (Scheme 12.6).

Such a catalyst in combination with zirconocene has a high initial activity (about 3000 g PE mmole·Zr^{-1}·hr^{-1}) [129]. A unique approach is used for forming nanosized Ti- and Zr-cenous catalysts, which is based on the dendronized (first or second generations) of cyclopentadienylic rings use (as the initial components) (Scheme 12.7) [131]. The catalytic properties of such systems in combination with MAO and the molecular mass characteristics of the obtained PE depend on the dendrimer generation degree.

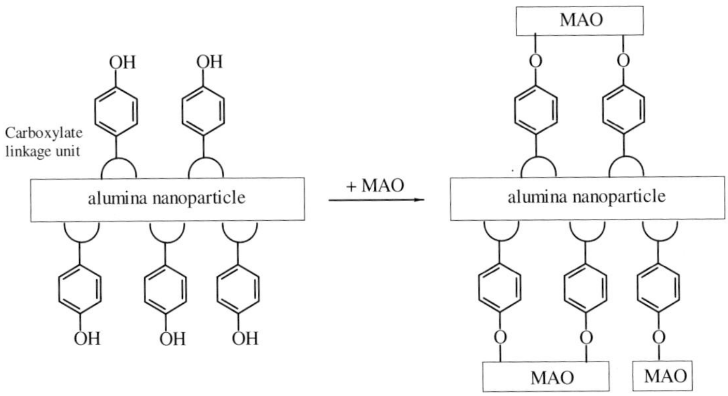
OH
OH
Carboxylate linkage unit
alumina nanoparticle
OH
OH
OH
+ MAO
MAO
O
O
alumina nanoparticle
O
O
O
MAO
MAO

Scheme 12.6

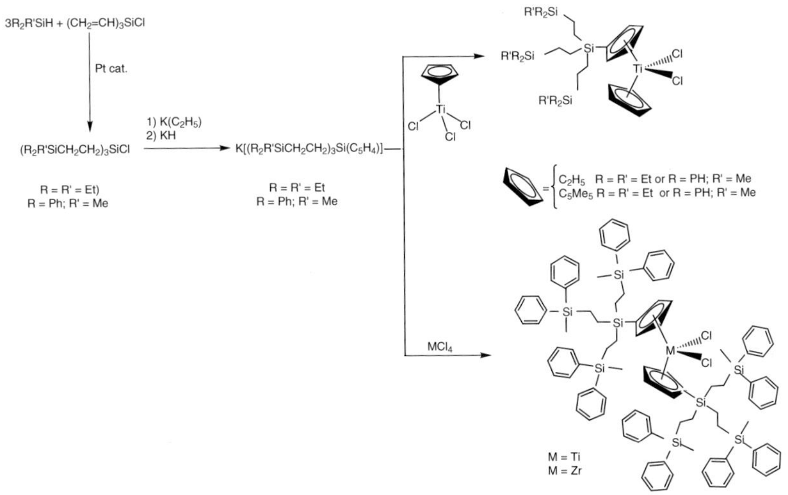
3R2R'SiH + (CH2=CH)3SiCl
Pt cat.
(R2R'SiCH2CH2)3SiCl
R = R' = Et)
R = Ph; R' = Me
1) K(C2H5)
2) KH
K[(R2R'SiCH2CH2)3Si(C5H4)]
R = R' = Et
R = Ph; R' = Me
MCl4
R'R2Si
R'R2Si
R'R2Si
C2H5 R = R' = Et or R = PH; R' = Me
C5Me5 R = R' = Et or R = PH; R' = Me
M = Ti
M = Zr

Scheme 12.7

Another peculiarity of metallocomplex catalysis is the increase in nucleation both at the stage of the formation of active centers and during the catalyzed process. This effect is very widespread in metallopolymer systems [18] and can be the reason for both their activation (for example, in hydrogenization reactions) and deactivation (for example, in ionic-coordinate polymerization). One of the substantial advantages of heterogenized complexes is the possibility of non-destructive control of the process intermediates. This can be illustrated by an example of hydrogenization reactions (see Sect. 12.2). An intermediate can practically always be isolated (sometimes even during the catalyzed reaction). Thus, one can study the transition state of a metal and find the correlation between the catalytic properties and its valence state, dispersion, and surfacial ratio. Some examples are given below for polymerization catalysis processes.

One of the best studied reactions in the field is the ethylene dimerization into isomeric butylenes under the action of so-called "nickel" catalytic systems (of type NiX_2–AlR_xCl_{3-x}) characterized by high efficiency (the total product yield achieves 3–5 kg/g of Ni). It is known that a homogeneous catalyst is practically deactivated over 1–2 hours and, according to [132], this is a result of the considerably great Ni_0 (colloidal) particles formation, which are not active in the reaction itself (Scheme 12.8):

$$Ni^{2+} + AlR_xCl_{3-x} \quad \rightarrow \quad Ni^{+} + Ni^{0} \quad \rightarrow \quad Ni_0(\text{colloidal})$$

Scheme 12.8

Part of the formed Ni^0 in such heterogenized "nickel" catalysts is stabilized and this prevents aggregation; another part is involved in a redox-cycle by Scheme 12.9:

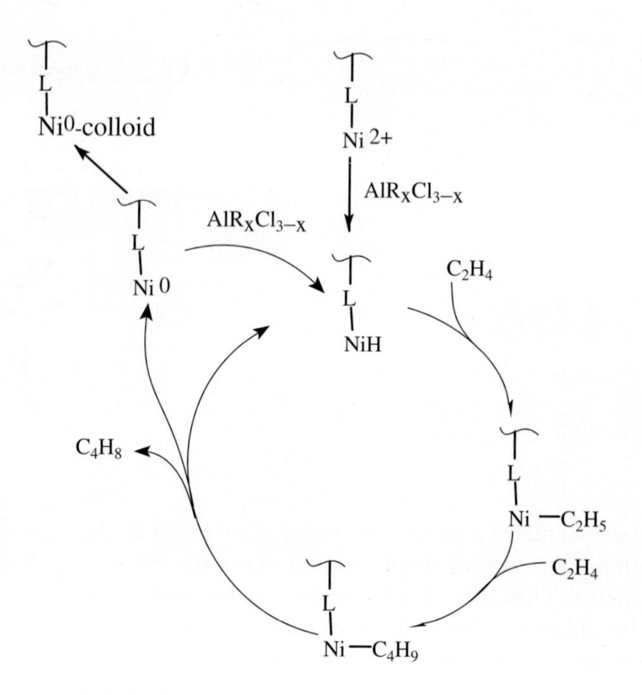

Scheme 12.9

The metamorphism of immobilized Ni^{2+} during ethylene dimerization was confirmed by magnetic permeability measurements in strong magnetic fields (up to 70 kOe) and at low temperature (4.2 K) [133, 134]. Under such conditions the susceptibility of different spin states depends non-linearly on the magnetic field strength and it allows one to get the distribution of the nickel states over the magnetic moment values (Table 12.6).

Table 12.6. Ferromagnetic nickel content (%) in interaction of products of polymer-immobilized $\cdot Ni(CH_3COO)_2$ (A*) and Ni (napht)$_2$ (B**) with $AlEt_xCl_{3-x}$ and C_2H_4

System	T, K		
	293	80	4.2
A + $AlEt_2Cl$	0.55	1.2	6.3
A + $AlEt_2Cl$ + C_2H_4	0.18	0.37	
B + $AlEt_2Cl$	27	34	37.5
B + $AlEt_2Cl$ + C_2H_4	34	53	
A + $AlEtCl_2$	0.34	0.31	2.3
A + $AlEtCl_2$ + C_2H_4	0.35	0.41	3.8
B + $AlEtCl_2$	31.2	31.6	51.4
B + $AlEtCl_2$ + C_2H_4	14.5	17.6	43.8

*A – heterogenized system; **B – homogeneous system

Particles with size ~1 nm (containing 10–30 Ni atoms) are formed, which are not ferromagnetic (supermagnetic) but present polymer-immobilized clusters with a ferromagnetic exchange. It is known [132] that the hybrid nickel complexes active in the ethylene dimerization are formed from nickel compounds of usual oxidation degree (+2) through a number of transformations with participation of organoaluminum cocatalysts (Scheme 12.9). Regeneration of active centers is provided by involving the formed Ni^0 in a reduction-oxidation cycle. Their deactivation runs through the colloidal nickel formation. So, prevention of the generation of nickel particles is an important condition of the efficiency of such catalytic systems [133].

Polymer-immobilized cobalt catalysts are used for butadiene polymerization. In the absence of a polymerizable monomer, a considerable part of Co^{2+} is reduced to a metallic state and at least 90% of all ferromagnetic cobalt is concentrated in considerably coarse particles with sizes ~10 nm [134]. At the same time under diene polymerization by the "cobalt" systems the monomer plays a substantial role in the formation of active centers. The cobalt quantitative content was analyzed in different magnetic states (ferro (Co^0)-, dia

(Co^+)- and paramagnetic (Co^{2+}) states) and it was shown that in interaction of the $CoCl_2$ macrocomplex with $Al(C_2H_5)_2Cl$ (in the absence of the monomer, so-called system "aging") half of the cobalt particles were ferromagnetic (and about 90% of it was concentrated in big particles, >10 nm) and the other half were diamagnetic (Co^+). In contrast to this system, only 6–8% of ferromagnetic cobalt is formed in the presence of isoprene and the main part of it remains diamagnetic, which can be explained by the stabilization of polymer-bonded Co^+ due to inclusion into a polymerization-active π-complex, which prevents the subsequent cobalt reduction. Probably, such a nucleation rise is a characteristic feature of many homogeneous catalysts and studies of this mechanism gives us a chance to control the catalyst activities.

12.5 Polymer-Protected Nanoparticles in Photocatalysis

In recent years success was achieved in the field of solar energy transformation into chemical energy and in modelling of molecular photocatalytic systems. Colloidal nanoparticle solutions have very interesting characteristics for photocatalysis, but their practical use is limited by the necessity to remove the nanoparticles from the solution after the photocatalytic process. Therefore use of thin nanoparticle layers on a carrier seems to be more promising. Mimicking the highly efficient energy transfer system is a subject to be tackled towards the ultimate goal of establishing solar energy conversion systems [135, 136]. Since only a small amount of nanoparticles are required to photosensitize the matrix, the nanoparticles are isolated from each other and are responsible for charge generation while the polymer is responsible for charge transport [137]. As a catalyst of the reducing semireaction, noble metal nanoparticles stabilized by polymers are often used, as well as iso- and heteropolycompounds of Mo and W able to take part in multielectron oxidative–reductive transformations. Such compounds are the analogue of the reaction centers in hydrogenase ferments. Colloidal platinum particles with sizes 1–2 nm in PVA, PAA, cellulose or other matrices are widely used for such purposes. The particle sizes are extremely important. If the heavy Pt particles are removed by a centrifugation (20 000 rev·min^{-1}) after catalytic reaction of hydrogen evolution, the remaining transparent solution containing about 20% of the initial platinum in very dispersed form becomes considerably more active [138]. These dimensions of Pt particles are critical for catalytic activity regardless of the nature of the protective colloid. Polymers based on micelle-protected Pt-clusters with mean size 1–2 nm were obtained by radical polymerization of unsaturated surfactants (such as sodium undecenoate $CH_2{=}CH(CH_2)_8COO^-Na^+$) and H_2PtCl_6 reduction by conventional methods. These polymers are catalytically very active for hydrogen evolution under the action of visible light in the system EDTA/$[Ru(bpy)_3]^{2+}$/$[MV]^{2+}$ (electron donor, photocatalyst and electron acceptor, correspondingly), as shown in Scheme 12.10 [139].

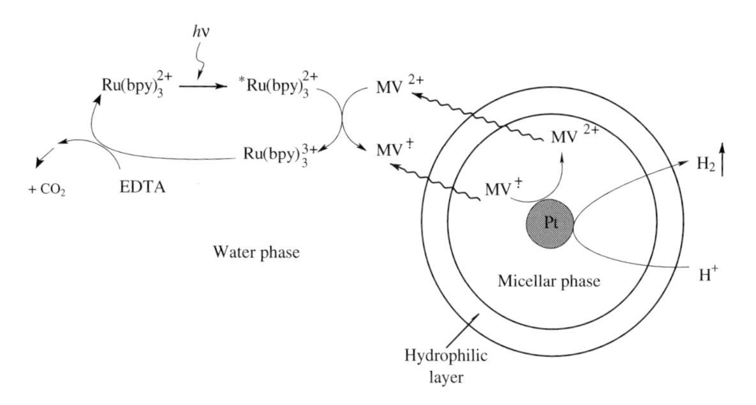

Scheme 12.10

It is important that a hydrophobic nucleus from such micelle-forming monomers is formed and a hydrophylic layer is generated on the stabilized nanoparticle surface, which assists in more effective charge division. The same principle was used to develop novel type semiconductors (metal sulfide or TiO_2, AgI, Ag) with nanoparticle-polyurea composites, using reverse micellar systems via diisocyanate polymerization [140], as well as RuS_2 nanoparticles, prepared in reverse micellar system onto thiol-modified PS [141]. They are effective as photocatalysts for the generation of H_2 from 2-propanol aqueous solution. These catalysts can be prepared also in the form of nanoparticle-containing transparent film by casting the solution onto a quartz sheet.

Charge separation and fabrication of effective photocatalysts were achieved using semicoductive nanomaterials, obtained by sol–gel synthesis. They present mainly TiO_2 of three known structural modifications (rutile, anatase and brookite). Anatase has the highest activity [142, 143]. The additional introduction of transient metals ions (in particular, copper) at the material-forming stage increases considerably the efficiency of the photocatalytic reactions [144, 145]. For example, the photocatalytic activity of TiO_2 was remarkably enhanced (from 0.1 μmole·h^{-1} to 2.4 μmole·h^{-1}) by the addition of Cu ion in TiO_2. The discharged H_2 amount depends on the various irradiation conditions and is maximal under the full arc irradiation of a Xe lamp [145] (Fig. 12.7). The approaches based on the immobilization of metal-containing chlorophyll (M = Mg, Zn) derivatives into mesoporous silica to construct an efficient energy transfer system between the chromophores (Fig. 12.8) [146, 147] and upon irradiating ($>$420 nm) an indium tin oxide electrode spin-coated with a mixture of TiO_2 nanoparticles/PVK and carotenoid [148], were elaborated. Photocatalytic systems were organized on the basis of lipidic vesicles and semiconductor nanoparticles [149].

The high photocatalytic activity of the heteronanoparticles TiO_2/SnO_2 was established earlier [150]. Now the colloidal solutions of semiconductive

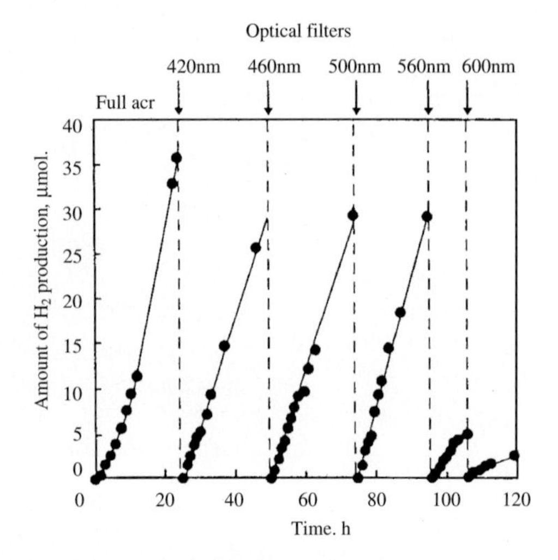

Fig. 12.7. Time course of photocatalytic H_2 production from aqueous methanol under various irradiation conditions

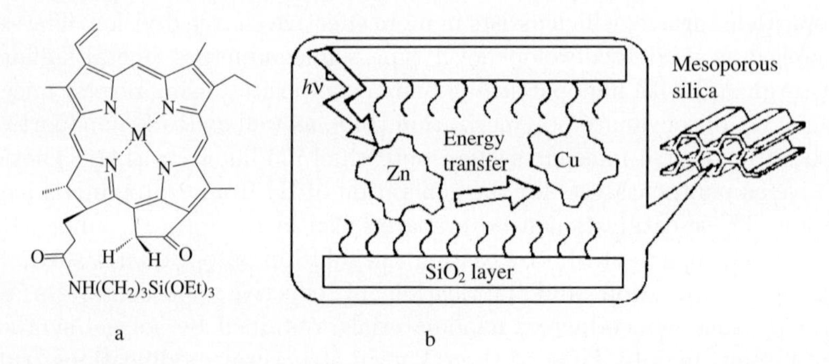

Fig. 12.8. Molecular structure of chlorophyllous pigments (**a**) and schematic representation of the energy transfer process in mesopores (**b**)

SnO_2 particles (with mean diameter ~3 nm) are produced commercially by "Johnson Matthey". Powder P25 (TiO_2 with 80% of anatase and 20% of rutile) is widely used in photocatalytic systems [151].

We do not analyze the series of other photochemical reactions catalyzed by metallopolymer catalysts, such as unimolecular isomerization, quenching of triplet or singlet states of other molecules through energy transfer, intramolecular charge-transfer, charge separation in synthetic molecular dyads and triads, photoinduced electron-transfer to solvent molecules, etc. A new direction is being intensively developed based on the possibility of singlet oxygen to oxidize alkenes, for example by polymer-immobilized clusters of Mo_6 [152] or platinum complexes incorporated in Nafion membrane [153]. In the last few years there has been a tremendous growth [154] in the utilization of semiconductors as catalysts in solar energy conversion, photochemical transformations of organic and inorganic compounds and photodegradation of organic pollutants. Therefore, many photocatalytic systems based on TiO_2 particles are being successfully used (TiO_2/UV-process) for waste cleaning (contaminated air, water, microorganisms contamination) [155].

Chapter 9 made accent on Mn polynuclear complexes as good models of oxygen evolution in fermentative-type photocatalytic systems. The physical properties of semiconductor nanoparticles are dominated by quantum confinement, and the widening HOMO–LUMO gap with decreasing particle size, which directly affects the photophysics of the material. Photochemical reaction induction is a very promising direction in the nanocatalysis field.

12.6 Metallopolymeric Nanocomposites in Catalysis

All the results presented above show that metal nanoparticles stabilized by polymers exhibit peculiarities, allowing us to unite and merge them into a new separate class of catalysts. In some systems they give the opportunity to combine high selectivity and high rates of processes. In many cases a polymer presents an inert stabilizer and works as a medium where many interactions between the nanoparticle and matrix take place.

Metallopolymer nanocomposites are very active as a bridge between heterogeneous catalysis and its homogeneous counterpart. Rational catalyst design can be proposed via templating polymeric nanostructured materials [156]. The areas of science and technology that have significantly benefited from advances in molecular recognition, separation or catalysis are very wide and diversified. It is supposed that the target of mimicking enzyme catalysis appears to be closer due to continued improvements in the synthesis of nanostructured materials.

The field of investigations and applications is permanently widening. In recent years some metal nanoparticles were used as catalysts in the production of carbon nanotubes (for example, palladium impregnated on graphite edges [157]) or together with nanotubes. Decomposition of ethane on nickel nanoclusters allows us to form carbon nanotubes [158]. The catalytic synthesis of nanotubes is carried out under mild conditions ($T \leq 920$ K), and the materials obtained had a relatively high surface area (100–200 $m^2 \cdot g^{-1}$), and mean diameter nanotubes 40–60 nm.

It is interesting that such approaches are not always readily possible until industrial process conditions are proved. Novel methods are possible as the polymer–nanoparticles catalysis with the dendrimers provides supports for recoverable catalysts [159].

Polymer-immobilized clusters and nanoparticles are intermediate objects of both homogeneous and heterogeneous catalysis that gives us a chance to establish the connections in the successive chain of homogeneous, heterogeneous and fermentative nanocatalysis [18,20]. On one hand, the nanoparticles can open a new stage of heterogeneous catalysis, but on the other hand, they are near to heterogeneous catalysts for many indicators. For example, they are obtained from usual soluble metal compounds due to a rise of nuclearity and the polymer plays the role of macroligand and stereochemical surrounding, they can functionize under mild conditions, etc. The main features of heterogeneous catalysts remains the same (the reactions run at an interface, the substrate is activated by adsorption, the catalysts can be easily separated from products and regenerated). In nanocatalysis conditions, a situation analogous to fermentative catalysis is easily modeled (the substrate enrichment, the structural correspondence, the favorable orientation of reacting molecules, and additional substrate activation due to multicenter catalysis). The principal relations between these two types of catalysts are given in Scheme 12.11.

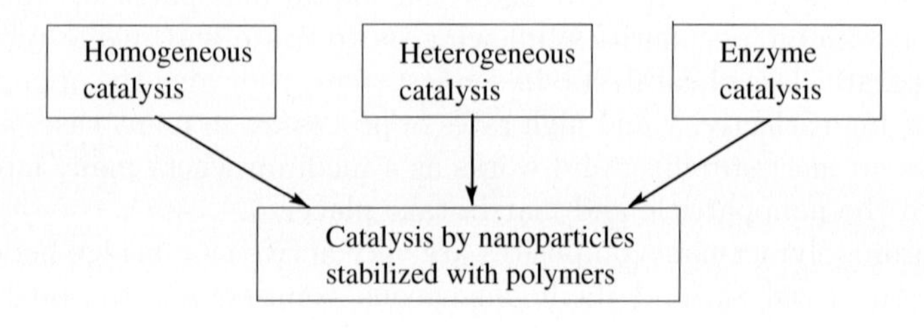

Scheme 12.11

It will be possible in the near future to form active centers for a given reaction, to control the selectivity and activity of the reaction by changing the cluster or nanoparticle nuclearity and/or the composition and to use proper stabilized carriers. The nanocatalysis by metallopolymers is now a rapidly developing field that give us a chance to create novel highly effective and selective catalysts, as well as novel processes based on them.

References

1. M. Boudart: J. Mol. Catal. **30**, 27 (1985)
2. D. Bazin: Topics in Catalysis **18**, 79 (2002)
3. A.K. Kakkar: Chem. Rev. **102**, 3579 (2002)
4. R.F. Khairutdinov, N. Serpone: Progr. React. Kinet. **21**, 1 (1996)
5. R.F. Khairutdinov, K.Ya. Burstein, N. Serpone: Chem. Phys. **16**, 2030 (1997)
6. J.H. Clark: *Catalysis of Organic Reactions by Supported Inorganic Reagents* (VCH, New York 1994)

7. P. Laszlo: *Preparative Chemistry Using Supported Reagents, vol.6* (Academic Press, New York 1987)
8. J.M. Thomas, W.J. Thomas: *Principles and Practice of Heterogeneous Catalysis* (VCH, Weinheim 1997)
9. A. Choplin, F. Quignard: Coord. Chem. Rev. **178-180**, 1679 (1998)
10. A. Sayari: Chem. Mater. **8**, 1840 (1996)
11. B.C. Gates: Chem. Rev. **95**, 511 (1995)
12. L.D. Rampino, F.F. Nord: J. Am. Chem. Soc. **63,** 2745; 3268 (1941)
13. K.E. Kavanagh, F.F. Nord: J. Am. Chem. Soc. **65**, 2121 (1943)
14. V.M. Varushchenko, S.A. Mikhalyuk, E.M. Natanson, V.D. Polkovnikov: Izv. AN USSR. Ser. Khim. 1346 (1971)
15. *Catalysis by Di- snd Polynuclear Metal Cluster Complexes* ed. by R.D. Adams, F.A. Cotton (Wiley-VCH, New York 1998)
16. D. Wöhrle, A.D.Pomogailo: *Metal Complexes and Metals in Macromolecules. Synthesis, Structures and Properties* (Wiley-VCH, Weinheim 2003)
17. *Metal Clusters in Catalysis* ed. by B.C. Gates, L. Guczi, H. Knösinger (Elsevier, New York 1986)
18. A.D. Pomogailo: *Catalysis by Polymer-Immobilized Metal Complexes* (Gordon and Breach Sci. Publ., Amsterdam 1998)
19. V.I. Bukhtiyarov, M.G. Slinko: Russ. Chem. Rev. **70**, 167 (2001)
20. A.D. Pomogailo: Kinet. Catal. **45**, 67 (2004)
21. D.C. Sherrington: Chem. Commun. 2275 (1998)
22. A.D. Pomogailo: *Polymer-Immobilized Metallocomplex Catalysts* (Nauka, Moscow 1988)
23. M. Haruta: Catalysis Today **36**, 153 (1997); CATTECH, **6**, 102 (2002)
24. M. Haruta, N. Yamada, T. Kobayashi, S. Iijima: J. Catal. **115**, 301 (1989)
25. K. Ruth, M. Hayes, R. Burch, S. Tsubota, M. Haruta: Appl. Catal. B **24**, L133 (2000)
26. A. Ueda, M. Haruta: Appl. Catal. B **18**, 115 (1998)
27. M. Haruta: J. Nanoparticle Research **5**, 3 (2003)
28. G. Pacchioni, N. Rosch: Acc. Chem. Res. **28**, 390 (1995)
29. G.A. Martin: Catal. Rev. Sci. Eng. **30**, 519 (1988)
30. A. Henglein: J. Phys. Chem. **97**, 5457 (1993)
31. H. Weller: Angew. Chem. Int. Ed. Engl. **32**, 41 (1993)
32. S.L. Lewis: Chem. Rev. **93**, 2693 (1993)
33. L. Pauling: J. Am. Chem. Soc. **62**, 2643 (1940)
34. F.H. Dickey: Proc. Nat. Acad. Sci. USA **35**, 227 (1949)
35. E. Bayer, W. Schumann: J. Chem. Soc. Chem. Commun. 949 (1986)
36. D.J. Bayston, J.L. Fraser, M.R. Ashton, A.D. Baxter, M.E.C. Polyvka, E. Moses: J. Org. Chem. **63**, 3137 (1998)
37. S. Sgorlon, F. Pinna, G. Strukul: J. Mol. Catal. **40**, 211 (1987)
38. A.A. Belyi, L.G. Chigladze, A.L. Rusanov, M.E. Volpin: Bull. Acad. Sci. USSR, Div. Chem. Sci. 1961 (1989)
39. A.D. Pomogailo: Russ. Chem. Rev. **66**, 750 (1997)
40. A.D. Pomogailo, A.S. Rozenberg, I.E. Uflyand: *Metal Nanoparticles in Polymers* (Khimiya, Moscow 2000)
41. E. Sulman, Yu. Bodrova, V. Matveeva, N. Semagina, L. Cerveny, V. Kurtc, L. Bronstein, O. Platonova, P. Valetsky: Appl. Catal. A. **176**, 75 (1999)
42. H. Hirai, N. Toshima: In: *Tailored Metal Catalysts* ed. by Y. Iwasawa (Reidel, Dodrecht 1986)

43. S.I. Pomogailo, G.I. Dzhardimalieva, V.A. Ershova, S.M. Aldoshin, A.D. Pomogailo: Macromol. Symp. **186**, 155 (2002)
44. A.M. Lyakhovich, S.S. Mikhailova, S.I. Pomogailo, G.I. Dzhardimalieva, A.D. Pomogailo: Macromol. Symp. **204**, 251 (2003)
45. J. Lieto, J. Rafalko, B.C. Gates: J. Catal. **62**, 149 (1980)
46. R. Mutin, W. Abboud, J.M. Bassel, D. Sinou: J. Mol. Catal. **33**, 47 (1985)
47. R.A. Sancez-Delgado, A. Andriollo, J. Pupa, G. Martin: Inorg. Chem. **26**, 1867 (1987)
48. J. Lieto, M. Wolf, B.A. Matrana: J. Phys. Chem. **89**, 991 (1985)
49. N.L. Pocard, D.C. Alsmeyer, R.L. McCreery, T.N. Neeman, M.R. Callstrom: J. Am. Chem. Soc. **114**, 769 (1992)
50. U. Schubert: New. J. Chem. **18**, 1049 (1994)
51. H. Hirai, N. Yakuda, Y. Seta, S. Nodoshima: React. Function. Polym. **37**, 121 (1998)
52. Ng Cheong Chan, G.S.W. Craig, R.R. Schrock, R.E. Cohen: Chem. Mater. **4**, 885 (1992)
53. J.D. Aiken, R.G. Finke: J. Mol. Catal. A **145**, 1 (1999); Chem. Mater. **11**, 1035 (1999)
54. Y. Lin, R.G. Finke: J. Am. Chem. Soc. **116**, 8335 (1994)
55. M.V. Seregina, L.M. Bronstein, O.A. Platonova, D.M. Chernyshov, P.M. Valetsky, J. Hartmann, E. Wenz, M. Antonietti: Chem. Mater. **9**, 923 (1997)
56. A.B.R. Mayer, J.E. Mark: J. Polym. Sci. Polym. Phys. B **35**, 1207 (1997); J. Polym. Sci. Polym. Chem. A **35**, 3151 (1997)
57. A.B.R. Mayer, J.E. Mark: Angew. Makromol. Chem. **52**, 268 (1999)
58. A. Warshawsky, D.A. Upson: J. Polym. Sci. Polym. Chem. A **27**, 2963 (1989)
59. N. Toshima, T. Teranishi, H. Asanuma, Y. Saito: Chem. Lett. 819 (1990)
60. C. Fajardo, A.L. Cabannes, C. Godinez, G. Villora: Chem. Eng. Comm. **140**, 21 (1996)
61. C. Milone, G. Neri, A. Donato, M.G. Musolino, L. Mercadante: J. Catal. **159**, 253 (1996)
62. H. Gao, Y. Xu, S. Liao, R. Liu, J. Liu, D. Li, D. Yu, Y. Zhao, Y. Fan: J. Mem. Sci. **106**, 213 (1995)
63. N. Toshima, K. Kushihashi, T. Yonezawa, H. Hirai: Chem. Lett. 1769 (1989)
64. N. Toshima, T. Yonezawa, M. Harada, K. Asakura, Y. Iwasawa: Chem. Lett. 815 (1990)
65. T. Teranishi, K. Nakata, M. Miyake, N. Toshima: Chem. Lett. 277 (1996)
66. T. Teranishi, N. Toshima: J. Chem. Soc. Dalton Trans. **20**, 2967 (1994)
67. Y. Wang, N. Toshima: J.Phys. Chem. **101**, 5301 (1997)
68. N. Toshima, T. Yonezawa: New J. Chem. 1179 (1998)
69. D. Bazin, C. Mottet, G. Treglia: Appl. Catal. A **200**, 47 (2000)
70. J.S. Bradley, E. Hill, M.E. Leonowicz, H. Witzke: J. Mol. Catal. **41**, 59 (1987)
71. V.M. Rudoi, B.G. Ershov, N.L. Sukhov, O.V. Dementeva, A.V. Zaitseva, A.F. Selivestrov, M.E. Kartseva, V.A. Ogarev: Colloid J. **64**, 832 (2002)
72. D.E. Bergbreiter, L. Zhang, V.M. Mariananam: J. Am. Chem. Soc. **115**, 9295 (1993)
73. C.-W. Chen, M. Akashi: J. Polym. Sci. Polym. Chem. A **35**, 1329 (1997)
74. A. Guner, M. Ataman: Colloid Polym. Sci. **272**, 175 (1994)

75. A. Zharmagambetova, B. Selenova, S. Mukhamedzhanova., L. Komashko: In: *The 7-th Int. Conf. on Polymer Supported Reactions in Organic Chem. POC-96 (Wroclaw, 1996)*
76. N. Toshima, P. Lu: Chem. Lett. 729 (1996)
77. A.D. Pomogailo, M.V. Kluev: Bull. Acad. Sci. USSR, Div. Chem. Sci. 1716 (1985)
78. H. Hirai, H. Wakabayashi, M. Komiyama: Bull. Chem. Soc. Jap. **59**, 545 (1986)
79. W. Yu, M. Liu, H. Liu, X. An, Z. Liu, X. Ma: J. Mol. Catal. A **142**, 201 (1999)
80. Y. Wang, H. Liu, Y.Y. Jiang: Chem. Commun. 1878 (1998)
81. S.A. Zavyalov, P.S. Vorontsov, E.I. Grigorev, G.N. Gerasimov, E.N. Golubeva, O.V. Zagorskaya, L.M. Zavyalova, L.I. Trakhtenberg: Kinet. Catal. **39**, 905 (1998)
82. M. Ohtaki, N. Toshima, M. Komijama, H. Hirai: Bull. Chem. Soc. Jpn. **63**, 1433 (1990)
83. N. Toshima, T. Takashi: Bull. Chem. Soc. Jpn. **65**, 400 (1992)
84. T. Teranishi, K. Nakata, M. Iwamoto, M. Miyake, N. Toshima: React. Function. Polym. **37**, 111 (1998)
85. H. Ohde, F. Hunt, C.M. Wai: Chem. Mater. **13**, 4130 (2001)
86. H. Ohde, C.M. Wai, H. Kim, M. Ohde: J. Am. Chem. Soc. **124**, 4540 (2002)
87. A.N. Collins, G.N. Sheldrake, J. Grosby: *Chirality in Industry: Developments in the Commercial Manufacture and Applications of Optically Active Compounds* (Wiley, New York 1997)
88. S.C. Stinson: Chem. Eng. News. **23**, 55 (2000)
89. C. Saluzzo, R. Halle, F. Touchard, F. Fache, E. Schulz, M. Lemaire: J. Organomet. Chem. **603**, 30 (2000)
90. P.A. Jacobs, I.F.J. Vankelecom, D. Vos: *Chiral Catalysts Immobilisation and Recycling* (Wiley-VCH, Weinheim 2000)
91. K. Fodor, S.G.A. Kolmschot, R.A. Sheldon: Enantiomer **4**, 497 (1999)
92. I.F.J. Vankelecom, P.A. Jacobs: Catal. Today **1**, 1905 (1999)
93. A. Wolfson, S. Janssens, I. Vankelecom, S. Geresh, M. Gottlieb, M. Herskowitz: Chem. Commun. 388 (2002)
94. F. Gelman, D. Avnir, H. Schumann, J. Blum: J. Mol. Catal. **146**, 123 (1999)
95. *Chiral Reactions in Heterogeneous Catalysis* ed. by G. Jannes, V. Dubois (Plenum Press, New York 1995)
96. J.U. Köhler, J.S. Bradley: Catal. Lett. **45**, 203 (1997); Langmuir **14**, 2730 (1998)
97. S. Bhaduri, K. Sharma: Chem. Commun. 207 (1996)
98. S.N. Sidorov, I.V. Volkov, V.A. Davankov, M.P. Tsyurupa, P.M. Valetsky, L.M. Bronstein, R. Karlinsey, J.W. Zwanziger, V.G. Matveeva, E.M. Sulman, N.V. Lakina, E.A. Wilder, R.J. Spontak: J. Am. Chem. Soc. **123**, 10502 (2001)
99. M. Okumura, S. Tsubota, M. Iwamoto, M. Haruta: Chem. Lett. 315 (1998)
100. E.A. Trusova, M.V. Tsodikov, E.V. Slivinskii, G.G. Hernandez, O.V. Bukhtenko, T.N. Zhdanova, D.I. Kochubey, J.A. Navio: Mendeleev Commun. 102 (1998)
101. V.D. Materra, D.M. Barnes, S. Chaudhuri: J. Phys. Chem. **90**, 4819 (1986)
102. T. Tabushi, K. Kurihara, K. Morimitsu: J. Synth. Org. Chem. Jap. **43**, 323 (1985)
103. S.N. Kholuiskaya, A.D. Pomogailo, N.M. Bravaya, S.I. Pomogailo, V.A.Maksakov: Kinet. Catal. **44**, 831 (2003)

104. Y. Li, M.A. El-Sayed: J. Phys. Chem. B **105**, 8938 (2001)
105. I. Komoto, S. Kobayashi: Chem. Commun. 1842 (2001)
106. R.F. Heck, J.P. Nolley: J. Org. Chem. **37**, 2320 (1972)
107. R.F. Heck: *Palladium in Organic Synthesis* (Academic Press, London 1985)
108. S. Klingelhöfer, W. Heitz, A. Greiner, S. Oestreich, S. Förster, M. Antonietti: J. Am. Chem. Soc. **119**, 10116 (1997)
109. A. Biffis: J. Mol. Catal. A **165**, 303 (2001)
110. M.M. Dell'Anna, P. Mastrorilli, F. Muscio, C.F. Nobile, G.P. Surannal: Eur. J. Inorg. Chem. 1094 (2002)
111. M.R. Buchmeister, T. Scharenia, R. Kempe, K. Wurst: J. Organomet. Chem. **634**, 39 (2001)
112. N. Arul Dhas, H. Cohen, A. Gedanken: J. Phys. Chem. B **101**, 6834 (1997)
113. L.K. Yeung, C.T. Lee, K.P. Johnston, R.M. Crooks: Chem. Commun. 2290 (2001)
114. L. Djakovitch, K. Koehler: J. Am. Chem. Soc. **123**, 5990 (2001)
115. D.E. Bergbreiter, P.L. Osburn, J.D. Frels: J. Am. Chem. Soc. **123**, 11105 (2001)
116. A. Deronzier, J. Moulet: Coord. Chem. Rev. **147**, 339 (1996)
117. H.H. Abou el Naga, A.E.M. Salem: Wear **96**, 267 (1984)
118. T. Werne, T.E. Patten: J. Am. Chem. Soc. **121**, 7409 (1999)
119. J. Pyun, K. Matyjaszewski, T. Kowalewski, D. Savin, G. Patterson, G. Kickelbrick, N. Huesing: J. Am. Chem. Soc. **123,** 9445 (2001)
120. S.C. Hong, K. Matyjaszewski: Macromolecules **35**, 7592 (2002)
121. S.L. Saratovskikh, A.D. Pomogailo, O.N. Babkina, F.S. Dyachkovskii: Kinet. Catal. **25**, 464 (1984)
122. A.M. Bochkin, A.D. Pomogailo, F.S. Dyachkovskii: Kinet. Catal. **27**, 914 (1986)
123. A.M. Bochkin, A.D. Pomogailo, F.S. Dyachkovskii: React. Polym. **9**, 99 (1989)
124. H.G. Alt, S.J. Palackal: J. Organomet. Chem. **472**, 113 (1994)
125. H. Rahiala, I. Beurroies, T. Eklund, K. Hakala, P. Trens, J.B. Rosenholm: J. Catal. **188**, 14 (1999)
126. Y.S. Ko, T.K. Han, S.I. Woo: Macromol. Chem. Phys. **17**, 749 (1995)
127. V.I. Costa Vaya, P.G. Belelli, J.H.Z. Santos, M.I. Ferreira, D.E. Damiani: J. Catal. **204**, 1 (2001)
128. S. Amigoni-Gerbier, S. Desert, T. Gulik-Kryswicki, C. Larpent: Macromolecules **35**, 1644 (2002)
129. S.J. Obrey, A.R. Barron: Macromolecules **35**, 1499 (2002)
130. R.L. Callender, A.R. Barron: Adv. Mater. **12**, 734 (2000)
131. R. Andres, E. Jesus, F.J. Mata, J.C. Flores, R. Gomez: Eur. J. Inorg. Chem. 2281 (2002)
132. F.K. Shmidt: *Catalysis of hydrogenation and dimerization reactions by first row transition metal complexes* (Irkutsk University Publ., Irkutsk 1986)
133. S.B. Echmaev, I.N. Ivleva, N.D. Golubeva, A.D. Pomogailo, Yu.G. Borodko: Kinet. Catal. **27**, 394 (1986)
134. I.N. Ivleva, S.B. Echmaev, N.D. Golubeva, A.D. Pomogailo: Kinet. Catal. **24**, 663 (1983)
135. D. Gust, T.A. Moore, A.L. Moore: Acc. Chem. Res. **34**, 40 (2001)
136. A. Maldotti, A. Molinari, R. Amadelli: Chem. Rev. **95**, 3811 (2002)
137. Y. Wang: Pure Appl. Chem. **68**, 1475 (1996)

138. J. Kiwi, M. Gratzel: J. Am. Chem. Soc. **101**, 7214 (1979)
139. N. Toshima, T. Takahashi, H. Hirai: J. Macromol. Sci.-Chem. A **25**, 669 (1988)
140. T. Hirai, T. Watanabe, I. Komasawa: J. Phys. Chem. B **103**, 10120 (1999)
141. T. Hirai, Y. Nomura, I. Komasawa: J. Nanoparticles Res. **5**, 61 (2003)
142. K.I. Zamaraev, M.I. Khramov, V.N. Parmon: Catal. Rev. **36**, 617 (1994)
143. J.M. Stipkala, F.N. Castellano, T.A. Heimer, C.A. Kelly, K.J.T. Livi, G.J. Meyer: Chem. Mater. **9**, 2341 (1997)
144. M. Hara, T. Kondo, M. Komoda, S. Ikeda, K. Shinohara, A. Tanaka, J.N. Kondo, K. Domen: Chem. Commun. 357 (1998)
145. Y. Sakata, T. Yamamoto, T. Okazaki, H. Imamura, S. Tsuchiya: Chem. Lett. 1253 (1998)
146. S. Murata, H. Furukawa, K. Kuroda: Chem. Mater. **13**, 2722 (2001)
147. H. Furukawa, T. Watanabe, K. Kuroda: Chem. Commun. 2002 (2001)
148. G. Gao, Y. Deng, L.D. Kispert: J. Chem. Soc. Perkin Trans. 2 1225 (1999)
149. O.V. Vasiltsova, V.N. Parmon: Kinet. Catal. **40**, 70 (1999)
150. K. Vinodgopal, I. Bedja, P.V. Kamat: Chem. Mater. **8**, 2180 (1996)
151. R.I. Bickley, T. Conzalez-Carreno, J.S. Less, L. Palmisano, R.J.D. Tilley: J. Solid State Chem. **82**, 178 (1991)
152. J.A. Jackson, M.D. Newsham, C. Worsham, D.G. Nocera: Chem. Mater. **8**, 558 (1996)
153. X.-H. Li, L.-P. Zhang, C.-H. Tung, C.-M. Che: Chem. Commun. 2280 (2001)
154. A. Mills, S.L. Hunte: J. Photochem. Photobiol. Chem. A **108**, 1 (1997)
155. R. Bauer, G. Waldner, H. Fallmann, S. Hager, M. Klare, T. Krutzler, S. Malato, P. Maletzky: Catal. Today **53**, 131 (1999)
156. M.E. Davis, A. Katz, W.R. Ahmad: Chem. Mater. **8**, 1820 (1996)
157. C. Pham-Huu, N. Keller, M.J. Ledoux, L.J. Charbonniere, R. Ziessel: J. Mol. Catal. Chem. A **170**, 155 (2001)
158. C. Pham-Huu, N. Keller, V.V. Roddatis, G. Mestl, R. Schlögl, M.J. Ledoux: Phys. Chem. Chem. Phys. **4**, 514 (2002)
159. van R. Heerbeek, P.C.J. Kamer, van P.W.N. Leeuwen, J.N.H. Reek: Chem. Rev. **102**, 3717 (2002)

13 Conclusion

The main objective of this book was to demonstrate that studies of nanosized metal-containing particles in a polymeric matrix have become in recent years a large and independent field of physics and chemistry of the ultradispersed substance state. This field can operate by its own widely used objects (polynuclear, cluster and nanosized particles), has its own methodology and specific methods of fabrication of hybrid polymer-inorganic nanocomposites. The methods are based on combinations of inorganic, elementoorganic, colloidal, physical and bioorganic chemistry, etc. In other words, the science of polymer-inorganic nanocomposites has achieved a high level of multidisciplinarity and integration, and this fact on one hand enriches the science, but on the other hand permanently demands the use of joint investigations (chemistry, physics, biology, etc.) to solve the problems arising.

In this monograph much attention has been paid to the various possible applications of the produced materials, from micro- (and even nanometallurgy) and laser treatment of surfaces to material sciences and the solution of specific catalytic problems. Practical needs will be the main driving force of the new science and it demands the creation of a new class of investigation instrumentation for specific nanophase experiments (for example, the scanning tunnel microscopes allow us to create nanosized tailored clusters on crystal surfaces).

Another motivation for the intensive studies of nanoparticles arises from the common nature of many processes running on nanoscales in both living organisms and inorganic systems. Therefore, much attention is focused on the so-called biomimetic approaches directed to the imitation modeling of organism various function and fabrication the novel biomaterial types (including the processes, connected with biomineralization).

The main successful results in the field are connected with the development of synthesis methods, which allow us to get stable nanosized particles in a wide range of compositions, dimensions, concentrations and polymeric matrices. On the other hand, the specifications of material usage must be always be considered and accounted for. The high art of synthesis is the main tool of successful design of the tailored nanocomposites. The search for new materials and catalysts remains a field of the chemist's private intuition, because until now we can't predict the correct connections between the novel chemical substance composition and their properties. In recent years our knowledge is considerably increasing and now we understand the general problems of the growth of nanoparticles in the polymer matrix and their role in the very complicated processes and characteristics of nanocomposites. Sometimes it is possible even to control particles size and dispersion during formation.

Broadly speaking, the prediction of the characteristics of hybrid nanocomposites is a fundamental problem of the near future. At present only the main mechanisms of particle stabilization by polymeric surfactants are known. The strength and stiffness of protective layers, their surface conformation and the methods of nanoparticle attachment to polymers are the main characteristics of the considered stabilization effects. At least the qualitative level of investigation has considerably increased in recent years. The atomic metal reaction abilities to the polymer surfaces (i.e. to its functional groups) are now studied intensively.

There are many other important theoretical problems, first connected with the physical and chemical peculiarities of the generation and growth of nanoparticles, as well as the influence of the polymer matrix composition and the special features of their structure (namely, fractal clusters and fiber formation, specificity of the cluster–cluster aggregation, and percolation processes). The investigations of conformational and dynamical properties of macromolecules now have not only theoretical value, but are of practical interest too, because they play a very important role near the interfaces (both in composites and biological structures).

The most important and intensively developing field is nanocatalysis. Metal colloid catalysts are very effective in many-electron processes because they present unusual electron "basins" where the electrons can easily move or escape from. Interesting studies relate to the interactions of clusters with buckminsterfullerenes and their attachment to *host* microporous channels, nanowires, dendrimer-templated nanocomposites.

The formation of nanoparticles under thermal transformation of the metal-containing precursors (being introduced into a polymer matrix by different methods) is a very simple and convenient method of nanocomposite production, though this method is not yet developed completely. Practical applications initiated the development of sol–gel methods and the approach of polymer intercalation into the interlayer space of the *host* lattices, including the various methods of in situ assembling, and without question they will be used and will develop in future. Special attention must be given to the action of pore orientation and the influence of interlayer space on the forming polymer crystallinity and stereoregularity (it is one of the fundamental problems of polymer and nanocomposite sciences in general). All these directions form the general problem of the description of chemical transformation in a systematic way. Now the main goal is to create well-based and informative models of the nature of nanostructured materials, because without such "intelligent" models we can't interpret the wealth of experimental evidence that is progressively increasing.

In the near future various methods will be developed to synthesize the interpenetrating polymer-inorganic networks of "inorganic nets into carbo- or heterogeneous-chain polymer nets" type and their structure will be studied in detail. This problem is of great general scientific interest, but at the same time it has considerable practical relevance (the applications problems probably would be dominating ones). Therefore the connection between experiments and technological tests must be well established. Great success is awaited in the field of nanobiomaterials and molecular self-organizing systems (including the above mentioned multilayer Langmuir–Blodgett films).

We had tried to present in this monograph a general survey of the mosaic picture, now containing the newest methods of composite and hybrid material synthesis, the continuous study of their characteristics and the widest range of their possible applications in different fields.

Nanoscience is based on the theoretical and experimental possibilities of direct material creation from atoms and molecules on the level of nanometric dimensions, studies of their properties or structure and their various (often very unusual) applications in the different fields of science and technology. These characteristics often can't be explained on the basis of classical physics, but they are unique and in the near future one can expect the development of novel materials with very high strength, coatings with specific properties, etc. The design and construction of nanomaterials becomes a very important problem for material science substances with special physical and chemical characteristics, especially for aviation and cosmic techniques, when strength and durability are the main parameters.

The usage of polymers (as the carriers of ultrafine metallic films or particles) to create coatings with specific characteristics is now widespread. Nanostructuration has critical importance as an important material base in the production of light and durable apparatus (for example, in aviation or cosmic technologies), novel information media materials, nonlinear optical active media, etc. In the foreseeable future such materials will be used in

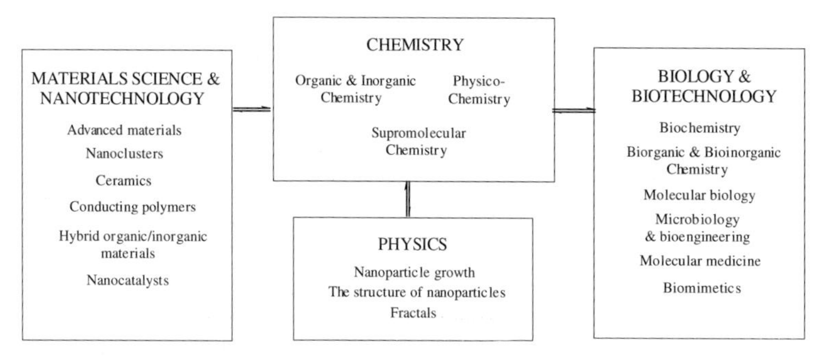

Fig. 13.1. The relations between chemistry, material science and biotechnology

tribomaterials, environmental control devices, solar energy conversion cells, targeted drug delivery systems, gene technology, electronic and optoelectronic devices, avionics, computers, and so on.

The connections between chemistry, material science and biotechnology (as well as with application aspects) are demonstrated in Fig 13.1.

Our future will be based on nanotechnologies, nanomaterials, nanodevices and nanosystems, which will be the key components of the technical and scientific progress of the millennium.

Index

Printing: Strauss GmbH, Mörlenbach
Binding: Schäffer, Grünstadt